Lecture Notes in Mechanical Engineering

Lecture Notes in Mechanical Engineering (LNME) publishes the latest developments in Mechanical Engineering—quickly, informally and with high quality. Original research or contributions reported in proceedings and post-proceedings represents the core of LNME. Volumes published in LNME embrace all aspects, subfields and new challenges of mechanical engineering.

To submit a proposal or request further information, please contact the Springer Editor of your location:

Europe, USA, Africa: Leontina Di Cecco at Leontina.dicecco@springer.com
China: Ella Zhang at ella.zhang@cn.springernature.com
India, Rest of Asia, Australia, New Zealand: Swati Meherishi
at swati.meherishi@springer.com

Topics in the series include:

- Engineering Design
- Machinery and Machine Elements
- Mechanical Structures and Stress Analysis
- Automotive Engineering
- Engine Technology
- Aerospace Technology and Astronautics
- Nanotechnology and Microengineering
- Control, Robotics, Mechatronics
- MEMS
- Theoretical and Applied Mechanics
- Dynamical Systems, Control
- Fluid Mechanics
- Engineering Thermodynamics, Heat and Mass Transfer
- Manufacturing Engineering and Smart Manufacturing
- Precision Engineering, Instrumentation, Measurement
- Materials Engineering
- Tribology and Surface Technology

Indexed by SCOPUS, EI Compendex, and INSPEC.

All books published in the series are evaluated by Web of Science for the Conference Proceedings Citation Index (CPCI).

To submit a proposal for a monograph, please check our Springer Tracts in Mechanical Engineering at https://link.springer.com/bookseries/11693.

Saman Halgamuge · Lifeng Ma · Weiguo Liang · Yongming Bian
Editors

The 10th International Conference on Advances in Construction Machinery and Vehicle Engineering

ICACMVE 2025

Springer

Editors
Saman Halgamuge
University of Melbourne
Melbourne, VIC, Australia

Weiguo Liang
President office
Taiyuan University of Science
and Technology
Taiyuan, Shanxi, China

Lifeng Ma
School of Mechanical Engineering
Taiyuan University of Science
and Technology
Taiyuan, Shanxi, China

Yongming Bian
College of Mechanical Engineering
Tongji University
Shanghai, China

ISSN 2195-4356 ISSN 2195-4364 (electronic)
Lecture Notes in Mechanical Engineering
ISBN 978-981-95-6910-6 ISBN 978-981-95-6911-3 (eBook)
https://doi.org/10.1007/978-981-95-6911-3

This Springer imprint is published by the registered company Springer Nature Singapore Pte Ltd.
The registered company address is: 152 Beach Road, #21-01/04 Gateway East, Singapore 189721, Singapore

Preface

From August 22–25, 2025, the 10th International Conference on Advances in Construction Mechanical and Vehicle Engineering (ICACMVE 2025) was grandly held in Taiyuan, Shanxi. The conference was organized by the China Construction Machinery Society and hosted by Taiyuan University of Science and Technology, the State Key Laboratory of Advanced Stainless Steel, the Collaborative Innovation Center for Heavy Machinery Equipment jointly established by provincial and ministerial authorities, the Shanxi Key Laboratory of Intelligent Technology and Systems for Heavy Equipment, and the Shanxi Society of Automotive Engineering. Centered around the core theme of "Low Carbon, Connectivity, Intelligent Innovation for the Future," the event brought together leading experts and scholars, industry leaders, and young researchers from around the world, building a high-level academic exchange platform focused on cutting-edge topics and future discussions. After ten successful editions, this conference not only summarizes the development achievements in the field but also deeply explores future breakthrough directions for the industry. Therefore, this proceedings volume is a precious crystallization of the intellectual collisions and refined outcomes from this academic feast.

This conference closely aligned with the core demands and cutting-edge trends in the field of construction machinery and vehicle engineering, featuring three main segments: keynote speeches, parallel sessions, and thematic interviews. Four parallel sessions accurately covered four major hotspot areas. First area is new designs, new structures, new processes, and reliability of construction machinery and vehicles; second is electrification, third is electro-hydraulic control, and new energy power; intelligence, automation, and novel sensors; and the last one is a Youth Scholars Forum. The content focused on key areas such as intelligent equipment design, new energy power systems, green manufacturing technologies, and intelligent perception and control, creating an academic dialogue in harmony with keynote speeches delivered by top experts including Academician Yushan Ma and Academician Zorin. Research teams, industry representatives, and university scholars from multiple countries and regions worldwide participated in in-depth discussions centered on industry challenges, technological bottlenecks, and future opportunities, injecting new ideas and vitality into the development of the field.

To ensure the high quality and innovation of academic achievements, this conference strictly adhered to a "double-blind peer review" process, conducting multiple rounds of rigorous evaluation on submitted papers from around the world. The assessment comprehensively considered dimensions such as academic rigor, technical innovation, practical application value, and logical integrity. The selected papers not only showcase the latest research developments in the field but also encompass interdisciplinary innovative results—including explorations of AI-based flexible manufacturing technologies, innovative solutions for new energy power systems under the dual-carbon framework, and studies on intelligent perception and reliability optimization, fully reflecting the development philosophy of "balancing technological innovation with sustainable development."

The successful publication of this proceedings volume would not have been possible without the strong support and dedicated efforts of all parties involved. Here, we would like to express our sincere gratitude to all the experts, scholars, and authors who contributed to the conference—especially to Academician Qingxue Huang, President of the Academic Committee and member of the Chinese Academy of Engineering, Academician Yushan Ma, along with the 13 keynote speakers and 51 parallel session presenters. Your insightful contributions laid a solid academic foundation for the conference. We also extend our heartfelt respect to the rigorous and responsible review panel—it is your professional scrutiny and impartial evaluation that ensured the academic quality and industry relevance of the proceedings. Our deep appreciation also goes to all members of the Academic Committee and the Executive Committee—your meticulous preparation and efficient execution guaranteed the smooth organization of the conference and timely compilation of the proceedings. Special thanks are also due to the host, organizer, and co-organizer institutions for their strong support; your collaborative efforts were instrumental in facilitating academic exchanges and disseminating research outcomes.

We firmly believe that the research results compiled in this proceedings volume will provide valuable academic reference for researchers and practitioners in the field of construction machinery and vehicle engineering, offering solid support for technological innovation and industrial upgrading. Furthermore, we look forward to these achievements driving deep cross-national and cross-disciplinary cooperation, helping the industry overcome technical barriers and achieve the development goals of "efficiency, intelligence, and sustainability," thereby contributing to the sustainable development of the global industry.

The Organizing Committee of the 10th International Conference on Advances in Construction Mechanical and Vehicle Engineering

Committee Member

Honorary Chair

Shi Laide — China Construction Machinery Society, China

Conference Chair

Weiguo Liang — Taiyuan University of Science and Technology, China

Conference Co-chairs

Yongming Bian — Tongji University, China
Lifeng Ma — Taiyuan University of Science and Technology, China
Marcus Geimer — Karlsruhe Institute of Technology, Germany

Publication Chairs

Yongming Bian — Tongji University, China
Weiguo Liang — Key Laboratory of In-situ Modified Mining of the Ministry of Education, China
Lifeng Ma — Taiyuan University of Science and Technology, China
Saman Halgamuge — 'University of Melbourne, Australia

Academic Committee Chairs

Zhihua Zhong — Chinese Academy of Engineering, China
Huayong Yang — Zhejiang University, China

Academic Committee Co-chairs

Xuedong Chen	Hefei General Machinery Research Institute, China
Hehua Zhu	Tongji University, China
Jianlong Zheng	Changsha University of Science and Technology, China
Xinping Yan	Wuhan University of Technology, China
Shirong Ge	China University of Mining and Technology, Beijing, China
Zongquan Deng	Harbin Institute of Technology, China
Jue Zhong	Central South University, China
Bangchun Wen	Northeastern University, China
Baojun Lin	University of China Academy Sciences, China
Guobiao Wang	Tianjin University, China
Pei'en Feng	Zhejiang University, China
Huiqiang Zheng	Tongji University, China
Jürgen Fleischer	German Academy of Engineering, Germany
Saman K. Halgamuge	The University of Melbourne, Australia
Michael Abramovici	Ruhr University Bochum, Germany

Members

Weihua Bao	Shanghai Automation Instrumentation Co., Ltd., China
Hui Cai	China Railway Engineering Equipment (Tianjin) Co., Ltd., China
Ming Cao	University of Shanghai for Science and Technology, China
Donghui Cao	SANY Heavy Industry Co., Ltd., China
Gengxiang Cao	CCCC Third Harbor Engineering Bureau Co., Ltd., China
Yuanwen Cao	Chongqing Jiaotong University, China
Li Chen	Tongji University, China
Haiying Cheng	Chang'an University, China
Wenming Cheng	Southwest Jiaotong University, China
Xiaoming Cheng	Ningbo University of Technology, China
Chi Feng	Shandong Lingong Construction Machinery, China
Jinling Cong	Shihezi University, China
Xichen Dong	Wuhan University of Technology, China

Wenjun Duan	China Railway Engineering Services Limited, China
Maohua Fu	Shanghai Zhenhua Heavy Industries Co., Ltd., China
Rongzhi Gao	Zoomlion Concrete Pumping Machinery Branch, China
Shunde Gao	Dalian University of Technology, China
Yanbei Gao	Chang'an University, China
Mingde Gong	Yanshan University, China
Rui Guo	Yanshan University, China
Chuanxin Guo	Beijing Institute of Construction Mechanization Co., Ltd., China
Wenwu Guo	Shijiazhuang Tiedao University, China
Xiaohong Guo	Chongqing Jiaotong University, China
Rujiang Hao	Shijiazhuang Tiedao University, China
Qin He	Shandong Jianzhu University, China
Lin Hu	Changsha University of Science & Technology, China
Yongbiao Hu	Chang'an University, China
Hui Huang	Huaqiao University, China
Ming Huang	Guangxi Liugong Machinery Co., Ltd., China
Yi Huang	China Communications Construction Company, China
Haibo Huang	Ningbo University, China
Heting Huang	Xiamen XGMA International Trading Co., Ltd., China
Qiufang Huang	Fujian Longma Sanitation Equipment Co., Ltd., China
Yunpeng Huang	Shenji (Shanghai) Intelligent System R&D & Design Co., Ltd., China
Ziming Kou	Taiyuan University of Technology, China
Peng Lan	Xi'an University of Architecture and Technology, China
Hong Li	Yunnan Construction Machinery Industry Association, China
Zheng Li	Inner Mongolia University of Science & Technology, China
Zong Li	Xuzhou Construction Machinery Group Co., Ltd., China
Anhu Li	Tongji University, China
Suoyun Li	Xuzhou Construction Machinery Group Co., Ltd., China
Taijie Li	Da Gang Holding Group Co., Ltd., China

Yunhua Li	Beihang University, China
Zhonggang Li	Lanzhou Jiaotong University, China
Deng Lin	Zhejiang Yongan Construction Machinery Co., Ltd., China
Mingzhi Lin	Guangxi Liugong Machinery Co., Ltd., China
Guangjun Liu	‘Tongji University, China
Xiangdong Liu	Inner Mongolia University of Technology, China
Xinhui Liu	Jilin University, China
Yuxin Liu	Zoomlion Construction Crane Branch, China
Yuanhong Liu	Beijing Institute of Construction Mechanization Co., Ltd., China
Zhengfu Liu	Kunshi Control Systems (Shanghai) Co., Ltd., China
Liang Long	Changsha Zoomlion Environmental Industry Co., Ltd., China
Liang Lu	Tongji University, China
Nianli Lu	Harbin Institute of Technology, China
Biao Ma	Shanghai Municipal Engineering Design Institute (Group) Co., Ltd., China
Xianjun Meng	China Railway Tunnel Bureau Group Company Limited Equipment Branch, China
Junying Min	Tongji University, China
Tao Ni	Yanshan University, China
Lisheng Qin	Shanghai Tongxin Electromechanical Control Technology Co., Ltd., China
Jiang Qiu	Shanghai Environmental Engineering Design & Research Institute Co., Ltd., China
Quan Long	Taiyuan University of Technology, China
Haibo Shang	Tianjin Construction Machinery Research Institute Co., Ltd., China
Jijun Shen	China Construction Machinery Co., Ltd., China
Jinliang Sheng	Tongji University, China
Xianxin Shi	XCMG Xuzhou Heavy Machinery Co., Ltd., China
Hong Sun	Shenyang Jianzhu University, China
Ruming Teng	Dalian University of Technology, China
Zhicheng Tian	CIMIC (Beijing) Vehicle Inspection Engineering Research Institute Co., Ltd., China
Anlin Wang	Tongji University, China
Caiying Wang	Huicong Construction Machinery Network, China
Dajiang Wang	Qinhuangdao Tianye Tolian Heavy Industry & Technology Co., Ltd., China

Jiru Wang	Hangzhou Huaxin Mechanical & Electrical Engineering Co., Ltd., China
Jiangbo Wang	Jiangsu Donghua Testing Technology Co., Ltd., China
Zhiyong Wang	Caofeidian Port Group Co., Ltd., China
Zhongbin Wang	China University of Mining and Technology, China
Chong Wu	Tongji University, China
Binxing Wu	Zoomlion Heavy Industry Co., Ltd., China
Renzhi Wu	Tongji University, China
Haibo Xie	Zhejiang University, China
Ligang Xin	Yantai Hyde Specialized Vehicle Co., Ltd., China
Bing Xu	Zhejiang University, China
Huaiyu Xu	Xuzhou Xugong Schwing Machinery Co., Ltd., China
Zhaohong Xun	Anhui Taiyuan Construction Machinery Co., Ltd., China
Bin Yang	Phoenix (China) Investment Co., Ltd., China
Wenbiao Yang	Anhui Taiyuan Construction Machinery Co., Ltd., China
Hong Yi	Lanzhou University of Technology, China
Chenbo Yin	Nanjing Tech University, China
Haoshan Yu	Chinese Academy of Geological Sciences, China
Ye Yuan	Zoomlion Heavy Industry Co. Earthmoving Machinery Company, China
Wenhui Yue	Hunan University of Science and Technology, China
Qingliang Zeng	Shandong Normal University, China
Hao Zhang	Tongji University, China
Li Zhang	Anhui Taiyuan Construction Machinery Co., Ltd., China
Qing Zhang	Tongji University, China
Yuan Zhang	LIBO Heavy Industries Science & Technology Co., Ltd., China
Daqing Zhang	Shanhe Intelligent Equipment Co., Ltd., China
Guosheng Zhang	Tianjin Construction Machinery Research Institute, China
Junhui Zhang	Zhejiang University, China
Shejun Zhang	China Railway Construction Heavy Industry Corporation Limited, China
Shengjun Zhang	Beijing Institute of Construction Mechanization Co., Ltd., China
Weiguo Zhang	Shanghai Maritime University, China

Qihuai Chen	Huaqiao University, China
Zhe Chen	Tongji University, China
Min Cheng	Chongqing University, China
Christopher Ehrmann	Tongji University, China
Ruqi Ding	East China Jiaotong University, China
Changhong Fu	Tongji University, China
Xiangyu He	Changsha University of Science & Technology, China
Zeran Hou	Tongji University, China
Jiahai Huang	Taiyuan University of Technology, China
Tianliang Lin	Huaqiao University, China
Changhui Liu	Tongji University, China
Liang Lu	Tongji University, China
Junying Min	Tongji University, China
Morteza Mohammadzaheri	Birmingham City University, UK
Muhammad Saleem	Imam Abdulrahman Bin Falsal University, Saudi Arabia
Youcheng Shi	Lanzhou University of Technology, China
Bo Sun	Tongji University, China
Yuantao Sun	Tongji University, China
Tang Tang	Tongji University, China
Yichao Tang	Tongji University, China
Valerian Croitorescu	University POLITEHNICA of Bucharest, Romani
Jianwei Wang	Yanshan University, China
Kun Wang	Tongji University, China
Liang Wang	Tongji University, China
Yajing Wu	Tongji University, China
Yao Xiao	Tongji University, China
Dongdong Xu	Tongji University, China
Meng Yang	Tongji University, China
Chunlong Yu	Tongji University, China
Zhicheng Yuan	Tongji University, China
Cuixiang Zhu	Tongji University, China

Contents

Mechanical Design and Analysis

Mechanical Design and Analysis

Bridge Parameters and Performance of Internal Recirculation Ball Screw

Zujun Xiong[1], Bo Dang[1], Lijun Meng[2], and Jun Song[3](✉)

[1] Department of Mechanical Engineering, Xijing University, Xi'an, Shaanxi, China
[2] Shandong Wantong Hydraulic Co., Ltd., Rizhao, Shandong, China
[3] Shanxi Dedao Equipment Manufacturing Group Co., Ltd., Taiyuan, Shaanxi, China
info@dedao.group

Abstract. The performance of the reverser affects the overall performance of the ball screw pair, and the high-performance reverser is the key to the design of the ball screw pair. When using the UG bridging curve method to design the internal reverser, it is necessary to accurately determine the influence of parameters G1, G2, and G3 on the performance of the inverter. By using ADAMS to three groups of different parameters (G1, G2, and G3) reverser performance simulation analysis, the ball is gotten through the impact of the reverser and nut running speed, and the impact of the ball is assessed through the reverser and nut running speed, to determine the best G3 reverser performance parameters. The research results provide an important guiding basis for engineering design.

Keywords: bridging curves · internal recirculation · performance study · ADAMS

1 Introduction

UG's application scope is expanding as a comprehensive CAD/CAE/CAM software [1]. The UG bridge curve method is usually used to design internal recirculation inverters. Compared with traditional design curves such as cubic curve, quintic curve, and sine curve [2], this method avoids complicated calculation and is simpler, faster, and more intuitive. In the bridge curve method, the connection parameters' choice of G1, G2, and G3 will affect the inverter performance. Therefore, studying the performance of the inverter under different parameters to determine the optimal performance has become the core task of the design of ball screw pairs. However, there is no relevant research data to discuss this in depth currently.

In this study, three three-dimensional models of three kinds of ball screw pairs were established by UG, and the parameters were set as G1, G2, and G3 respectively, and the other parameters remained the same. Subsequently, these models were imported into ADAMS for simulation analysis to evaluate the performance of the inverter under different parameters. The specific analysis includes the impact force generated by the ball as it passes through the reverser and the speed at which the nut runs. Through the comprehensive analysis of impact force and velocity indexes, the optimal performance

S. Halgamuge et al. (Eds.): ICACMVE 2025, LNME, pp. 3–12, 2026.
https://doi.org/10.1007/978-981-95-6911-3_1

parameters of the reverser were determined. These research results have certain generality, can be applied to the design of all internal circulation inverters, and provide an important design basis for engineering design.

2 Establish a Three-Dimensional Model

The model of ball screw pair established in this study is 6310, the nominal diameter of the screw is φ63 mm, the lead Ph is 10 mm, and the contact Angle α is 45°. The design of the bead return curve of the inverter is very important. The bridge curve method is used in UG to generate different bead curves by adjusting the parameters G1, G2, and G3 to obtain the inverters with different characteristics. These parameters directly affect the shape and performance of the inverter, as shown in Fig. 1. The three-dimensional model of the ball screw pair finally constructed is shown in Fig. 2, which lays a foundation for subsequent performance analysis.

The model in Fig. 2 has been simplified to save computational resources and shorten the simulation time. Specifically, some of the reversers and balls were removed, leaving only one reverser and ball for a recirculating loop. Subsequently, the G1, G2, and G3 inverters were replaced with the inverter positions in the simplified model for subsequent dynamic simulation analysis. This method not only improves computational efficiency but also enables the focus on the performance evaluation of key components, providing strong support for optimal design.

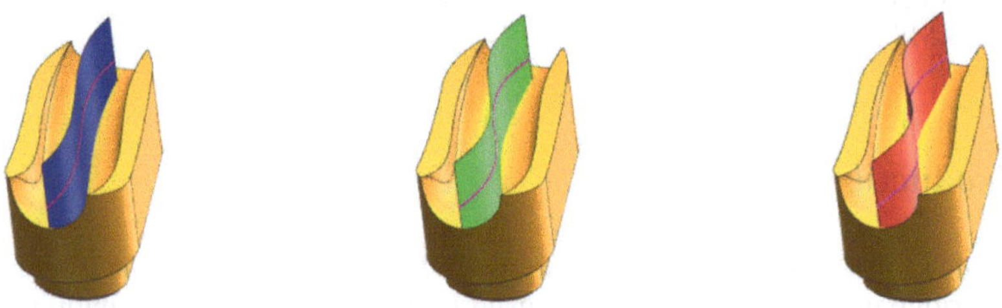

(a) G1 (b) G2 (c) G3

Fig. 1. Reverse curve.

Fig. 2. Ball Screw Pair.

3 ADAMS Dynamics Simulation

The model was exported to Parasolid format of UG10.0 version, and the dynamics simulation analysis was performed in ADAMS. The specific steps include setting the lead screw as a rotating motion pair, the nut as a moving pair, and setting a fixed pair between the reverse device and the nut. Contact force models between ball and screw, ball and reverse, ball and nut, and ball to ball are then added. Finally, according to the need to set the simulation parameters, and start the process of dynamic simulation [3, 4].

3.1 Add a Joint Pair

According to the movement of the ball screw pair, the lead screw is used as the driving part, and the lead screw drive is given 1 r/s. According to the screw rotation, the nut moves along the axis, respectively, adding a rotating pair, a moving pair, and a drive to the screw, and adding a fixed pair to the reverse device and the nut [5], as shown in Fig. 3 below.

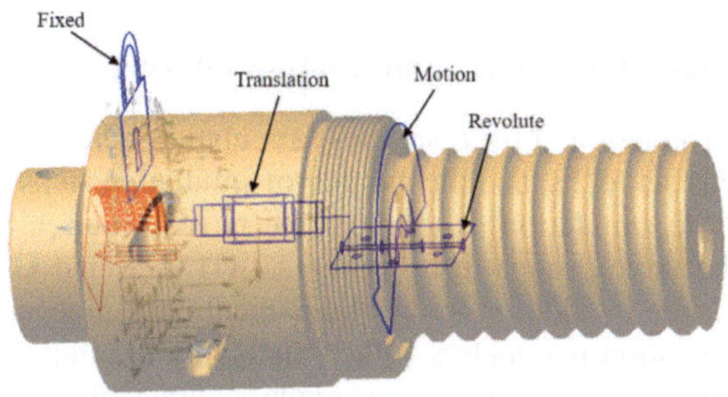

Fig. 3. Schematic diagram of the joint.

3.2 Add Contact Force

The contact parameters are set as follows: the contact force model selects the impact model, the contact stiffness can be set as the system default, the force index is set to 1.5, generally 1.5 when the two members are steel, the penetration depth is set to 0.1 mm, and other parameters can be set as system default. Set ball and screw, ball and inverter, ball and nut, and ball and ball, for a total of 100 contact pairs. Contact parameters are shown in Fig. 4 below.

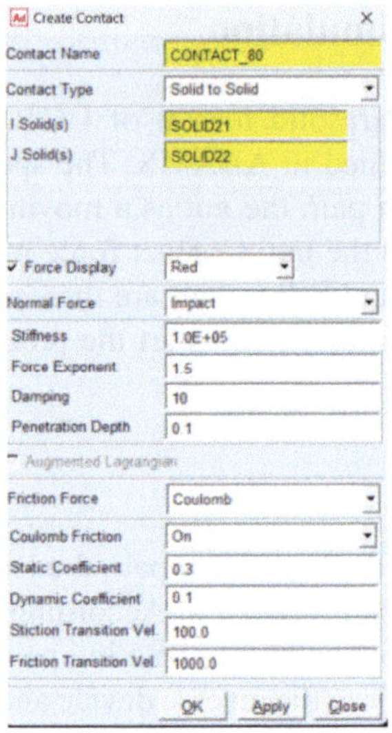

Fig. 4. Contact parameters.

3.3 Set the Simulation Parameters and Analysis Objects

To facilitate the interpretation of the results, the simulation time is set to an integer, and the simulation time is set to 2 s, and the number of simulation steps is 400.

To ensure that the simulation experiment meets the requirements of the control variable method, the three groups of experiments, G1, G2, and G3, should keep the analysis object the same in addition to the above-mentioned motion pair, contact force parameters, and simulation parameters. When analyzing the impact force, the impact force of the same ball on the inverter can ensure the accuracy of the simulation results to the greatest extent. As shown in Fig. 5 below, the red balls in the figure are the objects of the impact force analysis.

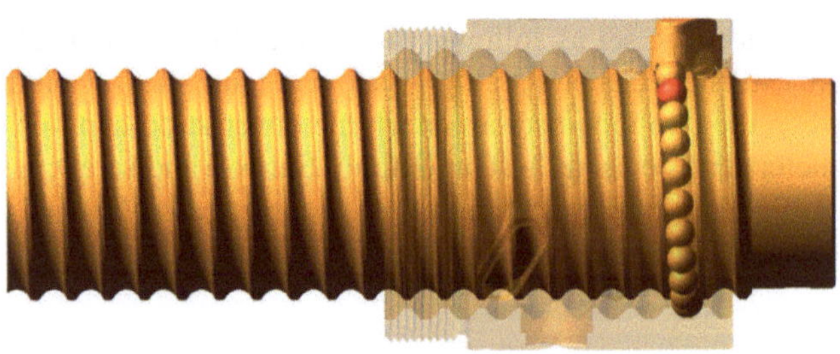

Fig. 5. Impact force analysis object.

4 Simulation Results and Post-Process

After the simulation, the impact force results and the nut running speed results are output in Fig. 6 below. The impact force curves of the G1, G2, and G3 model balls on the inverter are given, respectively. As shown in Fig. 7 below, the running speed curves of G1, G2, and G3 models are given, respectively.

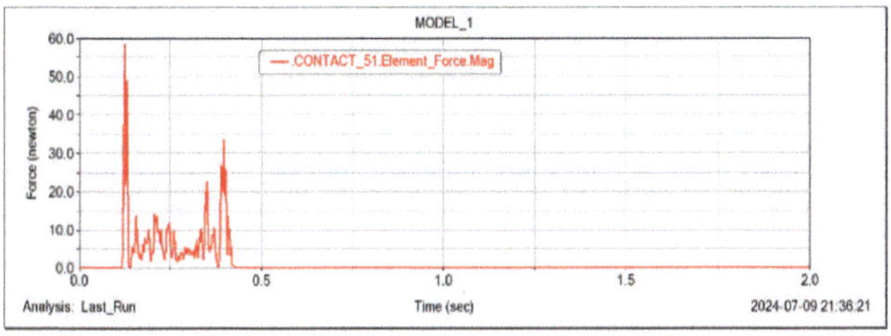

(a) G1

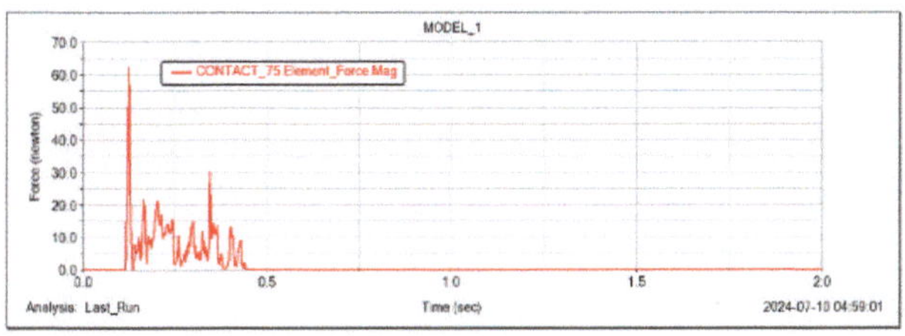

(b) G2

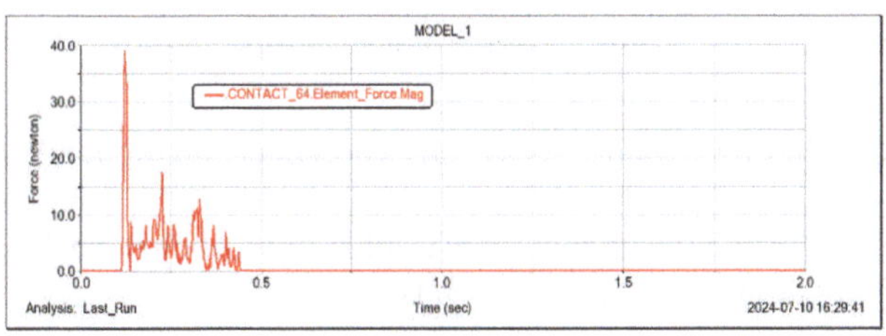

(c) G3

Fig. 6. Impact force curve.

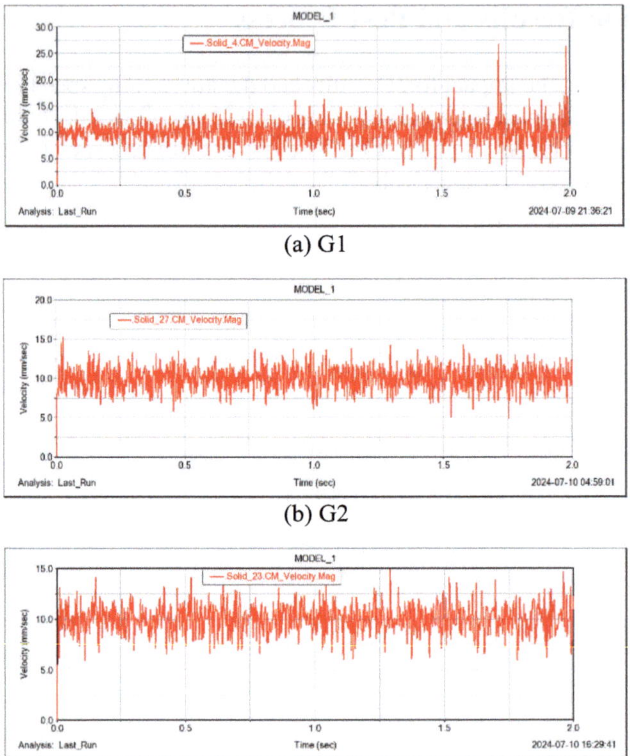

(a) G1

(b) G2

(c) G3

Fig. 7. Curve of nut velocity.

Using ADAMS in Figs. 6 and 7 above, all export data and digital data are convenient for data analysis.

From Fig. 6 (a)–(c) that the impact force is maximum when the ball enters the reverse device. The magnitude of the load here can also be used as a reference index to evaluate the performance of the reverser.

From Fig. 7 (a)–(c), the nut operates at a speed of approximately 10 mm/s, which is consistent with the nut lead (10 mm) and the screw speed (1 r/s). This shows that the simulation results meet the design requirements and agree with the actual situation.

5 Data Processing and Analysis

5.1 Impact Force Analysis

The derived impact force data were averaged to evaluate the mechanical properties of the ball passing through the reverser.

As shown in Fig. 8, the difference between the average impact forces of G1 and G2 is small, with a difference of 0.46 N. In contrast, the average impact force of the G3 is significantly lower than that of the G1 and G2, with a maximum difference of 2.77 N, showing a clear trend of performance improvement. From the average analysis of the impact force, the reverser corresponding to G3 shows the best performance in terms of mechanical properties.

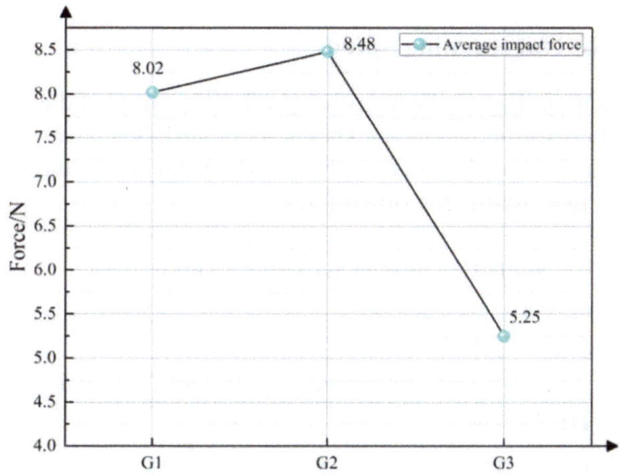

Fig. 8. Average impact trend chart.

5.2 Impact Force Analysis at the Entrance

The impact at the reverse entrance is one of the biggest, because the average impact is easily affected by the most value. Therefore, we compared the load at the entrance to further validate the analysis, as shown in Fig. 9 below.

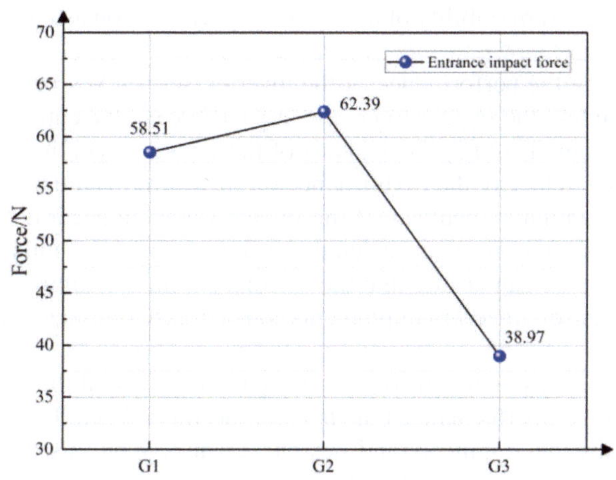

Fig. 9. Impact force at the entrance.

As shown in Fig. 9, the distribution of the impact force at the entrance is like that of the average impact force, and the impact force of G3 is the smallest, which is only 38.97 N. This further verifies the superiority of the inverter corresponding to the G3 parameter in terms of performance.

5.3 Analysis of the Nut Running Speed

According to the data in Fig. 7, the average speed of the nut and the standard deviation were calculated; the specific results are shown in Table 1. The average operating speed of the nut is 10 mm/s, and the smaller the standard deviation, the smaller the speed fluctuation and the more stable the operation.

Table 1. Nut Running Speed Analysis Table.

Type	Calculated value	
	Average velocity mm/s	Standard deviation
G1	9.98	1.84
G2	9.98	1.44
G3	10	1.54

After analyzing the data in Table 1, it can be concluded that although the average speed of G1 and G2 is the same, they are both slightly lower than the theoretical 10 mm/s. The average speed of the G3 is the same as the theoretical speed. In terms of standard deviation, the values of G2 and G3 are almost identical, with a difference of only 0.1, showing a small velocity fluctuation. Relatively speaking, the standard deviation of G1 is significantly larger, indicating that its velocity fluctuates greatly. In summary, G3 has the best performance from stability of nut running speed and consistency with theoretical speed.

When evaluating the performance of the inverter, several key parameters are taken into account, including the impact force, the impact force at the entrance, and the operating speed of the nut. Through the analysis of these parameters, it is found that when the parameter G3 is selected, the performance of the inverter is the best. This discovery was based on a quantitative analysis of the impact force, an accurate measurement of the impact force at the entrance, and a detailed examination of the nut's operating speed. With this approach, we can ensure that the optimal parameter configuration is selected during the design and engineering implementation phases to optimize the overall inverter performance.

In addition, the analysis results of this study not only reveal the direct influence of parameter selection on performance but also provide valuable guidance for engineering design. This means that engineers and designers can rely on this data to make more precise system designs to ensure that the expected performance standards are met in real-world applications. Through this data-based approach, the number of trials and errors can be reduced, the design efficiency can be improved, and the cost effectiveness can be maximized.

6 Conclusions

In this study, the UG modeling and ADAMS dynamic simulation technology are used to analyze G1, G2, and G3 inverters deeply. Through the simulation, the impact force and nut speed were plotted in detail, paying special attention to the impact force at the entrance as well as the nut speed, which are critical parameters for evaluating the inverter performance. The simulation results reveal the differences in the key performance indices of different types of inverters, which provides an important reference for engineering design and performance optimization.

The simulation results show that the inverter with G3 parameter configuration has the best performance. This finding not only has guiding significance for the current study of 6310 ball screw pair, but also the conclusion and method can be applied to other types of internal circulation reversers. Through this analysis, we can provide the design team with directions to improve existing products and develop new models of reversers, thus ensuring that products can meet higher performance standards at the design stage. These results will greatly promote the optimization of engineering design, making the design of reversers and ball screws more efficient and reliable, and ultimately improving their overall performance in practical applications.

References

1. Wang, L., Bi, C.: Application and development of UG technology. Machinery **37**(07), 51–54 (2010)
2. Feng, H.: Fundamentals of Ball Screw Pair Dynamics and Design. Machinery Industry Press, Beijing (2015)
3. Li, X., Wang, K., Hao, D., et al.: Modeling and simulation of ball screw pair dynamics based on ADAMS. Mach. Tool Hydraul. **43**(13), 142–144 (2015)
4. Zhang, X., Tang, W., Wang, A.: Dynamic simulation of ball screw auxiliary ball circulation system based on ADAMS. Mach. Manuf. Autom. **39**(02), 110–113+119 (2010). https://doi.org/10.19344/j.cnki.issn1671-5276.2010.02.033
5. Sun, L., Liu, G., Ma, S., et al.: Inner loop ball screw pair of reverser optimization design and simulation. J. Mech. Sci. Technol. (01), 13–18 (2018). https://doi.org/10.13433/j.CarolCarroll Nki.1003-8728.2018.0103

Design and Analysis of a Viscous Damper with Quasi-Zero Stiffness

Yi Zhang[1], Jiale Peng[2], Dongbo Tian[3], and Zhihui Gao[1](✉)

[1] School of Mechanical Engineering and Automation, Beihang University, Beijing, China
gaozhihui@buaa.edu.cn

[2] Beijing Institute of Spacecraft System Engineering, The China Academy of Space Technology, Beijing, China

[3] Beijing Institute of Spacecraft Environment Engineering, The China Academy of Space Technology, Beijing, China

Abstract. Owing to the flexible structures, the mechanical vibration excited during the operation of the flexible arm impacts the control precision and stability of the flexible arm. To eliminate vibrations, the viscous damper can be installed in a flexible arm. Damping and mass are introduced into the system. However, without a stiffness element to support the mass, the damper cannot effectively function against gravitational force. Given this, a design of a viscous damper based on the principle of viscous damping and quasi-zero stiffness is proposed. The rhombic mechanism, which possesses quasi-zero stiffness, can counteract the gravitational impact on the mass. The structural parameters of the damper are optimized based on the optimal damping coefficient obtained from the frequency response function. Modal tests demonstrate a significant reduction in the amplitude of the host structure due to the viscous damper.

Keywords: viscous damper · Quasi-zero Stiffness · Vibration · Frequency Response

1 Introduction

Compared with the rigid arm, the flexible arm possesses the advantages, such as lightweight construction, high-speed operation, and low energy consumption. However, its insufficient rigidity makes it prone to deformation and elastic vibrations. External interference can deteriorate this deformation, resulting in increasingly severe vibrations that impede the flexible arm's capability to achieve high speed and precision.

Incorporating structural damping into the main structure effectively reduces vibration in the flexible arm and enhances dynamic performance. Attaching a subsystem, such as a dynamic vibration absorber, is a common method to attenuate forced vibration responses at specific frequencies. The vibration absorber is widely utilized for its straightforward design and substantial vibration suppression capabilities. Since its invention by FRAHM [1] in 1909, extensive research has been conducted on dynamic vibration absorbers over nearly a century. Den Hartog [2] initially proposed the "fixed-point" theory to reduce the

S. Halgamuge et al. (Eds.): ICACMVE 2025, LNME, pp. 13–20, 2026.
https://doi.org/10.1007/978-981-95-6911-3_2

amplitude of the main structure and subsequently derived the analytical solution for the minimum amplitude criterion of the undamped primary structure using a single degree of freedom damper. Bonsel et al. [3] developed a linear dynamic absorber, and by comparing the study with and without damping, it was found that the damped dynamic absorber not only effectively suppresses the formant peak, but also simultaneously reduces other peaks. Based on the aforementioned research, Brock [4] derived the formula for optimal damping ratio in various scenarios. To eliminate the multi-modal vibration of the main structure, Rice [5] studied the parameter optimization design problem of two single-degree-of-freedom dynamic vibration absorbers for suppressing vibrations in a cantilever beam.

However, the primary function of the dynamic vibration absorber is to effectively reduce vibrations at a particular frequency. Consequently, it is highly influenced by the targeted mode and may not function properly under minor disturbances due to its narrow operating frequency band. The dynamic vibration absorber has a natural frequency. When the vibration frequency of the main structure approaches the natural frequency, resonance phenomena occur, causing the vibration of the controlled object to diverge. To mitigate the wide-band vibration of the main structure. Lanchester [6] invented a torsional vibration damper for the crankshaft of an internal combustion engine in 1914. Mohammad et al. [7] found that linear viscous dampers perform better than nonlinear dampers both in terms of attenuation and robustness. To reduce the vibration during the finishing process of thin-walled parts, Gubanov [8] proposed a viscous damper that uses air viscosity and a freely movable mass. Li et al. [9] proposed a composite viscous damper based on the principles of fluid viscous damping and eddy current damping.

The viscous damper can effectively reduce vibration for objects in rotation or horizontal motion. However, the lack of stiffness prevents the damper from adequately supporting mass when the object moves against gravity, thereby rendering it ineffective. Gatti et al. [10] proposed a design scheme for a viscous damper with a quasi-zero stiffness device, which effectively supports mass and disregards additional stiffness during operation. However, its complexity, challenging analysis, and large size have hindered its widespread adoption. In this paper, a viscous damper with a quasi-zero stiffness device is designed with a simple structure and straightforward analysis. Optimal structural parameters for effective vibration mitigation are determined through model building and numerical simulation.

2 Design of the Damper

A viscous damper based on quasi-zero stiffness is proposed to achieve vibration suppression in a wide frequency range. The device can prevent failure caused by unsupported mass under the influence of gravity.

2.1 Damper Principle Design

As shown in Fig. 1, the viscous damper consists of a cylindrical shell, a mass block, and a uniformly filled fluid medium. There is a specific radial gap between the mass block and the shell. As the shell and object move together, the relative movement between the mass block and the shell causes the fluid medium in Region A to be forced to flow to Region B through the radial gap. The viscous medium generates a viscous force acting on the mass block, impeding its relative motion. When the viscous damper is installed on the vibrating controlled object, the kinetic energy of the mass block is partly converted into thermal energy dissipation, thereby dissipating energy of the controlled object to suppress vibrations.

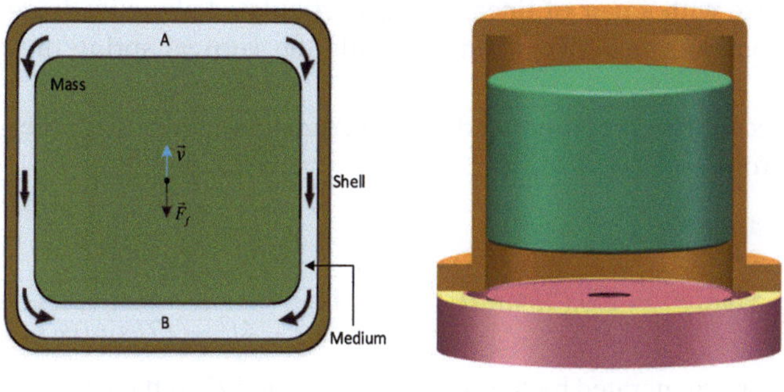

(a) structure (b) 3D View

Fig. 1. The viscous damper principal design.

It can be known from the relevant theories of fluid gap flow that the flow rate Q will be derived from the following equation when the fluid flows in the annular gap.

$$Q = \frac{h^3 a}{12\mu_a l}\Delta P \tag{1}$$

where h is the gap width of the damper, $a = \pi D$ is the perimeter of the circular gap, D is the diameter of the mass block, μ_a is the aerodynamic viscosity of the fluid medium, l is the length of the mass block, and ΔP is the pressure difference. v is the flow speed of the fluid medium in the gap, and is also the relative speed of the mass block and the shell. The gap width of the damper is half of the radial clearance g. The flow rate Q can also be obtained as follows:

$$Q = v\frac{\pi D^2}{4} \tag{2}$$

The resistance F_f caused by the pressure difference on the mass block is given as follows:

$$F_f = \frac{\pi D^2}{4}\Delta P \tag{3}$$

Therefore, the damping coefficient of the damper can be obtained as follows:

$$c = \frac{F_f}{v} = 6\pi\mu_a\frac{lD^3}{g^3} \tag{4}$$

2.2 Quasi-Zero Stiffness Mechanism Design

To counteract the gravity acting on the mass of the viscous damper, a quasi-zero stiffness device is integrated into the viscous damper. This device ensures effective vibration suppression across a wide range of frequencies, particularly in the direction of gravity. When the viscous damper is operating, the device exhibits low stiffness characteristics. The device will effectively support the mass when the viscous damper is stationary.

As depicted in Fig. 2(a), four equal-length rods are connected by pins to form a diamond structure, with a pair of linear springs articulated along its diagonal. The horizontal spring generates negative stiffness, while the vertical spring provides positive stiffness.

When the mechanism is compressed in the vertical direction, the force generated can be determined from

$$\begin{cases} F_a = k_a\delta \\ F_b = 4k_b x\tan(\theta) = 2k_b(2l\sin(\theta_0) - \delta)\left[1 - \dfrac{2l\cos(\theta_0)}{\sqrt{4l^2 - (2l\sin(\theta_0) - \delta)^2}}\right] \end{cases} \quad (5)$$

where F_a is the force generated by the vertical spring, and F_b is the vertical component of the force generated by the combination of the horizontal spring and four-bar linkage. As is illustrated in Fig. 2(a). Where δ is the amount of vertical spring compression, x is the amount of horizontal spring elongation, k_a and k_b represent the stiffness coefficients of vertical and horizontal springs, respectively. θ_0 is the initial angle between the bar and the horizontal direction. l is the length of the linkage. Rewriting Eq. (5) in non-dimensional form results in

$$\overline{F} = \frac{F_k}{k_a l} = \Delta + 2r_k(2\sin(\theta_0) - \Delta)\left[1 - \frac{2\cos(\theta_0)}{\sqrt{4 - (2\sin(\theta_0) - \Delta)^2}}\right] \quad (6)$$

where $\overline{F}$ is the non-dimensional force, $F_k = F_a + F_b$, $\Delta = \delta/l$ is the non-dimensional displacement, and r_k is the ratio of the vertical to the horizontal spring stiffness are. According to Eq. (6), the relationship between the non-dimensional force and the non-dimensional displacement is shown in Fig. 2(b) when $r_k = 1.5$ and $\theta_0 = 60°$. It can be seen that an overall linear-like quasi-zero stiffness behavior is achieved in the approximate range where $\Delta \in [0.5, 0.9]$, which is indicated in Fig. 2(b) by a shaded region. The structural parameters of the quasi-zero stiffness device can be obtained according to the working range and mass of the desired viscous damper.

3 Dynamic Analysis of the Device

To suppress the vibration of the controlled object, a viscous damper with quasi-zero stiffness can be conveniently installed on the controlled object as shown in Fig. 3(a). To study this system's dynamic characteristics, the dynamic model is illustrated in Fig. 3(b).

According to Fig. 3(b), the equations of motion of the system can be obtained as follows:

$$\begin{cases} M\ddot{x}_1 + c_1\dot{x}_1 + k_1 x_1 - c_2\dot{z} - F_k = F\cos(\omega t) \\ m\ddot{x}_2 + c_2\dot{z} + F_k = 0 \end{cases} \quad (7)$$

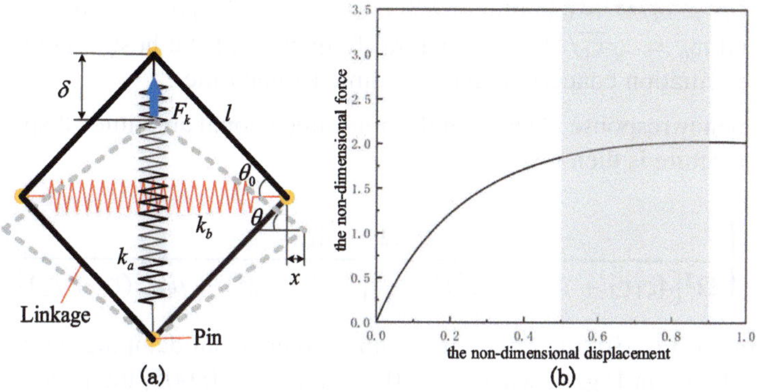

Fig. 2. Quasi-zero stiffness mechanism design.

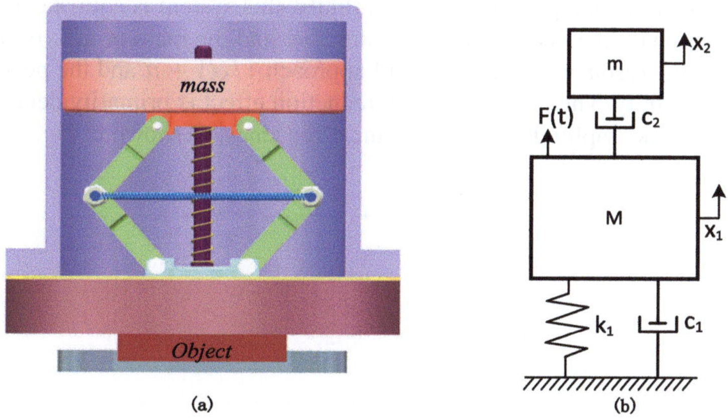

Fig. 3. Viscous damper based on quasi-zero stiffness.

where M, k_1 and c_1 are the mass, stiffness, and damping of the host structure, respectively. m and c_2 are the mass and damping of the viscous damper, respectively. F is the excitation force, and F_k is the force generated by the quasi-zero stiffness mechanism. $z = x_2 - x_1$ is the relative displacement between the mass of the viscous damper, x_2, and the mass of the host structure, x_1. Due to $F_k = mg$ caused by the quasi-zero stiffness device, F_k can be regarded as a constant, and Eq. (7) can be rewritten in non-dimensional form as follows:

$$\begin{cases} (1+\mu)\widehat{x}_1'' + \mu\widehat{z}'' + 2\zeta_1\widehat{x}_1' + \widehat{x}_1 = \cos(\Omega\tau) \\ \widehat{x}_1'' + \mu\widehat{z}'' + 2\zeta_2\widehat{z}' = 0 \end{cases} \tag{8}$$

where $\widehat{x}_1 = x_1/x_0$ and $\widehat{z} = z/x_0$ are the non-dimensional displacement and $x_0 = F/k_1$ is the static deflection due to the excitation force; $\zeta_1 = c_1/\sqrt{k_1 M}$ and $\zeta_2 = c_2/\sqrt{k_1 M}$ are the non-dimensional damping coefficients of the host structure and viscous damper, respectively; $\mu = m/M$ is the ratio of mass; $\Omega = \omega/\omega_n$ is the non-dimensional forcing frequency and $\omega_n = \sqrt{k_1/M}$ is the natural frequency of the host structure; $\tau = \omega_n t$ denotes differentiation concerning the non-dimensional time.

The frequency response in terms of the non-dimensional amplitude displacement $\widehat{X}_1$ of the host structure is then given by

$$\left|\widehat{X}_1\right| = \left\{ \frac{\mu^2\Omega^2 + 4\zeta_2^2}{\Omega^2\left[4\zeta_1\zeta_2 + \mu(1-\Omega^2)\right]^2 + 4\left[\zeta_2(1-\Omega^2) - \mu\Omega^2(\zeta_1+\zeta_2)\right]^2} \right\}^{\frac{1}{2}} \tag{9}$$

According to Eq. (9), the effect of the non-dimensional damping on the frequency response as shown in Fig. 4, when $\mu = 0.57$ and $\zeta_1 = 0.005$, the peak amplitude of the host structure decreases by installing a viscous damper with smaller damping, and the change in resonance frequency is relatively small. When the damping increases, the resonance frequency will decrease, and the peak amplitude will decrease first and then increase. When the damping approaches infinity, this system is equivalent to an oscillating system, with the mass of the main structure and the mass of the viscous damper. In this case, the resonance frequency will approach $1/\sqrt{1+\mu}$ and the peak amplitude will tend to infinity. The prime vibration reduction effect is primarily demonstrated by the minimal peak amplitude at the resonance frequency, as shown in Fig. 4.

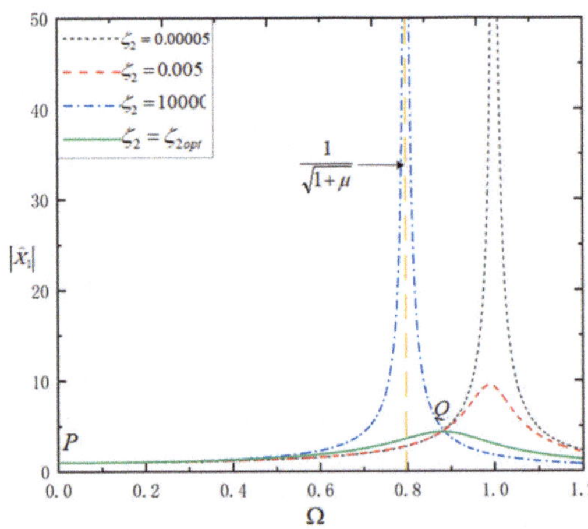

Fig. 4. Effect of the non-dimensional damping on the frequency response.

To achieve the best vibration reduction, obtaining the optimal structural parameter of the damper is essential. Because the damping of the host structure is relatively small and its impact on the frequency response curve is minimal, Eq. (9) can be approximated in the proximity of $\zeta_1 = 0$ as

$$\left|\widehat{X}_1\right| \approx \left\{ \frac{\mu^2\Omega^2 + 4\zeta_2^2}{\Omega^2\mu^2(1-\Omega^2)^2 + 4\left[\zeta_2(1-\Omega^2) - \mu\Omega^2\zeta_2\right]^2} \right\}^{\frac{1}{2}} \tag{10}$$

As shown in Fig. 4, the curves of the response for different values of the damping ratio pass through two fixed points, P and Q. Equation (10) can be rewritten as

$$\left|\widehat{X}_1\right| = \sqrt{\frac{A\zeta_2^2 + B}{C\zeta_2^2 + D}} \tag{11}$$

where A, B, C, D are coefficient equations. Because the coordinates of the fixed points in the response curves do not change with the damping ratio, $A/C = B/D$ is always established. According to Eq. (10), the equation can be obtained as follows:

$$\left(\frac{1}{1 - \Omega^2 - \mu\Omega^2}\right)^2 = \left(\frac{1}{1 - \Omega^2}\right)^2 \tag{12}$$

The horizontal coordinates of two fixed points can be obtained to give.

$$\Omega_P = 0, \Omega_Q = \sqrt{\frac{2}{2 + \mu}} \tag{13}$$

If the resonant peak is at point Q, it indicates the best damping effect, and the corresponding damping is the optimal damping of the viscous damper. In this case, $\left|\widehat{X}_1\right|'_Q = 0$ and solved for the optimal damping coefficient to give

$$\zeta_{2opt} = \frac{\mu}{\sqrt{2(1 + \mu)(2 + \mu)}} \tag{14}$$

Equations (4) and (14) are solved for the optimal structural parameters of the device. Meanwhile, the viscous damper designed based on these parameters can effectively reduce the vibration amplitude of the controlled object along the direction of gravity.

4 Conclusions

In this study, a viscous damper with a quasi-zero stiffness device is proposed. This viscous damper can effectively suppress broadband vibration. However, the lack of stiffness prevents the damper from adequately supporting mass when the controlled object moves against gravity, rendering it ineffective. Additionally, the quasi-zero stiffness device can counteract the gravitational influence of the mass block within a specific travel range, ensuring effective vibration suppression by the damper along the direction of gravity. And it will not affect the vibration suppression effect in other directions. Model building and numerical simulation reveal the impact of device parameters on dynamic behavior. Based on the optimal damping coefficient, the device's optimal structural parameters can be determined.

Acknowledgment. This project is supported by the National Natural Science Foundation of China (No. 52075014).

References

1. Hermann, F.: Device for Damping Vibrations of Bodies, 989958 (1909)
2. Den Hartog, J.P.: Mechanical Vibration. McGraw-Hill, New York (1940)
3. Bonsel, J.H., Fey, R.H.B., Nijmeijer, H.J.N.D.: Application of a dynamic vibration absorber to a piecewise linear beam system. Nonlinear Dyn. **37**(3), 227–243 (2004)
4. Brock, J.: A note on the damped vibration absorber. J. Appl. Mech. **13**(4), A284–A284 (1946)
5. Rice, H.J.: Design of multiple vibration absorber systems using model data. J. Sound Vib. **160**(2), 378–385 (1993)
6. Lanchester, F.W.: Damping torsional vibrations in crank shafts, no. 1085443 (1914)
7. Vakilinejad, M., Grolet, A., Thomas, O.: A comparison of robustness and performance of linear and nonlinear Lanchester dampers. Nonlinear Dyn. **100**(1), 269–287 (2020)
8. Gubanov, G.: Broadband pneumatic mass damper for the elimination of workpiece vibrations. CIRP J. Manuf. Sci. Technol. **30**, 184–194 (2020)
9. Li, D., Xu, R., Dong, Q., Xu, Y.: Design and application of composite lanchester damper for milling thin-walled part. In: Journal of Physics: Conference Series, vol. 2760, no. 1, p. 012070. IOP Publishing (2024)
10. Gatti, G., Shaw, A.D., Gonçalves, P.J.P., Brennan, M.J.: On the detailed design of a quasi-zero stiffness device to assist in the realisation of a translational Lanchester damper. Mech. Syst. Signal Process. **164**, 108258 (2022)

Study on the Core Deformation of Super-Thick Steel Plate by Single Roller Heavy Reduction in Continuous Casting Process

Yang Liu[1,2](✉), Weihua Wang[1], Ning Hao[1], Yunfei Wei[3], Xin Li[1], Wei Zhang[1], Yonglin Kang[4], and Guobiao Di[1]

[1] Research Institute of Technology, Shougang Group Co., Ltd., Beijing 100043, China
365410529@qq.com
[2] Beijing Key Laboratory of Green Recyclable Process for Iron and Steel Production Technology, Beijing 100043, China
[3] Shougang Jingtang United Iron and Steel Co., Tangshan 063200, Hebei, China
[4] University of Science and Technology Beijing, Beijing 100083, China

Abstract. The technology of single roll heavy reduction in a continuous casting process of more than 20 mm at the end of the slab solidification is an important technical means to improve the qualification rate of flaw detection and other comprehensive performance of the heavy plate. In this paper, the processes of the slab direct rolling and the slab first by single point heavy reduction then rolling were simulated by the finite element software ABAQUS. The equivalent strain and the cumulative deformation at the full plate thickness during the two production processes were calculated and analyzed. The results show that the heavy reduction is beneficial to increasing the strain and the cumulative deformation in the core area of the plate. In industrial production, the 252 mm plate with good performance was successfully produced by the 400 mm slab through single roll heavy reduction and rolling, and the compression ratio during the rolling process was only 1.6.

Keywords: Continuous casting slab · Single point heavy reduction · Super-thick steel plate · Deformation penetration · Numerical simulation · Low compression ratio

1 Introduction

In order to solve the internal defects such as central segregation of slab, different types of heavy reduction technology at the solidification end of continuous casting have been developed in Japan, Korea, and China, and good results have been achieved [1–5]. However, due to the relatively small reduction amount of single-point reduction, the effect of pressure penetration on the center of the thick slab is limited, and the problem of center porosity and microstructure density of the slab can not be further improved. The single roll heavy reduction technology developed by Shougang Jingtang makes full use of the natural large temperature gradient by the surface and the center of the slab in the secondary cooling zone of continuous casting. The large diameter casting roll installed

S. Halgamuge et al. (Eds.): ICACMVE 2025, LNME, pp. 21–31, 2026.
https://doi.org/10.1007/978-981-95-6911-3_3

in the secondary cooling zone of the continuous caster implements a more than 20 mm reduction through one pass so that the center of the slab can be efficiently deformed, which greatly improves the porosity and density of the slab center, thereby improves the performance of the plate.

In this paper, the process of 400 mm slab is directly rolled to 252 mm and the process of 400 mm slab is first subjected to single roll heavy reduction of 20 mm then rolled to 252 mm were calculated by the finite element numerical simulation, the equivalent strain and the cumulative deformation at the full slab thickness under the two processes are comparatively analyzed.

2 Model Establishment

The finite element software ABAQUS was used to establish the model of single-roll heavy reduction and the model of multi-pass rolling, as shown in Fig. 1. The models were discretized by the CPE4RT four-node plane strain thermal structure coupling unit. The heavy reduction roll and the mill rollers were all regarded as rigid material, they were discretized by the explicit rigid unit R2D2. The two models were divided into 1,600 units and 1,701 nodes. The temperature field of the slab by the heavy reduction was calculated by the solidification heat transfer module.

2.1 Model Construction

1) The slab is set as an isotropic elastic-plastic material model, and the upper and lower surface states of the slab are completely consistent during casting or rolling.
2) Compared with the large deformation of the slab during the casting or rolling process, the deformation of the casting roll or hot roll can be ignored, and the casting roll or hot roll is defined as a rigid body.
3) The Coulomb friction model is used in casting or rolling.
4) Since the length and width of the slab are much larger than the thickness, the rolling model can be defined as a plane strain model.

2.2 Contact and Boundary Conditions

1) Temperature boundary condition

   When simulating the casting or rolling process, the casting roll or hot roll is set to 20 °C, and the temperature field of the slab is distributed according to the actual temperature. Before the large reduction, the temperature field in the whole thickness direction has a temperature gradient difference of high center and low edge. The temperature field in the full thickness direction of the slab before rolling after large reduction and before direct rolling is set as a uniform temperature value without a temperature gradient in the full thickness direction.
2) Displacement and velocity boundary conditions

   During the casting process, the X and Y displacement degrees of freedom of the casting roller are set to zero, and only the rotation degree of freedom in the Z direction is set to zero. The rotation angular velocity of the casting roller is set to 0.024 rad/s;

during the rolling process, the X and Y direction displacement degrees of freedom of the hot roll are set to zero, only the Z direction rotation degree of freedom, and the rotation angular velocity of the hot roll is set to 2 rad/s; the displacement boundary of the hot roll between passes is the X-direction freedom constraint, and the Y-direction displacement is assigned according to the pass reduction. The specific value is shown in Sect. 2.3. The degree of freedom of the slab is not constrained, and only the initial velocity is given, that is, $V_0 = 0.083$ m/s, and the initial velocity is loaded as the predefined field of the coupling model.

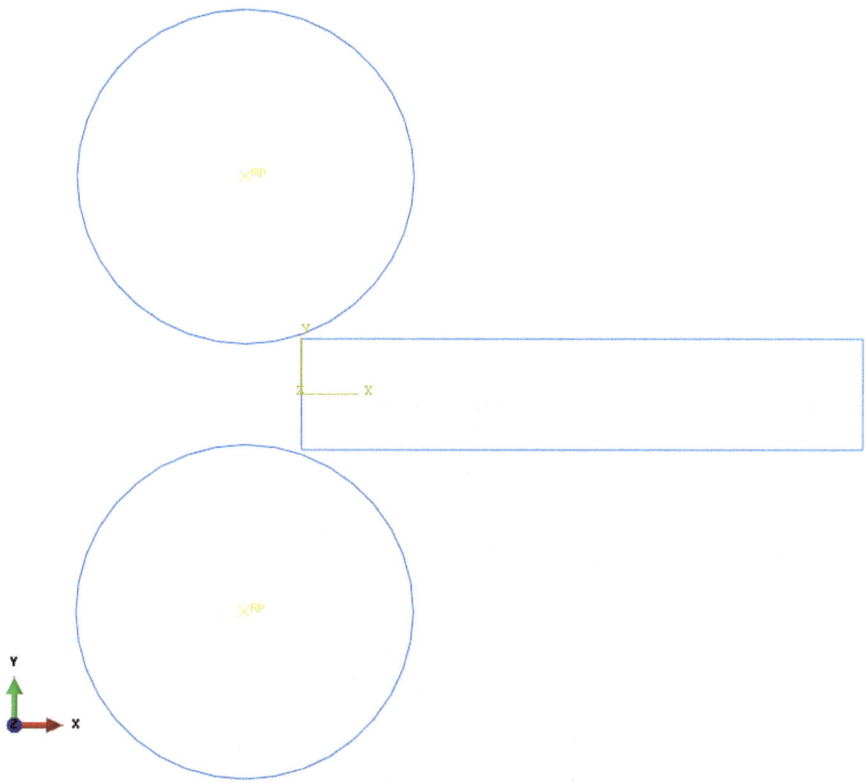

Fig. 1. A schematic of a two-dimensional model with high pressure.

2.3 Simulated Processes

1) Direct rolling

   Direct rolling is to reduce the 400 mm slab to 252 mm by the rolling mill.

2) Single roll heavy reduction and rolling

   The 400 mm slab casting online is pressed to 380 mm by the single roll heavy reduction, the reduction amount is 20 mm. Then the 380 mm plate is rolled at the temperature of 1,200 °C. During the rolling process, the plate is rolled from 380 mm to 252 mm, and the total rolling amount is 128 mm. Table 1 shows the two kinds of rolling specifications.

Table 1. The two kinds of rolling specifications.

Rolling pass	Direct rolling		Single roll heavy reduction and rolling	
	Plate thickness before rolling/mm	Reduction ratio/%	Plate thickness before rolling/mm	Reduction ratio/%
1	400	9.18	380	9.50
2	363.3	12.03	343.9	12.30
3	319.6	13.95	301.6	13.49
4	275	3.93	260.9	1.95
5	264.2	2.31	255.8	1.37
6	258.1	1.39	252.3	0.75
7	254.5	1.06	–	–

3 The Model of the Slab Temperature Field

3.1 Analysis of Temperature Field Model

The accurate temperature field is an important basis for simulating the deformation of the reduction process. Figure 2(a) shows the temperature in the full thickness of the 400 mm slab casting online before the heavy reduction. Both the upper and the lower surface temperatures are 860 °C, the core temperature is 1,450 °C, higher than the surface. Figure 2(b) shows the full thickness temperature of the 380 mm slab by 20 mm heavy reduction before rolling. Figure 2(c) shows the full thickness temperature of the 400 mm slab before direct rolling. Both of them are 1,200 °C.

Figure 3 compares the temperatures of the 400 mm slab before heavy reduction and the 400 mm slab before direct rolling. It shows that all the temperatures at the full thickness of the 400 mm slab before direct rolling are 1,200 °C. However, the temperature difference between the surface and the core of the slab before heavy reduction is about 600 °C, which gives the natural advantage to penetrating the reduction pressure into the core of the slab.

3.2 Model Verification

In this paper, the temperature of the slab at the outlet of the continuous casting machine is measured in order to indirectly verify the rationality and accuracy of the temperature field distribution of the slab before casting and rolling. The calculated temperature of the model and the measured temperature are shown in Fig. 4.

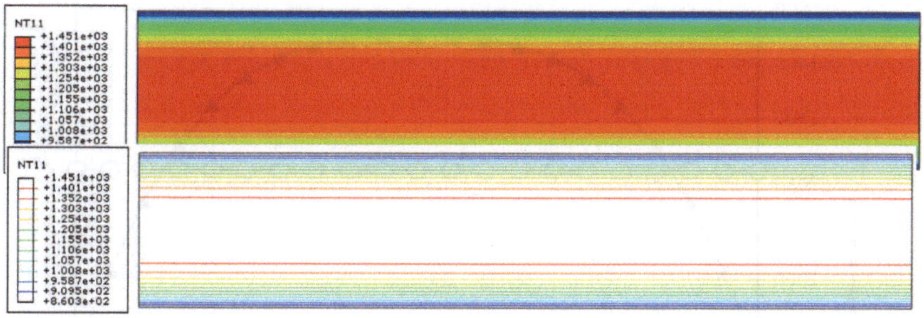

(a)

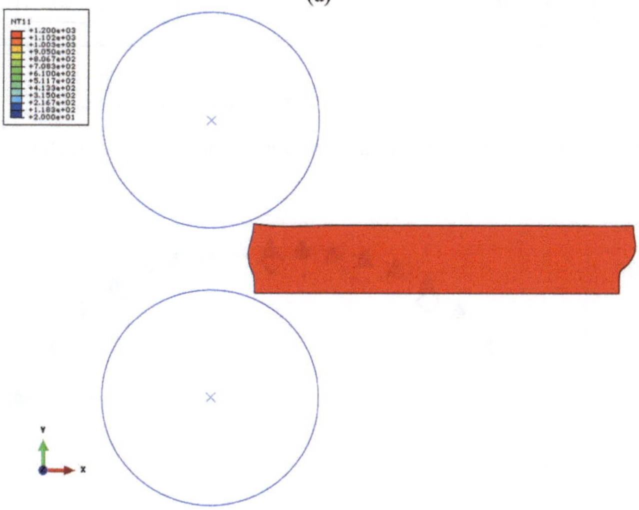

(b)

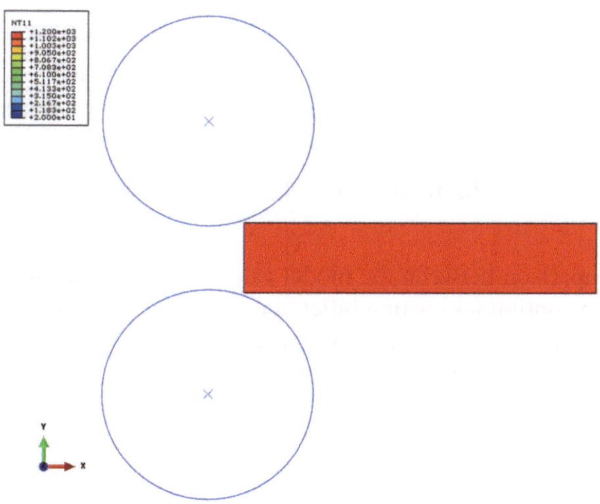

(c)

Fig. 2. Temperature field cloud map. (a) Before heavy reduction (b) 380 mm slab before rolling (c) 400 mm slab before rolling.

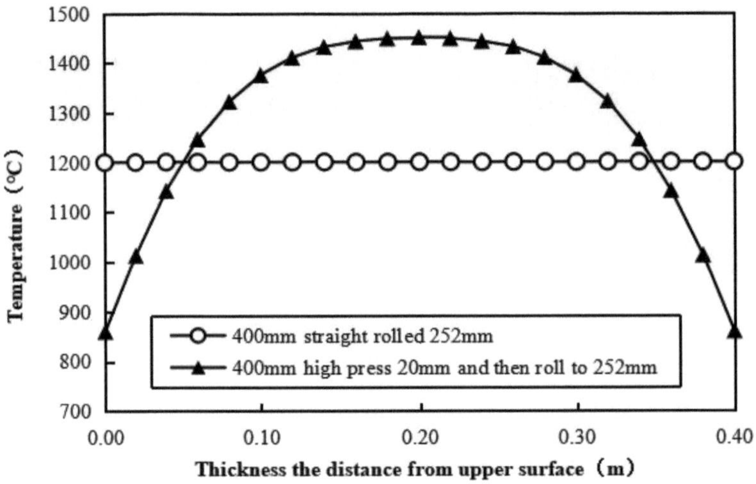

Fig. 3. The full thickness temperature distribution of the slab under different conditions.

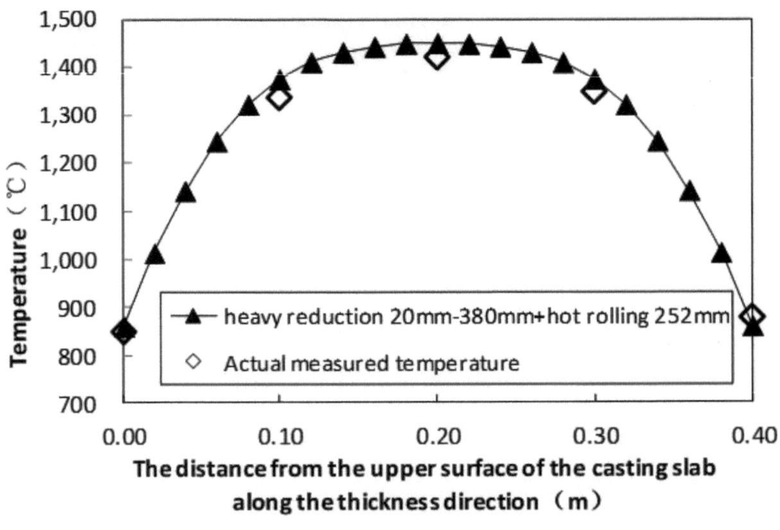

Fig. 4. Model calculation verification.

From the comparison between the model calculated temperature and the measured temperature of the continuous casting billet, it can be seen that the average temperature difference between the center and 1/4 of the casting billet is 20.9 °C, and the model error is within a reasonable range, which has high calculation accuracy.

4 Analysis of the Simulation Results

Figure 5 shows the cloud maps of the equivalent strain at each rolling pass of slab direct rolling from 400 mm to 252 mm. The equivalent strain near the surface of the plate is significantly higher than that in the central area, which is mainly because the compression ratio of the whole rolling process and the reduction ratio by each pass are both small.

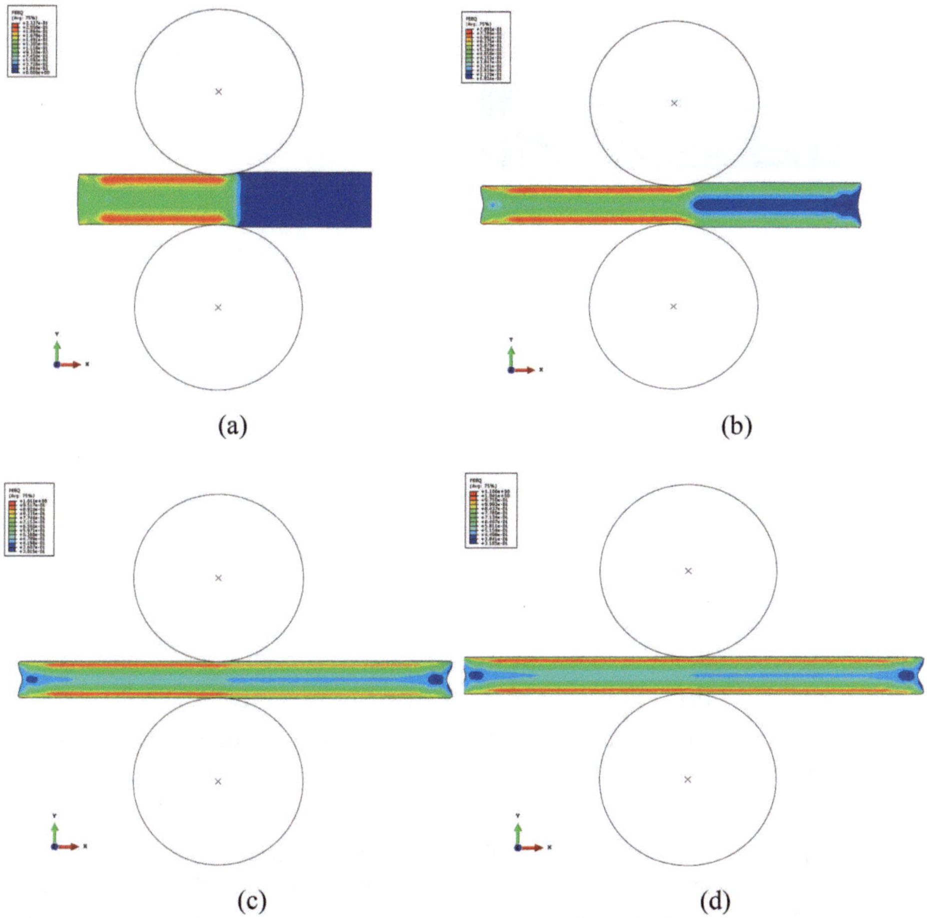

Fig. 5. The cloud maps of the equivalent strain at each rolling pass during the direct rolling. (a)–(d): The first rolling pass ~ the seventh rolling pass.

Figure 6 shows the equivalent strain cloud maps of the slab at the process of single roll heavy reduction and each pass of rolling. The equivalent strain near the core of the slab is obviously higher than that near the upper and the lower surface. This is because there is a temperature gradient of about 600 °C between the surface and the core of the online casting slab, which ensures that the reduction pressure fully penetrates from the surface to the core of the slab during the process of the single-point heavy reduction.

The large strain effect near the core of the slab has been accumulated into the subsequent rolling process, which can be clearly seen from Figs. 6(b)–(g).

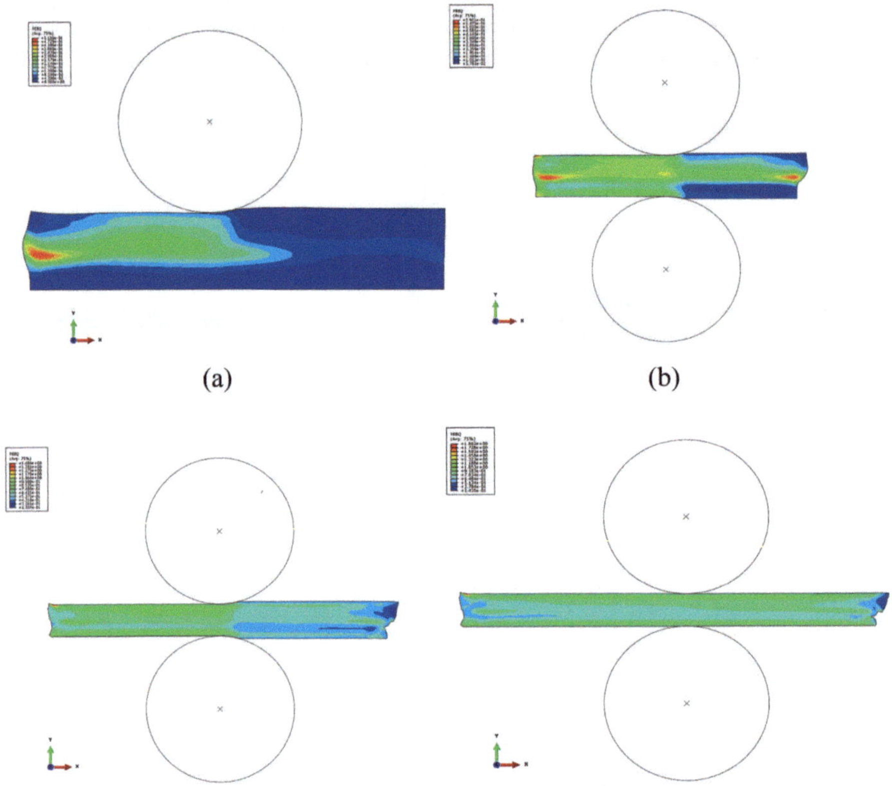

(a) (b)

(b) (d)

Fig. 6. The equivalent strain cloud maps of the slab at the process of single roll heavy reduction. (a): the process of heavy reduction (b) –(d): the first ~ the sixth rolling pass.

Figure 7 shows the calculation results of the equivalent strain at the full thickness of 252 mm plate. In the plate by heavy reduction and rolling, the equivalent strain is mainly concentrated in the central area, and has good uniformity along the whole thickness direction. But in the plate by the direct rolling, the equivalent strain decreases from the surface to the central area and is the lowest at the center of the plate. So the heavy reduction promotes the pressure penetration and the strain generation in the central area. Strain is the basis for obtaining good microstructure of materials.

Figure 8 shows the simulation results of the cumulative deformation of each node at the full thickness of the plates. The 400 mm slab is divided into 20 grids along the thickness direction, and the height of each grid is 20 mm. To produce the 252 mm plate by 400 mm slab, the total cumulative reduction is 148 mm. The cumulative deformation of each grid is an important index for measuring the reduction of permeability.

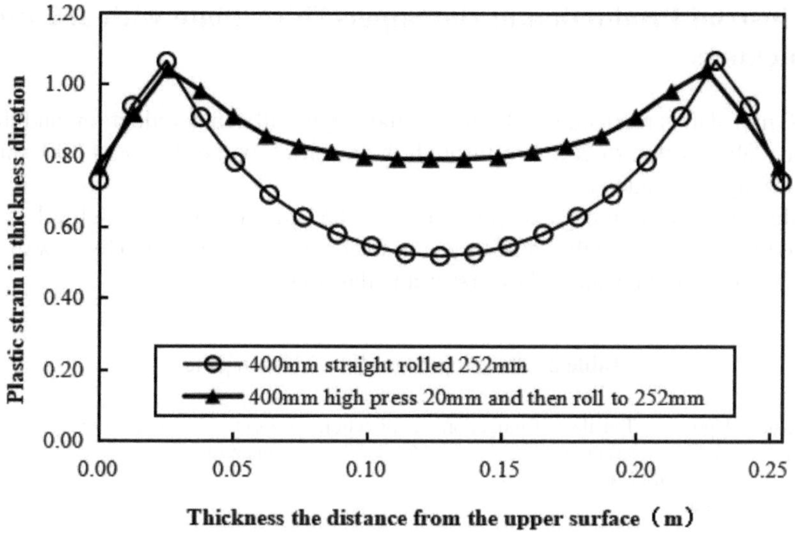

Fig. 7. The equivalent strain at the full thickness of 252 mm plate.

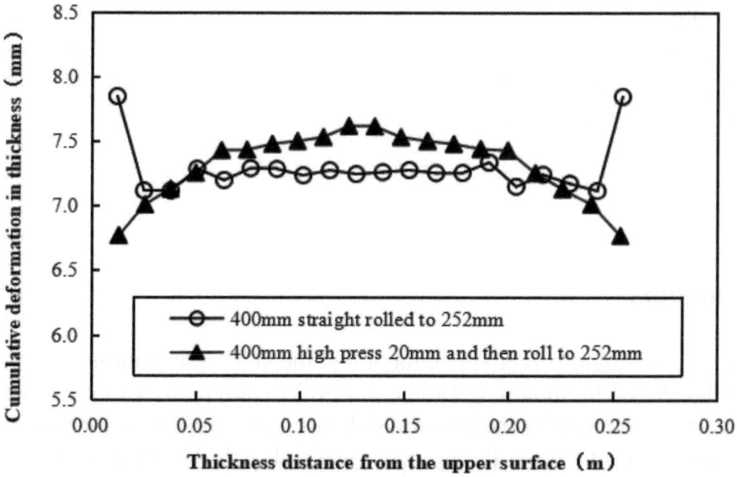

Fig. 8. The cumulative deformation along the thickness direction of the plate.

It can be seen that the accumulated deformation of the directly rolled plate is mainly concentrated in the area near the upper and lower surface, and is almost the same at other thickness positions. The accumulated deformation of the plate by single roll heavy reduction and rolling is mainly concentrated in the central area, and is smaller near the surface of the plate, which shows that the reduction deformation along the thickness direction by the heavy reduction has obvious accumulation and genetic effect, and also has a good effect on improving the performance and uniformity of the super-thick plate.

5 Industrial Production of the Super-Thick Plate with 252 mm Thickness

The 400 mm slab is reduced to 380 mm by the single roll heavy reduction, and then the 380 mm slab is rolled to 252 mm through the 6 rolling passes by the 4,300 mm plate mill in Shougang Jingtang.

Table 2 shows the measured properties of the 252 mm plate, which all meet the standard requirements. At the same time, the steel plate has been tested by flaw detection and meets the national standard of first-grade flaw detection.

Table 2. The properties of the 252 mm plate.

Performance	Upper yield strength. MPa	Tensile strength. MPa	Elongation. %	Z-direction. %	Impact temperature, °C	Impact energy, J			
						1	2	3	Average
Standard	≥275	430–590	≥22	≥25	0	27			
Upper surface	456	563	25	58					
Up to 1/4	308	463	35.5		0	331	339	343	338
Center	295	459	31.5		0	339	331	340	337
Down to 1/4	307	457	36.5		0	332	342	342	339
Lower surface	450	552	25						

6 Conclusions

The technology of single-roll heavy reduction at the end of the continuous slab solidification was developed in Shougang Jingtang. The 252 mm plate with good performance was successfully produced by the 400 mm slab through heavy reduction and rolling, The compression ratio during the rolling process was only 1.6. The main conclusions are as follows:

1) In the plate produced by the heavy reduction and rolling, the equivalent strain is mainly concentrated in the central area and almost equal along the whole thickness direction. But in the plate by direct rolling, the equivalent strain decreases from the surface to the central area and is lower than that in the plate through the heavy reduction.
2) The accumulated deformation of the directly rolled plate is mainly concentrated in the area near the upper and lower surface and is almost the same at other thickness positions. The accumulated deformation of the plate by single roll heavy reduction and rolling is mainly concentrated in the central area and is smaller near the surface of the plate.

3) The temperature difference between the surface and the core of the slab at the end of solidification is about 600 °C. The single roll heavy reduction to the slab makes the high-temperature metal in the core fully deform. This large strain effect near the core has been accumulated. Then the strain and cumulative deformation in the plate core are improved after rolling. So the penetration effect is much higher than that of direct rolling.

References

1. Hiraki, S., Yamanaka, A., Shirai, Y., et al.: Development of new continuous casting technology (PCCS) for very thick plates. Materia Japan **48**(1), 20 (2009)
2. Yin, C.-H., Won, Y.-M., Park, J.-K., et al.: Continuous Cast Slab and Method for Manufacturing the Same: US, 2010/0296960 (2010)
3. Yim, C.-H., Seo, J.-D.: Advanced steelmaking technologies for CO_2 emission reduction and slab quality improvement. In: The 5th International Congression the Science and Technology of Steelmaking (ICS 2012), Dresden, pp. 1–6 (2012)
4. Zhu, M.: Some considerations for the new generation of high-efficiency continuous casting technology development. Iron and Steel. **54**(8), 21–28 (2019)
5. Kang, Y., Tang, D.: Advanced Technology of Energy Saving and Emission Reduction During the Rolling Process. Metallurgical Industry Press, Beijing (2020)

Research and Optimization of Dynamic Performance of Grinding Wheel Trimmer

Menghui Cui[1,2], Gongxue Zhang[1(✉)], and Shaoqiang Cao[2]

[1] School of Mechanical and Electrical Engineering, Shaanxi University of Science and Technology, Xi' 710021, China
zhanggx@sust.edu.cn
[2] Qinchuan Machine Tool Co., Ltd., Baoji 721009, China

Abstract. SolidWorks is used to create a three-dimensional model of the grinding wheel dresser, and then the completed 3D model is imported into ANSYS Workbench software. Utilizing ANSYS Workbench, structural stiffness and modal analysis of the dresser are conducted. A series of analyses are performed, including defining boundary conditions, specifying material physical properties, setting constraint loads, meshing, and solving. The resulting stress, strain, and modal vibration patterns are analyzed to identify the weak areas of the dresser. Structural optimization is then applied to regions with insufficient stiffness. By comparing the deformation and stress before and after the structural optimization, the rationality of the optimization is demonstrated, ensuring the performance of the dresser and enhancing its stiffness.

Keywords: grinding wheel dresser · structural stiffness · modal analysis · structural optimization

1 Introduction

Currently, the development trends of gear grinding machines are mainly reflected in high-speed operation, high precision, intelligence, multifunctionality, and environmental sustainability. As a core component of gear grinding machines, the grinding wheel dresser directly participates in the grinding process, thus imposing higher requirements on its structural design and performance. Most existing dressers use a diamond stylus or a single roller for dressing. During the grinding process, the diamond stylus or single roller bears the grinding force, and prolonged use may lead to a decline in stiffness. The stiffness of the dresser significantly impacts the machining accuracy and operational stability of the machine tool. Insufficient stiffness can cause significant deformation of the dresser under force, thereby affecting the machining accuracy of the machine tool, leading to workpiece dimensional deviations or substandard surface quality. This, in turn, affects machining stability and surface finish, and may result in fatigue damage to the dresser under long-term force, shortening its service life and impairing its dynamic response characteristics, thereby reducing its performance under high-frequency or high-speed conditions.

S. Halgamuge et al. (Eds.): ICACMVE 2025, LNME, pp. 32–44, 2026.
https://doi.org/10.1007/978-981-95-6911-3_4

Therefore, conducting a finite element analysis on the structural stiffness and modal characteristics of the dresser is essential. Based on the analysis results, the overall performance of the dresser can be improved. Additionally, by employing topology optimization technology, excess material from the front-end housing can be removed while ensuring structural strength, achieving lightweight design and enhancing structural stiffness and dynamic performance. This reduces weight, lowers inertial forces, minimizes vibration, increases natural frequency, avoids resonance phenomena, optimizes stress distribution, and extends service life, providing technical support for the development of high-performance, low-cost dressers.

2 Structural Stiffness Analysis of Grinding Wheel Dresser

2.1 Establishment of Finite Element Model

Based on the existing two-dimensional drawings, a three-dimensional model of the grinding wheel dresser was constructed using SolidWorks. The main components of the dresser include the rear housing, sliding plate, front housing, motor mounting seat, motor, and roller, among others. The simplified model was then converted into a compatible format and imported into ANSYS Workbench software for subsequent analysis. During the meshing process, attributes such as element type and material properties were assigned to each component of the dresser. Since the structure of the dresser is relatively complex, with numerous features such as bolts, chamfers, and fillets that could affect mesh quality, these structures were simplified during the finite element model establishment to enhance the mesh quality. Generally, selecting an appropriate mesh density is crucial - ensuring computational accuracy while avoiding excessively prolonged calculation times due to overly dense meshing. Therefore, tetrahedral elements were chosen for meshing the dresser structure. The overall element size was set to 5 mm. The meshing result is shown in Fig. 1.

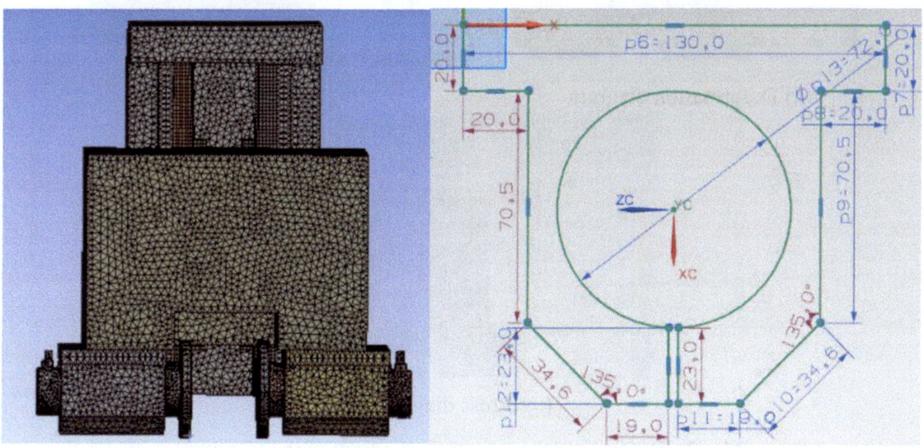

Fig. 1. Finite Element Model and Key Component Dimensions.

2.2 Definition of Materials and Boundary Conditions for the Dresser Model

To reasonably simplify the mechanical model and facilitate calculations, all components of the dresser in the structural design are made of structural steel. The dresser structure is fixed through the rear housing, and in the analysis process, the boundary constraints can be simplified to a fully fixed condition at the bottom of the housing. During operation, the dresser is primarily subjected to its own weight and the force exerted by the grinding wheel on the roller. The dressing operation is achieved by the motor driving the roller to interactively engage with the grinding wheel [1]. For the dresser, the reaction force exerted by the grinding wheel on the roller is its primary external load. This external force is analogous to the cutting force during the grinding of a workpiece by the grinding wheel, and its magnitude depends on factors such as the dressing amount, the material of the grinding wheel, and its grinding characteristics.

2.3 Calculate the Stress and Strain of the Conditioner

After defining the material, setting the boundary conditions, performing mesh division, and applying loads for the dresser, the post-processing phase can be initiated. During the finite element analysis process, once the solver's calculations are completed, the ANSYS post-processing module is utilized to extract and analyze the results. Through this module, the overall and localized deformation, stress, and strain distributions of the dresser can be obtained. Specific results are illustrated in the accompanying figures.

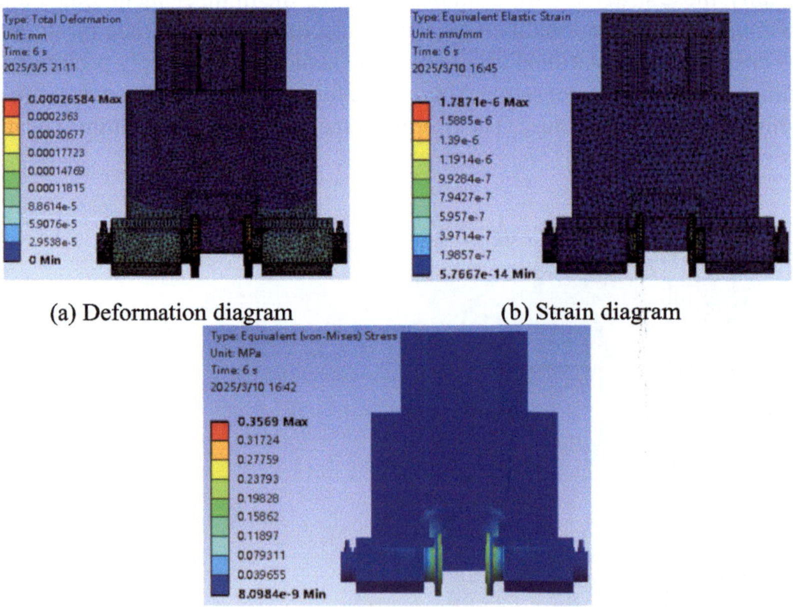

(a) Deformation diagram (b) Strain diagram

(c) Stress diagram

Fig. 2. Deformation stress-strain diagram of the trimmer.

As shown in Fig. 2(a), the areas with the greatest deformation are primarily concentrated on the two rollers and the motor mounting seat. The likely causes for this are the reaction force exerted by the grinding wheel on the rollers or excessive deformation due to the large diameter of the rollers. From Figs. 2(b) and 2(c), it can be observed that the stress is mainly concentrated at the bottom of the sliding plate, which is likely caused by the self-weight of the dresser. Therefore, structural optimization is required for the rollers, motor shaft, and the self-weight of the dresser.

3 Modal Analysis of the Trimmer

3.1 Selection of ANSYS Modal Analysis Methods

The dresser is an elastic structure composed of multiple components, with a continuous mass distribution and theoretically possessing infinite degrees of freedom. In a multi-degree-of-freedom system, the vibration of the system is the superposition of vibrations at multiple natural frequencies, indicating that the vibration of such a system is not characterized by a single frequency but rather by a combination of multiple frequencies. Therefore, in practical analysis, it is only necessary to focus on the low-order natural frequencies and their corresponding mode shapes that significantly influence the dynamic characteristics of the structure.

The modal analysis methods in ANSYS [2] primarily include the Subspace Method, Block Lanczos Method, Reduced Method, Unsymmetric Method, Damped Method, Power Dynamics Method, and QR Damped Method. Yin Hong et al. [3] enhanced the stratification efficiency of Variational Mode Decomposition (VMD) by integrating it with the Stochastic Subspace Identification (SSI) method for modal parameter identification, while also applying Singular Value Decomposition (SVD) technology for noise reduction, significantly improving the accuracy of modal identification. Xia Zunping [4] proposed using a spectral kurtosis identification tool to distinguish between the modal and harmonic frequency components of rotating machinery structures, achieving accurate elimination of harmonic frequency components by removing them from measured signals. Qin Chao [5] addressed the periodic harmonic frequency characteristics generated during cutting processes and introduced a cepstrum editing technique to remove periodic harmonic components, applying this method to the processing of milling machine cutting vibration data.

In modal analysis, the Block Lanczos method is efficient, stable, and flexible in modal analysis, making it particularly suitable for handling large-scale systems and extracting multiple modes. Therefore, the Block Lanczos method is typically used to solve the eigenvalue problem of the system. The natural frequencies of the dresser reflect its free vibration characteristics. To ensure stable system operation, it is essential to maintain a certain separation between the dresser's natural frequencies and its operating frequencies to avoid resonance. Natural frequencies are inherent properties of the system, primarily determined by the material properties, structural characteristics, boundary conditions, and stiffness of the components. Higher elastic modulus and greater structural stiffness generally result in higher natural frequencies, whereas higher density may lead to lower natural frequencies. The constraint conditions of the structure (e.g., fixed, free, or elastic support) influence its vibration characteristics, thereby affecting the natural frequencies, and are unrelated to external forces or loads applied during operation. Therefore, in modal analysis, it is unnecessary to consider the influence of external loads, and the finite element model established during static analysis can be directly used.

3.2 Analysis of Solution Results

Before conducting the modal analysis of the dresser, it is necessary to preset the number of modal orders and the convergence percentage. Typically, there is no need to calculate all the natural frequencies and mode shapes of the system; extracting the first few modes is sufficient to meet the analysis requirements. This is because low-order modes have a more significant impact on the vibration system, and the lower the order, the greater the degree of influence. By adjusting the deformation display scale and animation frame rate, different output forms can be selected to optimize the visualization of the results, thereby facilitating a better observation and analysis of the outcomes. The deformation and natural frequencies correspond to each mode and can be obtained from the analysis results. The first six natural frequencies from the modal analysis are: 388.44 Hz, 545.42 Hz, 713.34 Hz, 1221 Hz, 1424.5 Hz, and 1540.6 Hz.

To minimize the interference of external excitation on the dynamic characteristics of the dresser, the excitation frequency of the external load should be significantly lower than its first natural frequency. The mode shapes corresponding to the first six natural frequencies are illustrated in the figures. Figure 3(a) shows that the first mode is primarily characterized by the dresser swinging up and down around the Y-axis. Figure 3(b) indicates that the second mode mainly involves left and right swinging around the Z-axis. Figure 3(c) reveals that the third mode is dominated by rotation around the X-axis. In Fig. 3(d), the fourth mode again exhibits the characteristic of swinging up and down around the Y-axis. Figure 3(e) displays that the fifth mode primarily involves left and right swinging around the Z-axis. Figure 3(f) shows that the sixth mode is mainly characterized by rotation around the X-axis. These deformation directions align with the Z-axis and X-axis directions of the dresser feed motion, indicating deficiencies in the dynamic performance of the dresser. To address this issue, two approaches can be considered: first, optimizing the structural design to enhance the stiffness of the dresser; second, selecting a motor model with a lower rated speed while meeting the power requirements.

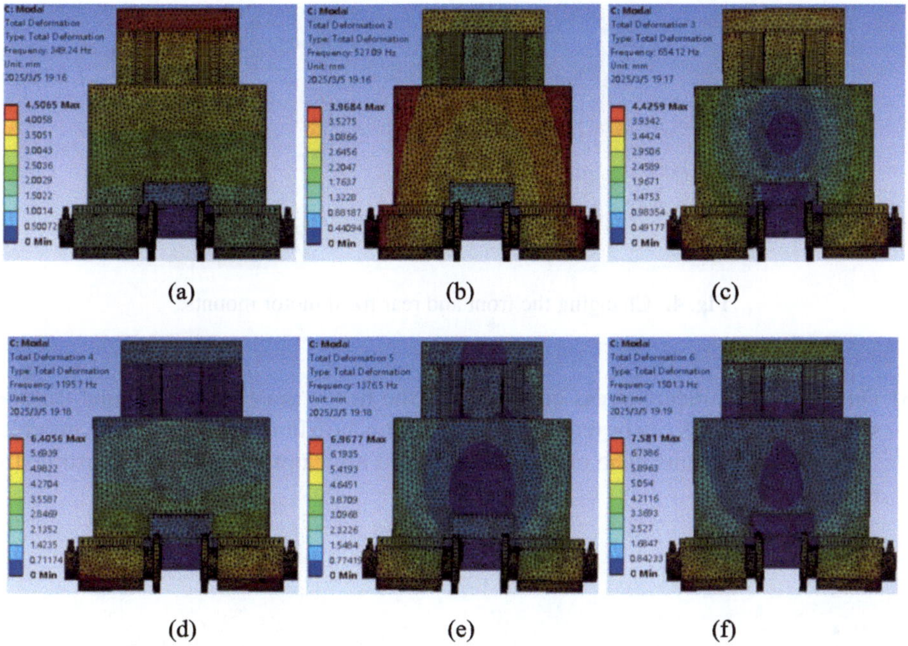

Fig. 3. Mode shapes of the first six modes.

4 Analysis and Optimization of Trimmer Structure

Based on a comprehensive consideration of the static structural characteristics, modal analysis results, and design requirements of the trimmer, and to ensure the effective realization of design functions and the stable installation of components, it is essential to strengthen the stiffness of weak areas while maintaining the overall structural framework. By analyzing the relationship between roller diameter and stiffness, the following structural optimization proposals are suggested:

(1) The diameter of the roller should not be excessively large, typically ranging between 60 and 150 mm [6]. The original structure in this context employs a roller diameter of 130 mm. Where the structure permits, reducing the roller diameter can decrease roller deformation, thereby indirectly enhancing roller stiffness.

(2) Adding support plates to the two fixed motor mounts can reduce the vibration of the motor spindle caused by the roller during the machining process, thereby minimizing deformation due to vibration. The structural optimization is illustrated in Fig. 4.

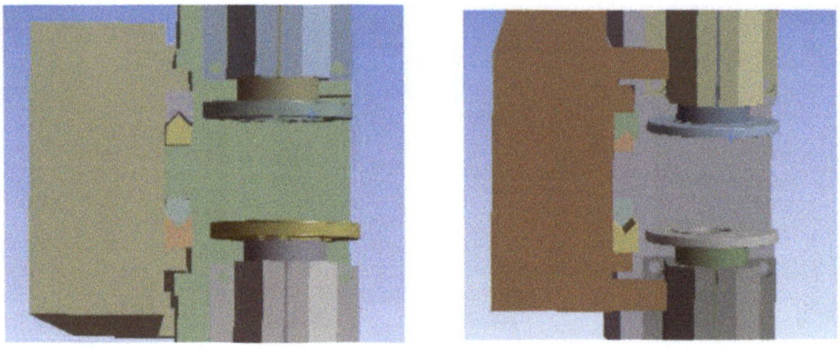

Fig. 4. Changing the front and rear fixed motor mounts.

(3) Due to the significant stress generated by the heavy mass of the original trimmer structure, the stress can be reduced by decreasing the trimmer's own weight, provided the structure permits. The changes before and after optimization are illustrated in Fig. 5.

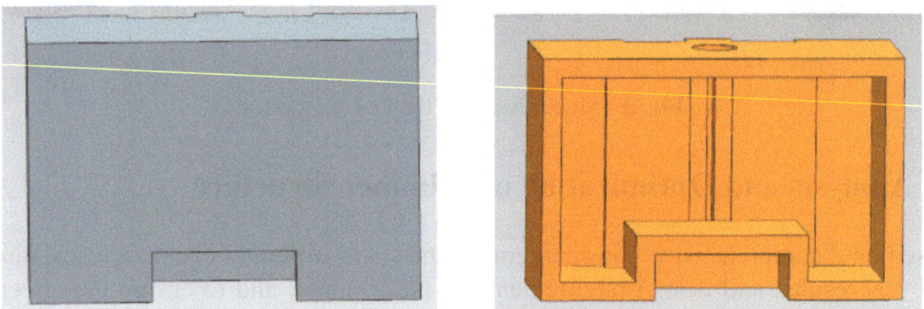

Fig. 5. Changing the shell of the front and rear trimmers.

5 Comparison of Deformation Before and After Optimization to Verify the Rationality of the Optimization

5.1 Comparison of Deformation Between the Original Structure and the Optimized Structure

Under a constant load, the deformation variation of the trimmer was observed by reducing the roller diameter before structural changes, as shown in Fig. 6. When the roller diameter was 130 mm, the maximum deformation of the trimmer was 0.00026584 mm, as illustrated in Fig. 6(a); When the roller diameter was 120 mm, the maximum deformation was 0.00018149 mm, as shown in Fig. 6(b); When the roller diameter was 110 mm, the maximum deformation was 0.00015255 mm, as depicted in Fig. 6(c); When the roller diameter was 100 mm, the maximum deformation was 0.0001304 mm, as presented in Fig. 6(d).

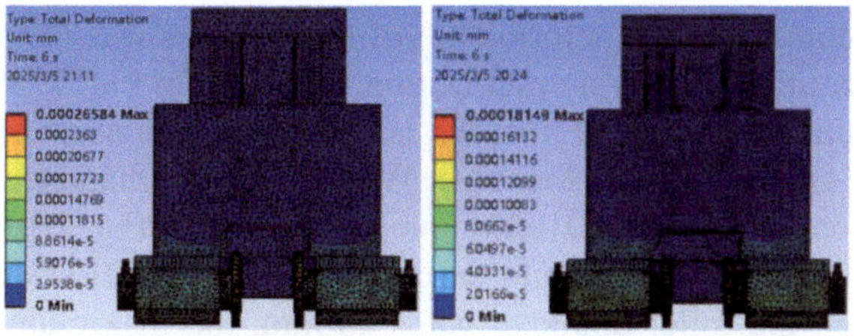

(a) 130 mm (b) 120 mm

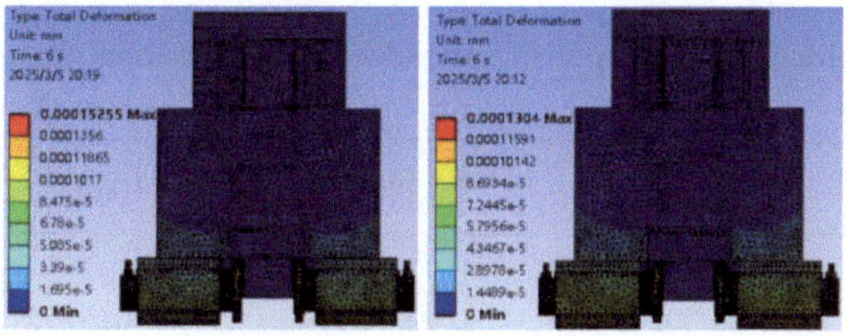

(c) 110 mm (d) 100 mm

Fig. 6. Deformation diagram of the original structure modifier.

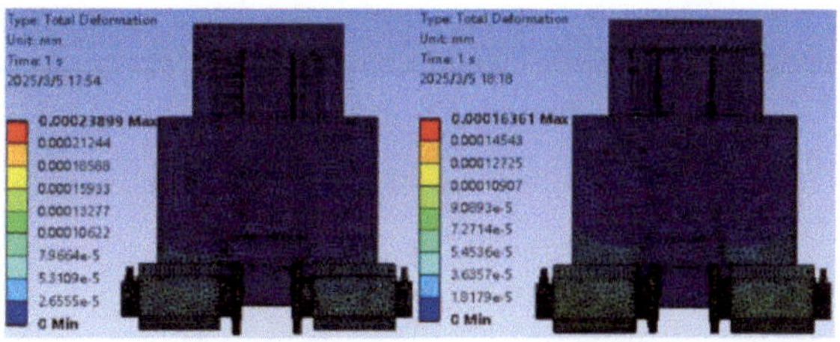

(a)130 mm (b)120 mm

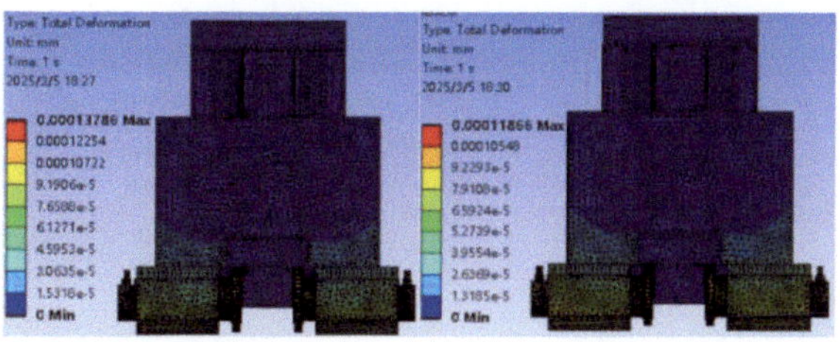

(c)110 mm (d)100 mm

Fig. 7. Deformation diagram of the modifier after structural optimization.

Under a constant load, the deformation of the trimmer was observed by reducing the roller diameter after structural optimization, as shown in Fig. 7. After optimization, when the roller diameter was 130 mm, the maximum deformation of the trimmer was 0.00023899 mm, as illustrated in Fig. 7(a); When the roller diameter was 120 mm, the maximum deformation was 0.00016361 mm, as shown in Fig. 7(b); When the roller diameter was 110 mm, the maximum deformation was 0.00013786 mm, as depicted in Fig. 7(c); When the roller diameter was 100 mm, the maximum deformation was 0.00011866 mm, as presented in Fig. 7(d).

Table 1. Deformation before and after structural optimization.

	130mm	120mm	110mm	100mm
Deformation before structural change	0.00026584	0.00018149	0.00015255	0.0001304
Deformation after structural changes	0.00023899	0.00016361	0.00013786	0.00011866
Deformation reduction	10%	9.8%	9.6%	9%

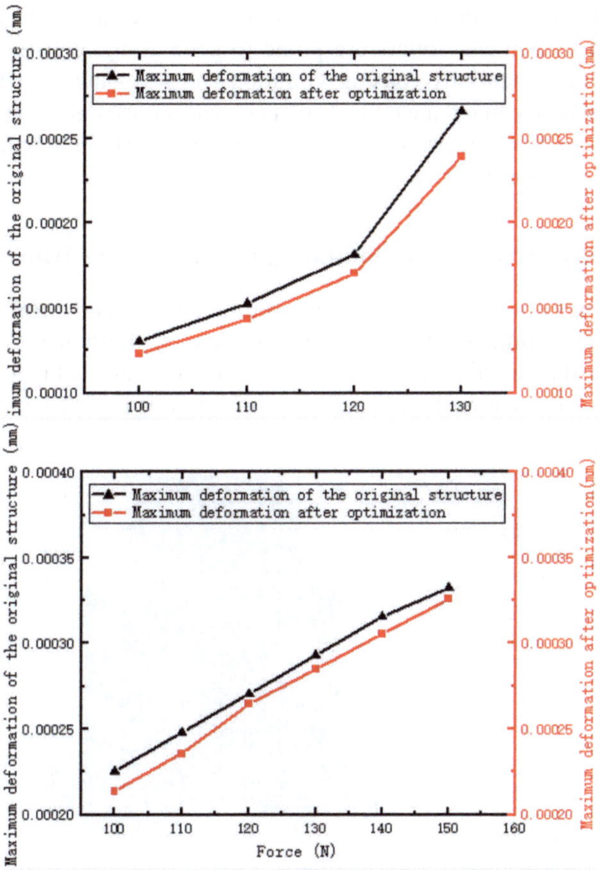

Fig. 8. Comparison of deformation before and after optimization.

The original structure is compared with the optimized results, as shown in Table 1. The deformation of the optimized roller with a diameter of 130 mm is reduced by 10% compared to the original 130 mm structure. Similarly, the deformation of the optimized roller with a diameter of 120 mm is reduced by 9.8%, the optimized 110 mm roller by 9.6%, and the optimized 100 mm roller by 9% compared to their respective original structures. Comparing the deformation trends under different loads with the same roller diameter, both before and after optimization, the deformation increases as the load increases. This is because the increase in load causes the roller to bear greater stress, resulting in larger elastic deformation. However, the deformation after optimization is reduced compared to that before optimization. It can be observed that the optimized roller diameter and the addition of a support plate to the fixed motor base have a significant impact on the overall deformation of the dresser, indirectly affecting its overall stiffness. As shown in Fig. 8, the influence mechanism of roller diameter and the addition of supports to the motor mounting base on the dresser is as follows: the larger the roller diameter, the lower the stiffness and the greater the moment of inertia. This may lead to a reduction in the response speed of the dresser and a decline in dynamic performance. Large-diameter rollers are prone to generating centrifugal force during high-speed rotation, potentially causing vibration and resonance issues. Adding supports to the motor base can enhance local stiffness and reduce deformation. Comparing the optimized design with the original, the reduction in deformation results in increased stiffness values and improved natural frequencies.

5.2 Comparison of Stress After Optimizing the Shell of the Trimmer

The maximum stress of the housing before optimization was 0.3569 MPa, as shown in Fig. 9(a); After optimization, the maximum stress of the housing was reduced to 0.32973 MPa, as illustrated in Fig. 9(b). The optimization resulted in a stress reduction of 7.6%. These results demonstrate that the optimization is justified and effective.

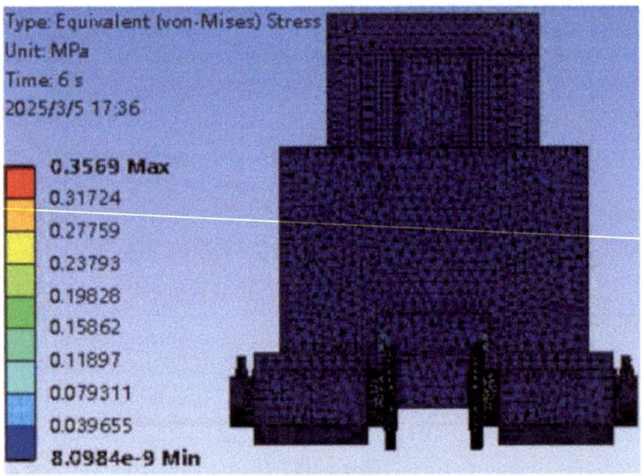

(a) Stress diagram before change

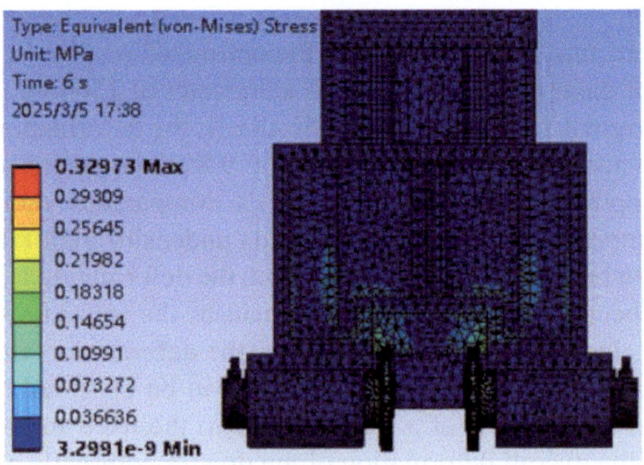

(b) Stress diagram after change

Fig. 9. Stress change diagram.

6 Conclusion

Using ANSYS software, a finite element simulation analysis was conducted on the structural stiffness and modal characteristics of the CNC forming wheel dresser. The results indicate that the roller diameter has a significant impact on the overall stiffness of the dresser, and its deformation directly affects the forming accuracy of the silicon carbide roller dressing grinding wheel. Therefore, to reduce deformation, it is essential to reasonably select the roller diameter and add support structures to the motor mount. To enhance the dynamic performance of the dresser, the following two measures can be adopted: first, improving the structural design to increase overall stiffness, and second, selecting a motor with a lower rated speed while ensuring power requirements are met. The structural calculations and optimization analysis demonstrate that reducing the roller diameter, decreasing the self-weight of the dresser, and reinforcing the motor mount support can effectively enhance the dresser's performance. This optimization method lays a solid technical foundation for developing dressers that combine high performance and cost-effectiveness. For future challenges, using high-performance materials in critical areas and economical materials (such as aluminum alloy) in non-critical areas can balance manufacturing costs. Installing sensors (e.g., accelerometers, temperature sensors) in key parts of the dresser enables real-time monitoring of parameters such as vibration, deformation, and temperature. Additionally, using equipment like surface roughness testers and profilometers to measure the surface quality and geometric accuracy of the dressed grinding wheel allows for the evaluation of the dresser's performance.

References

1. Zhu, X.: Optimization of CNC forming grinding wheel dresser. Precis. Manuf. Autom. (02), 6–8 (2012)
2. Long, K., Li, Z.: Mastering ANSYS 2022 Finite Element Analysis from Beginner to Advanced. Posts & Telecom Press (2022). 12.331
3. Yin, H., Dong, K., Peng, Z.: Structural modal parameter identification based on VMD-SSI. J. Vibr. Shock **39**(10), 81–91 (2020)
4. Xia, Z., Wang, T.: Harmonic mode detection method based on spectral kurtosis. Eng. Mech. **30**(12), 255–258+266 (2013)
5. Qin, C.: Research on Modal Parameter Identification Methods of CNC Machine Tools under Periodic Cutting Excitation. Huazhong University of Science and Technology (2018)
6. Li, J., Zhang, F.: Research Status of Forming Grinding Wheel Dressing Technology. Mech. Eng. (10), 19–22 (2006)

Research on the Operating Principle and Design Analysis of Mud Pump

Fangceng Wang[1(✉)] and Junpeng Cui[2]

[1] School of Mechanical Engineering, Shandong Huayu Institute of Technology, Dezhou, Shandong, China
502581253@qq.com

[2] Shandong Dezhou Mobile Company, Dezhou, Shandong, China

Abstract. The mud pump is the heart of geological drilling equipment and also one of the key equipment in operations such as cementing, fracturing, and acidification. It plays a role in various industries such as petrochemicals, coal gasification engineering, power plants, and mining. This paper first introduces the working principle of the mud pump, draws the overall 3D graphics of the mud pump, and separately draws the 3D graphics of each component, introducing the parts contained in each component. Finally, the checks that need to be done before the operation of the mud pump were introduced.

Keywords: Mud pump · Structure · Design analysis · Hydraulic end

1 Introduction

A mud pump is a reciprocating piston pump that converts the rotational motion of the power source into the reciprocating linear motion of the crosshead and piston through a crank connecting rod mechanism, compressing low-pressure mud into high-pressure mud. The working process of a mud pump is divided into two steps: the suction process and the discharge process. They are separated at the crosshead, with one end being the hydraulic end of the pump and the other end being the power end. When fluid flows from the hydraulic end to the power end, negative pressure will be generated in the pump body, and the liquid will enter the pump cylinder through the suction valve. When fluid flows from the power end to the hydraulic end, it is compressed, generating strong pressure that is discharged from the cylinder through the discharge valve [1]. The working principle is shown in Fig. 1.

S. Halgamuge et al. (Eds.): ICACMVE 2025, LNME, pp. 45–51, 2026.
https://doi.org/10.1007/978-981-95-6911-3_5

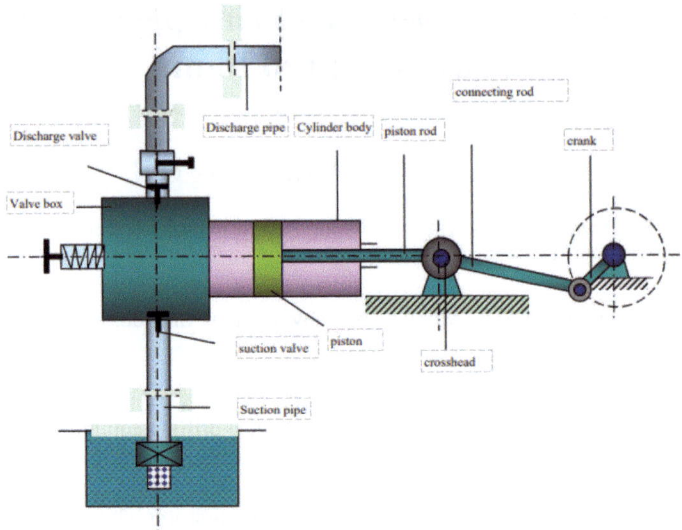

Fig. 1. Schematic diagram of the working principle.

2 Basic Structure of a Mud Pump

The main parameters of the design include: maximum cylinder diameter of 170 mm, rated power of 735 kW, stroke length of 254mm, rated stroke of 140SPM, and valve chamber API16 #.

The power end is used to transmit power and convert motion modes and speeds. Provide appropriate power for the hydraulic end. The hydraulic end is used to convert mechanical energy into liquid internal energy for transporting mud. The power end includes: base, frame, small gear shaft assembly, crankshaft connecting rod device, crosshead assembly, lubrication assembly, spray pump assembly. The hydraulic end includes: valve box, suction valve, discharge valve, cylinder head, valve cover, cylinder liner, piston, piston rod, suction pipe, suction air bag (to reduce water hammer phenomenon), discharge pipe, discharge filter screen assembly, discharge air bag, and safety valve [2]. The three-dimensional structure is shown in Fig. 2.

The main power of the mud pump relies on the crank connecting rod device, which converts the rotational motion of the crank into the reciprocating linear motion of the crosshead and piston, thereby converting low-pressure mud into high-pressure mud. The three-dimensional diagram of the crank connecting rod device is shown in Fig. 3. The crank connecting rod device mainly includes the main bearing end cap, main bearing sleeve, sealing gasket, inner retaining ring, outer retaining ring, connecting rod, locating ring, hollow crankshaft, large gear ring, eccentric wheel bearing, etc. [3].

The small gear assembly includes key, small gear shaft, wear-resistant sleeve, end cover, bearing sleeve, retaining ring, sealing gasket, disc pump flange and other main components, as shown in Fig. 4.

The hydraulic end assembly is shown in Fig. 5, including the valve box assembly, cylinder head flange, cylinder head, cylinder head plug, discharge pipeline, valve cover,

Fig. 2. Three-dimensional diagram of a mud pump.

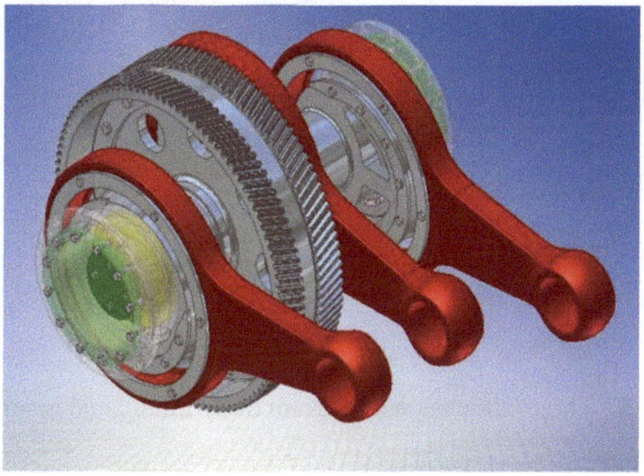

Fig. 3. Three-dimensional diagram of crank connecting rod mechanism.

valve stem guide, spring guide, valve assembly, piston sealing ring, spray pipe assembly, cylinder liner, suction pipeline and other main components [4].

The valve box assembly is shown in Fig. 6. Remove the valve cover, cylinder head, cylinder head plug, and locating plate, extract the insert plate assembly, and take out the valve stem guide (lower), valve spring, and valve body (including valve rubber and valve nut, etc.). Use a randomly configured hydraulic valve puller tool to pull out the upper discharge valve seat and lower suction valve seat, respectively. Clean the mating surfaces of the upper and lower valve seat holes (using high-quality cleaning agents), and confirm that each valve seat hole is very clean and dry (free of dirt, grease, rust inhibitors, etc.). Use diamond sandpaper to remove all burrs on the surface, and the sandpaper should circulate along the valve seat hole. Thoroughly clean and dry the valve seat, install the suction and discharge valve seats into the valve chamber hole, install the valve body, and

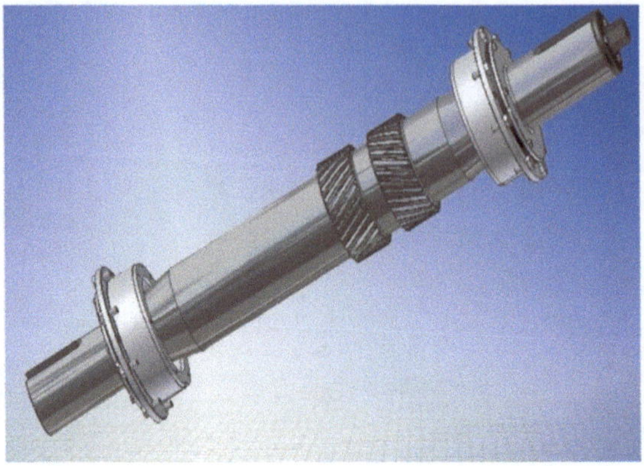

Fig. 4. Three-dimensional diagram of the small gear assembly.

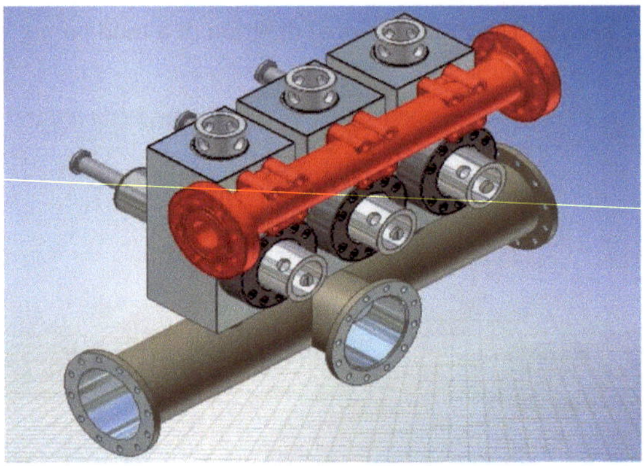

Fig. 5. Three-dimensional diagram of hydraulic end assembly.

use a copper rod hammer to strike the upper part of the guide rod to make the valve seat in place. Install the valve spring and other components. As a vulnerable component, the cylinder liner is made of ceramic material.

The air bag, as shown in Fig. 7, can effectively reduce pressure fluctuations in the discharge system, thereby obtaining a more uniform liquid flow. The airbag can only use compressed nitrogen or compressed air, and cannot use flammable or explosive gases such as oxygen and hydrogen. When maintaining the airbag, the air pressure inside the airbag must be zero. At this time, the pressure of the pump must also be zero. It cannot be judged solely by observing the reading of the pressure gauge. The pressure cap of the exhaust valve must be opened to release the pressure. When the residual pressure is small, the pressure gauge cannot display it, but this low pressure may also lead to accidents [5].

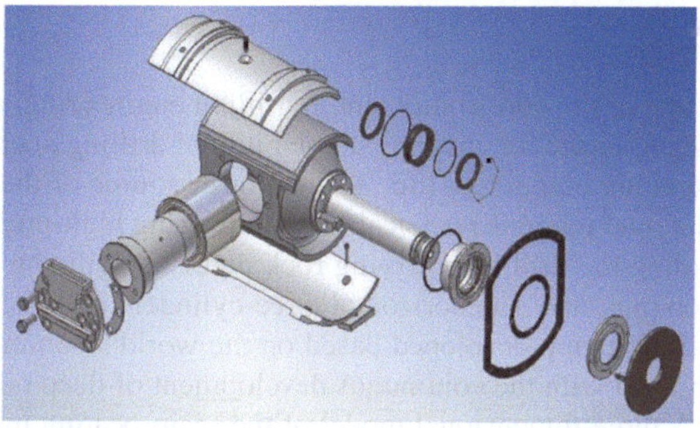

Fig. 6. Three-dimensional diagram of valve box assembly.

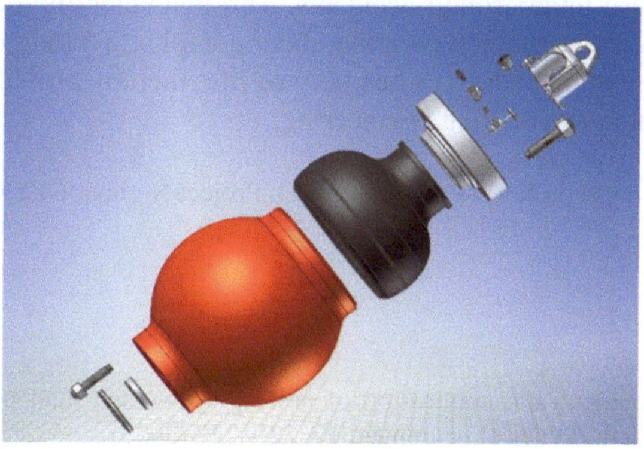

Fig. 7. Three-dimensional diagram of the airbag assembly.

3 Operational Checks

Check if there is any abnormal noise from the piston rod and intermediate rod clamp. Check all nuts on the valve box, as well as the valve cover and cylinder head, for looseness. Check if there are any leaks at the high-pressure seals and if there are any piercing sounds at the pump valves and cylinder liners. Pay attention to changes in pump pressure and handle any abnormal situations properly. Pay attention to whether the water supply of the sprinkler pump is normal. Pay attention to whether the temperature of each bearing part, crosshead guide plate, and other parts with relative motion is too high or abnormal. The general oil temperature should not exceed 80 °C. Is the pressure gauge reading of the power end lubrication circuit at the rear of the rack normal. It is strictly prohibited to use cylinder liners and pistons with two different diameters in a three-cylinder pump. The basic elements for the normal operation of a drilling pump are: moderate power, moderate pump pressure, medium speed operation, no fluctuation in pump pressure, normal temperature, and normal sound [6].

4 Conclusion

The mud pump is one of the core equipment in drilling engineering, and its reasonable design is of great significance for ensuring the stability of drilling operations, improving efficiency, and extending equipment life. As the power source of the drilling fluid circulation system, the mud pump is responsible for delivering high-pressure drilling fluid to the drill bit, cooling the drill bit, carrying rock debris, stabilizing the wellbore, and balancing formation pressure. The horizontal three-cylinder single-acting reciprocating mud pump is a drilling pump developed based on the world's commonly used drilling pump design, combined with the continuous development of deep oil and gas resource drilling in China. A three-dimensional model of the reciprocating mud pump is established to evaluate its design rationality. If the design is not reasonable, it may lead to a decrease in pump efficiency, an increase in energy consumption, intensified component wear, and even cause underground accidents. Through scientific design, mud pumps can maintain efficient and stable operation under complex working conditions, significantly reducing downtime and extending service life, thereby providing guarantees for the safety and economy of drilling engineering.

Acknowledgments. Dezhou R&D Program Project, Project Number DZSKJ202426.

References

1. Gong, C.: Development and application of new spray welded mud pump cylinder liners. Petroleum Petrochem. Mater. Procurement **17**, 225–227 (2023)
2. Ji, W., Zhen, F., Xu, F.: Reciprocating sealing with V-shaped sealing ring performance analysis. Off road World **17**(15), 203 (2022)
3. Ma, L., Zhang, W., Kang, X.: The development of petroleum drilling technology and high power mud pumps. Petroleum Mining Mach. **29**(3), 15–16 (2000)
4. Wang, D., Sun, J., Zhang, J., et al.: Confucius' oil drilling equipment discussion on the current status and development direction of backup technology. Petroleum Mach. **49**(1), 47–52 (2021)
5. Liang, Q., Wang, T., Zhang, Y., et al: Hydraulic performance analysis of 3200 HP five-cylinder drilling mud pump Mech. Electr. Eng. Technol. **50**(01), 53–58+206 (2021)
6. Zhao, H.: Progress has been made in the application testing of key seals for drilling pumps. Synthetic Rubber Indust. **45**(04), 347 (2022)

Parametric Design of Boiler Reducing Tee Elbow Based on SolidWorks

Juye Wan[1], Yixin Zhou[2](✉), Xunyang Zheng[2], and Junting Cheng[1]

[1] School of Electrical and Mechanical Engineering, Chuzhou University, Anhui, China
[2] School of Mechanical Engineering, Liaoning Technical University, Fuxin, China
1431686203@qq.com

Abstract. Boiler reducing tee elbows feature complex structures with numerous interrelated dimensions. To address the inefficiency and repetitive modeling tasks in their design process, this study utilizes secondary development on the SolidWorks platform through VB.NET under the Visual Studio environment. By establishing modular assembly modeling and parametric design for critical dimensions, a parametric design system for boiler reducing tee elbows is developed. A human-machine interface enables parametric modeling of different elbow types. This modular assembly approach significantly enhances design efficiency, accuracy, and flexibility. The proposed parametric system has increased the design efficiency of elbows by more than 90% and effectively shortened the product development cycle. And optional forms such as ComboBox and CheckBox have been added to the design system, enhancing the diversity of the design.

Keywords: Secondary development · SolidWorks · Parametric design · Reducing tee elbow

1 Introduction

The boiler reducing tee elbow is a device connected to the boiler burner to withstand high-temperature combustion. The inner layer of the elbow needs to be coated with silicon carbide industrial ceramic material. Meanwhile, to ensure dimensional correlation and fundamental similarity between the silicon carbide layer and the steel structural shell, parametric design can significantly improve work efficiency [1].

Given the complex structural variations and prolonged design cycles of boiler reducing tee elbows, it is crucial to utilize SolidWorks secondary development technology for elbow design. Parametric design refers to establishing variable parameters for critical dimensions during modeling, enabling technicians to modify parameters according to required dimensions for flexible parametric modeling. This approach effectively reduces development cycles and labor costs. Technicians can rapidly complete model construction based on dimensional requirements, including 3D models of individual components, 2D drawings, and fully assembled 3D models with corresponding 2D documentation.

To meet the combustion and operational requirements of industrial boilers, the elbow's inner wall requires a silicon carbide ceramic layer for high-temperature resistance, while the cast iron shell differs from the ceramic in dimensions and geometry.

S. Halgamuge et al. (Eds.): ICACMVE 2025, LNME, pp. 52–61, 2026.
https://doi.org/10.1007/978-981-95-6911-3_6

Therefore, during elbow component processing, separate 3D models and 2D machining drawings must be created for both the shell and ceramic layer. Different application scenarios require varying elbow types, with port positions being non-fixed and port quantities ranging from single to triple configurations. Common characteristics among different elbow models provide convenience for parametric design. Using SolidWorks 3D modeling software, parametric processing is implemented according to structural and dimensional features of different components, with reserved interfaces for customization. Through VB.NET API calls, an optimized human-machine interface is developed, ultimately realizing elbow designs of various dimensions and types [2].

Common parametric modeling methods include the complete programming method and the template parameter modification method. The complete programming method adopts a bottom-up approach and realizes the modeling of parts from scratch through API calls. The template parameter modification method adopts the top-down idea. First, the model is pre-modeled. The equation function in SolidWorks is used to associate the dimensions between parts, and the corresponding interfaces are reserved for the program to call, thereby achieving rapid parameterization of the model.

The boiler reducer tee elbow belongs to a large assembly. If the fully programmed method is adopted, it will lead to an increase in programming volume and is not easy to achieve. Therefore, in this study, the template parameter modification method is adopted to conduct the secondary development of the boiler reducer tee elbow.

2 Structural Division and Parametrization of the Elbow

2.1 Structural Division

The main structure of the boiler reducing tee elbow includes: bent pipe, silicon carbide ceramic, inlet flange, outlet flange, thick flange, stiffening plates, and stops (Fig. 1). The bent pipe primarily consists of an inlet and outlets, with the outlet divided into a main outlet and two branch outlets. The inclusion of branch outlets depends on practical requirements, resulting in diverse and complex configurations. Parametric design can significantly reduce workload. By decomposing the structure, complex models can be modularized, simplifying model creation and providing enhanced convenience for relevant personnel [3].

2.2 Parametric Design of the Elbow

The parametric design process for elbows is a technology based on software like SolidWorks, aimed at enhancing design efficiency and accuracy. First, SolidWorks is extended using secondary development tools to support parametric design functionalities. Next, key parameters of the elbow—such as bend angle, radius, etc.—are defined, along with parameter constraints. Then, based on design requirements and limitations, the parametric modeling function is utilized to create an initial model. Adjusting parameter values allows real-time modification of the geometric shape to meet specifications. Subsequently, the elbow model undergoes verification and optimization through simulation analysis or experimental testing. Feedback from these evaluations guides fine-tuning

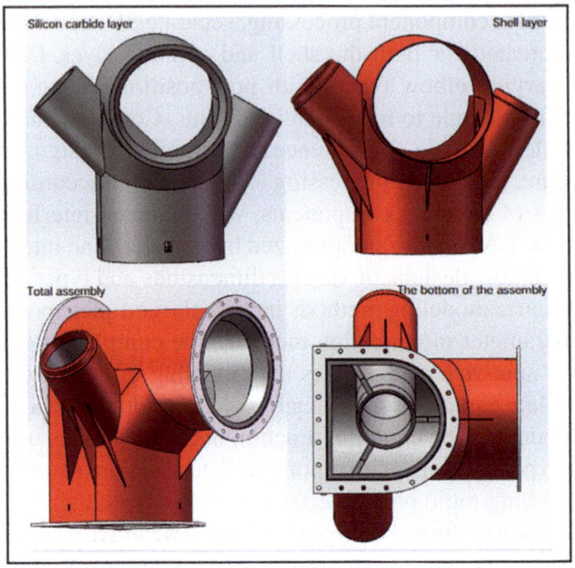

Fig. 1. The main structure of the boiler reducing tee elbow.

and optimization of parameters to further refine the design. Finally, the parametrically designed elbow model is finalized, and engineering drawings are exported [4]. The specific process is shown in Fig. 2:

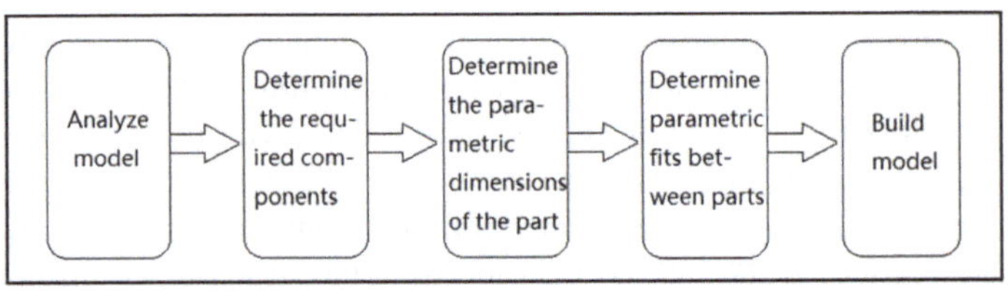

Fig. 2. Parametric design process of elbows

When programming, developing the code for bend angles presents significant challenges, as it requires accounting for the extension of pipe angles connected to circular flanges and ensuring compliance with dimensional requirements. Additionally, proper sizing between the elbow and flange connections must be guaranteed to prevent dimensional mismatches. During 3D model generation, sketches and sketch commands in SolidWorks are employed, with constraints added for dimensional parameters [5].

The parametric design workflow enables rapid generation of compliant elbow models, reducing repetitive modeling efforts and human errors while improving design efficiency and precision. It offers designers flexibility and convenience, allowing them to swiftly adapt to design changes and optimization demands [6].

3 Development System Design

3.1 System Design Process

To parametrically establish models, we need to create a human-machine interaction interface. Windows Forms (WinForm) is a client-side development technology based on the.NET Framework platform. WinForm offers simplicity, user-friendliness, and rapid development, along with diverse control combinations. By rationally arranging various controls, a parametric modeling interface for elbows can be constructed. Controls such as Labels, ComboBoxes, Buttons, CheckBoxes, TextBoxes, and FolderBrowserDialogs are added to build the main interface for elbow parametric modeling. Finally, VB.NET is used to program these controls in WinForm to realize model creation [7].

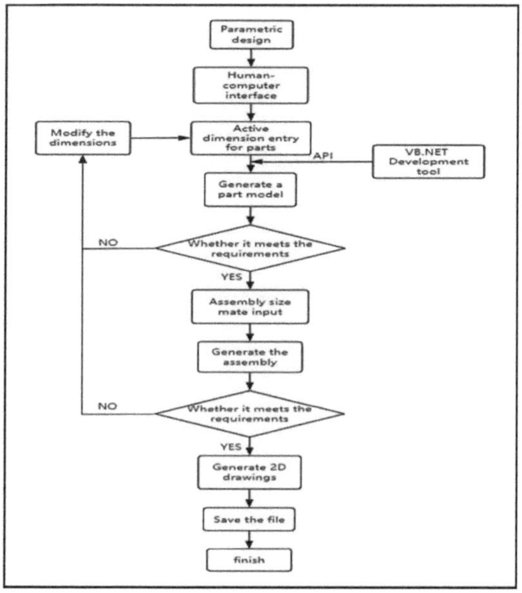

Fig. 3. System design process.

The model design interface is divided into multiple pages based on different design steps. First, users input required dimensional parameters following the page instructions to generate compliant component models. Next, assembly fit dimensions are entered to complete the overall assembly of parts. Finally, 2D production-ready drawings are automatically generated for both individual components and the assembled structure. During modeling, files are automatically saved to the selected storage path. If generated models fail to meet specifications, dimensions can be modified to reconstruct the models [8]. The specific process is shown in Fig. 3.

3.2 Establish a Human-Computer Interaction Interface

Taking the reducing tee elbow as an example, the system development process begins by entering the development interface. The main interface is divided into three modules: Create Part Model, Create Assembly Model, and Generate 2D Part Drawings. Designers input dimensional parameters based on requirements and determine the number and positions of ports according to the model type. After design completion, the model is fully assembled, and 2D drawings are generated. Before model generation, users must first select a storage location to ensure data retention. Compared to manual design methods, this system significantly reduces modeling time and avoids repetitive tasks. It also allows flexible adjustments to port positions and even reduction of port quantities based on practical needs [9]. Figure 4 is the parametric design page diagram of the flange:

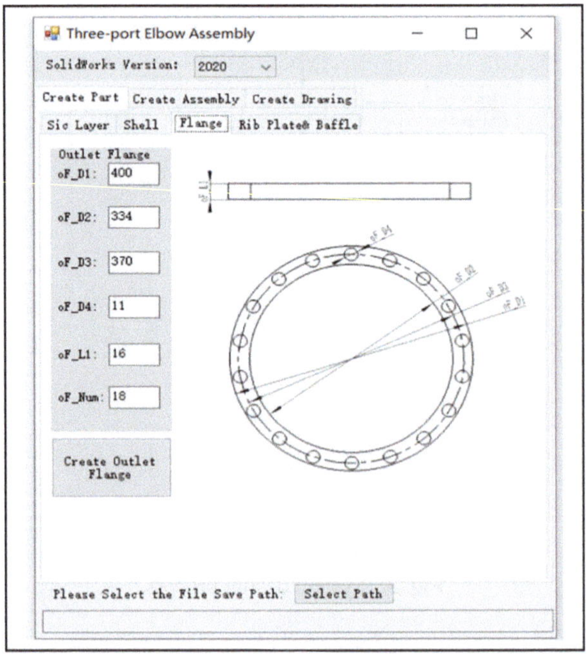

Fig. 4. Parametric modeling interface of the Flange.

Compared with the traditional manual design method, the proposed three-way elbow design system of this research has more advantages, which are specifically reflected as follows: Through the design time comparison table (Table 1), it can be seen that this system significantly shortens the design time of the product and improves the design efficiency. Among them, the design time of the shell module is now 0.1 h, which is 90% shorter than before. The time consumed by the silicon carbide module has been shortened by 92.86% and is now 0.05 h. The final assembly is reflected in the design time of 0.08 h, and the efficiency has increased by 92%.

Table 1. Design time.

Design project	Time for system design	Time for manual design
Shell Module	0.1 h	1 h
Sic Module	0.05 h	0.7 h
Assembly	0.08 h	1 h

To ensure dimensional accuracy, constraint functions are integrated into the code. These functions verify that input values fall within valid ranges and comply with fundamental geometric relationships, preventing errors. For instance, during part design, input dimensions must be numerical values, dependent dimensions automatically calculate valid ranges, and if inputs exceed limits, the program triggers alerts to prompt corrections, ensuring model compliance and preventing operational errors.

The following is a sample program that demonstrates the implementation and application of the constraint box. A constraint box is a tool used to define and manage the numerical range of an object, aiming to limit its value. The sample program demonstrates how to use constraint boxes to validate data input, limit the range of variables during simulation and optimization processes. By defining appropriate constraint boxes, ensure the accuracy and consistency of the data, and avoid invalid values that exceed the expected range. The example is as follows:

```
If Is Numeric(A_R1.Text) And Val(A_R1.Text) > 0 Then.
Else.
NumError = True.
MessageBox.Show("Please enter the correct A_R1!","The format is incorrect", MessageBoxButtons.OK, MessageBoxIcon.Warning).
Exit Sub.
End If.
If Val(CD_D1.Text) < 2 * Val(B_D1.Text) Then.
Else.
NumError = True.
MessageBox.Show("CD_D1 Need to be less than 2*B_D1!", "The value doesn't meet the requirements!", MessageBoxButtons.OK, MessageBoxIcon.Warning).
Exit Sub.
End If.
```

After completing all component designs, users enter the assembly interface (Fig. 5), select saved part models, input fit dimensions, and click Generate Assembly to finalize the assembly structure.

To ensure that the development program can be correctly connected to SolidWorks, it is necessary to add references to the library files of SolidWorks before development, thereby completing the setup of the development environment. After adding references is completed, the API in SolidWorks needs to be debugged. The code for calling the API in SolidWorks is as follows:

```
Imports SwConst.
Imports SldWorks.
```

The code for opening the model and performing size-driven operations is as follows:

swApp = CreateObject("SldWorks.applicatin").

swApp.Visible = True.

swApp.NewDocument("C:\ProgramData\SolidWorks\SOLIDWORKS"& SolidworksEdition& "\templates\gb_part.prtdot", 0, 0, 0).

part = swApp.ActiveDoc.

After connection, the model can be drawn through code, and the assembly needs to be completed by selecting parts. The code example for selecting the part path is as follows:

If OpenSicFilePath.ShowDialog = DialogRe-sult.OK Then TextBoxSicPath.Text = OpenSicFilePath.FileName.

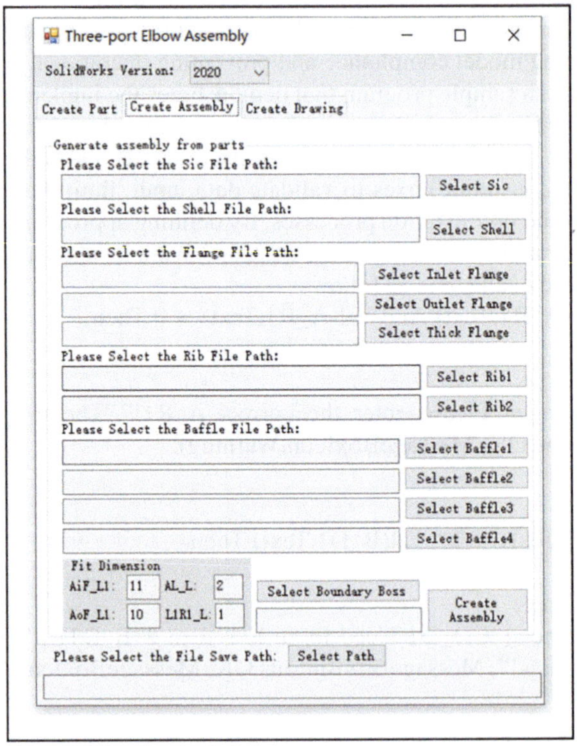

Fig. 5. Assembly mating interface.

3.3 Generate Engineering Drawings

After completing the modeling and assembly of a series of parts, the 3D model can be converted into a 2D engineering drawing through SolidWorks. Each established part and the overall assembly can generate a 2D engineering drawing with one click through the interactive interface (Fig. 6).

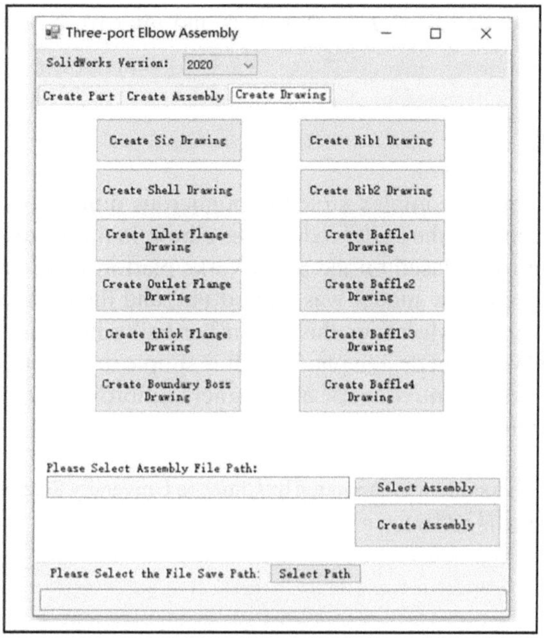

Fig. 6. 2D drawing generation interface.

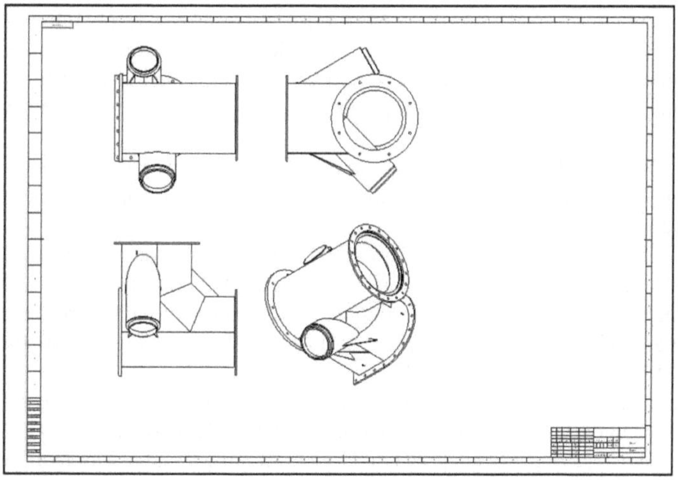

Fig. 7. 2D engineering drawing of an elbow.

In addition, SolidWorks can also adjust the scale and position of the views in the engineering drawing according to actual needs. Meanwhile, users can manually add the required dimension annotations to better express the design intent. The final generated engineering drawings can be saved in the required format files, such as PDF, DWG, etc., to facilitate subsequent editing of the drawings and improve processing efficiency, as shown in Fig. 7.

4 Conclusions

To solve the problems of complex structure, numerous dimensional correlations, and low modeling efficiency of the boiler reducer tee elbow, in this study, VB.NET was used as the development tool. Based on the SolidWorks platform, the parametric design of the boiler reducer tee elbow model was carried out, and the engineering drawing was generated simultaneously. Moreover, this system has built a human-computer interaction interface using the WinForm design platform. It is simple to operate, runs smoothly, can effectively reduce the repetitive work of designers, improve work efficiency, and has certain reference significance.

Acknowledgment. This study was supported by Chuzhou University Scientific Research Start-up Fund, project number: 2022qd043.

References

1. Hasan, A.B., Wikander, J., Onori, M.: Assembly design semantic recognition using solidworks-API. Inter. J. Mech. Eng. Robot. Res. **5**(4), 280–287 (2016)
2. Aritz, A., Camba Jorge D., Daniel, J.S., Manuel, C.:An Improved Explicit Reference Modeling Methodology for Para-metric Design.Computer-Aided Design, vol. 161 (2023).
3. Jorro-Aragoneses, J.L., Díaz-Agudo, B., Recio-García, J.A., Jimenez-Díaz, G.: Recolibry suite: a set of intelligent tools for the development of recommender systems. Autom. Softw. Eng. **27**(1), 1–27 (2020)
4. Shuai, W., Wufa, L., Peng, Z.: Research on parametric design of mortise and tenon furniture components based on secondary development of solidworks Comput. Appli. Softw. **40**(03), 28–32 (2023)
5. Zhang,X., Tao, J., Yang, H., et al.: Parametric design of fish vaccine syringe base based on secondary development of soildworks. Anhui Agricult. Sci. **48**(22), 209–212+216 (2020)
6. Liu, S.: Application of solidworks Secondary Development in Parametric Modeling Horizon of science and technology, vol. (7), pp. 178–179 (2019). https://doi.org/10.19694/j.carolcarrollnki, ISSN: 2095–2457.2019.07.073
7. Li, H.: Research on parametric design method of coal preparation equipment supported by solidworks secondary development. Shandong Coal Sci. Technol. **43**(03), 93–97 (2025)
8. Fang, Q., Zhao,Z., Ding, H.: et al.: Secondary Development and research on low-temperature circulating rapeseed dryer design based on solidworks. Ploidy J. Anhui Agricult. Univ. (6), 1093–1098 (2024). https://doi.org/10.13610/j.carolcarrollnki.1672-352-x.20241224.016
9. Zhang, J., Lin, X., Hong, R., et al.: Parametric modeling system of milling gear cutter head based on solidworks secondary development. Machinery Manufact. Autom. **3**(02), 134–138+152 (2024). https://doi.org/10.19344/j.cnki

An Integrated Dynamic Model of Helicopter Considering Beating Vibration

Peihui Shen[1(✉)] and Lijuan Chen[2]

[1] General Aviation Industry College, Fujian Chuanzheng Communications College, Fuzhou, China
751659875@qq.com

[2] School of Safety and Environment, Fujian Chuanzheng Communications College, Fuzhou, China

Abstract. According to the common phenomenon of beating vibration in helicopter coupled systems caused by the rotation of the main rotor, tail rotor, and operating engine, which might lead to structural fatigue, electronic equipment failure, and passenger discomfort, a three-degree-of-freedom concentrated mass model considering the impact of beating vibration is established. Selecting representative vibration parameters based on existing literature experimental platform, vibration responses are analyzed from time-domain, phase diagrams and frequency-domain characteristics. It was explained that as the difference in beat frequency increases, the amplitude value and extreme difference of beating decrease, and at the same time, the response and amplitude of the beating vibration are significantly higher in the hovering state than in the forward flight state. Dynamically adjusting the damping coefficient could weaken the beating vibration response. This provides a good theoretical basis for the vibration reduction structure and frequency design of the main rotor, and tail rotor.

Keywords: dynamic model · beating vibration · phase diagram · time-frequency characteristics

1 Introduction

Engineering vibration is widely present in various fields such as machinery, construction, and aerospace, and is an inevitable phenomenon in energy transfer and daily systems. Helicopter vibration is mainly caused by the rotation of the main rotor, tail rotor, and operating engine. High frequency periodic loads can easily lead to structural fatigue, electronic equipment failure, and passenger discomfort [1, 2]. The current research on helicopter vibration focuses on the integration of intelligent and multi-dimensional control technologies. The core purpose of these studies is to drive the development of vibration reduction technology towards the combination of active control and structural optimization [3, 4]. In terms of helicopter dynamics modeling, the main methods include three types: multi-body dynamics theory, finite element modeling and Computational Fluid Dynamics (CFD) coupling structural dynamics. Certainly, the latest

S. Halgamuge et al. (Eds.): ICACMVE 2025, LNME, pp. 62–69, 2026.
https://doi.org/10.1007/978-981-95-6911-3_7

progress focuses on real-time modeling under the digital twin framework, integrating flight control system parameters with composite material anisotropy characteristics to achieve collaborative simulation of flight dynamics, vibration, and noise.

However, there is limited research on the common phenomenon of beating vibration in helicopter coupled systems. Beating vibration refers to the phenomenon of periodic fluctuations in amplitude generated when two or more multi-source exciting vibrations with similar frequencies are superimposed. In a helicopter, if the speed of the main rotor and tail rotor is designed to be close to integer multiples, it may cause low-frequency flapping effects, resulting in pulsating loads on the body, causing dizziness or instrument reading jitter. This article used the concentrated mass stiffness damping method to establish the overall dynamic model of the helicopter system. Through analysis, the main purpose was to eliminate vibration through frequency active isolation technology, so that the excitation frequency difference changed avoiding the sensitive frequency band of the human body. Simultaneously, changing damping or other system parameters to improve comfortable driving provides a theoretical basis for the design of helicopters.

2 Modeling and Analysis of Helicopter System Considering Beating Vibration

Helicopter dynamics analysis currently focuses on interdisciplinary and high-precision numerical simulation techniques, particularly in the fields of rotor systems, power transmission, and structural response, where significant progress has been made. The laboratory seat test system from [5] is shown in Fig. 1(a), and another laboratory seat test system from [6] is shown in Fig. 1(b). The iterative verification of numerical simulation and experimental data significantly improved the accuracy of dynamic parameter prediction.

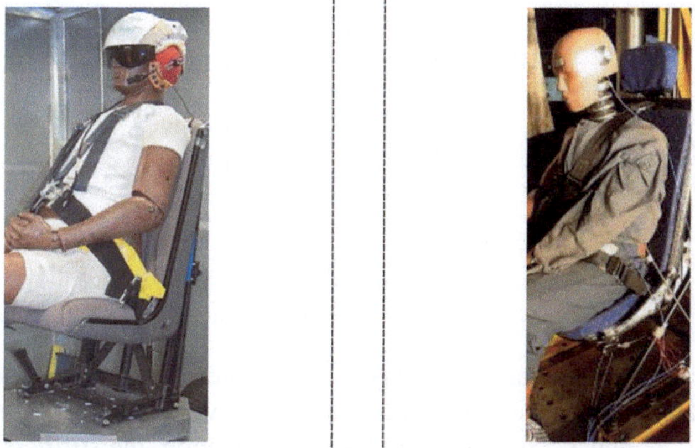

(a) (b)

Fig. 1. Existing literature experimental platform

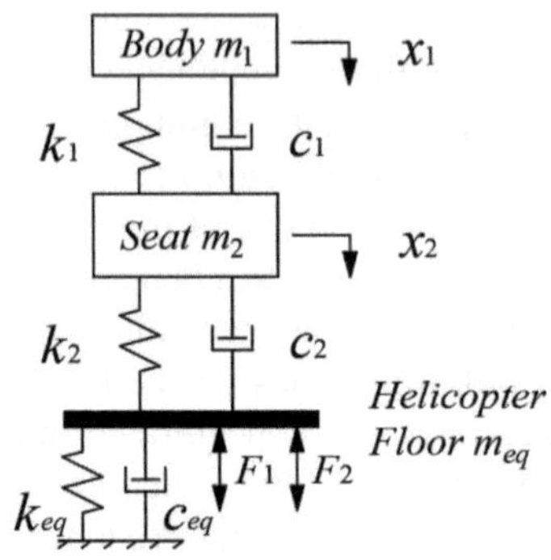

Fig. 2. The integrated dynamic model

Based on the existing literature experimental platform, a three-degree-of-freedom concentrated mass model considering the impact of beating vibration is representative established, as shown in Fig. 2. The dynamic response equation analysis is shown in Eq. (1) and Eq. (2).

$$\begin{cases} m_1\ddot{x}_1 + c_1(\dot{x}_1 - \dot{x}_2) + k_1(x_1 - x_2) = m_1 g \\ m_2\ddot{x}_2 + c_2(\dot{x}_2 - \dot{x}_3) + k_2(x_2 - x_3) - c_1(\dot{x}_1 - \dot{x}_2) - k_1(x_1 - x_2) = m_2 g \\ m_{\mathrm{eq}}\ddot{x}_3 + c_{\mathrm{eq}}\dot{x}_3 + k_{\mathrm{eq}}x_3 - c_2(\dot{x}_2 - \dot{x}_3) - k_2(x_2 - x_3) = F_{\mathrm{e}}(t) + m_{\mathrm{eq}} g \\ F_{\mathrm{e}}(t) = F_1\sin(\omega_1 t + \varphi_1) + F_2\sin(\omega_2 t + \varphi_2) \end{cases} \tag{1}$$

It can be written in matrix form as:

$$[M]\{\ddot{x}\} + [C]\{\dot{x}\} + [K]\{x\} = \{p\}$$

$$[M] = \begin{bmatrix} m_1 & 0 & 0 \\ 0 & m_2 & 0 \\ 0 & 0 & m_{eq} \end{bmatrix}, [C] = \begin{bmatrix} c_1 & 0 & -c_{eq} \\ 0 & c_1 + c_2 & -c_{eq} \\ -c_{eq} & -c_{eq} & c_2 + c_{eq} \end{bmatrix},$$

$$K = \begin{bmatrix} k_1 & 0 & -k_{eq} \\ 0 & k_1 + k_2 & -k_{eq} \\ -k_{eq} & -k_{eq} & k_2 + k_{eq} \end{bmatrix},$$

$$x = \begin{Bmatrix} x_1 \\ x_2 \\ x_3 \end{Bmatrix}, p = \begin{Bmatrix} m_1 g \\ m_2 g \\ m_{\mathrm{eq}} g + F_1 \sin(\omega_1 t + \varphi_1) + F_2 \sin(\omega_2 t + \varphi_2) \end{Bmatrix} \tag{2}$$

where the concentrated masses m_1, m_2 and x_{eq} represent the driver, the seat, and the helicopter platform respectively. Displacement in the vertical direction relating to the cab floor are x_1, x_2 and x_{eq} respectively. For simplification, the stiffness and damping parameters for first-order vibration reduction between the pilot and operating chair are k_1 and c_1. The stiffness and damping parameters for second-order vibration reduction between the seat and helicopter floor are k_2 and c_2. The stiffness and damping parameters between the helicopter platform and external airflow environment are equivalently simplified to k_{eq} and c_{eq}. The vibration of a helicopter mainly comes from the vibration coupling of the main rotor, tail rotor, transmission system, and engine. For the convenience of analysis, only two representative vibration sources are considered for vibration sampling here, $F_e(t)$ is the periodic coupling excitation. ω_1 and ω_2 are the angular velocity of the exciter, F_1 and F_2 are the centrifugal excitation force from main rotor and tail rotor, φ_1 and φ_2 are the initial phase of these two forces.

3 Numerical Analysis

According to the integrated dynamic model in Fig. 1 and the derivation method of Eq. (1) and Eq. (2), computer simulation is used.

Originating from existing studies [7–10], the following parameters have been selected as $m_1 = 61$ kg, $m_2 = 13.5$ kg, $m_{eq} = 13.5 \times 10^3$ kg, $k_1 = 37.7 \times 10^3$ N/m, $c_1 = 159$ Ns/m, $k_2 = 22.6 \times 10^3$ N/m, $c_2 = 750$ Ns/m, where the mass of the pilot is the total sum of the mass of the head, upper torso, viscera and pelvis. F_1 and F_2 are both selected within the range of [100, 500] N. The frequency difference ($\Delta = \omega_1 - \omega_2$) is selected within the range of [1, 10] 2π, which is sensitive to drivers. It is assumed that $\varphi_1 = \varphi_2 = 0$. The parameter k_{eq} is respectively selected within the range of [10^4, 10^5] N/m or the range of [10^5, 10^6] N/m according to different flight states of hovering or forward flight. The parameter c_{eq} is also selected within the range of [10^2, 10^3] Ns/m or the range of [10^3, 10^4] Ns/m according to different flight states of hovering or forward flight.

Selecting the relatively small frequency difference, the responses of the helicopter vibration system are displayed in Figs. 3, 4, 5 and 6. From the displacement response in Fig. 3 to the velocity response in Fig. 4, it can be seen that the beating vibration phenomenon is very obvious. From Fig. 5, it can be found that phase diagrams of pilot, seat and helicopter floor have similar characteristics. Figure 6 shows the amplitude characteristics of these three in the frequency domain, clearly indicating the two excitation frequency energy branches.

The harm of beating vibration to drivers could not be ignored. Long-term exposure might cause fatigue, decreased attention, musculoskeletal pain, and even affect driving safety. Therefore, reducing the impact of flapping vibration and improving driving comfort and safety are crucial. From the time-domain characteristics in Fig. 7(a) and frequency-domain characteristics in Fig. 7(b), it can be seen that as the difference in beat frequency increases, the amplitude value and extreme difference of beating decrease, which provides a reference for the frequency design of the main rotor, and tail rotor.

When selecting the stiffness and damping parameters according to different flight states such as hovering and forward flight, the helicopter system responses are shown in Fig. 8. From the time-domain characteristics in Fig. 8(a) and frequency-domain characteristics in Fig. 8(b), it can be found that the response and amplitude of the beating vibration are significantly higher in the hovering state than in the forward flight state. Dynamically adjusting the damping coefficient, such as magnetorheological or electrorheological fluid dampers, could weaken the beating vibration response. Through controlling by doubling the damping coefficient c_2 as shown in Figs. 9(a) and 9(b), the vibration amplitudes are obviously reduced, which is beneficial for the equipment and pilots.

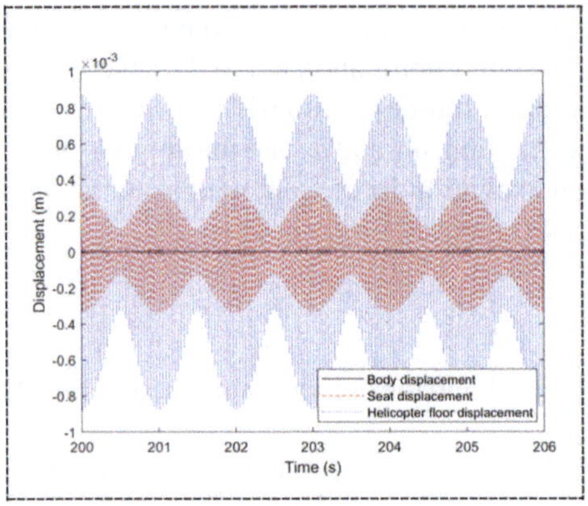

Fig. 3. Beating vibration response characteristics of displacement.

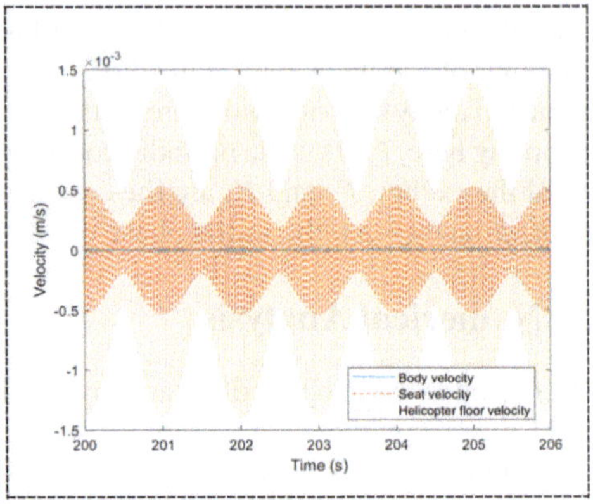

Fig. 4. Beating vibration response characteristics of velocity.

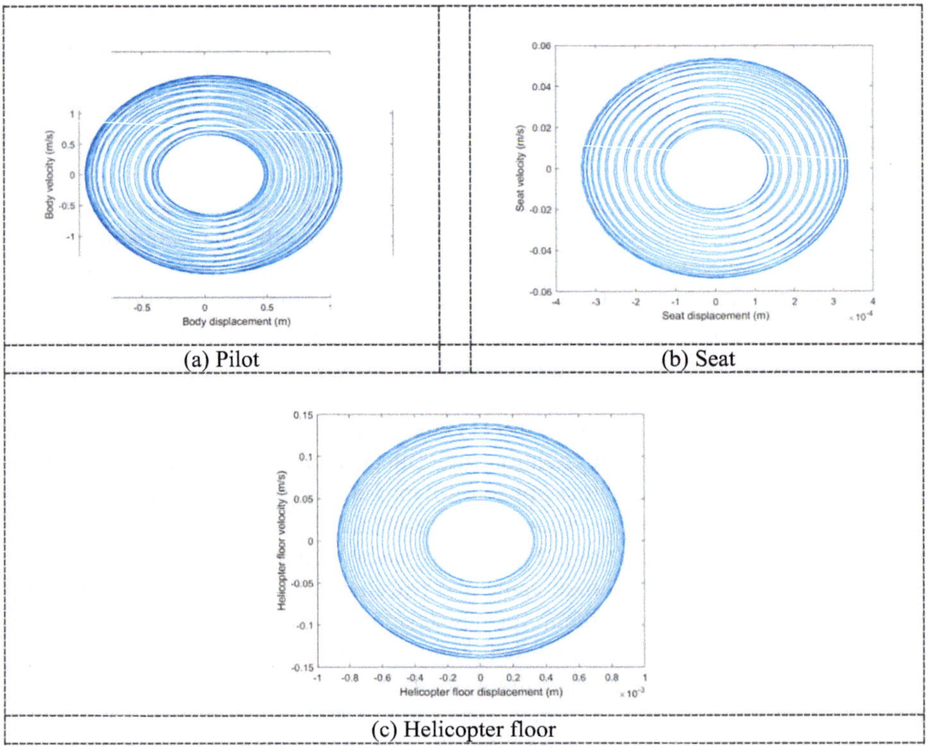

(a) Pilot (b) Seat

(c) Helicopter floor

Fig. 5. Phase diagram of beating vibration system.

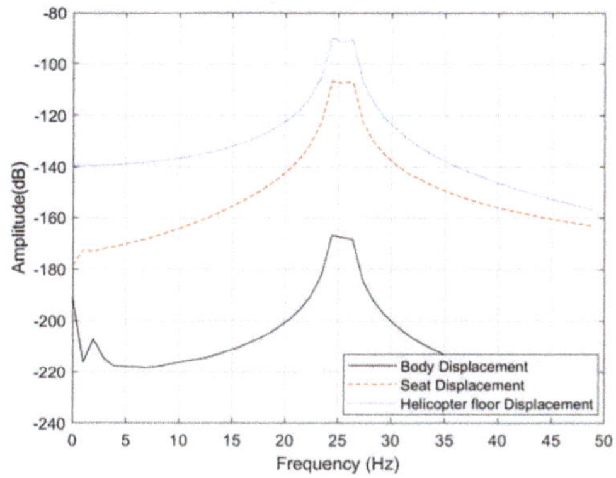

Fig. 6. Amplitude frequency characteristics of beating vibration.

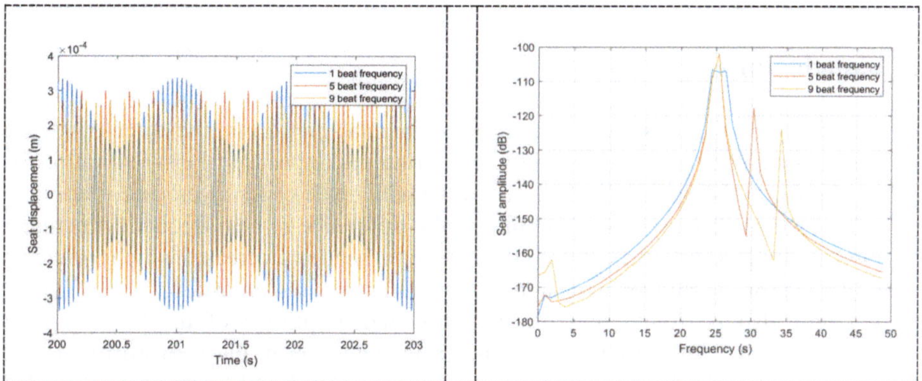

(a) Time domain (b) Frequency domain

Fig. 7. Amplitude characteristics under different beat frequency.

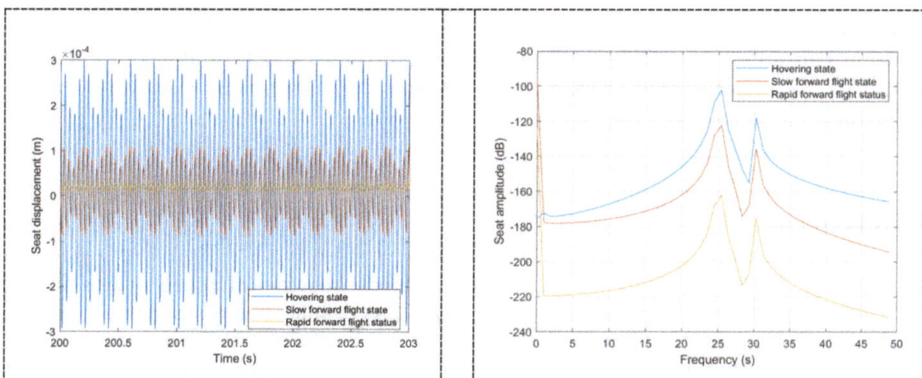

(a) Time domain (b) Frequency domain

Fig. 8. Amplitude characteristics under different flight state.

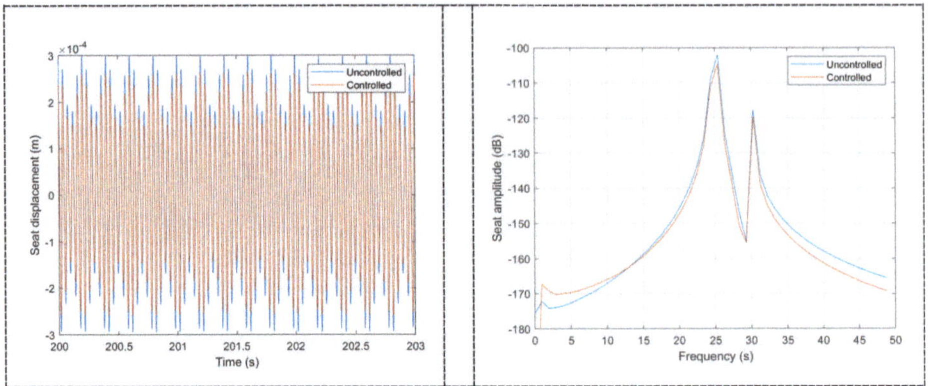

(a) Time domain (b) Frequency domain

Fig. 9. Comparison of different responses after damping control.

4 Conclusions

(1) According to the existing literature experimental platforms, the three-degree-of-freedom concentrated mass model considering the impact of beating vibration was put forward. The dynamic responses of the helicopter system were analyzed.
(2) After investigation, it was found that as the difference in beat frequency increases, the amplitude value and extreme difference of beating decrease, which provides a reference for the frequency design of the main rotor, and tail rotor.
(3) It could also be found that the response and amplitude of the beating vibration are significantly higher in the hovering state than in the forward flight state. Dynamically adjusting the damping coefficient could weaken the beating vibration response. Sensing flight state parameters in real-time and adaptively adjusting damping characteristics could be designed beneficially for the equipment and pilots.

Acknowledgment. These works have been financially supported by the Natural Science Foundation of Fujian Province of China with Grants no. 2022J01389, A Team Project for Research on General Aviation Relative Application Technologies and Industrial Services with Grant no. 20200208.

References

1. Ma, X.Y., Li, Y.J., Cao, Y.Y., et al.: A nonlinear biodynamic model of helicopter seat: ATD system exposed to vertical impact for predicting the risk of injury in airworthiness certification. Nonlinear Dyn. **10**, 1–37 (2024)
2. Oliveira, C.G.D., Nadal, J.: Transmissibility of helicopter vibration in the spines of pilots in flight. Aviat. Space Environ. Med. **76**(6), 576–580 (2005)
3. Feng, Z.Q., Chen, Z.B., Xing, X.D.: An adaptive shock mitigation control method for helicopter seat system with magnetorheological energy absorbers. J. Intell. Mater. Syst. **33**(8), 987–1001 (2022)

4. Bang, S.W., Hong, S.B., Lee, Y.B., et al.: Active Airframe vibration control study using a small-scale model for lift-offset compound helicopter. Inter. J. Aeronaut. Space Sci. **24**, 77–91 (2023)
5. Chen, Y., Wickramasinghe, V., Hu Mzimcik, D.: Development of adaptive seat mounts for helicopter aircrew body vibration reduction. J. Vib. Control **15**(12), 1809–1825 (2009)
6. Hu, D.Y., Yang, J.L., Hu, M.H.: Full-scale vertical drop test and numerical simulation of a crashworthy helicopter seat/occupant system. Int. J. Crashworthiness **14**(6), 565–583 (2009)
7. Tamer, A., Zanoni, A., Cocco, A., et al.: A Generalized index for the assessment of helicopter pilot vibration exposure. Vibration **4**, 133–150 (2021)
8. Choi, Y.T., Wereley, N.M.: Biodynamic response mitigation to shock loads using Magnetorheological Helicopter Crew Seat Suspensions. J. Aircr. **42**(5), 1288–1295 (2005)
9. Tamer, A., Muscarello, V., Masarati, P., et al.: Evaluation of vibration reduction devices for helicopter ride quality improvement. Aerosp. Sci. Technol. **95**, 105456 (2019)
10. Xie, J.G., Zhong, X.Y., Shi, X.P., et al.: Research on lumped parameter model of aviation seat cushion. J. Civil Aviation Univ. China **42**(5), 9–14 (2024)

Design of Wheel-Shaped Hook Anti-derailment Device for High-Speed Trains

Jinqi Wang(✉)

China Changsha Central South University, Changsha 410075, China
1731986985@qq.com

Abstract. High-speed train, with its advantages such as large passenger capacity, high speed, and low energy consumption, has become an important part of the modern transportation system. However, derailment accidents are still the major hidden danger threatening the operational safety of high-speed trains. To this end, this paper proposed a design of wheel-shaped hook anti-derailment device for high-speed trains and carry out numerical simulation analysis to verify the effectiveness of the device. This device can limit the lifting of wheel set through the hook-shaped structure when the train is lifted, thus effectively reducing the risk of derailment. Moreover, it can ensure the train pass through curves by installing small wheels on the hook-shaped anti-derailment device. The application of the device is of great significance for ensuring the operational safety of the high-speed train.

Keywords: Anti · derailment · High · speed train · Wheel · shaped hook

1 Introduction

With the rapid development of global economy and the acceleration of urbanization process, high-speed railway, as an efficient, convenient and safe mode of transportation, has been widely constructed around the world. By the end of 2023, the length of China's railway in operation has reached 159,000 km, including 45,000 km of high-speed railways, ranking first in the world in terms of scale [1].However, despite the remarkable progress has been made in terms of the safety of high-speed trains, derailment accidents still remain a major hidden danger threatening the operational safety of high-speed trains. Once a derailment accident occurs, it often causes serious casualties and property losses. In 2020, the 'Frecciarossa' high-speed train in Italy derailed unexpectedly, causing at least 2 deaths and 31 injuries [2] In 2022, the derailment accident on Guiyang-Guangzhou Railway line led to the death of the driver of Train D2809, 1 train attendant and 8 passengers were injured [3] (as shown in Fig. 1). These incidents have all attracted great attention from all sectors of society to the operational safety of high-speed trains.

In order to effectively reduce the probability of derailment accidents and ensure the safety of passengers' lives and property, it is of vital significance to research, develop and apply anti-derailment devices for high-speed trains. The anti-derailment device can

S. Halgamuge et al. (Eds.): ICACMVE 2025, LNME, pp. 70–77, 2026.
https://doi.org/10.1007/978-981-95-6911-3_8

(a) Guiyang-Guangzhou Railway line (b) 'Frecciarossa' high-speed train

Fig. 1. Derailment accidents of high-speed trains

monitor the interaction status between the wheels and track in real time during the operation of the train. When abnormal situations occur, it can promptly take corresponding measures to prevent the train from derailing, providing a reliable technical guarantee for the safe operation of high-speed trains.

The Railway Technical Research Institute of Japan has developed a new type of bogie [4], which can prevent the reduction of wheel pressure and reduce the lateral force between the wheel and the track. In addition, Kajitani Y et al. from the Japanese National Railways [5] proposed a L-shaped anti-derailment device and an anti-deviation limiter. Masaki Seki [6] proposed an anti-derailment guard rail through a large number of experimental studies. The device was arranged on the inner side of the track and achieved the prevention and control of derailment by restricting the lateral displacement of the wheel set. Germany has developed an advanced anti-derailment early warning system. By installing a large number of sensors on the bogies, wheels and tracks, it can monitor the operation status of the train in real time, the contact situation between the wheels and the tracks, as well as the geometric parameters of the tracks and other information, ensuring that the wheels always stay on the track. South Korea has developed a technology for evaluating the running safety of trains on floating bridges [7]. They have grasped the vibration characteristics of ocean waves by utilizing the existing statistical data, and derived the derailment safety standards for the wheels when trains run on the tracks of floating railway bridges with low-frequency swaying. It helps to detect potential derailment hazards in a timely manner.

In recent years, with the continuous innovation and development of China's high-speed railway technology, a series of important breakthroughs have been achieved in the research of anti-derailment devices for high-speed trains in China, and a variety of anti-derailment devices have been developed. Hou et al. [8] developed a lateral limiting device for high-speed trains. Jiang et al. [9] invented an active anti-derailment method for high-speed trains based on electromagnetic-driven clasping of the rails. Jing et al. [10] invented a derailment blocking system for high-speed trains. These devices are innovative in terms of structural design, monitoring methods, and control strategies, etc., and are able to adapt to the complex operation environment and line conditions of China's high-speed railways.

During the operation of train, although the wheel flanges of the bogie can prevent the bogie from derailing to a certain extent, the number of restraint points is limited. When the train is running at a high speed, the contact force will increase with the increase in speed, and the acting time is short. In collision accidents, jumping derailment is the main form of train derailment. According to the analysis based on the derailment evaluation criteria, the lateral offset and the lifting amount of the wheel set are the key indicators for evaluating collision-induced jumping derailment. Therefore, in order to prevent the bogie from derailing, the anti-derailment device needs to effectively restrain the vertical and lateral displacements of the wheels. Based on the above requirements, some teams have preliminarily proposed the design of a hook-shaped anti-derailment device for railway vehicles. The inner side of the device closely fits the surface of the rail, so as to restrict the relative movement between the bogie and the track in both lateral and vertical directions. However, when the train enters a curve line, this device is highly likely to generate a certain amount of friction with the track, which may affect the operation of the train. To address this issue, a row of small wheels is added to the hook-shaped anti-derailment device to ensure the curve pass ability of train will not affect by the anti-derailment device.

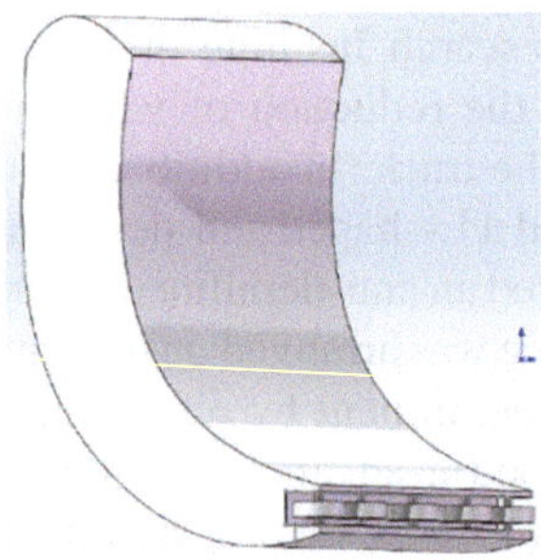

Fig. 2. The initial anti-derailment device

2 Design of the Initial Anti-derailment Device

A three-dimensional model of the initial anti-derailment device was established using a three-dimensional modeling software (as shown in Fig. 2). Then the finite element pre-processing software Hypermesh was employed to perform geometric processing and meshing, and the model was imported into Workbench for simulation analysis. The material of the anti-derailment device is Q235. Set the bottom surface of the track as a fixed constraint, and apply a vertical force to the anti-derailment device. Since the mass of a high-speed train is about 55t, and a single car body contains 2 bogies, there are 4 anti-derailment devices in one car. Therefore, a vertical force of 137.5 kN was applied to a single anti-derailment device, and the force-bearing point was set at the top of the device, which is the position connected to the bogie frame (as shown in Fig. 3). However, due to the small structural dimensions, with a width of 40 mm and a thickness of approximately 15.6 mm, the calculation results failed to converge.

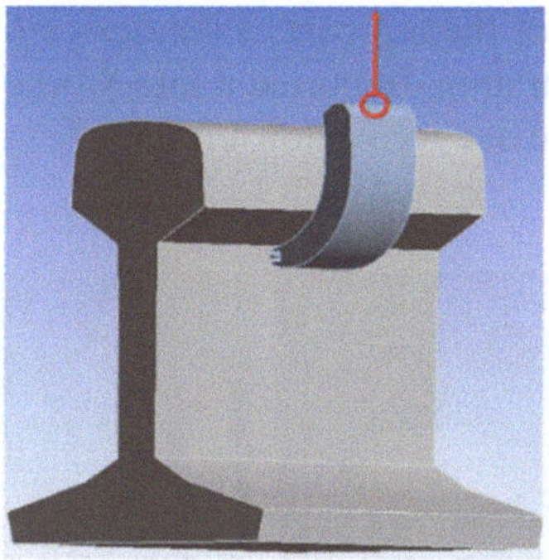

Fig. 3. Assembly drawing of the anti-derailment device and the track

3 Improvement of the Anti-derailment Device

On the premise of meeting the requirements of the bogie's structural dimensions, the dimensions were enlarged by 2.5 times, and the width is about 100mm, the thickness is about 40mm. Then changed the top of the device from a curved shape to a flat shape, which is more conducive to the connection with the bogie. Meanwhile, considering that the axles of the small wheels on the device might be too thin to bearing stress, the number of small wheels was reduced from 5 to 3. Additionally, the axles of the small wheels were thickened, and the thickness of the small wheels was increased. As a result, an improved anti-derailment device was obtained (as shown in Fig. 4).

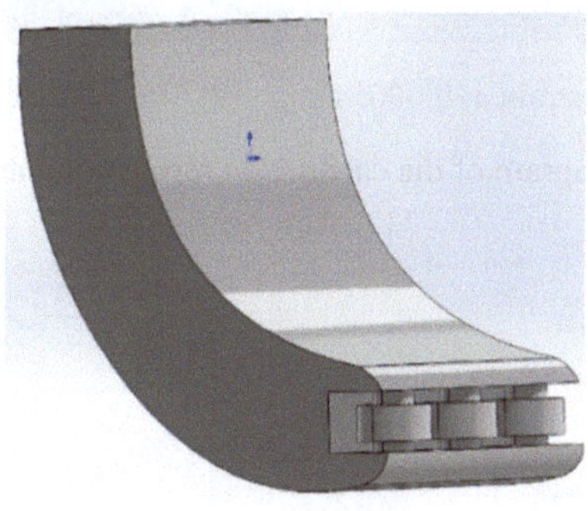

Fig. 4. The improved anti-derailment device

Imported it into Workbench, and the stress and strain of the device were shown in Fig. 5. The calculation results showed that the stress is mainly concentrated in the part connected to the bogie, and the strain is too large, reaching 0.6262, meaning that the anti-derailment device has been fractured.

In order to solve this problem, we consider to further improve the device from two aspects:

(1) Increased the curvature of the top part of the device (as shown in Fig. 6), so as to reduce the stress concentration. Imported it into Workbench for calculation, and the stress and strain of the device were shown in Fig. 7.

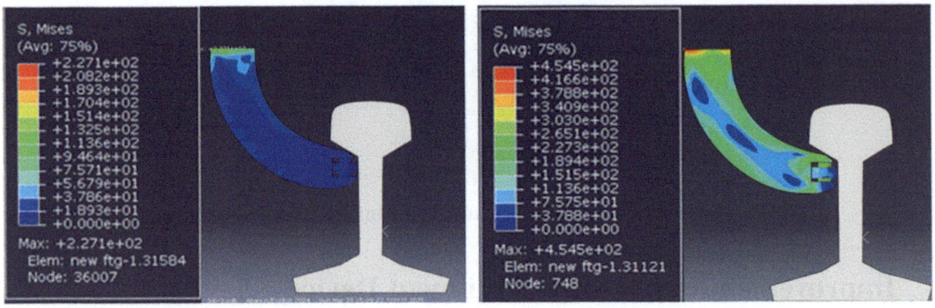

(a) The stress at 0.0003s (b) The stress at 0.0007s

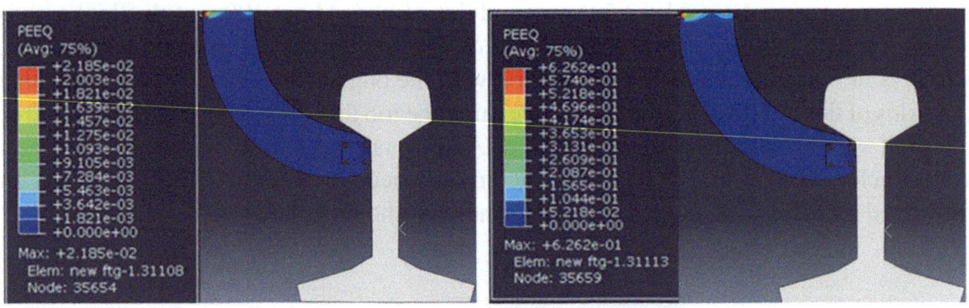

(c) The equivalent plastic strain at 0.0005s (d) The equivalent plastic strain at 0.001s

Fig. 5. Schematic diagram of the calculation results for anti-derailment device 1

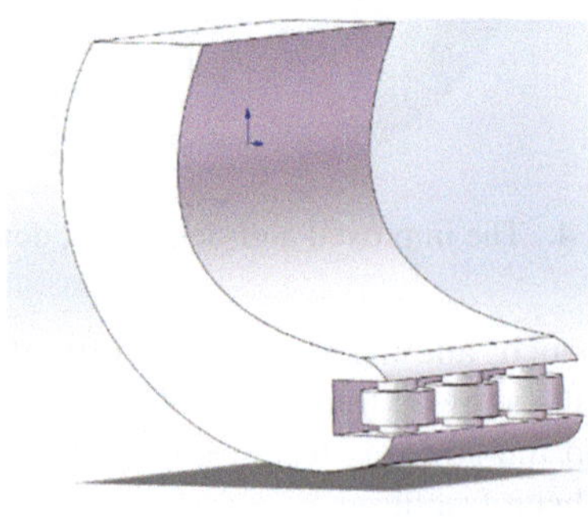

Fig. 6. The anti-derailment device after curvature improvement

It can be seen that the position where the main equivalent strain occurs has shifted from the top of thedevice to the wheel axle, and the strain value was only 0.03171.

(2) Replaced the material of the device with Q345, imported it into Workbench for calculation, and the stress and strain of the device were shown in Fig. 8.

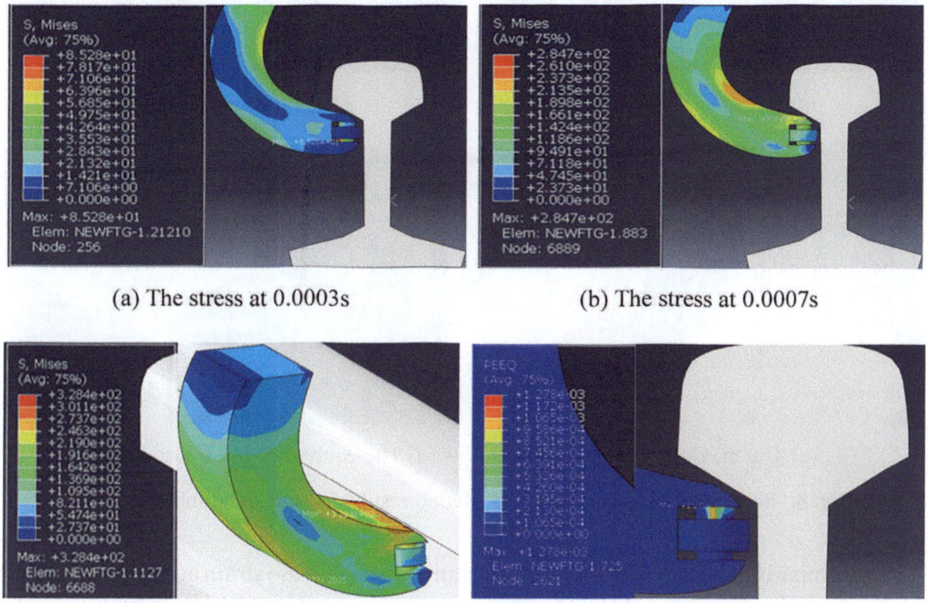

(a) The stress at 0.0003s

(b) The stress at 0.0007s

(c) The maximum stress

(d) The equivalent plastic strain at 0.0006s

Fig. 7. Schematic diagram of the calculation results for anti-derailment device 2

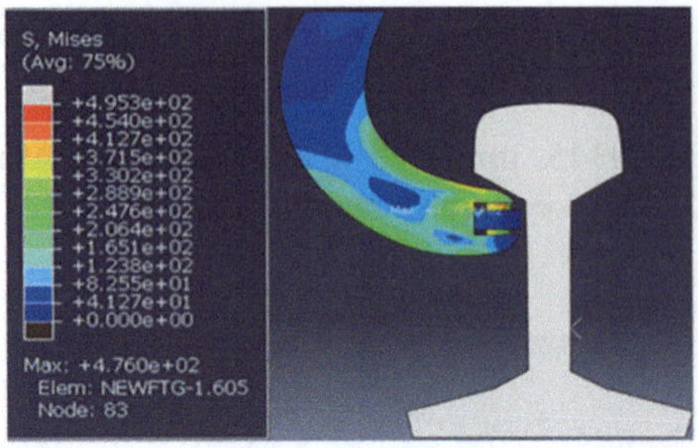

(a) The stress at 0.0006s

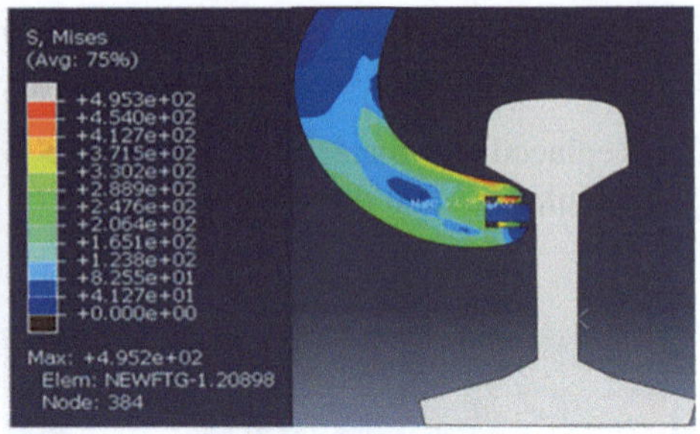

(b) The stress at 0.0014s

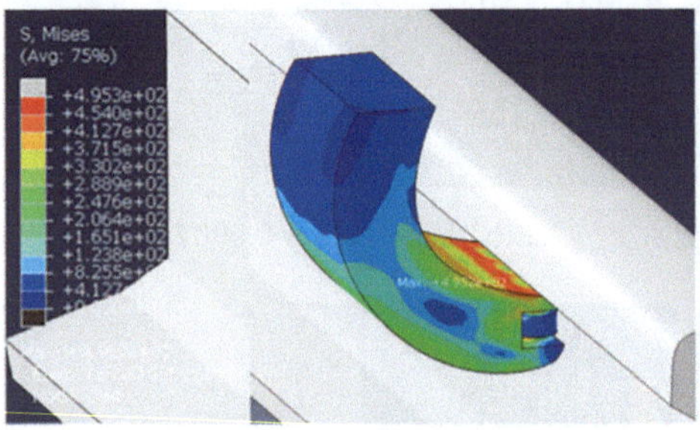

(c) The maximum stress

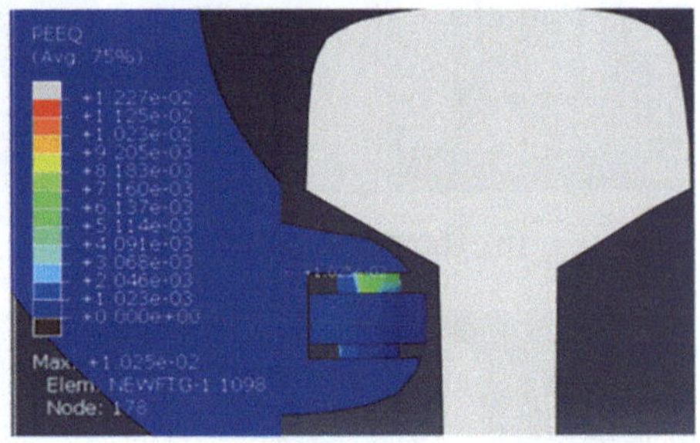

(d) The equivalent plastic strain at 0.001s

Fig. 8. Schematic diagram of the calculation results for anti-derailment device 3

By comparing the calculation results, it can be seen that the strain and stress are better when select the material Q345. The maximum strain is only 0.01448, which indicates that the anti-derailment device can limit the lifting of the wheel set, effectively reducing the risk of derailment, and the device will not suffer fracture damage.

4 Conclusion

This paper proposed a hook-shaped anti-derailment device with small wheels and conducted numerical simulation calculations to analyze the practicality of the designed device. The structural strength of the initial anti-derailment device does not meet the strength requirements for actual use. After improving the design of the structure, the calculation results showed that the device has large deformation. Subsequently, the device was further optimized in terms of structure and material. The results showed that the optimized anti-derailment device can limit the lifting of wheel set through the hook-shaped structure when the train is lifted, thus effectively reducing the risk of derailment.

References

1. Qi, H.: The Acceleration of Modern Railway Infrastructure Construction. Economic Daily, 2024–01–11(006)

2. Track World, Super Structure. A high - speed train derailed in Italy, killing 2 people and injuring 31[EB/OL]. https://www.sohu.com/a/371573215_661255, 2020–02–08
3. Beijing Youth Daily. The National Railway Administration introduced the cause of the derailment accident of Train D2809[EB/OL]. https://news.sina.com.cn/o/2022-06-10/doc-imizmscu6080022.shtml, 2022–6–10
4. Railway Technical Research Institute. RTRI Develops a New-Type Bogie for Reducing the Risk of Derailment[EB/OL]. http://www.rtri.or.jp/eng/press/2016/nr20160721_01_detail.html, 2016–11–09
5. Kajitani, Y., Kato, H., Koji, A.: Development of an L-Shaped Guide to Prevent Deviation from Rails **15**, 53–56 (2010)
6. Seki, M., Konagai, K., Muramatsu, H., et al.: Research and development for preventing derailment by large-scale earthquakes. J. Japan Soc. Civil Eng. Ser. F6 Safety Problem **69**(1), 1–18 (2013)
7. Seo, S., Park, J.H.: Safety criteria against derailment for train running on vibrating track at low frequencies. J. Korean Soc. Railway **23**(2), 165–172 (2020)
8. Hou, S., Huo, X., Zhang, X.: et al. Study on the dynamic characteristics and structural strength of the lateral limiting device of high-speed trains under collision derailment accidents. J. Railway Sci. Eng. **18**(03), 588–595 (2021)
9. Jiang, L., Duan, H., Liu, J., et al.: An Active Anti - derailment Method for High - speed Railway Trains Based on Electromagnetic - driven Rail - clamping Device. China: CN202211348193.0, 2025–01–28
10. Jing, L., Zhang, S., Liu, K., et al.: A Derailment Blocking System for High - speed Trains and Its Structural Design Method. China: CN202310206175.7, 2023–07–04

Design and Experimental Investigation of Vibration-Damping Alloy Noise Reduction Pads for Metro Wheels

Fufei Wang, Chenghao Lv(✉), Qi Wang, and Hailong Yin

Jiangsu Tianqi New Materials Application Research Institute Co., Ltd., Zhenjiang 212100, China

lvchenghao12138@163.com

Abstract. To achieve vibration and noise reduction in metro wheels, a novel metro noise reduction plate was developed by integrating the vibration suppression principle of advanced vibration-damping alloy materials with the advantages of damped dynamic vibration absorbers. ANSYS finite element simulations were first conducted to demonstrate that the noise reduction plate exhibits significant vibration suppression above 1000 Hz, with the frequency difference ratio between the plate and wheel maintained within 10%. Subsequent semi-anechoic chamber tests were performed to evaluate the vibroacoustic radiation characteristics of freely suspended metro wheels with and without the vibration-damping alloy plate. Experimental results revealed full-band noise reduction of 11.2 dB(A) under radial excitation and 12.9 dB(A) under axial excitation. Notably, noise reduction exceeding 12 dB(A) was achieved in the frequency range above 3000 Hz, consistent with simulation predictions demonstrating superior high-frequency vibration absorption characteristics. These findings demonstrate that the vibration-damping alloy plate effectively suppresses metro wheel vibrations.

Keyword: Vibration and Waves · Vibration and Noise Reduction · Vibration-Damping Alloy

1 Introduction

While the continuous development of urban rail transit systems provides convenient mobility solutions, concomitant vibration and noise pollution pose challenges to urban living environments. Extensive testing data from global railway networks reveals that railway noise primarily comprises three components: wheel-rail interaction noise, traction system noise, and aerodynamic noise [1]. Notably, when operational speeds remain below 120 km/h, wheel-rail noise dominates the acoustic emission spectrum, accounting for over 68% of total noise energy [2]. This establishes wheel vibration and noise reduction designs as critical optimization targets for improving rail transit environmental performance.

Present mitigation approaches for wheel vibration and noise primarily involve enhanced wheel damping and dynamic vibration absorber (DVA) installations. Experimental results from Hamburg Metro wheels demonstrate that resilient wheels achieve

S. Halgamuge et al. (Eds.): ICACMVE 2025, LNME, pp. 78–86, 2026.
https://doi.org/10.1007/978-981-95-6911-3_9

20–22 dB(A) noise reduction in curved sections [3, 4]. Despite their effective noise control and improved passenger comfort, resilient wheels face challenges including high manufacturing costs, elevated pricing, and complex maintenance, leading researchers to prioritize DVA investigations. In Germany, the VICON-RASA DVA installed on ICE2 high-speed trains reduces total radiated noise by 5–8 dB(A) [5]. Domestic research on wheel DVAs started relatively late. Chang et al. engineered an elastic wheel noise-reduction panel featuring a hybrid rubber-metal composite structure, which achieved a mere 2 dB(A) reduction in radial wheel modal noise; this design proved fundamentally constrained by inherent limitations including rubber deterioration and temperature susceptibility [6]. Conversely, Ding et al. implemented sandwich-type damping patches comprising rubber and stainless steel layers on Shanghai Metro Line 6, with experimental verification demonstrating a substantial 12.7 dB(A) average noise attenuation during curved track operations performance that sharply diminished to functionally negligible levels in linear track segments [7].

This study addresses the limitations of existing wheel-mounted dynamic vibration absorbers—specifically bolt fractures, inadequate straight-track noise reduction, and damping-layer degradation—by proposing a novel high-damping alloy noise-reduction plate for metro wheels. Utilizing the alloy's broad vibration-damping frequency range, low mass, high damping capacity, and superior mechanical properties, we conducted vibration simulations on wheels equipped with this component and performed impact-induced acoustic testing in semi-anechoic chambers. The high-damping alloy plate demonstrated significant noise reduction performance.

2 Principle and Structure

As a novel metallic material for vibration and noise reduction, damping alloys function through the following mechanism: Under external stress, the internal twin boundaries within the material undergo microscopic elastic deformation [8]. With increasing stress intensity, these twin boundaries experience progressive growth and expansion while generating new twin structures. Upon removal of the external force, the twin boundaries automatically revert to their original configuration. This reversible motion of twin boundaries under cyclic loading effectively dissipates vibrational energy through hysteresis loss. Notably, damping alloys not only maintain the mechanical properties characteristic of conventional metals but also exhibit exceptional damping characteristics comparable to elastomers, demonstrating a logarithmic decrement consistently around 0.3 with peak values reaching 0.72 [9]. Furthermore, these alloys maintain operational effectiveness across an exceptionally broad frequency spectrum ranging from 0.01 Hz to 5 MHz.

Based on the vibration-damping mechanism of the developed alloy, we designed a metro wheel noise reduction washer illustrated in Fig. 1. The washer comprises eight units uniformly distributed around the wheel circumference. Adopting a laminate configuration, the structure consists of two flat plates and three mass layers. The mass layers and plates are joined using semi-tubular hollow rivets. This three-layer mass design enhances the washer's vibration amplification capability, thereby absorbing greater amounts of wheel vibration energy.

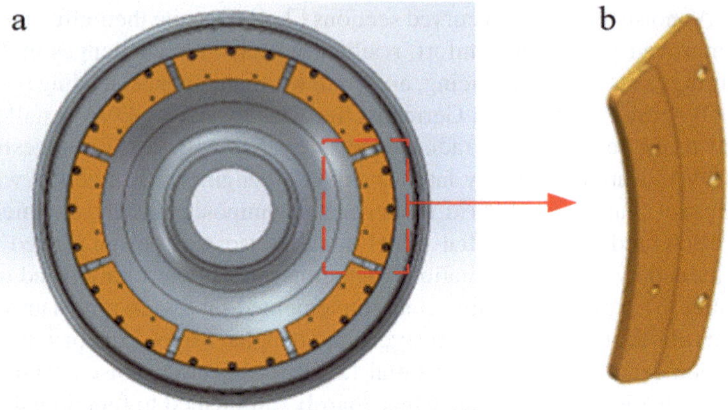

Fig. 1. A) Subway wheel and (b) noise reduction pad.

3 Finite Element Analysis

3.1 Computational Modal Analysis

To evaluate the effective vibration absorption frequency range of the vibration-damping alloy noise reduction pad on subway wheels—defined by the optimal ratio of the pad's natural frequency (ω) to the wheel's fundamental natural frequency (ω_n)—finite element analysis (FEA) was conducted using ANSYS [10]. The computational framework incorporated the material parameters specified in Table 1, with the mesh configuration demonstrating the following discretization characteristics:

- Noise reduction pad: 14, 231 quadratic hexahedral elements (SOLID186) with 103, 460 nodes.

Table 1. Material parameters.

Component	Elastic Modulus (Pa)	Poisson's Ratio	Density (kg/m^3)
Noise Reduction Pad	8×10^{10}	0.338	7250
Subway Wheel	2×10^{11}	0.300	7850

As detailed in Table 2, the modal analysis reveals that the damper maintains a frequency deviation ratio below 10% relative to the wheel across all modes above 1000 Hz. The operational bandwidth of vibration absorption exhibits a positive correlation with frequency. For instance, the damper's 4513 Hz mode effectively suppresses eight consecutive wheel natural frequencies spanning 4149.2–4858.3 Hz. Beyond 5000 Hz, the damper demonstrates multi-modal coupling capabilities, simultaneously attenuating over five wheel modes per octave. This multi-mode coupling mechanism enables broadband energy dissipation through superposition effects, particularly enhancing suppression efficiency in the mid-to-high frequency range.

Table 2. Modal analysis results of noise reduction sheets and subway wheels.

ω (Hz)	ω_n (Pa)	ω/ω_n (%)	ω (Hz)	ω_n (Pa)	ω/ω_n(%)	ω (Hz)	ω_n (Pa)	ω/ω_n (%)
306.42	458.05	33.10	3335.7	3091	7.92	4513	4435.8	1.74
599.62	458.05	30.19		3471	3.90		4691.8	3.81
997.09	1105.8	9.83		3589.5	7.07		4707.6	4.13
1692.5	1704	0.67		3471	6.34		4816.6	6.30
1968.1	1892.2	4.01	3691.1	3471	6.34		4858.3	7.11
	2028.3	2.97		3589.5	2.83	5411.6	5055	7.05
	2086.3	5.67		3709.4	0.49		5198.7	4.10
2143.8	2028.3	5.69		3743.5	1.40		5235.6	3.36
	2086.6	2.74	3985.8	3709.4	7.45		5631	3.90
2350.7	2406.2	2.31		3743.5	6.47		5762.1	6.08
2700	2658	1.58		4149.2	3.94		5970.4	9.36
	2801.6	3.63		4317.2	7.68	5474.1	5055	8.29
	2805.5	3.76		4385.2	9.11		5198.7	5.30
	2815.6	4.11	4513	4149.2	8.77		5235.6	4.56
3124.5	3091	1.08		4317.2	4.54		5631	2.79
	3471	9.98		4385.2	2.91		5970.4	8.31

3.2 Vibration Characteristics Analysis

To further verify the damping efficacy of the noise reduction pad, harmonic response analyses were conducted under radial and axial excitations for wheels with and without the damper. As demonstrated in Fig. 2a, the radial excitation tests revealed that the damper exhibited superior attenuation in the 2000–6400 Hz frequency range, with a maximum reduction of 72.98 dB at 5632 Hz. This high-frequency performance is attributed to the superposition effect of six natural frequencies above 5411.06 Hz within the damper structure. Conversely, limited suppression below 1500 Hz corroborates the frequency ratio discrepancies identified in modal analysis, confirming the damper's reduced efficacy in low-frequency regimes.

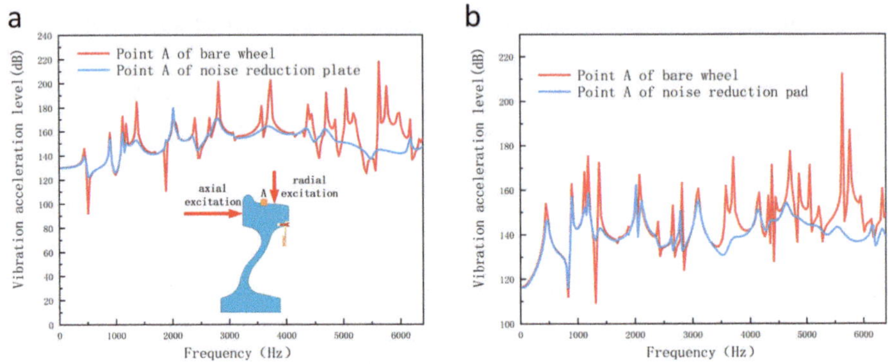

Fig. 2. (a) Vibration acceleration level of the radial excitation and (b) axial excitation.

For axial excitation (Fig. 2b), the total root-mean-square (RMS) acceleration of the damped wheel decreased by 3.58 dB across the 0–6400 Hz spectrum. The comparatively lower attenuation arises from two synergistic factors: (1) reduced overlap between the damper's natural frequencies and the wheel's inherent frequencies (tread, rim, and web) in the low-frequency range, and (2) incomplete energy absorption due to mismatched frequency compatibility in the 0–1500 Hz regime. These results underscore the frequency-dependent performance characteristics of the damper, where mid-to-high frequency vibrations are preferentially suppressed through multi-mode coupling mechanisms.

4 Experimental Analysis

To evaluate the vibration and noise mitigation performance of the vibration-damping alloy noise reduction pad on metro wheels, this study conducted drop-weight impact tests in the Acoustics Laboratory of Changzhou Southwest Jiaotong University Rail Transit Research Institute. The experiments were performed in a semi-anechoic chamber to isolate external acoustic interference.

Table 3. Noise reduction in the 0–6400 Hz frequency band.

Excitation	Bare wheel (dB(A))	Noise reduction pad wheel (dB(A))	Noise reduction amount (dB(A))
Radial	97.4	84.5	12.9
Axial	100.1	88.9	11.2

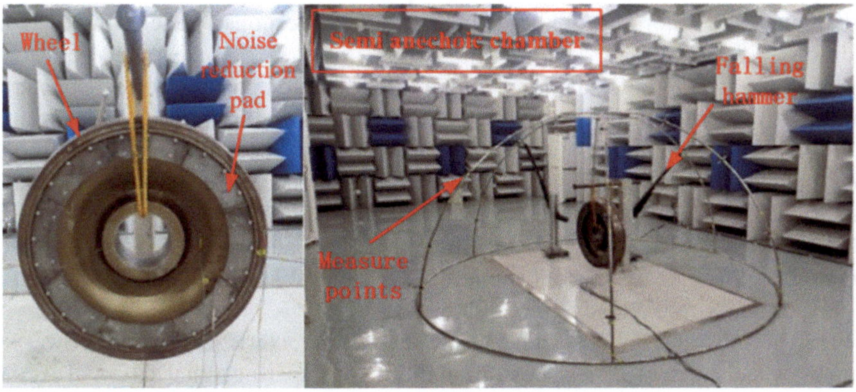

Fig. 3. Experimental test of noise reduction plate wheel.

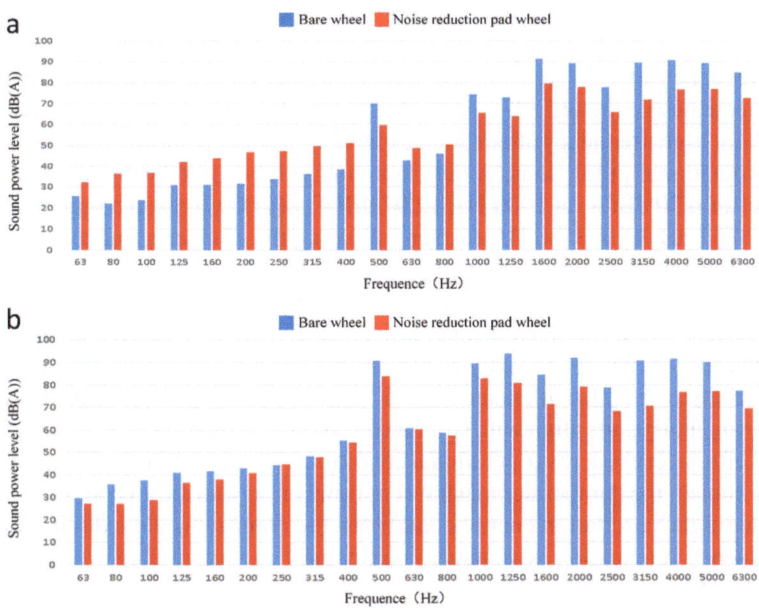

Fig. 4. (a) Sound power level of the radial excitation and (b) axial excitation.

As illustrated in Fig. 3, the test setup involved impacting freely suspended wheels—both with and without the damping pad—using a 25 mm steel ball dropped from a standardized height. Axial excitation targeted the wheel rim, whereas radial excitation focused on the nominal rolling circle. A 20-point spherical envelope method was implemented to capture the sound power radiation characteristics, with 20 calibrated acoustic sensors uniformly distributed over a hemispherical measurement surface. Each impact event triggered synchronized data acquisition across all sensors, ensuring a comprehensive recording of the wheel's vibratory response under controlled input conditions.

The impact of vibration-damping alloy noise reduction pads on wheel acoustic radiation was investigated through drop-weight impact tests under axial and radial excitation conditions. As shown in Table 3, the total radiated sound power level of the bare wheel under radial excitation was 97.4 dB(A), while the damped wheel achieved a 12.9 dB(A) reduction to 84.5 dB(A). For axial excitation, the TRSPL decreased by 8.9 dB(A) (from 98.6 dB(A) to 89.7 dB(A)). These experimental results align with the trend observed in harmonic response analysis, confirming the superior noise attenuation performance of the damper under radial loading conditions.

Figure 4 illustrates the 1/3-octave band spectral characteristics of sound power levels for metro wheels under free-suspension conditions, comparing radial and axial excitation scenarios. The vibration-damping alloy demonstrates significant noise attenuation above 1000 Hz, with peak reductions aligning with resonant frequency coupling between the damper and wheel structures. Under radial excitation, the total radiated sound power level decreases by 8.88 dB(A) at 1000 Hz, 11.6 dB(A) at 2000 Hz, and reaches a maximum attenuation of 17.6 dB(A) at 3150 Hz. This enhanced suppression at 3150 Hz is attributed to two synergistic mechanisms: the steel ball impact-induced resonance at the wheel's natural frequency of 3091 Hz and the simulated coupling of the damper's 3124.5 Hz (1.08% frequency ratio) and 3335.7 Hz (7.92% ratio) modes. Experimental results confirm superior high-frequency attenuation exceeding 12 dB(A) above 3000 Hz, consistent with computational predictions. Axial excitation exhibits similar trends, with TRSPL reductions of 6.70 dB(A) at 1000 Hz, 12.97 dB(A) at 2000 Hz, and a maximum attenuation of 20.02 dB(A) at 3150 Hz, further validating the damper's frequency-dependent performance characteristics.

The drop-weight impact tests demonstrated that the vibration-damping alloy noise reduction pad exhibits significant suppression effects on the acoustic radiation noise of metro wheels. The damper not only reduces rolling noise during straight-line operation (radial excitation) but also attenuates squeal noise in curved sections (axial excitation). The experimental results align with simulated resonance coupling trends, confirming the damper's effectiveness across operational conditions. Due to the energy constraints of the steel ball impact, the damper demonstrates optimal energy absorption at 3150 Hz. However, under higher excitation energy conditions, the damper would exhibit even greater high-frequency attenuation advantages, particularly in suppressing wheel resonances above 3000 Hz.

5 Conclusions

This study developed a novel noise reduction pad integrating vibration-damping alloy and dynamic vibration absorber (DVA) principles, validated through computational simulations and semi-anechoic chamber acoustic tests. The key findings are as follows:

- Modal calculations revealed that the vibration-damping alloy noise reduction pad exhibits natural frequencies within a 10% deviation ratio from the wheel's resonant frequencies above 1000 Hz. Notably, a single natural frequency of the pad demonstrated the capability to simultaneously absorb vibration energy from multiple wheel frequencies.
- The simulations demonstrated superior vibration suppression under radial excitation compared to axial excitation. The maximum vibration acceleration level reduction of 10.38 dB was observed at the wheel web, with absorption frequency bands aligning closely with modal analysis results. High-frequency attenuation was particularly effective.
- Semi-anechoic chamber tests confirmed total RMS noise reductions of 12.9 dB(A) and 11.2 dB(A) under radial and axial excitations, respectively. The peak attenuation at 3150 Hz corresponds to the optimal performance of the noise reduction pad, which is constrained by the limited impact energy of the steel ball used in testing.

References

1. Thompson, D.J., Jones, C.J.C.: A review of the modelling of wheel/rail noise generation. J. Sound Vib. **231**(3), 519–536 (2000)
2. Zhong, S., et al.: A parametric study on the effect of ring damper on wheel damping. J. Vib. Control **31**(5–6), 1034–1043 (2025)
3. Wang, B., Zhao, Q.: Application of rubber elastic wheels in light rail vehicles. Urban Rail Transit Research **1**(4), 37–40 (1998)
4. Liu, Y., et al.: Study on the vibration and noise reduction performance of dynamic vibration absorbers for train wheels. Noise Vibrat. Control **35**(3), 1 (2015)
5. Donders, S., et al.: A reduced beam and joint concept modeling approach to optimize global vehicle body dynamics. Finite Elem. Anal. Des. **45**(6–7), 439–455 (2009)
6. Chang, L., et al.: Design and performance analysis of noise reduction plates for tram elastic wheels. Noise Vibrat. Control **43**, 200–207 (2023)
7. Ding, J., et al.: Experimental study on vibration control of single-span rotor using multiple dynamic vibration absorbers. J. Mech. Electr. Eng. **36**(1), 5 (2019)
8. Chen, J., et al.: Noise reduction measures and effects of damped wheels on shanghai metro line 6 vehicles. Electric Locomotives Urban Rail Veh. **34**, 59–60 (2011)
9. Dong, L., et al.: Preparation and research on the structure and properties of high damping Mn-Cu alloy. Vacuum **229** (2024)
10. Wang, X., et al.: Vibration characteristics analysis of wheel-damper systems for rail vehicles. J. Dyn. Control **2** (2025)

Investigation into the Mechanical Behavior of the Crankshaft-Connecting Rod System in a V-Shaped Marine High-Pressure Air Compressor

Jun Feng[1,2] and Jing Liu[1](✉)

[1] Country School of Marine Science and Technology, Northwestern Polytechnical University, 710072, ChinaXi'an
jliu@cqu.edu.cn, jliu0922@nwpu.edu.cn

[2] Hengyuan Science and Technology Co., Ltd., China State Shipbuilding Corporation (CSSC), Beijing 410400, China

Abstract. As a critical auxiliary machinery in naval vessels, the V-type marine high-pressure air compressor exhibits significant dependence of its overall operational stability on the mechanical performance of the crank-connecting rod system. This study investigates the crank-connecting rod system of a V-type marine high-pressure air compressor, simulating the motion of pistons and connecting rods at all stages and deriving their inertial forces and moments. Using ADAMS software, a comparative simulation analysis was conducted on the crank-connecting rod system, incorporating flexible components, frictional kinematic pairs, and joint clearances. The results demonstrate that the flexible crankshaft absorbs a portion of the piston gas forces. Friction significantly influences the X-direction bearing load of the crankshaft in the low-frequency range, while the system exhibits enhanced dynamic stability in the mid-to-high frequency range. The clearance magnitude and different connecting-rod clearance configurations significantly affect the excitation forces on the crankshaft bearing in the load direction. Notably, smaller clearances induce larger excitation forces in the low-frequency range. This study provides a theoretical foundation and design reference for the crank-connecting rod system of differential-type marine high-pressure air compressors.

Keywords: V-type marine high-pressure air compressor · Crank-connecting rod · Kinematic and dynamic · Simulation

1 Introduction

As a core auxiliary system in naval vessels, the marine high-pressure air compressor not only serves as a critical enabler for ship propulsion systems but also forms the operational backbone of submarines—sustaining life-support functions and combat readiness. Its functionalities span multiple mission-critical naval operations: from supplying startup power for main/auxiliary engines, to precision pneumatic control of weapon launch

S. Halgamuge et al. (Eds.): ICACMVE 2025, LNME, pp. 87–114, 2026.
https://doi.org/10.1007/978-981-95-6911-3_10

systems, and even auxiliary command execution (e.g., whistle signaling), all relying on its stable compressed air output [1]. Current research on air compressors primarily focuses on health management, fault detection, and diagnostic methodologies [2–5].

The primary moving components of air compressors are crank-connecting rod mechanisms, which have been extensively studied by numerous scholars. Jiang [6] proposed a novel method for diagnosing wear faults in connecting rod small-end bearings through multibody simulation modeling and angular domain analysis of vibration signals. Experimental validation confirmed its capability to accurately identify faults and assess severity levels, providing a foundation for online monitoring. Godara [7] designed an internal combustion engine connecting rod using high-strength low-alloy (HSLA) steel, modeled it in Fusion 360, and conducted finite element analysis (FEA) in ANSYS 2021. The results validated compliance with operational safety standards, providing a reference for lightweight and high-efficiency connecting rod design. Peng [8] established a torsional vibration model of a shaft system incorporating variable inertia and nonlinear friction, analyzing the effects of connecting rod ratio, rotational speed, reciprocating mass, and damping. Experimental validation confirmed the model's accuracy, demonstrating that parameter variations exacerbate high-frequency resonance, while optimized damping improves vibration characteristics. Xiao [9] investigated the NL9340 diesel engine crankshaft by establishing free-mode and rigid-flexible coupled models to analyze natural frequencies, vibration modes, and dynamic characteristics. Simulations revealed a peak load of 30,000 kN at the connecting rod journal, with maximum stress and deformation concentrated at the fillet transition of the second cylinder's journal. Sakhraoui [10] developed a geometric and kinematic model of the GBCM engine, demonstrating that its connecting rod kinematics are equivalent to those of conventional engines. The results indicate that under identical crank-to-connecting-rod ratios, the kinematic behavior of the GBCM engine can be directly derived from classical engine theory. Albaghdadi [11] proposed a novel optimization method for vibration analysis of a double-crank rocker mechanism, with a detailed discussion on the optimized results. To minimize the influence of the aforementioned factors on the crank-slider mechanism, Wang [12] proposed, for the first time, a hybrid evaluation method combining entropy weighting and Grey Relational Analysis based on TOPSIS (GrT-TOPSIS) to achieve the optimal combination of process parameters. Gregory [13] investigated the frictional performance of hydrodynamic bearings in crank-connecting rod mechanisms under specific lubrication conditions. By employing the Elastohydrodynamic (EHD) motion equation and Lagrange's equations, a hydrodynamic lubrication mathematical model was established, yielding more realistic dynamic behavior results of the system. Matekar [14] designed a slider-crank mechanism with adjustable joint clearances at both ends of the connecting rod. Additionally, two models were developed in ADAMS/View to conduct simulation-based validation and analysis of this adjustable joint-clearance slider-crank mechanism. Lombardero [15].

proposed a mathematical model that accounts for mechanical interactions affecting stress states and material degradation processes, providing more accurate fatigue damage estimation. When applied to fatigue analysis of a planar slider-crank mechanism, the results demonstrated that joint clearance significantly influences the fatigue damage of the connecting rod. Cheng [16] investigated the influence of revolute joint clearance on the motion of reciprocating compressors. The results demonstrated that clearance significantly affects the kinematic parameters of the connecting rod. Finally, a nonlinear analysis was conducted using Lyapunov exponents.

Currently, research on the crank-connecting rod system of high-pressure air compressors in V-type marine vessels remains relatively limited. To address this gap, this paper first employs kinematic simulation to analyze the motion characteristics of the multi-stage pistons and connecting rods, and subsequently derives the inertial forces and moments of the crank-connecting rod mechanism. Building upon this foundation, a rigid-flexible coupled dynamic model of the crank-connecting rod system is established to analyze its dynamic response under ideal gas forces, while further investigating the influence of friction pairs on the main shaft load characteristics. Furthermore, incorporating the influence of kinematic joint clearances, this study systematically investigates the dynamic behavior of the crank-connecting rod mechanism in V-type marine high-pressure air compressors, aiming to provide theoretical foundations for the optimal design of such compressor systems.

2 Dynamic Modeling Method for Crank-Connecting Rod System of V-Type Marine High-Pressure Air Compressor Considering Clearance and Friction Effects

2.1 Calculation Method for Internal Forces in the Crank-Connecting Rod System of a V-Type Marine High-Pressure Air Compressor

To optimize space utilization, marine air compressors commonly adopt a V-type configuration featuring two piston banks arranged at a 60° included angle between cylinder axes. This compact design approach effectively minimizes the overall system dimensions while maintaining operational performance requirements. As the fundamental structural component of the compressor, the crank-connecting rod system facilitates the progressive pressurization of ambient air from low to high pressure states. Figure 1 illustrates the configuration and working principle of this system in V-type marine high-pressure air compressors.

Driven by the electric motor, the crankshaft (1) executes rotational motion at an angular velocity of ω. Through the Tier I-II-III connecting rod (2) and Tier I-II-IV connecting rod (9), this rotary motion is kinematically transformed into linear reciprocating motion of the Tier I-II-III piston (3) and Tier I-II-IV piston (8), respectively.

As illustrated in the enlarged structural diagram of the piston, distinct cross-sectional areas formed by regions of varying diameters generate differential pressure zones during reciprocating motion. When the piston moves along the green arrow direction. When the piston moves along the red arrow, the high-pressure and low-pressure zones accomplish gas compression, while the intermediate-pressure zone achieves gas expansion and intake.

During the operation of an air compressor, the primary forces acting on the crank-connecting rod system include the gas force, reciprocating inertia force, rotational inertia force, frictional force, piston lateral thrust, and crankshaft bearing reaction force. The gas force, generated by gas pressure, constitutes the primary driving force in the compressor. For computational convenience, the following sign convention is adopted: Positive gas force corresponds to a tensile load on the connecting rod. Negative gas force corresponds to a compressive load on the connecting rod. Accurate analysis and calculation of the various forces acting on the crankshaft-connecting rod system constitute a fundamental aspect of air compressor research. Therefore, it is essential to conduct detailed computational studies on these forces.

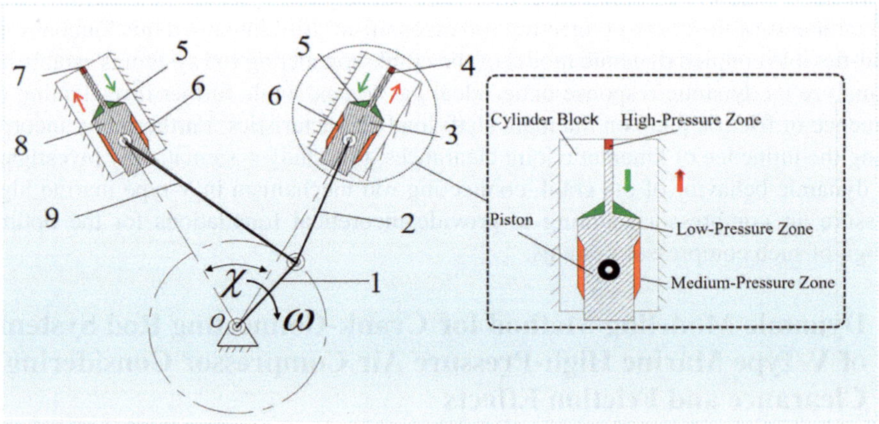

1—Crankshaft;2—Stage I-II-III connecting rod;3—Stage I-II-III piston;4—Stage III cylinder chamber;5—Stage I cylinder chamber;6—Stage II cylinder chamber;7—Stage IV cylinder chamber;8—Stage I-II-IV piston;9—Stage I-II-IV connecting rod;ω—Rotational speed of the crankshaft; χ —Phase angle between two piston columns, with a value of 60°.

Fig. 1. Working principle of the stepped-type marine air compressor crankshaft-connecting rod system

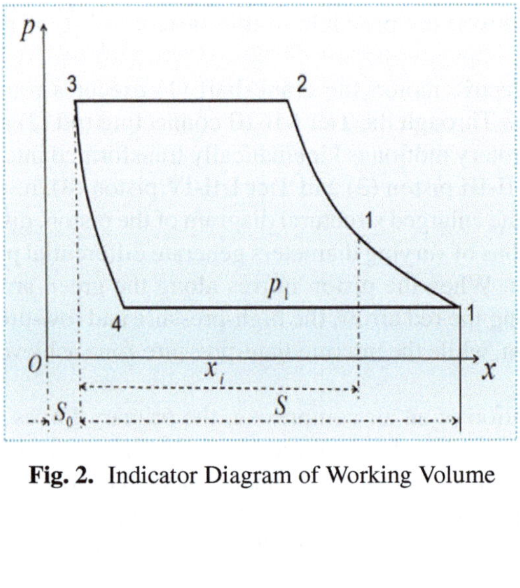

Fig. 2. Indicator Diagram of Working Volume

Calculation of Gas Force. In V-type marine high-pressure air compressors, the two cylinder banks (designated as BankI-II-III and BankI-II-IV) perform periodic multistage gas compression through their coordinated operation as the crankshaft rotates. A complete working cycle consists of four processes: compression, exhaust, expansion and intake, with the corresponding pressure-volume indicator diagram shown in Fig. 2. Let, P_1, P_2, P_3 and P_4 represent the gas forces acting on the respective cylinders during each stage (compression, exhaust, expansion, and intake). The governing equations are formulated as follows:

$$\begin{aligned} F_{g1} &= p_i A = \left(\frac{S + S_0}{x_i + S_0}\right)^n p_1 A, F_{g2} = p_i A, \\ F_{g3} &= p_i A = \left(\frac{S_0}{x_i + S_0}\right)^m p_3 A, F_{g4} = p_i A = p_1 A \end{aligned} \tag{1}$$

where, p_i is the pressure of the gas inside the cylinder at the i-th point during the compression process. S represents the piston stroke, S_0 is the equivalent stroke of the clearance volume, x_i is the displacement of the piston at the corresponding point i. p_1, p_2, p_3 and p_4 are the actual pressures inside each cylinder at the end of the intake process. m is the expansion coefficient, and A is the cross - sectional area of the cylinder.

For optimal vibration control in the compressor, the gas forces between the two piston banks are designed with a 60° phase difference and equal magnitude. Therefore, only the gas force calculation for Bank I-II-III is required during the design process. The crankshaft rotation angles for compression, exhaust, expansion, and intake processes are 0°, 127°, 180°, 210°, and 360° respectively. With reference to the relevant standard [17], the actual intake pressures for Bank I-II-III pistons are determined as follows, Primary stage actual intake pressure $p_{11} = 0.1$ MPa, Secondary stage actual intake pressure $p_{12} = 0.45$ MPa; Tertiary stage actual intake pressure $p_{13} = 2.6$ MPa. Polytropic Process Indices for Multi-stage Compression, Primary compression process index $n_1 = 1.35$, Secondary expansion process index $n_2 = 1.26$, Tertiary compression process index $n_3 = 1.4$. The ideal gas force acting on the piston as a function of crankshaft angle, calculated from the gas pressure and piston area of a single cylinder, is plotted in Fig. 3. For Bank I-II-III, the primary and tertiary pistons initiate gas compression at 0° crankshaft angle, while the secondary piston begins gas expansion at the same angular position. When the primary and tertiary pistons complete partial compression, the discharge valves open under differential pressure, during which the gas force acting on the pistons remains constant while the pistons continue moving toward the cylinder heads. As the piston approaches top dead center (TDC), the discharge valve closes under differential pressure. Subsequently, during the downward stroke, the gas force undergoes an abrupt decline while the secondary cylinder initiates compression of the crank-end gas volume. Due to the presence of high-pressure gas in the clearance volume, the intake valves of the primary and tertiary stages open with a deliberate time delay. Following the opening of the primary and tertiary intake valves, the air compressor initiates the gas intake phase, during which the gas force acting on the pistons remains constant until the subsequent compression cycle begins.

The principal forces acting on the crankshaft-connecting rod mechanism include the gas force, reciprocating inertia force, reciprocating friction force, and lateral force

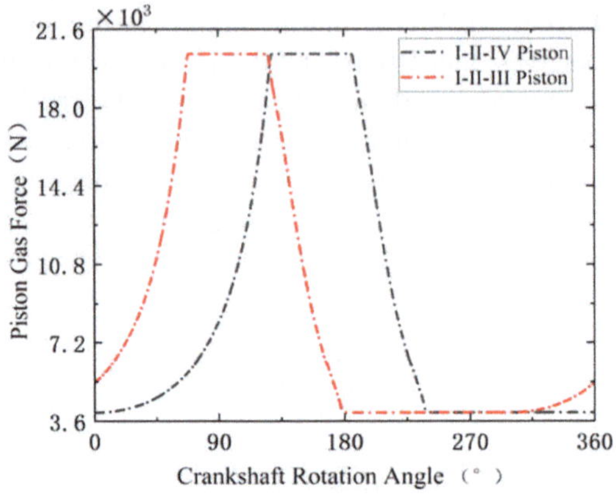

Fig. 3. Curves of Combined Gas Forces for I-II-IV and I-II-III Pistons as a Function of Crankshaft Angle

acting on the piston; the rotational inertia force and connecting rod force acting on the crankpin; as well as the driving force and rotational friction force acting on the main bearings [17]. The gas force constitutes the primary driving force in compressors, with the magnitudes of all other dynamic forces being intrinsically correlated to its variation. The reciprocating inertia forces and rotational inertia forces of moving components are determined by their masses and kinematic velocities, while the lateral thrust is provided by the cylinder liner and primarily depends on the crank-connecting rod geometry.

Calculation of Reciprocating Inertia Forces and Moments. The V-type marine high-pressure air compressor is a two-bank configuration where the resultant reciprocating inertia force of the crank-connecting rod mechanism equals the vector sum of inertial forces generated by both piston assemblies and their corresponding connecting rod small ends, as illustrated in Fig. 4. For multi-bank compressors, the reciprocating inertia force for any individual bank is calculated by:

$$\begin{cases} F_{Isi}^{I} = m_{si} r\omega^2 \cos(\theta - \phi_i + \delta_i) \\ F_{Isi}^{I} = m_{si} r\omega^2 \cos(\theta - \phi_i + \delta_i) \end{cases} \tag{2}$$

where, F_{Isi}^{I} denotes the primary (first-order) reciprocating inertia force for any bank; F_{Isi}^{II} the secondary (second-order) reciprocating inertia force; m_{si} the reciprocating mass of any bank; r the crank radius; θ the rotational angle of the I-II-IV crankshaft relative to the y-axis; ϕ_i the angle between the cylinder centerline of any bank and the y-axis; and δ_i the angular offset of any crankshaft relative to the first bank in the rotational direction.

To calculate the synthetic reciprocating inertia force of a V-type high-pressure air compressor on a ship, the first - and second-order reciprocating inertia forces of each row are decomposed into components along the x-axis and y-axis directions. The calculation formulas for the reciprocating inertia forces and inertia torques of multi-row reciprocating air compressors [17] are shown in Table 1. Where, F_{si}^{ab} is the a-th order reciprocating inertia force of any row i in the b-axis direction, m_{si} represents the reciprocating motion

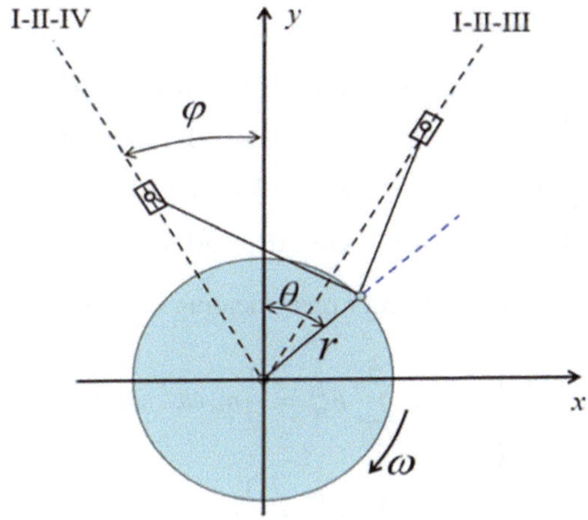

Fig. 4. Schematic Diagram of the Crank-Connecting Rod Mechanism Motion in a V-Type Marine High-Pressure Air Compressor

Table 1. Calculation Formulas for Reciprocating Inertia Forces and Inertial Moments in Multi-Column Air Compressors

	The component force of any row in the x-axis direction	The component force of any row in the y-axis direction
First-order reciprocating inertia force	$F_{si}^{Ix} = m_{si}r\omega^2\cos(\theta - \phi_i + \delta_i)\sin\phi_i$	$F_{si}^{Iy} = m_{si}r\omega^2\cos(\theta-\phi_i+\delta_i)\cos\phi_i$
First-order reciprocating inertia torque	$M_{si}^{Ix} = F_{si}^{Ix}a_i$	$M_{si}^{Iy} = F_{Isi}^{Iy}a_i$
Second-Order Reciprocating Inertia Force	$F_{si}^{IIx} = \lambda m_{si}r\omega^2\cos 2(\theta - \phi_i + \delta_i)\sin\phi_i$	$F_{si}^{IIy} = \lambda m_{si}r\omega^2\cos 2(\theta - \phi_i + \delta_i)\cos\phi_i$
Second-order reciprocating inertia torque	$M_{si}^{IIx} = F_{si}^{IIx}a_i$	$M_{si}^{IIy} = F_{si}^{IIy}a_i$

mass of any row i, M_{si}^{ab} represents the a-th order reciprocating inertia torque of any row i in the b direction, and a_i represents the distance from the inertia force to the symmetry center.

To mitigate excessive vibration during operation of V-type marine high-pressure air compressors, the crank-connecting rod system must be designed to ensure balanced reciprocating inertia forces between both cylinder banks. Therefore, the two sets of piston assemblies and the two sets of connecting-rod assemblies need to ensure equal mass through different materials and structural dimensions, that is $m_{s1} = m_{s2} = m_s$. Meanwhile, due to the limitation of the ship's space, $\varphi = 30°$, and the resultant forces of the first-order and second-order reciprocating inertia forces of the two rows on the

$x - x$ and $y - y$ axes can be expressed as:

$$\begin{cases} F_{s1}^{Ix} = m_s r\omega^2 \cos(\theta + 30^\circ)\sin(-30^\circ) \\ F_{s1}^{IIx} = \lambda m_s r\omega^2 \cos 2(\theta + 30^\circ)\sin(-30^\circ) \\ F_{s2}^{Ix} = m_s r\omega^2 \cos(\theta - 30^\circ)\sin 30^\circ \\ F_{s2}^{IIx} = \lambda m_s r\omega^2 \cos 2(\theta - 30^\circ)\sin 30^\circ \end{cases} \tag{3}$$

The resultant force of the first-order reciprocating inertia force:

$$\begin{cases} F_s^{Ix} = \sum_{i=1}^{2} F_{si}^{Ix} = \frac{1}{2} m_s r\omega^2 \sin\theta \\ F_s^{Iy} = \sum_{i=1}^{2} F_{si}^{Iy} = \frac{3}{4} m_s r\omega^2 \cos\theta \end{cases} \tag{4}$$

$$\frac{\left(F_s^{Ix}\right)^2}{\left(\frac{1}{2} m_s r\omega^2 \sin\theta\right)^2} + \frac{\left(F_s^{Iy}\right)^2}{\left(\frac{3}{4} m_s r\omega^2 \cos\theta\right)^2} = 1 \tag{5}$$

The resultant force of the second - order reciprocating inertia force:

$$\begin{cases} F_s^{IIx} = \sum_{i=1}^{2} F_{si}^{IIx} = \frac{\sqrt{3}}{2} \lambda m_s r\omega^2 \sin 2\theta \\ F_s^{IIy} = \sum_{i=1}^{2} F_{si}^{IIy} = \frac{\sqrt{3}}{4} \lambda m_s r\omega^2 \cos 2\theta \end{cases} \tag{6}$$

$$\frac{\left(F_s^{IIx}\right)^2}{\left(\frac{\sqrt{3}}{2} \lambda m_s r\omega^2 \sin 2\theta\right)^2} + \frac{\left(F_s^{IIy}\right)^2}{\left(\frac{\sqrt{3}}{4} \lambda m_s r\omega^2 \cos 2\theta\right)^2} = 1 \tag{7}$$

$$F_s^x = \sum_{i=1}^{2} F_{si}^x = \frac{1}{2} m_s r\omega^2 \sin\theta + \frac{\sqrt{3}}{2} \lambda m_s r\omega^2 \sin 2\theta \tag{8}$$

First-order reciprocating inertia torque:

$$M_s^{1x} = \frac{a}{4} m_s r\omega^2 \sin\theta, M_s^{1y} = \frac{3a}{4} m_s r\omega^2 \cos\theta \tag{9}$$

$$M_s^1 = \sqrt{\left(M_s^{1x}\right)^2 + \left(M_s^{1y}\right)^2} = \frac{a}{4} m_s r\omega^2 \sqrt{1 + 8\cos^2\theta} \tag{10}$$

Second - order reciprocating inertia torque:

$$\begin{cases} M_s^{2x} = \frac{\sqrt{3}a}{4} \lambda m_s r\omega^2 \sin 2\theta \\ M_s^{2y} = \frac{\sqrt{3}a}{8} \lambda m_s r\omega^2 \cos 2\theta \end{cases} \tag{11}$$

$$M_s^2 = \sqrt{\left(M_s^{2x}\right)^2 + \left(M_s^{2y}\right)^2} = \frac{\sqrt{3}a}{8}\lambda m_s r\omega^2\sqrt{3\sin^2 2\theta + 1} \tag{12}$$

The above analysis demonstrates that the first and second-order reciprocating inertia forces in the V-type marine high-pressure air compressor's crank-connecting rod system form elliptical trajectories, while the corresponding first- and second-order inertia moments vary periodically with crankshaft rotation angle. Due to the relatively small crank web spacing a and crank-connecting rod ratio λ,, the second-order inertia moment remains negligible. While higher-order (n $\geq$ 3) reciprocating inertia forces and moments exist, conventional analysis typically considers only the first and second-order components for engineering purposes.

Calculation of Rotational Inertia Forces and Moments. Within the realm of mechanical system dynamics, the calculation of rotational inertia forces demonstrates relative simplicity compared to the computationally intensive analysis required for reciprocating inertia forces, which exhibit complex harmonic variability. When the system's angular acceleration is explicitly defined, established theoretical frameworks confirm the absence of higher-order rotational inertia moments during rotary motion, thereby significantly simplifying the associated dynamic analysis procedures. The rotational inertia force F_r generated by rotating mass m_r is given by:

$$F_r = m_r r\omega^2 \tag{13}$$

The connecting rod converts the crankshaft's rotational motion into the piston's linear reciprocating motion. Consequently, during the operation of the crank-connecting rod mechanism, the connecting rod's movement comprises both reciprocating and rotational components. For computational simplification, the total reciprocating mass is equivalently concentrated at the piston pin center, while the unbalanced rotating mass is entirely transferred to the crankpin center [17]. The total reciprocating mass per bank is given by:

$$m_s = m_p + (0.3 \sim 0.4)m_1 \tag{14}$$

where, m_p is the mass of the piston and m_1 is the mass of the connecting rod.

The unbalanced rotating mass is defined as:

$$m_r = m_c + (0.7 \sim 0.6)m_1, m_c = m_{c1} + m_{c2}\frac{\rho}{r} \tag{15}$$

where, m_c is the unbalanced mass transferred to the center of the crankpin and mc_1, m_{c2} are the two parts of the assembled mass in the crankshaft, in kg, as shown in Fig. 5.

The rotational inertia force is consequently determined as:

$$F_r = \left(m_{c1} + m_{c2}\frac{\rho}{r} + (0.7 \sim 0.6)m_1\right)r\omega^2 \tag{16}$$

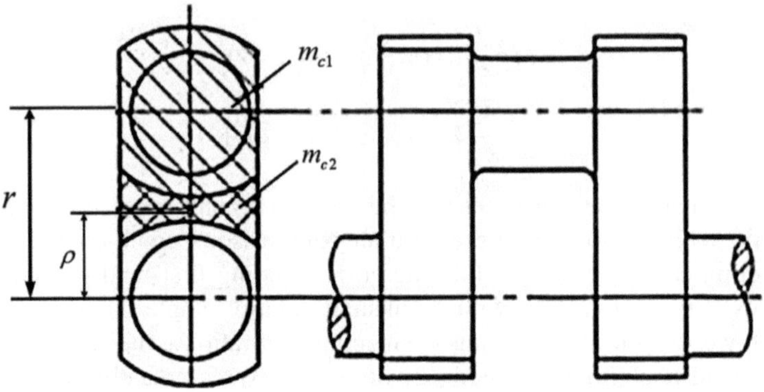

Fig. 5. Conversion of Crankshaft Mass

Calculation of Friction Forces and Piston Forces. During the working process of each row of pistons, it is assumed that the reciprocating frictional force F_f is a constant force and is always opposite to the direction of piston movement. The indicated power of each row is denoted as N_i, and then:

$$N_i = \frac{n}{60} P_{si} V_{si} \lambda_{vi} \frac{n_i}{n_i - 1} (\varepsilon_s^{\frac{n_i - 1}{n_i}} - 1) \tag{17}$$

$$\lambda_{vi} = 1 - \alpha_i (\varepsilon^{\frac{1}{m_i}} - 1) \tag{18}$$

where, n is the rotational speed of the compressor, ε_s is the actual pressure ratio of intake and exhaust, P_{si} is the nominal intake pressure, V_{si} is the working volume, n_i is the compression process index, α_i is the relative clearance volume, ε is the nominal pressure ratio, and m_i is the polytropic process coefficient.

$$F_f = 0.6 \times \frac{N_i \left(\frac{1}{\eta_m} - 1 \right) \times 60}{2Sn} \tag{19}$$

where, η_m is the mechanical efficiency. Referring to relevant data [17], $\eta_m = 0.85$ is taken.

In the compressor, the gas force F_g, the reciprocating inertia force F_{Is}, and the frictional force F_f all act along the direction of the cylinder center-line. Their algebraic sum is called the piston force F_p, that is:

$$F_p = F_g + F_{Is} + F_f \tag{20}$$

The lateral thrust force F_h acting on the piston can be expressed as:

$$F_h = F_p \frac{\lambda \sin \theta}{\sqrt{1 - \lambda^2 \sin^2 \theta}} \tag{21}$$

The variation curves of piston force and lateral thrust force as functions of crankshaft rotation angle are illustrated in Fig. 6 and Fig. 7. The piston force exhibits phase-dependent fluctuations during both intake and exhaust strokes when reciprocating inertia forces and frictional effects are accounted for in the analysis. Both the resultant piston force and lateral thrust force simultaneously attain their peak magnitudes upon discharge valve opening. When the intake valve opens, the resultant piston force reaches its negative peak, while the lateral thrust force exhibits a phase-delayed reversal to its maximum negative value.

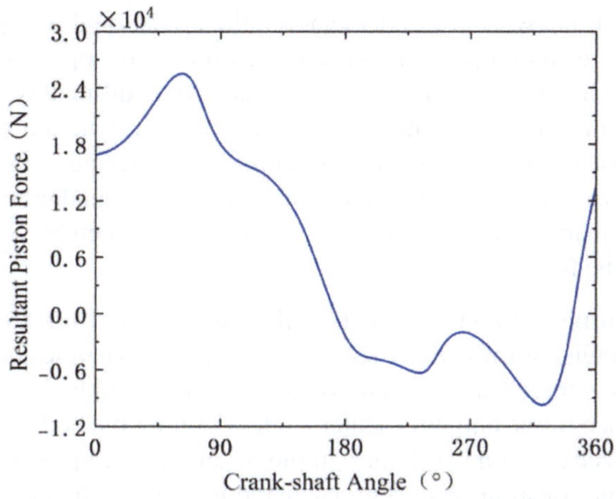

Fig. 6. Variation of Resultant Piston Force with Crankshaft Angle

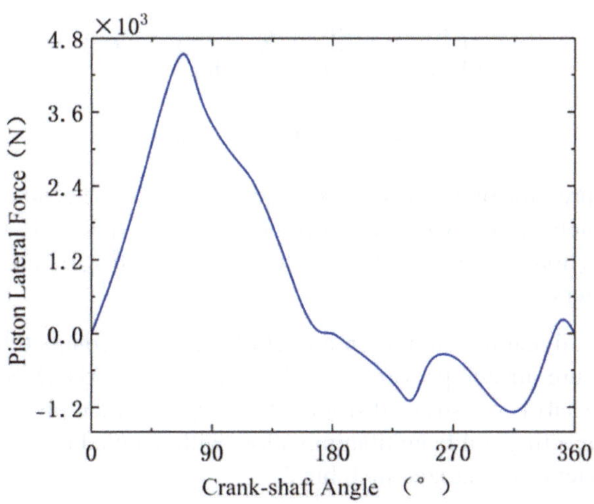

Fig. 7. Variation of Piston Lateral Force with Crankshaft Angle

2.2 Dynamic Model of the Crank-Connecting Rod System for V-Type Marine High-Pressure Air Compressors

Significance of Developing a Dynamic Model for the Crank-Connecting Rod System of V-Type Marine High-Pressure Air Compressors in Research and Practice. This model not only enables rigorous analysis of the kinematic behavior of system components and their interactions with gas pressure, inertial forces, and frictional effects--thereby establishing a fundamental framework for understanding the operational mechanisms of V-type marine high-pressure air compressors—but also facilitates the optimization of materials and structural design to reduce component loads and wear, ultimately enhancing overall compressor performance and reliability. Furthermore, the dynamic model serves as a virtual platform for exploring novel design concepts, fostering deep interdisciplinary integration, and providing both theoretical and practical foundations for technological innovation in marine high-pressure air compressors—thereby driving continuous advancements in the field.

Rigid-Body Dynamic Model of the Crank-Connecting Rod System. ADAMS employs the Lagrangian formulation to derive the system's dynamic equations of motion. The software selects three Cartesian coordinates of each rigid body's center of mass in the inertial reference frame and three Euler angles defining the body's orientation as the Cartesian generalized coordinates. It then employs Lagrange's equations with multipliers to handle holonomic constrained systems (with redundant coordinates) or nonholonomic constrained systems, ultimately deriving the kinematic equations expressed in terms of these Cartesian generalized coordinates.

$$\frac{d}{dt}\left(\frac{\partial K}{\partial \dot{q}_j}\right) - \frac{\partial K}{\partial q_j} + \sum_{i=1}^{n} \frac{\partial \psi_i}{\partial q_i}\lambda_i = F_j \tag{22}$$

$$\psi_i = 0(i = 1, 2, \cdots, m) \tag{23}$$

where, K is the expression of the system's kinetic energy, q_i is the generalized coordinates describing the system, ψ_i is the constraint equations of the system, F_i is the generalized forces in the directions of the generalized coordinates, and λ_i is an $m \times 1$ matrix of Lagrange multipliers.

In this study, a dynamic simulation model of the crank-connecting rod system for marine high-pressure air compressors was developed using ADAMS, as illustrated in Fig. 8. The crankshaft is constructed from 40Cr alloy steel, the piston from aluminum alloy, and the connecting rod from titanium alloy, with detailed material properties and geometric parameters summarized in Table 2.

Dynamic Model of the Crank-Connecting Rod System with Flexible Components. The equations of motion for flexible bodies are derived from the following Lagrange's equation:

$$\begin{cases} \frac{d}{dt}\left(\frac{\partial L}{\partial \dot{\xi}}\right) - \frac{\partial L}{\partial \varsigma} + \frac{\partial \Gamma}{\partial \varsigma} + \left[\frac{\partial \psi}{\partial \varsigma}\right]^T \lambda - Q = 0 \\ \psi = 0 \end{cases} \tag{24}$$

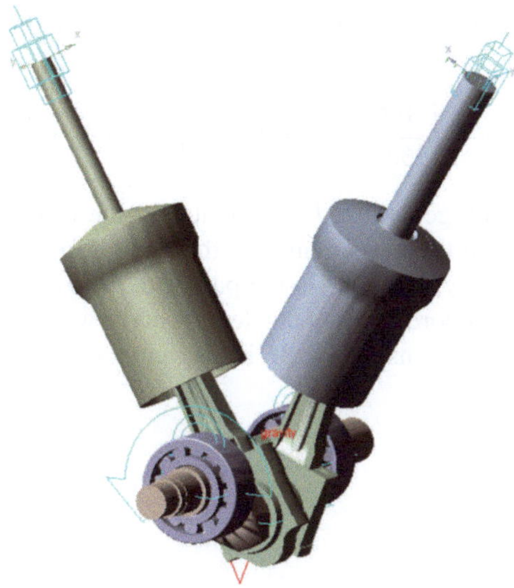

Fig. 8. Rigid Body Crank-Slider Mechanism Modeled in ADAMS.

Table 2. Material Properties of Main Components.

Name	Material	Density(kg/m^3)	Elastic Modulus (N/m^2)	Poisson's Ratio	Dimensions/mm	Mass/kg
Piston	2A12	2770	7.22E + 10	0.33	/	7.2
Two-stage connecting rod	TC4	4510	1.13E + 11	0.342	265	2.4
Two - stage piston pin	38CrMoAlA	7850	1.9176E + 11	0.272	/	/
Bearing	GCr15	7827	2.01E + 11	0.277	/	/
Crankshaft	40Cr	7870	2.11E + 11	0.277	50	12.67

$$\zeta = \left[x, y, z, \psi, \theta, \varphi, q_i(i = 1, \cdots, M)\right]^T = [r, \psi, q]^T \tag{25}$$

where, ψ is the constraint equation, λ is the Lagrange multiplier corresponding to the constraint equation, ζ is the generalized coordinate, Q is the generalized force projected onto ζ, L is the Lagrangian term, defined as $L = T - W$, where T and W are the kinetic energy and potential energy respectively, and Γ is the energy dissipation function.

Substitute the obtained T, W and Γ into Eq. (32), and the final differential equation of motion is:

$$M\ddot{\zeta} + \dot{M}\dot{\zeta} - \frac{1}{2}\left[\frac{\partial M}{\partial \zeta}\dot{\zeta}\right]^T \dot{\zeta} + K\zeta + f_g + D\dot{\zeta} + \left[\frac{\partial \psi}{\partial \zeta}\right]^T \lambda = Q \tag{26}$$

Among them, $\ddot{\zeta}$ is the generalized coordinates of the flexible body and their time derivatives; M and $\dot{M}$ are the mass matrix of the flexible body and its derivative with respect to time respectively; $\partial M / \partial \zeta$ is the partial derivative of the mass matrix with respect to the generalized coordinates of the flexible body, which is a tensor of $(M + 6) \times (M \times 6) \times (M + 6)$ dimensions, where M is the number of modes.

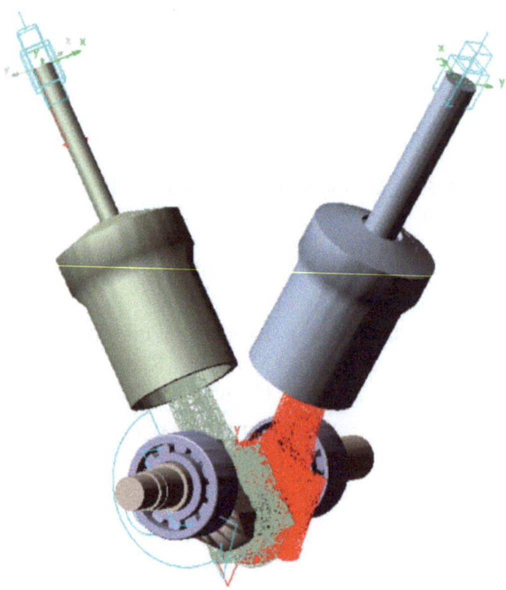

Fig. 9. Flexible Connecting Rod System in ADAMS

Traditional multibody system dynamics theories have predominantly focused on the motion analysis of rigid bodies, whereas rigid-flexible coupling research incorporates deformation effects of flexible components, representing a critical yet challenging problem in the field of flexible multibody system dynamics. The crank-connecting rod system of V-type marine high-pressure air compressors constitutes a multibody dynamic system, for which rigid-flexible coupling analysis based on theoretical formulations presents greater complexity compared to finite element simulations. This study employs ADAMS simulation software to perform rigid-flexible coupling analysis on the crank-connecting rod system, with tetrahedral elements utilized to model the flexible behavior of both the crankshaft and connecting rod components. The crankshaft was discretized into 11,650 finite elements, while the connecting rods for Banks I-II-III and I-II-IV were meshed with 6,354 elements each. The dynamic models of the flexible connecting rod and crankshaft are illustrated in Fig. 9 and Fig. 10, respectively.

Fig. 10. Flexible Crankshaft System in ADAMS

Clearance Model of the Crank-Connecting Rod System. The clearance model between the crankshaft and connecting rod big end is illustrated in Fig. 11, where R denotes the inner radius of the connecting rod big end and r represents the crankshaft journal radius.

As derived from Fig. 11, the radial clearance between the crankshaft journal and connecting rod big end bearing is quantified as c

$$c = R - r \tag{27}$$

The vector diagram illustrating the clearance between the crankshaft journal and connecting rod big end is presented in Fig. 12. In the global XOY coordinate system, $\mathbf{p}$ is the vector from the center of the crankshaft to the center of the big end of the connecting rod, $\mathbf{A}$ is the position vector of the big end center in the XOY frame, and $\mathbf{B}$ is the position vector of the crankshaft in the XOY coordinate system.

As illustrated in Fig. 11,

$$\mathbf{p} = \mathbf{A} - \mathbf{B} \tag{28}$$

The clearance magnitude $|\mathbf{p}|$ between the crankshaft journal and connecting rod big end can be expressed as:

$$|\mathbf{p}| = \sqrt{\delta_x^2 + \delta_y^2} \tag{29}$$

When the crankshaft comes into contact and collides with the big end of the connecting rod, the contact depth δ can be calculated using Eq. (38).Where $\delta \geq 0$ indicates contact collision between the crankshaft and connecting rod big end, while $\delta < 0$ signifies no contact collision occurs.

$$\delta = |\mathbf{p}| - c \tag{30}$$

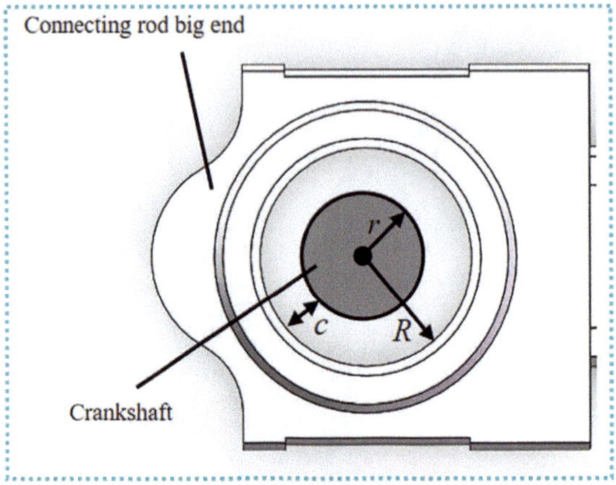

Fig. 11. Clearance Model of Crankshaft and Connecting Rod Big End

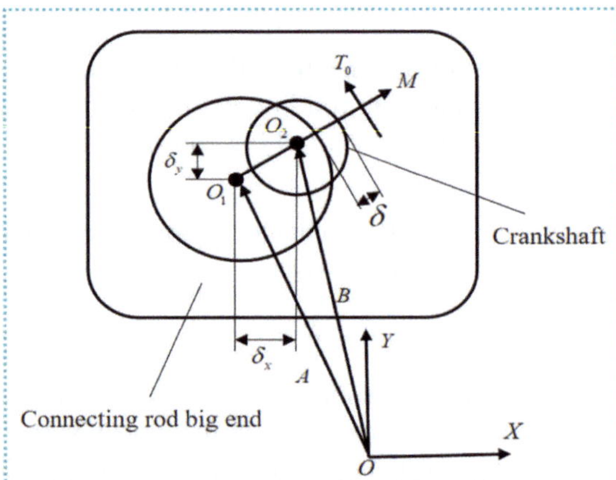

Fig. 12. Vector Diagram of Clearance Between Crankshaft and Connecting Rod Big End

When contact collision occurs between the crankshaft and the connecting rod big end, the tangential velocity $\mathbf{V_{T0}}$ and normal velocity $\mathbf{V}_M$ at the contact point can be expressed as:

$$\begin{cases} \mathbf{v}_M = \left(\dot{\mathbf{A}} - \dot{\mathbf{B}}\right)^T \mathbf{M} \\ \mathbf{v}_{\mathbf{T}_0} = \left(\dot{\mathbf{A}} - \dot{\mathbf{B}}\right)^T \mathbf{T}_0 \end{cases} \tag{31}$$

where, $\mathbf{M}$ is the unit normal vector, $\mathbf{T}_0$ is the unit tangential vector, and $\mathbf{M} \cdot \mathbf{T}_0 = 0$.

Contact Model of the Crank-Connecting Rod System. In ADAMS, there are two primary methods for defining contact forces [18]: the restitution-based approach (compensation method) and the impact function method. Comparatively, the parameters in the restitution-based method are more difficult to set accurately. Therefore, the impact function method is generally preferred for calculating contact forces. The impact function method computes the contact force between two components based on the Impact function. The contact force consists of two components: an elastic force resulting from the mutual penetration between the two bodies, and a damping force induced by their relative velocity. The description of the IMPACT function is illustrated in Fig. 13.

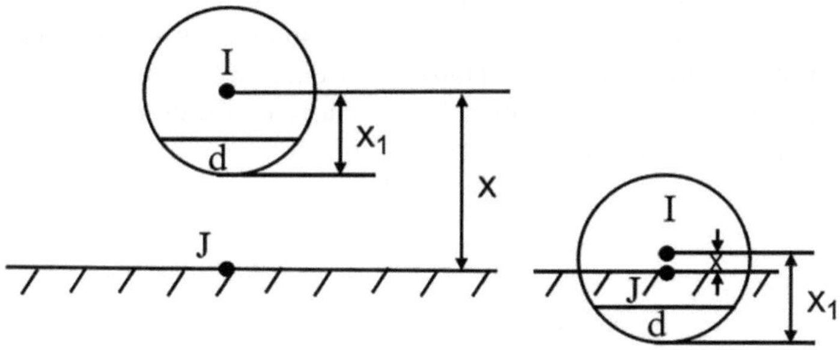

Fig. 13. Illustration of the IMPACT Function in ADAMS

Define the contact between the big ends of the I-II-IV and I-II-III connecting rods and the crankshaft using the built-in impact function method in ADAMS, while keeping other kinematic pairs in an ideal state. Simultaneously, the Coulomb friction model parameters are selected, and the contact force is calculated using the following formula:

$$IMPACT = \begin{cases} 0 & x \geq x_1 \\ MAX\left(0, k(x_1 - x)^e - STEP(x, x_1 - d, c_{\max}, x_1, 0) \cdot \dot{x}\right) & x < x_1 \end{cases} \tag{32}$$

where, x_1 is the initial distance between the two colliding bodies, x is the instantaneous distance during collision, $\dot{x}$ is the time derivative of the inter-body distance, k is the stiffness coefficient, e is the exponent of the contact force law, $c_{\max}$ is the maximum damping coefficient, and d is the penetration depth at which the damping force reaches its maximum value.

The contact stiffness coefficient k is calculated as follows [19]:

$$k = \frac{4}{3\pi(h_i + h_j)} R^{1/2}, \quad R = \frac{R_i R_j}{R_i + R_j}, \quad h_k = \frac{1 - \nu_k^2}{\pi E_k}, \quad k = i, j \tag{33}$$

where R_i,ν_i and E_i represent the radius, Poisson's ratio and elastic modulus of the component respectively. The elastic force exponent ranges from 1.3 to 1.5.

To prevent the discontinuity of the damping force during the collision process, the *STEP* function is adopted in the equation. Its form is $STEP(x, x_0, h_0, x_1, h_1)$, which is specifically as follows:

$$STEP = \begin{cases} h_0 & x \leq x_0 \\ h_0 + a \cdot \Delta^2 \cdot (3 - 2\Delta) & x_0 < x < x_1 \\ h_1 & x \geq x_1 \end{cases} \tag{34}$$

where, $a = h_1 - h_0$, $\Delta = (x - x_0)/(x_1 - x_0)$.

3 Results and Discussions

3.1 Comparative Analysis of Dynamic Models

Comparative Simulation Analysis of the Crankshaft-Connecting Rod System in a V-Type Marine High-Pressure Air Compressor with Different Flexible Components. The gas force acting on the pistons was constructed as a spline curve and applied to the corresponding piston centroids in the model, with the force direction aligned along the piston's motion path during gas expansion. By default, the gas forces on both piston stages were assumed equal in magnitude but with a phase difference of 60°, as reflected in the simulation results shown in Fig. 14 and Fig. 15.

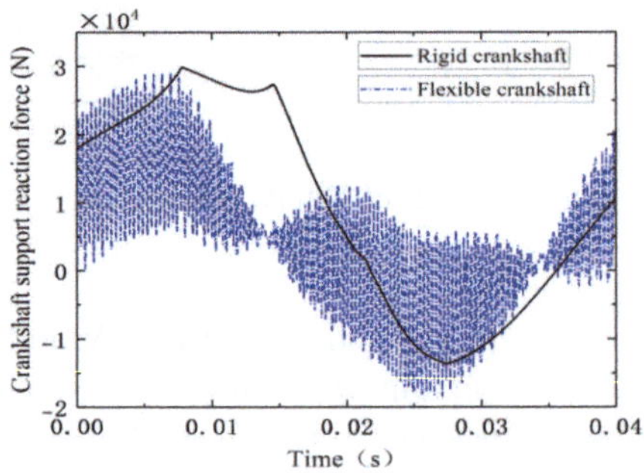

Fig. 14. Comparison of the crankshaft support reaction forces and flexible state

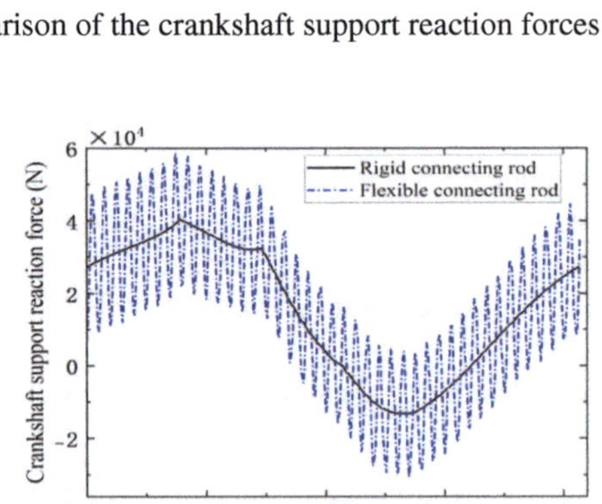

Fig. 15. Crankshaft support reaction forces with the connecting rod in rigid state between the rigidified and flexibilized crankshafts

As evident from Fig. 14, when only the crankshaft is treated as flexible, a comparative analysis of the crankshaft support reaction forces between the flexible and rigid crankshaft configureations reveals that the reaction forces in the flexible crankshaft case are significantly reduced compared to the rigid case. This demonstrates that the elastic deformation of the flexible crankshaft can effectively absorb a portion of the piston gas forces. When the connecting rod is modeled as flexible, the mean value of the crankshaft support reaction forces remains essentially unchanged compared to the rigid connecting rod case. This observation indicates that in the V-type marine high-pressure air compressor during operation, the crankshaft exhibits more significant flexible-body characteristics than the connecting rod, as illustrated in Fig. 15.

Systematic analysis reveals that neither the introduction of crankshaft flexibility nor connecting rod flexibility fundamentally alters the excitation force trends at the crankshaft support. For V-type marine high-pressure air compressors, the dynamic analysis typically considers only crankshaft flexibility. However, given the crankshaft's inherently high stiffness, rigid-body dynamics generally suffice to meet routine computational requirements.

Analysis of Friction Effects on Excitation Forces in the Crankshaft-Connecting Rod System. By appropriately configuring the contact model parameters, as listed in Table 3. The gas forces acting on the pistons in Fig. 3 were applied to the centroids of the I-II-III and I-II-IV stage pistons based on their phase relationships. Under the assumption that both piston stages bear equal loads with a 60° phase difference, multibody dynamics simulations were conducted to obtain the x-direction of the crankshaft support forces on the main shaft for both the I-II-III and I-II-IV stage connecting rod big ends, with and without friction effects. The results are presented in Fig. 16 and Fig. 17, respectively.

Table 3. Parameter Settings for the Impact Function and Coulomb Friction

Parameters	Value	Parameters	Value
Collision Stiffness $k(N/m)$	1.2E + 10	Kinetic Friction Coefficient μ_d	8.5E-002
exponent of the contact force law e	1.5	Static Friction Coefficient μ_s	7.0E-002
Maximum Damping Coefficient $c_{\max}$	18500	Friction-Velocity Relationship in Kinetic Friction v_d	0.1
Penetration Depth $d(m)$	5.58E-004	Static Friction Velocity v_s	1.0E-002

As evidenced in Fig. 16 and Fig. 17, both configurations exhibit pronounced force peaks in the low-frequency range, with amplitudes approaching 1.2×10^4 N. This observation indicates significant crankshaft bearing loads along the x-direction under low-frequency excitation. As the frequency increases from 0 Hz to 1000 Hz, the force amplitude exhibits an overall decreasing trend, indicating enhanced dynamic stability and reduced susceptibility to external disturbances at higher frequencies. When friction is not considered, the force amplitude approaches zero beyond 250 Hz, indicating that the dynamic force in the x-direction becomes negligible when the excitation frequency exceeds 250 Hz. When friction is considered, small amplitude fluctuations persist beyond 250 Hz, indicating that the system exhibits dynamic force variations in the medium-to-high frequency range. This demonstrates that the presence of friction can modify the frequency spectrum of excitation forces in the x-direction of the crankshaft bearing.

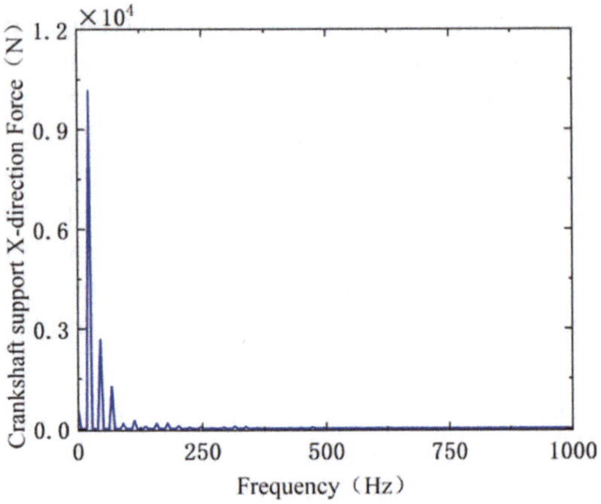

Fig. 16. Frictionless Condition at the Connecting Rod Big End

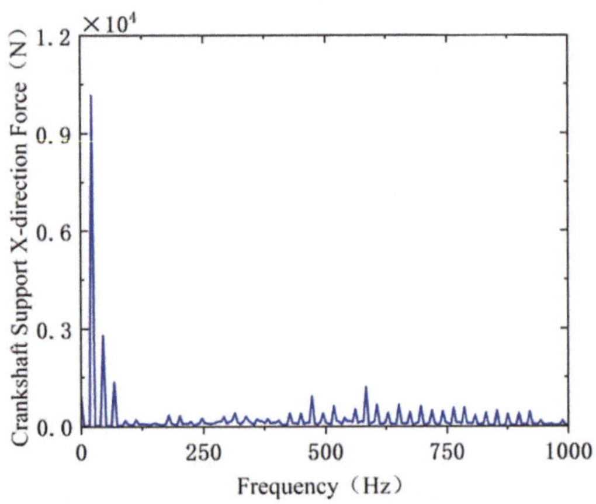

Fig. 17. Frictional Condition at the Connecting Rod Big

3.2 Analysis of Connecting Rod Big End Clearance Effects on Crankshaft-Connecting Rod System Contact Loads

For clearance adjustment purposes, the crankshaft-connecting rod system of the V-type marine high-pressure air compressor was simplified, with the schematic shown in Fig. 18. The simulation results are presented in Fig. 19, where both curves exhibit distinct periodic fluctuations. The curves repeat similar variation patterns at approximately 0.044 s intervals. The curves reach their minimum values at 0.022 s and 0.066 s, while peaking at 0.044 s. The high degree of similarity in curve trends between the original and simplified models indicates that all critical characteristics have been preserved, validating the use of the simplified model for subsequent investigations.

The V-type marine high-pressure air compressor experiences challenges in maintaining stable operational speeds due to the combined effects of unbalanced inertial forces,

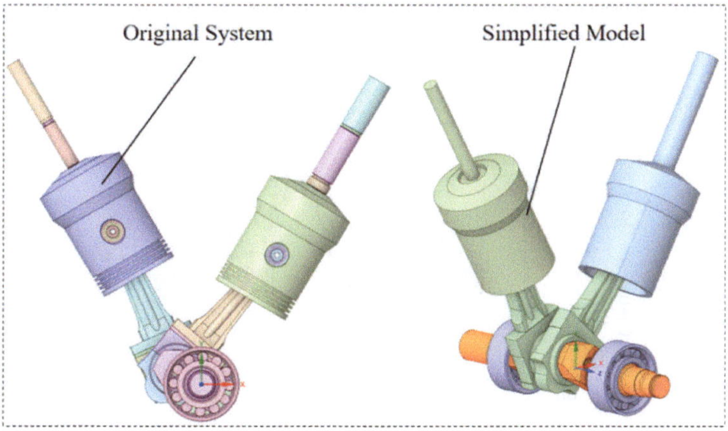

Fig. 18. Model Processing Results

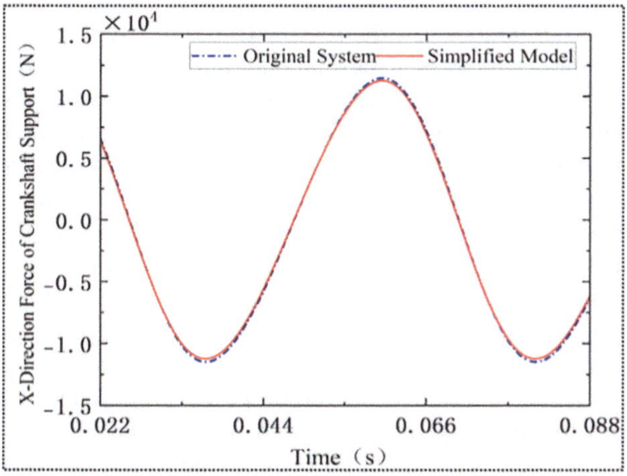

Fig. 19. Comparison of Original &Simplified Model Performance

frictional resistance, gas forces, and kinematic joint clearances. Unstable operational speeds induce significant whole-body vibrations, compromising the vessel's stealth performance and ultimately degrading its combat effectiveness. Given the nonlinearity and complexity of gas forces, friction, and clearance variations, a high-moment-of-inertia flywheel is typically connected to the compressor crankshaft. The flywheel mitigates speed fluctuations by storing excess kinetic energy during power surges and releasing it during power deficits, thereby stabilizing operational dynamics. When the driving load fluctuates—particularly during load increases—the flywheel releases its stored kinetic energy to maintain stable compressor operation, thereby ensuring the vessel's combat readiness. To accurately quantify the influence of joint clearances on the driving power of the V-type marine high-pressure air compressor's crank-connecting rod mechanism—and thereby provide theoretical support for flywheel design—this study employs Adams

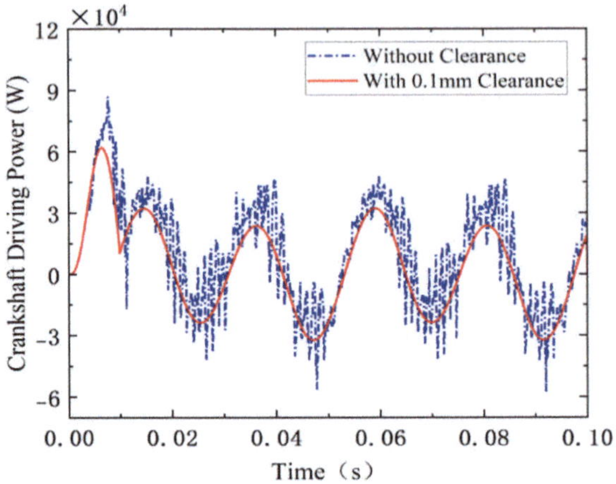

Fig. 20. Impact of Clearance on Crankshaft Drive Power

to conduct dynamic simulations of the mechanism with explicit clearance modeling. In the simulation, the compressor's startup condition is replicated by constraining the rotational speed to a fixed value of 1350 RPM without applying gas forces, thereby isolating the driving torque variations attributable solely to inertial and frictional effects. The big-end bore diameter of both connecting rod stages was designed to exceed the corresponding crankshaft journal diameter by 0.1 mm. A STEP function was applied to ramp the rotational speed to 1350 RPM within 0.1 s, enabling comparative analysis of driving torque and power under both ideal (zero-clearance) and clearance conditions. The simulation results are detailed in Fig. 20. The shorter the startup time, the greater the required starting torque. In particular, a startup period of 0.01 s demands a substantially higher torque, after which the driving power converges to a steady-state value. Under the zero-clearance condition, the torque power amplitude ultimately stabilizes with fluctuations within the range of ± 32,282 W. When radial clearance exists between the crankshaft journal and connecting rod big end, the torque power exhibits transient fluctuations that significantly exceed those observed under ideal joint conditions. Without the stabilizing effect of a flywheel, the instantaneous torque power amplitude can reach 86,971 W—a 170% increase compared to the zero-clearance case (±32,282 W).

In V-type marine high-pressure air compressors, the clearance range between the crankshaft and the connecting rod big end is influenced by multiple factors, including working pressure, rotational speed, lubrication conditions, and manufacturing precision. Generally, an appropriate clearance ensures effective lubrication and smooth operation, whereas excessive or insufficient clearance can adversely affect compressor performance. Excessive clearance may intensify impact and vibration between mating surfaces, accelerating component wear and potentially causing abnormal noise. Conversely, insufficient clearance can lead to overheating, seizure, and other operational failures. To investigate the influence of clearance on the operational performance of the V-type marine high-pressure air compressor, the rated rotational speed of the compressor was set at 1350 RPM. According to the relevant technical specifications for marine air compressors, the wear clearance of kinematic pairs generally should not exceed

0.1 mm, while the installation clearance between the crankshaft and connecting rod is specified as 0.1 mm. Therefore, three distinct clearance values—0.10 mm, 0.15 mm, and 0.20 mm—were selected for the crankshaft and connecting rod big-end interface to establish a dynamic simulation model for computational analysis, yielding the time-domain acceleration diagrams in the *x*-direction for the corresponding pistons under different clearance conditions, as illustrated in Figs. 21a-c. The peak acceleration transients in the X-direction for Stage I-II-IV pistons consistently occur at top/bottom dead centers. Increased clearance leads to reduced contact frequency but higher acceleration amplitudes. Corresponding velocity-acceleration phase diagrams under varying clearance conditions (Figs. 21d-f) demonstrate progressive concentration of phase trajectories near top/bottom dead centers regions with larger clearances.

To systematically investigate the influence of rotational speed on piston dynamic characteristics, computational simulations were performed with a fixed clearance of 0.1 mm at three distinct speed conditions (450, 900, and 1350 r/min). The resultant time-domain acceleration profiles along the *x*-direction under varying rotational speeds are comparatively presented in Figs. 21a-c. At 450 r/min, the zero-clearance condition exhibits smooth acceleration profiles with minimal fluctuations, whereas the clearance case demonstrates pronounced oscillatory behavior characterized by frequent high-frequency spikes and troughs in the time-domain response. At 900 r/min, both congfigurations exhibit increased oscillation amplitudes. Notably, the clearance cause higher peak accelerations. At 1350 r/min, the peak value increases significantly. The curve with clearance shows an extremely wide and irregular fluctuation range. In summary, the analysis demonstrates that increasing rotational speed significantly amplifies both the peak magnitudes and oscillation intensity of acceleration profiles. Furthermore, the presence of clearance exacerbates these dynamic instabilities.

Under varying rotational speeds with 0.1 mm clearance, the velocity profiles of Stage I-II-IV pistons exhibit progressively intensified instability as speed increases. This clearance-induced dynamic deterioration not only compromises piston kinematic performance but may also adversely affect the operational integrity of the entire mechanical system. The time-domain acceleration profiles in the *x*-direction for Stage I-II-IV pistons across different rotational speeds are presented in Figs. 22a-c, while their corresponding velocity-acceleration phase portraits are shown in Figs. 22d-f.

3.3 Analysis of Connecting Rod Big-End Clearance Effects on Crankshaft-Linkage Mechanism Dynamics

To investigate the relationship between excitation frequency and the crankshaft housing's X-direction excitation force under different connecting rod big-end clearances, the crankshaft speed was set to 1350 r/min. The big-end clearances of the I-II-III and I-II-IV stage connecting rods were increased to 0.1 mm, 0.15 mm, and 0.2 mm, respectively, while all other kinematic pairs were modeled as rigid (zero-clearance) connections. The resulting X-direction excitation forces at the crankshaft bearing location are presented in Fig. 23. At low excitation frequencies, the excitation forces under all three clearance conditions rapidly reached a peak value. When the connecting rod big-end clearance is 0.1 mm, the peak excitation force reached its maximum value of approximately 6.5 kN. As the clearance increased to 0.15 mm and 0.2 mm, the peak values decreased

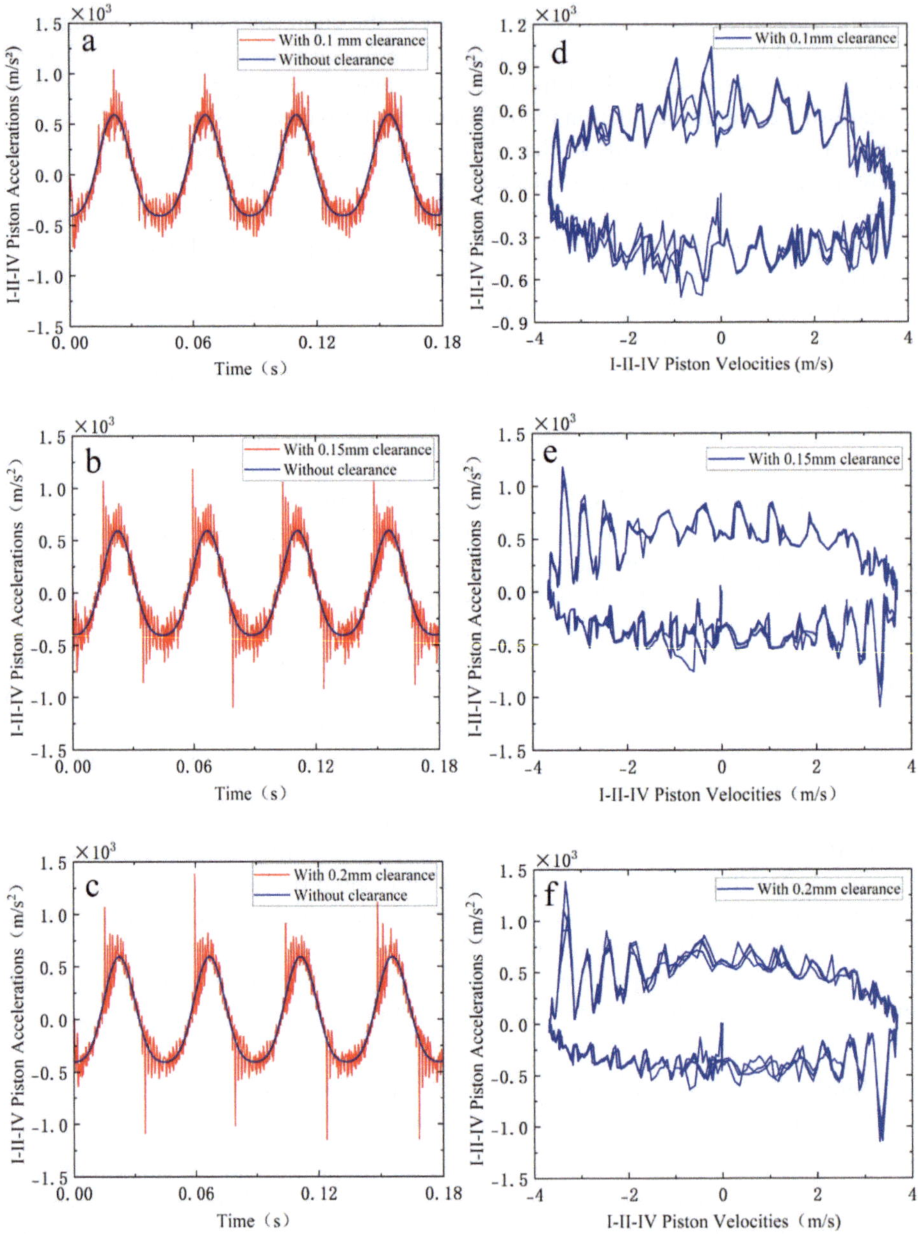

Fig. 21. Comparative analysis of dynamic characteristics for I-II-IV pistons under varying clearance dimensions

to 5 kN and 2 kN, respectively. This inverse correlation clearly indicates that smaller bearing clearances generate significantly higher excitation forces in the low-frequency range. Beyond the peak, the excitation force decreased rapidly with increasing excitation

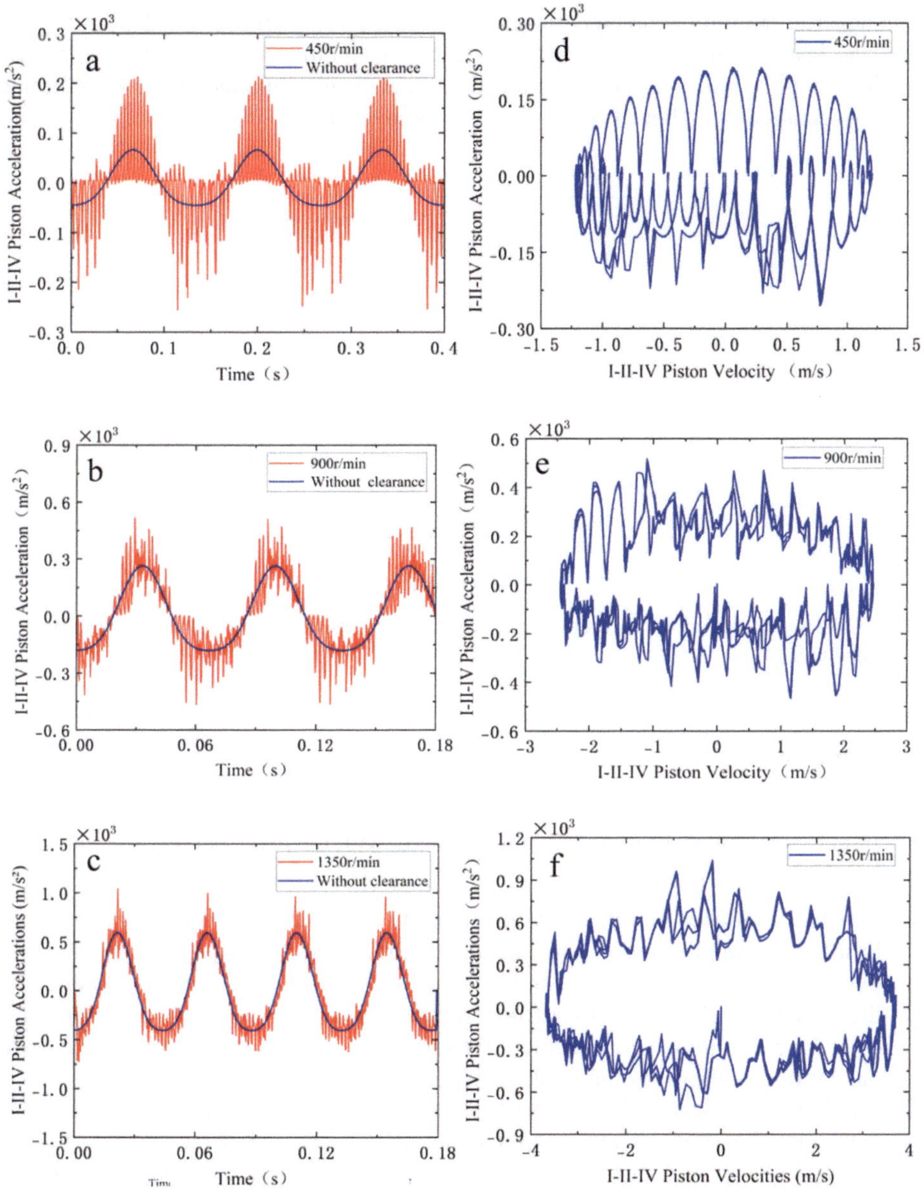

Fig. 22. Comparative analysis of dynamic characteristics for I-II-IV pistons under varying velocity dimensions

frequency, subsequently entering a fluctuating state. During the oscillatory phase, the excitation force for the 0.1 mm clearance consistently exceeded that of the other two clearance configurations, while demonstrating relatively larger amplitude fluctuations. Conversely, the 0.2 mm clearance case exhibited the lowest excitation force magnitude with superior stability characteristics. In summary, under identical excitation frequencies, the bearing clearance size exerts a significant influence on the excitation force direction at the crankshaft mounting location. Specifically, smaller clearances correlate with systematically higher overall excitation force levels.

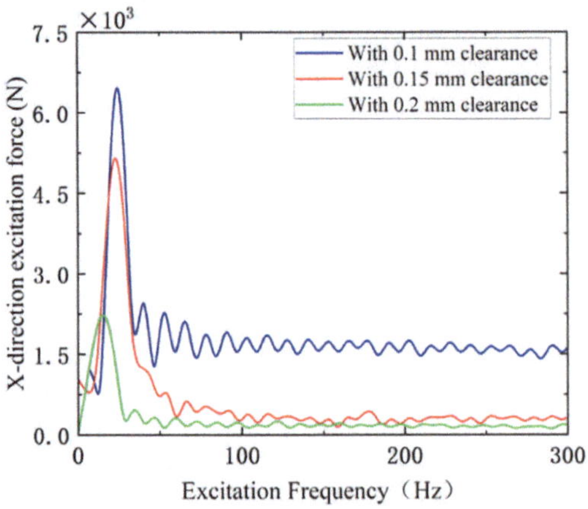

Fig. 23. *x*-direction excitation forces at the crankshaft housing versus connecting rod big-end clearance variations

4 Conclusion

This study conducts a comprehensive investigation into the V-type marine high-pressure air compressor crankshaft-connecting rod system, with the principal research outcomes summarized as follows:

(1) Flexibility-induced modifications in either the crankshaft or connecting rod primarily affect the amplitude domain rather than the fundamental frequency characteristics of main bearing excitation forces. For V-type marine high-pressure air compressors, the typically high crankshaft stiffness renders rigid-body dynamics modeling sufficient for conventional computational requirements, without necessitating excessive pursuit of flexible-body modeling complexity.
(2) Frictionless models predict complete attenuation of excitation forces beyond 250 Hz at the main bearing, whereas frictional interactions generate persistent spectral content up to 800 Hz, confirming friction's role in broadening the effective frequency response bandwidth by 2.2 times.
(3) The startup time exhibits a negative correlation with the required starting torque, particularly pronounced for ultra-short startups (<0.01 s). Under zero-clearance conditions, the driving power remains relatively stable, fluctuating within ± 32,282 W. In contrast, systems with clearances exhibit instantaneous torque power peaks reaching 86,971 W—significantly exceeding ideal clearance-free values. This unequivocally demonstrates that clearances amplify dynamic impacts. Consequently, integrating flywheels for inertial balancing should be prioritized in system design to mitigate such fluctuations.

(4) The peak X-direction acceleration transients of the I-II-IV stage pistons are triggered at top/bottom dead centers, with larger clearances reducing contact collision frequency but increasing acceleration amplitudes.
(5) Increasing rotational speed exacerbates piston velocity instability induced by connecting rod big-end clearance, which not only degrades piston-specific kinematic performance but may also adversely affect overall mechanical system operation.

In summary, this study provides comprehensive and in-depth theoretical foundations for the design of V-type marine high-pressure air compressor crankshaft-connecting rod systems. The findings contribute significantly to enhancing the overall performance and reliability of such compressors, offering substantial guidance for engineering practices in related fields. Future research could incorporate additional operational factors, such as complex working environments and material fatigue characteristics, to further refine the understanding of this system.

Acknowledgment. This work was Sponsored by Natural Science Foundation of Chongqing, China. Consortium: CSTB2023NSCQ-LMX0031.

References

1. Kim, B.-G., Kim, H.-R., Yang, C.-J., Kim, J.-H.: An Experimental Study on Reducing Condensation in Marine Air Compressors. J. Korean Soc. Marine Environ. Safety. **21**, 3 (2015)
2. Abhay Nambiar, Naveen Venkatesh S., Aravinth S., Sugumaran V., Sangharatna M. Ramteke, Max Marian, Prediction of air compressor faults with feature fusion and machine learning, Knowledge-Based Systems, **304**, 112519 (2024)
3. Tran, V.T., AlThobiani, F., Ball, A.: An approach to fault diagnosis of reciprocating compressor valves using Teager-Kaiser energy operator and deep belief networks. Expert Syst. Appl. **41**, 9 (2014)
4. Tran, V.T., et al.: Single and combined fault diagnosis of reciprocating compressor valves using a hybrid deep belief network. Proc. Institut. Mech. Eng. Part C: J. Mech. Eng. Sci. **232,** 20 (2018)
5. Gribbestad, M., Hassan, M.U., Hameed, I.A.: Transfer Learning for Prognostics and Health Management (PHM) of marine air compressors. J. Marine Sci. Eng. **9**, 1 (2021)
6. Jiang, Z., Mao, Z.; Yao, Z., et al,.: A Diagnosis method of the small end fault on reciprocating compressor connecting rod, IOP conference series. Mater. Sci. Eng. **90**, 1 (2015)
7. Godara, S.S., Brenia, V., Soni, A.K.et al.: Design & analysis of connecting rod using ANSYS software. Mater. Today: Proc. **56** (2022)
8. Peng, F., Huang, Y.: Torsional vibration behavior of compressor shaft system considering variable inertia characteristics of crank-connecting rod mechanisms. Inter. J. Non-Linear Mech. **170** (2025)
9. Xiao, Y., Zhao, H.; Wu, J., et al.: Simulation and analysis of the crankshaft system of the high-power marine engine. Mechanika (Kaunas, Lithuania: 1995), **30**, 6 (2024)
10. Sakhraoui, A., Saggar, M., Ayari, F., et al.: Kinematics modeling of the gear-based crank mechanism engine regardless of the compressions ratio variations. Sci. Rep. **14**, 1 (2024)
11. Albaghdadi, A.M., Baharom, M.B., Sulaiman, S.A.: Tri-planar balancing optimization of a double crank-rocker mechanism for shaking forces and shaking moments reduction. Proce. Estonian Acad. Sci. **70,** 3 (2021)

12. Wang, C., Le, T.D.M., Huynh, N.T.: Optimal rigid-flexible dynamic of space slider-crank mechanism with clearance joints. Sādhanā, **48**, 44 (2023)
13. Daniel, G.B., Cavalca, K.L.: Analysis of the dynamics of a slider–crank mechanism with hydrodynamic lubrication in the connecting rod–slider joint clearance. Mech. Mach. Theory **46**, 10 (2011)
14. Matekar, S.B., Fulambarkar, A.: M, Displacement analysis of slider in slider-crank mechanism with joint clearance. Aust. J. Mech. Eng. **20**, 4 (2020)
15. López-Lombardero, M., Cuadrado, J., Cabello, M, Martinez, F., Dopico, D., López Varela, A:. Simulating fatigue damage on planar mechanisms considering wear in revolute joints with clearance. Mech. Mach. Theory **203**, 105805 (2024)
16. Cheng, S.: The influence of the clearance on the motion of the reciprocating compressor connecting rod. In: Deng, K., Yu, Z., Patnaik, S., Wang, J. (eds) Recent Developments in Mechatronics and Intelligent Robotics, ICMIR 2018, AISC, vol. 856(2019)
17. Piston Compressor Design Writing Group. Piston Compressor Design. China Machine Press, Beijing, China (1974)
18. Li , Z., Li, B.: ADAMS Detailed Introduction and Examples. Tsinghua University Press (2021). (in Chinese)
19. Khemili, I., Romdhane, L.: Dynamic analysis of a flexible slider–crank mechanism with clearance. Eur. J. Mech. A. Solids **27**, 5 (2008)

Design and Analysis of Bidirectional Openable Hatch Cover

Xuan Wang[1,2](✉)

[1] State Key Laboratory of Deepsea Manned Vehicle, China Ship Scientific Research Center, Wuxi 214082, China
275489205@qq.com
[2] TaiHu Laboratory of Deepsea Technological Science, Wuxi 214082, China

Abstract. This paper designs a hatch cover for a manned submersible. The hatch cover is located at the top of the manned capsule. It should have a two-way opening and closing. Also, it must be as light as possible while meeting the pressure resistance requirements. Through the analysis of the flat plate, spherical cap butterfly cover plate, the butterfly cover plate is selected for the design, and the structure is calculated and checked by empirical formula and finite element method. Finally, it is proved that the hatch meets the requirements through air tightness test, pressure cylinder test, and opening tests.

Keywords: hatch cover · bidirectional openable · manned submersible

1 Introduction

Cabin hatch cover is a general term for the cover. Its opening and closing devices are used to close various hatches on underwater work platforms. It is a closed device that ensures the water-tightness and internal safety of the platform. Its main function is to resist external pressure and ensure that waves/seawater do not invade the interior of the cabin. As a necessary piece of equipment for various underwater work platforms, hatch covers have developed into different series of product types according to different work tasks after decades of development. According to the different depths of underwater work, it can be divided into wind- and rain-tight ventilation hatch covers with a depth of 5 m or less, ship hatch covers with a depth of 20 m or less, hundred-meter-level underwater hatch covers, and kilometer-level deep manned hatch covers [1–3].

This article designs a hatch cover installed on a manned submersible. The hatch cover is located at the top of the manned cabin and serves as a passage for personnel to enter and exit the cabin. The manned hatch cover is crucial for the safe and efficient entry and exit of personnel. According to the function of the manned cabin, the hatch cover should not only have the characteristics of reliable sealing, operability, and lightweight flexibility of traditional hatch covers, but also ensure that the hatch cover can be opened and closed in both directions underwater. If there is a sealing failure or an opening failure, the operation task will fail. Thus, it is necessary to conduct design research and experimental verification for hatch covers.

S. Halgamuge et al. (Eds.): ICACMVE 2025, LNME, pp. 115–126, 2026.
https://doi.org/10.1007/978-981-95-6911-3_11

To achieve the design goal of bidirectional opening and closing of personnel underwater, reducing the weight of the hatch cover and designing a balance mechanism are the key points. In terms of reducing the weight of the hatch cover, the design of the cover plate is optimized through the comparison of multiple schemes. We also innovatively designed a new type of balancing mechanism to achieve bidirectional torque balance. Based on a comprehensive comparison of various hatch opening forces, a single-person onshore opening force of no more than 15 kg is designed (due to underwater buoyancy, the underwater opening force is slightly smaller than onshore) for bi-directional single-person opening and measurability.

2 Composition of the Hatch Cover

The hatch cover that can be opened in both directions underwater consists of a surrounding wall, a cover plate, a locking device, a balancing mechanism, and a sealing element (Fig. 1). Among them, the enclosure wall is used to reinforce the opening of the manned cabin and serves as the installation base for the hatch cover. The cover plate is a component that carries pressure and ensures sealing, located at the upper part of the surrounding wall, and can rotate around a fixed axis. The end face of the cover plate is generally grooved for installing seals. The balancing mechanism is used to balance the weight of the hatch cover during the opening and closing process. After personnel enter and exit the manned cabin through the hatch cover, they use the hatch cover locking device to lock the hatch cover to prevent seawater from entering. In order to design the bi-directional opening hatch cover underwater, it is necessary to conduct design analysis on the hatch cover plate, locking device, and balancing mechanism. The detailed process is as follows.

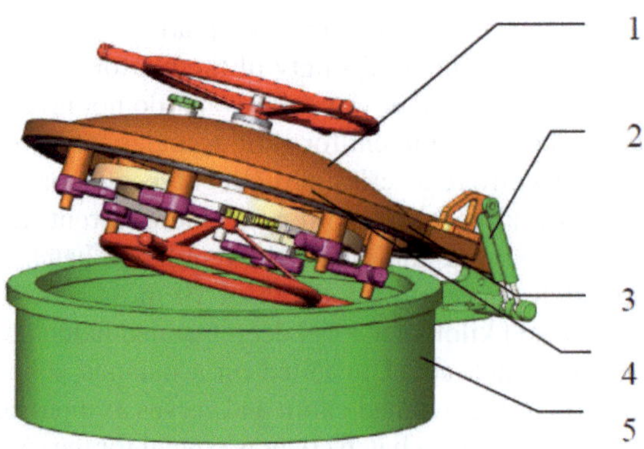

Fig. 1. Schematic diagram of hatch cover composition.

The operation process of the hatch cover is as follows. By rotating the small hand wheel on the transmission device, the hand wheel and gear ring are driven to rotate, causing the connecting rod to swing. This clamps or rotates the rolling head into or out of the flange (welded to the surrounding wall). The handwheel is connected by a connecting rod and located on both the inner and outer sides of the hatch cover, allowing for operation of the hatch cover. By combining the transmission ratio of small gears, transmission wheels, and large gears, it is easy to open the hatch cover.

3 Hatch Design

3.1 Cover Plate Design

Scheme Selection. The hatch cover requires bi-directional opening underwater, which improves the flexibility of bi-directional locking operation and enables the operability of the hatch cover opening and closing process. Cover plate design is the focus of hatch cover design. In the initial design stage, three options were chosen to carry out the cover plate design.

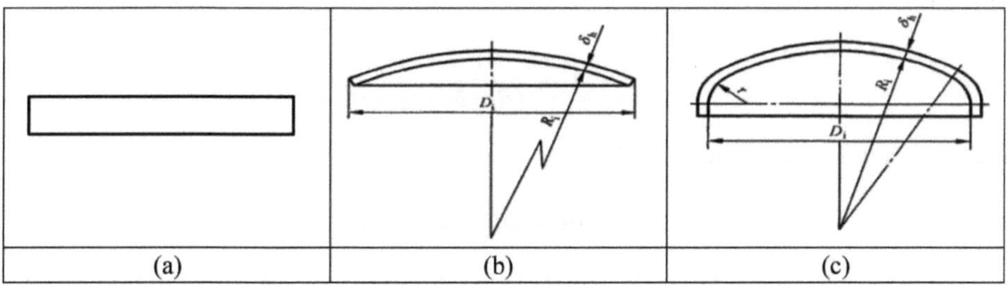

Fig. 2. Cover plate scheme.

Option 1 adopts a flat form (Fig. 2a). The flat form cover plate is flat and does not protrude from the line shape. However, this form has poor pressure resistance. Under the same working conditions, the weight of the cover plate is the heaviest. It is generally used for surface ships or shallow-depth hatch covers.

Option 2 adopts the form of a ball crown (Fig. 2b). This form has the best pressure resistance and the lightest weight of the cover plate under the same working conditions. However, its outer diameter is relatively high and will protrude in a linear shape, usually used on deep hatch covers.

Option 3 adopts a disc-shaped form (Figure 2c). This form has better pressure resistance, and the weight of the cover plate is lighter under the same working conditions. Among the three outer diameter schemes, it is centered and will protrude some lines.

Q345 was initially selected as the material for the hatch cover plate, and the weights of the flat plate, spherical cap, and butterfly cover plate were compared according to the design requirements. Simultaneously, we evaluate the resistance effects of different cover plates, as shown in Table 1.

In summary, considering the pressure resistance and linearity, the optimal design for the cover plate is the disc-shaped form of option three.

Calculation of Specifications. Zhu Ying, Zhao Qian, and Li Youjie carried out the analysis and research of the structural strength of surface and underwater hatch covers

Table 1. Comparison of the cover plate scheme.

Cover plate scheme	Pressure resistance/weight	Line type
Flat form	Worst/106.8 kg	Best
Ball crown	Best /42.2 kg	Worst
Disc-shaped form	Better /50.6 kg	Better

[4–6]. According to the GJB150 relevant provisions, the calculation formula for the convex surface of the butterfly head under pressure is as follows.

$$[p] = \frac{B}{(Ro/\delta e)} \tag{1}$$

where B is looked up from a table of coefficients A,

$$A = \frac{0.125}{(Ro/\delta e)} \tag{2}$$

where δe is the effective wall thickness.

$$\delta e = \delta n - C \tag{3}$$

The materials selected are alloy steel Q345, titanium alloy TA2, and aluminum alloy 6061-T6, and the thickness of the disc-shaped head is calculated separately.

Alloy steel Q345

If we suppose δn is 10 mm, and additional thickness C is 2 mm, we can obtain the following values.

A = 0.001761

B = 159

$[p] = 2.25 > p_c$

Ttanium alloy TA2

If we suppose δn is 11 mm, and additional thickness C is 1 mm, we can obtain the following values.

A = 0.00183

B = 145

$[p] = 2.13 > p_c$

Aluminum alloy 6061-T6

If we suppose δn = 20 mm, and additional thickness C is 2 mm, we can obtain the following values.

A = 0.0037

B = 69

$[p] = 1.83 > p_c$

Based on the preliminary design results mentioned above, we also include material and processing costs. The preliminary calculation results are shown in Table 2. Considering both weight and economy, we select alloy steel Q345 for this design.

Table 2. Calculations and analysis of cover plate processing costs.

Material	Theoretical weight/kg	Material cost	Processing fee	Total price
Alloy steel Q345	62	700	2,000	2,700
Titanium alloy TA2	38	1,520	6,000	7,520
Aluminum alloy 6061-T6	35	700	3,500	4,200

Finite Element Verification. The strength and stability of the hatch cover's cover plate structure are calculated and verified using finite element analysis. The static external water pressure and boundary constraints are applied to the finite element model of the cover plate structure, and the stress level state can be obtained through static calculation analysis. Figure 3 shows the stress levels of the cover plate structure model, with a maximum stress of 267 MPa. It is less than the yield strength of Q345, and meets the design requirements.

Figure 4 gives the deformation displacement cloud map of the cover plate structure. Its maximum displacement is 0.2 mm, which meets the performance requirements.

3.2 Balance Mechanism Design

Scheme Selection. Option 1 is in the form of a torsion spring (Fig. 5a), which is widely used in ship hatch covers and is located outside the hatch cover. Its performance varies with angle, and corrosion issues need to be considered for underwater use.

Option 2 is a gas spring (Fig. 5b). Compared with ordinary springs, the advantages of gas springs are that the dynamic force does not change much (generally within 1:1.2). It is easy to control, and the performance is stable. Unlike mechanical springs, gas springs have an almost linear elastic curve.

Considering the stability of performance and corrosion issues, we selected the gas spring method for the design of the balancing mechanism.

Gas Spring Design. This balancing device is designed using gas springs. The spring is composed of a pressure cylinder, a piston rod, a piston, a sealing guide sleeve, filling material, and so on. Compared with ordinary springs, gas springs have many advantages, including minimal dynamic force changes (generally within 1:1.2), easy control, and stable performance. Unlike mechanical springs, gas springs have an almost linear elastic curve. Its appearance is shown in Fig. 6.

Based on the optimal design scheme and the characteristics of the gas spring, we conduct a preliminary design for its layout. We optimize the position and force of the spring based on the weight moment curve of the hatch cover, as shown in Fig. 7.

Combining the installation method, we present the curves of weight torque, spring torque, and human torque as shown in Fig. 8.

From Figure 8, it can be seen that the maximum value of the heavy torque is 410 Nm, which occurs at the starting position. The maximum spring torque is 390 Nm, which

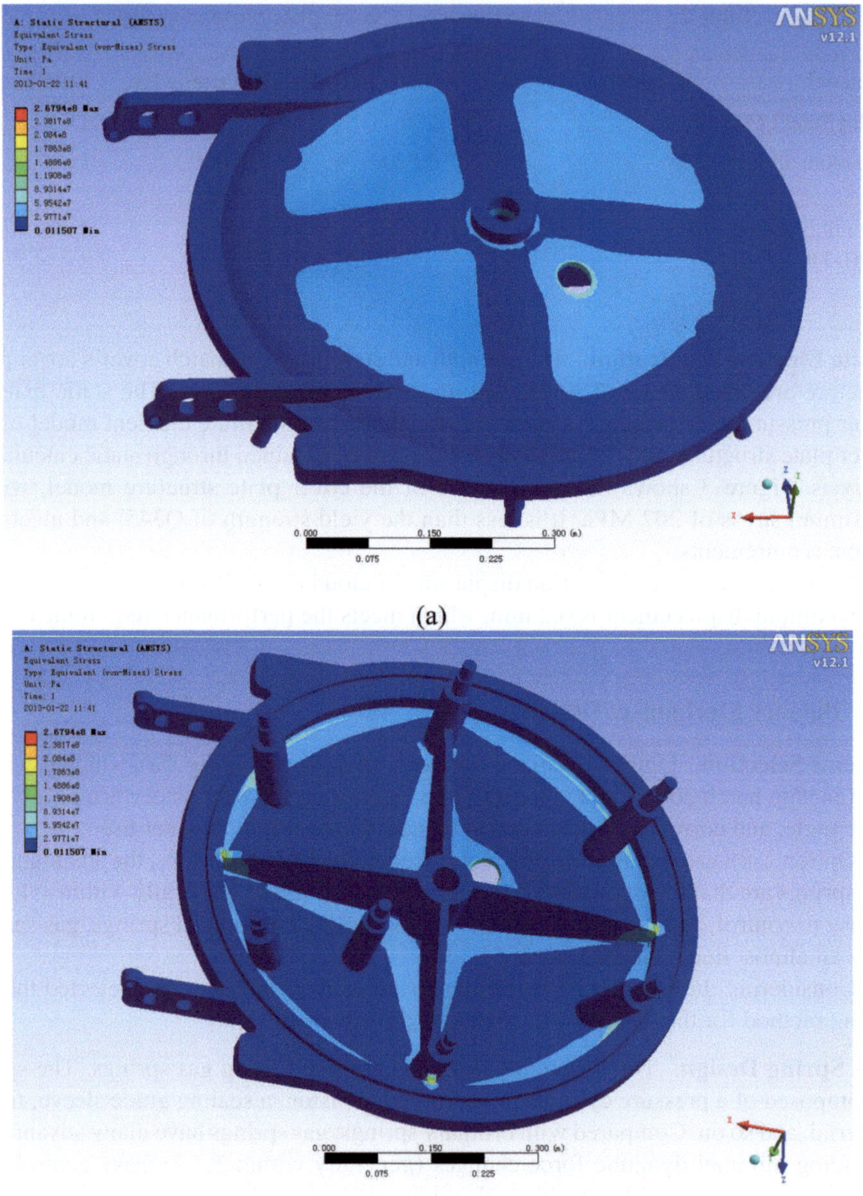

Fig. 3. Stress level of the cover plate structure model (convex and concave surface).

occurs at the beginning. Figure 9 shows the human work curve. The maximum force required for human work during the opening process is 11 kg, while the maximum force required for human work during the closing process is about 2.5 kg.

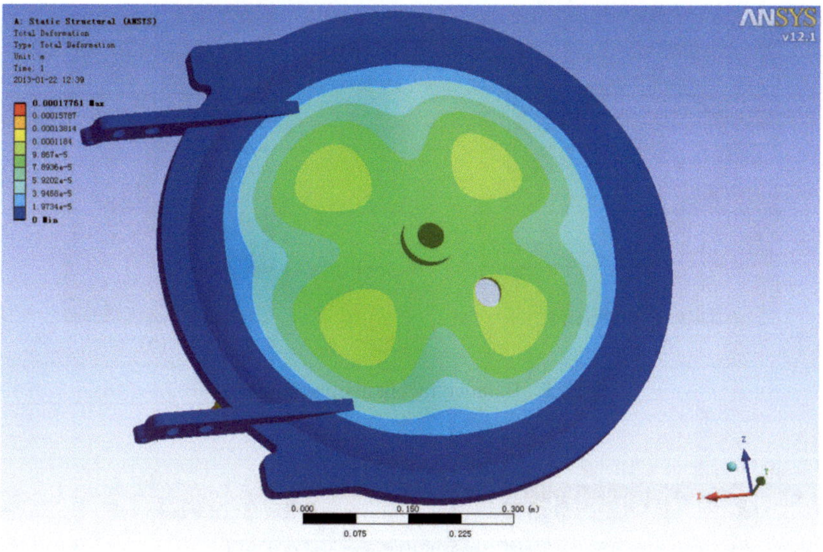

(a)

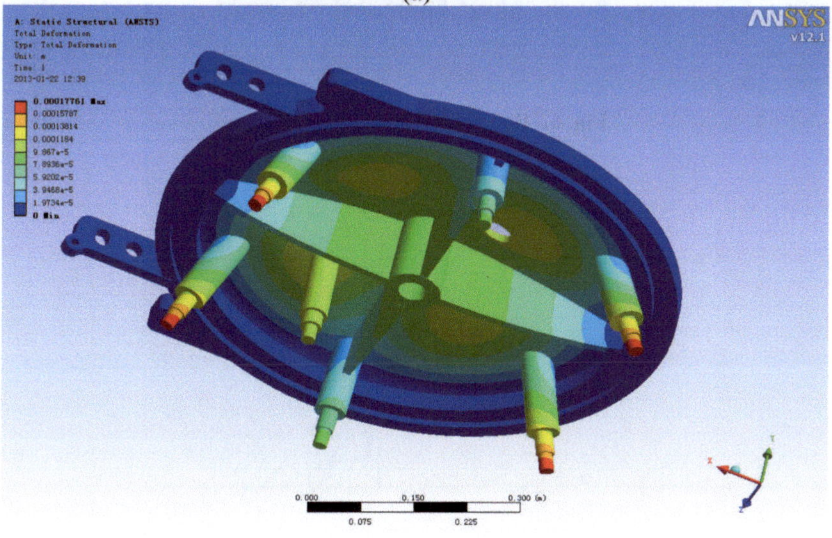

(b)

Fig. 4. Displacement contour map of the cover plate structure model (convex and concave surface).

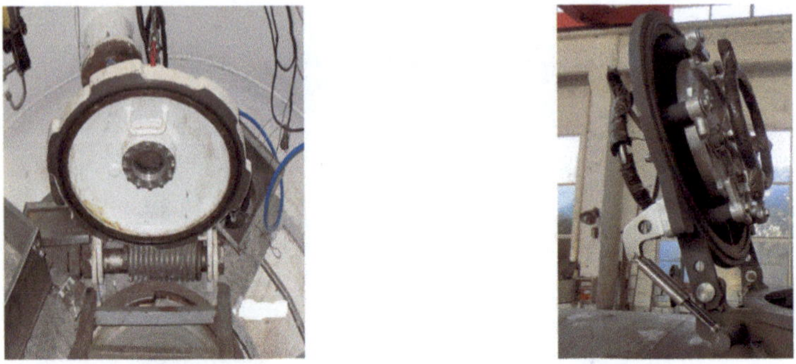

(a) (b)

Fig. 5. Balance mechanism scheme.

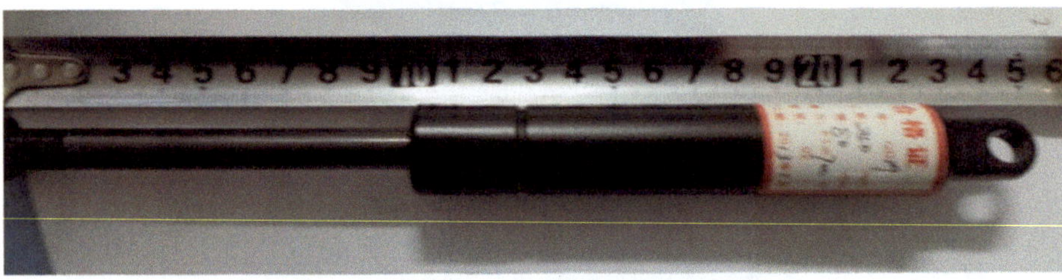

Fig. 6. Balance mechanism appearance.

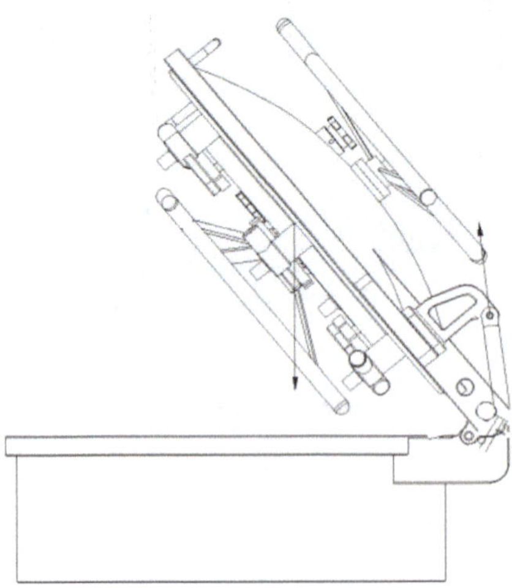

Fig. 7. Force analysis diagram of a gas spring.

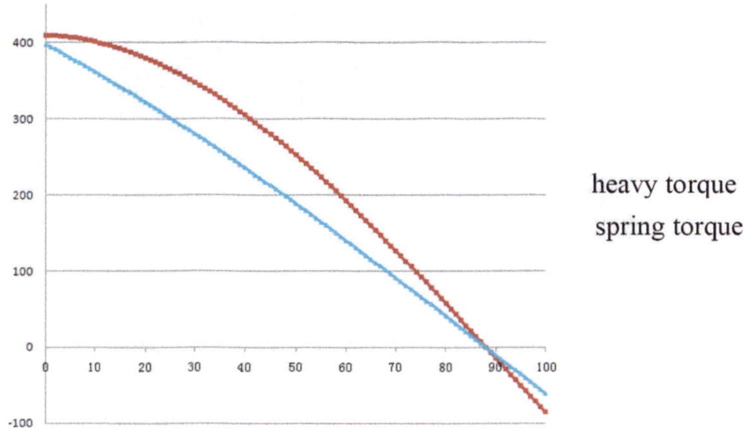

Fig. 8. Relationship curve between heavy torque and spring torque.

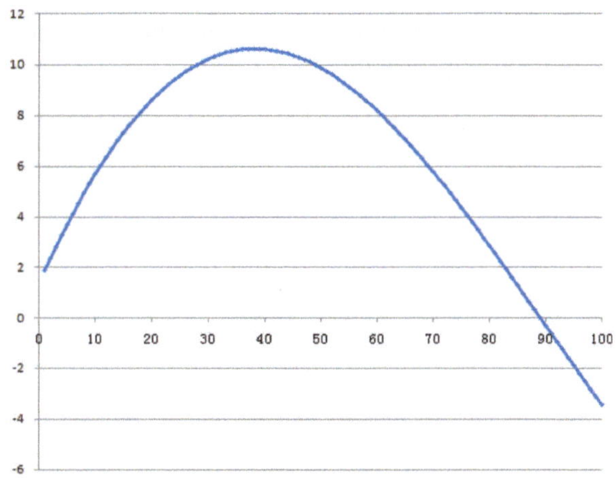

Fig. 9. Human resources curve.

4 Test Verification

We carried out the hatch cover test in Fig. 10. The test content includes the air-tightness test of the hatch cover, the pressure cylinder, and the opening test. Firstly, the auxiliary fixture was used to seal the opening end of the hatch cover reinforcing wall, and the seal was ensured. During the test, gas was filled and pressurized to 0.1 MPa. The air tightness of all welded joints and connections was checked with soapy water without leakages. The hatch cover was assembled on the head for the pressure cylinder test. Its test pressure was 0.75 MPa, and the pressure was maintained for 0.5 h. After the test, the strength and sealing performance of the hatch cover were checked. Its seal was leak-free, and the hatch cover was not deformed.

To assess the opening force of the hatch cover, multiple attempts were made on land without any issues, meeting the requirements for use. The maximum opening force

Fig. 10. The hatch cover pressure test.

is 11 kg, meeting the requirement of a maximum opening force of 15 kg or less. The specific measurement data are shown in Figs. 11 and 12.

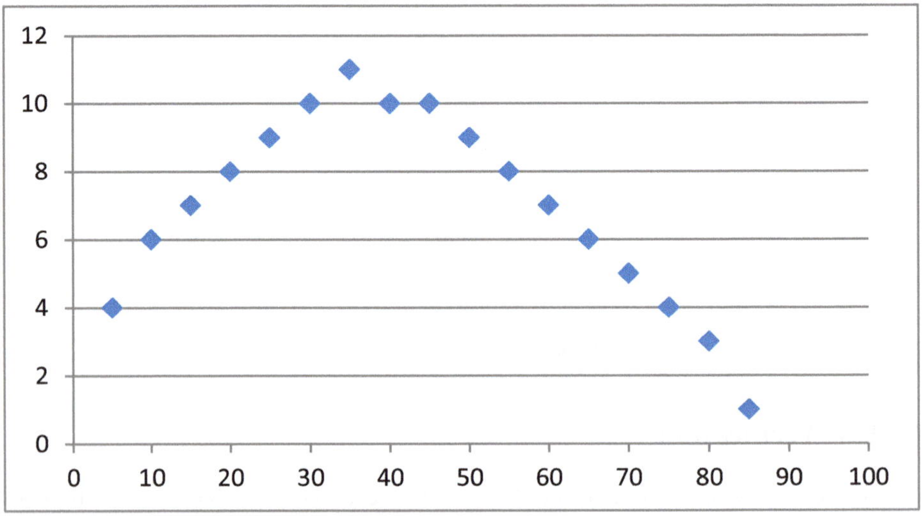

Fig. 11. Opening force measurement results.

Fig. 12. Opening test experiment.

5 Conclusion

This article aims to achieve the design goal of bidirectional opening and closing for underwater personnel. We use two measures, which are reducing the weight of the hatch cover and designing a balance mechanism. On the one hand, by comparing and analyzing the flat plate form, spherical cap form, and disc-shaped head, the implementation of adopting the disc-shaped head was determined. The design of the cover plate was optimized through standard calculation, finite element analysis, and other methods to reduce the weight of the hatch cover. On the other hand, a gas spring was innovatively designed as a balancing mechanism to achieve bidirectional torque balance. In order to assess the opening force of the hatch cover, it was opened multiple times on land without any issues. It can also be opened at times by a single person, which meets the operational requirements.

References

1. Xu, K., Zhao, F.H., Luo, S., et al.: Design and experiment for Deep-Sea automatic opening and closing hatch cover. Ship Eng. **4**(10), 37–41 (2022)
2. Yu, S.: Study on Hatch Cover and Drive Mechanism for Full-Ocean-Depth Manned Submersible. Shanghai Ocean University, Shanghai (2018)
3. Fan, H.T., Qi, X., Song, Q., et al.: Design and experiment for automatic opening and ClosingHatch cover of wet-dry-conversion compartment. Ship Eng. **39**(S1), 19–22 (2017)
4. Zhu, Y.: Exploration of the structural strength of container port openings. Mar. Equipment Mater. Mark. **31**(11), 19–21 (2023)
5. Zhao, X., Zhu, T.Y., Gao, X.: Exploration of the structural strength of container port openings. Ship Ocean Eng. **52**(03), 113–116 (2023)
6. Li, Y.J.: Research on the Ultimate Bearing Capacity of Fully-Transparent Manned Submersible Spherical Cabin. Shanghai Ocean University, Shanghai (2023)

Structural Analysis of a Mobile Aero-Engine Test Platform Chassis Based on Finite Element Method

Zebing Fan[1], Jingyu Zhao[2,3,4], Ran Li[1], Rui Guo[2,3,4](✉), Hao Wang[1], Dianmin Chen[2,3,4], and Shuaiqi Dou[2,3,4]

[1] AECC SiChuan Gas Turbine Research Establishment, Chengdu, Sichuan 610500, P. R. China
[2] State Key Laboratory of Crane Technology, Yanshan University, Qinhuangdao, Hebei 066004, P. R. China
guorui@ysu.edu.cn
[3] Hebei Key Laboratory of Special Carrier Equipment, Yanshan University, Qinhuangdao, Hebei 066004, P. R. China
[4] School of Mechanical Engineering, Yanshan University, Qinhuangdao, Hebei 066004, P. R. China

Abstract. The structural stiffness and strength of chassis frames significantly impact the operational reliability and testing accuracy of mobile aero-engine test platforms. To ensure structural integrity under representative load conditions, a comprehensive finite element analysis (FEA) was conducted on a typical chassis of a mobile aero-engine test platform, focusing on front and rear axle torsion and static vertical load conditions. Key performance indicators, such as torsional stiffness, structural deformation and stress distribution, were evaluated quantitatively to demonstrate compliance with safety and stability requirements. Furthermore, parametric sensitivity analysis and response surface-based multi-objective optimization were employed to explore lightweight design possibilities. The optimization results showed a mass reduction of approximately 15%, achieving an optimal balance between structural weight and torsional stiffness. These results confirm the validity of the analytical and optimization approaches proposed, providing practical guidance and theoretical support for designing lightweight, high-performance chassis structures for mobile aero-engine testing applications.

Keywords: Aero-engine · Mobile Test Platform · Lightweight Design · Chassis Structure Optimization · Torsional Stiffness

1 Introduction

The rapid advancement of new-generation aero-engines, which feature higher thrust-to-weight ratios and stricter reliability requirements, has led to a substantial increase in the need for advanced and flexible ground testing systems [1–3]. Modern aero-engine testing demands increasingly modular and adaptable platforms. Although traditional fixed aero-engine test stands offer high-precision measurement capabilities, their limited mobility and adaptability often renders them unsuitable for decentralised or rapidly deployed testing environments [4].

S. Halgamuge et al. (Eds.): ICACMVE 2025, LNME, pp. 127–134, 2026.
https://doi.org/10.1007/978-981-95-6911-3_12

In contrast, mobile aero-engine test platforms have attracted increasing attention for their advantages in structural integration, operational versatility, and the ability to support full-scale engine tests in decentralized locations [5, 6]. Within these systems, the chassis frame is the main load-bearing component supporting the aero-engine, suspension and measurement devices. The mechanical integrity of the chassis frame, especially with regard to torsional stiffness and strength, is critical for ensuring platform stability, vibration isolation, and test accuracy, as previously discussed by Liu et al. [3] and Zhang et al. [7].

A substantial body of research has explored structural stiffness analysis, fatigue life assessment and design optimisation in the context of heavy-duty transport equipment, vehicle frames. This research includes studies by Li et al. [4], Zhang et al. [7] and Zhou et al. [8]. These studies have made significant contributions by establishing reliable methodologies for load-bearing analysis and structural optimisation under complex loading conditions. More recently, Xu and Li [9], Shen et al. [10] and Zhao et al. [11] have introduced advanced multi-objective optimisation techniques for chassis structures, improving torsional rigidity while reducing weight.

Building upon these achievements, the present study focuses on the structural analysis and lightweight design of the world's first mobile aero-engine test platform chassis. this study utilizes the finite element method (FEM) following the GB/T 5620–2020 standard to comprehensively evaluate structural stiffness and strength performance of a mobile aero-engine test platform chassis under representative load conditions. Through detailed finite element modeling, critical performance metrics were quantitatively assessed, and parametric optimization based on response surface methodology (RSM) was performed to achieve chassis lightweight design. This work provides theoretical guidance and practical support for advancing aero-engine test platforms from fixed to modular, integrated, and mobile configurations. The overall framework is shown in Fig. 1.

2 Chassis Structure Simulation

2.1 Finite Element Simulation Setting

The chassis frame adopts a dual-longitudinal beam and multi-crossbeam configuration, designed to achieve low weight, high load-bearing capacity, and superior torsional stiffness. Key geometric and material parameters are summarized in Tables 1.

Q345 steel was used for all structural parts. Simulating three representative load cases: front axle torsion, rear axle torsion, and static vertical load. For torsional analysis, a forced vertical displacement of 1 mm was applied to specific suspension hard points, and the resulting reaction forces were used to compute the torsional moments and stiffness. Mesh density and constraint types were determined to ensure convergence and physical relevance.

2.2 Static Load Stiffness Analysis

The hard points of the front and rear suspensions on the fixed loading platform were defined as fixed supports, and a vertical load of 30,000 N was applied to the load-bearing

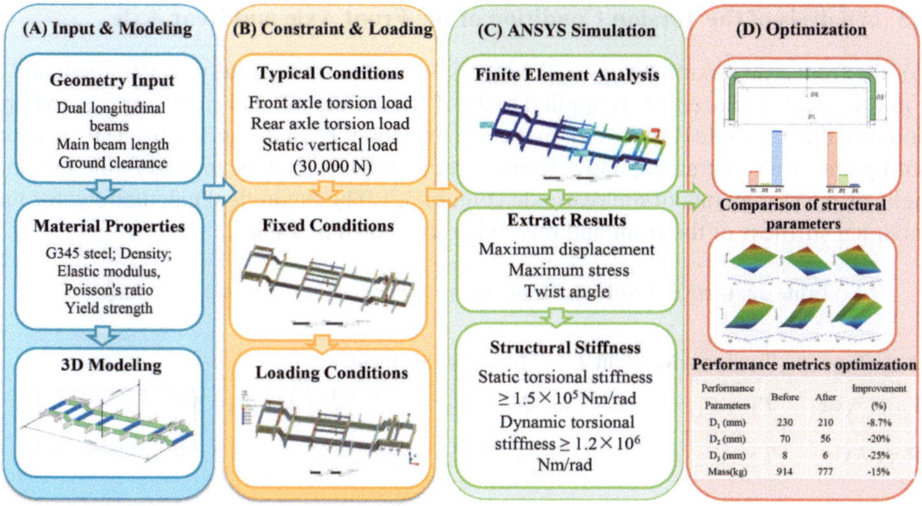

Fig. 1. Research structure and flowchart

Table 1. Material properties of the transportation platform

Materials	Density / (t/mm^{-3})	Elasticity modulus/ (MPa)	Poisson's ratio	Yield strength/ (MPa)
Q345	7.85×10^{-9}	2.1×10^{5}	0.3	345

surface. The boundary conditions are shown in Fig. 2(a); The stress cloud map is shown in Fig. 2(b).

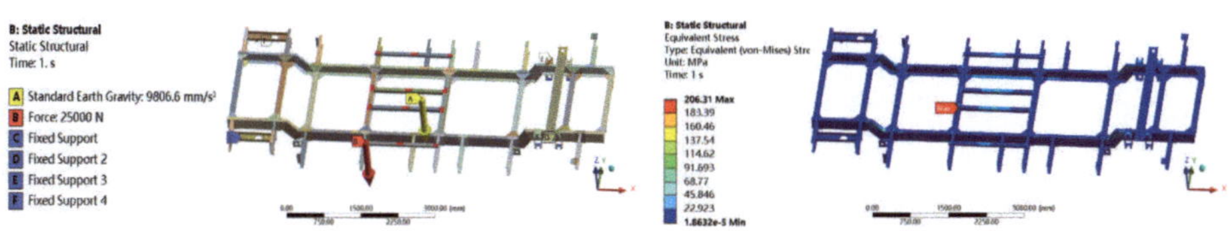

(a) Static load boundary conditions **(b)** Static stress cloud map

Fig. 2. Static Stiffness Analysis Diagram

The simulation results show that the maximum stress is 206.31 MPa.

The analysis indicates that the selected material has an allowable stress of 345 MPa. With a safety factor of 1.5, the permissible limit is 230 MPa, which exceeds the maximum simulated stress of 206.31 MPa. Therefore, the structural strength satisfies the design requirements.

2.3 Analysis of the Torsion Condition of the Front Axle and Rear Axle

For the front axle torsion analysis, the rear axle hard points are fully constrained, and a 1 mm vertical displacement is applied to the Z-axis of the left and right suspension hard points of the front axle (Figure 3a). In the rear axle torsion condition, the front suspension hard points are constrained, and a 1 mm vertical displacement is applied to the left and right suspension points of the rear axle (Figure 3c). The theoretical calculations for the torsional stiffness of the front and rear axles are shown in Table 2.

Table 2. Torsional Stiffness Calculation Formulas for Front and Rear Axles

Axle Type	Torsional Moment T	Torsional Stiffness
Front Axle	$T_1 = F_1 \times L_1$	$K_{t1} = \frac{T_1}{\theta_1} = \frac{10294 \times 2.25}{\arctan(1/(L_1/2))} = 4.54 \times 10^5 \mathrm{N} \cdot \mathrm{m}/^\circ$
Rear Axle	$T_2 = F_2 \times L_2$	$K_{t2} = \frac{T_2}{\theta_2} = \frac{8104.9 \times 2.212}{\arctan(1/(L_2/2))} = 3.448 \times 10^5 \mathrm{N} \cdot \mathrm{m}/^\circ$

3 Structure Parametric Optimization

To achieve structural weight reduction while ensuring sufficient torsional stiffness and structural reliability, the critical geometric parameters influencing the torsional stiffness of the chassis frame were identified through preliminary finite element simulations. The height of the main longitudinal beams (D_1), cross-beam spacing and dimensions (D_2), and thickness of beam plates (D_3) were selected as design variables. Parametric Modeling Variables for the Chassis are illustrated in Fig. 4.

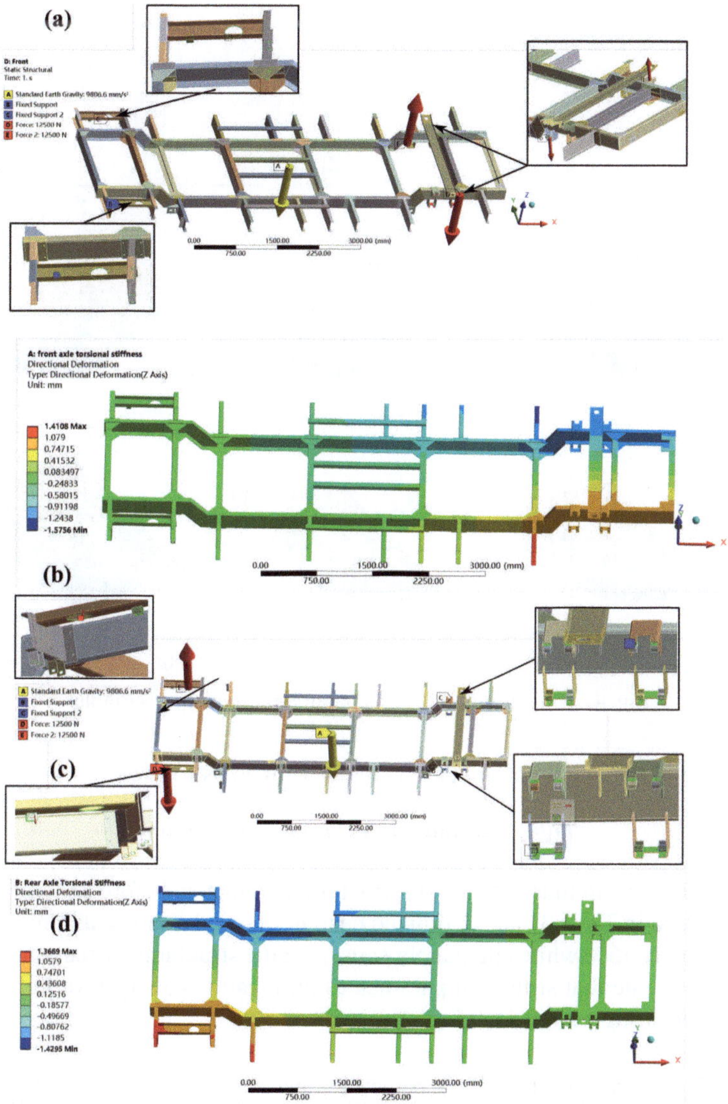

Fig. 3. Torsional analysis of the chassis: (a) Front axle boundary condition, (b) Z-displacement (front); (c) Rear axle boundary condition, (d) Z-displacement (rear).

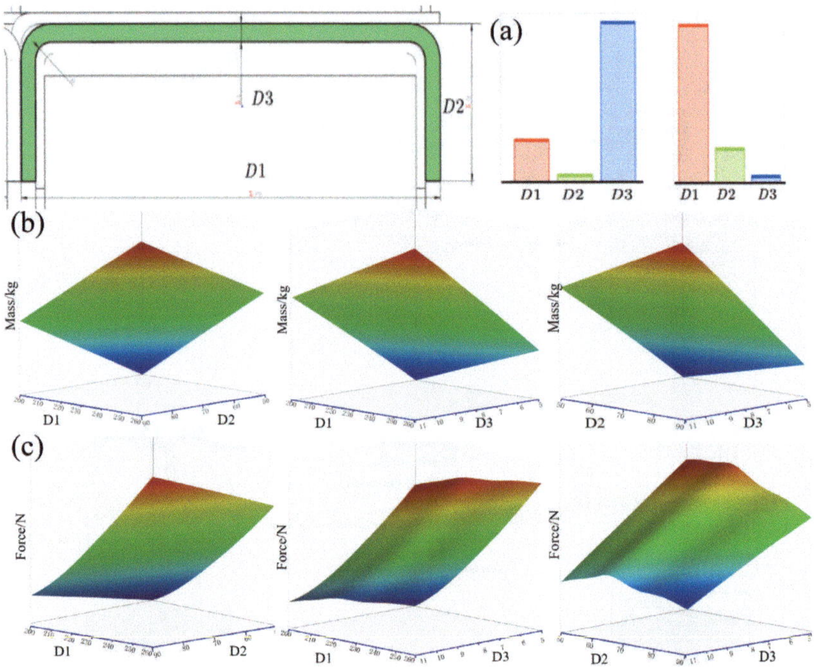

Fig. 4. Illustration of chassis parametric modeling variables

The optimal structural parameters were determined as follows:

$$D_1 = 210\text{mm}, D_2 = 56\text{mm}, D_3 = 6\text{mm}$$

A comparative summary of chassis structural parameters before and after optimization is presented in Table 3. The optimized design achieved a substantial mass reduction of approximately 15% while adequately satisfying the structural stiffness requirements. The results indicate that structural parameter optimization can achieve light weighting goals while ensuring performance.

Table 3. Comparison of structural parameters and performance metrics optimization

Performance Parameters	Before Optimization	After Optimization	Improvement (%
D_1 (mm)	230	210	−8.7%
D_2 (mm)	70	56	−20%
D_3 (mm)	8	6	−25%
Mass(kg)	914	777	−15%

4 Conclusion

This study assessed the chassis performance of a mobile aero-engine test platform under three loading scenarios: front/rear axle torsion and vertical static load. The key findings are:

(1) Torsional stiffness: The chassis met design requirements under both axle conditions, ensuring structural stability during operation.
(2) Strength: Under a 30,000 N static load, stress levels remained within safe limits, confirming adequate structural strength.
(3) Light weight: Parametric refinement reduced mass by 15% (from 914 kg to 777 kg) while maintaining stiffness and safety, demonstrating the effectiveness of the lightweight design.

Future work will focus on integrating modal and fatigue analysis to further enhance structural reliability.

Acknowledgments. Supported by the National Natural Science Foundation of China (Grant No.12173054, 52405073)

Disclosure of Interests. The authors have no competing interests to declare that are relevant to the content of this article.

References

1. Arghode, V.K., Gupta, A.K.: Development of a flexible aeroengine test platform. Appl. Therm. Eng. **66**(1–2), 538–546 (2014)
2. Wang, Z., Zhang, Y., Sun, X.: Design and verification of thrust measurement system for aero-engine test rigs. Measurement. **159**, 107788 (2020)
3. Liu, Y., Zhang, H., Wu, J., et al.: Research on structure optimization of aero-engine test stand based on finite element analysis. J. Mech. Eng. **53**(5), 89–96 (2017)
4. Li, B., Zhang, W., Liu, Q.: Modal characteristics and measurement decoupling design of thrust test platform. Sensors. **18**(11), 3825 (2018)
5. Chen, H., Zhao, Y., Li, D.: Lightweight design of chassis based on multi-objective optimization. Eng. Struct. **188**, 495–505 (2019)
6. Wang, S., Lin, J., Guo, M.: Fatigue life prediction of welded chassis based on FEA and experimental validation. Int. J. Fatigue. **143**, 105998 (2021)
7. Zhang, L., Sun, X., Wang, C.: Experimental verification of structural vibration in engine test rig. J. Vib. Control. **22**(16), 3543–3555 (2016)
8. Zhou, L., Han, Y., Chen, Z.: Response surface-based sensitivity analysis for chassis structural performance. Thin-Walled Struct. **173**, 108947 (2022)
9. Xu, Q., Li, Y.: Optimization of cross-beam structures in heavy-duty vehicle frames. Structure. **34**, 2058–2068 (2021)
10. Shen, Y., Wei, Y., Fang, J.: Multi-objective design of vehicle frames under vibration and load coupling. Mech. Mach. Theory. **145**, 103716 (2020)
11. Zhao, L., Zhang, M., Yu, C.: Design optimization of modular transport platforms based on FEM. Struct. Multidiscip. Optim. **57**(3), 1011–1023 (2018)

A Unified Index Method for Safety Assessment of Crawler Crane Boom Structure

Gurko Valerii, Hongsheng Zhang, Shuo Li, and Yue Wang(✉)

School of Mechatronics Engineering, Harbin Institute of Technology, Harbin, Heilongjiang, China 150001

wang_yue@hit.edu.cn

Abstract. Crawler cranes operate under highly variable loads and configurations, making rapid, reliable safety assessment of boom structure essential. This study proposes a Unified Safety Index (USI) that consolidates three critical indicators: strength (maximum equivalent stress), stability (linear buckling coefficient), and stiffness (displacement) into a single quantitative score. Gaussian membership functions transform each indicator into fuzzy safety grades ("Safe", "Warning", "Danger") using thresholds drawn from national and international standards. An Improved Combination Weighting method based on game theory (ICWGT) combines the Analytic Hierarchy Process and entropy weights to balance expert judgment and data dispersion, resulting in an optimized vector. The fuzzy relationship matrix and ICWGT weights are synthesized to obtain a categorical USI and a continuous score. Validation of finite-element simulations covering various configurations demonstrates accurate categorization across "Safe", "Warning", and "Danger" states. Sensitivity analysis, in which each weight varied by $\pm 5\%$, resulted in a change of less than 3% in the score and did not alter the safety class, confirming the robustness of the proposed method.

Keywords: Crawler Crane · Safety Assessment · Unified Safety Index · Fuzzy Logic · Game Theory

1 Introduction

Crawler cranes are pivotal in construction and transportation, handling complex lifting tasks across diverse operational scenarios. Prior studies show that structural failures, instability, or excessive deformation can lead to catastrophic accidents [1, 2]. Current safety assessment methods often focus on individual metrics — strength (maximum equivalent stress), stability (linear buckling coefficient), and stiffness (maximum nodal displacement) — in isolation.

To overcome these shortcomings, this study proposes a Unified Safety Index (USI) that integrates the three critical indicators into a comprehensive quantitative metric. The methodology leverages fuzzy logic to address uncertainties in safety state transitions, where conditions may not be strictly "safe" or "unsafe" but exist in intermediate zones [3–6]. To assign appropriate weights to each indicator, an Improved Combination Weighting method based on game theory (ICWGT) is employed [7, 8]. This approach

S. Halgamuge et al. (Eds.): ICACMVE 2025, LNME, pp. 135–142, 2026.
https://doi.org/10.1007/978-981-95-6911-3_13

optimally balances objective weights derived from data variance via the entropy method [9] with subjective weights informed by expert judgment through the Analytic Hierarchy Process (AHP) [10]. Safety thresholds are established based on national and international standards [11, 12], ensuring compliance and applicability. The USI model is validated using finite element analysis (FEA) across various crane configurations, demonstrating its accuracy in classifying safety states and its robustness through sensitivity analysis.

The paper is structured as follows: Sect. 2 outlines the methodological basis for the USI, defining its two-level conceptual model. The determination of safety levels using Gaussian fuzzy sets and calibrating membership functions. The weighting of indicators based on FEA results and synthesis, including the fuzzy relationship matrix and ICWGT optimization. Obtain the final classification, the maximum membership principle is applied, assigning a USI to the state with the highest membership value. For a more detailed assessment, this paper introduces a quantitative Unified Safety Score. The method's robustness is tested through sensitivity analysis. Section 3 concludes with key findings and implications for enhancing crawler crane safety management.

2 Methodological Basis for the Unified Safety Index (USI)

2.1 Determining Security Levels Using Gaussian Fuzzy Sets

To structure the assessment, a two-level method is created, serving as the basis for the entire system. Target level (USI): The goal of the method is a single, comprehensive indicator representing overall structural safety. Indicator level comprises direct, quantitatively measurable indicators prescribed by safety standards and obtained through analysis.

In this study selected three critical indicators: Strength (U_1), assessed by maximum equivalent stress ($\sigma_{\max}$) to evaluate proximity to material yield and plastic deformation risk. Stability (U_2), assessed by the linear buckling coefficient (λ) indicating the safety margin against compressive failure. Stiffness (U_3), assessed by maximum nodal displacement ($\delta_{\max}$) at the boom end. To address ambiguity in safety state transitions beyond simple "safe" or "unsafe", a fuzzy evaluation method is developed using fuzzy sets. These sets assign membership degrees between 0 and 1, allowing partial membership to represent gradual safety boundaries.

The next step is to introduce clear quantitative thresholds ($C_{\min}$ and $C_{\max}$) that define the boundaries between safe and unsafe states. These thresholds are based on the national standard TSG 51–2023. The method based on a set of factors (U_i) comprising of indicators and a set of assessment levels (V_i) consisting of linguistic safety terms $\{V_1, V_2, V_3\} = \{\text{Safe, Warning, Dangerous}\}$.

The introduction of a 10% threshold for the strength and stiffness parameters is a risk management strategy. This threshold is based on the "safety factor" principle. For example, for strength, the yield strength of FGS9OWV steel is 583 MPa. Setting the "Warning" limit to 525 MPa ($583 \times 0.9 \approx 525$ MPa) creates a warning zone. The boundaries of the danger zone are defined from the maximum permissible value of the material to the maximum obtained value of the indicator based on FEA. According to state standards, the stiffness parameter has a warning threshold of 20% of the sum of the length of the main boom (L_{main}) and the jib (L_{jib}) of the crane $\delta_{\max} > 20\% \times \sum$

$L_{main} + L_{jib}$. The Danger grade boundaries are selected on the basis of maximum FEA results.

The buckling coefficient λ = 1.34 [13] represents a sharp critical limit defined by safety standards. Values below this threshold indicate a high and imminent risk of buckling. Therefore, the "Warning" state for this parameter is defined as $\lambda \in [1.34, 1.47]$, indicating that the crane is precisely at the stability limit, and any value below this value is immediately classified as "Dangerous". Table 1 presents the quantitative thresholds that serve as the basis for further fuzzy classification.

Table 1. Corresponding quantitative thresholds.

Indicator	Grade: Safe (V_1)	Grade: Warning (V_2)	Grade: Danger (V_3)
Strength (σ_{max}) MPa	$0.1 \leq \sigma_{max} < 525$	$525 \leq \sigma_{max} \leq 582$	$582 < \sigma_{max} < 800$
Stability (λ)	$5.0 \geq \lambda > 1.47$	$1.34 \leq \lambda \leq 1.47$	$0.1 \leq \lambda < 1.34$
Stiffness (δ_{max}) cm	$0.1 \leq \delta_{max} < 194$	$194 \leq \delta_{max} \leq 216$	$216 < \delta_{max} < 430$

Parameter Calibration for Gaussian Membership Functions. To transform the crisp thresholds from Table 1 into a flexible fuzzy representation, Gaussian membership function are employed due to their smooth transitions and physically realistic modeling of continuous parameters. Each safety level (V_j = Safe, Warning, Danger) for every indicator requires two calibrated parameters:

Mathematical expectation (E_x): Represents the midpoint of the safety levels range, ensuring the core of the fuzzy set aligns with the most typical value for that state.

$$E_x = (C_{min} + C_{max})/2 \tag{1}$$

where C_{min} and C_{max} are explicit boundaries.

Entropy (E_n): Quantifies the fuzziness of the transition between safety levels. For a Gaussian function, ensures membership at boundaries C_{min} and C_{max} creating a smooth and decisive transition. This parameter controls how quickly the degree of membership decreases as values move away from the center, similar to the standard deviation.

$$E_n = (C_{max} - C_{min})/2\sqrt{-2\ln(\beta)} \tag{2}$$

Parameter's stress and buckling change continuously, for the Gaussian function, the chosen value of β = 0.01 for calculations avoids abrupt changes while ensuring boundaries are practically. An example of calculation from Table 1 for parameters in each condition represented in Table 2.

2.2 Weighting of Indicators and Synthesis

Fuzzy Relationship Matrix. This matrix provides a quantitative snapshot of the current state of the crane, indicating the degree to which each indicator aligns with each safety

Table 2. Parameters calibration.

E_i	σmax V_1 (MPa)	λ V_1	δmax V_1(cm)	σmax V_2 (MPa)	λ V_2	δmax V_2(cm)	σmax V_3 (MPa)	λ V_3	δmax V_3(cm)
E_x	262.55	3.235	97	553.5	1.405	205	691	0.72	323
E_n	86.5	0.582	32	9.4	0.02	3.62	35.91	0.204	35,3

category. The matrix R is populated by calculating the degree of membership (r_{ij}) for each indicator ($i = U_1, U_2, U_3$) to each safety level ($j = V_1, V_2, V_3$). A Gaussian membership function is used, chosen for its smooth transition curve, which accurately reflects the continuous nature of physical parameters such as stress and displacement.

$$r_{ij} = \exp\left(-\frac{\left(x_i - E_{xj}\right)^2}{2\left(E_{nj}\right)^2}\right) \tag{3}$$

where x_i represents the actual measured value of the indicator for a given operating scenario. E_{xj} and E_{nj} are the expected value and entropy parameters previously calibrated for the j-th safety level. Each element (r_{ij}) in the matrix denotes the degree of membership of indicator i to safety level j, ranging from 0 (no membership) to 1 (full membership).

This process results in a matrix for each operational scenario, serving as the core input for the final weighted synthesis. For instance, for a scenario where the measured stress is 528.48 MPa (Warning), the buckling factor is 1.77 (Safe), and the displacement is 205.58 cm (Warning). Once the matrix is obtained, normalization is performed to ensure each row sum to 1. This turns absolute membership degrees into relative ones, showing the distribution of "confidence" for one indicator across all safety categories. The resulting Fuzzy Relationship Matrix would be:

$$R = \begin{bmatrix} r_{11} & r_{12} & r_{13} \\ r_{21} & r_{22} & r_{23} \\ r_{31} & r_{32} & r_{33} \end{bmatrix} = \begin{bmatrix} 0.35 & 0.64 & 0.03 \\ 0.99 & 0 & 0.01 \\ 0.05 & 0.893 & 0.057 \end{bmatrix} \tag{4}$$

Optimized Weighting of Indicators Using Game Theory. Rationally weighing the influence of indicators is crucial for the creation of unified index. It should rely not only on objective weights based on data (w_{obj}) but also on subjective expert weights (w_{sub}). To address this issue, an improved combined weighting method based on game theory (ICWGT) applied to find the optimal balance between weights derived from data variance (entropy method) and weights derived from expert judgment (AHP). A single combined vector W_{ICWGT}, which is a linear combination of the base vectors consisting of the weight vector $w_1 = w_{\text{sub}}$ (from the AHP). It is assumed that the influence of each indicator has an equivalent weight contribution and $w_2 = w_{\text{obj}}$ (from the entropy method).

$$w_{\text{obj}} = \frac{d_j}{\sum_{j=1}^{n} d_j} \tag{5}$$

$$E_j = -\frac{1}{\ln(m)} \sum_i r_{ij} \ln\left(r_{ij}\right); \tag{6}$$

where $d_j = 1 - E_j$, m is the number of FEA calculations.

$$W_{\mathrm{ICWGT}} = \sum_{k=1}^{2} \alpha_k w_k^{\mathrm{T}} \tag{7}$$

where α_k are the combination coefficients subject to optimization, satisfying the conditions $\sum \alpha_k = 1$ and $\alpha_k > 0$.

The optimal coefficients ($\alpha_1 = 0.474$, $\alpha_2 = 0.526$) minimize the deviation between the final vector W and each of the initial vectors w_k. This game-theoretic objective is formulated as:

$$\min \left\| \sum_{j=1}^{2} \alpha_j w_j^{\mathrm{T}} - w_j^{\mathrm{T}} \right\|_2, \text{ for } i = 1, 2 \tag{8}$$

The final, optimized weight vector is calculated using these coefficients:

$$W_{\mathrm{ICWGT}} = \alpha_1 w_{\mathrm{sub}} + \alpha_2 w_{\mathrm{obj}} \tag{9}$$

The resulting weights used in this study are presented in Table 3.

Table 3. Calculated Weights.

Indicator	Objective Weight (w_{obj})	Subjective Weight (w_{sub})	Optimized (W_{ICWGT})
Strength (U_1)	0.199	0.333	0.264
Stability (U_2)	0.407	0.333	0.368
Stiffness (U_3)	0.394	0.333	0.362

2.3 Calculation of the Unified Safety Assessment Index

The final step synthesizes the fuzzy relation matrix R and the combined weights W_{ICWGT} using the fuzzy synthesis operation $B = W_{\mathrm{ICWGT}} \otimes R$. This yields a score vector $B = [b_1, b_2, b_3]$, where b_1, b_2, and b_3 are the final membership grades for the Safe, Warning, and Dangerous states, respectively.

To obtain the final classification, the maximum membership principle is applied, assigning a USI to the state with the highest membership value. For a more detailed assessment, this paper introduces a quantitative Unified Safety Score (USS):

$$\mathrm{USS} = \frac{(b_1 \times 0.2) + (b_2 \times 0.4) + (b_3 \times 0.9)}{b_1 + b_2 + b_3} \tag{10}$$

This rating translates the security status onto a continuous scale from 0 to 1, where a rating close to 0 corresponds to "Safe," and a rating closer to or more than 1 corresponds to "Dangerous." For practical visualization in the monitoring system, these ratings are mapped to warnings: a USS rating equal to or less than 0.3 indicates a "Safe" status, a rating greater than 0.3 and less than 0.6 indicates a "Warning" status, and a rating of 0.6 or higher indicates a "Dangerous" status. To validate the method, it applied to a dataset of FEA simulations. Each "Scenario" in the dataset represents a unique operating condition of the crane, defined by parameters such as main boom length, jib configuration, and operating radius. Table 4 presents the detailed results for 6 representative scenarios.

Table 4. Detailed Safety Evaluation Results and Comparison for Selected Scenario.

Scenario Config	Stress (MPa)	Buckling (λ)	Disp. (cm)	Individual Assessment	Vector B	USS (Score)	Final USI
1	259.37	2.79	161.43	Safe / Safe / Safe	[0.88, 0.0, 0.07]	0.24	Safe
2	250.77	2.96	161.18	Safe / Safe / Safe	[0.89, 0.0, 0.07]	0.24	Safe
3	509.32	1.62	200.71	Safe / Safe / Warning	[0.51, 0.38, 0.27]	0.5	Warning
4	528.48	1.77	205.37	Warning / Safe/ Warning	[0.51, 0.51, 0.27]	0.55	Warning
5	696.16	1.93	322.56	Danger / Safe / Danger	[0.33, 0.0, 0.66]	0.67	Danger
6	771.31	1.64	394.52	Danger / Safe / Danger	[0.23, 0.05, 0.51]	0.66	Danger

To test the robustness of the evaluation method, a sensitivity analysis is conducted on the ICWGT weights. The objective is to determine whether minor changes in the calculated weights result in significant changes in the resulting safety classification. . Each of the three indicator weights (W_1 for Strength, W_2 for Buckling, and W_3 for Stiffness) changed by $\pm5\%$. When one weight changed, the other two were adjusted proportionally so that their sum remains equal to 1. The results presented in Table 5 indicate that the final classification remains "Danger" in all test cases.

Table 5. Sensitivity Analysis.

Adjusted Weight	W_{ICWGT} Vector	Vector B	USS	Changes	Final USI
W_1 (+5%)	[0.277, 0.368, 0.362]	[0.302, 0.0, 0.67]	0.68	+1.47%	Danger
W_1 (−5%)	[0.251, 0.368, 0.362]	[0.288, 0.0, 0.64]	0.66	−1.5%	Danger
W_2 (+5%)	[0.264, 0.386, 0.362]	[0.31, 0.0, 0.65]	0.68	+1.47%	Danger
W_2 (−5%)	[0.264, 0.35, 0.362]	[0.28, 0.0, 0.65]	0.69	+2.9%	Danger
W_3 (+5%)	[0.264, 0.368, 0.38]	[0.3, 0.0, 0.67]	0.68	+1.47%	Danger
W_3 (−5%)	[0.264 0.368, 0.344]	[0.3, 0.0, 0.64]	0.67	0%	Danger

3 Conclusion

This study has successfully developed and validated a Unified Safety Index (USI) for boom structures of crawler cranes, integrating strength, stability, and stiffness into a quantitative safety measure. By employing Gaussian membership functions and an Improved Combination Weighting method based on game theory, the USI effectively captures the nuanced safety states of crawler cranes, providing a more reliable and rapid assessment tool. Validation against finite element simulations demonstrated the USI's accuracy in classifying safety states across various crane configurations, given the impracticality of physical stability tests due to their prohibitive cost and significant danger, with no public reports of such tests currently available. Furthermore, sensitivity analysis has revealed that the method maintains consistent safety classifications even with minor variations in indicator weights, underscoring its robustness. The USI offers a practical tool for enhancing safety management in crawler crane operations, potentially reducing the risk of accidents and improving operational efficiency.

Acknowledgements.. This work was Supported by the Longjiang Science and Technology Talent Spring Goose Support Program (CYCX24011), Key R&D Program Project of Heilongjiang Province (No. 2022ZX01A03).

References

1. Shepherd, G.W., Kahler, R.J., Cross, J.: Crane fatalities — a taxonomic analysis. Saf. Sci. **36**(2), 83–93 (2000)
2. Aneziris, O.N., et al.: Towards risk assessment for crane activities. Saf. Sci. **46**(6), 872–884 (2008)
3. Zadeh, L.A.: Fuzzy sets. Inf. Control. **8**(3), 338–353 (1965)
4. Aihua, L.: Risk assessment of crane operation hazards using modified FMEA approach with Z-number and set pair analysis. Heliyon. **10**(9) (2024)
5. Ross, T.J.: Fuzzy Logic with Engineering Applications, 3rd edn. Wiley, Hoboken (2010)
6. Shi, H., Liu, Z.: Hu-Chen Liu.: a new linguistic preference relation-based approach for failure mode and effect analysis with dynamic consensus reaching process. Inf. Sci. **610**, 977–993 (2022)

7. Zhu, Z., Luo, Y., Xiao, H., Xiong, C., Chen, C.: Research on the crane safety assessment method based on the cloud model and ICWGT. Appl. Sci. **12**(9), 4399 (2022)
8. Aihua, L.: Human error risk prioritization in crane operations based on CPT and ICWGT. PLoS One. **16**(1) (2024). https://doi.org/10.1371/journal.pone.0297120. Accessed 11 July 2025
9. Zhu, Y., Tian, D., Yan, F.: Effectiveness of entropy weight method in decision-making. Math. Probl. Eng., 1–5 (2020)
10. Saaty, R.W.: The analytic hierarchy process—what it is and how it is used. Math. Model. **9**(3–5), 161–176 (1987)
11. Standardization Administration of China (SAC): GB/T 1591–2018: High Strength Low Alloy Structural Steels. SAC, Beijing (2018)
12. State Administration for Market Regulation (SAMR): TSG 51–2023: Regulation on Safety Technology for Lifting Appliances. SAMR, Beijing (2023)
13. Zhang, H., Lu, N.: Comparative study on strength safety factor and structural stability coefficient of finite element method. Chin. J. Constr. Mach. **18**(4), 310–313 (2020)

Influent Factors Analysis and Stress Prediction of V-Type Structure Based on RSM

Chu Jiang[1], Xin Wang[1(✉)], Guohua Wang[2], Yang Teng[2], and Yan Wu[2]

[1] Dalian University of Technology, Dalian 116024, China
wangxbd21@163.com
[2] Dalian Boiler and Pressure Vessel Inspection & Testing Institute CO. Ltd., Dalian 116024, China

Abstract. As the research object, a V-type structure in fixed leg of gantry crane applied in handling bridge sections is analyzed, and its FEA model is built. According to the result of the typical load case, there is stress concentration at the V-type structure. Three factors affecting the stress concentration are analyzed based on response surface methodology (RSM), including V-type angle, corner radius and plate thickness of V-type structure. The results show that the ascending sequence of influence is the corner radius, the plate thickness and the V-type angle. Then, the trends of the maximum stress as it varies with a single factor are analyzed. As a result, within a certain range, the maximum stress decreases with the increase of V-type angle in a linear relationship, and decreases rapidly with the increase of the corner radius and then tends to level off, and decreases with the increase of the plate thickness in an inversely proportional function relationship. Furthermore, according to multi-factor coupling influence analysis, the agent model reflecting the relationship between influencing factors and stress mapping is established and its accuracy is verified these provide the technical basis for the subsequent stress-life prediction.

Keywords: V-type structure · agent model · stress prediction · RSM

1 Introduction

V-type structure is widely used in the products in the field of construction machinery, which is used as gantry crane for bridge construction [1], and its fixed leg is a typical representative of V-type structure, as shown in Fig. 1. With the development of national construction, a large number of cranes are put into production. The structural failure occurs frequently, and has a base growth trend. The V-type structure of fixed leg for a gantry crane is subjected to variable loads all year round, so fatigue damage becomes one kind of failures. Since surface cracks often appear at the sharp corners of the V-type structure, where the radius of curvature is often very small so as to lead to stress concentration, it is very important to analyze the factors affecting the stress of the V-type structure.

S. Halgamuge et al. (Eds.): ICACMVE 2025, LNME, pp. 143–150, 2026.
https://doi.org/10.1007/978-981-95-6911-3_14

Currently, structural stress analysis is mainly an agent model approach. Romdhane Ben Khalifa used numerical simulation, response surface methodology (RSM) and artificial neural networks (ANNs) to predict the critical weld results with less deviation [2]; Narges Mokhtari processed the mesh space of the reinforcing plate through CNN, and successfully predicted the von Mises stress after the plate strengthening [3]; For key positions of the turbine,Cheng Muwei constructed and trained a temperature and stress prediction model based on Long Short-Term Memory (LSTM) neural network, and evaluated the generalisation ability of the model through cross-validation and noise implantation [4]; Xing Xinlei performed finite element simulation calculations using COMSOL software, extracted the pipeline stress critical points of the stress distribution data, established stress dataset, and the stress prediction using the gradient boosting tree algorithm got better results [5].

At present, there are not many in-depth studies on the stress and geometric relationship of this V-type structure. Therefore, the V-type structure is taken as the research object, the finite element analysis method is adopted for exploring the stress concentration location, fits the stress data with the RSM model. The corner radius, plate thickness, and the V-type angle are taken as the influencing factors. The influence trend of single factor and multi-factor coupling on stress are analyzed, and the agent model between multi-factors and maximum stress is established to realize the fast stress prediction.

2 Finite Element Static Analysis

The V-type structure from the fixed leg structure of 900 t gantry crane, is one kind of box structures, which consists of the top and down flange plates, the webs, and inner rib plates. So shell element is chosen to build the finite element model. The material is Q345B, and the modulus of elasticity and Poisson's ratio are respectively 200GPa and 0.3. Considering that the gradient of the stress at the concentration of the stress varies greatly,, taking the tip point as the center, the V-type structure is radially divided into 4 areas, which are gradually transitioned from the center subdivided meshes to the coarse ones. The mesh size of the four regions are 80 mm, 40 mm, 20 mm and 10 mm respectively, as shown in Fig. 2(b). This results in a smooth transition from coarse to fine mesh, with 34806 nodes and 34795 cells in the model.

In this paper, one section of the leg is selected, according to the structural connection, fixed constraints are applied to the bottom of the leg, and the load shared to this leg by its upper self-weight and the external load of the lift is applied to the top of the leg, which is 6123.5kN. The load is applied in the form of linear load to the four upright plates formed by the leg. The boundary conditions are shown in Fig. 2(a).

For the above model, boundary conditions and loads, ANSYS software is used for static analysis and the stress cloud is obtained as shown in Fig. 3.

It can be seen that there is a stress concentration at the weld seam of the V-type structure corner, with a maximum stress of 158.78 MPa. From the structure, it is easy to see that the stress at this place is related to the corner radius, V-type angle, V-type plate and bending plate thickness (the following text referred to as "plate thickness"). Figure 4 gives the sensitivity analysis result of the three factors influence on the stress, and the order of influence is corner radius > plate thickness t > V-type angle.

Fig. 1. V-type structure.

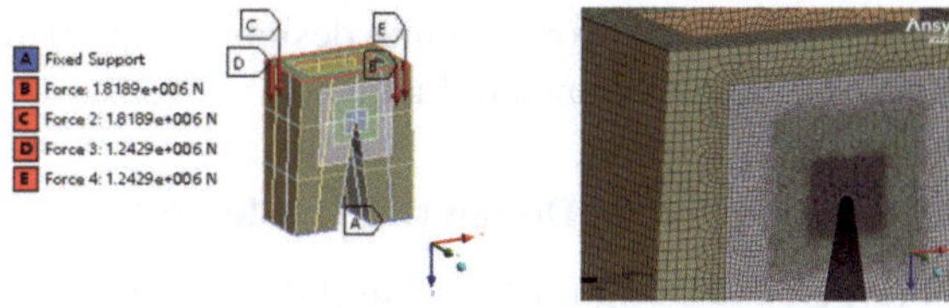

a) Boundary conditions b) Meshes

Fig. 2. Finite element model.

a) Overall stress diagram b) Local stress diagram

Fig. 3. Stress diagram.

In order to analyze the effect of each single factor on the stress, a single factor is changed while keeping other parameters constant to get the curve of this factor with respect to stress. Considering the large amount of work involved in finite element modelling, parametric modelling is implemented with the help of response surface methodology to obtain the stress variation curve quickly.

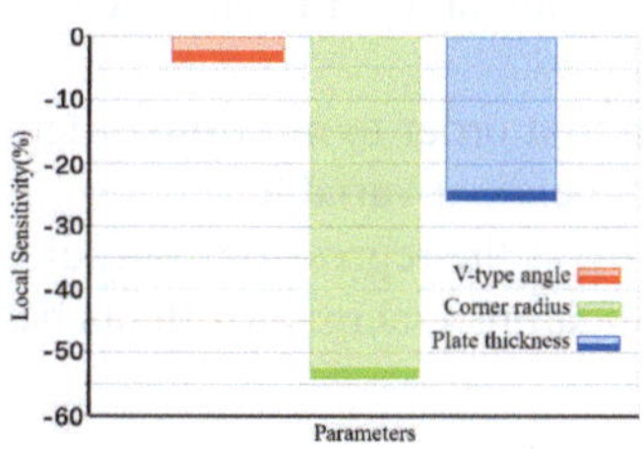

Fig. 4. Sensitivity analysis.

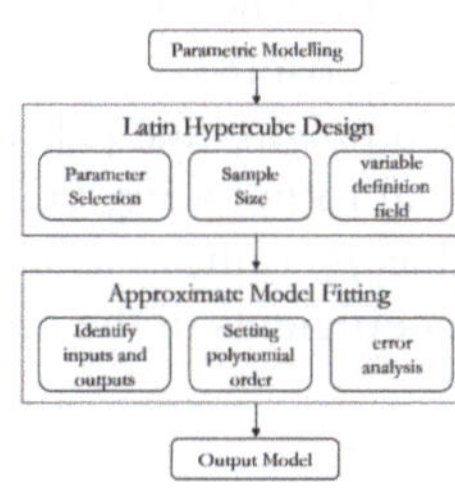

Fig. 5. Work flow.

3 Response Surface Methodology (RSM) and Calculation Flow

In response surface method, multiple quadratic regression equations is used to fit the functional relationship between variables and response values. An experimental design was made to study the effect of each variable and its coupling on the response values, and multivariate optimization through the regression equations to ultimately determine the relationship between the variables and the response values. The calculation process is mainly divided into experimental design (sampling data extraction) and approximate simulation fitting, as shown in Fig. 5.

3.1 Latin Hypercube Design (LHD) Method

Latin Hypercube Design (LHD) was proposed by McKay in 1979 [6] and belongs to the stratified sampling techniques. As one of the main methods of Design of Experiments (DOE), LHD especially suitable for sampling data with high dimensional variables. The working principle is to divide each dimension of the variable dimensions into multiple equal probability mutually exclusive intervals, and take a point at random in each interval. The order of the dimensions is then disrupted to eliminate inter-dimensional interference. The resulting samples can uniformly cover each variable dimension, and the combination of RSM model can represent the complex mapping relationships of the parameters in the system. Compared with the center composite design (CCD), uniformly distributed sampling and other experimental design methods, the spatial coverage is uniform. The steps are:

(1) The number of parameter levels is specified as 10, setting the range of parameter variations, the range of parameter values are divided in each part into 10 equal parts, then the 3-dimensional parameter space will be divided into 1,000 subintervals, each subinterval called a hypercube.
(2) A 10×3 dimensional matrix is defined, each column of which consists of random rational numbers obeying a uniform distribution on the interval [0, 1], so the sample point's specific locations are determined [7].

3.2 Response Surface Model (RSM)

The essence of RSM is a multivariate generalization of Taylor's expansion, which approximates the true function by a polynomial model, and completes the optimization at a small trial cost [8].

The fitted polynomial [9], which can be from first order to multiple orders, is given in polynomial form by Eq. (1). Where, y is the dependent variable, $x_1, x_2,..., x_M$ is the input variable, β_0 is the constant term, $\beta_1, \beta_2,..., \beta_M$ is the regression coefficient, β_{ii} and β_{ij} are the coefficients of the interaction terms indicating synergistic effects between the variables.

$$y = \beta_0 + \sum_{i=1}^{m} \beta_i x_i + \sum_{i=1}^{m} \beta_{ii} x_i^2 + \sum_{i}^{m} \sum_{\substack{j \\ (i<j)}} \beta_{ij} x_i x_j \tag{1}$$

4 Factor Impact Analysis

4.1 Single Factor Impact Analysis

The basic parameters of each influencing factor of this structure are external load 6132.5kN, V-type angle for 14°, corner radius for 62 mm, plate thickness for 12 mm. Firstly, only the V-type angle factor is changed, and 20 sets of sample data are extracted in the range of V-type angle [1°, 40°] through the LHD method, mapping relationship is fitted through the RSM model, which is shown in Eq. (2), and the curve of maximum stress variation with V-type angle is shown in Fig. 6. It can be seen that the stress varies basically linearly and decreasingly with the V-type angle, and the fitted polynomial at the primary level, which also shows a linear trend.

$$\sigma = -1.4\alpha + 179.29 \tag{2}$$

Where σ is the maximum stress and α is the V-type angle.

Secondly, only the corner radius is changed, and 20 sets of sample data are extracted in the range of corner radius [1 mm, 120 mm] through LHD method, mapping relationship is fitted through the least squares method, which is shown in Eq. (3), and the curve of maximum stress with corner radius is shown in Fig. 7. It can be seen that the stress is decreasing with the corner radius and then tends to be stable, the fitted equation shows the inverse proportional function trend.

$$\sigma = 775.05r^{-0.585} \tag{3}$$

Where σ is the maximum stress and r is the corner radius.

Finally, only the plate thickness is changed, and 20 sets of sample data are extracted in the range of plate thickness for [4 mm, 30 mm] through LHD method, mapping relationship is fitted through the least squares method [10], which is shown in Eq. (4), and the curve of maximum stress with the thickness is shown in Fig. 8. It can be seen that the stress with plate thickness is decreasing nonlinear change, the fitted equation shows the inverse proportional function trend.

$$\sigma = 1249.4t^{-0.845} \tag{4}$$

Where σ is the maximum stress and t is the plate thickness.

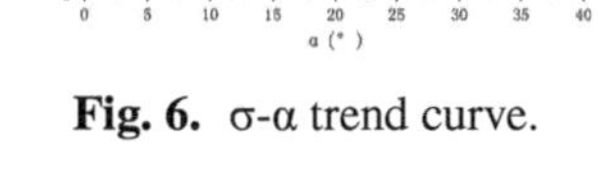

Fig. 6. σ-α trend curve.

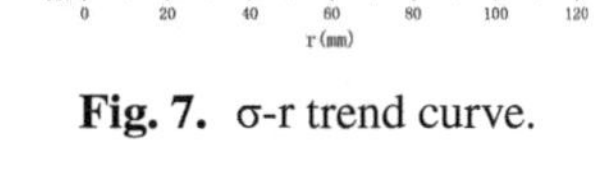

Fig. 7. σ-r trend curve.

4.2 Multi-factor Coupling Impact Analysis

The stress effects of the three single factors were analyzed previously under the basic parameter conditions. If multiple factors produce changes at the same time, it is possible to cause different stress trends, which indicates that there is a coupling relationship

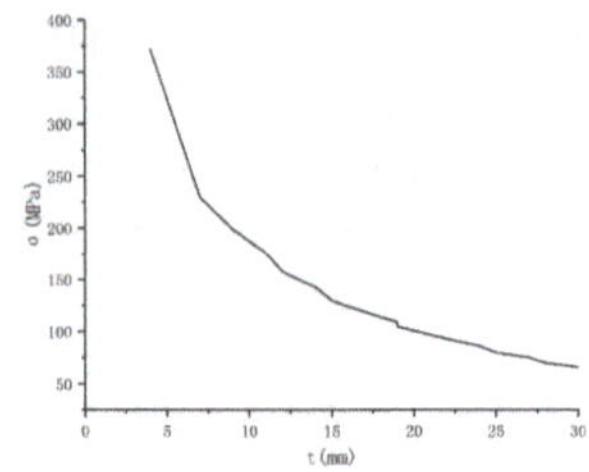

Fig. 8. σ-t trend curve.

among the factors. For this reason, the stress trend analysis is carried out for three factors in the value range: V-type angle for [1°, 40°], corner radius for [1 mm, 120 mm], and plate thickness for [4 mm, 30 mm], as shown in Fig. 9. The three axes represent the V-type angle, corner radius, plate thickness, and the color value is stress. For the plane formed by the corner radius and the plate thickness, it is easy to see that the variation of each *r-t* in-plane stress under different V-type angle, but the overall decrease with the increase of V-type angle. For the other two planes, different stress changes also occurred. From this, it can be seen that there is a coupling relationship among the three factors.

For this purpose, 1000 sets of sample data were taken by LHD method, and the mapping relationship among the three factors and stress by RSM model is shown in Eq. (5), and the two-factors multiplication relationship among the three factors can also be seen in the equation, which indicates the coupling relationship among the three factors.

$$\begin{aligned}\sigma &= -0.001\alpha^4 + 3.007r^4 + 0.002t^4 + 0.042\alpha^3 - 0.008r^3 - 0.183t^3 - 0.343\alpha^2\\ &+0.788r^2 \quad + 5.549t^2 - 0.033\alpha r + 0.202\alpha t - 0.015rt\\ &-3.92\alpha - 29r - 79.2t + 931.138\end{aligned} \tag{5}$$

where σ is the maximum stress, r is the corner radius, and t is the plate thickness.

The variable fitting accuracy of this polynomial is shown in Fig. 10. 50 of the 1000 sets of samples are selected for accuracy verification, and the coefficient of determination R^2 is applied for accuracy analysis. It can be seen that $R^2 = 0.90303$, which indicates that this polynomial has a high fitting accuracy, and can be used as an agent model to achieve fast stress prediction.

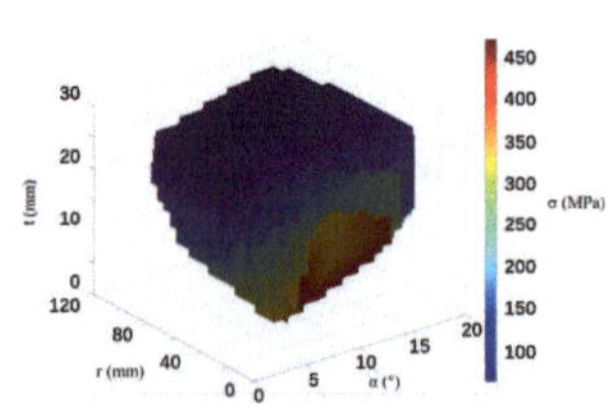

Fig. 9. Three-variable coupling relationship to stress.

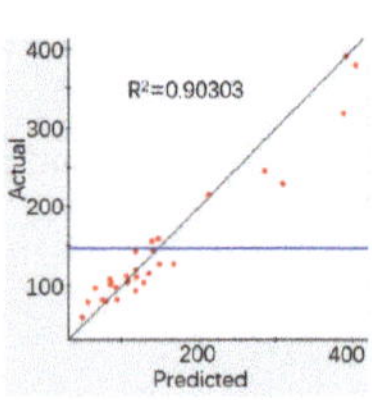

Fig. 10. Error analysis.

5 Conclusion

(1) The finite element model of V-type structure is established. According to the structural stress distribution the stress concentration location is at the structural corner.
(2) The sensitivity of the three factors affecting the stress, namely, V-type angle, corner radius, plate thickness is analyzed. It is indicated that the ascending influence of the three factors is the corner radius, the plate thickness, and the V-type angle.
(3) Through the RSM, the relationship between the factors and the stress is obtained, and the trends of single factor and multi-factor coupling on the stress is analyzed.
(4) Through the accuracy analysis, it is shown that the polynomials fitted to multi-factors and stresses can perform the mapping relationship better, which can achieve fast stress prediction and provide support for the subsequent stress-fatigue prediction.

Acknowledgements. This project is supported by the Dalian Inspection Testing Certification Group Scientific Research Project (Project No.: J-2404-001-GJ-Q).

References

1. Yang, Q.X.: Research on Lightweighting of Key Components of 1000T Tyre-Type Gantry Crane. Yanshan University (2020)
2. Khalifa, B.R., Toumi, O.: Prediction of the friction stir welding temperature and ultimate tensile stress for 6061 T6 aluminum. Proc. Inst. Mech. Eng. C J. Mech. Eng. Sci. **239**(14), 5622–5635 (2025)
3. Mokhtari, N., Cai, Y., Jelovica, J.: Comparison of deep learning techniques for prediction of stress distribution in stiffened panels. Thin-Walled Struct. **215**(PA) (2025)
4. Cheng Muwei, Yang Xiaoguang, Fan Yongsheng, Jiang Renjie, Shi Duoqi. A Fast prediction method for local service temperature and stress of aero-engine turbine discs based on LSTM J. Aerosp. Power, 1–15.
5. Xinlei, X.I.N.G., Guangyuan, W.E.N.G., Haoran, W.A.N.G., et al.: A machine learning model for stress prediction of oil and gas pipelines based on magnetomechanical simulation. Tech. Supervision Pet. Ind. **41**(03), 1–5 (2025)
6. Chen, C.: Research on Comprehensive Performance of Crane Telescopic Boom and its Multi-Objective Optimisation Design. Hunan University (2013)
7. Zi-Xu, J.: Research on Parameter Calculation of Super High-Rise Structural Engineering and its Visualisation Based on Digital Twin. Harbin Institute of Technology (2024)
8. Liang, A.: Research on Lightweighting and Experimental Verification of Construction Machinery Rims. Yanshan University (2024)
9. Yang, D.: Research on Large Tonnage Double Slewing Railway Rescue Crane Turntable System. Southwest Jiaotong University (2014)
10. Yan, L., Xiang, D., Qiwen, L., et al.: Research and application of rapid lightweight design system for cranes. J. Mech. Eng. **54**(09), 205–213 (2018)

Stress Influence Analysis of Right-Angled Structure Based on ANSYS

Muhao Wang[1], Xin Wang[1(✉)], Yan Wu[2], Yanchuan Liu[2], and Wenjing Wang[2]

[1] Dalian University of Technology, Dalian 116024, China
wangxbd21@163.com

[2] Dalian Boiler and Loads Vessel Inspection & Testing Institute CO., Ltd, Dalian 116024, China

Abstract. For the problem of fatigue damage that occurs easily in the right-angled structure of bridge cranes, the stress influence of the right-angled structure is analyzed with Ansys. In this paper, the right-angled structure of bogie for the bridge crane is as the research object, and its FEA model is built with plate elements. After sensitivity analysis, the factors influencing stress are determined. According to the stress influence sequence, the trends between these factors and the stress are studied respectively by response surface method (RSM). Then, an agent model, that considers the overall structure and integrates the multi-factors effects, is built to enhance the efficiency of stress calculation for right-angled structures.

Keywords: Right-angled structure · agent model · finite element analysis

1 Introduction

The right-angled structure is commonly at the end beam and the bogies of bridge cranes and gantry cranes, as shown in Fig. 1. For assembling the gearbox, it is inevitable to exit the right-angled opening where has usually an arc transition. Due to the existence of right-angled, when the wheels bear the supporting load, the stress flow lines within the structure change direction suddenly, which causes stress concentration and lead to fatigue failure, as shown in Fig. 2. Therefore, it is of great significance to analyze the factors influencing stress and relationship of stress-fatigue life for the right-angled structure.

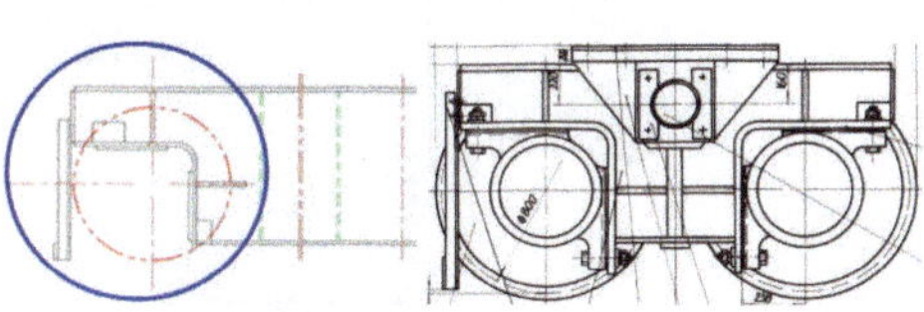

(a) At the beam (b) At the bogie

Fig. 1. Right-angled structure.

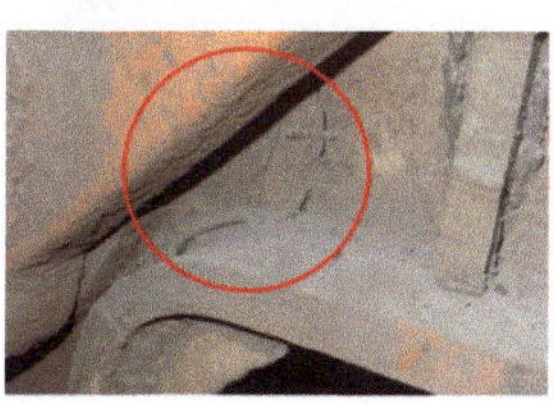

Fig. 2. crack at right-angled structure.

S. Halgamuge et al. (Eds.): ICACMVE 2025, LNME, pp. 151–158, 2026.
https://doi.org/10.1007/978-981-95-6911-3_15

Wei, Guo conducted a study using a multi-layer model and finite element analysis method, and based on the stress spectrum pattern, the fatigue life of the crack area was predicted [1]. Liu, D. K. established a finite element model of large-scale overhead cranes in hydropower stations. Actual loads were applied to the finite element model for stress time history calculation. The safe service life of the structure is estimated [2]. Muratovic, Enis conducted analytical calculations on single-girder overhead cranes via the finite element method. The maximum bending stress and deflection of the bridge were determined [3]. Abdulnaser M. Alshoaibi predicted the crack propagation trajectory under constant amplitude loading using Ansys software [4]. Zhang Jingyong proposed three types of cross-bar plate arrangements for the frame of the traveling crane bogie. Through finite element stress analysis using Ansys, the optimal structural form was determined [5]. Wan Zhengxi established a finite element model of bogie with a specific structural form using Ansys. The stress distribution of the bogie under static loading was analyzed [6]. Zheng Yun adopted a method combining FEA and experimental research. The fatigue resistance of traditional steel crane girders with right-angle abrupt change supports was improved [7]. Li Deming optimized the bogie structure of bulldozers using virtual simulation technology. The safety factor of the structure was increased [8].

The current researches mostly analyze the overall stress of the right-angled structures. However, there is not much detailed research on the influence factors of the stress. In this paper, ANSYS software is used to carry out finite element analysis on the right-angled structure of the bogie. The sensitivity of each parameter is analyzed at first. And then the response surface method is used to find out the trends of parameters and stresses. Furthermore, the agent model of the right-angled structure is built through RSM.

2 Finite Element Model of the Bogie

The object of this study is the right-angled structure of bogies for a bridge casting crane. Its material is Q355B, with an elastic modulus of 200 GPa and a Poisson's ratio of 0.3. The overall structure of the finite element model is established as plate elements. The size of the model grid is 12 mm, with 26,346 nodes and 25,924 elements.

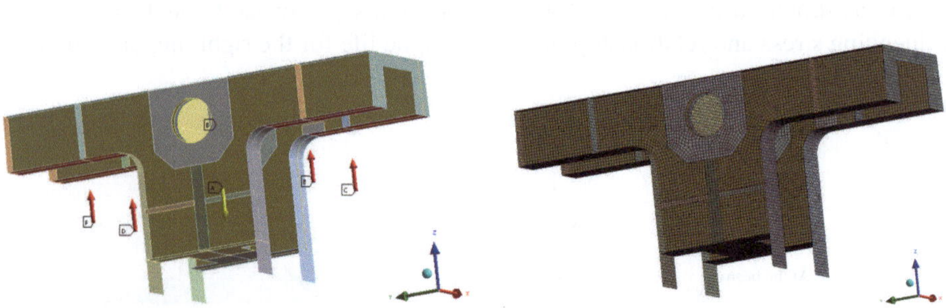

(a) Boundary conditions (b) Mesh generation

Fig. 3. The FEA model.

The wheel load is applied to the lower half of the circular hole in the middle of bogie. The boundary condition at the horizontal padding plates under the lower flange plate of bogie is UZ, ROTX, and ROTY. And the boundary condition at the vertical padding plates is UX, UY, and ROTZ, as shown in Fig. 3.

The results of different load cases are shown in Table 1. As can be seen from Fig. 4, the maximum stress on the bogie occurs at the arc of the right-angled structure. When the wheel load reaches maximum value of 89 tons, the maximum stress in the right-angled structure of the bogie stands at 347 MPa. This value is below the material's yield strength of 355 MPa.

Table 1. Wheel load of the bogie under different load case.

Load case	Wheel load/t	Max. stress/MPa	Load case	Wheel load/t	Max. stress/MPa
Deadweight in midspan	23.5	91	Deadweight in span end	33.5	130
Rated load in midspan	44.5	173	Rated load in span end	72.5	283
dynamic rated load in midspan	50	195	dynamic rated load in span end	82	320
Travel dynamic load in midspan	55	214	Travel dynamic load in span end	89	347

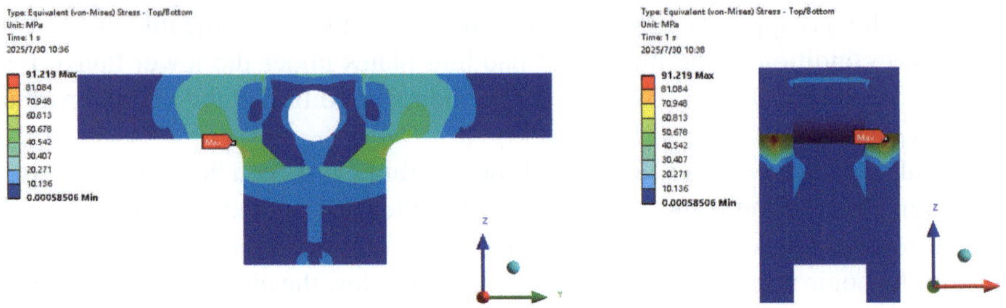

(a) 23.5t wheel load

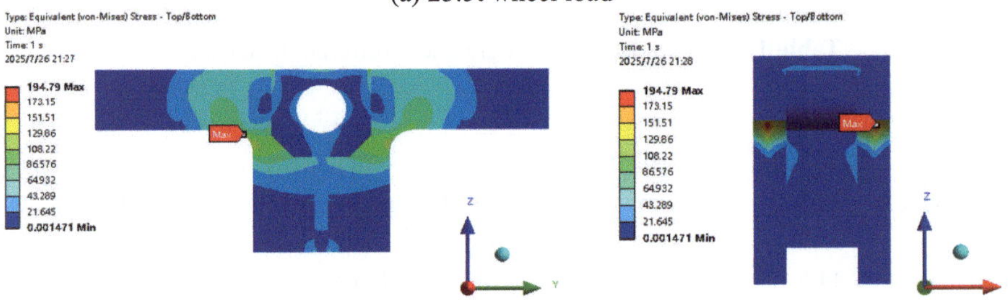

(b) 50t wheel load

Fig. 4. Stress diagram of bogie

3 Sensitivity Analysis of Stress Influencing Factors

The stress at the arc is influenced by factors such as lower flange plate thickness, arc radius, web thickness, top flange plate thickness, and applied load. From sensitivity analysis, the ascending sequence of the factors influencing stress is: load > lower flange plate thickness > arc radius > web thickness > top cover plate thickness, as shown in Fig. 5. The effect of load on stress is linear. The top flange plate thickness has a relatively minor effect on stress. Thus, the lower flange plate thickness t_1, arc radius r, and web thickness t_2 are selected for analysis.

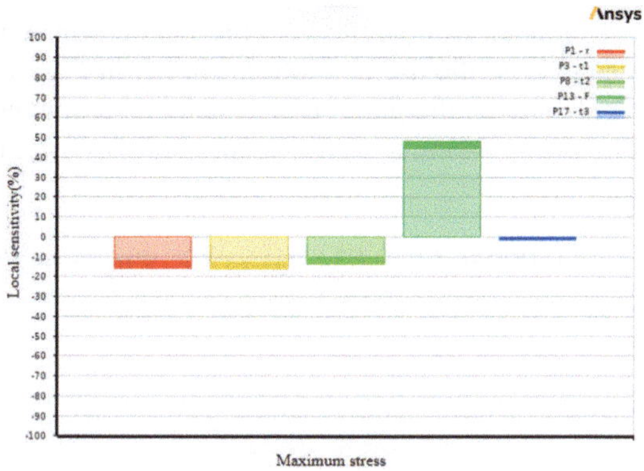

Fig. 5. Parameter sensitivity analysis

4 Analysis of Factors Influencing Stress

For the basic model of bogie, its lower flange plate thickness is 30 mm, arc radius is 85 mm, and web thickness is 24 mm. The single-factor analysis is based on the basic model. RSM comprises experimental design and data fitting. This method is applied to system analysis and optimization [9]. Latin hypercube sampling (LHS) is employed to extract experimental sample data within the parameter ranges. The reliability of the surrogate model is enhanced by the stratified random strategy [10]. The Kriging model, as a parameter-stress agent model, is used to fit the sample data results. At the same time, the stress variation trends with each factor under 10 t load of wheel load as a unit load are analyzed.

(1) The factor of lower flange plate thickness

Lower flange plate thickness is taken as a single-variable factor, whose range is 15-40 mm. The correspond stress curve at the arc position of right-angled structure is shown in Fig. 6. It can be seen that the stress at the arc gradually decreases with the increase in plate thickness, which is close to a linear relationship. The fitting formula, as the agent model of σ-t_1, is obtained by RSM and LHS.

$$\sigma = -0.0007{t_1}^3 + 0.078{t_1}^2 - 3.6497t_1 + 95.338 \tag{1}$$

(2) The factor of arc radius

The arc radius is taken as a single variable factor, whose range is 50-150 mm. The correspond stress curve at the arc position is shown in Fig. 7. It can be seen that the stress gradually decreases as the arc radius increases, which is also close to a linear relationship. The fitting formula, as the agent model of σ-r, is obtained by RSM and LHS.

$$\sigma = -0.000003r^3 + 0.0019r^2 - 0.4506r + 65.438 \tag{2}$$

(3) The factor of web thickness

The web thickness is taken as a single variable factor, whose range is 15-40 mm. The correspond stress curve at the arc position is shown in Fig. 8. It can be seen that the stress gradually decreases as the plate thickness increases, which is also close to a linear relationship. The fitting formula, as the agent model of σ-t_2, is obtained by RSM and LHS is as follows.

$$\sigma = -0.0003{t_2}^3 + 0.0341{t_2}^2 - 1.7199t_2 + 64.09 \tag{3}$$

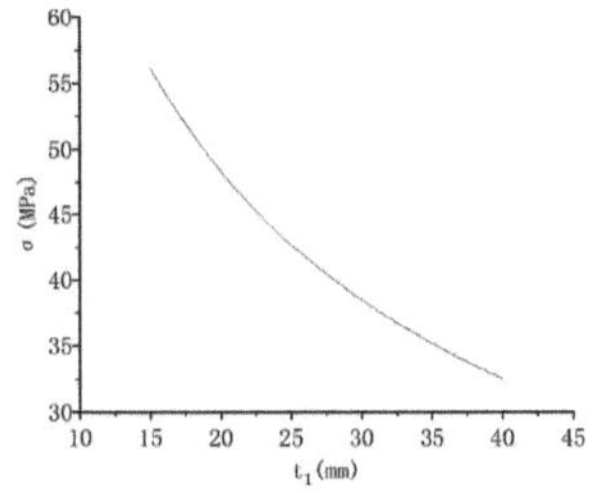

Fig. 6. $t_1 - \sigma$ curves.

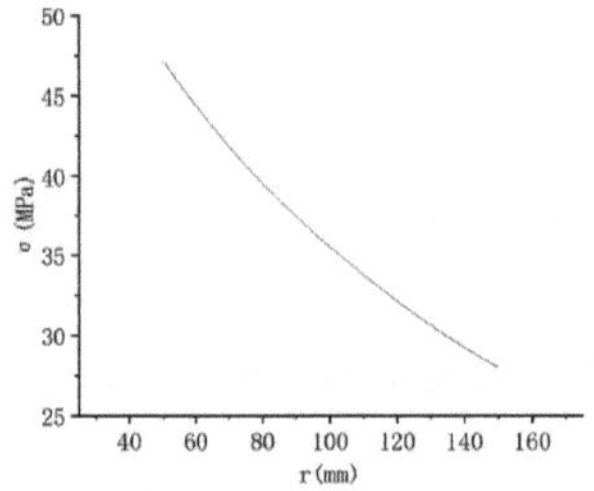

Fig. 7. $r - \sigma$ curves.

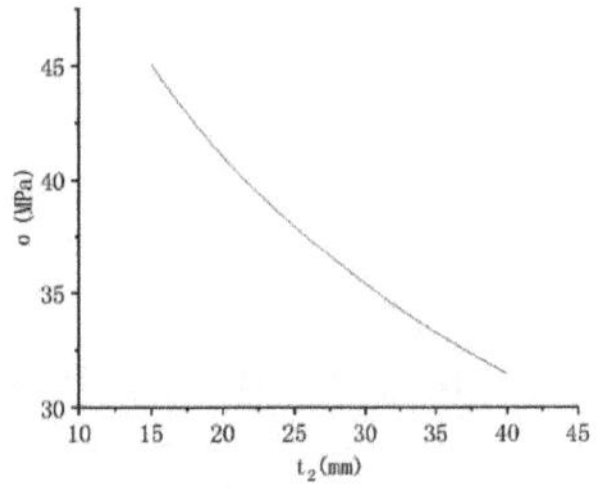

Fig. 8. $t_2 - \sigma$ curves.

(4) Comprehensive analysis of multiple factors

Given that factors in practical operations are not independent and instead interact with each other. In order to describe the actual trend of maximum stress at arc more accurately, the comprehensive multi-factor analysis of the right-angled structure is required.

by RSM, the mapping relation between multi-factors and stress is shown in Formula (4). It can be seen that there exist pairwise multiplicative relationships among the three factors, which indicates their mutual influence.

$$\begin{aligned}\sigma = {} & -0.002{t_1}^3 + 0.003r^2 + 0.185{t_1}^2 + 0.04{t_2}^2 + 0.001rt_1 + 0.004rt_2 + \\ & 0.015t_1t_2 - 0.688r - 7.182t_1 - 2.722t_2 + 204.888\end{aligned} \tag{4}$$

The coefficient of determination (R^2) is as the criterion for analyzing the accuracy of the agent model. The closer R^2 is to 1, the higher the accuracy of the agent model. As sample points for accuracy verification, 100 sets of data are extracted from the 1,000 sets. The fitting accuracy is as shown in Fig. 9. R^2 is 0.99015, which shows high fitting accuracy of the agent model.

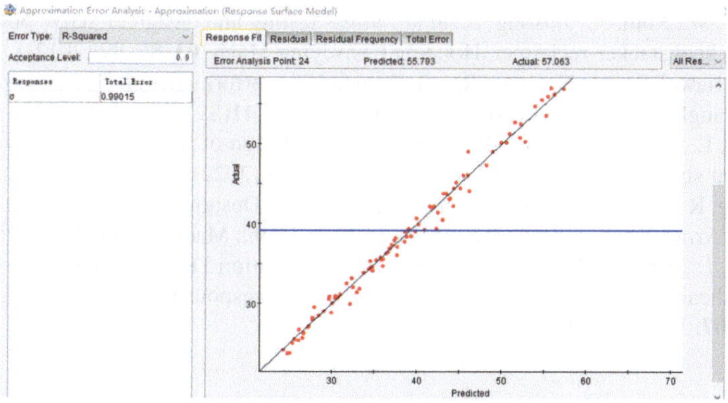

Fig. 9. Accuracy analysis.

5 Conclusion

(1) In this paper, a FEA model of the right-angled structure is established. According to the stress distribution, the stress concentration is at the arc of the lower flange plate.
(2) Parametric analysis of different factors shows their ascending influence, that is lower flange plate thickness, arc radius, and web thickness.
(3) Within a certain range, the maximum stress at the right-angled structure gradually decreases as the thickness of the lower flange plate increases, and also gradually decrease as the arc radius increases, and also gradually decrease as the thickness of the web plate increases. These curves basically show a linear relationship.
(4) An agent model for multiple factors and stress is built through RSM, which can quickly and accurately obtain the stress, and provides a technical foundation for the rapid calculation of subsequent stress-fatigue predictions.

Acknowledgements. This project is supported by the DALIAN INSPECTION TESTING CERTIFICATION GROUP Scientific Research Project (Project No.: J-2404-001-GJ-Q).

References

1. Wei, G., Yin, Q., Zhao, Y., et al.: Fatigue failure analysis and life prediction of wheel load local area of ladle crane. Meas. Sci. Technol. **35**(12), 126120 (2024)
2. Liu, D.K., Jiang, J.F., Shao, X.Y., et al.: Study on fatigue life estimation method of welded structure of lifting machinery. J. Phys. Conf. Ser. **2680**(1), 012013 (2024)
3. Muratovi E, Oli M, Muminovi A, et al.: Analytical calculation and FEM analysis of single girder bridge crane made out of hot-rolled profiles. Springer, Cham 233, 54–61 (2021)
4. Alshoaibi, A.M., Fageehi, Y.A.: 3D modelling of fatigue crack growth and life predictions using ANSYS. Ain Shams Eng. J. **13**(4), 101636 (2021)
5. Zhang, J., Zeng, G.: Finite element analysis of traveling bogie based on ANSYS workbench. Mech. Eng. Autom. **06**, 96-97+ 104 (2016)

6. Zhengxi, W., Man, X., Jinsong, Z., et al.: Stress testing and design of active bogie frame for bucket wheel stacker-reclaimer. Hoisting Conveying Mach. **03**, 85–89 (2012)
7. Yun, Z., Jiawei, S., Xin, Z., et al.: Research on fatigue performance and strengthening methods of right-angled abrupt steel crane girders. Eng. Mech. **41**(S1), 275–281 (2024)
8. Deming, L., Tao, G., Caifeng, W., et al.: Optimal design of bulldozer bogie structure based on virtual simulation technology. Welding. **10**, 47–53 (2022)
9. Wenlong, R., Xie Gaomin, X., Dehong.: Optimization Design of Truss Structure for spherical antenna Array based on response surface method. China Mach. **13**, 11–15 (2025)
10. Yanmin, L., Jingyu, W., Jinggan, S., et al.: Optimization Design of the Crossbeam on the 60MN shear-compression testing machine based on response surface method. Mech. Des. Manuf. **07**, 173–177 (2024)

Design and Simulation of Unpowered Dust Suppression Structure Installed on Grab Ship Unloader

Peng Zhang and Rumin Teng(✉)

Dalian University of Technology, Dalian 116003, China
tengrumin@163.com

Abstract. To address dust escape during the unloading process of grab ship unloaders, this study endeavors to develop an efficient unpowered dust suppression structure. Firstly, the formula for dust suspension velocity is derived to clarify the conditions governing dust diffusion. Subsequently, four two-dimensional models of the dust suppression structure are designed, and comparative simulation analyses are performed to optimize the structure—with a focus on parameters such as guide plate width and inlet width. Finally, a three-dimensional model is constructed based on the optimal two-dimensional configuration (Structure IV), and its rationality is validated using the FLUENT-DPM model to observe particle movement trajectories. The results demonstrate that the unpowered dust suppression method is feasible, with Structure IV—characterized by filleted corners and widened channels—exhibiting the optimal dust suppression performance, reducing the outlet airflow velocity to approximately 0.95 m/s. Two-dimensional simulations enable preliminary evaluation of structural accuracy, while three-dimensional simulations realistically replicate actual operating conditions. The FLUENT-DPM model clearly reveals that large particles tend to settle after rebounding, whereas small dust particles may escape with the airflow.

Keywords: Grab Ship Unloader · Dust Suppression · Chamber Design · Fluid Simulation

1 Introduction

As shown in Fig. 1 is the funnel structure of the grab ship unloader, and Fig. 2. Dust escape generated during the unloading process of grab ship unloaders causes environmental pollution. Existing dust removal and suppression methods mainly include bag dust removal, ventilation dust removal, spray dust removal, foam dust removal, electrostatic dust removal, etc. Ventilation dust suppression has a wide application range and strong control over dust, but its dust removal efficiency is low in open environments. Electrostatic precipitators have high dust removal efficiency, but they are greatly affected by the physical properties of dust [1]. The cost of chemical dust suppressants is very high. Spray dust removal devices are easily blocked by dust or impurities in water, and the water mist increases the humidity of the workplace, leading to conveyor slippage and

S. Halgamuge et al. (Eds.): ICACMVE 2025, LNME, pp. 159–169, 2026.
https://doi.org/10.1007/978-981-95-6911-3_16

easy icing at low temperatures. Based on the shortcomings of the above dust removal and suppression methods, this paper takes the unpowered dust suppression device as the research object, and achieves the effect of unpowered dust suppression through the structural design of the windbreak wall and the structure of the airflow channel.

Fig. 1. Grab ship unloader hopper.

Fig. 2. Grab ship unloader funnel unloading.

Zhu Yong and Qian Hao handled two different types of materials, namely wood chips and grain/fertilizer, during the unloading process of ship unloaders. After adopting conventional water mist dust removal and vacuum dust removal methods, they applied dust-proof funnel technology and used microporous baffle windbreak wall technology.

They conducted simulation analyses on different windbreak wall structures under the conditions where the wind is parallel to the track and perpendicular to the track.Zhang Hanzhong, Meng Wenjun, and Wang Beibei conducted a study on the gas-solid two-phase flow field during the grab unloading process. The purpose was to reflect the movement of dust through the velocity and trajectory of the air flow field, and further derive the suspension velocity of dust particles under different particle sizes and air flow angles [2].

2 Two-Dimensional Simulation Analysis

2.1 Correlation Analysis Between Dust Suspension Velocity and Dust Diffusion of Grab Ship Unloader

Shows the dust emission during the material unloading process of the grab ship unloader. When the upward force of the air flow carrying dust and the gravity of the dust are equal in magnitude and opposite in direction in the vertical direction, reaching an equilibrium state, the air flow velocity corresponding to the upward force is called the suspension velocity. When the escape velocity of the air flow field is greater than the suspension velocity, dust, as a part of the air flow field, will diffuse. The formula for the suspension velocity is derived by simplifying factors with minor impacts. The Magnus force, Saffman force, virtual mass force, buoyancy, pressure gradient force, and thermophoretic force have very low weights and can all be neglected; only the influence of gravity (F_9) and drag force (F_D) is mainly considered. When dust is in a suspended state in the air flow, the forces acting on it in the vertical direction reach an equilibrium state. According to Newton's second law

$$m\frac{dV_y}{dt} = F_D \bullet \cos\theta - F_g = 0 \tag{1}$$

In the formula:m is the mass of dust, in kg;Vy is the velocity in the vertical direction, in m/s;θ is the angle between the direction of air flow movement and the positive direction of the y-axis. Assuming the dust is spherical, the gravity of the dust can be obtained as follows:

$$F_g = \frac{\pi}{6} d_p^3 \rho_p g \tag{2}$$

In the formula:dp is the radius of dust, in m;ρp is the dust density, in kg/m^3.The Reynolds number is related to the dynamic viscosity coefficient of air flow (μ), the absolute value of the relative velocity between dust and air flow (|uf − up|), the air flow density (ρf), and the dust particle size (dp). It can be expressed as follows:

$$R_e = \frac{\rho_f d_p |u_f - u_p|}{\mu} \tag{3}$$

In the formula:uf is the air flow velocity, in m/s;up is the dust movement velocity, in m/s.Based on Eqs. (1), (2), and (3), the force balance of dust in the vertical direction can be obtained, and the relative velocity between the air flow and dust is as follows:

$$u_f - u_p = \begin{cases} \frac{\rho_p d_p^2 g}{18\mu \cos\theta}, R_e \le 2 \\ 0.1527\left[\frac{\rho_p g d_p^{1.6}}{\rho_f^{0.4}\mu^{0.6}\cos\theta}\right]^{0.714}, 2 \le R_e \le 500 \\ 1.74\sqrt{\frac{\rho_p d_p g}{\rho_f \cos\theta}}, 500 \le R_e \le 2\times 10^5 \end{cases} \quad (4)$$

The dust movement velocity (up) is much smaller than the air flow velocity (uf). Therefore, the air flow velocity corresponding to the suspended state of dust can be obtained as follows:

$$u_f \approx u_f - u_p \quad (5)$$

Taking dust with a density of 1023 kg/m^3 as the research object, and the dynamic viscosity coefficient of gas (μ) at room temperature is set to 1.81×10^{-5}, the air flow suspension velocity corresponding to different particle sizes and air flow angles can be calculated by Eq. (5). Similarly, if the suspension velocity and air flow angle are known, the particle diameter under this suspension velocity can also be calculated [3].

2.2 Analysis on Dust Flowability and Escape Law during Unloading Process of Grab Ship Unloader

When materials fall into the funnel, a vertical material column is initially formed. Due to the shearing effect of the air flow, small dust particles are separated from the material column, forming the first stage of dust escape. After the materials fall to the bottom of the funnel, they collide with the funnel surface, rebound, and diffuse, forming the second stage of dust escape [4]. Subsequently, large dust particles will eventually settle freely. However, small dust particles are easily affected by the velocity of the air flow field and shearing effect, and are separated from the particle flow. Meanwhile, small dust particles have a low suspension velocity and strong followability in the air. Therefore, the movement of dust can be reflected through the velocity and trajectory of the air flow field, thereby determining the amount of dust escape [5].

Considering the discontinuity of grab unloading, there is a certain interval between two unloading operations, during which no materials or air flow enter the silo. Thus, as long as dust escape can be suppressed during the unloading process, the dust will have settled to the bottom of the silo before the next unloading operation [6].

2.3 Influence of Inlet Airflow Width and Width Between Guide Plates on Escape Velocity

The surface plot of the maximum upward airflow velocity on the cross-sections of the gaps on both sides of the inlet airflow under the baffle, under different widths between guide plates and different inlet widths, is shown in Fig. 3.

Based on the above data, the surface plot of the maximum upward airflow velocity in the dust suppression chamber under the baffle, under different baffle opening widths and different inlet widths, is shown in Fig. 4.

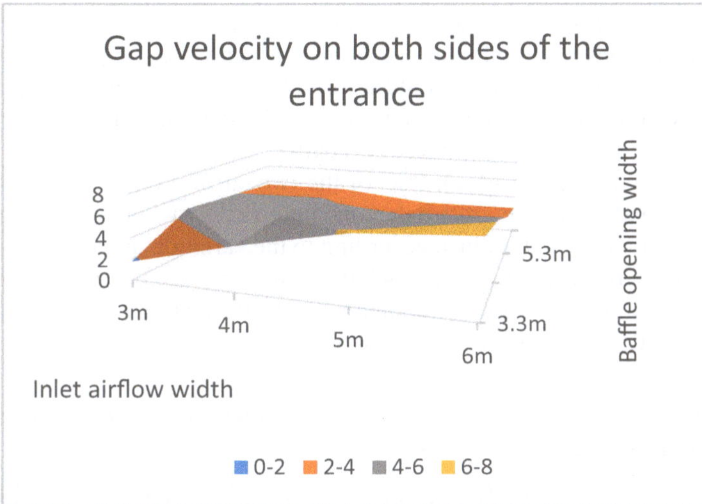

Fig. 3. Surface Plot of Maximum Upward Airflow Velocity on Cross-sections of Gaps on Both Sides of Inlet Airflow Under Baffle.

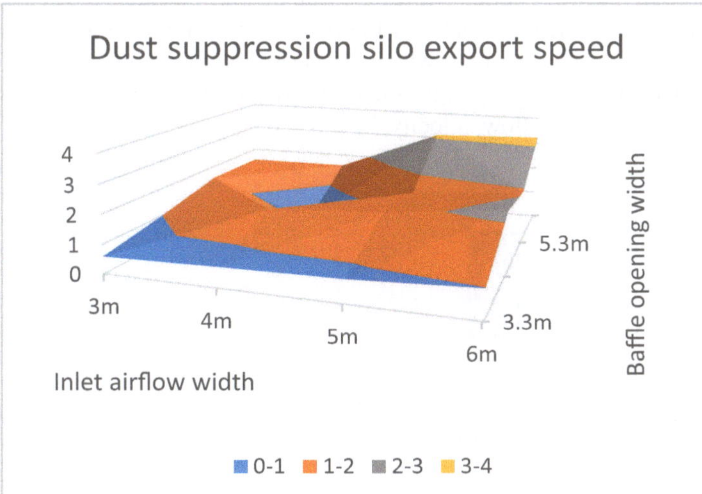

Fig. 4. Surface Plot of Maximum Upward Airflow Velocity in Dust Suppression Chamber Under Baffle.

The above two surface plots respectively present the data acquired from sixteen combinations, where the inlet airflow width ranges from 3 m to 6 m and the width between guide plates ranges from 3.3 m to 6.3 m. As observed from the two surface plots, when the inlet airflow width and guide plate width increase gradually—with the guide plate distance remaining consistently greater than the inlet airflow width—the velocity of the airflow escaping from the inlet decreases progressively. Specifically, when the guide plate width reaches 6.3 m and the inlet airflow width reaches 5 m, the airflow velocity drops to a minimum of 2.69 m/s. Meanwhile, the velocity at the outlet of the dust suppression chamber should also be taken into account; thus, the optimal airflow inlet width and guide plate width need to be determined by comprehensively evaluating the escape velocities of these two parts.

Through the comprehensive analysis of the two surface plots, it is confirmed that the guide plate width must be greater than the inlet airflow width. On this premise, the optimal width should be determined according to the variation in outlet escape velocity induced by the increase in inlet airflow width. However, since the inlet airflow width is essentially a dynamically changing parameter during the material unloading process, no precise combination conclusion can be drawn.

2.4 Iterative Optimization of 2D Model for Unpowered Dust Suppression Chamber of Grab Ship Unloader

Analysis on Flow Guidance Effect and Resistance Problem of Initial Flow Guide Structure: As can be seen from Fig. 5, The velocity contour of Structure I reveals that after gas enters through the inlet, its velocity surges at the corner of the guide plate as a result of changes in the flow channel. Blocked by the wall and guide plates, most of the gas flows out through the lower outlet, while only a small portion enters the dust suppression chamber. Due to the narrow channel of the dust suppression chamber, the velocity of the incoming gas increases further, which gives rise to a potential problem: "when the inlet gas velocity is excessively high, the outlet of the dust suppression chamber still maintains a relatively high flow velocity."

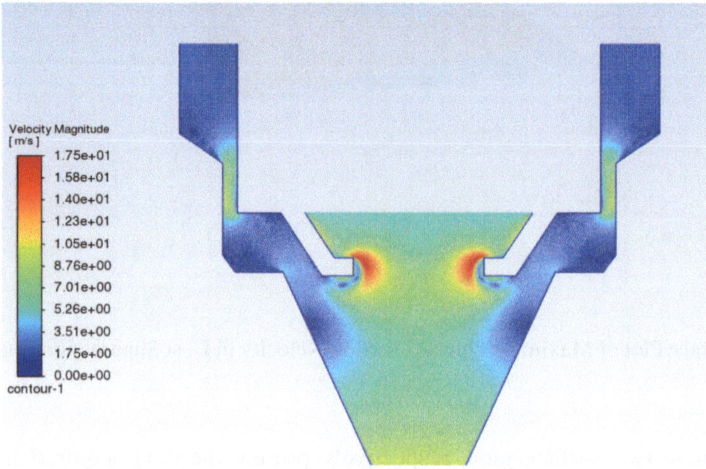

Fig. 5. Velocity Contour of Structure I

Rationality Verification of New Silo Structure: As can be seen from Fig. 6, The original design intention of this structure is to reverse the dust flow via the guide plates, enabling the dust to pass through the pipes of the dust suppression chamber upon entering the chamber, and ultimately guiding it to the lower inlet—thereby achieving dust recovery and control. However, as indicated by the velocity contour results, in practice, only a limited volume of fluid (carrying dust) enters the dust suppression chamber through the guide plates. This phenomenon is presumably attributed to the excessively long path of the dust suppression chamber and the excessively high frictional resistance along the path. Eventually, the maximum velocity of the airflow escaping from the outlet reaches 7.9 m/s, rendering this dust suppression chamber structure unreasonable.

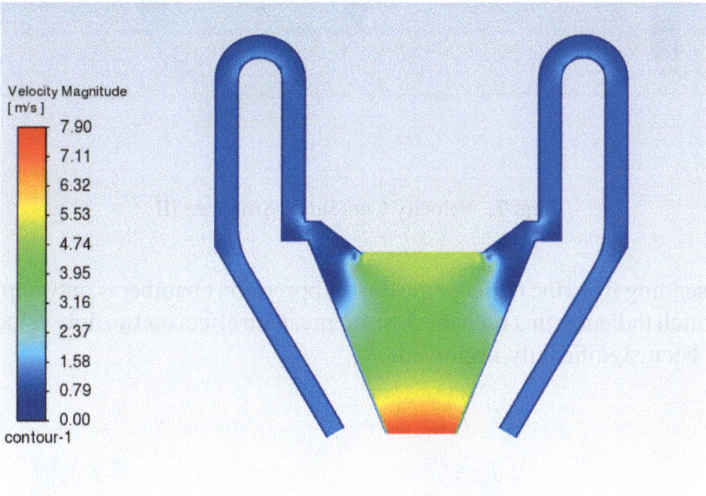

Fig. 6. Velocity Contour of Structure II

Simplification of Dust Suppression Chamber Structure and Analysis on Influence of Inlet Dust Escape on Airflow in Dust Suppression Chamber: As can be seen from Fig. 7, In Structure II, the flow-guiding effect of the guide plates on dust has initially manifested. As shown in the velocity contour, the escape velocity of dust from the outlets on both sides of the inlet has decreased to 2.56 m/s. Under the guidance of the guide plates, part of the dust enters the tall and long dust suppression chamber, where the gas flow velocity gradually reduces, reaching 1.53 m/s at the outlet of the dust suppression chamber. Meanwhile, the dust discharge velocity at the lower outlet has also decreased to 3.06 m/s, which indicates that the dust suppression effect has been improved in comparison with the original structure.

Analysis on Airflow Velocity Improvement of Dust Suppression Chamber After Adding Fillets and Optimizing Channels: As can be seen from Fig. 8, Structure IV is optimized to address the defects of previous structures: it modifies the sharp corners between the dust suppression chamber and guide plates into fillets and increases the channel width of the dust suppression chamber. As shown in the velocity contour, the gas is smoothly guided into the dust suppression chamber through the guide plates, and the airflow velocity inside the chamber is significantly reduced. Eventually, the velocity

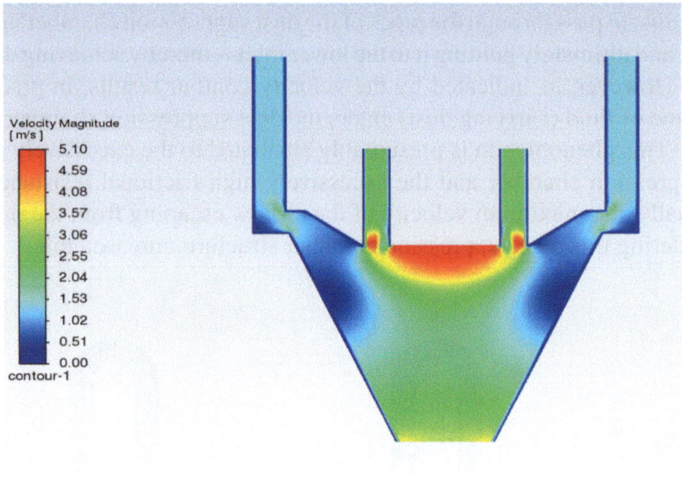

Fig. 7. Velocity Contour of Structure III

of the gas escaping from the outlet of the dust suppression chamber is only approximately 0.95 m/s, which indicates that both the dust suppression effect and airflow velocity control effect have been significantly improved.

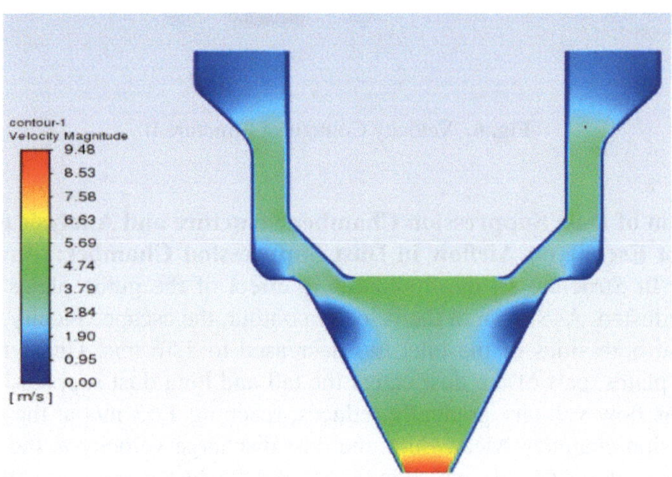

Fig. 8. Velocity Contour of Structure IV

Iterative Optimization Logic of Structures I to IV and Progressive Law of Dust Suppression Effect: Different models from Structure I to Structure IV are established based on the core logic of gradually resolving the dust escape problem through "problem-oriented iterative optimization". Specifically, targeting the specific defects exposed by previous models (such as excessively high flow velocity, flow guidance failure, and local turbulence), structural parameters are adjusted in a targeted manner, ultimately achieving an improvement in the dust suppression effect. The specific logic can be divided into a progressive process of "identifying problems → implementing targeted optimization → verifying effects" [7].

3 3D Model FLUENT-DPM Simulation

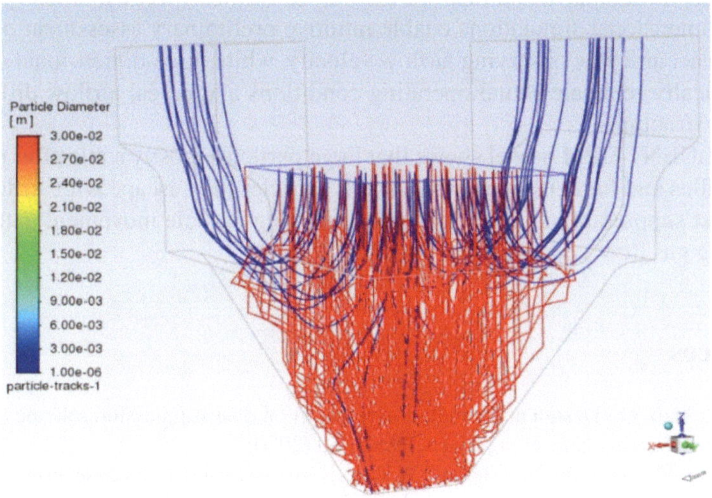

Fig. 9. Distribution diagram of particles with two different diameters

As can be seen from Fig. 9,This 3D model is established based on Structure VI, which showed the optimal effect in the 2D model simulation. In the figure, the red lines represent the movement trajectories of large particles (30 mm in size), while the blue lines represent those of small dust particles (0.001 mm in size). It can be observed from the figure that large particles collide with the inner wall of the funnel. Since the funnel wall is set to "reflect" by default in the simulation, the particles rebound after colliding with the wall and eventually settle in the flow field. Tiny dust particles move with the air in the flow field, and some of them adhere to large particles and settle together. Additionally, it is evident that in the fluid domain at the set inlet, dust does escape directly when materials are fed.

From the simulation results, it can be seen that by establishing the fluid domain at the inlet to realistically simulate the unloading process, some dust is also observed to escape from the inlet. Therefore, it is necessary to consider dust suppression methods such as water spraying at the inlet. However, the simulation results after adding the inlet fluid domain are more ideal than those of previous structures. Meanwhile, establishing the inlet fluid domain is more reasonable and closer to the actual situation compared to the previous approach—where particle generation started directly from the funnel inlet, ignoring the process of materials falling freely.

4 Conclusion

The following conclusions are drawn from the simulation:

(1) The unpowered dust suppression method for grab unloading is feasible, with Structure IV—featuring optimal design—exhibiting the best dust suppression effect. Key design criteria include: the inlet baffle opening width must exceed the grab's opening width to prevent material escape; the dust suppression chamber should have a widened inlet channel (with greater width reducing airflow velocity) and an uppermost chamber wider than the channel to further lower exit velocity.
(2) Two-dimensional simulations enable intuitive preliminary assessment of structural design accuracy by observing airflow velocity, while three-dimensional simulations realistically replicate actual operating conditions and reveal airflow diffusion laws in real structures.
(3) The FLUENT-DPM model shows that large particles mostly settle after rebounding off baffles and funnel slopes, while small particles may escape with airflow through the dust suppression chamber or inlet, clarifying particle movement patterns under varying gas flow velocities.

References

1. Yong, Z., Hao, Q.: Design and simulation research on dust suppression scheme for grab ship unloaders. Hoisting Conveying Mach. **09**, 47–50 (2020)
2. Hanzhong, Z., Wenjun, M., Beibei, W.: Study on gas-solid two-phase flow field of grab unloading based on CFD-DEM coupled simulation. Min. Res. Dev. **42**(04), 166–172 (2022)
3. Dongxue, W.: Study on the Flow Law of Airflow Induced by Free Falling of Bulk Materials. China University of Mining and Technology (2020)
4. Hongfa, S.: Dust Falling and Dispersion Mechanism and Dust Suppression Methods Based on Industrial Production Processes. Xi'an University of Architecture and Technology (2020)
5. Karwat, B., Rubacha, P., Stańczyk, E.: Simulational and experimental determination of the exploitation parameters of a screw conveyor. Eksploatacja Niezawodność Maintenance Reliab. **22**(4), 741–747 (2020)
6. Elaswad, R., Sheimy, E.N., Mohamad, A.: Trajectory dynamics of particles in accelerated toroidal pipe: A computational study using CFD-DPM simulations. J. Comput. Sci., 78102285 (2024)
7. R. E, N. E, A. A M.: Review—basic and advanced inertial navigation fluid-based technology. ECS Sens. Plus. **2**(2) (2023)

Study of the Seismic Influence on Analysis of Nuclear Polar Cranes by Different Geological Conditions

Benxin Zhang[1], Rumin Teng[1](✉), and Zhisheng Xu[2]

[1] Dalian University of Technology, Dalian 116024, China
tengrumin@163.com

[2] Dalian Shipbuilding Industry Equipment Manufacturing Co., Ltd., Dalian 116012, China

Abstract. Seismic analysis of nuclear polar cranes is a core verification method to ensure their functional integrity under safe shutdown earthquake (SSE) conditions. To address the massive computational challenges brought by multiple geological conditions and different holding positions, a dynamic simulation of nuclear polar cranes is conducted considering nonlinear factors. Then the multi-point excitation time history analysis is completed by inputting SA/MA/HA site displacement time history data, and the stresses of key components are compared to deeply explore the influence of three geological conditions on seismic performance. The study reveals that the peak responses occurred in different geological conditions and there is not a certain condition envelops all dangerous results. So, it is suggested that corresponding analyses should be conducted depending on the actual different site conditions. It is not feasible to adopt a single scenario as the overall reference. Only in this way can the analysis results be more reliable.

Keywords: Nuclear Polar Cranes · Seismic Analysis · Time history analysis · Geological condition

1 Introduction

The nuclear polar crane is a key piece of equipment in nuclear power plant construction. Its full-lifecycle functions span the construction, operation, and decommissioning phases. The crane performs critical tasks such as high-precision lifting of hundred-tonne objects, remote maintenance, and leak-tight disassembly in sealed high-radiation environments. As seismic Class I (C-I) equipment, the nuclear polar crane must withstand a safety shutdown earthquake (SSE/SL-2) in accordance with international and domestic standards [1, 2].

Seismic analysis of nuclear polar cranes primarily employs spectrum analysis and time-history analysis. Li Yongliang validated the reliability of the nuclear polar crane-steel containment vessel coupled system under SSE conditions through spectrum analysis and time-history analysis [3]; Yaroslav Dubyk et al., revealed the conservatism of the response spectrum method and identified structural weaknesses based on nonlinear time-history analysis of wheel-rail contact [4]; Feau Cyril et al. proposed a simplified

S. Halgamuge et al. (Eds.): ICACMVE 2025, LNME, pp. 170–176, 2026.
https://doi.org/10.1007/978-981-95-6911-3_17

modeling method to accurately characterize the seismic response of crane main beams [5]; Wang Zhiyun et al. confirmed through scaled model tests (1:16) that earthquake wave types, input directions, and time-series similarity significantly affect dynamic responses [6]; Huang Maosong et al. conducted centrifugal tests on soft soil sites, demonstrating that soil properties and underground structures significantly influence earthquake wave energy transmission [7].

The above research has laid a foundation for seismic analysis of nuclear polar crane structures, but research on the impact of geological conditions remains insufficient. Sometimes it is assumed that SA soft soil seismic effects are the most dangerous, or multiple site full-condition analysis are applied to pursue enveloping properties, resulting in resource waste. Therefore, it is necessary to investigate the influence of geological conditions to ensure crane safety based on seismic analysis.

2 Building Dynamic Simulation Model Including Nonlinear Factors

2.1 Layout of the Nuclear Polar Crane

The nuclear polar crane, as shown in Fig. 1, is mainly composed of main bridge girders, main and auxiliary trolleys, traveling mechanisms, a maintenance gantry, polar rails, and electrical control systems, among other components. It is installed on the polar rails at the top of the nuclear island containment vessel. The bridge girders span the polar rails; the main traveling system enables circular movement along the polar rails, while the trolleys move radially along the bridge girders. These two systems operate in tandem to ensure comprehensive lifting coverage over the entire area.

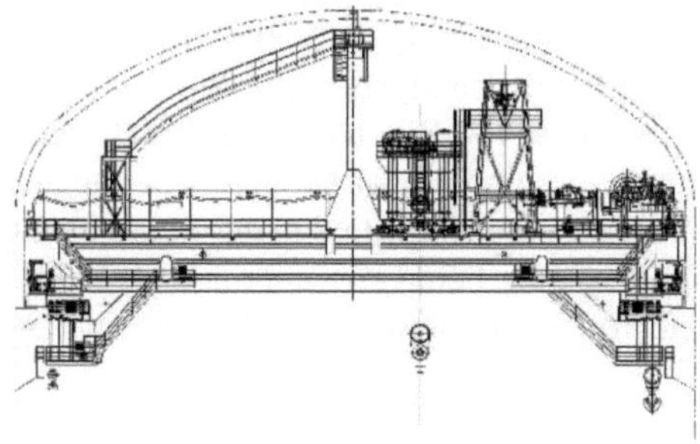

Fig. 1. The drawing of nuclear polar crane.

2.2 Selection of the Element Types to Simplify the Model

The simplified system, as shown in Fig. 2, includes selected element types such as Beam, Mass, Link, Combin39, and Contact elements. Definition of the coordinate system: The origin is the intersection of the rotational centerline and the top plane of the polar rails.

The x-axis is parallel to the bridge girders, the y-axis is perpendicular to the bridge girders, and the z-axis is perpendicular to the polar slewing plane. The experimental modal frequencies of the physical model of the nuclear polar crane are consistent with those of the finite element model established using this modeling method; furthermore the acceleration response waveforms and phases are are in good agreement in most operating conditions, thus validating the effectiveness of the model settings [8].

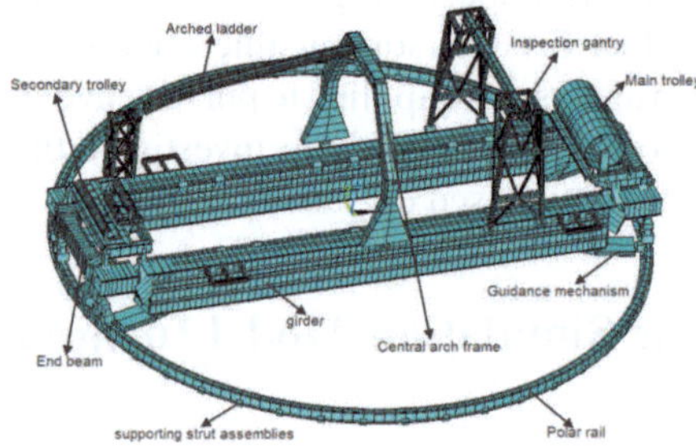

Fig. 2. finite element simulation model

Commonly used methods for seismic analysis include spectral analysis and nonlinear time-history analysis. Response spectrum analysis in the frequency domain or the mode superposition method in the time domain can only be applied to linear dynamic systems or to a limited number of nonlinear dynamic systems. Therefore, nonlinear time-history dynamic analysis often becomes the only solution for most nonlinear dynamic structural systems with nonlinear characteristics. Trolleys, traveling mechanisms, buffers, and steel ropes of the nuclear polar crane exhibit nonlinear behaviors such as friction and collisions during earthquakes; thus, the time-history analysis method that accounts for nonlinearity offers higher accuracy in seismic analysis of the crane [8]. This study employs the multi-point excitation time-history analysis method.

The commercial software ANSYS is used to solve the nonlinear dynamic system of the nuclear polar crane, where nonlinear behaviors include rope systems and contact pairs. The main steel structures are simplified using Beam elements; ropes are modeled as link elements with tension-only behavior; and mass elements are assigned at specific positions to compensate for the total weight, ensuring it matches that of the actual equipment.

3 Applying Boundary Conditions

3.1 Setting of Restraint Conditions

As shown in Fig. 3(a), contact pairs are established between the wheels and rails to simulate nonlinear contact using Element 178. For the slewing traveling system, only z-direction contacts are established between the wheels and rails because there are no wheel flanges. For the trolley's wheel-rail contact pairs, contacts are established in two directions (namely the z-direction and y-direction) since the wheels are equipped with double flanges. The normal contact force at the wheel-rail interface is calculated based on contact stiffness and damping. Frictional damping coefficients are assigned different values based on whether the wheels are braking or non-braking: 0.2 for braking wheels and 0.017 for non-braking ones. The configuration of contact pairs enables simulation of wheel-rail squeezing and separation behaviors. The guidance mechanisms are simplified using Combin39 elements, which can simulate both spring characteristics and contact behavior. The polar rails are bolted to the polar box beam, which is mounted on 45 columns of the main supporting strut assemblies. Full restraints are applied at the strut assembly points, as shown in Fig. 3(b).

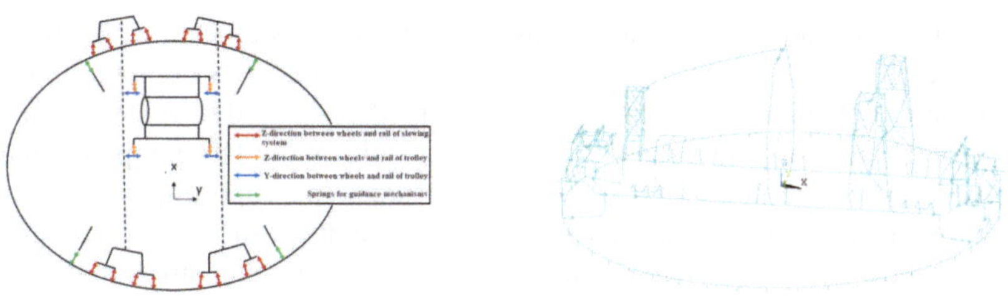

(a) Wheel-rail contact and horizontal wheel spring schematic

(b) Fully restraints at the main supporting strut assemblies

Fig. 3. Wheel-rail contact and horizontal wheel spring schematic.

3.2 Selection of Load Conditions

Referring to GB/T 3811–2008, NB/T 20234–2013 "Design Guidelines for Cranes Specialized for Nuclear Power Plants", and ASME NOG-1-2020 "Manufacturing Standards for Bridge and Gantry Cranes (Uplink Bridge, Multi-Girder Type)", under SSE conditions, the no-load condition and lifting hazardous loads are combined with four relative positions of the trolley along the bridge girders (i.e., trolley located at the end span, mid span, and left/right quarter span), and the hook block at its upper/lower limit positions. These combinations result in a total of 12 load cases.

3.3 Selection of Geological Conditions

According to GB 50011–2010 "Code for Seismic Design of Buildings", building sites are classified into four categories based on seismic conditions: favorable, general sites encompass all other areas not classified as favorable, unfavorable, or hazardous. Favorable sites refer to stable bedrock, hard soil, or open, flat, stable medium-hard soil; general sites encompass all other areas not classified as favorable, general, or unfavorable; unfavorable sites include soft soil, liquefiable soil, and unfavorable topographies such as elongated protruding mountain ridges, isolated high hills, steep slopes, and steep cliffs; hazardous sites are areas prone to earthquake-induced landslides, collapses, and surface displacements along active fault zones.

The site categories for nuclear polar crane seismic analysis are typically SA (soft soil), MA (medium-hard soil), and HA (hard soil). In this study, the design response spectra for the aforementioned three site categories were used, with corresponding geological seismic time-history data applied to 12 different load cases for comparative analysis.

Furthermore, there are 3 excitation acceleration histories per site. Combined with the 12 load cases, these result in 36 calculation conditions. This constitutes a substantial computational effort. However, comprehensive calculations that account for site categories and load positions facilitate more informed crane design, as a single analysis can encompass all conditions.

4 Analysis of Seismic Results

This computational analysis applies seismic time-history data for three geological conditions to each of the 12 load cases, resulting in a total of 108 computational cases. The multi-point excitation time-history analysis method is used to evaluate the seismic response of the nuclear polar crane. Based on the results of the 108 computational cases, the analysis examined horizontal wheel forces, forces in supporting strut assemblies, hoisting rope forces, and stresses in the main components of the polar crane (including girders, end beams, and the maintenance gantry). The maximum stresses are presented in Fig. 4, while the maximum force response diagrams for vertical forces in supporting strut assemblies and hoisting ropes are shown in Fig. 5.

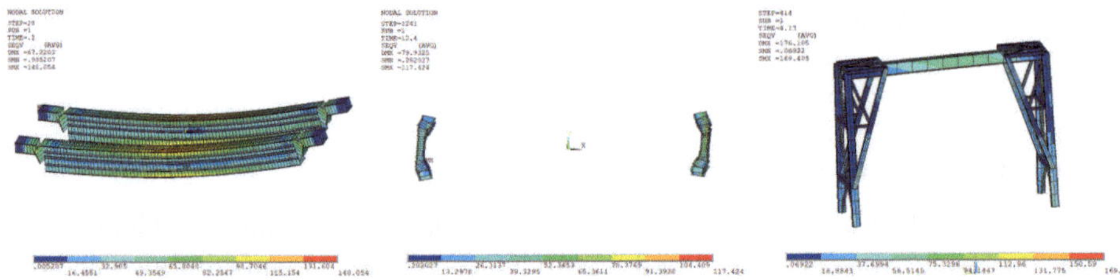

(a) Maximum response stress cloud of girder (under MA seismic loading, case2)

(b) Maximum response stress cloud of end beam (under SA seismic loading, case5)

(c) Maximum response stress cloud for inspection gantry (HA seismic loading, case5)

Fig. 4. Maximum response stress cloud for girder, end beam, and inspection gantry.

When comparing the results of seismic analyses under different geological conditions, although peak Von Mises stresses and peaks of key structural internal force responses are higher under SA (soft soil) conditions, some peaks still occur under MA (medium-hard soil) and HA (hard soil) conditions.

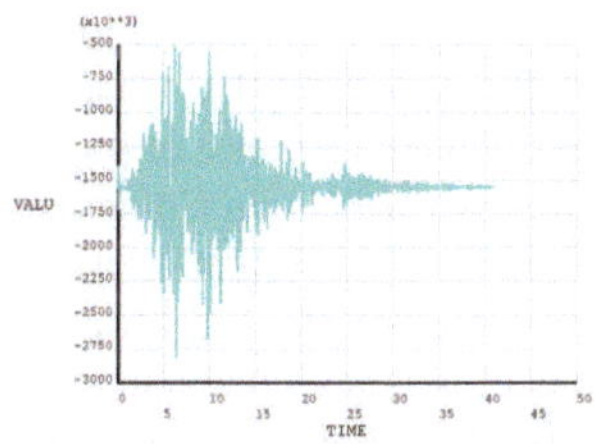

(a) Maximum force response of supporting strut assemblie in vertical direction (SA seismic load, case5)

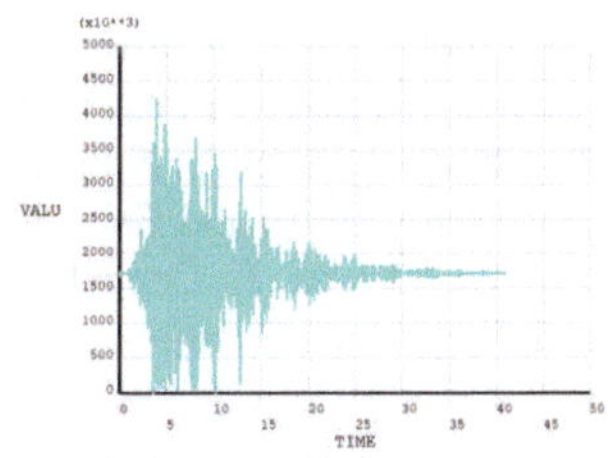

(b) Maximum force response of steel wire rope (HA seismic load, case5)

Fig. 5. Vertical direction of supporting strut assemblie, the maximum force response diagram of steel wire rope.

5 Conclusion

A dynamic simulation model for the nuclear polar crane that accounts for nonlinear behaviors has been established. Under safety shutdown earthquake (SSE) loading, 12 load cases were established. These cases were then analyzed using three sets of geological seismic time-history data for comparative purposes. The analysis indicates that peak Von Mises stresses in critical components and peak structural internal force responses are highest under SA (soft soil) conditions; however, some response peaks under MA (medium-hard soil) and HA (hard soil) conditions still exceed those under SA conditions. Seismic analysis results for different site conditions do not exhibit enveloping characteristics, indicating that a single site condition cannot encompass all extreme responses.

In engineering applications, to achieve enveloping seismic results, it is typically necessary to perform full-condition analyses for SA, MA, and HA site conditions, thereby increasing computational workload. This study reveals that peak responses under different site conditions do not exhibit enveloping characteristics. It is recommended that when the operational site condition of the nuclear polar crane is clearly defined, seismic analysis should be prioritized according to actual geological conditions. This conclusion provides a theoretical foundation for optimizing seismic computational strategies.

References

1. NB/T20234-2013: Design Guidelines for Cranes Specialized for Nuclear Power Plants, Beijing: China Planning Press, (2013). in Chinese
2. ASME NOG-1—2020: Rules for Construction of Overhead and Gantry Cranes (Top Running Bridge, Multiple Girder). New York: American Society of Mechanical Engineers, (2020)
3. Yong Liang Li: Research on seismic analysis of circulating cranes in nuclear power plants. Dalian University of Technology, (2018). in Chinese
4. Dubyk, Y., Antonchenko, V.: Seismic analysis of Vver-1000 polar crane. In: Proceedings of the ASME Pressure Vessels and Piping Conference 2021: PVP., Virtual, Online, volume 5. Operations, Applications, and Components; Seismic Engineepolar; Non-Destructive Examination: American Society of Mechanical Engineers, 1: V00 5T08A02-V005T08A02 (2021)

5. Feau, C., Politopoulos, I., Kamaris, G.S., et al.: Experimental and numerical investigation of the earthquake response of crane bridges. Engineepolar Struct. **84**, 89–101 (2015)
6. Zhiyun W, Shouju L, Shen Y: Numerical analysis for earthquake response of polar crane. IOP Conference Series: Earth and Environmental Science, 719 (4) (2021). in Chinese
7. Huang, M., Yao, J., Yu, J.: Centrifugal experimental simulation of seismic response of pile-beapolar underground structures in soft clay. J. Geotech. Eng. **46**(S1), 1–5 (2024) in Chinese
8. Huali Liu: Characterization of seismic response of nuclear polar suspension under multi-point excitation. Dalian University of Technology, 001202 (2024). DOI: https://doi.org/10.26991/d.cnki.gdllu. in Chinese

Study of Simulation Result of the Boom System for Heavy Load Aerial Work Platform in Different Modeling Way

Mingjian Li[1], Rumin Teng[1](✉), and Linkai Zhou[2]

[1] Dalian University of Technology, Dalian 116024, China
tengrumin@163.com
[2] Dalian Shipbuilding Industry Co., Ltd, Dalian 116024, China

Abstract. For a certain type of heavy load aerial work platform, in the design process of the telescopic boom system, in addition to ensuring that the strength meets the requirements, it is also necessary to ensure good rigidity. This is to provide a better working experience for the operators and enable them to carry more materials. Different models were established for the boom system: one was a model based on shell elements with contact non-linearity considered as the main body, and the others was simplified beam element model. The results were compared and analyzed under typical working conditions to determine a more optimal modeling method, in order to find an efficient computational analysis model for calculating the variation law of end displacement.

Keywords: Aerial work platform · Boom system · Comparison of modeling method · Finite element analysis

1 Introduction

Aerial work platforms play a significant role in fields such as municipal construction and engineering projects [1]. During the design process, in addition to ensuring that the boom system meets the strength requirements, it is also necessary for it to have good rigidity. For the boom system, the main factor affecting its stiffness is the deformation of the boom [2]. Therefore, this paper conducts a stiffness analysis of the typical operating conditions of the boom system.

Yin Jun proposed a negative feedback matrix control strategy for the 2°-of-freedom vibration model, which reduced the amplitude of the boom rotation vibration by more than 80% [3]. Wang Sai pointed out that the telescopic arm exhibited geometric non-linearity under the condition of long working length of the upper arm. Therefore, he proposed a method for non-linear calculation and correction of the overall deformation of the telescopic arm [4]. Woo C L and his colleagues designed an optimized boom mechanism to prevent the lifting machine's boom from deforming, and verified the structural safety of the mechanism [5]. J DANIEL GLAD STEPHEN took the cross-sectional area of the boom as the objective function and used the Lagrange multiplier

S. Halgamuge et al. (Eds.): ICACMVE 2025, LNME, pp. 177–185, 2026.
https://doi.org/10.1007/978-981-95-6911-3_18

method to obtain the optimal dimensions of the telescopic boom's cross-sect [6]. Wen Shuang derived the dynamic equations of the hybrid arm system based on its dynamic characteristics, and was able to accurately predict the vibration characteristics of the arm and frame system [7]. Xing Wei optimized the geometric parameters of the boom section to reduce the weight of the telescopic arm by 5.77%, and applied the research results to the actual product [8].

The non-linear contact of shell elements not only involves a large amount of modeling work but also takes a long time to solve. Therefore, it is proposed to use beam elements for modeling. The differences in calculation results are compared and analyzed to prepare for the subsequent work.

2 Overview of Heavy Load Aerial Work Platforms

The heavy load aerial work platform is mainly composed of the platform, telescopic boom, fly jib, travelling system, rotary table, counterweight, etc. Its main structure is shown in Fig. 1. The transportation capacity of the heavy load aerial work platform studied in this paper is 2.4 t. The single telescopic arm structure mode can no longer meet the requirements of the transportation capacity. Therefore, the heavy load aerial work platform studied in this paper is planned to adopt a double telescopic arm structure form.

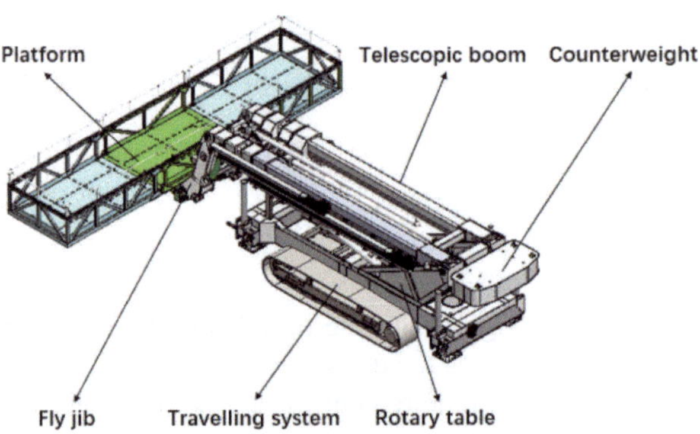

Fig. 1. Heavy load aerial work platform.

The specific technical parameters of the heavy load aerial work platform are shown in Table 1.

Table 1. Parameters of heavy load aerial work platform.

technical parameters	value
maximum operating height	26.7 m
maximum height of the bottom surface of the work platform	25 m

(*continued*)

Table 1. (*continued*)

technical parameters	value
maximum working range ability	13.5 m
maximum extent of the work platform	12.9 m
maximum cargo capacity	2400 kg
maximum passenger capacity	1800 kg
maximum allowable side load	680 N

3 Simplified Model Adopted by Different Element Types

3.1 Finite Element Analysis Based on Shell Elements Considering Non-Linear Contact

The materials selected for the heavy load aerial work platform are BS700, with an elastic modulus of 210,000 (N/mm^2) and a Poisson's ratio of 0.3 [9]. To ensure the mechanical properties of the original structure, multiple elements need to be used for modeling. The main structure is established using the shell element, the slider is established using the solid element, and the oil cylinder is established using the beam element. Different elements are directly connected using the contact element or coupling method, the specific elements selected are shown in Table 2.

Table 2. The specific elements selected and their attributes.

element name	element nature
Shell181	It is generally composed of four nodes and is suitable for replacing large flat surfaces.
Solid185	It is generally composed of eight nodes, one of which is used for the mesh establishment of complex models.
Bean188	Composed of two nodes, it is highly suitable for replacing slender beam structures.
Conta173	Used for setting the three-dimensional structure contact. When there is penetration with the target unit, contact is generated.
Targe170	The target surface is generated by the target entity or the planar grid and is matched with the corresponding contact unit surface.

The model constraints are to limit the three degrees of freedom (UX, UY, UZ) at the base of the telescopic boom and the hinge points of the luffing cylinder. The loading positions are at the head of the main boom and the nearby articulated luffing hinge points of the fly jib. The specific loading points and values are shown in Fig. 2, and the model with the constraints applied is shown in Fig. 3.

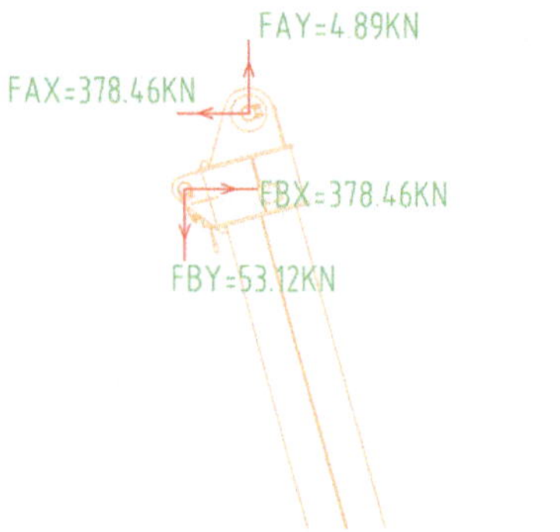

Fig. 2. Forces applied on the top of booms.

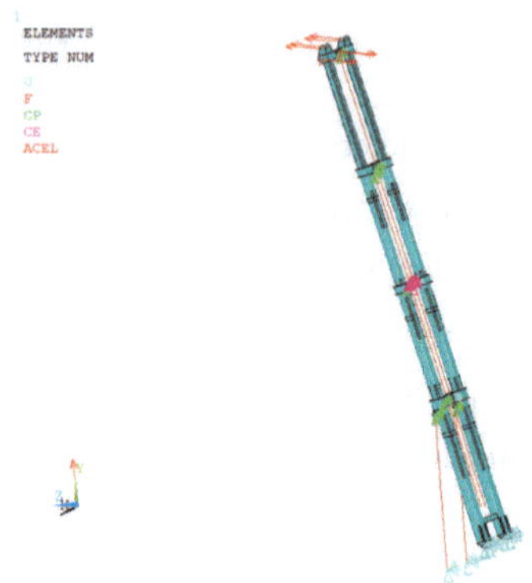

Fig. 3. FEA model by shell elements.

The vertical displacement obtained through calculation is 159.501 mm. The displacement is shown in Fig. 4.

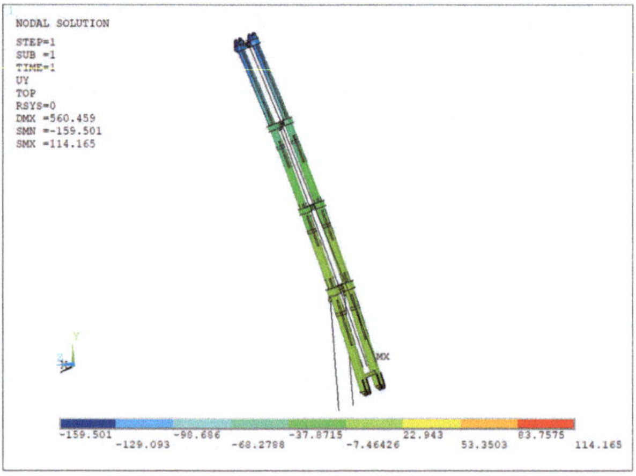

Fig. 4. Displacement based on shell elements considering non-linear contact.

3.2 Finite Element Analysis of Bench-Shaped Beam

The main boom is mostly of thin-walled structure. The ratio of its length to cross-sectional width meets the conditions for beam element modeling. Since using beam element modeling for finite element modeling and analysis can significantly increase the calculation efficiency [10], it is proposed to adopt bench-shaped beam for the analysis. The bench-shaped beam model does not take into account the overlapping relationship between adjacent segments. At the overlapping areas of the boom, a model is established using the more reliable beam tube section. Since it is consistent with the shell element condition, the model constraints and loads are the same as those in the previous text.

The typical cross-section is shown in Fig. 5, the detail is shown in Fig. 6, and the model after applying the constraints and loads is shown in Fig. 7.

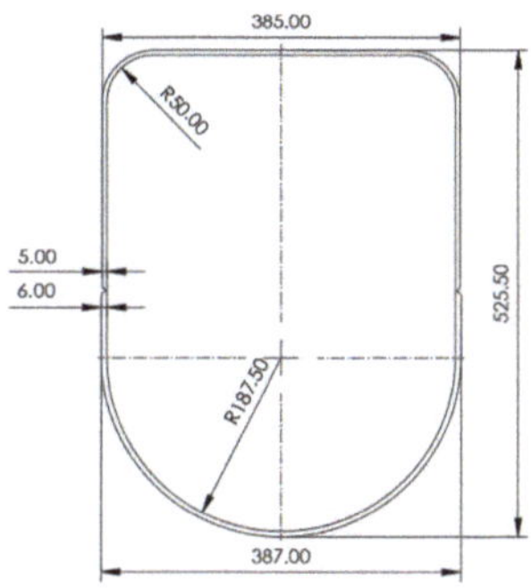

Fig. 5. Typical cross-section.

Fig. 6. Model details of bench-shaped beam.

Fig. 7. FEA model by shell bench-shaped beam.

The vertical displacement obtained through calculation is 154.446 mm. The displacement is shown in Fig. 8.

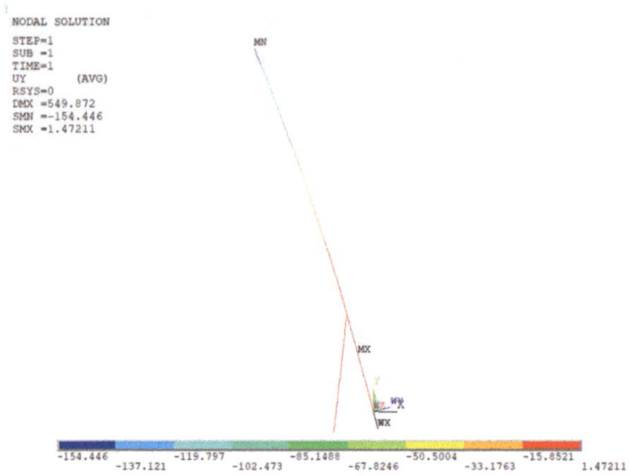

Fig. 8. Displacement of bench-shaped beam.

3.3 Finite Element Analysis of Overlapping Beam

The overlapping beam is different from the bench-shaped beam in that it takes into account the overlapping part at the two arm tubes. Two beam elements are established in this overlapping area, and the coupling command is used to couple the two beams together. The cross-section, load and constraints selected for the model are the same as those in the previous section. The details of the connection part of the model are shown in Fig. 9. The model after applying the constraints and loads is shown in Fig. 10.

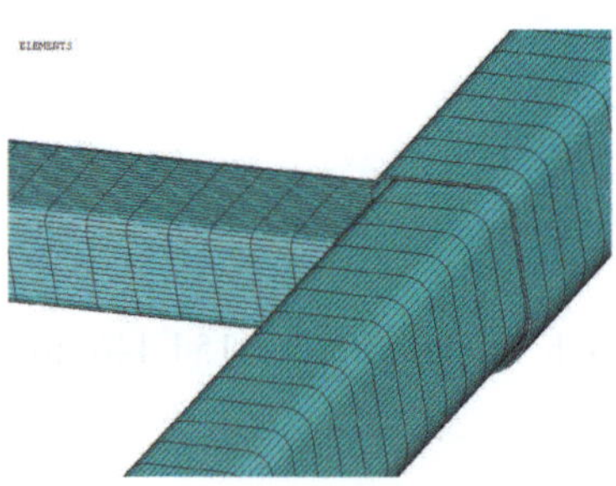

Fig. 9. Model details of overlapping beam.

Fig. 10. FEA model by overlapping beam.

The vertical displacement obtained through calculation is 138.011 mm. The displacement is shown in Fig. 11.

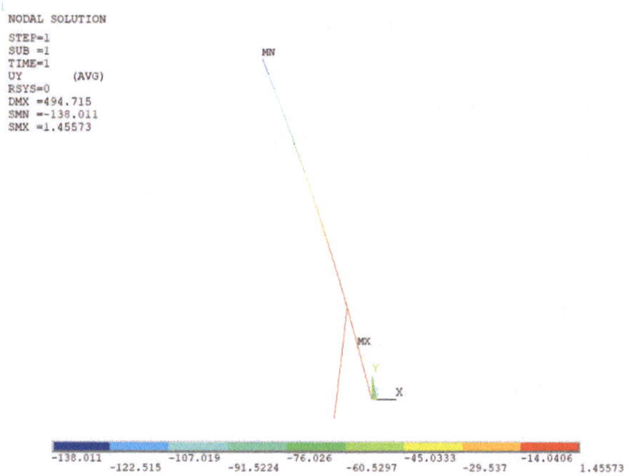

Fig. 11. Displacement of overlapping beam.

4 Comparison and Analysis of Results from Different Models

The deformation results obtained from the three models are relatively close. The model considering non-linear contact, which is based on shell elements, established an extension oil cylinder. This part is not the subject of our research, so it does not affect our comparison and analysis of the results. The vertical displacement results of the overloaded aerial work platform obtained through different modeling methods are shown in Table 3.

Table 3. Displacements at the ends of different modeling methods.

modeling method	displacement(mm)
model based on shell elements considering non-linear contact	−159.501
the bench-shaped beam model	−154.446
the overlapping beam model	−138.011

In order to obtain the relative error between the beam element model and the shell element-based model that takes into account non-linear contact, we established Eq. (1). Where, Δ is the relative error between the model based on shell elements considering non-linear contact and model of beam element, D_Q is the end displacement of the model based on shell elements considering non-linear contact, D_B is the end displacement of the model of the beam element.

$$\Delta = \frac{| D_Q - D_B |}{D_Q} \times 100\% \tag{1}$$

The relative error of the bench-shaped beams and the overlapping beams are shown in Table 4:

Table 4. Relative error.

modeling method	value
the bench-shaped beam model	3.17%
the overlapping beam model	23%

From Table 4, we can see that the relative error of the bench-shaped beam model is relatively small, and the modeling method of the bench-shaped beam is also simpler. Therefore, in comparison, we can conclude that using the bench-shaped beam to establish the model is a better solution.

5 Conclusion

Through simulation analysis of the heavy load aerial work platforms in different elements, the relative errors between the bench-shaped beam element model and the overlapping beam element model and the model mainly based on shell element considering non-linear contact were calculated. The relative errors were compared, and it was determined that modeling with the bench-shaped beam element can efficiently and accurately obtain the deformation law at the end.

References

1. Rumin, T.: Research on Rapid Design and Motion Planning of Aerial Work Platform Boom System. Dalian University of Technology (2012)
2. Feifan, R., Chuangang, F., Fuzhong, S., et al.: Lightweight Design of Aerial Work Platform Boom Based on multi-condition topology optimization. J. Mach. Des. Res. **41**(03), 80-87+94 (2025). https://doi.org/10.13952/j.cnki.jofmdr.2025.0200
3. Jun, Y.: Research and application of active control for multi-joint long flexible arm frame rotation vibration. Chin. J. Constr. Mach. **23**(03), 520–525 (2025). https://doi.org/10.15999/j.cnki.311926.2025.03.034
4. Sai, W., Shuai, S., Rumin, T., et al.: Precise calculation and experimental verification of lifting performance of long-extension boom full-terrain cranes. Mech. Des. **42**(06), 1–6 (2025). https://doi.org/10.13841/j.cnki.jxsj.2025.06.002
5. Woo, C.L., Y, J.T., Bin, W.O., et al.: Structural analysis according to the thickness of an aquaculture crane boom. Trans. Korean Soc. Mech. Eng. A. **45**(5), 387–392 (2021)
6. Stephen, J.D.G., Banerjee, A., Lahiri, A.: Optimization of cross section of mobile crane boom using Lagrange multiplier's method. In: 2nd International Conference on Advances in Mechanical Engineering., Part 2 of 2: ICAME 2018: Kattankulathur, India, 22-24 March 2018.: Institute of Physics, pp. 936–941 (2018)
7. Shuang, W.: Spider-Type High-Altitude Work Platform Design and its Structural Analysis and Optimization. Shandong University (2021)
8. Wei, X.: Analysis and Optimization of the Boom Structure of a Certain Type of Ladder-Type High-Altitude Work Platform. Yanshan University (2019)
9. Xin, C.: Research on Lightweighting of the Vehicle Structure of the Straight Arm Type High-Altitude Work Vehicle. Jilin University (2022)

10. Guanghui, G.: Research on structural design and analysis of super large portal cranes. China Heavy Equipment. **01**, 53-56 + 68 (2025)

Analysis of Lubrication Characteristics of Marine Stern Bearings under Wear Area Conditions

Jianhui Zhou[1], Zidu Zheng[1], Zhenguo Song[1], Feng Sun[1], Zehao Huang[1], Zhengxiang Li[1], Zixin Mei[2(✉)], and Shilong Zhang[2]

[1] China Shipbuilding R&D Center, Wuhan, China
{359828,313009,299579}@whut.edu.cn
[2] School of Marine and Energy Power Engineering, Wuhan University of Technology, Wuhan, China
313009@whut.edu.cn

Abstract. This research employs a simulation method to explore the lubrication properties of marine stern bearings in wear scenarios. Initially, an accurate mechanical model of the stern bearing and journal was constructed to characterize their original configuration. The model was subsequently imported into specialized simulation software to conduct computational fluid dynamics (CFD) and tribological simulations. The simulation process particularly examined how geometric deformations induced by wear (such as reduced effective contact length) influence lubrication behavior. By quantifying wear severity through variations in bearing length (representing differing contact areas), realistic operational conditions were effectively simulated. The research successfully established quantitative relationships between wear levels, fluid film parameters, and bearing stiffness, clearly revealing the degradation mechanisms of lubrication performance under various wear states. The findings hold significant engineering value, providing robust support for optimizing stern bearing structural design, improving lifespan prediction accuracy, and enhancing the reliability of marine propulsion systems. This advancement contributes substantially to promoting safer and more efficient navigation practices in the maritime industry.

Keywords: sliding bearing · lubrication performance · Wear · aspect ratio

1 Introduction

Sliding bearings, as crucial components in modern power systems, are extensively utilized across diverse sectorsh such as marine engineering, automotive manufacturing, and aerospace. Functioning as key supporting elements in power units, they provide critical support for ship propulsion shafting systems. However, their operational reliability is compromised by multiple factors, with bearing wear identified as a primary failure mechanism. Prolonged ship operation leads to progressive deterioration of bearing wear. As wear areas expand, the lubrication performance degrades significantly, shortening the

S. Halgamuge et al. (Eds.): ICACMVE 2025, LNME, pp. 186–194, 2026.
https://doi.org/10.1007/978-981-95-6911-3_19

service life of stern bearings. This not only compromises ship navigation safety but also risks stern shaft seizure and propeller detachment—incidents that may trigger serious accidents, severely impair the reliability of marine power systems, and potentially cause catastrophic maritime disasters. Hence, a thorough study on the lubrication properties of marine stern bearings is vital to ensuring the safe and dependable operation of ships.

Lü et al. [1] proposed a lubrication model incorporating wall slip and turbulence effects and established its corresponding numerical computation procedure. Zhang T. et al. [2] found that the critical speed for the deterioration of bearing lubrication conditions increases with rising external load, surface roughness, and radial clearance. Leng et al. [3] developed a dynamic bearing analysis model that accounts for the deformation of flexible bearing bushes and the effects of surface roughness, specifically considering the roughness and dynamic characteristics of non-metallic liners. Prajapati et al. [4] investigated the influence of non-Gaussian surface topography on the tribological performance of sliding bearings operating under mixed elastohydrodynamic lubrication conditions. Frost et al. [5] discussed potential differences in bearing characteristics arising from the use of different types of lubricants with the same viscosity grade. Liu et al. [6]. studied the effects of eccentricity ratio, shaft tilt angle, and bearing structural parameters on lubrication and dynamic characteristics. Shetty et al. [7] observed that, during the initial operation phase of bearings, the substantial flow of grease from unswept regions forming grease reservoirs does not affect the lubricating film thickness.

Based on a comprehensive analysis of existing literature, systematic research on the dynamic evolution behavior of wear area in sliding bearings and its impact mechanisms on system performance remains notably insufficient. As the bearing bush and journal undergo continuous frictional contact during sustained operation, the effective load-bearing area at their interface exhibits a progressive geometric reduction. This physical process directly triggers the degradation of critical mechanical properties of the lubricating oil film, manifested as: distortion of the oil film pressure distribution, significant attenuation of the oil film stiffness coefficient, and instability in damping characteristics. The degradation of these mechanical properties further induces a holistic deterioration in the dynamic performance of the shafting system. The core objective of this paper is to establish a quantitative correlation model between the wear area and the mechanical parameters of the oil film, thereby revealing the inherent mechanisms governing the dynamic response of the lubricating film during the gradient variation of wear area.

2 Methodology

For marine sliding bearings, the analysis of lubricating oil film bearing capacity is typically carried out using the Reynolds equation. In this paper, the Reynolds equation will be non-dimensionalized, followed by further discrete derivation using the finite difference method and the successive over-relaxation (SOR) iterative method. The wear cross-sectional view of the sliding bearing is as shown in Fig. 1.

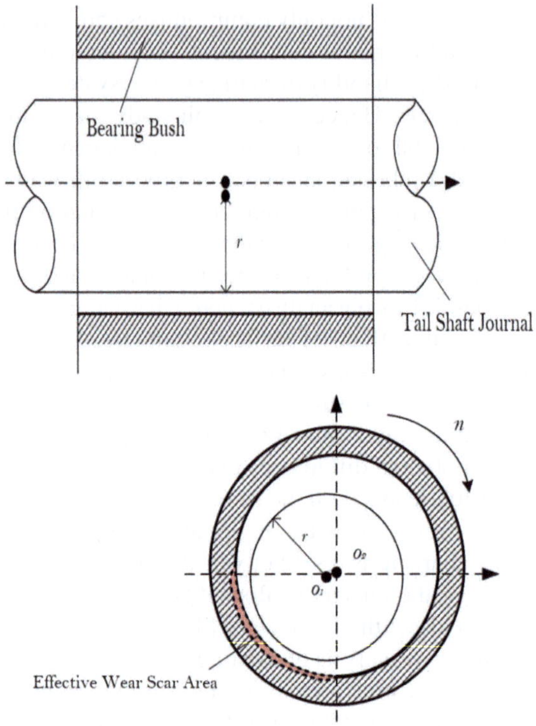

Fig. 1. Simplified Stern Bearing Model.

The Reynolds equation is derived from the Navier-Stokes equations in fluid mechanics. The equation is as follows:

$$\frac{\partial}{\partial x}\left(\frac{\rho h^3}{\mu}\frac{\partial p}{\partial x}\right)+\frac{\partial}{\partial y}\left(\frac{\rho h^3}{\mu}\frac{\partial p}{\partial y}\right)= \\ 6(V_1-V_2)\frac{\partial(\rho h)}{\partial x}+6\rho h\frac{\partial}{\partial x}(V_1+V_2)+12\frac{\partial(\rho h)}{\partial t} \tag{1}$$

where p represents the lubricant film pressure; define a Cartesian coordinate system with mutually perpendicular axes x, y. ρ and μ denote the density and viscosity of the lubricant respectively, h is the lubricant film thickness; V_1 and V_2 correspond to the tangential velocities of the upper and lower boundaries.

For liquids, the variation range of their density is usually so small that it can even be neglected. When the squeezing effect is ignored, the control equations based on hydrodynamic lubrication theory can be simplified accordingly. By eliminating the transient terms containing the normal velocity component, the original equations are transformed into a steady-state flow model, thereby focusing on the analysis of the wedging effect and the extension effect dominated by shear action. This simplified treatment is of significant engineering importance in the study of low-speed or quasi-static lubrication conditions, as it effectively reduces the complexity of equation solving while retaining the core dynamic characteristics of the lubricating film's load-carrying capacity. The modified formula is as follows:

$$\frac{\partial}{\partial x}\left(\frac{h^3}{\mu}\frac{\partial p}{\partial x}\right)+\frac{\partial}{\partial y}\left(\frac{h^3}{\mu}\frac{\partial p}{\partial y}\right)=6U\frac{\partial h}{\partial x} \tag{2}$$

The process of calculating the force of oil film lubrication using MATLAB is as follows: The process of calculating the oil film pressure of a sliding bearing requires parameter initialization first, that is, setting the number of axial and circumferential divisions, defining the wear area value, setting the eccentricity, bearing diameter, bearing width and relaxation factor, and calculating the axial step size. Then, initialize the two-dimensional film thickness array and calculate the liquid film thickness at each circumferential division point through traversal to construct a complete film thickness distribution matrix. Next, based on the grid division parameters and lubrication theory model, derive the discretization coefficient matrix including the pressure gradient term. Then initialize the pressure boundary conditions, set the pressure at both ends of the bearing and the circumferential outlet area to the ambient pressure, and use the successive over-relaxation iteration algorithm to iteratively calculate the pressure field until the pressure values at each grid node meet the preset convergence accuracy requirements. Then introduce the Reynolds boundary condition to correct the pressure distribution in the oil film rupture area, making the calculation model more consistent with the actual lubrication conditions. Finally, based on the converged pressure field data, draw the oil film pressure distribution cloud chart or contour chart to visually present the pressure variation rules in the circumferential and axial directions of the bearing. This paper studies the influence on the lubrication characteristics of sliding bearings by changing the wear area of the sliding bearings.

3 Result

Following reciprocating sliding motion between the bearing and journal, the effective load-bearing area at the interface progressively diminishes. This degradation directly compromises bearing lubrication characteristics, evidenced by altered hydrodynamic pressure distribution patterns and reduced lubricant film stiffness. Collectively, these changes detrimentally impact rotor dynamic behavior. Consequently, wear severity can be quantified using geometric parameters of the contact length, establishing a wear assessment model based on contact area reduction. Variations in contact length correlate with discrete states of interface load-carrying capacity, providing a quantitative framework for analyzing lubrication characteristic evolution throughout the wear progression lifecycle. The 3D liquid film pressure and film thickness are as shown in Figs. 2 and 3, while the 2D liquid film analysis is as shown in Figs. 4 and 5. The trend analysis chart can be seen in Fig. 6

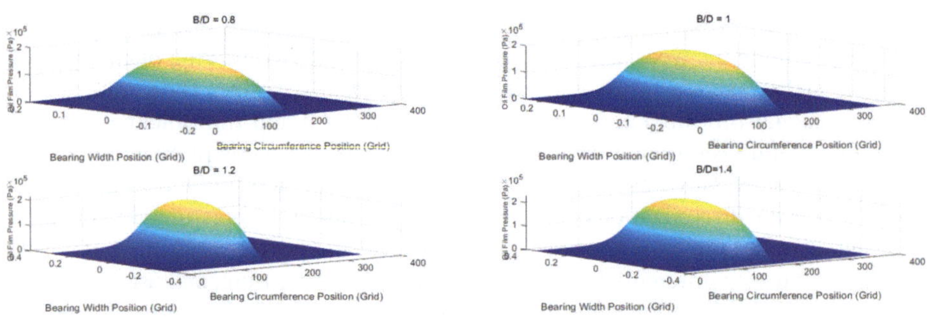

Fig. 2. 3D Oil Film Pressure Distribution.

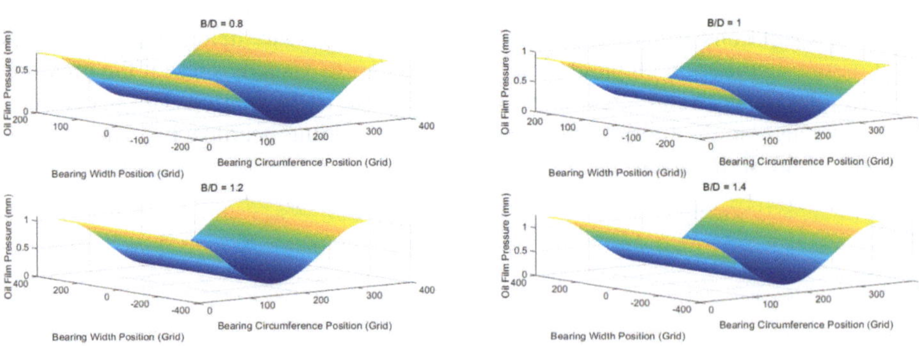

Fig. 3. Oil Film Thickness Distribution.

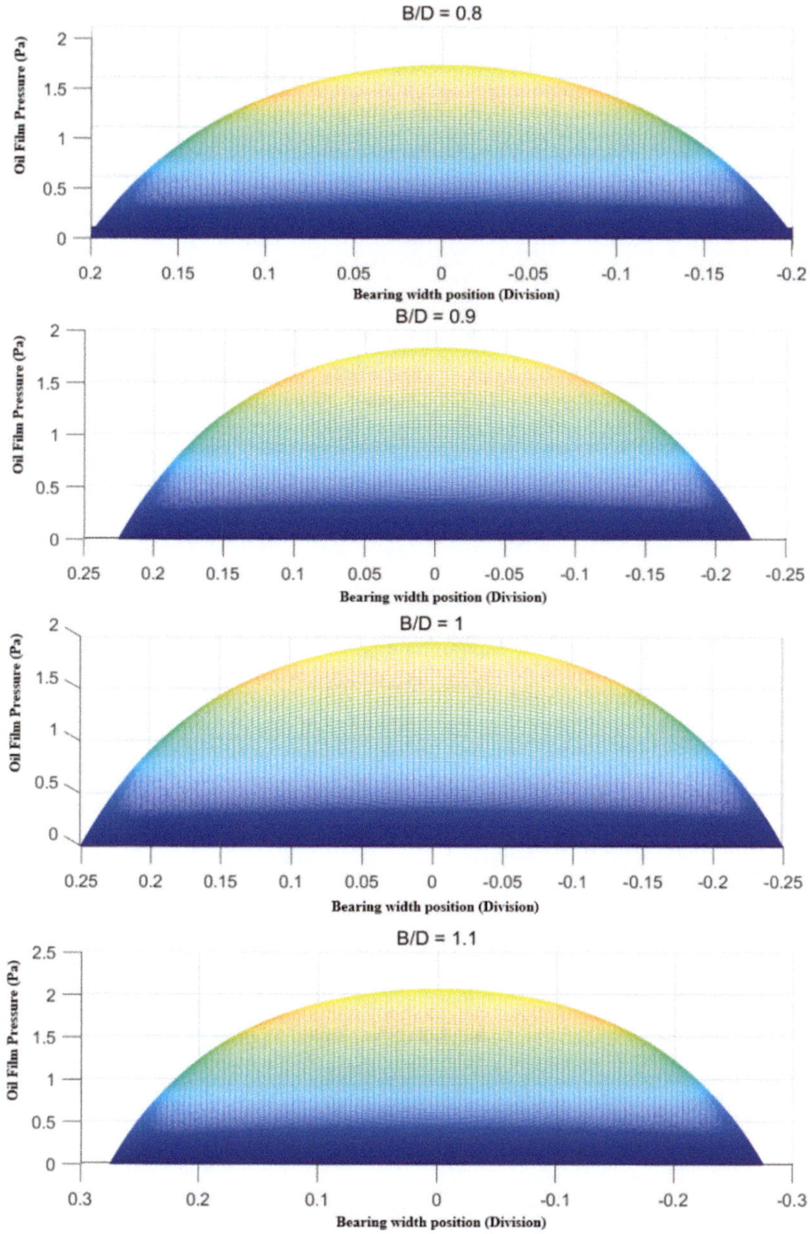

Fig. 4. Oil Film Pressure Distribution.

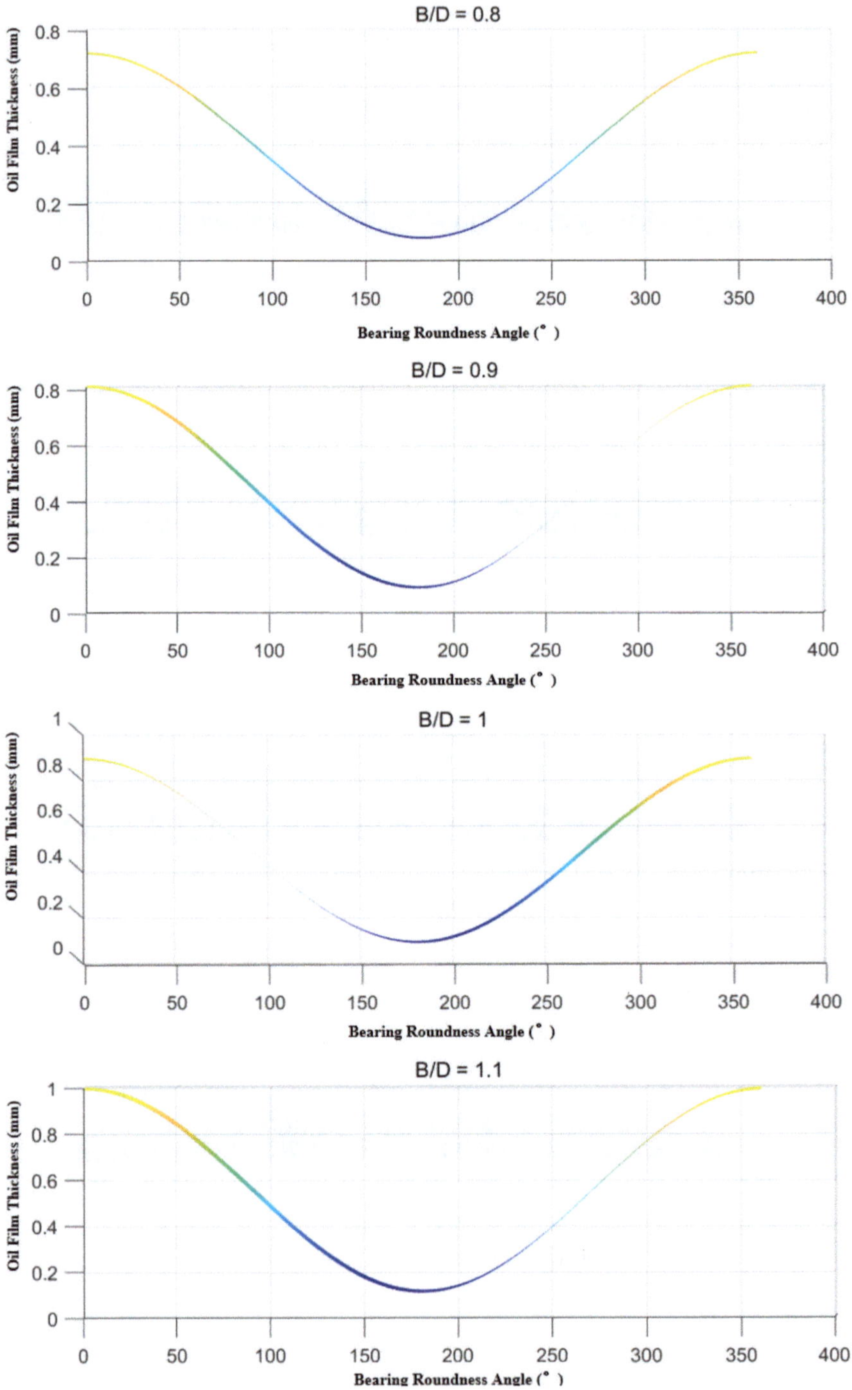

Fig. 5. Oil Film Thickness Distribution.

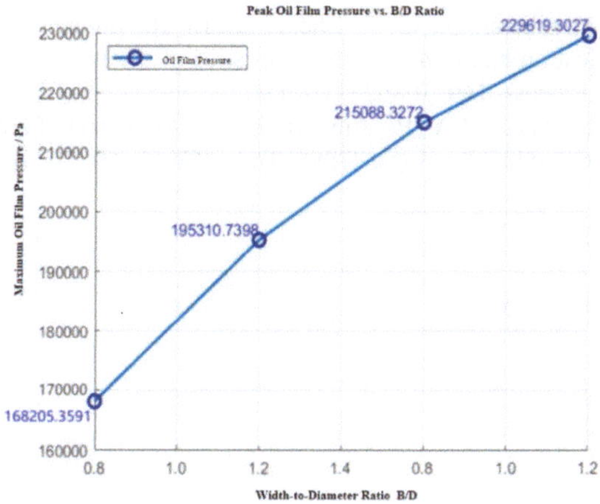

Fig. 6. Oil Film Pressure Profile

4 Conclusion

The continuous expansion of wear area induces a progressive increase in load-bearing stress on the lubricant film. The abrupt escalation of frictional power consumption may potentially trigger lubrication failure, consequently compromising the operational reliability of mechanical systems.

After repeated friction with the journal, the effective contact area between the bearing and the journal decreases, significantly impacting the lubrication of the bearing. As the contact area continues to diminish, the film thickness will correspondingly decrease.

An analysis of the liquid film pressure distribution in the composite bearing reveals that wear does not alter the distribution area but only causes a change in the peak value, which still occurs at the same location. This is because a variation in the aspect ratio does not affect the distribution of the liquid film; it merely changes the maximum load that the bearing can withstand.

References

1. Lü, F.-R., Zhang, X.-J., Rao, Z.-S., Zheng, S.-Y., Shi, J.-J.: Lubrication characteristics of marine stern tube bearings considering wall slip and local turbulence. Chuan Bo Li Xue/J. Ship Mech. **27**, 1304–1314 (2023)
2. Zhang, T., Gao, H., Yu, F., Yang, K.: Performance decay of stern bearing based on lubrication numerical model and state parameters. Chin. J. Ship Res. **17**, 133–140 (2022)
3. Leng, Z., et al.: Study of dynamic coefficients of water film in marine stern tube bearing considering roughness and deformation. Proc. Inst. Mech. Eng. Part J: J. Eng. Tribol. **235**, 2257–2271 (2021)
4. Prajapati, D.K., Ramkumar, P.: Surface topography effect on tribological performance of water-lubricated journal bearing under mixed-EHL regime. Surf. Topogr. Metrol. Prop. **10**, 045022 (2022)
5. Frost, J., Frycz, M., Kowalski, J., Wodtke, M., Litwin, W.: Environmentally acceptable lubricants (EAL) compared with a reference mineral oil as marine stern tube bearing lubricant – experimental and theoretical investigations. Tribol. Int. **189**, 109001 (2023)

6. Liu, Q., Ouyang, W., Cheng, Q., Li, J., Cheng, Q., Li, R.: Influences of bidirectional shaft inclination on lubrication and dynamic characteristics of the water-lubricated stern bearing. Mech. Syst. Signal Process. **169**, 108623 (2021)
7. Shetty, P., Meijer, R.J., Osara, J.A., Pasaribu, R., Lugt, P.M.: Effect of grease filling on the film thickness in deep groove ball bearings. Tribol. Trans. **67**, 62–69 (2024)

Operating Characteristics and Energy Efficiency of a Heavy-Duty Robotic Arm Featuring Chamber-Independent Pump-Valve Coordinated Control with Hydraulic-Pneumatic Energy Storage

Zhixin Yan[1,2], Lianpeng Xia[1,2], Tao Liang[1,2], and Long Quan[1,2](✉)

[1] Key Lab of Advanced Transducers and Intelligent Control System of Ministry of Education and Shanxi Province, Taiyuan University of Technology, Taiyuan, China
yanzx163@163.com

[2] College of Mechanical Engineering, Taiyuan University of Technology, Taiyuan, China

Abstract. Hydraulic systems remain the preferred actuation method for heavy-duty robotic arms under heavy-load and high-impact working conditions, owing to their high power density and excellent shock resistance. However, significant throttling losses at valve orifices and the dissipation of gravitational potential energy have long hindered energy efficiency improvements. To address these challenges, this study proposes a novel hydraulic–pneumatic energy storage system incorporating chamber-independent pump–valve coordinated control. In the proposed design, the driving chambers of a three-chamber hydraulic cylinder are independently metered and integrated with a coordinated pump–valve control strategy, which reduces throttling losses while enhancing motion controllability. The energy storage chamber of the three-chamber hydraulic cylinder is directly connected to a hydraulic accumulator to balance the gravitational load of the working device and overcome inertial forces, enabling efficient recovery and utilization of gravitational potential energy. Experimental results demonstrate that, based on the three-chamber hydraulic cylinder, the throttling losses of the proposed system are reduced by approximately 70%, and the overall energy consumption is reduced by about 30% compared with the load-sensing system.

Keywords: Heavy-duty robotic arm · Three-chamber hydraulic cylinder · Independent metering control · Pump-valve coordinated control · Hydraulic-pneumatic energy storage

1 Introduction

Valve-controlled hydraulic systems are widely employed in heavy-duty robotic arms, yet their average energy efficiency is only about 20% [1]. In particular, during lowering operations, a large portion of gravitational potential energy is dissipated as heat through

S. Halgamuge et al. (Eds.): ICACMVE 2025, LNME, pp. 195–204, 2026.
https://doi.org/10.1007/978-981-95-6911-3_20

valve throttling, representing a critical bottleneck that limits overall machine performance and cost-effectiveness. The root cause lies in the mechanical coupling of the inlet and outlet ports in conventional four-way spool valves, which results in substantial throttling losses and restricts the independent and coordinated control of pressures and flow rates in the two actuator chambers. This limitation not only reduces controllability but also causes irreversible conversion of gravitational potential energy into heat during lowering, leading to energy waste and increased thermal management demands [2].

To improve control performance and energy efficiency, independent metering control systems decouple the actuator's inlet and outlet flow control, enabling multivariable independent regulation. This approach effectively reduces throttling losses while enhancing response speed and control accuracy [3]. Existing research has explored various system architectures and control strategies. System architectures include cartridge valves [4], amplified valves with differential pressure compensation [5], multi-position center functions [6], and pilot-operated high-speed switching valves [7]. Control strategies include pump–valve coordination [8], active fault tolerance [9], fault detection [10], and synchronization control [11], all of which have achieved notable progress in energy saving and performance enhancement. However, multi-orifice coordinated control increases algorithmic complexity and imposes higher computational and hardware demands on the controller, leading to system complexity and higher costs, which limit large-scale engineering applications.

Meanwhile, recovering the gravitational potential energy released during the lowering process is another key to improving system energy efficiency. Current energy recovery methods can be classified into hydraulic, electrical, and mechanical types. Among them, hydraulic energy recovery offers the advantages of direct energy conversion and high power density. In particular, hydraulic–pneumatic energy storage has gained wide attention due to its compact structure and rapid response [12]. Several innovative approaches have been proposed, such as multi-chamber accumulators [13], multi-chamber hydraulic cylinders [14], and hydraulic transformers [15], achieving potential energy recovery efficiencies of up to 76% [16]. However, the energy density and nonlinear charging characteristics of hydraulic accumulators still constrain further efficiency improvements, highlighting the need for more efficient and compact structural and control solutions.

To address these issues, this study proposes a hydraulic circuit for heavy-duty robotic arms featuring chamber-independent pump–valve coordinated control with hydraulic–pneumatic energy storage. In this system, the two driving chambers of a three-chamber hydraulic cylinder are connected to an independent metering valve system, enabling efficient and precise control during both lifting and lowering operations. Simultaneously, the energy storage chamber is directly connected to a hydraulic accumulator for high-efficiency recovery and reuse of gravitational potential energy. Based on this configuration, a coordinated control strategy is developed, and the system's operational characteristics and energy efficiency are experimentally evaluated.

2 System Working Principle

As shown in Fig. 1, the proposed hydraulic circuit for a heavy-duty robotic arm integrates chamber-independent pump–valve coordinated control with hydraulic–pneumatic energy storage. The system consists of an electro-hydraulic power source, an independent metering valve group, a hydraulic accumulator, and a three-chamber hydraulic cylinder. The independent metering valve group is connected to the electro-hydraulic power source, the tank, and the two driving chambers (A and B) of the three-chamber hydraulic cylinder, enabling decoupled control of the inlet and outlet flow paths. The hydraulic accumulator is directly connected to the energy storage chamber (C) of the three-chamber hydraulic cylinder to balance the self-weight of the robotic arm and overcome inertial loads, while also allowing direct recovery and utilization of the gravitational potential energy.

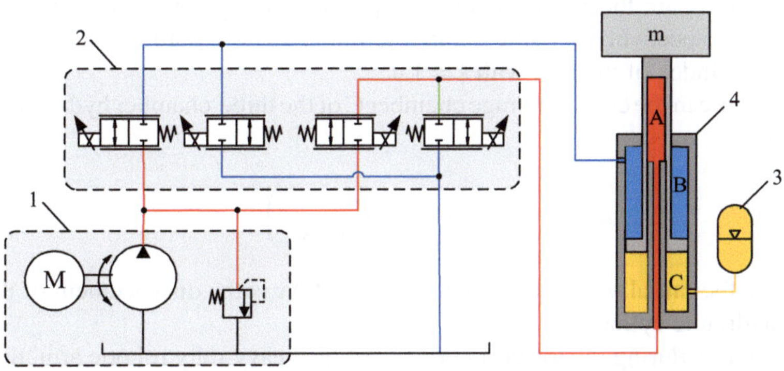

Fig. 1. Schematic of the hydraulic circuit for a heavy-duty robotic arm with chamber-independent pump–valve coordinated control and hydraulic–pneumatic energy storage 1-Electro-hydraulic power source; 2-Independent metering valve group; 3-Hydraulic accumulator; 4-Three-chamber hydraulic cylinder.

The dynamic force balance equation of the three-chamber hydraulic cylinder can be expressed as:

$$A_a p_a - A_b p_b + A_c p_c = m\dot{v} + Bv + F_f + F_L \tag{1}$$

where A_a, A_b and A_c are the piston areas of the driving chambers A and B, and the energy storage chamber C of the three-chamber hydraulic cylinder, respectively; p_a, p_b and p_c are the corresponding chamber pressures; m is the mass of the piston rod; v is the piston velocity; B is the damping coefficient; F_f and F_L represent the friction force and external load force of the cylinder, respectively. The combined forces from driving chamber A and energy storage chamber C drive the piston rod in the extension direction, while the force from driving chamber B drives it in the retraction direction.

Since the hydraulic accumulator is directly connected to the energy storage chamber C of the three-chamber hydraulic cylinder, and pipeline losses are neglected, the pressure in chamber C is assumed to be equal to the hydraulic oil pressure in the accumulator. During the lowering process of the heavy-duty robotic arm, the hydraulic fluid in chamber C is compressed into the accumulator, converting the gravitational potential energy of the arm directly into hydraulic energy stored in the accumulator. Conversely, during the lifting process, the high-pressure fluid in the accumulator flows into chamber C, driving the arm upward and directly converting the stored hydraulic energy back into gravitational potential energy for reuse. Throughout this process, energy conversion occurs only between mechanical energy and hydraulic energy, involving minimal conversion stages and achieving a high recovery efficiency.

Since the lifting and lowering processes of the heavy-duty robotic arm are relatively short, the gas inside the accumulator does not have sufficient time to exchange heat with the environment during the state change. Therefore, its state equation can be expressed as:

$$p_x {V_x}^k = p_0 {V_0}^k \tag{2}$$

where p_x and V_x are the instantaneous gas pressure and gas volume in the hydraulic accumulator, respectively; p_0 and V_0 are the initial gas pressure and gas volume; k is the polytropic index of the gas, with $k = 1.4$.

The pressure in the energy storage chamber C of the three-chamber hydraulic cylinder is given by:

$$p_c = p_{c0}\left(\frac{V_0}{V_0 - A_C \Delta x}\right)^n \tag{3}$$

where p_{c0} is the initial pressure in chamber C, and Δx is the displacement of the three-chamber hydraulic cylinder.

Furthermore, during the lowering process of the heavy-duty robotic arm, the energy stored in the hydraulic accumulator can be expressed as:

$$E_s = \int A_c p_c v dt \tag{4}$$

where E_s denotes the energy stored in the hydraulic accumulator.

3 Control Strategy

Depending on the load and motion state during lifting and lowering, the system operates in four typical quadrants: resistive extension, overrunning retraction, resistive retraction, and overrunning extension, as illustrated in Fig. 2.

First quadrant - Resistive extension:

The pressures in driving chamber A and energy storage chamber C of the three-chamber hydraulic cylinder are high, while the pressure in driving chamber B is low. In this condition, the hydraulic pump and the hydraulic accumulator jointly drive the cylinder to extend, with both the inlet and outlet valves fully open.

Second quadrant - Overrunning retraction:

Driving chamber A and energy storage chamber C are at high pressure, and driving chamber B is at low pressure. The inlet valve is fully open, and the hydraulic pump

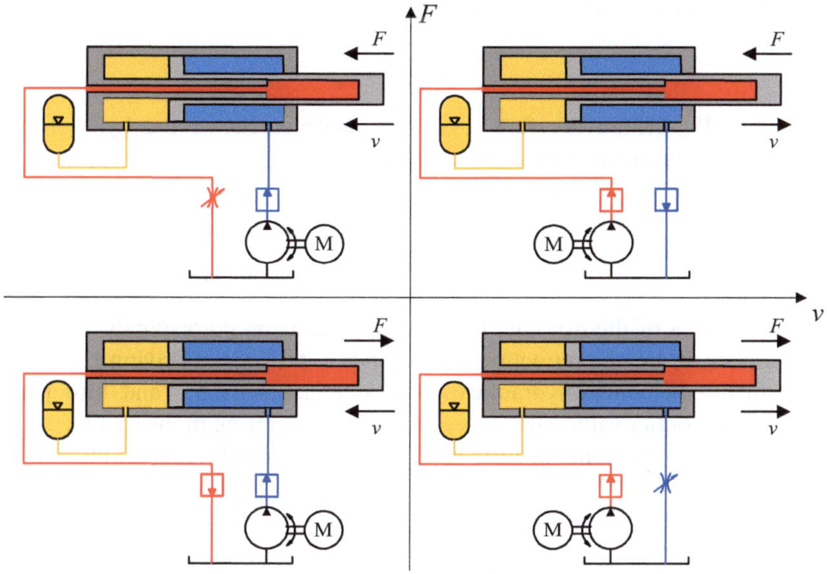

Fig. 2. Control Strategy for Four-Quadrant Operating Conditions.

supplies oil to chamber B. The operating speed of the cylinder is regulated through throttling at the outlet valve.

Third quadrant - Resistive retraction:

Driving chamber B is at high pressure, while driving chamber A and energy storage chamber C are at low pressure. The hydraulic pump drives the cylinder to retract, with both the inlet and outlet valves fully open.

Fourth quadrant - Overrunning extension:

Driving chamber B is at high pressure, and driving chamber A and energy storage chamber C are at low pressure. The inlet valve is fully open, and the hydraulic pump supplies oil to chamber A. The operating speed of the cylinder is regulated through throttling at the outlet valve.

In all four operating conditions, the inlet ports of the three-chamber hydraulic cylinder are controlled in a fully open state to minimize throttling losses. Under resistive conditions, the outlet ports are also kept fully open. Under overrunning conditions, the outlet ports are throttled to control the motion speed of the cylinder. By adopting a coordinated pump–valve control approach, the proposed strategy maintains precise motion control while significantly improving overall system energy efficiency.

The flow continuity equations for driving chambers A and B of the three-chamber hydraulic cylinder are given by:

$$q_A = A_A \mathrm{v} \tag{5}$$

$$q_B = A_B \mathrm{v} \tag{6}$$

where q_A and q_B represent the flow rates of driving chambers A and B, respectively.

The flow rate of the hydraulic pump is expressed as:

$$q_p = nV_p \tag{7}$$

where q_p and V_p denote the flow rate and displacement of the hydraulic pump, respectively; n is the rotational speed of the motor.

The motor speed is given by:

$$n = \frac{Av}{V} \tag{8}$$

where A is the area of the driving chamber; the extension corresponds to the area of driving chamber A, and the retraction corresponds to the area of driving chamber B.

When the three-chamber hydraulic cylinder operates in the first and third quadrants, both the inlet and outlet valves are fully open. When operating in the second and fourth quadrants, the inlet valve remains fully open, while the control signal for the outlet valve is:

$$\mathrm{u} = \frac{vA}{Q_n}\sqrt{\frac{\Delta p_n}{\Delta p}} \tag{9}$$

where u is the control signal of the outlet valve; Q_n is the rated flow of the outlet valve at the rated pressure difference Δp_n; Δp is the actual pressure difference across the outlet valve.

Under resistive conditions, the system operates in pump-controlled mode, where the hydraulic pump output flow matches the system demand in real time, greatly reducing throttling and overflow losses. Under overrunning conditions, the system operates in valve-controlled mode. On one hand, the energy storage chamber recovers energy through the hydraulic accumulator; on the other hand, the outlet valve throttles the flow to the driving chamber to control the operating speed of the three-chamber hydraulic cylinder in real time, while compensating for the nonlinear charging characteristics of the hydraulic accumulator. The inlet valve supplies oil to the driving chamber on demand via the hydraulic pump.

4 Experimental Study

The working device of a hydraulic excavator, including the boom, arm, and bucket, is a typical heavy-duty robotic arm. Taking a 6-ton hydraulic excavator as the research object, the operational characteristics and energy efficiency of a traditional load-sensing system and a chamber-independent pump–valve coordinated system based on independent metering valves were studied on the basis of a three-chamber hydraulic cylinder. The constructed experimental setup is shown in Fig. 3.

Figure 4 illustrates the operational characteristics of the chamber-independent pump–valve coordinated hydraulic–pneumatic energy storage system. As shown in Fig. 4(a), under the control of the joystick, the three-chamber hydraulic cylinder first retracts and then extends, completing the lowering and lifting of the boom. During operation, the

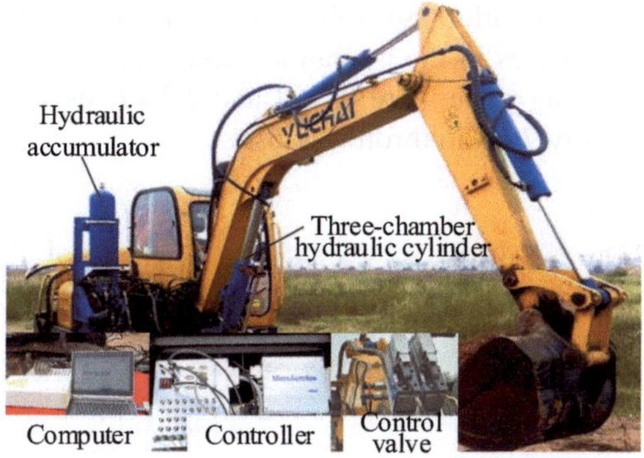

Fig. 3. Hydraulic Excavator Experimental Setup.

average velocity of the three-chamber hydraulic cylinder is approximately 100 mm/s, exhibiting smooth and controllable motion. Figure 4(b) shows that no cavitation occurred in any chamber of the three-chamber hydraulic cylinder. During boom lowering, under the action of the excavator's self-weight and the pressure in driving chamber B of the cylinder, fluid is charged from the energy storage chamber C into the hydraulic accumulator, causing the pressure in chamber C to gradually increase, thereby directly storing gravitational potential energy as hydraulic energy. During boom lifting, the hydraulic accumulator supplies fluid into chamber C of the three-chamber hydraulic cylinder, which together with driving chamber A drives the cylinder to extend. Throughout the process, efficient recovery and reuse of gravitational potential energy is achieved.

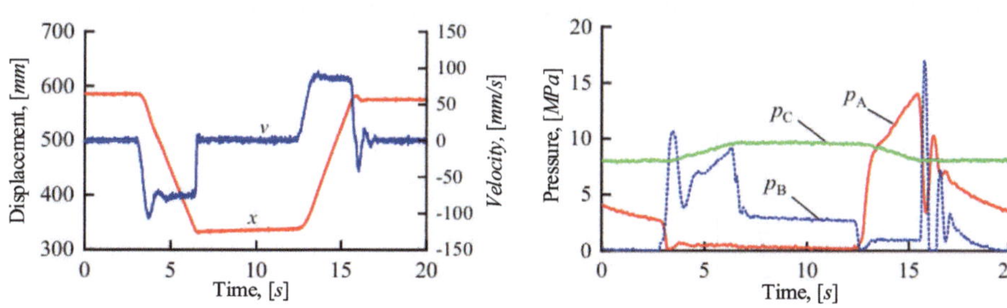

(a) Displacement and velocity of the three-chamber hydraulic cylinder

(b) Pressure in each chamber of the three-chamber hydraulic cylinder

Fig. 4. Operational Characteristics of the Chamber-Independent Pump–Valve Coordinated Hydraulic–Pneumatic Energy Storage System.

Figure 5 shows the loss power of hydraulic valves in the chamber-independent pump–valve coordinated system and the load-sensing system. The comparison indicates that the new system, by adopting the independent metering valve control principle and decoupling the inlet and outlet ports of the three-chamber hydraulic cylinder, effectively reduces throttling losses in the hydraulic valves. The energy losses for the new system and the load-sensing system are 0.5 *kJ* and 2.3 *kJ* respectively, representing a reduction of approximately 70% in throttling loss.

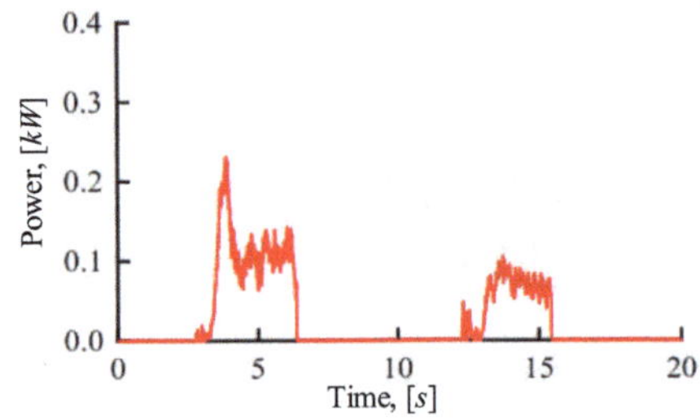

(a) Chamber-independent pump-valve coordinated system

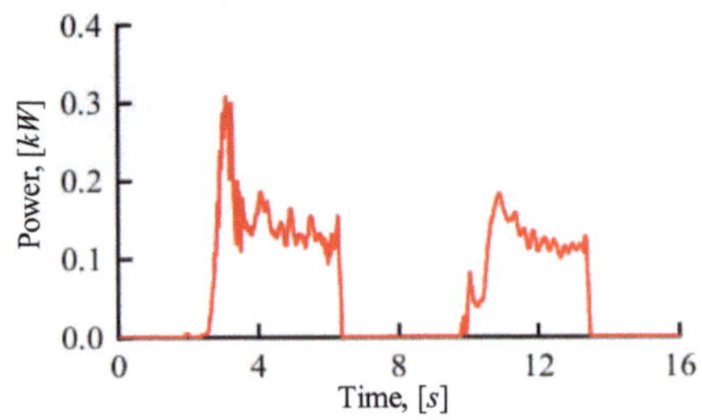

(b) Load-sensing system

Fig. 5. Loss Power of Hydraulic Valves.

Figure 6 illustrates the power profiles of the hydraulic pump and hydraulic accumulator in the chamber-independent pump–valve coordinated system. During boom lowering, energy is first stored in the hydraulic accumulator, and during boom lifting, the hydraulic accumulator assists the hydraulic pump in lifting the boom. The total output energy of the hydraulic pump throughout the process is approximately 10.9 *kJ*. Compared with the load-sensing system, the proposed configuration reduces total energy consumption by about 30%, demonstrating its effectiveness in improving overall efficiency.

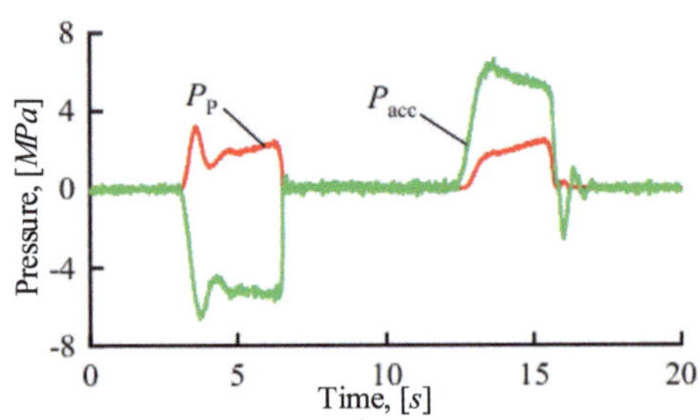

Fig. 6. Power of Hydraulic Pump and Hydraulic Accumulator in the Chamber-Independent Pump-Valve Coordinated System.

5 Conclusions

This study addressed the low energy efficiency of hydraulic systems in heavy-duty equipment by developing a chamber-independent pump–valve coordinated control system with hydraulic–pneumatic energy storage. The main conclusions are as follows:

a. The application of independent metering control enables decoupled regulation of pressure and flow in each actuator chamber, effectively reducing throttling losses and enhancing controllability compared with conventional four-way spool valves.
b. Directly connecting the energy storage chamber of the three-chamber hydraulic cylinder to a hydraulic accumulator allows efficient capture and reuse of gravitational potential energy, eliminating intermediate conversion stages and associated losses.
c. Integrating the independent metering valve group with a coordinated pump–valve control strategy for the driving chambers achieves significant energy savings. Experimental results indicate that throttling losses are reduced by approximately 70% and overall energy consumption by about 30% relative to a load-sensing system.

These findings verify the feasibility and effectiveness of the proposed approach and provide a technical foundation for enhancing the energy efficiency of heavy-duty hydraulic systems in practical engineering applications.

Acknowledgments. This work was supported by the National Key Research and Development Program of China (2023YFB3406601) and the Youth Program of the Shanxi Provincial Basic Research Plan (202403021222049).

Disclosure of Interests. The authors have no competing interests to declare that are relevant to the content of this article.

References

1. Love L J, Lanke E, Alles P: Estimating the impact (energy, emissions and economics) of the US fluid power industry. Oak Ridge National Laboratory (ORNL), Oak Ridge, TN (United States) (2012)
2. Quan, L., Z, Y., S, Z.: New progress in energy-efficient electrical drive and hydraulic transmission technologies for non-road mobile equipment. J. Mech. Eng. **59**(20), 385–400 (2023)
3. R, D., W, X., Cheng, M.: Safety performance evaluation of the intelligent independent-metering electro-hydraulic control system. J. Mech. Eng. **60**(4), 101–112 (2024)
4. Backé, W.: Design systematics and performance of cartridge valve controls. In: Proceedings of the International Conference on Fluid Power, pp. 1–48 (1987)
5. Jansson, A., Palmberg, J.-O.: Separate controls of meter-in and meter-out orifices in mobile hydraulic systems. SAE Trans. **377-383** (1990)
6. Hu, H., Zhang, Q.: Multi–function realization using an integrated programmable e/h control valve. Appl. Eng. Agric. **19**(3), 283–290 (2003)
7. Zhong, Q., Xu, E., Jia, T.: Dynamic performance and control accuracy of a novel proportional valve with a switching technology-controlled pilot stage. J. Zhejiang Univ.-SCI. A. **23**(4), 272–285 (2022)
8. Xu, B., Ding, R., Zhang, J.: Pump/valves coordinate control of the independent metering system for mobile machinery. Autom. Constr. **57**, 98–111 (2015)
9. Ding, R., Cheng, M., Jiang, L.: Active fault-tolerant control for electro-hydraulic systems with an independent metering valve against valve faults. IEEE Trans. Ind. Electron. **68**(8), 7221–7232 (2020)
10. Fischer E, Weber J: Data analysis for the evaluation and design of a model-based fault detection based on an independent metering system for mobile hydraulic drives. Universitätsbibliothek der RWTH Aachen (2023)

11. H, H., H, C., Li, L.: Design and control of four corner leveling system based on independent metering for high-speed hydraulic presses. J. Mech. Eng., 1–13 (2024)
12. Lin, T., Chen, Q., Ren, H.: Review of boom potential energy regeneration technology for hydraulic construction machinery. Renew. Sust. Energ. Rev. **79**, 358–371 (2017)
13. Yang, C., Zhou, L., Wang, J.: Research on energy saving system of hydraulic excavator based on three-chamber accumulator. J. Energy Storage. **72**, 1–19 (2023)
14. Xia, L., Quan, L., Cao, D.: Research on energy saving characteristics of large hydraulic excavator boom driven by dual hydraulic-gas energy storage cylinder. J. Mech. Eng. **55**(20), 240–248 (2019)
15. Gong, J., Zhang, D., Liu, C.: Power control strategy and performance evaluation of a novel electro-hydraulic energy-saving system. Appl. Energy. **233**, 724–734 (2019)
16. Ge, L., Dong, Z., Quan, L.: Potential energy regeneration method and its engineering applications in large-scale excavators. Energy Convers. Manag. **195**, 1309–1318 (2019)

Route Planning and Motion Simulation of Backhoe Hydraulic Excavator Based on Slewing Lifting Boom Process

Xiaowei Chen, Qinglu Shi(✉), Jinning Zhi, and Longfei Li

School of Mechanical Engineering, Taiyuan University of Science and Technology, Taiyuan 030024, Shanxi, China
shiqinglu@tyust.edu.cn

Abstract. In the cyclical operation cycle of excavator, the rotation of the rotary table occupies a lot of time, so the route planning including the rotation process has more engineering application value to shorten the cycle time, improve work efficiency and reduce energy consumption, based on the detailed theoretical analysis of the common operation cycle conditions of backhoes including the rotation process, the RRT* algorithm is used to plan a relatively short, collision-free and high-efficiency bucket tip path,the simulation is carried out by using SOLIDWORKS, and the motion parameters of each component of the working device are obtained, which verifies the feasibility of the proposed path planning.

Keywords: Hydraulic excavator · Route planning · Rotating unloading operation · RRT* algorithm

1 Introduction

In order to achieve the dual carbon goal, it is urgent to study the energy-saving technology of hydraulic excavators. In the typical cycle of hydraulic excavators, the rotary loading process has become one of the key links restricting its work efficiency and fuel economy due to its high repeatability, time and high energy consumption. According to statistics, in conventional earthmoving operations, the rotary loading action can account for 50%–70% of the single cycle time, [1] and the energy consumption accounts for more than 40%. However, hydraulic excavators still have the problems of large rotation start-stop impact, low trajectory control accuracy, and low collaborative operation efficiency, which not only leads to a decrease in fuel economy, but also causes a series of problems such as fatigue damage and operational safety hazards of structural parts. In order to improve the operation efficiency of excavators, reduce energy consumption, and realize their automated and intelligent operations, path planning is one of the main problems.

Fa Su [2] used the bidirectional RRT (Bi-RRT) algorithm for route planning, Seied Mohammad Langari[3] improved route planning performance by embedding heuristics. However, the above literature does not consider the problem of bucket filling rate during the movement of the excavator, resulting in an increase in the number of rotation operations. Han Su et al.[4] carried out the shortest route planning based on the artificial fish

S. Halgamuge et al. (Eds.): ICACMVE 2025, LNME, pp. 205–218, 2026.
https://doi.org/10.1007/978-981-95-6911-3_21

school algorithm, Alessandro Cupo[5] used the Pareto optimization control algorithm for path planning. However, the models built in the above literature are relatively ideal and difficult to reflect the actual situation. Jing Bai[6] established a joint simulation model of excavator using AMESim simulation software and ADAMS dynamic simulation software. Bin Wang[7] proposed an intelligent loading mode through AMESim and simulated and tested it. Jianxin Zhu[8] proposed a segmented acceleration scheme for rotary motors to solve the problem of loading coordination. However, the above literature does not consider the problem of obstacle avoidance when there is an obstacle. Shoulei Ma et al.[9] identified the loading conditions of the excavator based on the boom angle signal and the pressure signal of the oil inlet and outlet of the rotary hydraulic motor, but the operation was cumbersome and not intelligent and efficient. Based on the above literature, the current research on the planning of excavator slewing loading path still has certain one-sidedness and limitations, and it is difficult to comprehensively and accurately reflect the overall movement of the excavator in the process of slewing loading. The research on the feasibility, accuracy and efficiency of the implementation of the planning path is also not in-depth enough.

In this paper, the working device and slewing motion parameters are theoretically analyzed from both forward and reverse aspects, and then the RRT* algorithm is used to study the route planning for the selected research objects, and the optimal or suboptimal slewing loading trajectory is given under the premise of controlling the attitude angle of the bucket. Finally, the planned trajectory is simulated to test the trajectory accuracy and operation efficiency of each stage of rotary loading.

2 Kinematic Analysis

2.1 D-H Parametric Model

Considering the rotational motion, the hydraulic excavator has 4° of freedom, and the D-H method is used to establish the spatial Cartesian coordinate system as shown in Fig. 1.[10]

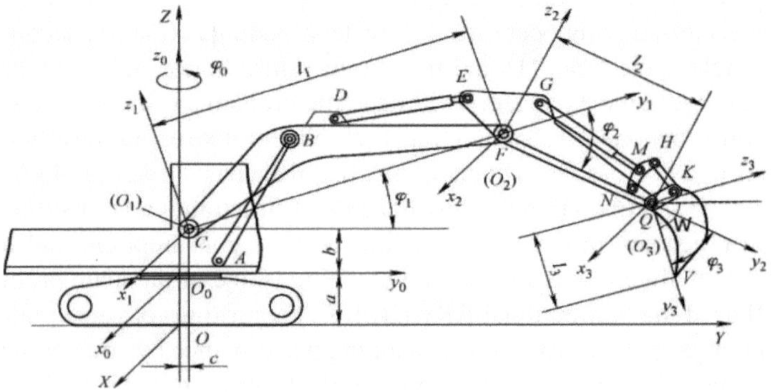

Fig. 1. Excavator D-H coordinate system.

The D-H parameters of the excavator corresponding to Fig. 1 are shown in Table 1. Among them, joint 1 is the swing mechanism, joint 2 is the boom, joint 3 is the stick, and joint 4 is the bucket.

Table 1. Excavator D-H parameters.

Joint	Connecting rod length /m	Offset /m	Deflection angle /°	Joint angle /°
1	0.4	1.44	90	−180 ~ 180
2	5.7	0	0	−49 ~ 58
3	2.7	0	0	−154 ~ 29
4	1.495	0	0	−152 ~ 36

2.2 Positive Kinematic Analysis

The correct kinematic solution can be understood as converting the joint space description $(\varphi_0, \varphi_1, \varphi_2, \varphi_3)^T$ into the Cartesian space description $(x_v, y_v, z_v, w)^T$, that is, the known swing joint angle $_0$, boom joint angle $_1$, stick joint angle $_2$, bucket joint angle $_3$, and the bucket tip V point relative to the basic coordinate system $\{O\}$ can be obtained the coordinate values of (x_v, y_v, z_v) fight with the stance w. Bucket tip coordinate system $\{V\}$ and base coordinate system $\{O\}$ the transformation matrix is[11]:

$$ {}_5^0T = \begin{bmatrix} C\varphi_0 C\varphi_{123} & -C\varphi_0 S\varphi_{123} & S\varphi_0 & C\varphi_0(c + l_1 C\varphi_1 + l_2 C\varphi_{12} + l_3 C\varphi_{123}) \\ S\varphi_0 C\varphi_{123} & -S\varphi_0 S\varphi_{123} & -C\varphi_0 & S\varphi_0(c + l_1 C\varphi_1 + l_2 C\varphi_{12} + l_3 C\varphi_{123}) \\ S\varphi_{123} & C\varphi_{123} & 0 & a + b + l_1 S\varphi_1 + l_2 S\varphi_{12} + l_3 S\varphi_{123} \\ 0 & 0 & 0 & 1 \end{bmatrix} \tag{1} $$

Among them, $C\varphi_i = cos\ \varphi_i$, $C\varphi_{ij} = cos\ (\varphi_i + \varphi_j)$, $S\varphi_{ij} = sin\ (\varphi_i + \varphi_j)$, $C\varphi_{ijk} = cos\ (\varphi_i + \varphi_j + \varphi_k)$, $S\varphi_{ijk} = sin\ (\varphi_i + \varphi_j + \varphi_k)$.

According to the following formula, the V point of the bucket tip can be obtained relative to the base coordinate system $\{O\}$ the coordinate values of (x_v, y_v, z_v):

$$ \begin{bmatrix} x_v \\ y_v \\ z_v \\ 1 \end{bmatrix} = {}_5{}^0T \begin{bmatrix} 0 \\ 0 \\ 0 \\ 1 \end{bmatrix} = \begin{bmatrix} C\varphi_0(c + l_1 C\varphi_1 + l_2 C\varphi_{12} + l_3 C\varphi_{123}) \\ S\varphi_0(c + l_1 C\varphi_1 + l_2 C\varphi_{12} + l_3 C\varphi_{123}) \\ a + b + l_1 S\varphi_1 + l_2 S\varphi_{12} + l_3 S\varphi_{123} \\ 1 \end{bmatrix} \tag{2} $$

The bucket attitude angle can be known from the geometric relationship:

$$ w = \varphi_1 + \varphi_2 + \varphi_3 \tag{3} $$

Therefore, the correct solution of the hydraulic excavator joint space description $(\varphi_0, \varphi_1, \varphi_2, \varphi_3)^T$ is the Cartesian space description $(x_v, y_v, z_v, w)^T$, and the specific relationship is as follows:

$$\begin{cases} x_v = C\varphi_0(c + l_1C\varphi_1 + l_2C\varphi_{12} + l_3C\varphi_{123}) \\ y_v = S\varphi_0(c + l_1C\varphi_1 + l_2C\varphi_{12} + l_3C\varphi_{123}) \\ z_v = a + b + l_1S\varphi_1 + l_2S\varphi_{12} + l_3S\varphi_{123} \\ w = \varphi_1 + \varphi_2 + \varphi_3 \end{cases} \tag{4}$$

2.3 Inverse Kinematic Analysis

The kinematic inverse is to convert the Cartesian space description $(x_v, y_v, z_v, w)^T$ into a joint space description $(\varphi_0, \varphi_1, \varphi_2, \varphi_3)^T$, the known bucket tip V point relative to the base coordinate system $\{O\}$ The coordinate values of (x_v, y_v, z_v) and the bucket attitude angle w can obtain the angle of the swing joint、 the angle of the boom joint、 the angle of the stick joint and the angle of the bucket joint. In this paper, the geometric method is used to perform the kinematic inverse solution of the excavator. The geometric relationship is shown in Fig. 2, and the kinematic inverse solution process is as follows:

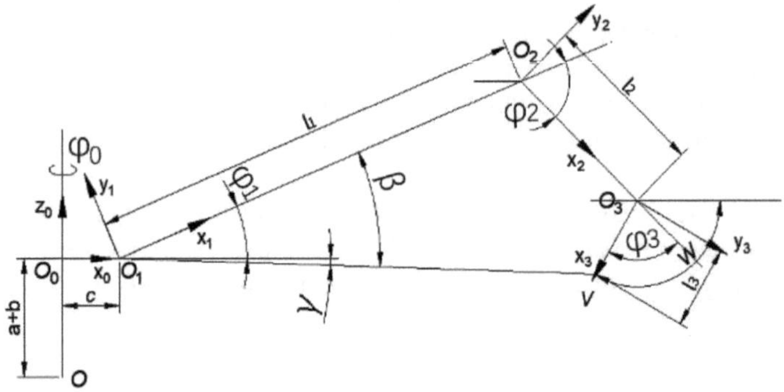

Fig. 2. Schematic diagram of kinematic inverse solution.

The angle of the swing joint $_0$ is as follows:

$$\varphi_0 = \tan^{-1}\frac{y_v}{x_v} \tag{5}$$

The coordinate values of the stick-bucket hinge point O_3 is as follows:

$$\begin{cases} x_{O_3} = x_v - \cos\varphi_0 l_3 \mathrm{cosw} \\ y_{O_3} = y_v - \sin\varphi_0 l_3 \cos w \\ z_{O_3} = z_v - l_3 \sin w \end{cases} \tag{6}$$

Swing - boom hinge point O_1 distance from stick - bucket hinge point O_3, $l_{O_1O_3}$ is as follows:

$$l_{O_1O_3} = \sqrt{\left(\frac{x_{O_3}}{\cos\varphi_0} - c\right)^2 + \left(z_{O_3} - a - b\right)^2} \tag{7}$$

The angle between the two lines of O_1O_2 and O_1O_3, is as follows:

$$\beta = \cos^{-1}\left(\frac{{l_1}^2 + {l_{O_1O_3}}^2 - {l_2}^2}{2l_1 l_{O_1O_3}}\right) \tag{8}$$

The angle between the two lines of O_1O_3 and O_0O_1, is as follows:

$$\gamma = \tan^{-1}\frac{z_{O_3} - a - b}{\frac{x_{O_3}}{\cos\varphi_0} - c} \tag{9}$$

The angle of the boom joint, φ_1 is as follows:

$$\varphi_1 = \beta + \gamma \tag{10}$$

The angle of the stick joint, φ_2 is as follows:

$$\varphi_2 = \cos^{-1}\left(\frac{{l_1}^2 + {l_2}^2 - {l_{O_1O_3}}^2}{2l_1 l_2}\right) - \pi \tag{11}$$

The angle of the bucket joint, φ_3 is as follows:

$$\varphi_3 = w - \varphi_1 - \varphi_2 \tag{12}$$

By geometry, the Cartesian space description $(x_v, y_v, z_v, w)^T$ is inversely solved into a joint space description $(\varphi_0, \varphi_1, \varphi_2, \varphi_3)^T$ and the specific relationship is as follows:

$$\begin{cases} \varphi_0 = \tan^{-1}\left(\frac{y_P}{x_P}\right) \\ \varphi_1 = \cos^{-1}\left(\frac{l_1^2 + {l_{O_1O_3}}^2 - l_2^2}{2l_1 l_{O_1O_3}}\right) + \tan^{-1}\left(\frac{z_{O_3} - a - b}{\frac{x_{O_3}}{\cos\varphi_0} - c}\right) \\ \varphi_2 = \cos^{-1}\left(\frac{l_1^2 + l_2^2 - {l_{O_1O_3}}^2}{2l_1 l_2}\right) - \pi \\ \varphi_3 = w - \varphi_1 - \varphi_2 \end{cases} \tag{13}$$

3 Analysis of Rotary Operations

3.1 Establish Simulation Model for Rotary Loading

In this paper, the commonly used 20-ton backhoe hydraulic excavator and the matching dump truck (length, width and height: 7 × 3 × 3.5 m) are taken as the research object. As shown in Fig. 3(a), the dump truck receiving and parking position should be parallel to the excavator as much as possible, and the rear of the truck should face the pit in order to avoid mutual influence. The distance between the excavator and the dump shop is 5.5 m, and the distance between the pit and the pit is 6.6 m.

Due to the special working conditions and few relevant references and data, by watching a large number of excavator rotary loading videos and communicating with practitioners in related fields, and finally using software SOLIDWORKS to build a detailed 3D model, as shown in the Fig. 3(a) below.

(a)SOLIDWORKS models.

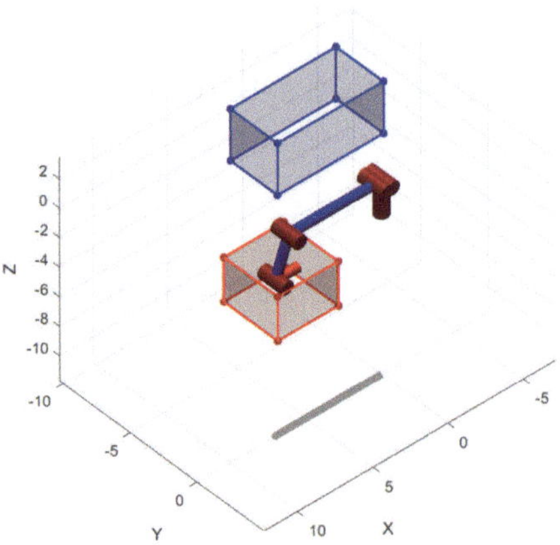

(b)MATLAB models.

Fig. 3. Rotating loading model.

3.2 Decompose the Process of Rotary Loading Action

As shown in Fig. 4, the rotary loading process can be divided into two stages: rotary lifting arm and unloading loading. In the stage of slewing lifting boom, to avoid spreading, keep the bucket mouth as horizontally up and parallel to the ground as possible during the slewing lift. In addition, special attention should be paid to the distance between the working device, especially the bucket, from the pit and dump truck to avoid collisions. Route planning through the three key poses shown in Fig. 4 can obtain a collision-free and relatively optimal path.

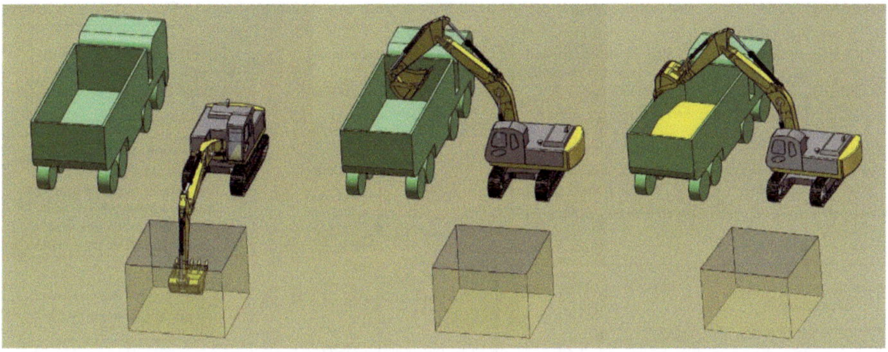

(a) The initial point of the slewing lifting boom. (b) The end point of the slewing lifting boom. (c) The discharge and loading.

Fig. 4. Schematic diagram of rotary loading.

4 Route Planning Based on RRT* Algorithm

The MATLAB robot toolbox is used to simplify the previously established D-H parametric model to the structure shown in Fig. 3(b), where the red wireframe represents the pit and the blue wireframe represents the dump truck.

Based on this model, the RRT* algorithm [12] is used for path planning. This algorithm is an important improved version of the classical RRT algorithm, [3] the core of this algorithm is to introduce two optimization mechanisms, "re-select parent node" and "rerouting", as shown in Fig. 5,while maintaining the exploration efficiency of the RRT algorithm, to gradually optimize the path quality, and finally converge to the optimal or near-optimal path.

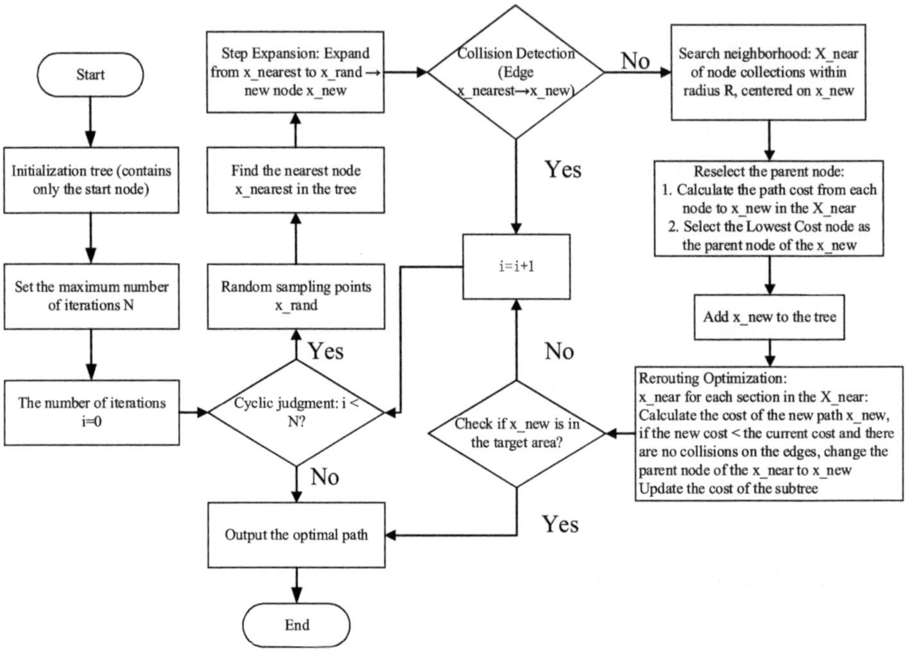

Fig. 5. RRT* Algorithmic flow diagram.

Set the maximum number of iterations to 500, the search radius is 0.1, the step size is 0.5, and the collision detection distance is 0.05. Write the code to make the robotic arm go through the three key positions mentioned above, and keep the end robotic arm representing the bucket as level with the ground as possible during the rotation of the boom, and finally obtain the motion trajectory and the angle change curve of each joint; Further, the joint angle is derived from time to obtain the joint angular velocity change curve.

There is a functional relationship between the joint angle and the length of the hydraulic cylinder,[13] then the cylinder length and flow change curve can be obtained. The final result is shown in Figs. 6 and 7:

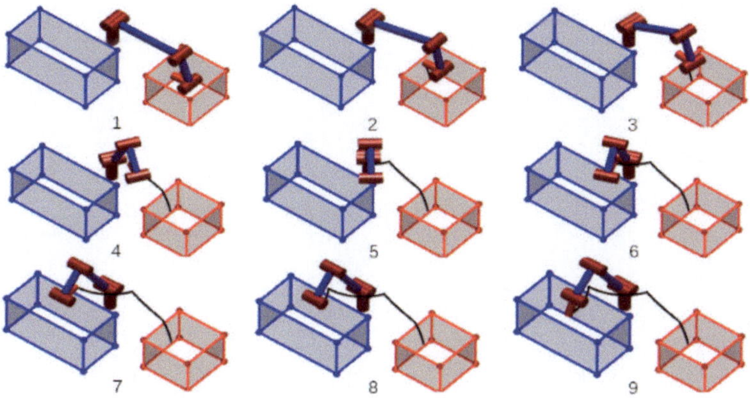

Fig. 6. Schematic diagram of the planning of the rotary loading route.

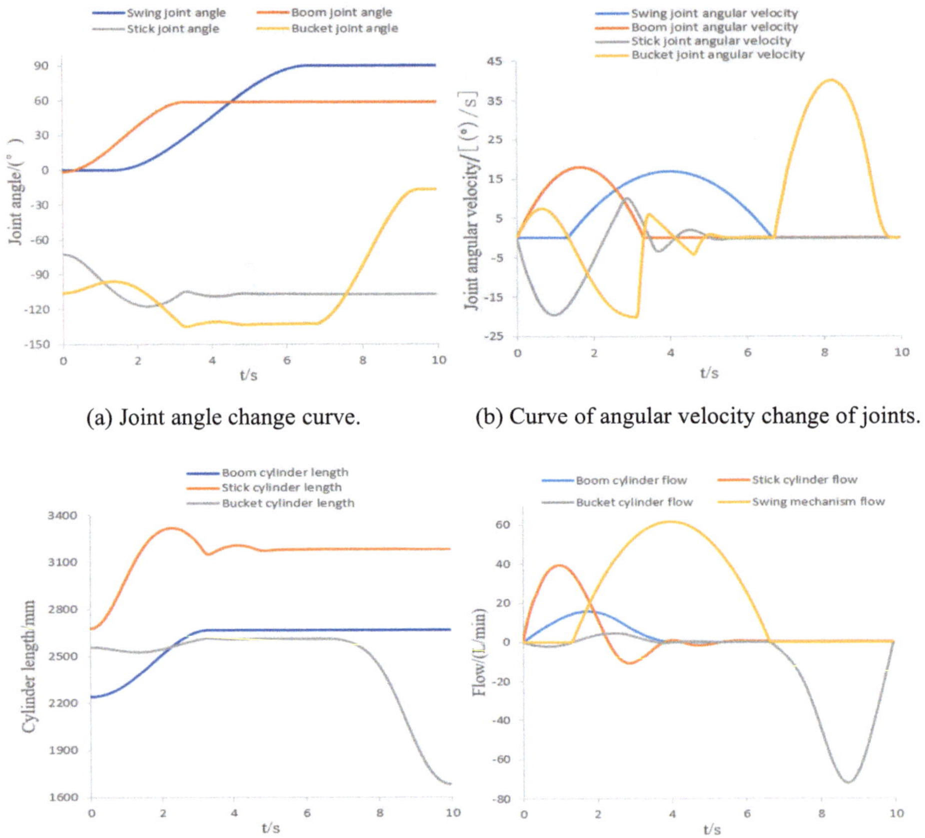

(a) Joint angle change curve.

(b) Curve of angular velocity change of joints.

(c) Cylinder length change curve.

(d) Flow change curve.

Fig. 7. Joint angle, hydraulic cylinder and flow change curve.

In Fig. 7, 0 ~ 3.4 s represents the process of the excavator avoiding obstacles from the initial point of the slewing lifting boom to the pit. 3.4 ~ 6.7 s represents the process of the excavator avoiding obstacles for the dump truck and reaching the end point of the slewing lifting boom.6.7 ~ 10 s represents the process of unloading and loading the excavator. In the time period of 0 ~ 6.7 s, due to the demand for obstacle avoidance, the curve fluctuates greatly. For the flow change curve, the flow rate required by the four components in the whole process is less than the maximum flow rate in the system of 429 L/min.

5 Simulation Results Analysis

The 3D model of the research object is established by using SOLIDWORKS, and the motion plug-in is used to simulate the above rotation process, and then the angle data generated by the RRT* algorithm is imported into it, and the simulation image as shown in Fig. 8 is obtained:

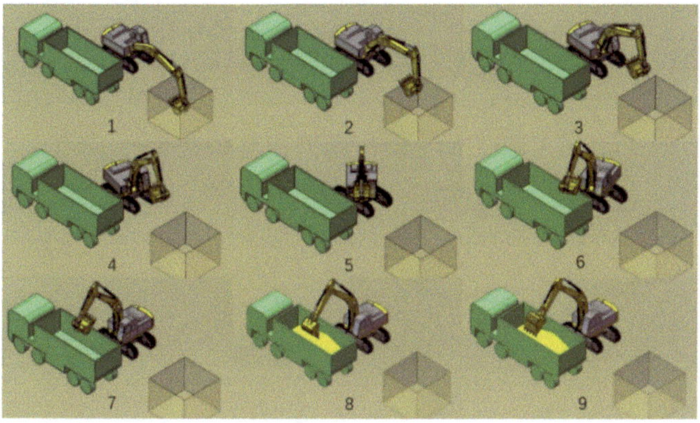

Fig. 8. Schematic diagram of simulation results.

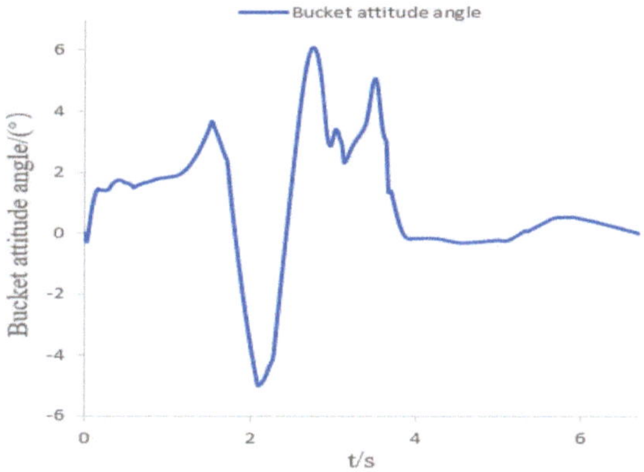

Fig. 9. Bucket attitude angle change curve.

Fig. 9 shows the change curve of the attitude inclination angle of the bucket mouth relative to the ground during the slewing lifting boom process. Also pay special attention to the distance between the excavator working device, especially the bucket, and the pit and dump truck to avoid collisions. Therefore, sensor commands are added to the bottom of the bucket and the tip of the bucket teeth of the 3D model, and the results are shown in Fig. 10.

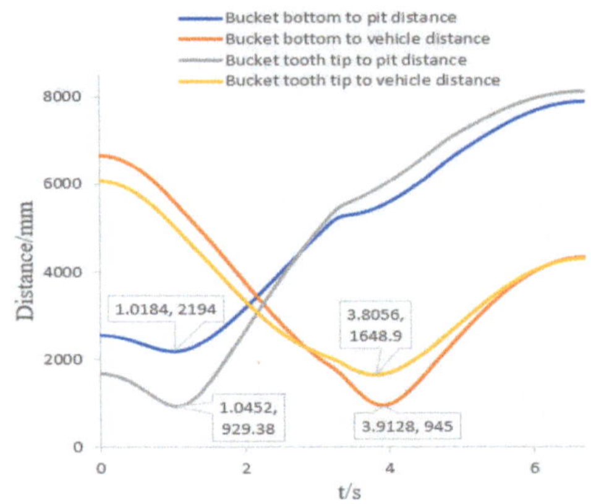

Fig. 10. Collision distance change curve.

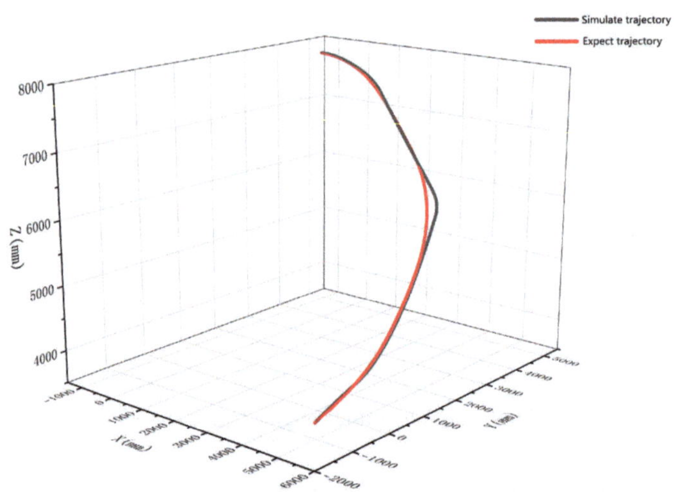

Fig. 11. Comparison diagram of the tip path of the bucket tooth.

From Fig. 9, it can be seen that the maximum deviation between the bucket mouth and the ground is about 6°. The curve fluctuates greatly within 0 ~ 3.7 s, which may be due to the low initial position of the boom and the fast process of rotating the boom, which limits the movement of the bucket in the pit and causes fluctuations. In 3.7 ~ 6.7 s, due to the detachment from the narrow pit and the slow movement of the boom, the obstacle avoidance process is relatively smooth, and the curve performance is relatively smooth.

Under normal circumstances, the time required for rotary loading is about 15 s, and Fig. 7 shows that the time after route planning is shortened to less than 10 s.Combined with Figs. 9 and 10, it can be verified that the excavator not only shortens the time on the planned slewing lifting boom trajectory, but also maintains the stability of the bucket and avoids collision with obstacles.

The simulated path of the tip of the bucket tooth and the expected path of the tip of the bucket tooth obtained by the RRT* algorithm are plotted in one figure, as shown in Fig. 11. The results show that the excavator can move according to the path derived by the RRT* algorithm well and is relatively optimal.

6 Conclusions

1. The kinematic model of the backhoe hydraulic excavator is established by using the D-H method, and the mutual transformation between the Cartesian space motion parameters and the joint space motion parameters can be realized through forward and reverse kinematic analysis, and the pose parameters of the end of the bucket can be accurately described.
2. By analyzing the rotary loading process, three key positions of the slewing lifting boom are determined: initial, termination and discharge, and a simplified model is established in MATLAB and the RRT* algorithm is used for route planning, and a collision-free and relatively optimal path is obtained, which provides a reference for the realization of excavator automation and intelligent operation.
3. With the help of 3D modeling technology, 3D simulation of the best trajectory after planning, it intuitively and clearly shows that the excavator can carry out efficient and collision-free operation according to the motion trajectory obtained by the RRT* algorithm.

The ideas and methods of this paper can be generalized to the path planning problem of the same type of backhoe hydraulic excavator under the same working conditions. Further research work can be carried out on the actual working conditions of the actual model, which can improve the efficiency of the operation and improve the stability and safety of the operation process.

Acknowledgments. This work is supported by the Shanxi Province Basic Research Program Project (Grant No. 202203021211196)

Disclosure of Interests. The authors declare no conflict of interest.

References

1. Tianliang, L., Jie, Y., Qiang, L., et al.: Simulation of new hydraulic excavator slewing drive system. J. Huaqiao Univ. (Nat. Sci. Ed.). **34**(03), 247–252 (2013)
2. Su Fa: Research on hydraulic excavator slewing loading trajectory planning and tracking control method. Jilin University, 2023. DOI: https://doi.org/10.27162/d.cnki.gjlin.2023.006510.
3. Langari, S.M., Vahdatikhaki, F., Hammad, A.: Improving the performance of RRT path planning of excavators by embedding heuristic rules. Adv. Eng. Inform. **62**, 102724 (2024)
4. Han, S., Wei, M., Chenbo, Y., et al.: Research on trajectory planning method under excavator rotary loading conditions. Manuf. Autom. **46**(06), 67-72+112 (2024)

5. Cupo, A., Cecchin, L., Demir, O., et al.: Energy-optimal trajectory planning for semi-autonomous hydraulic excavators. IFAC-PapersOnLine. **58**(28), 450–455 (2024)
6. Jing, B.: Research on the slewing compound action of medium-sized backhoe hydraulic excavator. Lanzhou University of Technology (2022). https://doi.org/10.27206/d.cnki.ggsgu.2022.001106.
7. Bin, W.: Research on intelligent loading mode of hydraulic excavator. Constr. Mach. Technol. Manag. **30**(12), 56–59 (2017). https://doi.org/10.13824/j.cnki.cmtm.2017.12.018
8. Zhu Jianxin, Liu Fuping, Wang Juan: Journal of Guangxi University (Natural Science Edition),(2013),38(02): 284–291. DOI: https://doi.org/10.13624/j.cnki.issn.1001-7445.2013.02.034.
9. Ma Shoulei, S., Han, Y.C., et al.: Research on working condition identification method based on hydraulic excavator operation cycle stage. Hydraul. Pneum. **47**(04), 19–26 (2023)
10. Qinglu, S., Muyi, L.: Hydraulic Excavator. Machinery Industry Press, Beijing (2011)
11. Xufeng, J., Qinglu, S., Hejing, S., et al.: Trajectory planning of slope condition of three-section arm excavator based on time optimization. Mech. Des. **41**(06), 130–138 (2024). https://doi.org/10.13841/j.cnki.jxsj.2024.06.028
12. Xiong, Z., Hou, W., Wang, C., et al.: Classified-RRT*: node classification RRT*-based algorithm with improved solution performance and convergence rate. Expert Syst. Appl., 129118 (2025)
13. Lina, P., Qinglu, S., Cong, Z., et al.: Three-section arm hydraulic excavator working device modeling and operation scope analysis. Constr. Mach. **52**(01), 53-59+9-10 (2021)

Design and Analysis of a Multi-functional Electric Tricycle Vehicle

Yi-huang Lian, Tian Tian(✉), Guo-biao Chen, Bo-tao Ding, Yu-heng Wu, and Hong-zhuo Peng

School of Automotive and Traffic Engineering, Guangzhou City University of Technology, Guangzhou, China
tiantian@gcu.edu.cn

Abstract. To meet the short-distance household travel needs, a multi-functional electric tricycle integrating cargo transportation, passenger carrying, and leisure functions has been designed. This study presents the overall vehicle design scheme, determines the selection of appropriate components and parameters, and conducts overall modeling of the electric tricycle for key components. Using ANSYS Workbench, finite element analysis (FEA) is applied to evaluate the frame structure under bearing and torsional loads. The results reveal that under a 450 kg load, the maximum total deformation of the tricycle frame is 1.5 mm with a maximum stress of 86.15 MPa; under a 900 N·m bending moment, the maximum deformation of the frame structure is 1.7 mm with a maximum stress of 97.77 MPa. These findings demonstrate that the strength and stiffness of the frame meet the service requirements.

Keywords: Multi-functional · electric tricycle vehicle · Structural Design · Finite Element Analysis

1 Introduction

For short-distance commuting, electric vehicles (EVs) have become the preferred choice for the majority of consumers [1]. Meanwhile, three-wheeled electric vehicles (TEVs) with multi-purpose capabilities are experiencing robust development in China and gaining significant global popularity, with their production volume, market demand, pricing, and market size continuing to grow [2]. Semi-enclosed TEVs, characterized by reduced size, lower manufacturing costs, and environmental friendliness, are well-suited for operation on narrow urban streets and offer easier parking. Furthermore, compared to two-wheeled electric vehicles (2WEVs), TEVs are equipped with an active roll compensation function that increases steering centripetal torque, enabling smoother turning, faster cornering speeds, and enhanced stability and maneuverability [3]. Consequently, TEVs not only achieve large cargo payloads but also mitigate safety risks during turning. The vehicle platform can accommodate larger-capacity battery packs, thereby extending the driving range.

S. Halgamuge et al. (Eds.): ICACMVE 2025, LNME, pp. 219–231, 2026.
https://doi.org/10.1007/978-981-95-6911-3_22

Cao Chi, Zhu Haiping, et al. designed a rear electric tricycle vehicle and presented a technical development plan for the entire vehicle [4]. Li Boqun investigated parameter identification of brushless DC motors and torque ripple suppression technology for 3WEVs to address operational stability [5]. Zhao Jianyong, taking a specific 3WEV model as a reference, developed a heavy-duty 3WEV by designing key performance parameters. Additionally, CAE technology was employed to analyze, improve, and optimize the frame design, achieving both sufficient strength and lightweight objectives [6]. Ye Yan invented an SOC (State of Charge) intelligent control system capable of precise self-diagnosis of various systems and components in 3WEVs, while incorporating safety preprocessing functions to enhance the intelligence level and performance of the vehicles [7].

Analysis of the above literature reveals the critical importance of technological development in multi-functional 3WEVs. In response to these findings, this paper presents the design of a multi-functional 3WEV integrating cargo transportation, passenger carrying, and leisure functions. It proposes an overall vehicle design scheme, determines appropriate component selections and parameter configurations, and conducts comprehensive modeling of key components. To validate safety, finite element analysis (FEA) using ANSYS Workbench was applied to optimize the frame structure and perform multi-directional simulations, verifying the safety of its structural design.

2 Electric Tricycle Structure Design

Combining the comprehensive functional requirements, safety, stability, and regulatory demands of the electric tricycle structural design scheme, a multi-functional electric vehicle is designed to align with the positioning and objectives specified in Table 1. The design adopts cost-effective lithium batteries and a differential motor directly connected to the rear axle, eliminating the traditional chain transmission structure. This optimization enhances energy conversion efficiency and improves overall power performance. Through these improvements, the new electric tricycle achieves a lighter curb weight, extended driving range, more convenient use and maintenance, and reduced noise levels, demonstrating significant advancements in both performance and user experience compared to conventional electric vehicles [6].

2.1 Dynamic Performance

Vehicle dynamics on flat roads are governed by longitudinal forces. The driving equation [8] is,

$$F_{\mathrm{t}} = F_f + F_w + F_i + F_j \tag{1}$$

where F_t is driving force; F_f is rolling resistance; F_w is air resistance; F_i is gradient resistance; F_j is acceleration resistance;

Driving force relates to motor torque T_m, transmission efficiency η_t, gear ratio I_0, and wheel radius r,

$$F_{\mathrm{t}} = \frac{T_m I_0 \eta_t}{r} \tag{2}$$

Table 1. I Design Scheme.

Design Orientation	Design Objectives
Functional Use	Passenger transport (short-distance commuting), cargo transport (daily goods)
Load Capacity	~450 kg (structural reinforcement required for >450 kg)
Range Requirement	30 Ah battery capacity +1500 W motor
Regulatory Compliance	Meets national 3WEV standards
Material Selection	High-strength steel (e.g., Q235)
Structural Form	Single longitudinal front beam + dual rear wheels (simple, low-load optimized)
Braking System	Dual disc brakes with E-ABS
Driving Comfort	Thickened sponge seats, adjustable positions, foot brake linkage, optional reverse gear
Load & Climbing	Front basket + under-seat storage; sustains 15° slopes at 450 kg load
Suspension	Front hydraulic fork + rear hydraulic shock absorption

The expression for rolling resistance is,

$$F_{\mathrm{f}} = f \cdot G = f \cdot mg \tag{3}$$

The expression for aerodynamic drag is,

$$F_w = \frac{C_D A u_a^2}{21.15} \tag{4}$$

The expression for grade resistance is,

$$F_i = G sin\alpha \tag{5}$$

The expression for acceleration resistance is,

$$F_j = m\delta \frac{du}{dt} \tag{6}$$

By substituting Eqs. (2), (3), (4), (5), and (6) into Formula (1), the dynamic equation for the maximum speed of the electric tricycle on flat roads can be derived as follows,

$$\frac{T_{\mathrm{m}} I_0 \eta_t}{r} = fmg + \frac{C_D A u_a^2}{21.15} \tag{7}$$

Based on the characteristics of the drive motor, the drive motor speed and vehicle travel speed satisfy the following relationship,

$$u_a = \frac{0.377 r \cdot n}{i_0} \tag{8}$$

Based on the technical requirements of national standards for the power performance, economic performance, and safety performance of pure electric vehicles (PEVs), the complete vehicle design parameters of the three-wheeled electric vehicle (TEV) are presented in Table 2. The gross vehicle mass (GVM) m is 450 kg, with a maximum speed of 40 km/h, an acceleration time from 0–30 km/h of less than 10 s, and a range exceeding 30 km. The overall model of the electric vehicle is illustrated in Fig. 1.

Table 2. Numerical values of vehicle design parameters.

Parameter	Value
Overall Dimensions (L × W × H)	3060 mm × 1300 mm × 1500 mm
Maximum Speed	40 km/h
Minimum Ground Clearance	160 mm
Maximum Front Wheel Turning Angle	45° (Left/Right)
Turning Radius	3 m
Payload Mass	450 kg
Vehicle Curb Mass	200 kg
Rolling Resistance Coefficient	0.02
Wheel Radius	360 mm
Transmission Efficiency (δ)	0.90
Transmission Ratio	10
Acceleration Time (0–40 km/h)	$<$10 s
Constant Speed Range (15 km/h)	$>$30 km

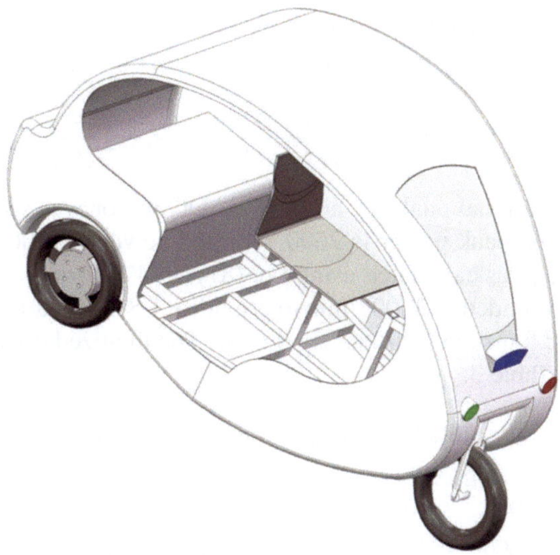

Fig. 1. The overall model of the electric tricycle vehicle.

2.2 Drive Motor Design Parameters for Electric Tricycle Vehicles

As one of the most critical components in pure electric vehicles, the drive motor is responsible for converting electrical energy stored in the power battery into mechanical energy, directly determining the vehicle's performance characteristics.

The power balance equation governing vehicle dynamics is expressed as [8],

$$P = \frac{1}{3600\eta_t} mg \cdot fcos\alpha \cdot u_a + \delta \cdot mu_a \frac{du}{dt} \tag{9}$$

The peak power of the drive motor must simultaneously satisfy the specific power demands for both maximum vehicle speed and acceleration performance [8]. This dual requirement constitutes a critical factor in ensuring the stable operation and energy-efficient performance of pure electric vehicles. Mathematically, this principle is expressed as follows,

$$Pmax \geq \max (P\,max1,\ Pmax2) \tag{10}$$

In Eq. (10), P_{max1} denotes the drive motor power required for a pure electric vehicle to operate at its maximum design speed under steady-state conditions. P_{max2} represents the drive motor power necessary to achieve the target acceleration performance (e.g., 0–40 km/h within a specified time).

The drive motor power required for a pure electric vehicle (PEV) to travel in straight-line motion at its maximum design speed umax on a smooth horizontal pavement within a closed test track is given by [8],

$$P_{max1} = \frac{1}{3600\eta_t} \cdot mgf \cdot u_{max} \tag{11}$$

The instantaneous peak power required for a pure electric vehicle to accelerate from standstill to terminal velocity ua is expressed as [8],

$$P_{max2} = \frac{1}{3600\eta_t}\left(\frac{mg \cdot f \cdot u_a \cdot t_m}{1.5} + \frac{\delta m u_a^2}{2\sqrt{t_m}}\right) \tag{12}$$

Through computational analysis, a brushless DC motor was selected with the following specifications: peak power is 2600 W; rated power is 1500 W; The maximum rotational speed n_{max} was calculated using Eq. (8) is 2947.2 r/min.

This value was rounded to 3000 r/min to provide a 1.8% operational margin for voltage fluctuations and transient overload conditions. The finalized drive motor parameters are summarized in Table 3.

Table 3. Motor Performance Parameters.

Parameter	Value	Unit	Design Basis
Peak Power	2600	W	Exceeds acceleration power demand (2.42 kW)
Rated Power	1500	W	Matches continuous operation at 40 km/h
Rated Speed	1500	r/min	Optimal efficiency point (89.7% per ISO 21782)
Maximum Speed	3000	r/min	Allows 1.8% overspeed tolerance

2.3 Technical Specifications for Power Battery Systems

The minimum operating voltage of the battery pack must be ensured to meet the minimum voltage requirement for the normal operation of the drive motor. Concurrently, the design of the power battery mandates that its maximum discharge current should not exceed 30 A. The energy output of a power battery pack is calculated using the following formula [8],

$$W_{ess} = \frac{U_{ess} \cdot C \cdot \varsigma_{soc} \cdot \eta}{1000} \tag{13}$$

In the formula, W_{ess} refers to the actual output energy of the battery pack (kWh); U_{ess} is the average voltage of the battery pack during normal operation; C represents the capacity of the battery pack (Ah), and ς_{soc} denotes the effective depth of discharge of the power battery, which is set to 0.8. The parameter setting premise for the power battery of electric tricycle vehicles is to meet the design requirement for driving range when the vehicle travels at a certain constant speed. It is assumed that the vehicle travels at a constant speed u_c in a closed area; at this time, the driving motor power should satisfy,

$$P_c = \frac{1}{3600\eta_t} \cdot mg \cdot f \cdot u_c \tag{14}$$

Then, the energy required for the electric tricycle vehicle to travel a distance S is,

$$W_c = P_c \cdot t = P_c \cdot \frac{S}{u_c} \tag{15}$$

Based on the vehicle parameters set in the previous text, the driving range of the vehicle under the 15 km/h constant-speed condition should be greater than 30 km. Through calculation, it is determined that only by connecting 16 battery cells in series with a battery pack capacity of 8.5 Ah can the energy provided by the power battery meet the power design requirements of the entire vehicle.

3 Modeling and Analysis of the Frame of the Electric Tricycle Vehicle

The designed electric tricycle frame adopts a side-beam welded frame structure, characterized by its modular front/rear sections and a central battery compartment. This configuration comprises five longitudinal beams, seven crossbeams, and multiple reinforcement struts, engineered to optimize structural integrity under dynamic loads. The main dimensional specifications of the frame are presented in the following Table 4, and the modeling is shown in Fig. 2.

Table 4. The Main Dimensional Specifications of the Frame.

Structural Element	Cross-Sectional Dimensions	Functional Significance
Rear Longitudinal Beams	40 mm × 100 mm	Maximized bending resistance for cargo/axle loads
Front Longitudinal Beams	40 mm × 60 mm	Balanced strength-to-weight ratio for steering dynamics
Crossbeams	40 mm × 60 mm	Enhanced torsional rigidity and crash energy absorption
Front-Rear Junction Struts	40 mm × 40 mm (×5 units)	Critical stress transfer points with uniform load distribution
Auxiliary Reinforcement Struts	40 mm × 30 mm	Local buckling prevention at high-stress nodes

Fig. 2. The model of the frame.

The material properties of the frame components are defined based on the model. According to the actual material used in the frame, its material property is defined as

Q235 steel, a type of metal. After defining the material properties, the relevant physical and geometric parameters of the frame components can be determined as shown in Table 5.

Table 5. Partial Physical and Geometric Parameters of the Frame

Parameter Name	Parameter Value	Parameter Name	Parameter Value
Total Material Volume	0.008 m^3	Centroidal Mass Moment of Inertia Ixx	30.627 kg·m^2
Material Density	7860 kg/m^3	Centroidal Mass Moment of Inertia Iyy	6.005 kg·m^2
Poisson's Ratio	0.266	Centroidal Mass Moment of Inertia Izz	33.302 kg·m^2
Young's Modulus	2 × 1011 Pa	Centroidal Mass Moment of Inertia Ixy	0.032 kg·m^2
Yield Strength	235 MPa	Centroidal Mass Moment of Inertia Ixz	0.002 kg·m^2
Mass	63.73 kg	Centroidal Mass Moment of Inertia Iyz	−1.639 kg·m^2
X-Coordinate of Centroid	0.978 mm	Principal Moment of Inertia $M1$	5.559 kg·m^2
Y-Coordinate of Centroid	−839.746 mm	Principal Moment of Inertia $M2$	30.627 kg·m^2
Z-Coordinate of Centroid	−285.39 mm	Principal Moment of Inertia $M3$	33.4 kg·m^2

To generate the finite element mesh, the "Finite Element Model Workspace" was initially accessed, followed by selecting the "Octree Tetrahedral Mesh" method. Key meshing parameters were set as follows: element size = 15 mm, absolute aspect ratio = 3 mm, and element type = linear elements. Upon completion, the model contained 37,564 nodes and 111,466 elements.

For the static load and torsional strength analysis of the tricycle frame, a coarser mesh size of 50 mm was adopted to balance computational efficiency and accuracy. The finite element mesh diagram is illustrated in Fig. 3. Material properties of Q235 steel were incorporated into the model to characterize its mechanical behavior under loading.

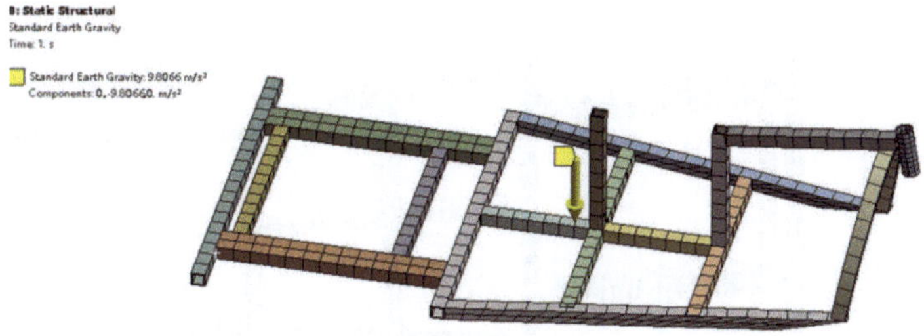

Fig. 3. Mesh Diagram of the Frame.

The static vertical load per wheel is specified as 150 kg, resulting in a total load of 450 kg for all three wheels. For torsional loading, each wheel experiences a torque of 300 N·m, with a total torsional load of 900 N·m across the three wheels. The application positions of the bearing loads are shown in Fig. 4, the application positions of the torsional loads are shown in Fig. 5, the positions of fixed constraints are shown in Fig. 6, and the addition of gravitational acceleration is illustrated in Fig. 7.

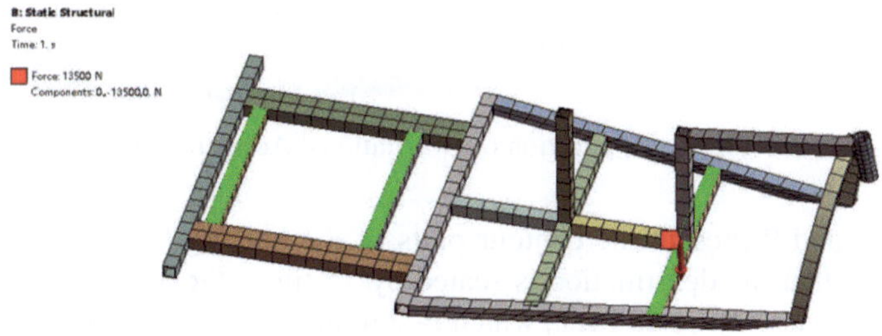

Fig. 4. Static Load Application Points.

Fig. 5. Torsional Load Application Points.

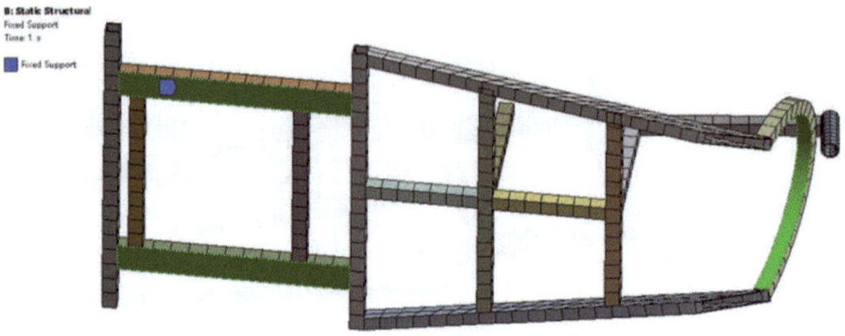

Fig. 6. Fixed Constraint Location.

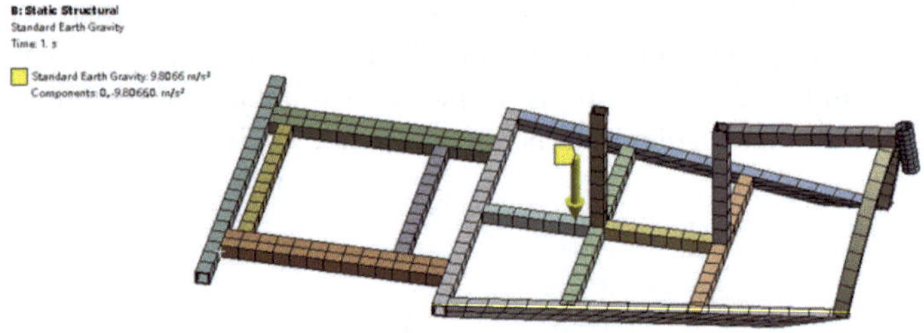

Fig. 7. Application of Gravitational Acceleration.

Figures 8 and 9 present the contour plots of the bearing load calculation results. It can be seen that the deformation is scaled by 75 times the actual deformation. The maximum stress is only 86.15 MPa, which is significantly lower than the yield strength of Q235 material (235 MPa). The maximum total deformation is 1.5 mm, indicating that the bearing strength under a 450 kg load is sufficient.

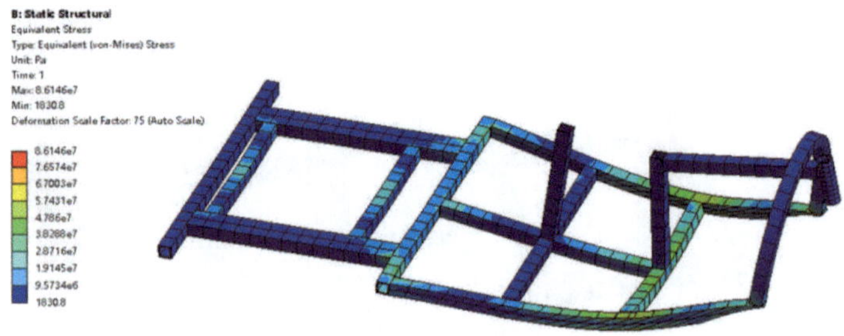

Fig. 8. Bearing Strength Stress Calculation Results.

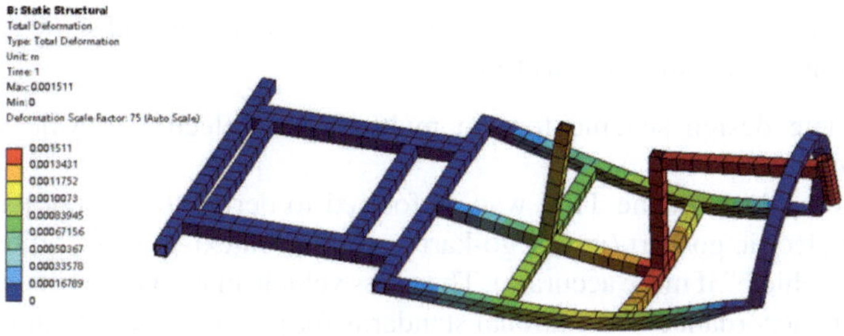

Fig. 9. Bearing Strength Deformation Calculation Results.

Figures 10 and 11 present the contour plots of torsional strength calculation results. It can be observed that the deformation is scaled by 66 times the actual deformation. The maximum stress is only 97.77 MPa, significantly lower than the yield strength of 235 MPa. The maximum deformation is 1.7 mm, indicating that the frame structure possesses sufficient strength under a 900 N·m bending moment.

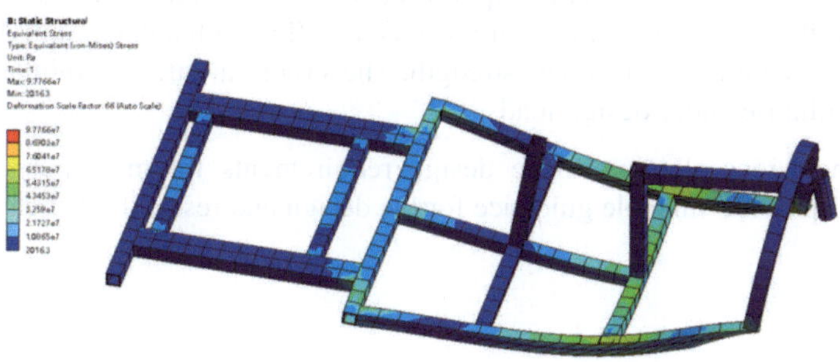

Fig. 10. Torsional Strength Stress Calculation Results.

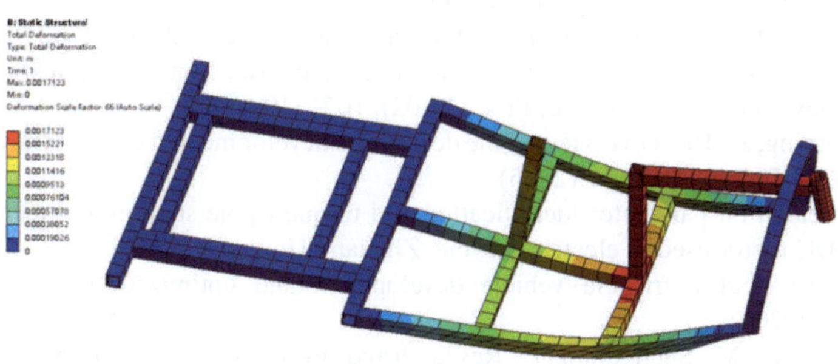

Fig. 11. Torsional Strength Deformation Calculation Results.

4 Conclusion

In this work, a multi-purpose electric tricycle vehicle (TEV) was designed, with matching of its powertrain parameters. Through computational analysis, the selection of the drive motor and battery module was determined. The overall design of the TEV's appearance and frame was conducted, followed by frame material selection and static analysis. The key conclusions are summarized as follows:

(1) A complete design scheme for the multi-purpose electric tricycle vehicle was developed.
(2) Dynamic analysis of the TEV was performed to derive the design parameters of the pure electric go-kart (note: "go-kart" may be context-specific; adjust to "three-wheeled vehicle" if more accurate). The gross vehicle mass (GVM) under full load is 450 kg. In accordance with national standards for pure electric vehicles' power performance, economic performance, and safety requirements, the designed TEV meets the following performance indicators: maximum speed ≤40 km/h; acceleration time from 0–30 km/h <10 s.
(3) Optimal parameters and selections for the drive motor and battery module were determined through computational calculations.
(4) Exterior design and frame modeling of the multi-purpose TEV were completed.
(5) The frame material was specified, and finite element analysis (FEA) was conducted. The results show: maximum total deformation under static load: 1.5 mm; maximum stress: 86.15 MPa (sufficient to support the 450 kg load);under a 900 N·m bending moment, the frame's maximum stress is 97.77 MPa, with a maximum deformation of 1.7 mm, confirming adequate strength. The structural safety is validated, with no risk of collapse under design loads.

The conclusions align with the design requirements for multi-purpose electric tricycles and provide valuable guidance for the design and research of such vehicles.

References

1. Xin, H.: Why domestically produced electric tricycles sell well overseas. Enterprise Observ. **06**, 118–119 (2024)
2. Zhu, R.: Chinese "Sanbengzi" gains popularity overseas: domestic electric tricycles must overcome these hurdles to go global. Mod. Logist. Daily (2024). 2024–12–02 (010)
3. Xiaofang, C.: Analysis and research on braking devices based on the application of electric tricycles. Dev. Innov. Mech. Electr. Prod. **37**(02), 107–110 (2024)
4. Chi, C., Haiping, Z.: Brief overview of the design and development of electric tricycle vehicles. Sci-Tech Dev. Enterp. **4**, 32–37 (2015)
5. Li, B.: Research on parameter identification and torque ripple suppression technologies for brushless DC motor used in electric tricycle. Zhejiang University (2022)
6. Zhao, J.: An electric tricycle vehicle development and optimization design. Shandong University (2020)
7. Ye., Yan, Jiankun, M., Shengqiang, Z.: Research and development of an intelligent SOC control system for electric tricycles. Inf. Syst. Eng. **01**, 16–19 (2024)
8. Zhisheng, Y.: Automobile Theory, 6th edn. China Machine Press, Beijing (2018)

Study on the Test Methods of Deicing Salt Corrosion for Automotive Components

Han-ying Luo[1], Zhao-tong Wang[1], Qing-xing Zheng[1], Wei-jian Luo[1], Tong Fu[1], and Jun Chen[1,2(✉)]

[1] CVC Testing Technology Co., Ltd., Guangzhou 510663, Guangdong, China
J-Chen@cei1958.com

[2] College of Chemistry and Chemical Engineering, Central South University, Changsha 410083, Hunan, China

Abstract. Given the significant environmental differences across different regions, automobile components are facing various corrosion failures. In extremely cold regions, the usage of deicing salt causes severe corrosive issues for the vehicles. To assess the corrosion effect of deicing salt accurately, the characteristics of common deicing salts and the corrosion mechanism of deicing salt on automobile components are comprehensively researched in this article. Furthermore, a systematic comparison of the cyclic corrosion test methods was conducted. Finally, the corrosiveness of the same deicing salt method to different components was compared through cyclic corrosion experiments.

Keywords: automotive components · deicing salt · corrosion mechanism · cyclic corrosion test

1 Introduction

According to the statistics from the General Administration of Customs of China, China's vehicle export volume exceeded 5.5 million in 2024, growing by approximately 12% year-on-year. With the rapid development of China's automobile manufacturing, vehicle export has become a current hot topic. However, due to the differences between domestic climate environments and those of exporting countries, China's vehicle export is still facing many problems. Exports to low-temperature regions like Russia make the cold resistance and corrosion resistance of automobiles key concerns for consumers. In these regions, deicing salt is a widely used chemical substance to melt ice and snow, mainly composed of NaCl, $CaCl_2$, $MgCl_2$, and KCl. Due to its simplicity of operation and low cost, chloride-based deicing salt has become a common method to melt ice and snow on roads [1]. It is estimated that global consumption of deicing salt exceeds 20 million tons annually. In Northern China, the annual usage of deicing salt amounts to 3 million tons, and this figure is increasing year by year. The extensive usage of deicing salt not only causes negative impacts on the environment and infrastructure but also brings serious corrosion problems to automobiles, especially posing multiple corrosion threats to metal parts, electronic systems, and coatings [2].

S. Halgamuge et al. (Eds.): ICACMVE 2025, LNME, pp. 232–243, 2026.
https://doi.org/10.1007/978-981-95-6911-3_23

Current research on deicing salts mainly involves ecology, environment, industry, transportation, materials, biology, and other fields [3]. As for corrosion research on deicing salts, domestic and international studies primarily focus on the corrosion mechanism and testing methods of chloride-based deicing salts. Li et al. [4] used dry-wet alternating and full-immersion corrosion tests to assess the corrosion resistance of bridge steel. Weathering bridge steels Q345qENH and Q420qENH exhibited significantly better corrosion resistance than the comparative specimen Q345qE in deicing salt corrosion environments. Moreover, different structural parts of the same bridge steel have undergone different degrees of corrosion under the corrosive effect of deicing salt. The phase structure and content of the rust layer also changed over time in the dry-wet alternating corrosion test. Shi et al. [5] used immersion and electrochemical methods to study the corrosion behavior of automotive galvanized steel sheets in simulated snow water solutions containing different chloride ion concentrations from deicing salts. Snow water solutions containing chloride-based deicing salts were more corrosive than pure snow water. As the Cl^- content increased, the average corrosion rate of automotive galvanized steel slowly increased. The corrosion rate was about half that of bare steel, indicating that corrosion resistance is significantly improved. Li et al. [6] analyzed the corrosion impact of different deicing salts on metals using the rotating coupon and dry-wet alternating periodic immersion methods. The results showed that both chloride-based and acetate-based deicing salts have certain corrosivity on metals, but acetate-based salts cause far less corrosion to carbon steel than chloride-based salts. Additionally, ambient temperature and changes between dry and wet environments significantly affect metal corrosion rates. Chongqing Changan Automobile conducted comparative analysis between outdoor exposure tests and under-shed tests at the Mohe Testing Station to study the impact of deicing salts on automotive component corrosion [7].

To provide technical insights and a reference basis for the deicing salt corrosion testing of automotive components, this study compares the characteristics of the main deicing salts and analyzes the corrosion mechanism of deicing salts. Then, the current deicing salt corrosion standards of the automotive industry are detailed. Furthermore, a comparative experiment is conducted to verify the corrosion difference of the same method to different components.

2 Characteristics Comparison of the Main Types of Deicing Salts

Deicing salts are diverse, mainly including chloride-based inorganic salts such as NaCl, $CaCl_2$, $MgCl_2$, and KCl, environmentally friendly organic deicing salts, and compound deicing salts. Deicing salts can melt snow by lowering the freezing point, disrupting ice crystal structure, and breaking vapor pressure equilibrium. Among them, chloride-containing deicing salts, due to their strong deicing ability and low cost, are widely used on roads. It is verified that rising ambient temperature can strengthen the snow-melting capability of deicing salts, and increasing the amount of deicing salt can increase the ice-melting rate. Moreover, calcium chloride demonstrates better snow-melting effects compared to others [8].

Given the widespread use of deicing salts, it is necessary to understand their characteristics in terms of snow-melting capability, cost, corrosivity, and environmental friendliness. The common deicing salts are classified and summarized in Table 1 to understand their corrosion impact.

Table 1. Characteristics and application field of common deicing salts

Type	Characteristics	Application field
Sodium Chloride (NaCl)	Low cost; freezing point can be lowered to −10 °C; strong corrosivity; harmful to roads, infrastructure, vegetation, and groundwater	Widely used for snow melting; usage can be reduced through mechanical snow removal
Calcium Chloride ($CaCl_2$)	Better snow-melting effect; freezing point can be lowered to −20 °C; lower corrosivity than sodium chloride; high cost	Suitable for low-temperature environments, bridges, airports, and other corrosion-sensitive areas
Magnesium Chloride ($MgCl_2$)	Freezing point can be lowered to −15 °C; low corrosivity; strong hygroscopicity (reduces dust); high cost	Often used in areas with high environmental requirements or in compound formulations
Potassium Acetate (CH_3COOK)	Freezing point can be lowered to − 30 °C; very low corrosivity; great environmental friendliness (biodegradable); very high cost	Mainly used for high-demand locations and environmentally sensitive areas, such as airport runways, golf courses, and bridges, as well as water source protection zones, urban green belts, etc

3 Analysis of the Corrosion Mechanism of Deicing Salt

Due to the complex vehicle operating environments, the hazards caused by deicing salt corrosion in winter cannot be ignored. Vehicles exported to extremely cold regions face the danger of low-temperature deicing salt corrosion. After melting ice and snow, deicing salts form strong electrolyte solutions containing Na and chloride ions. Chloride ions possess extremely strong penetration ability, and they are capable of penetrating metal passive films and causing electrochemical corrosion with metal materials, which accelerates pitting and intergranular corrosion in steel, aluminum alloys, etc. Metal materials play an important role in automobile manufacturing. The metal materials include high-strength steel, aluminum alloy, titanium alloy, magnesium alloy, etc. [9].

Q235 steel has good plasticity, toughness, and weldability, and is commonly used in automobile manufacturing for parts like chassis, frames, and body panels. In a salt-spray environment, its corrosion products mainly consist of α-FeOOH, β-FeOOH, γ-FeOOH, and Fe_2O_3 [5]. Chloride ions penetrate the Fe_2O_3 passive film to form soluble $FeCl_2{\cdot}4H_2O$.

In snow water containing deicing salt, the corrosion rate of galvanized steel is about half that of bare steel [10]. The zinc coating initially forms the corrosion product $Zn(OH)_2$ in Cl^- solution, which has very low solubility and can deposit on the material surface. The deposit continues to decompose, so it can be anticipated that after a certain period of corrosion, the zinc coating forms a thin, passive-like film composed of $Zn(OH)_2$ and ZnO. Due to its high resistance, this film hinders further corrosion of the zinc coating, which is the main reason for its lower corrosion rate compared to bare steel. Because chloride ions have a small radius and strong adsorption capacity, they continue to react with $Zn(OH)_2$ to form the soluble corrosion product $Zn(OH)_xCl_{(2-x)}$. The corrosion reaction equations for galvanized steel are as follows:

$$\text{Anodic reaction: } Zn \rightarrow Zn^{2+} + 2e^- \tag{1}$$

$$\text{Cathodic reaction: } 1/2O_2 + H_2O + 2e^- \rightarrow 2OH^- \tag{2}$$

$$Zn^{2+} + 2OH^- \rightarrow Zn(OH)_2 \tag{3}$$

$$Zn(OH)_2 \rightarrow ZnO + H_2O \tag{4}$$

$$Zn^{2+} + xOH^- + (2-x)Cl^- \rightarrow Zn(OH)_xCl_{(2-x)} \tag{5}$$

Aluminum alloy materials undergo oxidation with oxygen in the atmosphere and generate aluminum oxides firstly. Then, in a Cl^--containing corrosive environment, Cl^- penetrates the oxide film and reacts with the aluminum substrate to form soluble aluminum chlorides. Since AlCl is soluble, the main final corrosion products of aluminum alloys are aluminum oxides such as Al_2O_3 and AlOOH [11].

4 Comparison of Current Test Methods for Corrosion of Deicing Salt on Automotive Components

To evaluate the degree of corrosion caused by deicing salts on vehicle components, corresponding corrosion testing standards have been established, such as ISO 9227 [12] and GB/T 10125 [13]. The salt-spray corrosion test is currently the primary standard method to evaluate the corrosion situation caused by deicing salts, including neutral, acetic acid, copper-accelerated acetic acid, and alternating salt spray methods.

Table 2. Comparison of the mainstream cyclic corrosion methods

Main parameter	GMW 14872	CEPT00.00 -L-467	ISO 11997-1 Cycle B	PV 1210	5100Z-TRO -6000	NES M0158
Salt solution	0.9% NaCl, 0.1% $CaCl_2$, 0.075% $NaHCO_3$	0.5% NaCl	5% NaCl	5% NaCl	5% NaCl	5% NaCl

(continued)

Table 2. *(continued)*

Main parameter	GMW 14872	CEPT00.00 -L-467	ISO 11997-1 Cycle B	PV 1210	5100Z-TR0 -6000	NES M0158
pH	No requirements	No requirements	6.5~7.2	6.5~7.2	6.5~7.2	6.5~7.2
Solution application method	salt rain	salt rain	salt fog	salt fog	salt fog or salt immersion	salt fog or salt immersion
Salt spray stage	25 ± 3 °C, 45 ± 10% RH	25 ± 2 °C, ≈100% RH	35 ± 2 °C, ≈100% RH	35 ± 2 °C, ≈100% RH	1) 35 ± 2 °C, ≈100% RH; 2) 40 °C, immersion	1) 35 ± 2 °C, ≈100% RH; 2) 50 ± 2 °C, ≈100% RH; 3) 60 ± 2 °C, ≈100%RH
Humidity stage	49 ± 2 °C, ≈100% RH	40 ± 2 °C, 95% RH	40 ± 2 °C, ≈100% RH	40 ± 2 °C, ≈100% RH	50 ± 2 °C, 95 ± 5% RH	1) 50 ± 2 °C, 95 ± 5%RH; 2) 60 ± 2 °C, 95 ± 5%RH
Drying stage	60 ± 2 °C, ≤ 30% RH	50 ± 0.6 °C, 70 ± 3% RH	23 ± 2 °C, 50 ± 20% RH	23 ± 2 °C, 50 ± 5% RH	60 ± 2 °C, <30% RH	1) 60 ± 2 °C, <30% RH 2) 80 ± 2 °C, <30% RH
Weekend storage phase	≤35 °C, ≤55% RH	50 ± 0.6 °C, 70 ± 3% RH	23 ± 2 °C, 50 ± 20% RH	23 ± 2 °C, 50 ± 5% RH	Not involved	Not involved
Longest test period	95d	84d	84d	84d	80d	60d
Corrosion coupons	Test together with the sample	No	No	No	No	No

However, the actual condition for deicing salt corrosion involves that when the vehicles are driving on low-temperature roads, snow water or mud slurry containing deicing salts splashes onto the chassis and lower body components. Then, these components experience alternating conditions of low temperature, high temperature, high humidity, or ambient temperature/humidity, thus accelerating corrosion. The above standards only aim to simulate the influence of saline solution wetting but do not account for low-temperature freeze-thaw cycles, temperature gradients, or the coupling effects of other corrosive media. Hence, to simulate the actual corrosion environment of vehicles accurately, cyclic corrosion testing methods are widely used by foreign automobile companies, which include salt spray, temperature, humidity, condensation water, high temperature and drying, and low temperature tests. However, each foreign automotive company has their own cyclic corrosion test standards. For example, General Motors establishes a cyclic corrosion laboratory test standard, GMW 14872 [14]. Ford Motor develops an accelerated cyclic corrosion standard, CEPT00.00-L-46 [15]. Mercedes-Benz and BMW quote ISO 11997-1 [16], while Volkswagen adopts recirculating corrosion standard PV1210 [17]. Honda Motor and Nissan Motor refer to 5100 Z-TR0-6000 [18] and NES M0158 [19], respectively. These standards generally include salt spray, drying, and wet heat tests, but the sequence of each stage, duration, spraying method, and temperature and humidity conditions are all different. Due to issues such as lack of foundational research and long development cycles for new methods, domestic automobile companies find it difficult to establish a cyclic corrosion standard that fits the local environment and operating conditions. Most companies generally refer to the above standards from foreign companies. For example, SAIC, GAC Passenger Cars, and Great Wall refer to GMW 14872 for cyclic corrosion testing, while Li Auto refers to Volkswagen's standards and Geely refers to Volvo's standards. Given that the testing methods are diverse and not unified, it is of great necessity to conduct a comprehensive analysis and comparison of the mainstream cyclic corrosion testing methods, shown in Table 2.

As shown in Table 2, most standards use 5% NaCl as the corrosive solution with a pH of 6.5~7.2. The solution application methods include salt rain, salt fog, and immersion, combined with alternating cycles of dry and wet conditions. It can be found that during the salt spray stage, American standards apply salt solution using the salt rain method, while German and Japanese standards use the traditional salt fog method. The salt rain method mainly simulates situations where the corrosive media, such as deicing salts, splashes onto vehicles when the vehicles are driving. Whereas, the salt fog method simulates the dispersed atmospheric environment around vehicles, especially marine environments. According to Chen et al. [20], different application methods of the salt solution significantly affect the results of cyclic corrosion tests. Compared to the salt fog method, the immersion and salt rain methods result in much more severe corrosion for bare substrate specimens. However, the corrosion levels caused by the salt rain and immersion methods are quite similar. Additionally, during the salt spray stage, the temperature in American standards is 25 °C, while it is 35 °C in German standards. As for Japanese standards, the maximum temperature is 60 °C. For every 10 °C increase in test temperature, the reaction rate increases by approximately 2–3 times. However, when the temperature becomes too high, the oxygen content involved in the reaction decreases, which further alters the corrosion mechanism, resulting in inconsistency with real-world conditions. As for the humidity phase parameter, German standards set them as 40 ± 2 °C and ≈100% RH, based on ISO 6270-2. It simulates dew formation on vehicle bodies or components when encountering significant environmental temperature differences. This standard is usually used to evaluate the moisture and heat resistance of metal or non-metallic substrate coatings. Under these conditions, the failure mode is blistering, characterized by the decrease of coating adhesion. However, American standards set the conditions of this phase as 49 ± 2 °C and ≈100% RH, while Japanese standards' conditions are set as 50 °C or 60 °C and 95 ± 5% RH. To simulate the situation of vehicle parking, different standards set varying temperature and humidity storage or drying conditions based on the local climate environment and parking scenarios. American automotive companies, such as GM, set storage conditions as 25 °C and 45% RH, while Mercedes-Benz, BMW, and Volkswagen in Germany set standard storage conditions as 23 °C and 50% RH. Whereas, Japanese standards do not include storage conditions.

Based on the above analysis, factors such as temperature and humidity, transition times between conditions, corrosive medium composition, application method, and salt solution pH value all affect the test results. However, the parameters and indicators for cyclic corrosion tests vary in these standards. Each standard has its unique characteristics and advantages. Parameters for cyclic corrosion tests should be tuned to better reflect the corrosion characteristics of the service environment. For example, the application of the corrosive medium aims to ensure sufficient contact between the salt solution and the component surface, allowing chloride ions to cause electrochemical corrosion with the metal material. As for different components, the saline solution application method should be chosen based on the actual usage environment, e.g., spraying is recommended for the upper interior cabin, rinsing is suitable for the lower cabin, while immersion is fit for body parts below the grille area. Moreover, through adding cycles of wet/dry conditions, the corrosion rate can be accelerated.

5 Results and Analysis

To verify the adaptability of the same testing method on different automotive components, three types of stainless steel were selected to differentiate their deicing salt corrosion effect using the same corrosion method through the cyclic corrosion test.

5.1 Experimental Materials and Equipment

In this experiment, three types of stainless steel, 409, 436, and 439, were selected to represent three different vehicle parts. The stainless steel was cut into dimensions of 75 × 150 × 1.2 mm. Among them, material A is the domestic target material, while material B is the imported control material. The chemical composition of the materials is shown in Table 3. The equipment used in the experiment includes the ATLAS CCX2000 cyclic corrosion test chamber, an analytical balance, and a double-point micrometer.

Table 3. Chemical composition of test materials (mass percentage, %)

Steel type	C	Si	Mn	P	S	Cr	Ni	N	Ti	Mo
409-A	0.0085	0.4440	0.1770	0.0145	0.0010	11.3180	0.0795	0.0060	0.2430	-
409-B	0.0065	0.4760	0.3420	0.0210	0.0010	10.9210	0.0720	0.0060	0.1480	-
436-A	0.0080	0.1480	0.0960	0.0180	0.0010	17.7910	0.0990	0.0075	0.2630	0.9845
436-B	0.0130	0.0530	0.1370	0.0260	0.0010	17.3780	0.1130	0.0140	0.2840	1.1730

(continued)

Table 3. *(continued)*

Steel type	C	Si	Mn	P	S	Cr	Ni	N	Ti	Mo
439-A	0.0095	0.2100	0.1125	0.0215	0.0010	17.5995	0.0825	.0115	0.2955	-
439-B	0.0050	0.0920	0.0730	0.0165	0.0015	17.2300	0.0940	0.0110	0.1950	-

5.2 Experimental Method

The experimental method refers to ISO 14993:2018 [21]. The test conditions are listed in Table 4.

Table 4. Test conditions in ISO 14993:2018

Stage	Item	Parameter	Duration
Salt water spraying	Temperature	35 ± 1 °C	2 h
	Concentration of salt	5 ± 1%	
Drying	Temperature	60 ± 1 °C	4 h
	Relative humidity	20~30% RH	
Wetting	Temperature	50 ± 1 °C	2 h

Before the test, clean the sample with acetone (or ethanol) to remove oil firstly, and then dry and weigh it. Finally, place it in the cyclic corrosion test chamber and begin the test. After the test, remove the sample and rinse it using flowing water with a temperature not exceeding 40 °C, and then dry it immediately. Then, conduct an appearance inspection, take photos, and then remove the corrosion products. Finally, measure the pitting depth, weigh it again, and record weight loss.

5.3 Experimental Results

As shown in Figs. 1, 2 and 3 and Table 5, stainless steel 409 exhibited uneven pitting corrosion, with rust continuously growing inward, as shown in Fig. 1. The maximum pitting corrosion depth of 409-A is 0.137 mm, and the maximum mass loss is 0.152 g. As for 409-B, the maximum pitting corrosion depth is 1.110 mm, and the maximum mass loss is 0.490 g. However, stainless steels 436 and 439 showed slight rusting after 4000 h of cyclic corrosion testing, with the insignificant pitting.

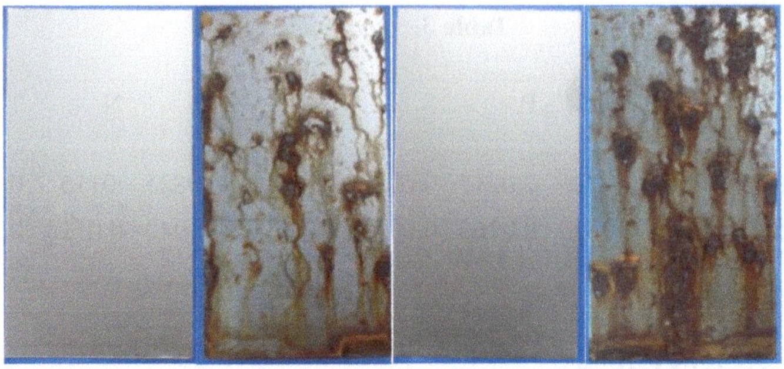

(a) (b) (c) (d)

Fig. 1. Cyclic corrosion test results (a) 409-A Before test; (b) 409-A After 800h; (c) 409-B Before test; (d) 409-B After 800h

(a) (b) (c) (d)

Fig. 2. Cyclic corrosion test results (a) 436-A Before test; (b) 436-A After 4000h; (c) 436-B Before test;(d) 436-B After 4000h

(a) (b) (c) (d)

Fig. 3. Cyclic corrosion test results (a) 439-A Before test; (b) 439-A After 4000h; (c) 439-B Before test;(d) 439-B After 4000h

Table 5. Test result of cyclic corrosion test

Steel type	Test time (h)	Max. Pit depth (mm)	Max. Mass loss(g)
409-A	800	0.137	0.152
409-B	800	1.110	0.490
436-A	4000	0.011	0.017
436-B	4000	0.014	0.037
439-A	4000	0.022	0.037
439-B	4000	0.024	0.104

6 Conclusion

In this study, the analysis of the characteristics and corrosion mechanisms of different deicing salts was conducted firstly. Then, the characteristics of different deicing salt corrosion test standards are compared. Finally, three types of stainless steel were selected to differentiate their deicing salt corrosion effect using the same corrosion method. The following conclusions can be drawn:

(1) Different deicing salts have significant differences in their ability to melt snow and ice. Currently, most corrosion studies focus on chloride-based deicing salts. The concentration should be determined based on the average annual amount of deicing salt spread in the specific region as a reference value.
(2) Temperature and humidity, transition times between conditions, corrosive medium composition, application method, and salt solution pH value all affect the test results. Parameters for cyclic corrosion tests should be tuned to better reflect the corrosion characteristics of the service environment.
(3) The three types of stainless steel demonstrate varying results under the same testing method. In addition to saline solution rinsing, temperature/humidity cycles, drying, and ambient storage conditions, the test cycle needs to include a low-temperature storage phase. Taking all of the above factors into consideration, the test cycle will better simulate the actual environment.

In the future, the correlation between laboratory accelerated corrosion test methods and actual operating conditions will be compared with low-temperature deicing salt road enhancement corrosion tests. Through intelligent corrosion detection methods and big data applications, a set of test standard methods with good effectiveness and low cost will be established.

Acknowledgement. This paper was supported by the CVC Testing Technology Co., Ltd. Class C Science and Technology Project "Research on Anti-Corrosion Methods for Exported Automotive Products in Specific Environments" (Project Number: SCRV2024031).

References

1. Liu, S., Hao, Y.: The use of de-icing salts in typical cities in china and north America and their impact on automotive corrosion. Shanghai Automob. **10**, 22–28 (2022)
2. Wang et al.: Corrosion mechanisms of automotive alloys in chloride environments. Nat. Mater. (2023)
3. Yang, F., Li, G., Cao, Y., Li, Y., Huang, Z., Han, Y.: Literature analysis of R&D and application of deicing agents at home and abroad in recent ten years. J. China Foreign Highw. **42**(08), 272–278 (2022)
4. Li, L., et al.: Study on corrosion resistance of bridge steel in de-icing salt environment. J. Chin. Soc. Corros. Prot. **40**(10), 448–453 (2020)
5. Shi, Y., Liang, P., Liu, F.: Effect of deicing salts on corrosion behavior of galvanized steel for automotive use. Mater. Prot. **47**(03), 53–55 (2014)
6. Li, H., Ren, X., Liu, Y., Yang, Y., Li, J., Cai, X.: Study on the corrosive effect of de-icing agents on different metals. Shanxi Chem. Ind. **43**(02), 13–15 (2023)
7. Huang, P., Yu, Y., Fan, M.: Research on test methods for corrosivity of deicing agents on automotive parts
8. Lei, Q., Shan, Q., Liao, J., Wang, Z.: Analysis of influencing factors on snow-melting effects of common de-icing salts and research on usage standards. Sichuan Cement **10**, 84–88 (2024)
9. Wang, C.: Discussion on the development trend of automotive metal materials. Auto Maintenance Technician **04**, 125–127 (2024)
10. Ning, S., Shi, Y., Liang, P., Lin, L.: Corrosion behavior of automotive Q255 steel in snow water containing deicing salt. J. Liaoning Shihua Univ. **35**(02), 58–66 (2015)
11. Zhao, P., Wen, L., Guo, W., Junsheng, W.: Accelerated corrosion behavior and mechanism of high-strength aluminum alloy under cyclic salt spray. Surf. Technol. **51**(10), 260–268 (2022)
12. ISO 9227: Corrosion tests in artificial atmospheres-Salt spray tests
13. GB/T 10125: Corrosion tests in artificial atmospheres-Salt spray tests
14. GMW 14872-2022: Cyclic corrosion laboratory test
15. CEPT_00.00-L-467: Global laboratory accelerated cyclic corrosion test
16. ISO 11997-1:2017: Paints and varnishes — Determination of resistance to cyclic corrosion conditions — Part 1: Wet(salt fog)/dry/humidity
17. PV 1210-2016: Body and Add-on Parts Corrosion Test
18. 5100Z-TR0-6000: Quality standard of coating for suspensions and the related parts of automobiles
19. NES M0158: Methods of compound corrosion test
20. Chen, L., Shen, Q.: Research on application of different saline solution application methods in cyclic salt spray test. In: Proceedings of 2019 SAE-China Congress
21. ISO 14993:2018: Corrosion of metals and alloys -- Accelerated testing involving cyclic exposure to salt mist, dry and wet conditions

Failure-Triggering Scenario Synthesis via Multiphysics Integration for Aircraft Seat Certification Under Dynamic Impacts

Yafeng Wang(✉) and Jijun Liu

National Key Laboratory of Strength and Structural Integrity, Aircraft Strength Research Institute of China, Xi'an, China
Wangyafeng623@163.com

Abstract. This research pioneers a multiphysics-fused critical scenario synthesis framework to resolve the certification challenge of identifying extreme dynamic loading conditions for aircraft seats during emergency landings. In compliance with FAA AC 25.562-1B and SAE AS8049B, we formulated a tri-axial parametric domain encompassing seat compliance characteristics, floor pre-distortion envelopes, and ATD deployment topologies. We develop a high-fidelity multibody FE model in Abaqus/Explicit to capture seat-track-ATD coupling dynamics, incorporating nonlinear material behavior and contact interactions. Through pre-deformation sensitivity screening and dynamic impact prognostics, the framework achieved a 62.5% reduction in computational burden (from 24 to 9 cases) while constraining predictive errors below 3.3%. Validation under dual-seat critical scenarios (10° right yaw + dual ATDs) reveals 913 MPa von Mises stress localization in leg-lock mechanisms, triggering incipient structural failure. This work establishes a certification-accelerated paradigm that synergizes failure mechanism prognostics with regulatory compliance, demonstrating 25.7% higher efficiency than conventional methods for next-generation seat safety validation.

Keywords: Multiphysics fusion · Incipient failure prognostics · Airworthiness certification acceleration · Seat-track-ATD coupling · Critical scenario synthesis

1 Introduction

Aircraft seats are vital safety-critical components for ensuring passenger safety during emergency landings of transport-category aircraft, as their dynamic protective performance directly influences both occupant safety and airworthiness certification efficiency. According to airworthiness regulations AC 25.562-1B [1, 2] and SAE AS8049B [3], dynamic impact tests must verify the structural reliability of seats under the most critical conditions. AC 25.562-1B mandates that seats must undergo vertical and longitudinal impact tests to demonstrate the continuity of load paths and compliance with human injury criteria (e.g., head injury criterion (HIC) and shoulder belt elongation). SAE AS8049B emphasizes that test configurations must be selected to maximize structural stress, ensuring critical load-bearing components withstand extreme conditions.

S. Halgamuge et al. (Eds.): ICACMVE 2025, LNME, pp. 244–255, 2026.
https://doi.org/10.1007/978-981-95-6911-3_24

Conventional methods for critical condition selection rely on static loading analysis [4, 5], which inherently introduces discrepancies and limits efficiency due to the inability to capture dynamic effects. To address these limitations, this study develops a coupled multi-parametric screening methodology, integrating finite element modeling, dynamic analysis, and multi-parametric extreme value screening. We developed an integrated seat-floor-ATD dynamic response model to systematically evaluate parameters such as seat deflection, floor pre-deformation, and ATD distribution. Sensitivity analysis and optimization algorithms were employed to identify the most critical test conditions.

The proposed methodology was validated through engineering practice. In a case study of a dual-seat configuration, the selected critical condition (right yaw of 10° combined with dual ATD loading) resulted in a maximum von Mises stress of 913 MPa in the seat leg lock, accurately capturing structural failure mechanisms. This study innovatively combines multi-parametric analysis with airworthiness validation, providing a systematic framework for aircraft seat dynamic impact testing. By enhancing the efficiency of condition screening and structural reliability assessment, the method significantly reduces certification cycles while ensuring compliance with regulatory requirements. The approach offers substantial engineering value for optimizing seat design and safety certification processes.

2 Airworthiness Regulations and Technical Standards Study

According to AC 25.562-1B, aircraft seats, restraint systems, and adjacent aircraft structures must undergo dynamic testing to verify their ability to protect occupants during impact. Specifically, two dynamic tests are mandated:

(1) Vertical Impact Test (Test 1): Evaluates the seat's performance under vertical acceleration, considering pitch deformation of the seat.
(2) Longitudinal Impact Test (Test 2): Simulates yawed landing conditions and floor deformation (pre-deformation of the seat on the floor by ±10° roll or pitch on either side) to assess structural integrity and occupant protection.

SAE AS8049B mandates selecting the most critical test configuration to ensure structural robustness. Key requirements include:

(1) Maximizing Structural Stress: For multi-seat configurations, ATD placement and seat yaw direction must be optimized to induce the highest stress in critical components (e.g., seat leg locks).
(2) Floor Pre-deformation: Deformation protocols shall maximize loads on primary load-bearing components (e.g., seat legs or tracks) to simulate extreme conditions.

The selection of critical test conditions must follow these principles:

(1) Multi-parametric coupling: For longitudinal impact, it must consider floor pre-deformation (±10° roll/pitch, ATD distribution (single/dual occupants), and yaw direction (left/right 10°). Vertical Impact: Requires ATD distribution analysis (single/dual occupants) without floor deformation.
(2) Theoretical basis: Interface loads between the seat and aircraft structure can be determined via static or dynamic seat/occupant analysis, serving as the foundation for selecting critical configurations.

3 Finite Element Modeling of Seat-Floor-ATD Coupled System

3.1 Modeling Principles and Component Discretization

In this section, we construct a multi-body coupled finite element model that integrates the seat assembly, floor structure, and ATD restraint system, accounting for the seat's complex configuration and impact scenarios. The modeling methodology adheres to well-established engineering principles [6, 7]:

(1) Structural Integrity: Preserve original assembly relationships and critical joints to ensure continuity of load paths;
(2) Hierarchical Precision Design: Use solid elements (e.g., C3D8, C3D10M) for high-stress regions (seat leg locks, slides) and shell elements (e.g., S4, S3) for non-critical areas (seat frames) to balance computational efficiency and accuracy;
(3) Consistency in boundary conditions: Accurately simulate component connections via joints, binding (Tie), and coupling (Coupling) constraints;
(4) Fidelity of load transmission paths: Match nodal constraints with actual mounting interfaces to prevent error accumulation from model simplifications.

As shown in Fig. 1, the dual-seat assembly is designed to withstand vertical (Test 1) and longitudinal (Test 2) impact loads. The finite element model employs a hybrid modeling strategy:

(1) Solid Elements: For load-critical components like seat leg locks (C3D8/C3D10M) and slides (C3D6) to capture localized stress concentrations;
(2) Shell Elements: For thin-walled structures (seat frames, seat pans) to optimize computational cost vs. accuracy trade-off.

Fig. 1. Finite element model of Aviation seat

3.2 Material Properties and Coordinate System

Material properties are summarized in Table 1.

Table 1. Material properties of seat components

Material	Model Type	Young's Modulus (E)	Poisson's Ratio (ν)	Density (ρ)
Aluminum Alloy	Bilinear Kinematic Hardening	70,000 MPa	0.33	2,700 kg/m^3
30CrMnSi Steel	Linear Elastic	210,000 MPa	0.33	7,900 kg/m^3

Baseline loads include structural self-weight and ATD masses (two 75 kg dummies applied as non-structural masses on seat pans). Impact loads comprise floor pre-deformation displacements (±10° roll/pitch) and vertical/longitudinal accelerations.

Origin: Intersection of the right-side seat foot and floor plane; X-axis: Aligned with the aircraft longitudinal direction; Y-axis: Vertical upward; Z-axis: Defined via the right-hand rule.

3.3 Boundary Conditions and Interaction Modeling

The coupled multi-body system is implemented through:

(1) Component interactions: Joint constraints simulate sliding contact between seat slides and floor tracks; Tie constraints: Enforce displacement continuity at interfaces (e.g., seat pan-leg connections); Coupling constraints (Coupling): Synchronize displacement at slide control points to ensure uniform load transmission.
(2) Boundary Conditions: Floor fixation: Full constraints (UX/UY/UZ = 0) applied to the floor model; ATD coupling: Rigid 6-degree-of-freedom (6-DOF) connections link ATD models to seat pans to capture occupant injury responses [8, 9].

3.4 Model Verification and Validation

Model validity is ensured through:

(1) Mesh Independence Verification: Grid convergence analysis was performed for the critical leg lock mechanism (referenced in Fig. 1). Von Mises stress under the representative longitudinal impact condition was evaluated across five progressively coarser mesh sizes ranging from 1.0 mm to 5.0 mm (Table 2). The results demonstrate that a mesh size of 3.0 mm achieves satisfactory convergence, exhibiting a von Mises stress of 707.1 MPa with a minimal deviation of only 1.4% compared to the 1.0 mm reference solution (717.2 MPa). Computational efficiency is significantly enhanced, with CPU time reduced by 68% (from 14.5 h to 4.6 h). Coarser meshes (4.0 mm and 5.0 mm) induced deviations exceeding 5% (3.9% and 6.2%, respectively), which violates typical engineering accuracy thresholds. This rigorous analysis confirms the selection of C3D8 element size ≤3 mm for critical regions as optimal for balancing computational cost and solution fidelity.;

Table 2. Mesh Convergence Analysis for Leg Lock Mechanism

Mesh Size (mm)	Von Mises Stress (MPa)	Deviation (%)	CPU Time (h)
1.0	717.2	Reference	14.5
2.0	712.5	0.7	6.2
3.0	707.1	1.4	4.6
4.0	689.3	3.9	3.1
5.0	672.8	6.2	2.4

(2) Static Benchmarking: Validate load paths and stress distributions under non-impact conditions;
(3) Dynamic Correlation: Compare simulation vs. test data for peak stress in seat leg locks under vertical/longitudinal impacts ($\leq$5% error) [10, 11].

4 Selection Methodology for the Most Critical Loading Cases in Aviation Seats

4.1 Two-Stage Screening Framework

To accurately identify the most critical loading cases for aviation seat dynamic impact testing (encompassing both vertical and longitudinal impacts), this study employs a staged Finite Element (FE) analysis [12, 13] approach. The core methodology involves two steps: first, determining the state inducing the most critical stress arising solely from floor pre-deformation (Stage 1); subsequently superimposing impact loads onto this pre-deformed state while considering the effects of yaw offset and occupant (ATD) configuration to screen for the ultimate most critical dynamic impact case (Stage 2). The evaluation criteria prioritize the peak load sustained by the rear chair leg locking mechanism under impact and the maximum Von Mises stress within the primary seat structure. These metrics are directly critical to the integrity of the seat-to-floor attachment and the survivability of the main load-bearing structure during impact, serving as key indicators for structural failure risk.

It is noteworthy that preliminary analysis revealed the stress level in the chair leg locking mechanism induced by ATD self-weight (75 kg) is relatively low (~20 MPa), significantly lower than stresses induced by floor pre-deformation and impact loads. Furthermore, its influence on stress comparison across loading cases is negligible. Therefore, a dual-occupant configuration was adopted in both stages. Stage 1 examined eight pre-deformation cases, and Stage 2 examined the impact cases, reducing the total workload.

The screening process for the most critical loading case comprises two distinct stages, with their objectives, variables, and total case counts summarized in Table 3:

Table 3. Two-Stage Screening Methodology for Aviation Seat Critical Conditions

Stage	Objective	Initial State	Variables	Cases	Metric	Critical Case
Stage 1	Identify max-stress pre-deformation state	Seat mounted, dual ATDs	Roll (±10°) & pitch (±10°) combinations on rails A/B (asymmetric)	8	Max σ_vm in leg lock	Ax + By +
Stage 2	Determine critical dynamic impact case	Fixed Ax + By + pre-deform	Longitudinal: Yaw offset (L/R 10°) Vertical: ATD config (dual only)	Long: 6 Vert: 3	• Max leg lock load • Max σ_vm in structure	Long: RD Vert: DDM

Ax + By +: *Seat Rail A first rotated +10° about the x-axis (roll), then Seat Rail B rotated +10° about the y-axis (pitch).*

4.2 Stage 1: Pre-deformation Analysis

Stage 1: Determination of Critical Pre-deformation State (as mandated by SAE AS8049B: 10° roll on one side +10° pitch on the other) that exerts the greatest influence on the stress of chair leg locks. Accounting for the structural asymmetry of the double seat assembly, 8 distinct cases analyzing different roll (±x) and pitch (±y) combinations on the left/right seat rails (A/B) were evaluated. A key finding was the minimal influence of ATD count (single/dual) on chair leg lock stress (~20 MPa). Consequently, all cases employed a dual-ATD configuration for nonlinear elastoplastic analysis, with the evaluation metric being the maximum Von Mises stress in the chair leg lock. Computational results indicated that case Ax + By + (in which Seat Rail A is rotated + 10° in roll followed by Seat Rail B rotated +10° in pitch) yielded the highest stress (913 MPa) [14]. We therefore established this case as the fixed pre-deformation state for subsequent impact analysis. Representative stress contours, displacement contours, and chair leg lock stress distributions for the 8 Stage 1 cases are illustrated in Figs. 2, 3 and 4. Calculated results are presented in Table 4.

Fig. 2. Von Mises stress contour of seat structure

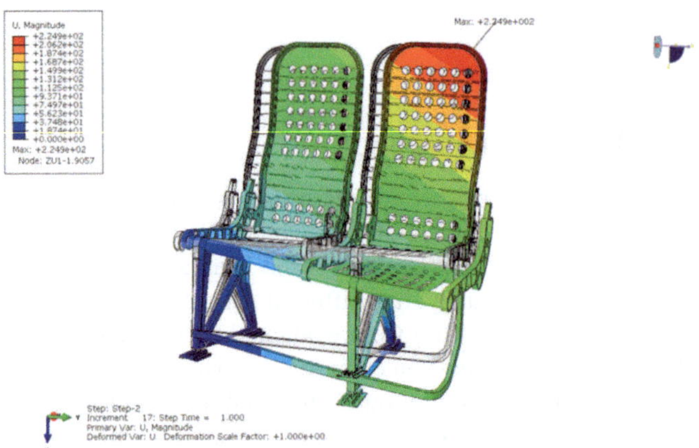

Fig. 3. Deformation contour plot of seat assembly

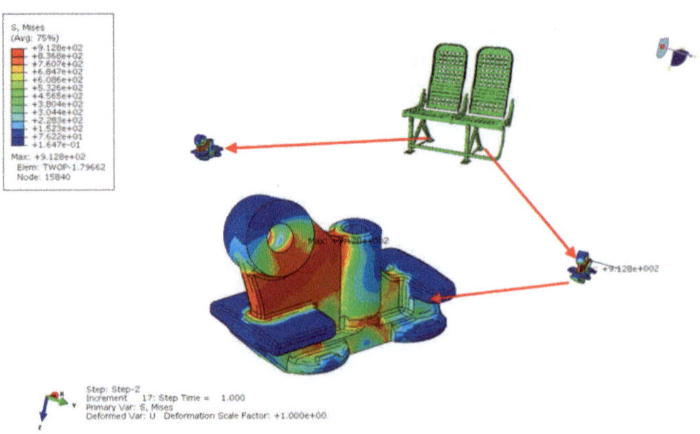

Fig. 4. Von Mises stress at leg lock mechanism

Table 4. Stage 1 Pre-deformation simulation results

Load Case ID	Von Mises/MPa
Ax-By+	856
Ax−By−	905
Ax+By−	858
Ax+By+	913
Bx-Ay+	909
Bx−Ay−	858
Bx+Ay−	901
Bx+Ay+	856

4.3 Stage 2: Dynamic Impact Analysis

Stage 2: Screening of critical dynamic impact cases based on the Ax+By+ pre-deformation state with dual ATDs. For longitudinal impact (Test 2), the variable was yaw offset direction (Left L/Right R), analyzed across 6 cases. For vertical impact (Test 1), the variable was ATD configuration (Left single/Right single/Dual), analyzed across 3 cases. Evaluation criteria involved a comprehensive evaluation of the maximum chair leg lock load and the maximum Von Mises stress in the primary seat structure. 2 cases are illustrated in Figs. 5 and 6. Calculated results are presented in Tables 5 and 6.

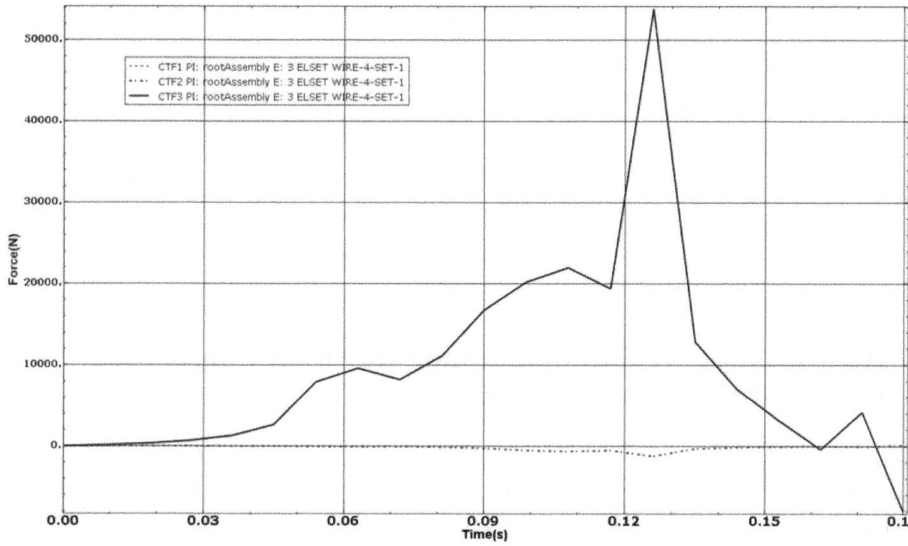

Fig. 5. Force time-history in aft leg lock mechanism

Fig. 6. Stress contour plot of the seat

Table 5. Stage 2 Longitudinal impact simulation results

Load Case ID	Von Mises/MPa	Peak Load of Leg Lock Mechanism/kN	Load Case ID	Von Mises/MPa	Peak Load of Leg Lock Mechanism/kN
LL	619	50	RL	693	56
LR	612	48	RR	686	52
LD	706	59	RD	717	62

Table 6. Stage 2 Vertical impact simulation results

Load Case ID	Von Mises/MPa	Peak Load of Leg Lock Mechanism/kN
LDM	242	3
RDM	369	6
DDM	582	10

5 Airworthiness Verification Testing Based on Conclusions from Multi-parametric Analysis

Airworthiness verification testing was conducted under the most critical loading conditions identified through the preceding multi-parametric coupled analysis (incorporating floor pre-deformation, yaw offset, and occupant distribution), as summarized in Table 7. The objective was to validate the effectiveness of the methodology and the structural reliability of the seat system.

Table 7. Critical Test Conditions Based on multi-parametric Analysis

Test Type	Yaw Offset	Floor Pre-deformation	Occupant Config	Dynamic Parameters
Longitudinal	Right 10°	Rail A: +10° Roll → Rail B: + 10° Pitch	Dual ATD	v ≥ 13.41 m/s, a ≥ 16 g
Vertical	None	None	Dual ATD	v ≥ 10.67 m/s, a ≥ 14 g

Test data (Table 8) demonstrated that the leg lock mechanism load served as the core compliance criterion. For the longitudinal impact (RD case), the measured maximum load was 64.1 kN, exhibiting a negligible deviation of only 3.3% from the predicted value of 62 kN. For the vertical impact (DDM case), the measured load was 10.2 kN (predicted: 10 kN), with an error of 2%. Regarding structural response consistency: post-impact inspection confirmed the absence of permanent deformation in the seat structure, and the load path remained intact, meeting the requirements stipulated in CCAR 25.562 [1, 2]. The test data validated the accuracy of the multi-parametric analysis [15] in predicting ultimate loads.

Table 8. Key Test Data vs. multi-parametric Predictions

Parameter	Longitudinal Impact (RD Case)	Vertical Impact (DDM Case)	Deviation
Peak Accel. (g)	16.31 (Req. ≥16)	14.4 (Req. ≥14)	≤1.9%
Leg Lock Load (kN)	64.1 (Pred. 62)	10.2 (Pred. 10)	≤4.0%
ATD Head Accel. (g)	386.3 (Right)/299.2 (Left)	---	---

Fatigue Life Assessment under Repeated Impacts: To address long-term structural integrity beyond the single dynamic impact event mandated by certification, a fatigue life assessment of the critical leg lock mechanism was conducted. Utilizing the stresslife (S-N) approach per SAE AIR6217 guidelines, the maximum von Mises stress amplitude (913 MPa) identified in the critical RD loading scenario (Section 4.2, Stage 1: Ax+By+ pre-deformation) was evaluated against the S-N curve of the 30CrMnSi steel material. By applying Miner's linear cumulative damage rule and incorporating a conservative safety factor of 2.0 (to account for material variability and load uncertainties), we predicted a fatigue life exceeding 10^5 loading cycles for the leg lock mechanism. This significantly surpasses the FAA AC 25.562-1B requirement for seats to withstand twice the design service life without failure, thereby providing assurance of the seat's durability under potential repeated emergency landing loads encountered over its operational lifespan.

6 Conclusions

This study establishes a methodology for screening the most critical loading conditions for aviation seat dynamic impact based on multi-parametric coupled analysis. Utilizing a two-stage (displacement pre-deformation followed by impact) finite element model, this approach accurately identifies critical failure modes. The key findings are summarized as follows:

(1) The developed two-stage displacement-impact analysis methodology, incorporating multi-parametric coupling, precisely identifies the risk of dynamic impact failure in aviation seats. Under the synergistic effect of a 10° right yaw offset and floor pre-deformation (Seat Rail A +10° roll/Seat Rail B +10° pitch), the peak load on the chair leg locking mechanism reached 62 kN (experimentally measured at 64.1 kN, deviation of 3.3%), with the associated Von Mises stress reaching 913 MPa, triggering a critical failure condition.
(2) The proposed methodology significantly optimizes the loading case screening process. Based on parameter sensitivity analysis, the original 24 potential loading cases were efficiently reduced to 9 critical ones. Finite element predictions demonstrated high fidelity, with errors between predicted and experimentally measured loads $\leq 2\%$, thereby validating the computational model's reliability.
(3) The established standardized workflow of "Pre-deformation Screening followed by Impact Verification" provides a robust framework for dynamic testing of various aviation seat types. This workflow offers a practical compliance solution for meeting the requirements stipulated in CCAR 25.562.

References

1. Civil Aviation Administration of China (CAAC): CCAR-25-R4: Airworthiness Standards for Transport Aircraft. Beijing: CAAC (2011). (in Chinese)
2. FAA: Dynamic evaluation of seat restraint systems and occupant protection on transport aviations: AC25.562-1B. Washington DC: FAA (2006)
3. SAE Technical Standards Board: performance standard for seats in civil rotorcraft, transport aircraft, and general aviation aircraft: AS 8049B. Warrendale: SAE Technical Standards Board (2005)
4. Astori, P., Ceresa, P., Morandini, M.: Numerical optimization of a seat energy absorber. In: 4th International KRASH Users Seminar (IKUS4), Amsterdam (2003)
5. Dhole, N.E.: Development and validation of a finite element model of a transport aircraft seat under part 25.562 dynamic test conditions. Wichita, Kansas: Wichita State University, pp. 2–6 (2010)
6. Bhonge, P., Lankarani, H.M.: Finite element modeling strategies for dynamic aircraft seats. In: Wichita Aviation Technology Congress & Exhibition, 2008-01-2272
7. Di Napoli, F., De Luca, A., Caputo, F., et al.: Mixed FE - MB methodology for the evaluation of passive safety performances of aeronautical seats. Int. J. Crashworthiness, 1–12 (2018)
8. Gowdy, V., Deweese, R., Beebe, M.S., et al.: A lumbar spine modification to the hybrid III ATD for aircraft seat tests. In: General, Corporate & Regional Aviation Meeting & Exposition. Wichita: SAE International (1999)
9. Taylor, A.M., Moorcroft, D.M., DeWeese, R.L.: Comparison of the hybrid II, FAA hybrid III, and THOR - NT in vertical impacts. In: Proceedings of the 73rd Annual AHS International Forum and Technology Display. Washington: American Helicopter Society, pp. 963–973 (2017)
10. Development and validation of an aircraft seat cushion component test volume 1: DOT/FAA/AR-05/5. Washington: Federal Aviation Administration (2005)

11. Embraer, S.A.: Use of dynamic simulation for certification and development. Wichita: FAA & National Institute for Aviation Research (2012)
12. Mengxiao, L., Jinwu, X., Yiru, R., et al.: Evaluation and analysis method of aircraft seat. J. Beijing Univ. Aeronaut. Astronaut. **42**(2), 383–390 (2016). (in Chinese)
13. Lamanna, G., Vanacore, A., Guida, M., et al.: Development of a head injury criteria - compliant aircraft seat by design of experiments. Aerospace (2019)
14. Lankarani, H., Mirza, M.G.: Parametric study of crashworthy bulkhead designs: DOT/FAA/AR-02/103. Washington, D. C.: Federal Aviation Administration (2002)
15. Federal Aviation Administration ANM -110: Transport airplane cabin interiors crashworthiness handbook: AC 25-17A. Washington, D. C.: Federal Aviation Administration (2009)

Intelligent Control and Dynamics

A Composite Control Strategy for ABS: Dynamic Threshold Adjustment Coupled with Anti-windup PID Compensation

Zhidong Zhang and Lu Zhang(✉)

Inner Mongolia University, Hohhot 010070, China
zzd3054476735@163.com

Abstract. Anti-lock braking systems (ABS) are crucial for maintaining vehicle stability during emergency braking, yet traditional ABS control methods are often limited by fixed slip ratio thresholds and insufficient adaptability to complex road conditions. These limitations can lead to increased braking distances and instability, particularly under varying surface friction. This study proposes a novel ABS control strategy integrating dynamic threshold adaptation with an anti-windup PID controller to enhance braking performance across diverse road conditions. The dynamic threshold algorithm adjusts the target slip ratio in real-time based on road surface information, while the anti-windup PID controller mitigates actuator saturation and ensures smoother braking. The proposed control strategy was evaluated using multi-scenario simulations on the Simulink platform. Results demonstrate an 8.6% reduction in braking time compared to conventional control methods, alongside improved slip ratio stability with deviations limited to ± 0.02. These findings indicate that the integration of dynamic threshold adaptation with anti-windup PID control significantly enhances braking stability and efficiency, particularly in scenarios involving sudden road surface changes. The study highlights the potential for this approach to be applied in intelligent vehicle braking systems, offering a robust solution to address the challenges posed by fluctuating road conditions. Future research should focus on further optimizing the dynamic threshold algorithm and exploring its implementation in real-world testing environments for broader application in autonomous driving systems.

Keywords: ABS · PID · slip ratio · adaptability · anti-windup

1 Introduction

With the advancement of intelligent vehicle technologies, braking control systems have become vital for safety, particularly during emergencies. The anti-lock braking system (ABS) plays a crucial role by preventing wheel lockup, preserving steering control, and enhancing stability during braking.[1] Traditional ABS strategies use fixed slip ratio thresholds to regulate braking torque. While effective under normal conditions, their performance drops significantly on complex or changing road surfaces, such as alternating wet and dry areas or sudden shifts in road friction, creating potential safety risks.

S. Halgamuge et al. (Eds.): ICACMVE 2025, LNME, pp. 259–267, 2026.
https://doi.org/10.1007/978-981-95-6911-3_25

Conventional methods like Bang-Bang control or rule-based logic are easy to implement but struggle with rapid changes in road conditions.[2] These static systems often cause excessive slip fluctuations and longer braking distances. As a result, researchers have turned to more adaptive and robust control methods. Approaches using fuzzy logic, sliding mode control, or neural networks have emerged. Fuzzy controllers improve adaptability but depend heavily on manual tuning.[3] Sliding mode control is robust to disturbances but causes high-frequency chattering, reducing actuator life. Neural networks can learn from data but require large datasets and lack interpretability and stability guarantees.[4]

Advanced slip ratio estimation techniques, such as those using recursive least squares with unscented Kalman filters, offer better accuracy but still face issues like parameter drift. Other methods, like dynamic models based on tire-road interaction,[5] improve adaptability but struggle with response delay and limited vehicle applicability. Moreover, many PID-based ABS systems lack effective anti-windup design.[6] Under persistent errors—like on slippery roads—integrator windup can cause actuator saturation and instability. Although anti-windup techniques exist, few achieve both quick response and overshoot suppression.[7]

To address these issues, this study proposes a composite ABS control strategy that combines dynamic slip threshold adaptation with an anti-windup PID controller. A dual-mode threshold adjustment algorithm recalibrates slip targets in real time based on road sensing and slip error feedback, reducing oscillations and improving effectiveness. Simultaneously, an integral reset mechanism in the PID controller prevents windup, enhancing stability and responsiveness. The overall system features an inner loop for adaptive threshold generation and an outer loop for pressure correction.

Simulations in Simulink across various road conditions show that this method significantly outperforms traditional strategies in braking time, slip ratio stability, and braking smoothness. The proposed approach offers a practical and scalable solution for future intelligent ABS design (Fig. 1).

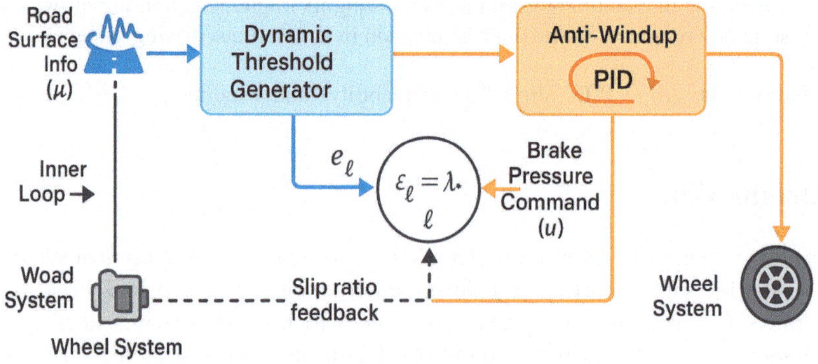

Fig. 1. Composite ABS Control Architecture

2 System Modeling and Controller Design

2.1 Vehicle and Tire Dynamics Modeling

This study adopts a single-wheel equivalent model to simplify the braking system dynamics analysis. The model's core consists of two parts: the vehicle longitudinal dynamics and the tire-road contact mechanics (Fig. 2).

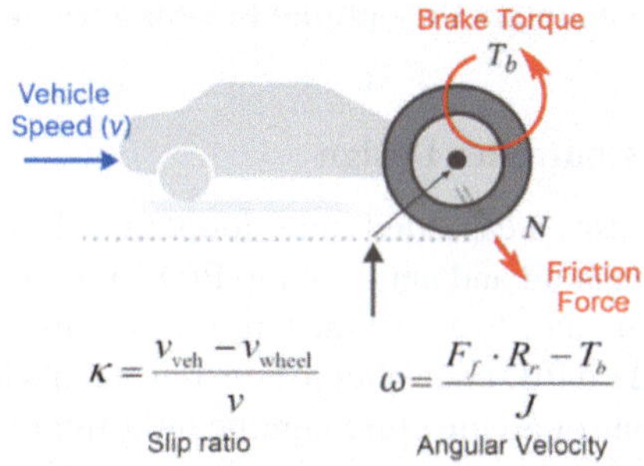

Fig. 2. Vehicle and Tire Dynamics Model for ABS Braking

Vehicle Longitudinal Dynamics. Assume the vehicle mass is m and the normal force on the ground is N (g is the gravitational acceleration). According to Newton's second law, the braking deceleration a_b can be expressed as:

$$a_b = \frac{F_f}{m} = \frac{\mu(k) \cdot W}{m} = \mu(k) \cdot g \tag{1}$$

where F_f is the tire longitudinal friction force, and $\mu(k)$ is the friction coefficient as a function of slip ratio.

By integrating the deceleration, the vehicle speed dynamics can be obtained as:

$$v_{\text{veh}} = v_0 - \int_0^t a_b(\tau)d\tau \tag{2}$$

where v_0 is the initial speed, and a_b is the braking deceleration as a function of time.

Single-Wheel Dynamics. The wheel's rotational dynamics are described by a torque balance. Assume the wheel rotational inertia is J, the effective rolling radius is R_r, and the brake torque applied is T_b. The wheel's angular acceleration a_w can then be expressed as:

$$\alpha_w = \frac{T_{\text{net}}}{J} = \frac{F_f \cdot R_r - T_b}{J} \tag{3}$$

where $F_f \cdot R_r - T_b$ is the net torque acting on the wheel (brake torque minus the friction torque).

This model neglects rolling resistance and aerodynamic drag, which is reasonable for analyzing emergency braking conditions.

Slip Ratio Calculation: Under braking conditions, the slip ratio κ is defined as:

$$\kappa = \frac{v_{\text{veh}} - v_{\text{wheel}}}{\max(v_{\text{veh}}, v_{\text{wheel}})} \times 100\% \tag{4}$$

This parameter reflects the extent of wheel slip: $\kappa = 0\%$ indicates pure rolling (no slip), while $\kappa = 100\%$ indicates the wheel is fully locked. As a key feedback variable for closed-loop control, the slip ratio is computed in real time from the difference between vehicle speed v_{veh} and wheel linear speed v_{wheel}. It drives the control strategy to adjust brake pressure in order to maintain the optimal tire-road friction state.

2.2 Anti-Windup PID Controller Design

The ABS control system uses a cascaded inner-loop/outer-loop structure to implement the proposed dynamic threshold and anti-windup PID strategy. The inner loop employs a Dynamic Threshold Generator to adjust the target slip ratio threshold in real time, and the outer loop uses a standard PID controller to generate a brake pressure correction. This coordinated dual-loop control architecture significantly improves the control accuracy and robustness of the ABS through dynamic threshold coupling and adaptive integration limiting (Fig. 3).

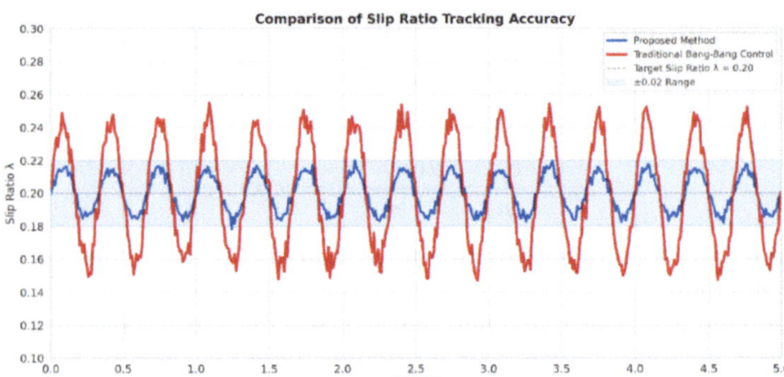

Fig. 3. Comparison of Slip Ratio Tracking Accuary

Inner-Loop Dynamic Threshold Adaptation Mechanism. To achieve optimal slip ratio tracking during braking, a dynamic threshold adaptive control strategy based on a dual closed-loop feedback is designed. The slip ratio error is defined as:

$$e = \kappa_{\text{actual}} - \lambda_{\text{target}} \tag{5}$$

where λ_{target} is the target slip ratio (threshold) adjusted in real time by the dynamic threshold generator according to the inferred road conditions, and κ_{actual} is the measured actual slip ratio. (Wheel angular velocity is obtained from wheel speed sensors, and vehicle speed is obtained via GPS fused with a Kalman filter.) The slip ratio error e evolves continuously, and the system triggers the threshold adaptation mechanism in real time based on this error:

1. **Error-driven adjustment:** If $e > 0.03$, decrease the target slip by increments of $\Delta_{dec} = 0.03$. If $e <$ -0.03, by increments of $\Delta_{dec} = 0.03$.

$$\lambda_{\text{target}}(k+1) = \begin{cases} \lambda_{\text{target}}(k) - \Delta_{dec},\ e > \delta_{\text{deadzone}} \\ \lambda_{\text{target}}(k) - \Delta_{dec},\ e < -\delta_{\text{deadzone}} \end{cases} \tag{6}$$

2. **Coordinated optimization mechanism:** A road-surface recognition module provides a baseline reference slip value w_{ref} for current conditions (for example, $w_{\text{ref}} = 0.20$ on dry asphalt), Define the target slip deviation as:

$$d = \lambda_{\text{target}}(k) - w_{\text{ref}} \tag{7}$$

If $|d| > 0.05$, it indicates the current threshold is mismatched with the environment. In this case, a warning is triggered and the target slip threshold is reset toward the reference value. A compensation step of 0.02 (absolute slip ratio) is used for the adjustment to reduce transients and ensure a smooth transition.

Outer-Loop Anti-Windup PID Control. The outer loop takes the slip ratio error from the inner loop and computes a brake pressure correction via a standard PID controller. The control law is given by:

$$\mu_{\text{PID}} = eK_p + K_i \int edt + K_d \frac{de}{dt} \tag{8}$$

where K_p is the proportional gain, K_i and K_d are the integral and derivative gains. These PID parameters are tuned via a discrete particle swarm optimization algorithm. To prevent integrator wind-up and actuator saturation, a saturation module is placed between the PID output and the hydraulic actuator, which clamps the control command within allowable upper and lower bounds. This anti-windup measure effectively suppresses hydraulic actuator saturation and avoids brake pressure overshoot caused by sustained integration.

3 Simulation Results and Analysis

To validate the effectiveness of the proposed ABS control strategy—combining dynamic threshold adaptation with an anti-windup PID controller—a series of simulations were conducted in MATLAB/Simulink across diverse braking scenarios. Key performance indicators include braking time, slip ratio tracking accuracy, and braking deceleration smoothness. Comparative analyses were performed against the conventional Bang-Bang control method to highlight the advantages of the proposed approach. Furthermore, visual analyses in the form of performance plots were generated to facilitate intuitive understanding of the improvements achieved.

3.1 Braking Time Optimization

The first performance metric analyzed is total braking time under dynamic road conditions (dry–wet alternation). As shown in Figure 4a, the conventional Bang-Bang control scheme results in a braking time of 14.14 s, whereas the proposed dynamic threshold method reduces this to 12.93 s—representing an 8.6% improvement. This enhancement is attributed to the real-time slip ratio recalibration capability of the inner-loop threshold generator, which rapidly aligns the target slip with current road adhesion characteristics. Moreover, the anti-windup PID controller's integral reset mechanism effectively limits brake pressure overshoot, reducing inertial lag and improving braking efficiency. The bar chart visualization not only quantifies the improvement but also includes annotations indicating the percentage gain, offering a clear comparative representation of system responsiveness.

3.2 Slip Ratio Stability Improvement

Slip ratio tracking performance is a critical determinant of ABS control accuracy. In Figure 4b, the time-history plots of slip ratio tracking reveal substantial differences in the control quality between the two methods. The conventional method demonstrates high-frequency oscillations within a $\pm$ 0.05 band around the target value (0.20), while the proposed method maintains deviations within $\pm$0.02 and achieves a standard deviation of 0.007. These results indicate a ~ 60% improvement in tracking accuracy. This performance gain is attributed to the memory-enhanced dynamic threshold logic, which applies a 0.03-deadzone and uses historical slip deviations to adaptively refine the threshold. The shaded tolerance bands and reference lines in the visualization clearly differentiate the control precision of the two methods, while road condition transitions (at 3 s, 7 s, and 11 s) are marked with vertical indicators, demonstrating the system's real-time adaptability to surface friction changes.

3.3 Braking Deceleration Smoothness

Deceleration smoothness directly influences ride comfort and safety, especially under emergency braking conditions. As depicted in Figure 4c, the proposed control method achieves faster transient convergence ($\approx$0.8 s) to steady-state deceleration compared to the conventional method ($\approx$1.2 s). Additionally, deceleration oscillations are substantially reduced—from $\pm$2.0 m/s^2 under the conventional scheme to $\pm$0.3 m/s^2 using the proposed controller, marking a ~ 97.5% reduction in fluctuation amplitude. This improvement is the result of the outer-loop anti-windup PID's integrator clamping and saturation logic, which mitigate control overshoot during torque regulation. The figure includes steady-state reference lines and labeled transient indicators, further emphasizing the superior stability and response speed of the proposed strategy.

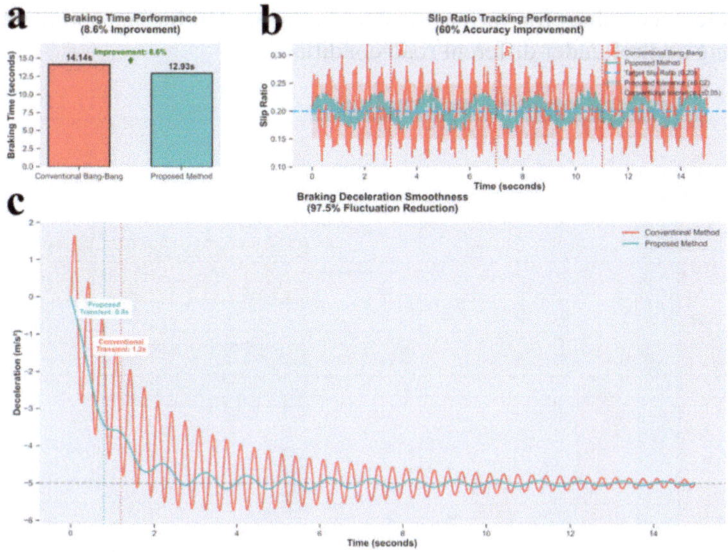

Fig. 4. Comparative Performance of Braking Time, Slip Ratio Tracking, and Deceleration Smoothness

4 Conclusion

This study tackled the adaptability and stability challenges of traditional ABS under complex road conditions by proposing a composite control strategy that integrates dynamic threshold generation with anti-windup PID control. Through theoretical modeling, controller design, and simulation validation, the following key outcomes were achieved:

1. **Dynamic threshold generation innovation:** Based on real-time road condition sensing and slip ratio error feedback, a dual-mode threshold adjustment algorithm was designed, achieving dynamic optimization of the target slip ratio in response to changing surfaces.
2. **Enhanced robustness via anti-windup PID:** By introducing conditional integration and a dynamic reset strategy, the hydraulic actuator's overshoot oscillation was effectively suppressed, demonstrating improved robustness of the ABS control.
3. **Simulation validation and engineering value:** In multi-condition Simulink tests, the proposed strategy shortened braking time from 14.14 s to 12.93 s (an 8.6% improvement), reduced braking distance by about 1.8 m, and lowered the slip ratio tracking standard deviation to 0.007(Fig. 5), relative to a conventional control method.

Overall, this research provides theoretical support and a practical solution for improving braking safety and environmental adaptability in intelligent vehicles. The dynamic threshold coupling framework can be extended to multi-physics coordinated control in by-wire chassis systems, supporting the high-precision control requirements of L4 autonomous driving. Future work will focus on more accurate identification of optimal slip ratio reference values for various road surfaces, and on further optimizing the slip ratio tracking control under different road conditions.

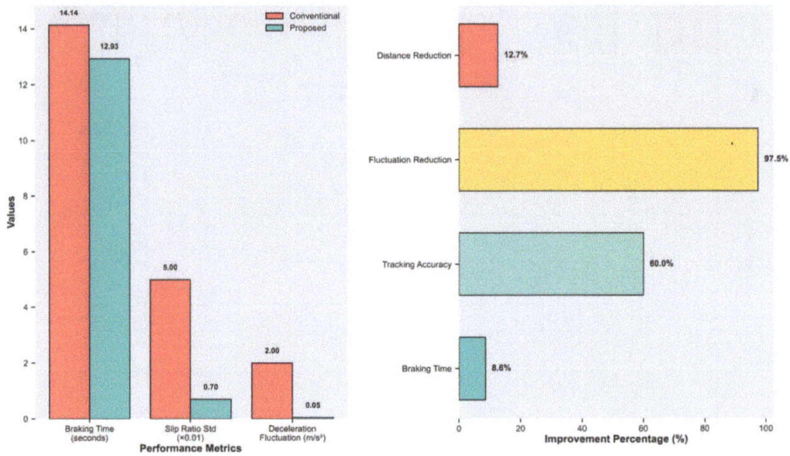

Fig. 5. Statistical and Overall Improvements of the Proposed Braking Control Strategy

Acknowledgments. Inner Mongolia Fund "Research on the Mechanism and Optimal Control of Multi-mode Regenerative Braking for Electric Heavy-duty Trucks" (2024MS05049)

Credit authorship contribution statement: ZhiDong Zhang: Methodology, Validation, Investigation, Writing an original draft. Lu Zhang: Writing review & editing, Supervision, Funding acquisition.

References

1. Aksjonov, A., Augsburg, K., Vodovozov, V.: Design and simulation of the robust ABS and ESP fuzzy logic controller on the complex braking Maneuvers. Appl. Sci. **6**(12) (2016)
2. Feng, X., Hu, J.: Discrete fuzzy adaptive PID control algorithm for automotive anti-lock braking system. J. Ambient. Intell. Humaniz. Comput. (2021)
3. Vošahlík, D., Hanis, T.: Real-time estimation of the optimal longitudinal slip ratio for attaining the maximum traction force. Control. Eng. Pract. **145**, 105876 (2024)
4. You, S., Gil, J., Kim, W.: Fixed-time slip control with extended-state observer using only wheel speed for anti-lock braking Systems of Electric Vehicles. IEEE Trans. Intell. Transp. Syst. **23**(7), 6368–6378 (2022)
5. Ghani, M.F., Athif Mohd Faudzi, A., Ghazali, R., Mohamaddan, S.: Application of fuzzy logic sliding mode control approach with PID structure for electro-hydraulic actuator tracking system. IEEE Access. **13**, 97818–97834 (2025)
6. Guo, L., Xu, H., Zou, J.: Acceleration slip regulation control strategy for four-wheel independent drive electric vehicles. IEEJ Trans. Electr. Electron. Eng. **14**(4), 630–639 (2019)
7. Tang, G., Yang, M., Liu, Q., Yang, S.: Design of an Adaptive Fuzzy Sliding Mode Controller for hydraulic position servo system. In: 2023 IEEE 12th Data Driven Control and Learning Systems Conference (DDCLS), pp. 1239–1244 (2023)

Development of a Novel Flat Labeling Machine Equipment

Yu Liu[1], Yiyao Wang[1], Dong Li[1], Yu Zhang[2], Yangxuan Ming[2], and Shenfeng Wu[1](✉)

[1] Enshi Cigarette Factory of Hubei Tobacco Industry Co., Ltd., 430070 Wuhan, China
2723871617@qq.com
[2] Wuhan University of Technology, 445000 Wuhan, China

Abstract. To address the shortcomings of traditional labeling machines in terms of labeling accuracy, efficiency, and stability, this paper proposes a novel fully automatic flat labeling machine. The equipment achieves an improvement in labeling accuracy from the traditional ±1.0 mm to ±0.5 mm, and the good product rate increased from 60% to 98%, while meeting the demand for high-speed labeling (up to 115 packs per minute), through innovative designs that include the introduction of a high-precision vision identification system and optimized cigarette box and label conveying systems. Additionally, to tackle issues related to unstable label transmission, a constant tension control and buffer zone have been designed, significantly enhancing the reliability and lifespan of the equipment. Experimental validation indicates that this device demonstrates superior labeling performance and strong industry applicability in practical applications.

Keywords: labeling machine · mechanical structure · visual recognition system

1 Introduction

The Industrial Revolution has driven the development of modern industrial technology. With increasing demands from consumers and businesses for product packaging and personalized customization, higher expectations are being placed on labeling technology and equipment. Currently, manual labeling machines face challenges such as high labor demand, low precision, and inefficiency [1]. The variety of labeled products has increased, and process quality requirements are improving year by year. Consequently, labeling equipment is constantly being updated, expanded, and improved, with increasing levels of automation and electromechanical integration.

Labeling machines are automated mechanical devices that use adhesives to affix labels, made from materials such as paper or metal, onto packaging containers [2]. Automatic labeling machines transport packaging containers to the labeling position via a conveyor belt. Using mechanical components such as label-pressing wheels, drive wheels, and take-up reels, they achieve precise label transfer and attachment. The drive wheels intermittently pull the label strip, separating the label from the roll and firmly

S. Halgamuge et al. (Eds.): ICACMVE 2025, LNME, pp. 268–289, 2026.
https://doi.org/10.1007/978-981-95-6911-3_26

attaching it to the container via the pressing wheel. The primary materials used are self-adhesive labels or self-adhesive films, and labels have adhesive on one side, allowing them to be applied directly after being peeled off the backing paper, without the need for additional glue [3]. Label material is clean, hygienic, non-molding, non-corrosive, and non-detaching, which effectively reduces the possibility of counterfeit products.

The international labeling machine market is highly competitive. The overall trend for foreign labeling machines is moving towards high speed, high efficiency, and high quality. The main development focuses include: compact structure, low energy consumption, lightweight design, high efficiency, and environmental friendliness, as well as flexibility and adaptability [4], while emphasizing low energy consumption, lightweight structures, compact designs, small footprints, multifunctionality, and environmental friendliness. Developed countries such as the United States, Germany, Italy, and Japan lead globally in labeling machine technology. Their products integrate advanced technologies such as microelectronics, Opto-mechatronics, new materials, biotechnology, and the industrial internet, achieving intelligent and energy-efficient development [5]. Cho and Bhushan [6] studied the effects of various polymers on friction and wear in labeling machines, investigating factors such as load, speed, and varnish coatings. The automation systems department in the United States, in collaboration with Italy, designed a series of rotary labeling machines for p-s label applications. The Sakmi KUBE labeling machine in the United States can achieve zero downtime and continuous operation when replacing materials for adhesive labels [7]. Currently, American labeling machines are transitioning from traditional processes to electromechanical-hydraulic integration, incorporating advanced technologies such as machine vision inspection and continuous improvement in control systems [8]. Chen [9] explored linear-time algorithms for image-labeling machines.

Currently, there are still many issues in the development of labeling machinery in China. First, there is a lack of macro-level regulation, leading to low-level and disorganized competition in product production and development. Second, insufficient funding results in slow equipment updates and iterations, making it difficult to develop a significant amount of advanced technologies. Third, there is a shortage of skilled professionals, which hinders the ability to independently develop products and innovate. Therefore, achieving automation or semi-automation in packaging machinery development holds great significance.

The primary issues with current labeling machines are: (1) a labeling accuracy of only ± 1 mm, which fails to meet high-precision requirements; (2) a product yield rate of only 60%, leading to significant economic losses; (3) an operating speed of just 80 labels per minute, coupled with poor stability, limiting production efficiency. To address these issues, this study designs a novel fully automated flat-surface labeling device, which includes a cigarette box conveying system, a label conveying system, and a vision recognition system. The main innovations of this paper include: (1) the adoption of a high-precision detection system based on visual recognition, which can correct labeling errors in real time and improve labeling accuracy; (2) the development of a cigarette box paging system based on a PPU manipulator and a vacuum conveyor, which addresses the issues of cardboard stacking jams and misalignment; (3) a constant tension control and buffer structure design in the label conveying system, which reduces the failure rate caused by uneven tension during the labeling process. These innovative measures collectively achieve comprehensive improvements in labeling accuracy, efficiency, and stability, providing new ideas for industry development.

2 Mechanical Structure Design of Labeling Machine

2.1 Overview of the Equipment

This paper focuses on the development and design of a labeling machine aimed at achieving the following research and development objectives:

(1) Labeling accuracy within ±0.5 mm;
(2) A yield rate of 98% at a speed of 80 packages per minute;
(3) A yield rate exceeding 85% at a speed of 100 packages per minute.

The structural components of the labeling machine are divided into the following main parts. The overall structural diagram of the equipment is shown in Fig. 1 below.

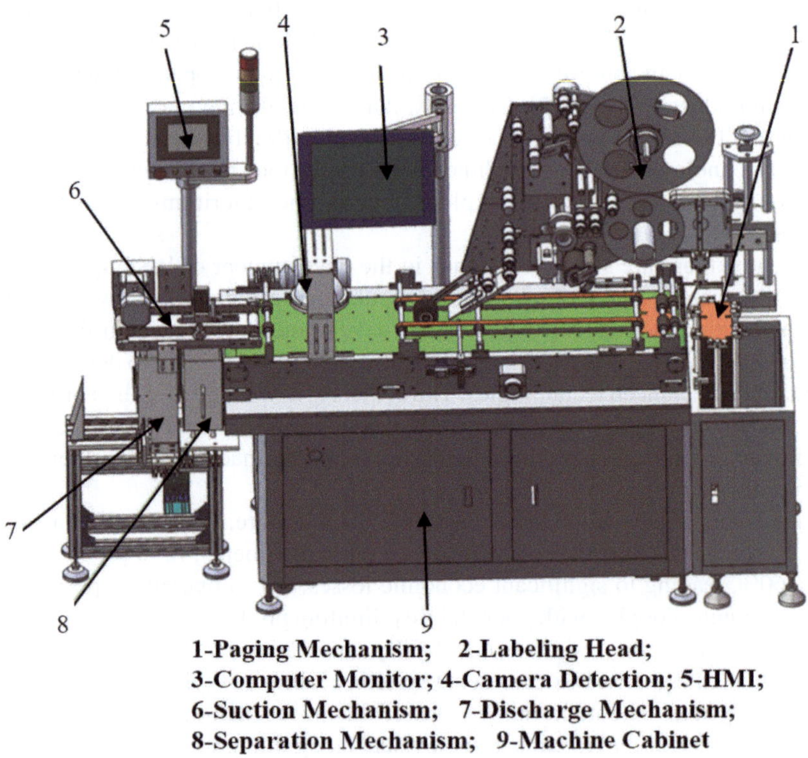

Fig. 1. Overall structural diagram of the equipment.

The system utilizes a suction cup and PPU robotic arm to transfer cigarette packs to the negative pressure conveyor belt. When the packs reach the designated position, the labeling unit performs labeling. The CCD inspection system detects label accuracy and grayscale values to ensure precision and consistency. If defective products are identified, the rejection mechanism automatically removes them from the production line, and the computer display updates the qualification rate in real time. Qualified products are then transported by the suction mechanism to the discharge unit, which operates automatically when the preset quantity is reached. The entire system is supported by a chassis unit to ensure stable operation.

2.2 Device Structure Design

Cigarette Box Conveying System Design. The cigarette box conveying system, consisting of a vacuum conveyor and paging mechanism, provides a stable and continuous carton supply to the labeling head. It automatically adjusts the paging speed and ensures cartons stay aligned on the conveyor belt, preventing adhesion and misalignment during start-up and speed changes. The cigarette box measures 238 mm in length and 95 mm in width, with a flap at the labeling position that may cause jamming when stacked.

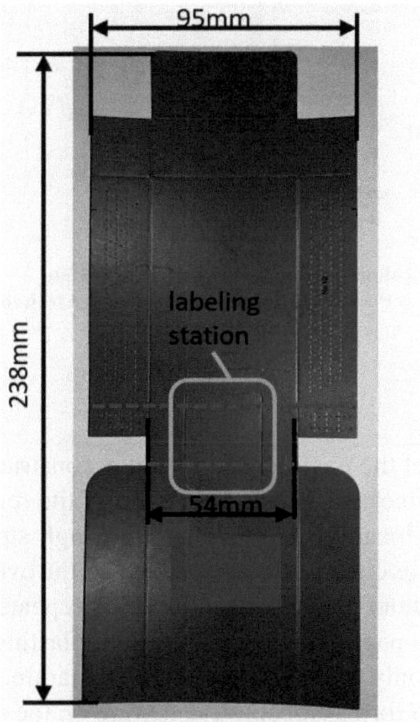

Fig. 2. Basic dimensions of the product.

Due to the raised portion at the labeling position of the carton, as shown in Fig. 2, stacking can easily cause the cartons to jam and prevent the use of friction for separation like traditional feeders. Therefore, this paper designs a feeding mechanism that uses a PPU robotic arm and suction cups for material handling. The PPU robotic arm functions as a cam mechanism, driven by the forward and reverse rotation of a servo motor to achieve reciprocating motion, thus completing the feeding process. Specifically, when the servo motor rotates forward, the PPU robotic arm drives the suction cup to lift the carton from below. Once the carton is securely attached, the servo motor reverses, and the PPU robotic arm precisely places the carton onto the vacuum belt. The structural diagram of the feeding mechanism is shown in Fig. 3.

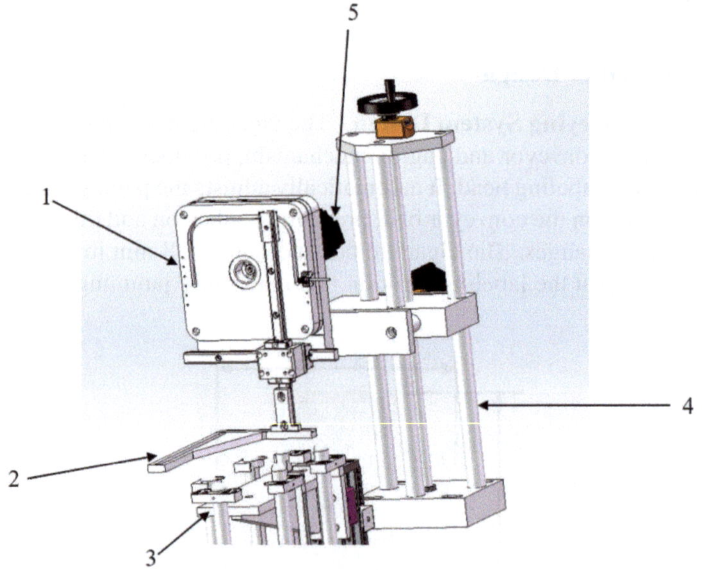

1-PPU Robotic Arm; 2-Suction Cup Position; 3-Carton Placement Area; 4-Adjustment Mechanism; 5-Servo Motor

Fig. 3. Paging mechanism.

The overall structure of the paging mechanism is constructed by using 304 stainless steel, with a surface roughness of Ra1.6. Considering the repetitive motion required at the suction cup placement location, a lightweight and high-strength aluminum alloy was selected as the material. This choice effectively reduces the overall weight while ensuring sufficient rigidity to withstand the stresses induced by repeated movements. To enhance the aesthetic appeal of the paging mechanism, a sandblasting treatment was uniformly applied. This process not only removes oxides and contaminants from the surface of the 304 stainless steel but also forms a decorative texture on the surface, contributing to the overall visual quality.

A vacuum conveyor is a device that uses negative pressure to achieve stable material transport. By installing a vacuum extraction device below the conveyor belt, a negative pressure environment is created, causing materials to adhere to the belt surface, thereby preventing sliding or falling during transport. However, since cigarette packs are evenly distributed on the vacuum belt, it is impossible to fill all the gaps on the belt. This results in air leakage during the suction process, reducing the actual vacuum pressure and making it difficult to firmly secure the cigarette packs on the belt. Consequently, misalignment often occurs during the labeling process. To address this issue, an upper-pressure belt was added to the vacuum belt, as shown in Fig. 4. The upper-pressure belt is secured with screws and springs, providing sufficient positive pressure to firmly fix the cigarette packs on the belt while allowing for flexible adjustments to ensure the packs enter the lower-pressure belt smoothly.

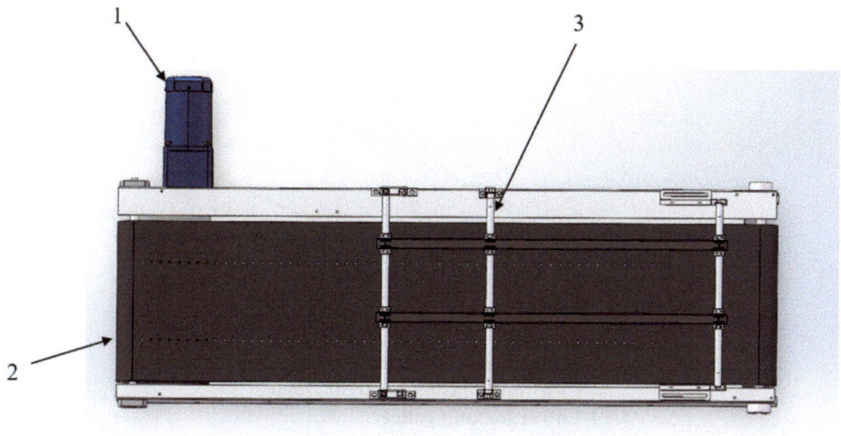

1-Stepper Motor; 2-Vacuum Belt; 3-Pressing Belt

Fig. 4. Vacuum conveyor.

In the early design phase, considerations were made regarding potential issues such as mislabeling, double labeling, and missed labeling during the overall debugging process. To prevent labels from adhering too tightly to the belt and becoming difficult to remove, a felt surface negative pressure belt was selected. This design effectively avoids overly tight adhesion of labels to the belt. Additionally, all other components are made of 304 stainless steel with a polished surface treatment, which not only enhances the aesthetics of the stainless steel but also improves its corrosion resistance to a certain extent.

Label Conveying System Design. The label conveying system consists of three main parts: the label collection area, the label supply area, and the labeling area. The design of the labeling head has been optimized. Compared to traditional labeling machines, a buffering area and locking rollers have been added to the label supply area, while a constant tension controller has been introduced in the label collection area. These improvements ensure constant tension during label winding, significantly enhancing labeling accuracy. These optimization measures work in synergy to improve the efficiency and quality of the entire labeling process, reducing labeling defects. The overall structure is shown in Fig. 5.

For the label supply area, a locking roller is designed to provide friction for transferring labels on the supply disc. This roller is controlled by an independent motor, and the spring-loaded swing arm has been modified into a movable form, incorporating a buffer zone and a photoelectric sensor for buffering. During the labeling process, the label paper on the left side of the buffer zone moves in the direction indicated by the arrow

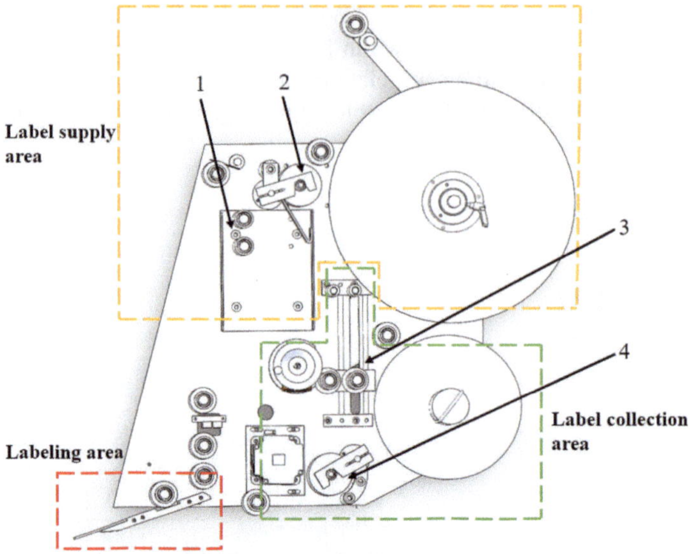

1-Buffer area; 2-Locking roller;
3-Constant Tension Control Device;
4-Lower locking roller

Fig. 5. An overall diagram of the modified label transport system plan.

Fig. 6. Brake mechanism structure diagram.

and gradually decreases. When the through-beam photoelectric sensor does not detect label paper, it sends a signal to the motor, prompting it to rotate, thereby controlling the rotation of the locking roller to drive the supplied disc and replenish the label paper in the buffer zone. Once the sensor detects the presence of label paper, the motor stops operating. This design effectively isolates the labeling area from the paper supply area while maintaining an appropriate margin between the two, ensuring that the radius of labels on the supplied disc does not affect the precision of labeling. Additionally, the movable swing arm design cushions the sudden start and stop of the motor, reducing machine impact and enhancing equipment stability and durability. The detailed structural diagram is shown in Fig. 6.

In the rewinding area, a constant tension control device is installed between the lower locking roller and the rewinding plate, with the label paper wound around the rollers of this device. A photoelectric sensor is positioned at the blue dashed line, obstructed by the rewinding plate. A motor is installed behind the rewinding plate to control its rotation. As the lower locking roller rotates, the label paper moves in the direction of the blue arrow, causing the slider to move downward due to gravity. When the slider enters the sensing range of the photoelectric sensor, the sensor sends a signal to activate the motor, which moves the slider upward. The motor stops when the slider exits the sensor's range. This design effectively isolates the labeling mechanism from the rewinding mechanism, maintaining an appropriate buffer zone between the two. It ensures that the tension on the label paper remains constant, so variations in the rewinding plate's size do not affect the labeling process. The detailed structure is shown in Fig. 7.

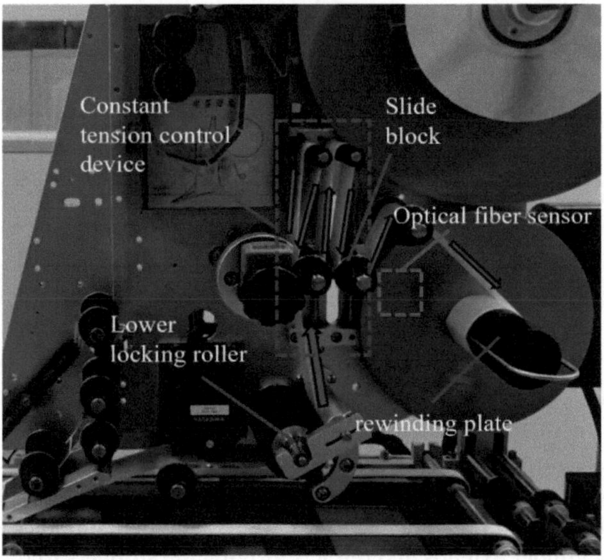

Fig. 7. Rewinding mechanism structure diagram.

The label applicator head is constructed entirely from 304 stainless steel and has been subjected to surface polishing treatment. The structure of the locking roller is shown in Fig. 8 and is mainly composed of components such as the support shaft, friction wheel, and eccentric roller. The function of the support shaft is to provide the necessary support; the axis of the eccentric roller is fixed at one end of the locking ring. When the locking ring tightly engages with the bearing on the axis of the friction wheel, the relative position between the eccentric roller and the friction wheel is locked. In this state, the motor drives the friction wheel to rotate, generating friction between the friction wheel and the eccentric roller, which propels the movement of the label paper. To enhance the frictional force between the friction wheel and the eccentric roller, the eccentric roller undergoes treatment.

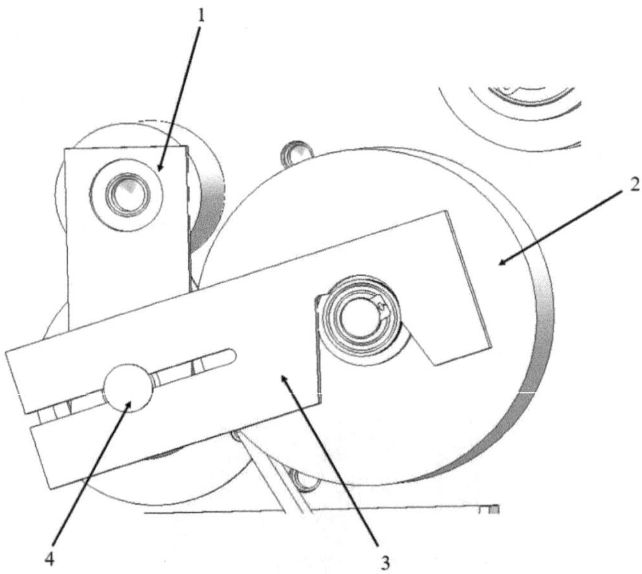

1-Support Shaf; 2-Friction Wheel;
3-Locking Ring; 4-Eccentric Roller

Fig. 8. Schematic diagram of the locking roller.

Design of the Rejection and Discharge Mechanism. The rejection and discharge mechanism mainly consists of two parts: the rejection belt section and the stacking section, as shown in Fig. 9 and Fig. 10, respectively. Cigarette packs with properly applied labels are inspected by the vision detection system. If no issues are detected, the packs are held beneath the belt by a vacuum belt and dropped into the staging area. Once a certain number of cigarette packs accumulate in the temporary storage area, the connecting rod cylinder extends and retracts, driving the stoppers to rotate, and causing the cigarette packs in the staging area to drop below. The baffle cylinders then retract to withdraw the baffle, and finally, the finished product ejection cylinder pushes the cigarette packs into the finished product stacking area.

In the design of the belt rejector, a plastic belt is primarily used due to its excellent elasticity and wear resistance, which allows it to maintain stable performance over long periods of operation. The support structure is made entirely of aluminum profiles to reduce overall weight, making it easier to transport and install. All other parts are made of 304 stainless steel with a polished surface treatment, further enhancing appearance quality and corrosion resistance.

If a cigarette pack is detected as defective by the vision detection system, the PLC sends a signal to the removing cylinder, which directly pushes the defective pack into the removing box, completing the rejection process.

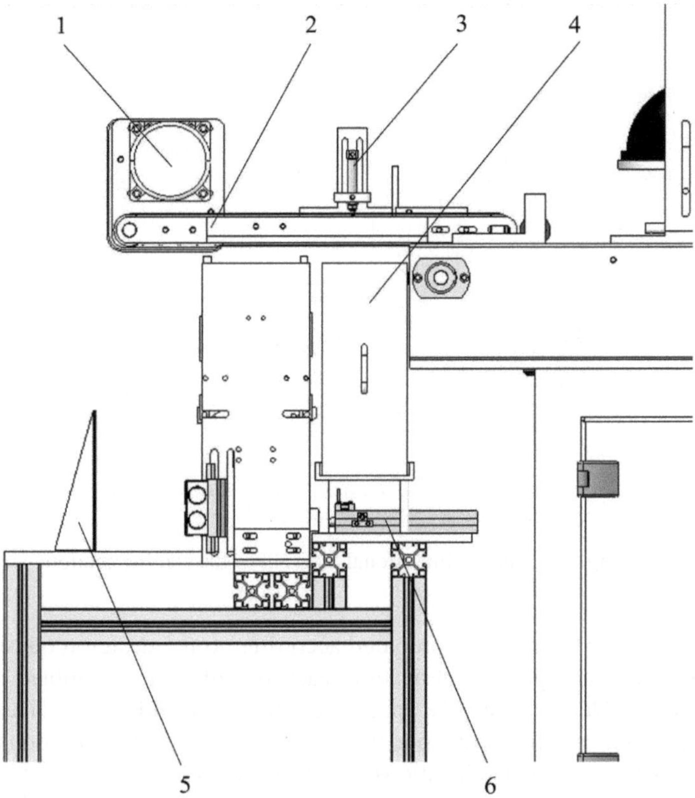

1-Motor of removing belt; 2-Removing belt conveyor; 3-Removing cylinder; 4-Removing box; 5-Finished product stacking area; 6-Finished product ejection cylinder

Fig. 9. Side view of rejection and discharge mechanism.

2.3 Processing and Assembly

After completing the overall design of the labeling machine equipment, three-dimensional (3D) models of each part were created. SolidWorks software was used to convert all 3D designs into two-dimensional (2D) manufacturing drawings, ensuring the machining accuracy and manufacturability of each component. Once the manufacturing drawings were finalized, the machining of the relevant components was arranged,

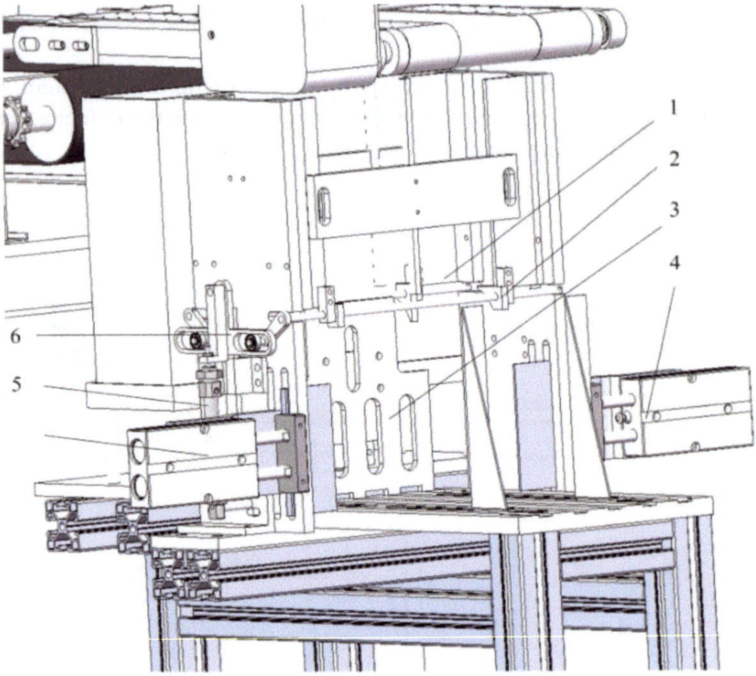

1-Staging area; 2-Stopper; 3-Pushing board; 4-Baffle cylinder;
5-Connecting rod cylinder; 6-Connecting rod

Fig. 10. Three-dimensional diagram of the staging section.

and various standard parts were procured according to the design requirements. Upon the completion of component fabrication, each module was assembled following the design drawings. During assembly, special attention was paid to critical aspects such as structural assembly accuracy, functional performance, and smooth motion to ensure that each part met the design specifications. After all modules were assembled, they were mounted onto a chassis made of aluminum alloy plates, completing the final assembly of the equipment. The assembly result is shown in Fig. 11. Practical operation and inspection confirmed that the equipment operates smoothly and functions properly.

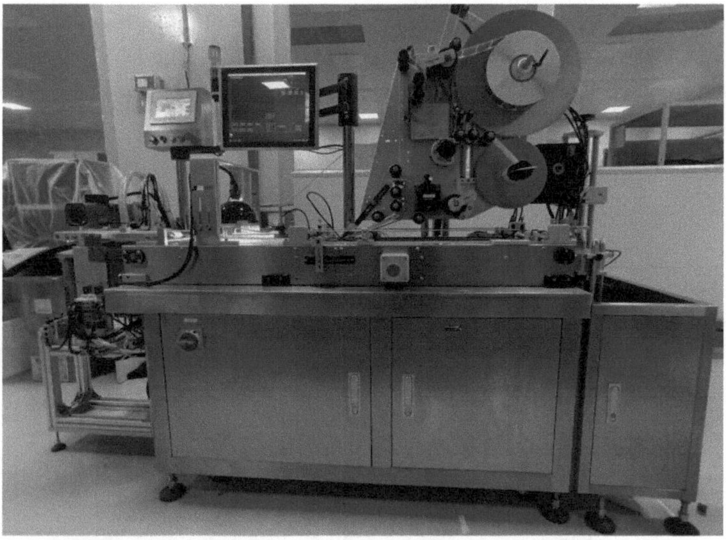

Fig. 11. Structural assembly diagram of the equipment.

3 Labeling Machine Control System Design

3.1 Visual Control System Solution

By utilizing machine vision technology to acquire and process label images, labeling equipment can adjust to the shape, size, and other characteristics of different products. This approach enhances production efficiency, reduces labor costs, and simultaneously improves the stability and reliability of the production line [10]. The hardware components of the vision system control are shown in Table 1.

Table 1. Hardware components of the vision system control.

Number	Product name	Model
1	Industrial cameras	MER-503-36U3M
2	Ring light	KM-DM20548
3	Senor	MC-FX501
4	Industrial computer	ARC-2300B-G2-S125R32
5	Display	APD-4117-C
6	PLC	SIMATIC S7–200

The specific equipment setup is shown in Fig. 12. The camera and light source are mounted on a bracket, allowing vertical and horizontal adjustments to achieve the optimal shooting position. The sensor is located above the conveyor belt, near the camera. The display is connected to the industrial control computer, which is linked to the 5 V input of a 5 V-to-24 V boost module via a USB-to-TTL interface. The 24 V output of the boost module is connected to a relay, which is further connected to the PLC's i2.7 interface for rejection control.

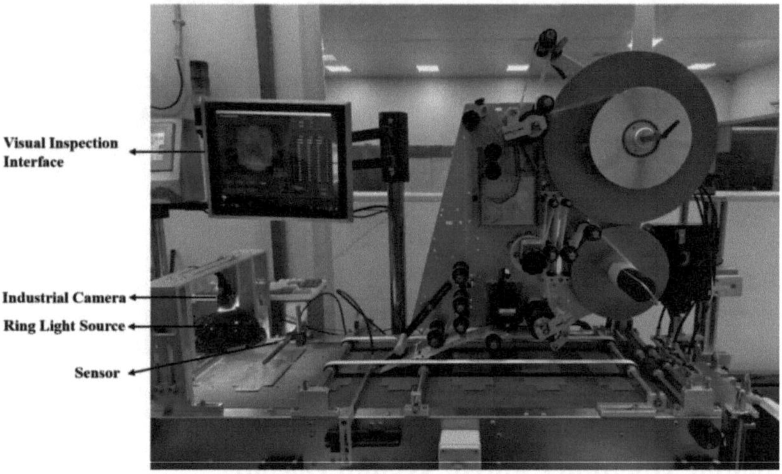

Fig. 12. Labeling system equipment diagram.

The control system flowchart of the machine is shown in Fig. 13. First, we reset the system, and all the mechanisms return to the start position. We set parameters like labeling velocity and conveyor velocity and press the start button. The vacuum conveyor starts to run, and the paging mechanism starts to page the cigarette boxes and put them on the vacuum conveyor. When the cigarette box passes through the optical fiber sensor, the optical fiber sensor sends a signal to PLC, making the label conveying system activate and perform a labeling action once. The visual inspection system will detect whether the label is properly attached. If so, the cigarette box will be sent to the stacking; If not, the removing mechanism will remove the defective one.

The electrical control diagram of the labeling machine is shown in Fig. 14.

We start the labeling machine system and open the visual inspection interface for image capturing. When the sensor above the conveyor belt near the camera detects the arrival of a cigarette pack label, it triggers the camera to capture an image. Various image processing techniques are then applied to the captured image of the cigarette pack label within the interface. The specific steps involved in the visual inspection and image processing are as follows: (1) Image Distortion Correction: We use the undistort_image function to correct image distortion. This step eliminates geometric distortion caused by the camera lens, making straight lines in the image appear closer to their real-world counterparts; (2) Grayscale Binarization: We use the cv2.inRange function to convert the grayscale image into a binary image. We apply median filtering to the binary image to remove noise and then invert the pixel values to reverse the black-and-white regions; (3) Contour Extraction and Denoising: We use cv2.findContours to extract contours and filter out small noise based on area. We retain contours with an area larger than 1, 400,

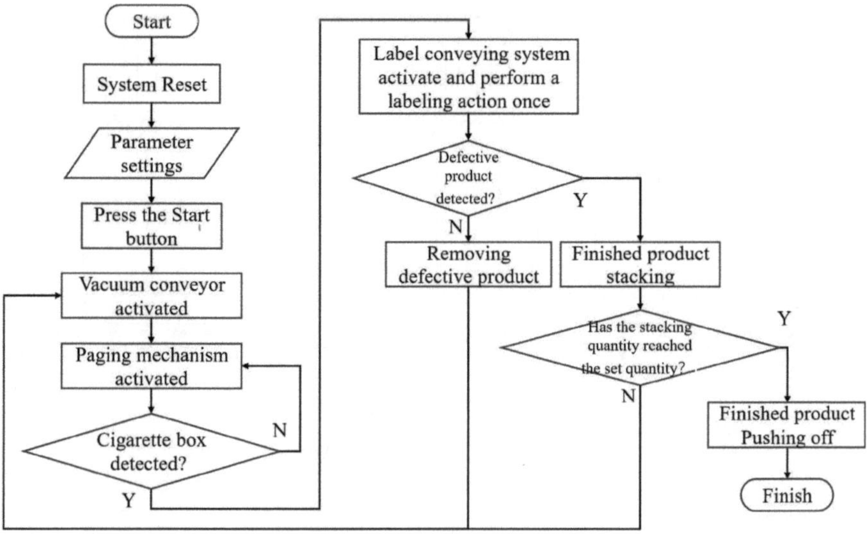

Fig. 13. Flowchart of control system.

removing insignificant noise; (4) Contour Point Interpolation: We apply the Bresenham algorithm to generate continuous pixel coordinates between contour points. This fills discrete contour points into continuous pixel points for subsequent processing; (5) Row and Column Pixel Statistics: We traverse each column and row of the image to count the number of white pixels, preparing for key region identification; (6) White and Black Region Extraction: We extract continuous intervals where the number of white pixels exceeds a certain threshold to determine the range of white regions. We use the white regions to infer black regions, providing references for key feature point extraction; (7) Line Fitting: We perform linear regression to fit lines to specific points. We fit the contour points to obtain linear equations and then calculate the distance and parallelism between lines; (8) Key Region Coordinate Calculation: We calculate the range of key columns and rows through the center points of regions. We extract and compute the key regions (column intervals and row intervals) in the image; (9) Distance Calculation: We use the fitted lines to compute distances between different regions and convert them into actual physical units. By calculating distances between key points or regions, we determine whether the inspection target meets the requirements.

After processing the captured cigarette pack label images, the resulting image displays the required lines for inspection. These lines are used to calculate relevant distances, which are then compared to predefined distance intervals. If the distances fall within the preset interval, the label is deemed acceptable; otherwise, it is considered defective. For defective cigarette pack labels, a rejection mechanism is initiated. First, the industrial computer sends a signal, which is transmitted through a USB-to-TTL communication cable to a 5 V-to-24 V step-up module (necessary because the PLC cannot directly recognize the high/low voltage levels from the industrial computer). After conversion, the signal is input into a relay, which then passes the signal to the PLC's I2.7 interface to

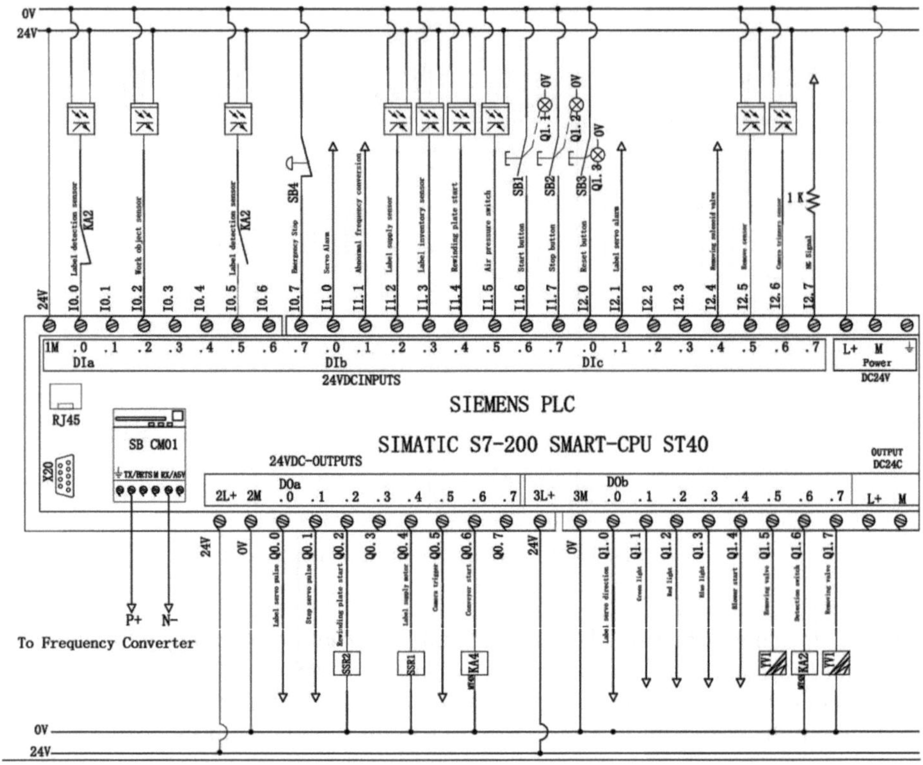

Fig. 14. Electrical control diagram of labeling machine.

trigger the rejection program. When the PLC receives the signal, it activates the rejection program, controlling the rejection mechanism to remove the defective label. The flowchart of the control system is shown in Fig. 15.

3.2 Human-Machine Interface Design and Operational Debugging

As shown in Fig. 16, the system mainly consists of seven interfaces: the user login interface, control interface, parameter setting interface, manual operation interface, production target setting interface, module input/output interface, and error log interface. The user login interface is designed to authorize different operation permissions based on user levels, ensuring the safety and controllability of the production equipment. The control interface is used to select the control state of each module. The parameter setting interface allows for the configuration of settings such as the number of pages to be separated, the number of separation cycles, and the labeling speed. The manual operation interface enables the control of servo motor start/stop, speed adjustments, resetting, and the communication status of the industrial control computer for each unit. The following figure shows the human-machine operation interface.

After confirming the correctness of the structural transmission components, the PLC and HMI programs developed on the upper-level PC were downloaded to the PLC and

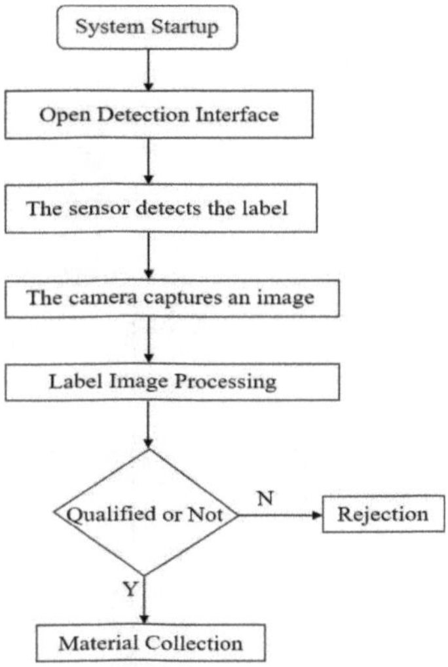

Fig. 15. Visual system control flowchart.

HMI via Ethernet communication. The variable table from the PLC was imported into the HMI program, enabling control of the device's various modules through the HMI interface. Comprehensive validation of the equipment's logical control, motion control, and other functionalities was performed, and the program was refined based on actual operational performance, ultimately resulting in a fully debugged and optimized control program.

Once the equipment control program was functioning correctly, the device was initially operated at a low speed to address any issues encountered during operation, primarily focusing on the stability of the paging mechanism and the label delivery system. Gradually, the operating speed of the equipment was increased, ensuring that it could operate smoothly at both low and high speeds and produce cigarette boxes meeting the required standards. Figure 17 shows the various stages of the labeling process. As illustrated, the cigarette boxes completed the labeling process, with the labeling accuracy controlled within ± 0.5 mm. This validates that the designed structure and con trol system function as intended, achieving the desired labeling functionality.

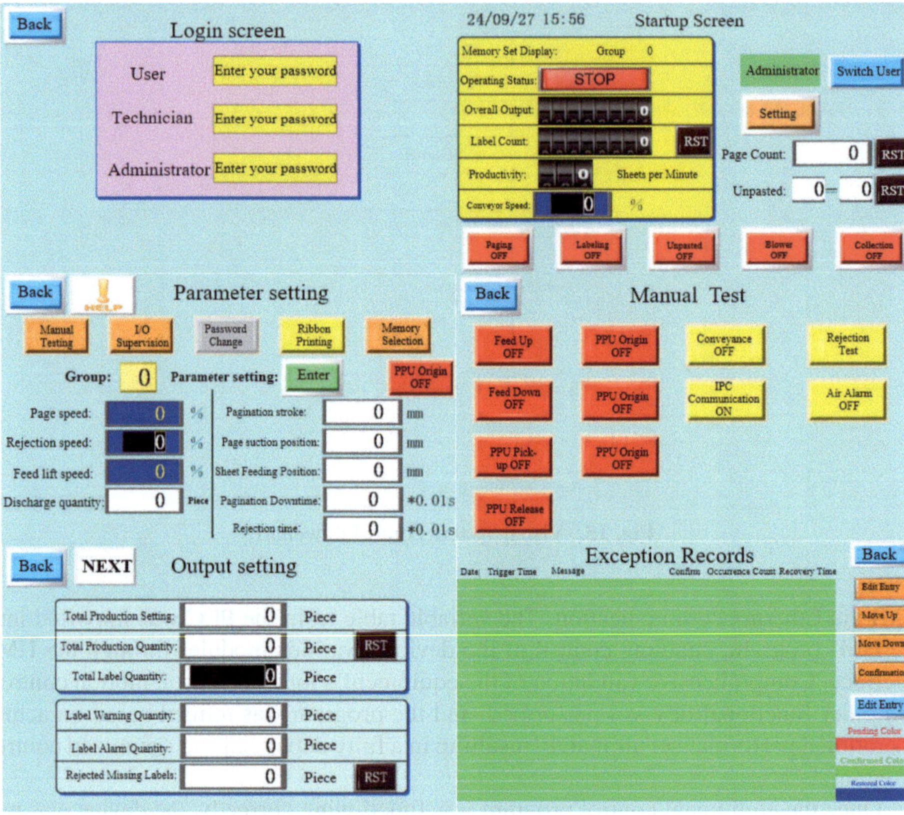

Fig. 16. Human-machine operation interface.

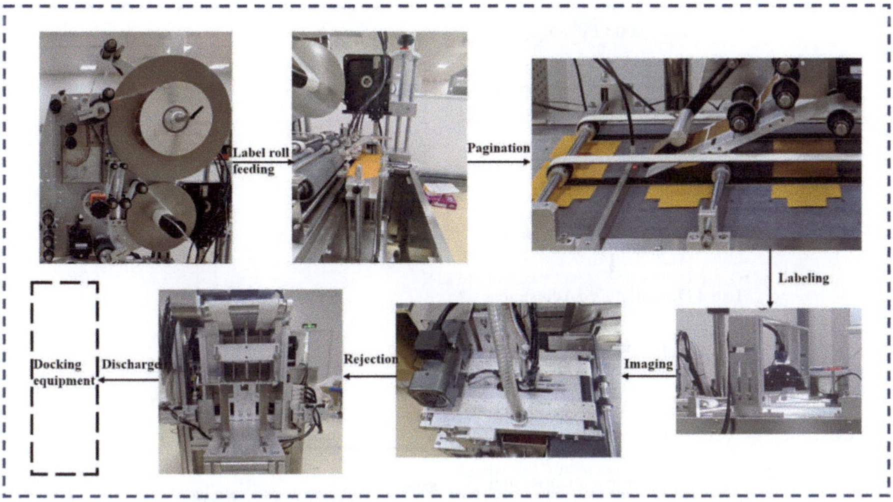

Fig. 17. Labeling stages.

4 Implementation Process and Technical Validation

4.1 Operating Procedure

The process of operating the labeling machine is shown in Fig. 18: First, we turn on the power and adjust the front and rear height of the labeling host. Next, we install the labels, adjust the peel board angle, and position the label detection eye and labeling eye. Then we conduct a test labeling, check the labeling effect, and perform accuracy correction. If any abnormalities occur, we will troubleshoot promptly. Once the labeling accuracy is confirmed to be correct, we begin official labeling. When the label roll is used up or needs replacement, we stop the labeling operation and replace the labels. Finally, after completing all labeling tasks, we turn off the power of the labeling machine.

First, we wrap the label along the path indicated in Fig. 19. Then, as shown in Fig. 20, we move the bottom guide belt ring to the same appropriate position. Finally, when the label is taut, we secure the tail end of the label's backing paper onto the label backing paper roll, as shown in Fig. 21.

Then, as shown in Fig. 22, we proceed to load the paper boxes. First, we loosen the movable handle and rotate the spacer notch to a position parallel to the rotating mechanism, allowing it to detach from the rotating mechanism. Next, we open the two cylindrical guardrails on the right side for the unloading operation. After loading is complete, as shown in Fig. 23, we return the two guardrails to their original position and secure the arc portion of the spacer onto the rotating mechanism. Finally, we tighten the movable handle to ensure that all components are stable and in place.

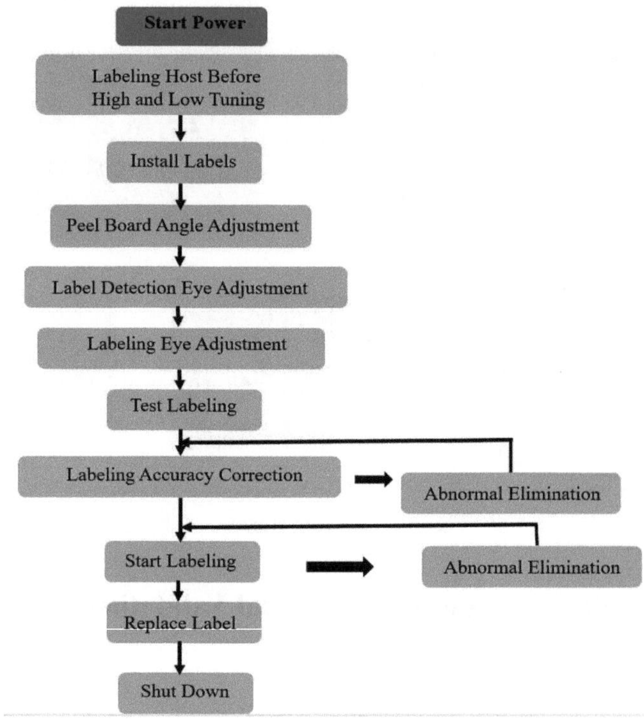

Fig. 18. Equipment operation flowchart.

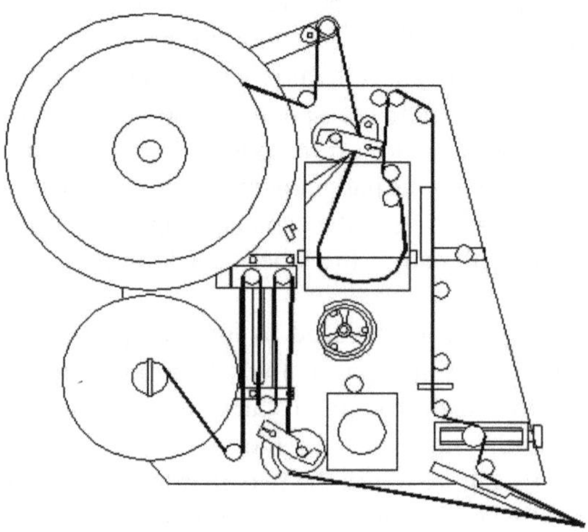

Fig. 19. Label wrapping path.

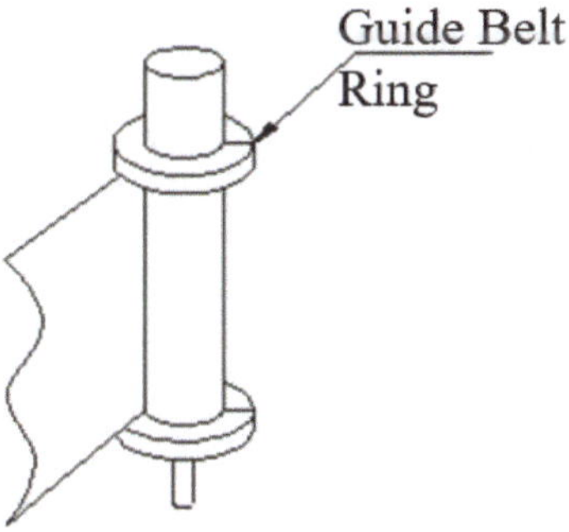

Fig. 20. Guide belt ring.

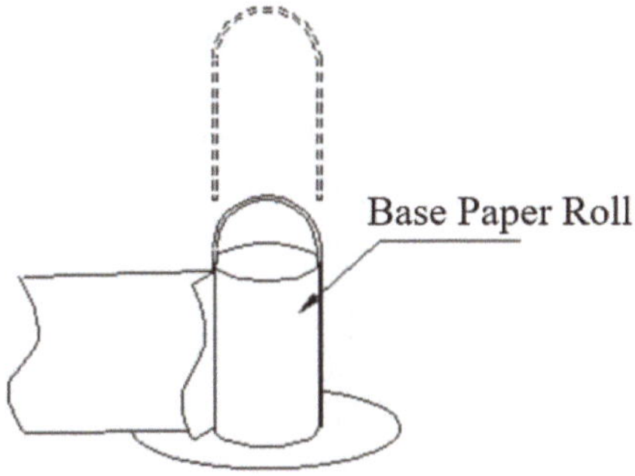

Fig. 21. Backing paper roll.

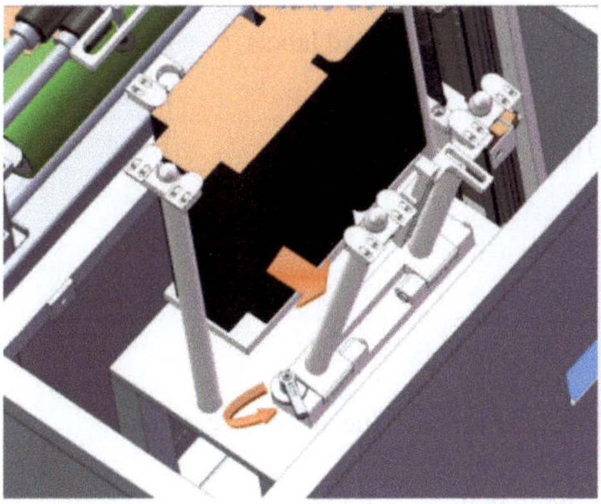

Fig. 22. Feeding process.

Fig. 23. Fastening process.

4.2 Labeling Effect Verification

After the verification of the cigarette box conveying system and visual detection system is completed, and the labeling machine has been debugged, the labeling conveying system is experimentally validated.

Experimental Method: We test the labeling success rate for different operational labeling speeds (80 packs/min, 100 packs/min, 115 packs/min) by recording once every 100 labels applied.

Experimental Materials: Cigarette boxes; labels (Table 2, 3 and 4).

From the experimental data in Tables 2, 3, and 4, it was found that the labeling success rate at 115 packs/min was the lowest. The labeling operations at speeds of 100 packs/min and 80 packs/min were relatively stable and met the expected requirements.

Table 2. Labeling quality rate of labeling machine at 80 packs/min.

Experiment No.	Total labels applied	Good labels	Bad labels	Good rate
1	100	100	0	100%
2	100	99	1	99%
3	100	99	1	99%
4	100	98	2	98%
5	100	98	2	98%

Table 3. Labeling quality rate of labeling machine at 100 packs/min.

Experiment No.	Total labels applied	Good labels	Bad labels	Good rate
1	100	99	1	99%
2	100	100	0	100%
3	0	98	2	98%
4	100	98	2	98%
5	100	98	2	98%

Table 4. The labeling quality rate of the labeling machine at 115 packs/min.

Experiment No.	Total labels applied	Good labels	Bad labels	Good rate
1	100	95	5	95%
2	100	93	7	93%
3	100	97	3	97%
4	100	98	2	98%
5	100	98	2	98%

5 Conclusion

This paper addresses the issues of poor labeling accuracy and low efficiency in current labeling machines by developing a new labeling machine. The design includes the cigarette box conveying system, label conveying system, rejection mechanism, discharge mechanism, and other structural components, as well as the driving elements for the labeling machine's vision system. By utilizing programming software to implement PLC and HMI programming for the labeling machine, the control speed and automation level of the equipment were improved. The designed components were verified to meet the requirements, and the equipment was confirmed to operate normally. Finally, experiments demonstrated the feasibility and accuracy of the new labeling machine.

References

1. Wang, R.X., Han, Z.H., Li, H.Y., Bian, L.T., Han, J.L., Jiang, J.: Design of a Labeling Machine System Based on PLC. Internet Things Technol. **14**(10), 73–74+77 (2024). https://doi.org/10.16667/j.issn.2095-1302.2024.10.019
2. Guo, K.N.: Design of a Fully Automatic Labeling Machine for bottled laundry detergent. Prog. Skill Maintenance. **06**, 103-106+153 (2022). https://doi.org/10.16184/j.cnki.comprg.2022.06.002
3. Xiao, P., Qin, A.P., Wu, L.L., Qin, H.L.: Design and research of a general-purpose self-adhesive labeling machine. Packag. Food Mach. **36**(05), 32–35 (2018) https://kns.cnki.net/kcms2/article/abstract?v=Ep7N7zfewySoUojTfSt9XRUD0BfiFXmo8tL1pk8PxYyI8oTcQmnmHC3ol8OgcduPl6rV_hRNkE6gilzyHlw2OfVN5CQ8jK6xihyXfzPqedTgj7EGXm1PGl3ipAHi8uSo4AHDd1VomFmzAV3Gq-oDaYagJnnxHtkIz7vli5isc5_huwd5Q82bPMGC70vOGxgbWzgTcT4T13U=&uniplatform=NZKPT&language=CHS
4. S. Wang. Design and research of an automated labeling machine for FRP tanks in water treatment. Wuhan University of science and technology, 2020. DOI: https://doi.org/10.27381/d.cnki.gwlgu.2020.001580.
5. Gurney, M.: Control system retrofit extends old labeling machine life, improving throughput by 25%. INTECH. **60**(1), 52 (2013) https://www.proquest.com/trade-journals/control-system-retrofit-extends-old-labeling/docview/1331780582/se-2
6. Cho, D.-H., Bhushan, B.: Friction and wear of various polymer pairs used for label and wiper in labeling machine, 10–19 (2016). https://doi.org/10.1016/j.triboint.2016.02.019
7. Sakmi, K.U.B.E.: Labeling machine running at high speed with zero shutdown. Wine Beverage Technol. Equipment. **04**, 43 (2017) DOI: CNKI: SUN: JYLZ.0.2017-04-025
8. C. S. Fahn and K. L. Lin. Method for labeling connected components and computer system using the method. 2013: 233–237. DOI: US20110158534 A1.
9. Chen, J.X.: Design and analysis of an automated labeling machine. Machinery. **59**(05), 20–21+49 (2021). https://doi.org/10.3969/j.issn.1000-4998.2021.05.007
10. Zhang Z., Xu M., and Chen S. Research on cylindrical labels unfolding algorithm based on machine vision 2021 IEEE 5th Advanced Information Technology, Electronic and Automation Control Conference (IAEAC). IEEE, 2021.5: 2, 568–2, 56572. DOI:https://doi.org/10.1109/IAEAC50856.2021.9390987.

Research on Digital Closed-Loop Circuit of Quartz Flexible Accelerometers

Mingyang Xiao[1], Yu Xia[1], Xinlei Yu[1], and Yongmeng Zhang[2](✉)

[1] College of Intelligence Science and Technology, National University of Defense Technology, Changsha 410073, China
[2] National Key Laboratory of Equipment State Sensing and Smart Support, National University of Defense Technology, Changsha 410073, China
zhangym@nudt.edu.cn

Abstract. Quartz flexible accelerometer is a kind of high-precision sensor widely used in inertial navigation system, the traditional quartz flexible accelerometer adopts analogue circuit to achieve the detection of differential capacitance and closed-loop feedback control. There are problems such as parameter calibration, error compensation process complexity, etc., which is difficult to meet the digital development needs of construction machinery and vehicle engineering, so there is an urgent need to develop digital circuits. This paper takes the existing accelerometer as the research object, on the basis of analyzing the domestic and foreign research dynamics, it designs a set of quartz flexible accelerometer digital closed-loop measurement and control circuits, completing the test of measurement and control circuit performance. The experimental validation shows that the linearity of differential capacitance detecting module reaches 0.998, with the fabricated accelerometer demonstrating 1679.491 LSB/g scale factor and 47 μg zero-bias instability.

Keywords: Quartz flexible accelerometer · Closed-loop measurement and control circuit · Differential capacitive detection · Incremental PID · Gravity field calibration

1 Introduction

As one of the core measurement elements of inertial navigation systems, the performance of high-precision accelerometers determines the accuracy and reliability of the navigation system [1]. As an important branch of high-precision accelerometers, quartz flexible accelerometers are widely used and play an important role in aerospace, equipment guidance, inertial navigation and other fields by virtue of their advantages such as small size, high measurement accuracy and strong anti-jamming ability [2].

In recent years, with the continuous advancement in intelligentization and automation of construction machinery, alongside the rapid development of intelligent connected vehicle (ICV) technologies in automotive engineering, the accuracy and stability requirements of accelerometers are increasing. The traditional quartz flexible accelerometers

S. Halgamuge et al. (Eds.): ICACMVE 2025, LNME, pp. 290–300, 2026.
https://doi.org/10.1007/978-981-95-6911-3_27

use analogue servo loops, which inherently have problems such as temperature drift, high noise, signal distortion, etc., which affect the control accuracy and stability of the servo loops, making it difficult to meet the requirements of high-precision applications [3]. In addition, in complex working environments, quartz flexible accelerometers are susceptible to a variety of interference and noise. The functions of the traditional analogue circuit need to be implemented through the chip, and the control function is usually relatively single, which makes it difficult to achieve advanced control strategies such as digital filtering, calibration and correction,[4] cannot flexibly cope with a variety of complex working environments.

Compared with the traditional analogue servo loop, digital servo loop shows its powerful advantages. Digital servo loop can adopt high-precision algorithms to process and analyze signals, such as digital filtering, calibration and correction, which can effectively suppress noise and interference and improve the measurement accuracy and stability of the system. Digital servo loops are usually implemented by editable logic devices or microprocessors, which can flexibly write control programs and algorithms according to the needs, realizing signal processing and control strategies in complex environments; Moreover, the digital servo loops have highly integrated design capabilities, which can effectively reduce costs and improve production efficiency. Therefore, it is of great practical significance to study the digital measurement and control circuit of quartz flexible accelerometer to improve the performance of inertial navigation system and promote the development of related fields.

At present, a number of domestic research institutions and universities have been engaged in the research and development of quartz flexible accelerometers, achieving a series of important results. However, there is still a gap with foreign research in terms of high precision, integration and intelligence.

The aim of this study is to design a set of high-performance quartz flexible accelerometer digital closed-loop measurement and control loop to improve the inadequacy of the existing technology and the accelerometer performance, contributing to the development of China's inertial navigation and other fields.

2 System Design and Methodology

2.1 Quartz Flexible Accelerometer Principle of Operation

The working principle of quartz flexible accelerometer is to achieve acceleration measurement through the offset of the quartz pendulum. When the accelerometer is subjected to external acceleration, the quartz pendulum will be under the action of the inertia force to occur a certain angle of deviation, so that the pendulum and the two sides of the pole plate constitutes the differential capacitance changes, differential capacitance detection module will be amplified by the small capacitance changes into a voltage transferred to the A/D converter module by the A/D converter module for the acquisition of the data by the digital controller to the processing and analysis of the digital signal, and finally through the D/A conversion module and the voltage to current module to convert the digital signal into a current signal to the force signal. Finally, through the D/A conversion module and the voltage-to-current module, the digital signal is converted into a current signal and transmitted to the torque coil, which pulls the quartz pendulum back to the equilibrium position in real time, reduces the nonlinearity, and improves the resolving power of the accelerometer.

The signal flow conversion schematic of the quartz flexible accelerometer designed in this study is shown in Fig. 1:

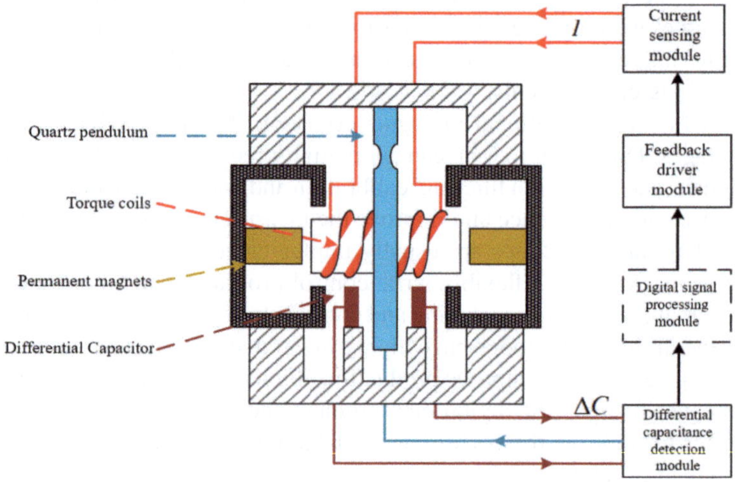

Fig. 1. Schematic diagram of quartz flexible accelerometer signal flow conversion.

2.2 Digital Closed-Loop Measurement and Control System Architecture

The digital servo loop of the quartz flexible accelerometer mainly consists of a differential capacitance detection module, an A/D conversion module, a controller module, a D/A conversion module, and a power amplifier module [4], as shown in Fig. 2:

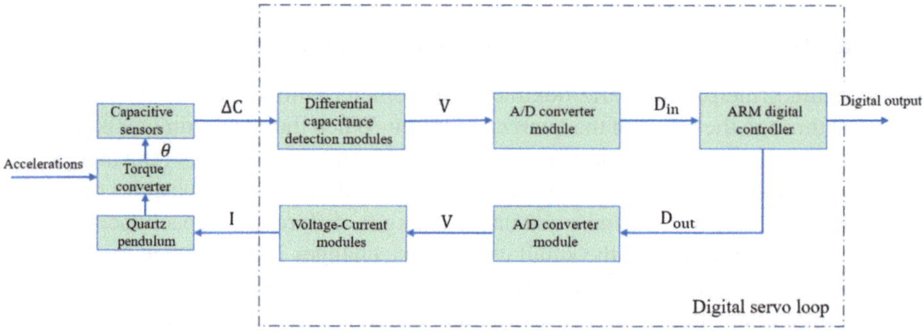

Fig. 2. Schematic diagram of digital servo loop composition of quartz flexible accelerometer.

A/D Converter Module. The A/D conversion circuit plays a key role in the system as a bridge connecting analogue and digital signals. In this project, an ADC data acquisition module is designed according to the experimental requirements, and the schematic diagram is shown in Fig. 3. This device uses a high-speed and high-precision successive approximation type analogue-to-digital converter, with up to ±20 V over-voltage protection circuit and a multiSPI digital interface to achieve SPI communication between the control module and ADC data acquisition module.

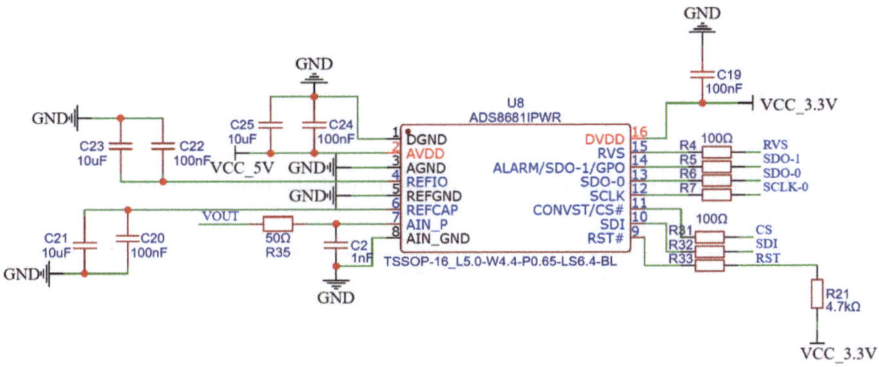

Fig. 3. Schematic diagram of A/D conversion circuit.

D/A Converter Module. The main function of the D/A conversion circuit is to convert the digital control signal into an analogue signal to achieve feedback control of the accelerometer. In the design, the choice of high-precision DAC8563 conversion device, the device can quickly and accurately convert digital signals into analogue voltage signals. After the D / A conversion access to the power amplifier circuit, the choice of OPA2363AIDGSR low-noise operational amplifier and 2 N1711 transistor to build the conversion circuit, the D / A converted analog signal for power amplification, to ensure that the output current signal can effectively drive the torque converter to achieve accurate control of the accelerometer pendulum, so as to maintain the system's closed-loop stable operation. The schematic diagram is shown in Fig. 4:

Differential Capacitance Detection Module. Differential capacitance detection circuit as the core part to achieve accurate acceleration measurement, the differential capacitance detection circuit designed in this study mainly consists of C/V conversion preamplifier, switched demodulation, switched capacitance low-pass filter, continuous low-pass filter, and differential to single-ended output module, in which except for the differential to single-ended gain constant 1, the rest of the levels of the gain can be adjusted. The block diagram of the system is shown in Fig. 5:

As can be seen from the above diagram the differential capacitance detection module system consists of three main modules:

The key module is mainly responsible for signal processing. The C/V converter preamplifier uses square wave excitation to convert the differential capacitance change into a voltage signal, which is demodulated by a cross-coupled switch, then filtered

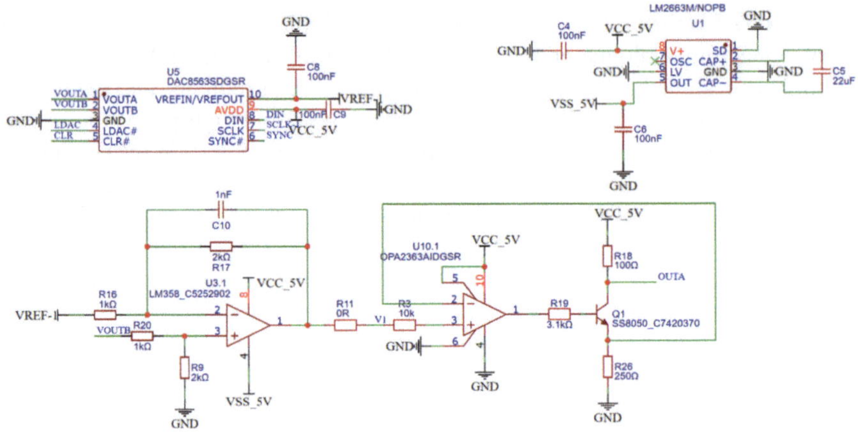

Fig. 4. Schematic diagram of D/A conversion circuit and power amplifier circuit.

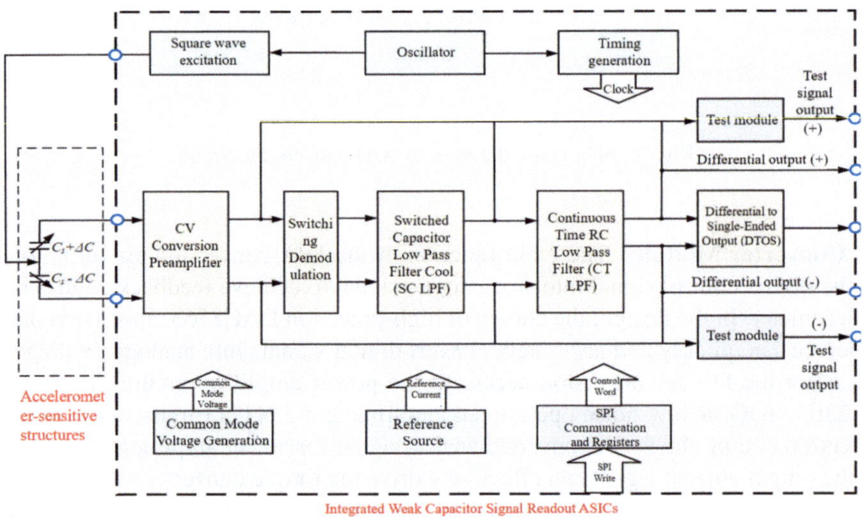

Fig. 5. Differential capacitance detection circuit system block diagram.

out by a two-stage low-pass filter to remove the high-frequency noise, and suppressed by a fully differential structure to suppress the common-mode interference, and finally adapted to the subsequent circuits by a differential-to-single-ended output.

The base module consists of an oscillator, a timing generation module, a square wave excitation module, a common mode voltage generation module and a reference source. The switched-capacitor detection circuit requires a clock signal for normal operation, so it needs an oscillator and a timing generator to generate a clock with a certain timing relationship. the implementation of C/V conversion requires a square-wave excitation on the differential capacitor pole plate, the square-wave excitation module is the oscillator to generate a clock signal for the frequency division process and adjust the square-wave amplitude after the application of the common pole plate, the above three modules are collectively referred to as clock The above three modules can be collectively referred to as the clock generation circuit. In switched capacitor detection circuits and DTOS circuits, it is necessary to set the common mode voltage, so the common mode voltage generation module is needed. The operational amplifier is the core of the ASIC for the differential capacitance detection system, and the bias current required for its normal operation is provided by the reference source.

In order to facilitate the debugging and testing of the whole system, some auxiliary modules are also designed, which mainly include SPI communication module, registers, and testing module. In order to realize the adjustment of the gain of the main path and the oscillator frequency and other module parameters, the SPI communication and register modules are designed, the registers are used to store the configuration words, and the SPI communication module communicates with the external (host computer). The output signals of each module on the main path can be output to the pins of the chip through the test module, and thus the waveforms of the key nodes can be observed during the debugging process.

With the above scheme, the detection of differential capacitance of the accelerometer head can be achieved with analogue differential output or analogue single-ended output. Good temperature characteristics can be achieved by internal reference current source. The ASIC can be configured by the upper computer to achieve gain and bandwidth adjustment, as well as the selection of internal test points and other functions. The configuration parameters of the ASIC can be solidified.

3 Experiments and Results

3.1 Differential Capacitance Detection Module Performance Test

The ratio of the output voltage of the differential capacitance detection module via the sampling resistor to the corresponding differential capacitance is the sensitivity of the capacitance detection section. The test experiment is carried out by configuring the trimming capacitance of the upper and lower plates in the differential capacitance detection module. After the experimental test, the relationship between the output voltage of the differential capacitance detection module and the differential capacitance curve is shown in Fig. 6 below:

According to the measured data, the linearity of the differential capacitance detection module is obtained as 0.998 [5]. Comparing the graphs, it is found that the relationship curve between the output voltage and the differential capacitance is basically a straight line, and the linearity of the differential capacitance detection part is high.

However, the relationship curve does not pass through the graph origin. After analysis, the reason for this situation is that the experimental test process uses the configuration of the trimming capacitance of the upper and lower plates in the differential capacitance test module to achieve the capacitance difference. The function of trimming capacitance is to ensure that the initial value of the connected differential capacitance sensor is zero, in the case of zero trimming capacitance of both the upper and lower plates does not guarantee that there is no capacitance difference, so there is still a voltage at the origin.

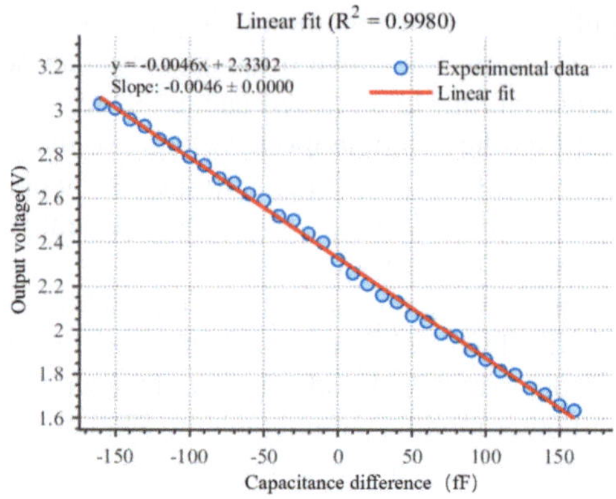

Fig. 6. Differential capacitance detection module fitting curve.

3.2 Gravity Field Tumbling Experiment and Calibration

Gravity field tumbling experiments are performed by connecting the head of a quartz flexible accelerometer to the designed digital closed-loop detection circuit and placing the accelerometer head on vibration-isolated precision indexing heads, turntables, and other test equipment in preparation for tumbling tests.

The gravity field tumbling test experiment is subdivided into two test modes, door state and pendulum state, the door state corresponds to the accelerometer PA axis coinciding with the equipment slewing axis, and the pendulum state corresponds to the accelerometer OA axis coinciding with the test equipment slewing axis, and both of them let the meeting axis rotate in the horizontal plane [6].

By rotating the test equipment, the orientation of the accelerometer relative to the gravitational acceleration is changed, which in turn changes the acceleration component on the input axis of the accelerometer, thus realizing the gravity field tumbling test. Use a multimeter to detect the voltage of the sampling resistor in the digital closed-loop detection loop, and record the data after the multimeter number is stabilized, and record the data once every 15° of rotation [7].

Since the sensitive axis of the quartz flexible accelerometer head is component to the acceleration of gravity during the experimental tests:

$$a_i = g \cdot \sin\theta_i \tag{1}$$

In the above equation a_i is the acceleration applied to the quartz flexible accelerometer head at the ith recording, g is the acceleration due to gravity, θ_i is the angle between the pendulum and the vertical direction.

From the above formula, the acceleration size will change with the quartz flexible accelerometer pendulum deflection angle showing sinusoidal law, and the sampling resistor voltage and accelerometer by the acceleration corresponding to the two terminal voltage should also be with the accelerometer inside the pendulum angle change showing sinusoidal changes, the experimental results corresponding to the image is as follows (Fig. 7):

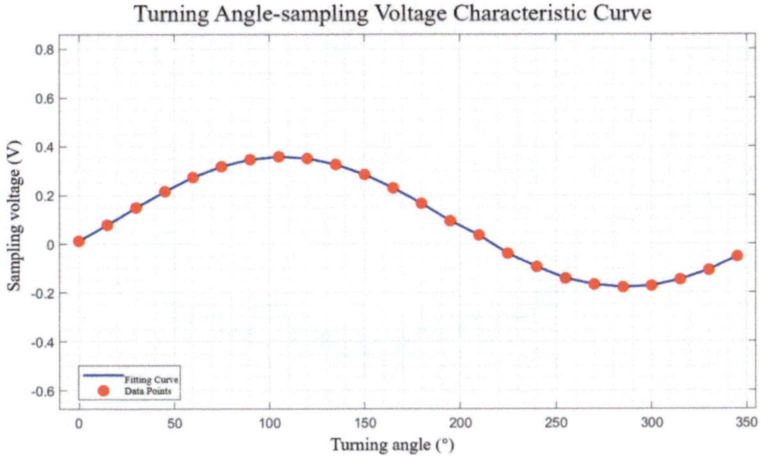

Fig. 7. Rotation angle-sampling voltage characteristic curve for gravity tumbling experiment.

Obviously, the voltage at the ends of the sampling resistor shows a sinusoidal pattern as the angle between the pendulum and the vertical direction changes.

The designed quartz flexible accelerometer was then calibrated using the four position experimental calibration method and the following mathematical model was used to test the quartz flexible accelerometer:

$$E = K_0 + K_1 a_1 + K_2 a_2^2 \tag{2}$$

In the above equation, E is the feedback voltage of the measurement and control loop of the quartz flexible accelerometer; K_0 is the bias value; K_1 is the scale factor of the accelerometer; and K_2 is the second-order nonlinearity coefficient [8].

Take the measured data at four positions of θ =0°, θ =90°, θ =180°, θ =270° and substitute them into the above equation, and use the least squares method to solve the best-fit solution, which solves the expression corresponding to the accelerometer as:

$$E = 0.0781 + 0.02615 a_1 + 0.0086 a_2^2 \tag{3}$$

The expression corresponding to the solved accelerometer yields a scale factor of 0.02615 V/ ms^2 for the accelerometer corresponding to the measurement, which is approximately 1679.491 LSB/g.

3.3 Zero-Bias Instability Test

When there is no acceleration input to the quartz flexible accelerometer, the closed-loop measurement and control loop of the accelerometer still has a corresponding feedback voltage output, defined as the zero bias of the accelerometer. When there is no external acceleration input to the accelerometer, the feedback voltage of the measurement and control loop captured by the designed digital measurement and control system can characterize the zero bias.

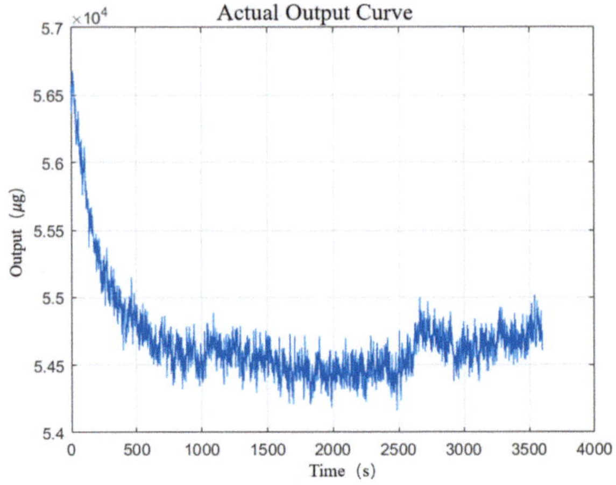

Fig. 8. Zero-bias test actual output curve.

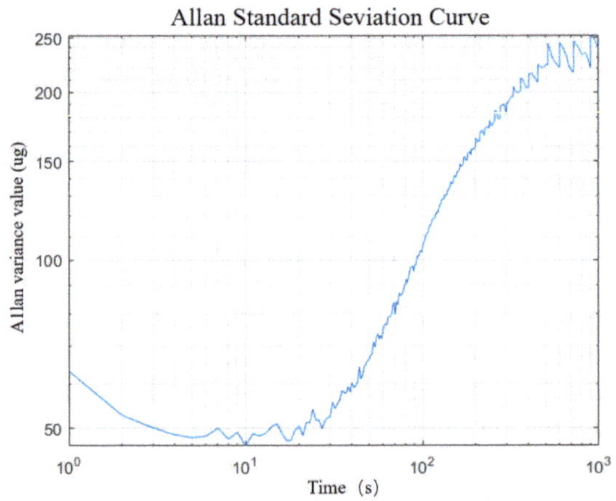

Fig. 9. Zero-bias instability test.

In this paper, the quartz flexible accelerometer is left at rest (so that the accelerometer input is zero), and the accelerometer zero bias data (shown in Fig. 8 above) is collected for 1 h, and the Allan variance is calculated, and finally the zero bias instability of the designed quartz flexible accelerometer is 47ug, as shown in Fig. 9 above. The zero-bias instability of the designed quartz flexible accelerometer is 47ug, as shown in Fig. 9 above.

4 Conclusions and Outlook

Aiming at the traditional analogue quartz flexible accelerometer's analogue rebalancing loop existing system noise, adjusting the system parameters must be redesigned system and other problems, this paper is committed to designing a quartz flexible accelerometer digital closed-loop measurement and control loop, which is mainly composed of differential capacitance detection module, A/D analogue-to-digital converter module, ARM controller module, D/A to digital analogue converter module, and voltage-to-current module several parts. The calibration of the designed accelerometer and the zero-bias non-stability test are completed by using the vibration isolation rotary table to perform the gravity tumbling experiment on the adapted pendulum accelerometer.

The experimental results show that the linearity of the differential capacitance detection circuit designed in this study reaches 0.998, which is capable of detecting weak signals accurately; the scale factor of the accelerometer is 1679.491 LSB/g after the calibration by the four-position experimental calibration method; and the zero-bias instability is 47 μg, which is a significant improvement in the accuracy and stability of measurement compared with the traditional scheme, according to the Alain variance calculation [9].

References

1. Yuan, Z.H.A.O., Yaliang, S.H.A.N.G., Yan, L.I.U., et al.: Construction of a performance analysis platform for quartz flexure accelerometers. J Bull. Guidance. **30**(05), 193–195+202 (2010). https://doi.org/10.15892/j.cnki.djzdxb.2010.05.005
2. Yan, J.: Research on Key Technology of Quartz Flexure Accuracy Improvement [D]. Zhejiang University (2020)
3. Chuan, C., Luyan, N., Liyong, L., et al.: Improved design of temperature stability of quartz flexure accelerometer. Chin. J. Inertial Technol. **30**(5), 658–665,673 (2022)
4. Ming, W., Wenbin, T.: Design of digital rebalancing loop for quartz flexure accelerometer. Sens. Microsyst. **3**, 103-105–103-109 (2015)
5. WöskeCA, F., Huckfeldt, M., Rievers, B.: Tailored accelerometer calibration by POD for thermospheric density retrieval with GRACE and GRACE-FO. Adv. Space Res. **74**(10), 4517–4542 (2024)
6. Wu, B., Tiantian, H., Haiyan, Z.: Study of dynamic modeling method for quartz flexible accelerometer. In: 2015 IEEE 12th International Conference on Electronic Measurement & Instruments (2015)
7. Silvestre, C., Cardeira, B., et al.: Accelerometer calibration and dynamic bias and gravity estimation: analysis, design, and experimental evaluation (article). IEEE Trans. Control Syst. Technol. **19**(5), 1128–1137 (2011)
8. Liu Chen, X., Zibin, L.Y.: Temperature characteristic test and modeling of quartz flexible accelerometer. J. Astronaut. Metrol. Meas. **40**(4), 69–74 (2020)
9. Wenhui, Y., Xiaodong, W., Bin, Q., et al.: Design of rebalancing loop for quartz flexible accelerometer based on digital controller. Navig. Control. **21**(1), 82–91 (2022)

Engineering Scene Environment Modeling Method Based on Improved FAST-LIVO

Yongming Bian[1], Linyuan Xie[1], Shubin Yu[1], and Li Chen[2(✉)]

[1] School of Mechanical Engineering, Tongji University, Shanghai 201804, China
[2] School of Automotive Studies, Tongji University, Shanghai 201804, China
lilychen@tongji.edu.cn

Abstract. As engineering scenes increasingly require higher precision, real-time performance and robustness from perception systems, traditional approaches are becoming inadequate for handling complex and dynamic environments. To tackle this issue, we present an enhanced multi-sensor collaborative mapping algorithm based on improvements to the FAST-LIVO framework. The initial step in this process is the optimization of the feature point extraction method. Furthermore, the Lucas-Kanade (LK) feature point optical flow method is introduced to replace the direct method, and the RANSAC algorithm is applied to optimize feature matching and eliminate mismatched points. Moreover, the error-state iterated Kalman filter (ESIKF) is constructed on the basis of reprojection error in order to enhance fusion accuracy. A novel visual map management scheme has been developed to enhance system operational efficiency. In order to validate the practical effectiveness of the algorithm, a hardware platform was established and a series of tests were conducted. The experimental results demonstrate that the mean absolute trajectory error of the proposed algorithm in a typical industrial park environment was reduced by 39%, with the map construction exhibiting greater consistency and continuity.

Keywords: Engineering Scene · Environment Modeling · Multi-sensor Fusion

1 Introduction

The laser-inertial-visual odometer integrates information from lidar, inertial measurement units and visual sensors to enable robust pose estimation and high-quality 3D map construction. This makes it a vital tool for achieving high-precision environmental perception [1]. Nevertheless, in the complex engineering scenarios examined in this paper, there are frequently multiple dynamic and static factors to be considered, including engineering machinery, on-site workers, fixed facilities (e.g. containers) and work objects [2]. These factors limit the traditional LIVO system's capabilities in feature point extraction and tracking, error modeling, and map maintenance strategies, making it difficult to meet the actual requirements of unmanned engineering machinery for long-term, high-intensity autonomous operations in complex industrial parks.

Therefore, this paper proposes a novel and improved environmental modelling technique for engineering scenarios that is based on FAST-LIVO [3], aiming to enhance the

S. Halgamuge et al. (Eds.): ICACMVE 2025, LNME, pp. 301–310, 2026.
https://doi.org/10.1007/978-981-95-6911-3_28

system's adaptability and mapping accuracy in complex engineering environments. The framework is illustrated in the Fig. 1 below.

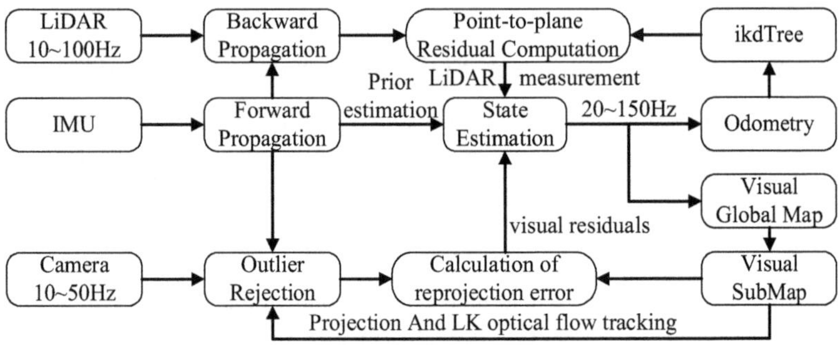

Fig. 1. Algorithm framework of this paper

2 Improvements to FAST-LIVO Algorithm

2.1 Optimize Feature Point Extraction Methods

In environmental perception and mapping tasks, high-quality feature point extraction is a prerequisite for achieving accurate feature matching and visual-inertial joint optimization. Traditional methods often employ feature extraction techniques based on prominent textures, such as SIFT [4], SURF [5] and ORB [6]. However, these methods are susceptible to failure when confronted with complex lighting conditions, low texture density and dynamic interference in industrial parks. Moreover, they impose substantial computational costs.

To enhance the system's robustness in real-world scenarios, this paper proposes a sparse visual feature extraction strategy that combines laser point cloud and image information based on the sparse direct method concept of the FAST-LIVO system. This strategy eschews reliance on conventional texture features such as corners or edges in images. Instead, it uses representative sparse points from a map constructed from laser point clouds, projects them onto image frames and extracts grayscale image patches at their corresponding positions to construct "visual map points".

2.2 Optimize Feature Point Association and Match Strategies

In laser-inertial-visual fusion systems, accurate and stable visual feature point association matching is crucial for improving pose estimation accuracy and mapping quality. In this section, based on the FAST-LIVO system framework, we introduce a visual feature association matching strategy that combines LK optical flow and IMU pre-integration, and use the RANSAC algorithm to eliminate mismatches.

Firstly, the LK optical flow method is introduced. In the laser-inertial-visual fusion odometer system, the incorporation of visual data has been shown to enhance the perception capabilities of lidar, particularly in scenarios characterised by sparse point clouds or degraded environments. Visual data has been shown to facilitate the acquisition of rich texture features, thereby enhancing the system's capacity for accurate pose estimation [7]. In this section, the LK optical flow method is introduced into the FAST-LIVO system with a view to replacing the traditional direct method. The aim is to enhance the system's real-time mapping and perception capabilities in complex construction environments within industrial parks. Optical flow is the term given to the study of the movement patterns of pixels over time in an image sequence. The objective of optical flow is to track the position changes of pixels between consecutive frames [8], as illustrated in Fig. 2.

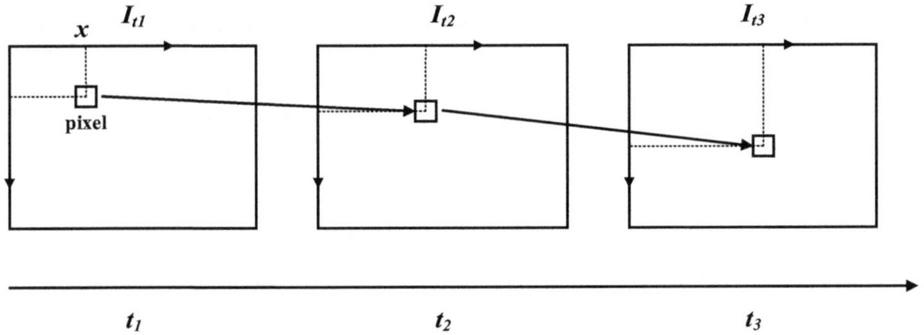

Fig. 2. LK optical flow method

Secondly, IMU pre-integration is introduced to improve the tracking performance of the LK optical flow tracking method. The system uses the sparse 3D point cloud constructed by the laser radar and selects stable 3D feature points P_w with semantic labels as the optical flow tracking objects. Between time points t and $t + \Delta t$, the IMU continuously provides acceleration and angular velocity data, and the system estimates the pose transformation through IMU pre-integration:

$$T_{t+\Delta t}^{t} = \exp\left(\hat{\xi}_{\Delta t}\right) \bullet T_t \tag{1}$$

Among them, $\hat{\xi}_{\Delta t}$ represents the motion increment calculated by IMU pre-integration, and exp(•) represents the exponential mapping from Lie algebra to Lie group.Subsequently, by using the initial pose $T_{t+\Delta t}^{t}$, the three-dimensional feature points are projected onto the current image frame to obtain the initial position of the optical flow tracking:

$$p_{t+\Delta t}^{init} = \pi\left(T_{t+\Delta t}^{t} \bullet P_w\right) \tag{2}$$

In the formula, $\pi(\bullet)$ represents the camera projection model. This method significantly reduces the drift error in the initial position estimation of traditional optical flow methods, significantly improves the success rate and convergence speed of optical flow matching and demonstrates stronger robustness, especially in cases of rapid movement or occlusion interference.

Finally, the RANSAC [9] algorithm is employed to eliminate mismatches, thereby ensuring the geometric consistency and estimation accuracy of the matching results. This process effectively suppresses mismatches that may occur in the optical flow method, thus improving the reliability of the visual front end and the system's overall estimation accuracy.

2.3 ESIKF Based on Reprojection Error

In order to further enhance the estimation of poses, this paper proposes the incorporation of observation residuals derived from reprojection within the observation update process of the ESIKF. ESIKF uses error states to define system state deviations and employs iterative optimization strategies to reduce the impact of nonlinear errors on estimation accuracy [10].

Under the error state filtering framework, the system state can be represented as x, including: position $p \in R^3$, velocity $v \in R^3$, rotation $R \in$ SO(3), and sensor bias (IMU gyroscope and accelerometer zero bias) $b \in R^6$.The error model for the system state is defined as follows:

$$\delta x = \begin{bmatrix} \delta p & \delta v & \delta \theta & \delta b \end{bmatrix}^{\mathrm{T}} \tag{3}$$

Among them, $\delta\theta$ is the rotation error, which is usually modeled using Lie algebra perturbation:

$$R = R_{\mathbf{nom}} \bullet \exp(\delta\theta) \tag{4}$$

Next, we proceed with observation residual modeling. In the ESIKF observation update process, we define the reprojection error as the observation residual. Among them, p_i^{abs} is the actual observed feature point position, and $\pi\left(X_i^C\right)$ is the theoretical projection point coordinate.

$$z_i = p_i^{obs} - \pi\left(X_i^c\right)\# \tag{5}$$

The observation equation can be expressed as Eq. (6). Here, $h(x) = \pi\left(X_i^c\right)$ is a nonlinear projection function, and v is Gaussian distribution measurement noise.

$$z_i = h(x) + v \tag{6}$$

Finally, in the ESIKF observation update equation, the Kalman gain is used for state correction:

$$K_t = P_t H_t^T \left(H_t P_t H_t^T + R_t\right)^{-1} \tag{7}$$

$$\delta x_t^+ = \delta x_t^- + K_t z_t \tag{8}$$

$$P_t^+ = (1 - K_t H_t) P_t^- \tag{9}$$

Among them, K_t is the Kalman gain matrix, P_t is the state covariance matrix, H_t is the observation Jacobian matrix, R_t is the measurement noise covariance matrix.The ESIKF optimization strategy enhances the weight of visual information in the state estimation process and significantly suppresses drift-caused errors.

2.4 Design Efficient Visual Map Management Scheme

In a laser-inertial-visual fusion odometry system, the core task of the visual map management module is to maintain stable tracking of visual feature points between consecutive frames, ensuring that visual observation information maintains good continuity and stability throughout the system's operational cycle [11].

In this study, high-precision sparse key points extracted from LiDAR data are leveraged as the foundational feature sources. These are combined with grayscale image patches to acquire corresponding visual information. The processes of feature point selection, tracking, and management are comprehensively optimized, enabling the robust construction and consistent maintenance of the visual map.

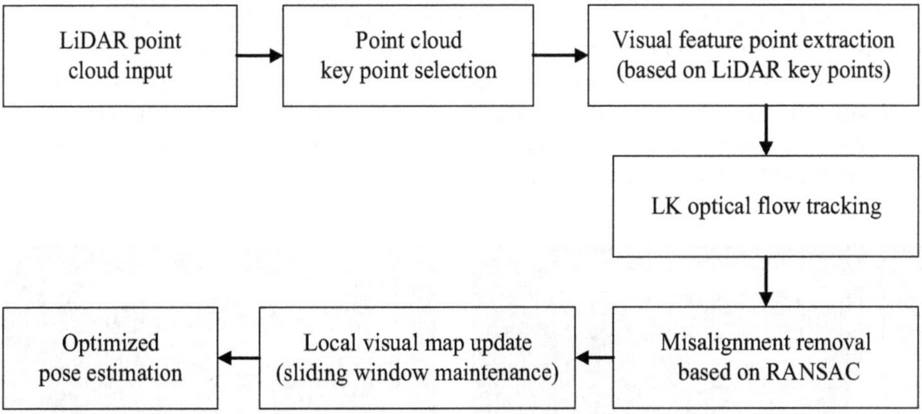

Fig. 3. Visual Map Management Flowchart

Fig. 3 presents the visual map management flowchart devised in this study. The structure of the text is as follows:

1. Feature point extraction: Based on the LiDAR point cloud map, sparse key points with structural stability and spatial representativeness are selected as the source of visual feature points. For related content, see Sect. 2.1.
2. Feature Point Tracking and Matching: Initial tracking employs the LK optical flow method guided by IMU pre-integration, combined with the RANSAC algorithm to remove mismatched points, ensuring the robustness and geometric consistency of the tracking process. Detailed content is provided in Sect. 2.2.
3. Local Map Management and Optimization: A sliding window mechanism is introduced to maintain a local visible map, and key frame filtering and dynamic point removal mechanisms are employed to enhance system efficiency and traceability.

3 Engineering Scene Environment Modeling Experiment

The experimental platform is equipped with a lidar system, an IMU, and a stereo depth camera. The lidar system uses the RS-LiDAR-16 which operates at the rotational speed of 600 rpm and has a point cloud scanning frequency of 10 Hz. The camera uses the

RS series stereo depth camera, which provides high-resolution RGB images and depth map information. Additionally, a 9-axis inertial measurement unit (model CH110) is employed, with angular velocity and acceleration output frequencies of 100 Hz. Prior to use, the IMU's internal parameters must be calibrated via Allan variance analysis and the calibration of the IMU used in this study was completed in advance. Furthermore, sensor preprocessing tasks have been completed in advance. The experiments for the improved algorithm were conducted on the Ubuntu 18.04-Melodic operating system. The test scenario is an industrial park scene from the NTU VIRAL dataset, which includes the dynamic and static environmental factors mentioned in this paper.

Fig. 4 shows the mapping results of FAST-LIVO and the algorithm proposed in this paper, allowing for a visual assessment of their mapping quality. Fig. 4(a) shows the point cloud map generated by FAST-LIVO. Overall, the map boundaries are relatively clear and the spatial structure exhibits a certain degree of continuity. However, after prolonged operation, the point cloud exhibits significant ghosting and distortion. Fig. 4(b) shows the point cloud map generated by the improved algorithm. By incorporating its key improvements, a noticeable enhancement in mapping quality is evident. The system now provides more precise details for static structures such as trees and obstacles, and is less affected by dynamic environmental factors.

(a)FAST-LIVO (b)Improved-LIVO

Fig. 4. Comparison of the effects of mapping

Next we conduct a comprehensive comparative analysis of the trajectory estimation accuracy of the original FAST-LIVO algorithm and the improved algorithm proposed in this paper in a park environment, focusing on two aspects: overall trajectory performance and absolute path error (APE). Based on this analysis, we will engage in an in-depth discussion, incorporating the improvements made in this paper.

First, compare the trajectory performance. Figure 5(a) shows the three-dimensional trajectory performance of the two algorithms in the complex environment of the industrial park, while Fig. 5(b) shows the projection of the trajectory on the XY plane.

A thorough comparison and analysis of the two figures above reveals that, in this particular industrial park scenario, both algorithms manifest a certain degree of trajectory offset in specific regions characterised by high error rates. In scenarios where the system position falls within the X-coordinate range of -5 to 0 and the Y-coordinate range of 5

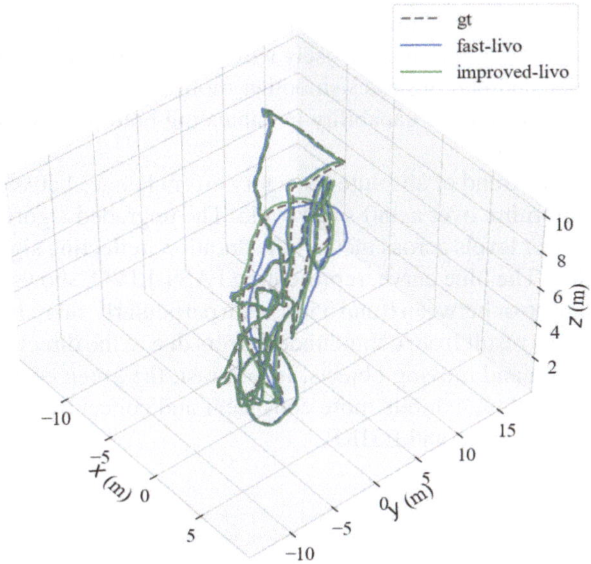

(a)In three-dimensional space

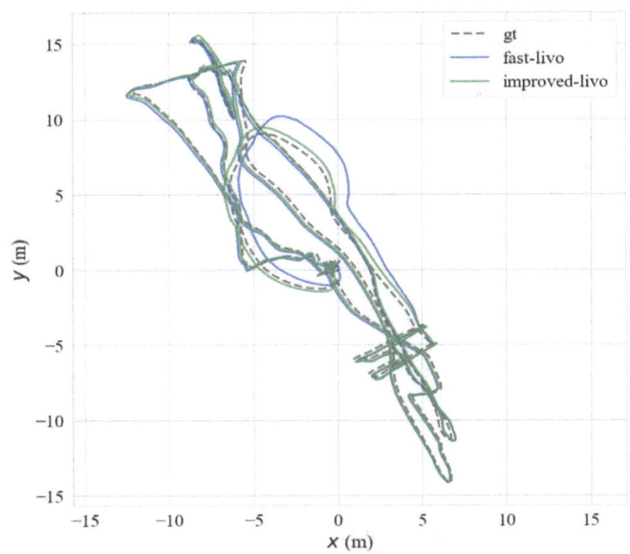

(b)On the XY plane

Fig. 5. Trajectory comparison

to 10, the enhanced algorithm has been shown to enhance the stability of feature point extraction in complex scenarios and the accuracy of multi-sensor fusion. This enhancement is achieved through the combined effects of the feature point optical flow method and the error state iterative Kalman filter model based on reprojection error. As a result, its estimated trajectory aligns more closely with the ground truth, exhibits narrower error fluctuation bands, and demonstrates smoother motion paths. These outcomes highlight the new algorithm's superior capability in enhancing both the accuracy and robustness of trajectory estimation.

Fig. 6 shows the trend of absolute trajectory error changes between FAST-LIVO and the improved algorithm over a 380-s test cycle. The upgraded algorithm maintains consistently lower error levels across most of the duration, reflecting significantly improved mapping stability. The blue curve, representing FAST-LIVO, shows frequent deviations above the median error between 0 and 350 s, with particularly sharp increases from 350 to 380 s. These spikes result from compounded errors due to the direct method's sensitivity to low illumination and moving objects. In contrast, the green curve, corresponding to the improved algorithm, exhibits more consistent and concentrated error values thanks to the optical flow method and ESIKF.

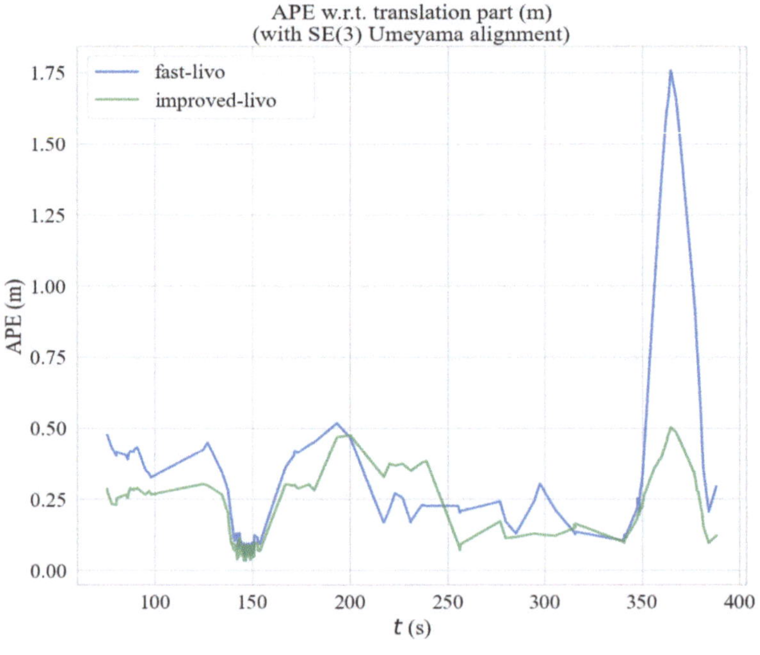

Fig. 6. Comparison of absolute trajectory error trends

Table 1 lists the statistical indicators of absolute trajectory error for the two algorithms in this scenario. The average error and median error of the improved algorithm were reduced by 0.097879 m and 0.036013 m respectively, and the RMSE decreased by approximately 52%, fully demonstrating its accuracy and stability in long-term trajectory estimation.

Table 1. Statistical indicators of absolute trajectory error

Algorithm	Average value(m)	Minimum value(m)	Median(m)	Standard deviation(m)	Root mean square deviation(m)
FAST-LIVO	0.250413	0.048882	0.130506	0.306402	0.395713
Improved-LIVO	0.152534	0.033969	0.094493	0.114081	0.190476

4 Conclusion

In summary, the selected industrial park scenario features wide roads, sparse features and diverse obstacles, which collectively present complex characteristics that impose extremely high demands on the robustness and accuracy of trajectory estimation algorithms. In such challenging environments, the algorithm proposed in this paper demonstrates exceptional trajectory estimation accuracy and mapping stability through the effective integration of feature point optical flow methods and the ESIKF model. Test results consistently demonstrate that its error control capability significantly outperforms FAST-LIVO, fully validating the improved algorithm's potential for application in low-light environments and feature-sparse regions. Future research could enhance the system's environmental adaptability under various operating conditions by incorporating larger, more comprehensive scene datasets and combining them with strategies such as transfer learning and data augmentation.

Acknowledgments. Supported by the National Key Research and Development Program of China (Grant No. 2023YFF0613200).

References

1. Lin, J., Zhang, F.: R 3 LIVE.: a robust, real-time, RGB-colored, LiDAR-inertial-visual tightly-coupled state estimation and mapping package. In: International Conference on Robotics and Automation, pp. 10672–10678. IEEE, Piscataway (2022)
2. Zhang, L., Zhao, J., Long, P.: An autonomous excavator system for material loading tasks. Sci. Robot. **6**(55), 12 (2021)
3. Zheng, C., Zhu, Q., Xu, W.: Fast-livo: Fast and tightly-coupled sparse-direct lidar-inertial-visual odometry. In: International Conference on Intelligent Robots and Systems (IROS), pp. 4003–4009. IEEE, Piscataway (2022)
4. Lowe, D.G.: Distinctive image features from scale-invariant keypoints. Int. J. Comput. Vis. **60**, 91–110 (2004)
5. Bay, H., Tuytelaars, T., Van Gool, L.: Surf: speeded up robust features. In: In: 9th European Conference on Computer Vision, pp. 404–417. ECCV, Berlin (2006)
6. Rublee, E., Rabaud, V., Konolige, K.: ORB: an efficient alternative to SIFT or SURF. In: International Conference on Computer Vision, pp. 2564–2571. IEEE, Piscataway (2011)
7. Cheng, J., Zhang, L., Chen, Q.: High precision and robust vehicle localization algorithm with visual-LiDAR-IMU fusion. IEEE Trans. Veh. Technol. **73**(8), 11029–11043 (2024)

8. Wang, X., Zhou, Y., Yu, G.: A lightweight visual Odometry based on LK optical flow tracking. Appl. Sci. **13**(20), 11322 (2023)
9. Derpanis, K.G.: Overview of the RANSAC algorithm. Image Rochester NY. **4**(1), 2–3 (2010)
10. Yu, C., Chao, Z., Xie, H.: An enhanced multi-sensor simultaneous localization and mapping (SLAM) framework with coarse-to-fine loop closure detection based on a tightly coupled error state iterative Kalman filter. Robotics. **13**(1), 2 (2023)
11. Bürki, M., Dymczyk, M., Gilitschenski, I.: Map Management for Efficient Long-Term Visual Localization in outdoor environments. Proc. 2018 IEEE Intel. Veh. Symp. (IV). **2018**, 682–688 (2018)

Research on Synchronous Control Strategy for Thousand-Ton Hydraulic Jacking System

Jianshen Wang, Xin Wang(✉), Panpan Wang, and Hongbiao Qian

Dalian University of Technology, Dalian 116024, China
wangxbd21@163.com

Abstract. For hydraulic synchronous jacking operations involving large-scale structures with weights exceeding 1,000 tons and large volumes, particularly when multiple modular units are coordinated for jacking, the poor asynchronous operations can easily lead to uneven force distribution, and then result in the object to tilt or overturn. So, it is important to research the hydraulic synchronous control technology. Based on analyzing the working principles of hydraulic jacking systems, a joint simulation model integrating hydraulic and control systems was constructed in AMESim and Simulink. For the problems such as slow response, large synchronization errors, and nonlinearity during system operation, the effectiveness of three hydraulic synchronization control strategies, including cross-coupled control, master-slave control, and parallel control, are analyzed. Taking a 400t*4-cylinder system as an example for co-simulation, the results shows that the cross-coupled control strategy outperforms the other two in terms of maximum dynamic error and time to reach stability, which can better meet the control requirements for smooth and rapid operation of the jacking system.

Keywords: Synchronous jacking · hydraulic control · co-simulation · cross-coupling control

1 Introduction

The increasing size and self-weight of large marine engineering structures bring significant challenges for installation and assembly, and due to the need to adapt to complex marine environments, the requirements for installation technology are extremely high. When jacking such structures, insufficient synchronization system accuracy can lead to uneven loading, off-center loading, or even structural damage or accidents [1]. Therefore, synchronization accuracy is critical for ensuring uniform stress distribution in the structure [2]. Currently, parallel control and master-slave control have their limitations: parallel control is susceptible to load disturbances, making it unsuitable for high-precision applications; master-slave control cannot eliminate load disturbances from slave cylinders, affecting overall precision. Cross-coupled control can avoid these shortcomings, and provide superior synchronization precision. Jia Shengchao of Harbin Engineering University compared three control methods and found that the synchronization error of cross-coupled control was the smallest [3]; Ye Xiangyang et al. of Taiyuan University

S. Halgamuge et al. (Eds.): ICACMVE 2025, LNME, pp. 311–318, 2026.
https://doi.org/10.1007/978-981-95-6911-3_29

of Technology showed that the maximum synchronization error of cross-coupled control was reduced by 81.25% compared with master-slave control [4]. Fan Diao et al. achieved high-accuracy position synchronization control of multiple hydraulic motors via a coupled controller that compensates for their position errors [5]. The synchronous control technology has already been applied in the actual products.

In this paper, a 400-ton * 4-cylinder jacking system is as the research object to study on the hydraulic synchronous control technology. By analyzing the hydraulic principle, a co-simulation model is built with AMESim and Simulink. In Simulink models, three control systems are established respectively, including master-slave, parallel, and cross-coupling. And then their synchronous effects are compared, so as to provide a more optimal control strategy for thousand-ton-class synchronous jacking systems.

2 Composition and Working Principle of a Thousand-Ton Jacking System

As the research object, a 1600 - ton jacking system with 400t*4 cylindersis shown in Fig. 1, which includes a top section, standard sections, a base support, and a conveying assembly. And its mechanisms cover a slewing mechanism, a jacking mechanism and a sprocket drive mechanism.

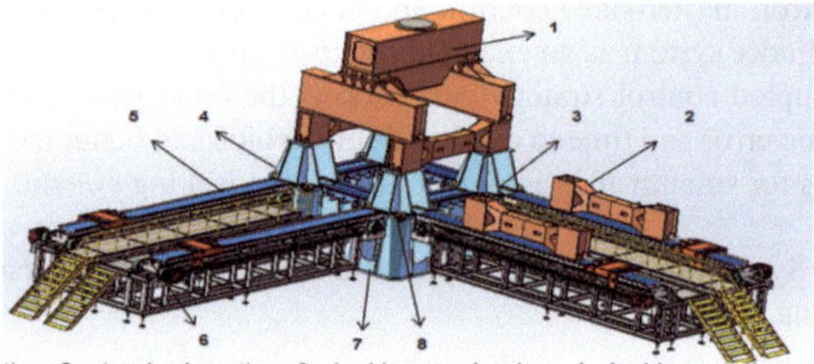

1-top section; 2- standard section; 3- Jacking mechanism; 4- Jacking rotating support; 5-Conveying assembly; 6-Sprocket drive mechanism; 7- Base support; 8-Jacking rotating support

Fig. 1. t jacking system

To jack large structures, the system applies 4 cylinders to jack synchronously the top and standard sections one-by-one until the required height The jacking process is as follow: a forklift alternately transports sections to the two right - angled conveying assemblies; the sprocket drive sends them into the base support. Once the position detection confirms the top section is on the 4 cylinder groups, the cylinders extend to jack up the sections. At the specified height, the jacking rotation mechanism rotates the support by 90°; the cylinder piston rods retract to set the standard section on the slewing support, then fully retract. Repeating this introduces more sections to lift the structure to the preset height.

During jacking, four cylinders work on the same heavy, large object. Asynchrony may cause uneven stress, leading to overload, tilting or overturning, and major accidents [6]. Thus, feedback control is needed to synchronize cylinder displacement. Since cylinder movement depends on the hydraulic system, a co - simulation model of the hydraulic and control systems is required.

3 Hydraulic System AMESim Model

The hydraulic principle of the jacking system is shown in Fig. 2. The four sets of hydraulic circuits for the cylinders are relatively independent, each driven by a servo motor to operate a fixed-displacement pump, supplying oil to the circuit with smooth speed control and high precision. The hydraulic fluid flows through the directional control valve into the rod chamber or non-rod chamber of the cylinder, enabling the extension and retraction of the cylinder piston rod. The extension speed of the hydraulic cylinder is controlled by servo speed regulation. The retraction is controlled using a switch combined with a throttle. The technical parameters of the hydraulic system are listed in Table 1. The AMESim hydraulic simulation model constructed based on this is shown in Fig. 3.

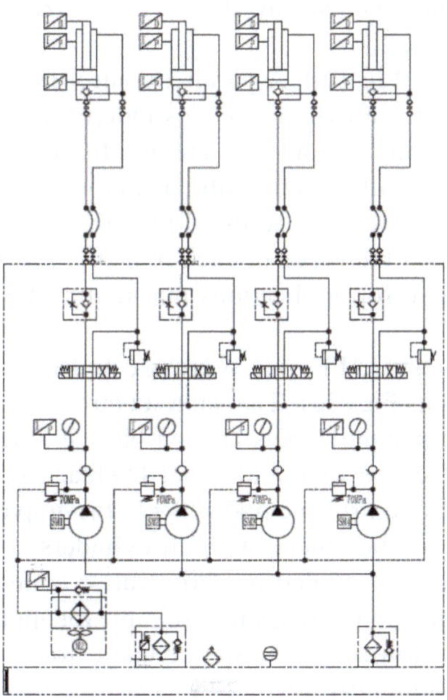

Fig. 2. Hydraulic principle diagram

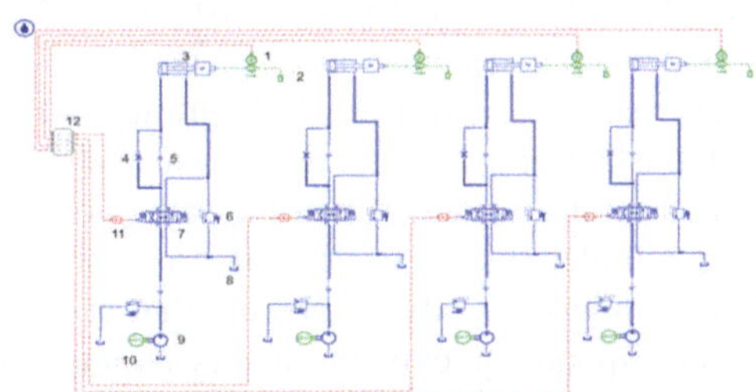

1-Displacement sensor; 2-Zero force source; 3-Hydraulic cylinder; 4-Throttle valve; 5-Check valve; 6-Relief valve; 7-Three-position four-way proportional directional control valve;8-fuel tank; 9-Pump; 10-Motor; 11-Gain; 12-Joint simulation interface

Fig. 3. Hydraulic model of the jacking system

Table 1. Hydraulic Component Parameters Table

Main Components	Parameters
System Flow/ Pressure	5.74L/min / 68MPa
Hydraulic Cylinder	Bore diameter 285mm, rod diameter 230mm
Directional Control Valve	Maximum flow rate 6.1 L/min
Pump	Displacement 4.3 ml/r, flow 6.1@1450 r/min,

4 Synchronous Control Strategy

From the hydraulic system principle, the four cylinder hydraulic circuits are relatively independent. Regulating displacement between groups to ensure cylinder synchronization and safety is key to control technology. Currently, there are three main synchronization control methods: master-slave, parallel and cross-coupling [7]. This paper will establish Simulink simulation models for each.

The master-slave control block diagram is shown in Fig. 4a). A master-slave control model is established in Simulink, as shown in Fig. 5a). In master-slave synchronous control, the control signal acts on the master cylinder, whose displacement output commands slave cylinders. Master cylinder signals or disturbances affect slaves, but slave commands or load disturbances do not affect the master or other slaves. This control is simple and stable but has persistent, hard-to-eliminate steady-state synchronization errors, reducing system control accuracy.

The parallel control block diagram is shown in Fig. 4b). A parallel control model is established in Simulink, as shown in Fig. 5b). Each cylinder moves independently and is not interrelated. When the same control signal is input, if there are differences in the cylinder system itself or if it is disturbed, it will not affect the displacement of other cylinders. Its advantages are that each cylinder can be adjusted flexibly and the system responds quickly, but its disadvantage is that the speed of each branch is easily affected by load disturbances, making it unsuitable for applications that require high speed accuracy.

The cross-coupled control block diagram is shown in Fig. 4c). A cross-coupled control model is established in Simulink, as shown in Fig. 5c). The same control signal is applied to each cylinder, and all output displacement signals are fed back into the cross-coupled controller. After cross-coupling, they become additional feedback signals to regulate the output of each cylinder. This control method reflects the impact of displacement output changes in one cylinder on the others, with all cylinders collectively suppressing synchronization errors [8]. Cross-coupled synchronization control achieves good results, producing smaller synchronization errors, and is suitable for applications requiring high synchronization precision. However, the cross-coupled structure is relatively complex, making tuning more challenging [9].

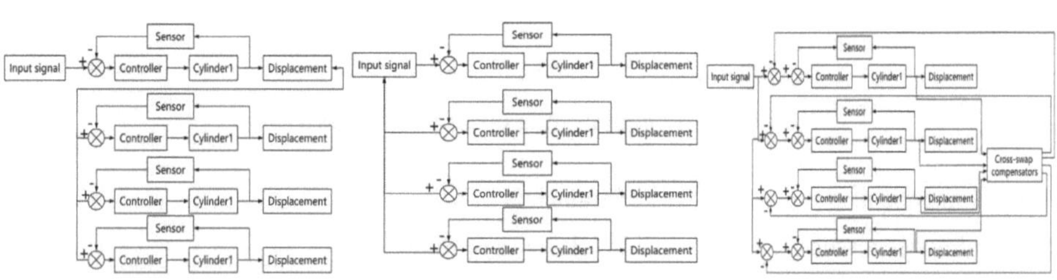

a) Master-slave control b) Parallel control c) Cross-coupling control

Fig. 4. Control schematic diagram

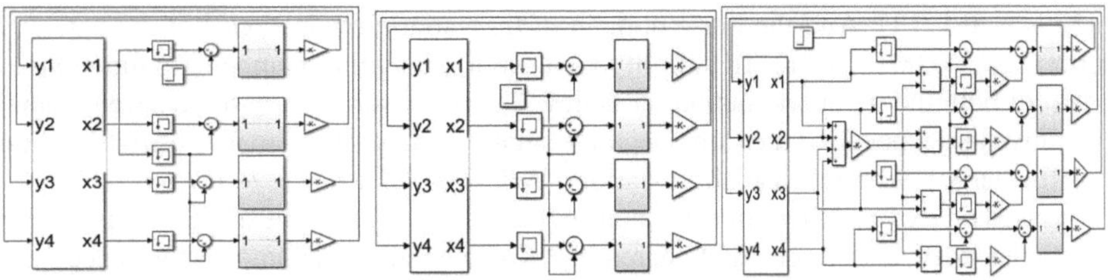

a) Master-slave control b) Parallel control c) Cross-coupling control

Fig. 5. Simulink simulation model diagram

5 Co-simulation of AMESim and Simulink Models Under Different Control Strategies

Co-simulation interfaces were established in AMESim and Simulink, respectively. The displacement signals from AMESim were output to Simulink, where they were converted into electrical signals via a control strategy and then output back to AMESim to control the valve opening. During actual jacking operations, the four hydraulic cylinders are individually controlled by four servo motors. During operation, inconsistencies in motor speeds may lead to discrepancies in jacking displacement. Therefore, in the joint simulation, different motor speeds are manually set to test the effectiveness of the control system's rapid synchronization. The motor speeds for the four sets of cylinders are set to 1305 r/min, 1400 r/min, 1450 r/min, and 1595 r/min, respectively.

5.1 Single-Cylinder Fuzzy PID Control

The main controller for each hydraulic cylinder is a fuzzy PID controller. The packaged fuzzy PID controller is shown in Fig. 6, with $KP = 500$, $KI = 450$, and $KD = 120$. A step input signal of 1 mm is applied, and the simulation time is 10 s.

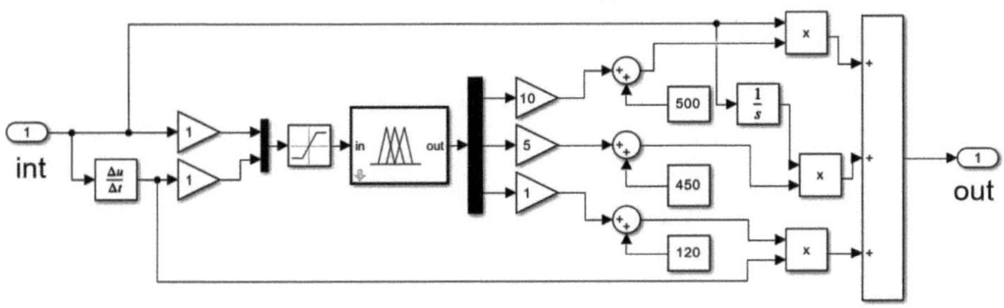

Fig. 6. Fuzzy PID controller

Set up the fuzzy controller in Simulink. The basic domain of the error e in the fuzzy controller is $[-1,1]$, the basic domain of the error change ec is $[-1,1]$, the basic domain of the proportional P change ΔKp is $[-1,1]$, and the basic domain of the integral I change

ΔKi is [−0.1,0.1]. And the basic domain of the variation value ΔKd of the derivative term D is [−0.001, 0.001]. To simplify the design, triangular membership functions are used for both the input and output variables. The fuzzy subsets are set to seven, namely [negative large, negative medium, negative small, zero, positive small, positive medium, positive large], with English abbreviations *[NB, NM, NS, Z0, PS, PM, PB]*.

Based on expert experience, the control parameter adjustment strategy is as follows: when the error e is large, first increase ΔKp to ensure tracking performance while limiting ΔKi and ΔKd; when e decreases to a moderate level, increase ΔKi and decrease ΔKp; when e is small, fine-tune ΔKp, ΔKi, and ΔKd to achieve control. Based on the control rules of the fuzzy PID controller, fuzzy rule tables for ΔKp, ΔKi, and ΔKd are established. Defuzzification is performed using the weighted average method [10].

6 Control System Simulation

For the master-slave control system, the displacements of the four cylinders are shown in Fig. 7a). As can be seen from the figure, during operation, the displacement difference between Cylinder 3 and Cylinder 1 is the largest. The maximum error curves for the two cylinders are shown in Fig. 8a), with a maximum error range of −0.186 to −0.185 mm. The time required to reach steady state is 7.0 s.

For the parallel control system, the displacements of the four cylinders are shown in Fig. 7b). The overshoot and settling time of parallel control are both smaller than those of master-slave control. During operation, the displacement difference between cylinder 2 and cylinder 1 is the largest. The maximum error curve is shown in Fig. 8b). The maximum error range is −0.065 to 0.141 mm, and the time required to reach steady state is 6.5 s.

For the cross-coupled control system, the displacements of the four cylinders are shown in Fig. 7c). The overshoot and settling time of the single-cylinder displacement under cross-coupled control are similar to those under parallel control and are also superior to those under master-slave control. During operation, the displacement difference between cylinder 2 and cylinder 1 is the largest. The maximum error curve is shown in Fig. 8c), with a maximum error range of −0.048 to 0.138 mm, and the time required to reach steady state is 4.4 s.

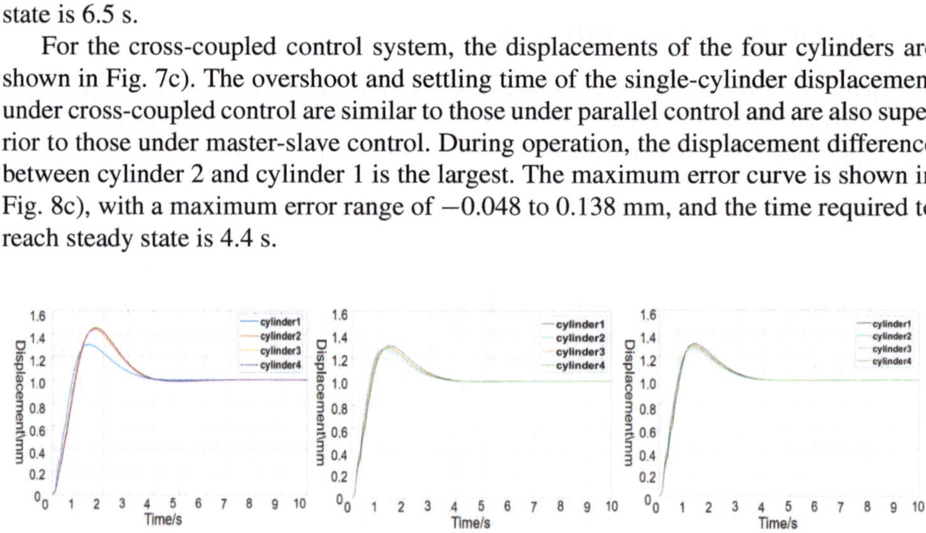

a) Master-slave control b) Parallel control c) Cross-coupling control

Fig. 7. Four-cylinder displacement curve

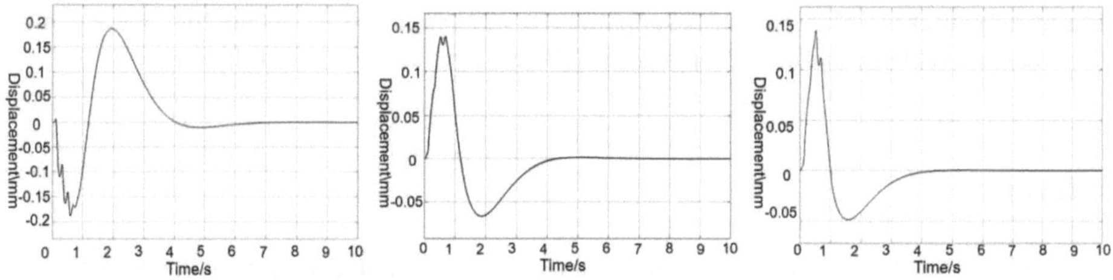

a) Master-slave control b) Parallel control c) Cross-coupling control

Fig. 8. Maximum error curve between cylinders

6.1 Simulation Comparison Analysis

Comparing the maximum error curves of the three control methods, cross-coupling control has the smallest error range and the fastest steady-state time, outperforming parallel and master-slave control. Specific data are given in Table 2 below. In cross-coupling control, the four cylinders' displacements interact, with dynamic errors adjusted in real time for rapid stabilization. Parallel control keeps cylinders independent-disturbances to one don't affect others. Master-slave control sees master cylinder disturbances or errors affect slaves, with poor anti-interference. Cross-coupling has stronger anti-interference and better adjustment in complex conditions.

Table 2. Data comparison table

Control Method	Maximum Error Range/mm	Stabilization Time/s
Master-Slave Control	−0.186–0.185	7.0
Parallel Control	−0.065–0.141	6.5
Cross-Coupling Control	−0.048–0.138	4.4

7 Conclusion

This paper addresses the problem that large structures may tilt or overturn due to asynchronous jacking of multiple modules during the jacking process. A co-simulation model is built by combining the hydraulic simulation software AMESim with the control simulation software Simulink, and modeling and simulation analyses are conducted for three synchronous control strategies: master-slave control, parallel control, and cross-coupling control. By comparing the maximum displacement error curves of the four cylinders, it can be seen that cross-coupling control has the smallest maximum error, with a range of −0.048 mm to 0.138 mm, and the optimal time to reach a steady state of 4.4 s, which can effectively meet the working requirements of the 1600-ton jacking system. Therefore, the cross-coupling control strategy is more advantageous.

Acknowledgements. This project is supported by the Qingdao Key Technology Research and Industrialization Demonstration Project Fund (Project No.: 24-1-3-hygg-6-hy).

References

1. Li, H.T.: Development trends of emerging marine engineering equipment. China Ship Surv. **09**, 43–48 (2024)
2. Yan Hongqin. Synchronous design and discussion of the hydraulic system for tower cranes. Construct. Mach. Technol. Manage. **38**(03), 24–25+28 (2025)
3. Jia, S.: Research on the jacking electrohydraulic servo system of a pressure-bearing operation device. Harbin: Harbin Institute of Technology (2015)
4. Ye, X., Meng, Q., Kou, Z.: Self-disturbance-resistant controller for single-track crane hydraulic system based on cross-coupling synchronous control strategy. Mech. Electr. Eng. (2025). 1–9 [2025–05–21]
5. Fan, D., Liu, H., Su, Y.: Integral robust cooperative control of hydraulic servo system based on cross-coupling. Mach. Tools Hydraulics **53**(04), 81–87 (2025). FAN D, LIU H, SU Y. Integral Robust
6. Chen, R., Sun, J., Pan, L.: Construction technology for overall jacking of large steel structure buildings. In: Proceedings of the National Conference on Innovation and Technology in Building Steel Structures (2023)
7. Defu, Z.: Key technologies of multi-hydraulic cylinder synchronous system for bridge jacking. Urban Construct. Theory Res. (Electron. Edn.) **32**, 69–71 (2024)
8. Lingzhi, T.: Research and application of multi-point synchronous hydraulic jacking systems. Crane Transport. Mach. **10**, 58–63 (2022)
9. Choi, Y.B., Lee, S.H., Jung, D.H.: Controller design based on a synchronous error compensator for wheeled cross-coupled systems. Int. J. Control. Autom. Syst. **22**(5), 1728–1738 (2024)
10. Phu, N.D., Hung, N.N., Ahmadian, A., et al.: A new fuzzy PID control system based on fuzzy PID controller and fuzzy control process. Int. J. Fuzzy Syst. (2020)

Design and Application of Health Prediction System for Port Hydraulic Equipment Based on Multi-source Data Fusion

Chun-Hui Zhang[1,2], Xiao-Ming Yuan[2], Rui Guo[2](✉), Hong-Xu Wei[1], Da-Wei Liu[1], Jing Sun[1], and Hong-yang Ye[1]

[1] Port Machinery Co., Ltd., Hebei Port Group, Qinhuangdao 066000, Hebei, China
[2] College of Mechanical Engineering, Yanshan University, Qinhuangdao, Hebei 066004, China
451094293@qq.com

Abstract. To achieve predictive maintenance for port hydraulic equipment, a multi-source data online detection system for port hydraulic equipment was designed, which can detect and analyze multiple state parameters in the system. Taking the hydraulic system of the port car dumper's hydraulic system as an example, based on the analysis of its working principle and fault problems, online detection and health prediction were carried out, and targeted maintenance and repair plans were proposed. The health status assessment and prediction results provide a basis for system fault diagnosis and maintenance, and have important reference and application value.

Keywords: Predictive maintenance · Multi-source data · Car dumper hydraulic system · Online detection · Fault diagnosis

1 Introduction

Various types of hydraulic equipment with complex structures have been there in ports. Once fault occurs, technicians need to make on-site judgments based on the fault phenomena, and such faults cannot be predicted in advance. For components that may have faults, they need to be disassembled and further judged on the test bench. The means of detection at the operational site are very limited. Therefore, the process of troubleshooting is complex and inefficient, affecting production efficiency.

At present, certain research achievements have been made in the detection of hydraulic system leakage. For instance, pressure signals can be collected and combined with technologies such as wavelet transform and neural networks for hydraulic system leakage testing. Additionally, portable hydraulic system leakage signal acquisition devices can be used to collect leakage signals. However, merely detecting the leakage volume cannot comprehensively reflect the health status of the hydraulic system. It is impossible to make precise health predictions for the system.[1, 2]

S. Halgamuge et al. (Eds.): ICACMVE 2025, LNME, pp. 319–328, 2026.
https://doi.org/10.1007/978-981-95-6911-3_30

Therefore, designing a multi-source data online detection system for port hydraulic equipment — which can realize online pressure resistance detection, leakage detection, contamination degree detection, online pressure detection, and oil temperature detection of port hydraulic equipment, and provide comprehensive references for predicting the health status of the hydraulic system — is of great significance for achieving predictive maintenance of port hydraulic equipment.

2 Design of Online Detection System for Multi-Source Data Fusion

The prediction of the health status in hydraulic systems often requires a comprehensive analysis of multiple state parameters such as cleanliness, pressure, flow rate, and temperature. An online detection system fusing multi-source data, as shown in Fig. 1, enables online detection of pressure resistance, leakage, contamination degree, pressure, and oil temperature for port hydraulic equipment. This system provides comprehensive data for fault diagnosis and prediction of the hydraulic system, and further supports the formulation of maintenance strategies.[3]

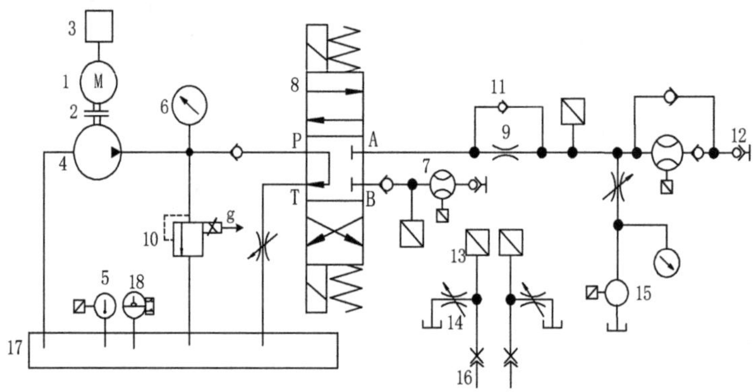

1- Servo motor, 2- Coupling, 3- servo driver, 4- Hydraulic pump, 5- temperature sensor, 6- pressure gauge I, 7- flow sensor I, 8- electromagnetic directional control valve, 9- fixed throttle hole, 10- proportional relief valve, 11- Check valve II, 12- quick Coupling II, 13- pressure sensor III, 14- Throttle valve II, 15- Oil contamination particles Counter, 16-pressure measuring pipeline I, 17-hydraulic oil tank, 18-liquid level sensor

Fig. 1. Schematic diagram of the online detection system

This data acquisition system adopts constant flow closed-loop control and consists of a controller, a touch human-machine interface, a flow transmitter, a pulse signal generator, an interface circuit, and a DC motor drive. Among these components, the controller, as the core of the system, exchanges data with peripheral devices.

Figure 2. shows the block diagram of the closed-loop control system. The flow signal set through the human-machine interaction interface is transmitted to the controller, causing the system to output a fixed flow. The main flow transmitter 1 installed on the oil outlet pipeline of the quantitative pump detects the output flow of the quantitative pump in real time, converts it into a 4 ~ 20 mA electrical signal, and feeds it back to the controller. The controller compares the feedback signal with the flow setting signal to keep the output flow of the quantitative pump stable within a certain range, unaffected by the oil temperature, test pressure, different types of hydraulic oil, and internal leakage of hydraulic components in the data acquisition system, thereby ensuring that the test accuracy during the data acquisition process remains unchanged within a specific range.

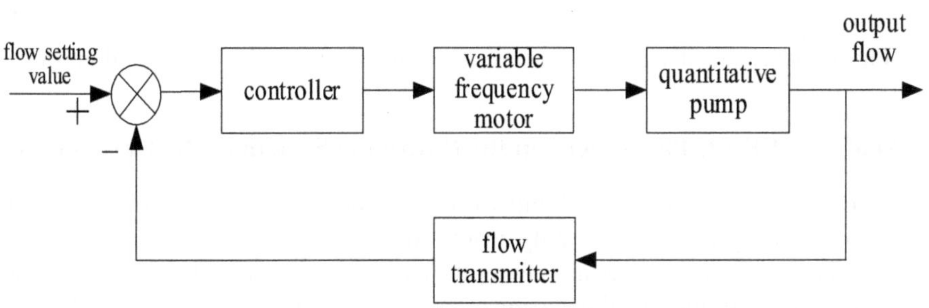

Fig. 2. Block diagram of the closed-loop control system

3 Analysis of Faults in the Hydraulic System of the Car Dumper

3.1 Introduction to the Principle of the Hydraulic System of a Car Dumper

Before the coal-carrying train carriages are flipped and unloaded by the car dumper, they are fixed by the pressure car beam cylinder and the leaning car plate cylinder. Both the pressure car and the leaning car of the car dumper are powered by the hydraulic system. The partial principle diagram of the pressure car system is shown in Fig. 3. The pressure oil controls the corresponding pressure car beam cylinder to be pressed down and lifted through the control valve group. The clamping of the carriage mainly relies on the pressure-holding effect of the hydraulic lock in the rod cavity of the pressure car beam and the corresponding branch circuit. The partial principle diagram of the leaning car system is shown in Fig. 4. When the leaning car plate extends and retracts, the pressure oil enters the rod cavity or rodless cavity of the corresponding leaning car plate cylinder through the leaning car control valve group. The holding of the car body during leaning mainly relies on the pressure-holding effect of the rodless cavity of the leaning car plate cylinder and the corresponding branch hydraulic lock.[4]

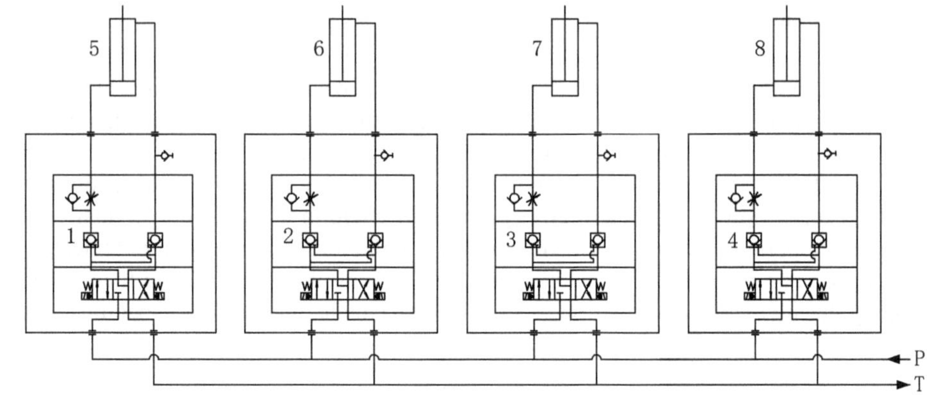

Fig. 3. Schematic diagram of the pressure car hydraulic system of the car dumper

3.2 Analysis of Fault Phenomena in the Hydraulic System of the Car Dumper

After the pressure beam of the car dumper is pressed down, a phenomenon occurs where the pressure in the cylinder's rod cavity drops rapidly until it reaches zero. During cyclic operation, the coal-carrying carriages enter the car dumper area. At this time, the leaning car plate extends and tightens the carriages. After extending to the designated position, the four pressure beams of the car dumper simultaneously press down on each carriage.

The control system detects whether the pressure signal of the oil cylinder reaches the set pressure value. If it does, the carriages are considered to be tightened, and the flipping action is allowed. If the pressure of the oil cylinder drops below the control system's set value at any moment during the flipping or returning operation, the car dumper will stop at the current position and the automatic cycle will halt. At this time, manual operation is required: the on-site patrol personnel should manually flip or return the car dumper. A poor pressure-holding effect causes the pressure in the rod cavity to drop rapidly below the set safety value, leading to the stoppage of the car dumper's automatic cycle operation and thus affecting unloading efficiency. This will hinder the smooth progress of production and increase the risk of the carriages falling off the track when they are flipped to the maximum angle.

Therefore, it is necessary to collect key data in the system, identify leaking components, assess the system's health status, and adopt targeted solutions to eliminate potential safety hazards in production.

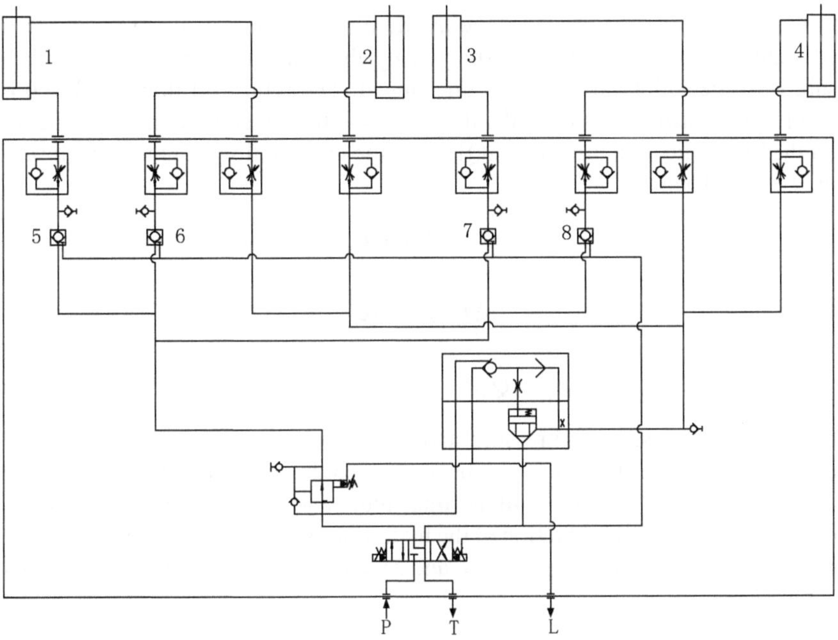

Fig. 4. Schematic diagram of the leaning car hydraulic system of the car dumper

4 Online Detection and Health Prediction

4.1 Principles of Health Prediction

Given the frequent production halts on-site caused by pressure-holding failures, initial analysis suggests the leakage originates from the hydraulic system. After repeated on-site inspections, no external leakage was detected. Therefore, it is necessary to inspect internal component leakage within the system, identify the leakage source, and on this basis assess the system's health status to implement targeted maintenance and repairs.

Since the pressure-holding state during the car pressing process in the pressure car beam system has the greatest impact on system safety, the focus is on inspecting the rod cavity of the pressure car beam cylinder and the hydraulic lock in the same branch. The system shown in Fig. 1 is connected to the car pressing hydraulic system and the car supporting hydraulic system of the car dumper respectively, in the manner shown in Fig. 5. After the system is started, the branch of the hydraulic cylinder is sealed with a flange cover. At this point, the output flow displayed by the main flow transmitter is Q, the flow displayed by the return oil flow transmitter is Q1, the flow entering the detected system is Q1x, and the leakage flow of the hydraulic lock branch is ΔQ1.Then

$$Q = Q_1 + Q_{1x} \tag{1}$$

$$\Delta Q_1 = Q_{1x} = Q - Q_1 \tag{2}$$

At the current test temperature, the hydraulic cylinder branch is restored to a connected state. The output flow Q displayed by the main flow transmitter and the set pressure P of the relief valve remain unchanged. The return oil flow displayed by the return oil flow transmitter is Q1a, the flow entering the tested system is Q2, the leakage flow of the hydraulic lock branch is ΔQ1a, and the leakage flow of the hydraulic cylinder branch is ΔQ2. Then

$$Q = Q_{1a} + Q_{2x} \tag{3}$$

$$\Delta Q_{1a} = \Delta Q_1 \tag{4}$$

$$\Delta Q_2 = Q_{2x} - \Delta Q_{1a} = Q_1 - Q_{1a} \tag{5}$$

Therefore, under specific pressure and temperature conditions, the ΔQ1 and ΔQ2 values for each branch of the pressure beam and leaning plate are obtained respectively. These values are then compared with normal values and sorted in descending order. This enables the diagnosis and prediction of the hydraulic system's health status, providing an important basis for system maintenance and repair as well as for predicting component service life.[5, 6]

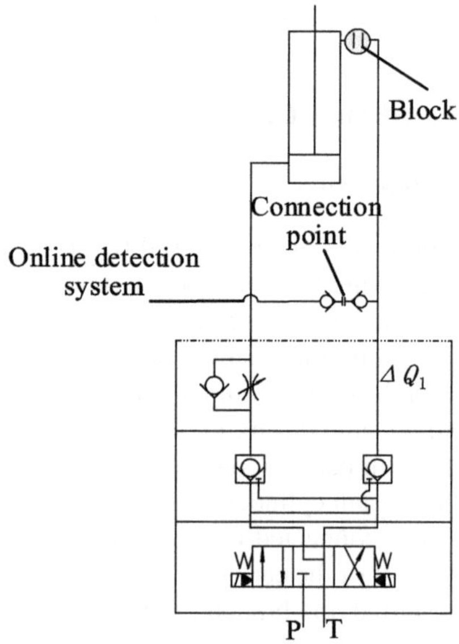

Fig. 5. Schematic diagram of data acquisition for the hydraulic system of the car dumper press

4.2 Online Detection

As shown in Fig. 6, at the online detection site, the oil cleanliness of the car dumper's hydraulic system is first collected and detected. The online detection system is connected to the port on the valve block of the car dumper's hydraulic system. The pump station of the car dumper system is started, and under no-load conditions, the cylinders of the pressure beam and leaning car plate are actuated for three to four cycles. This allows the oil in the car dumper system to flow through the contamination degree sensor into the data acquisition system. The cleanliness grade of the hydraulic oil in the system is NAS12.

Fig. 6. Online detection site

Based on the above principles and methods, and in combination with the hydraulic systems for the car dumper's pressure car and leaning car (shown in Figs. 3 and 4), the branch circuits of the No. 1 to No. 4 pressure car beam cylinders and the No. 5 to No. 8 leaning car plate cylinders were respectively blocked as illustrated in the figures. The pressure setting of the relief valve in the data acquisition system was referenced to the operating pressure of the car dumper's pressure car and leaning car hydraulic systems, with test pressures set to 5.6 MPa and 4 MPa respectively. The test temperature varied slightly according to the actual temperature detected by the temperature sensor, and the test duration was set to 60 s. The system parameter settings and collected data are shown in Table 1.[7]

Table 1. Data Acquisition Parameters for Hydraulic Lock Branches

Collection location (Hydraulic lock)	Test temperature (°C)	Test pressure (MPa)	Output flow$q$ (mL/min)	Return oil flow rate q_n (mL/min)
1#	7	5.6	1497	1486
2#	7	5.6	1497	1489
3#	8	5.6	1502	1496
4#	9	5.6	1502	1499
5#	10	4	1497	1497
6#	12	4	1502	1500
7#	12	4	1502	1497
8#	13	4	1502	1499

Remove the blocking flange from the cylinder branch, restore the corresponding cylinder branch to a connected state, keep the test pressure and temperature unchanged, and re-collect the system data. At this point, the return oil flow decreases because part of the oil leaks through the cylinder, as shown in Table 2.

Table 2. Data Acquisition Parameters of Hydraulic Cylinder Branch

Collection location (Oil cylinder)	Test temperature (°C)	Test pressure (MPa)	Output flow$q$ (mL/min)	Return oil flow rate q_n (mL/min)
1#	7	5.6	1497	1367
2#	7	5.6	1497	1450
3#	8	5.6	1502	1452
4#	9	5.6	1502	1472
5#	10	4	1497	1477
6#	12	4	1502	1483
7#	12	4	1502	1466
8#	13	4	1502	1481

4.3 Analysis of Health Prediction Results

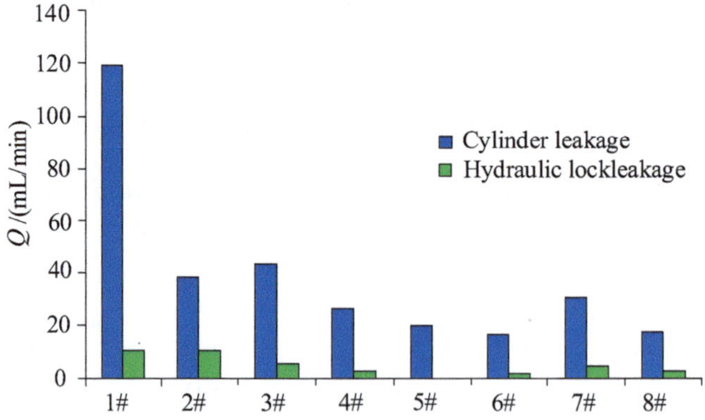

Fig. 7. Bar comparison chart of component leakage

Based on the flow data collected in Tables 1 and 2, a bar comparison chart was generated as shown in Fig. 7. It can be seen from the chart that due to system contamination after long-term operation, all system components have experienced varying degrees of wear. The leakage of the cylinders, in descending order, is as follows: 1#, 3#, 2#, 7#, 4#, 5#, 8#, 6#. The leakage rates of the hydraulic locks, from highest to lowest, are 1#,

2#, 3#, 7#, 4#, 8#, 6#, 5#. Among them, the leakage rate of the 1# pressure car beam cylinder is the highest at 119 mL/min, which is 2 to 3 times that of the other cylinders. The overall leakage of the pressure car system is better than that of the leaning car system. Therefore, in a relatively clean oil environment, with reference to the current actual production and usage frequency as well as the remaining life prediction theory, it is estimated that the remaining service life of the other hydraulic cylinders (excluding the problematic ones) will be 1 to 2 times their current service life. Except for the 1# and 2# hydraulic locks, the remaining service life of the other hydraulic locks is expected to be twice their current service life.

4.4 Maintenance and Repair Strategies for the Hydraulic System of the Car Dumper

Since the cleanliness grade of the system's hydraulic oil has exceeded the initial design requirement of NAS 9, it is necessary to replace the oil immediately and monitor the oil cleanliness on a monthly basis.

Due to significant internal leakage in both the No. 1 pressure car beam cylinder and the No. 1 hydraulic lock, they need to be replaced. To ensure the safety of the carriage during flipping, it is necessary to guarantee reliable pressure-holding performance of at least two pressure car beams. Therefore, the No. 2 hydraulic lock was replaced, while the No. 3 and No. 7 hydraulic locks were disassembled and cleaned; they will be replaced later based on usage and inspection conditions. The on-site improvement is shown in Fig. 8.

(a) Hydraulic oil replacement and system cleaning (b)Disassembly and cleaning of hydraulic locks

Fig. 8. On-site improvement situation

5 Conclusion

1. A multi-source data online detection system for port hydraulic equipment was designed to realize online pressure resistance testing, leakage detection, contamination degree detection, pressure detection, and oil temperature detection of port hydraulic equipment. This system provides comprehensive references for predicting the health status of hydraulic systems.

2. Online detection of the car dumper's hydraulic system enables rapid analysis of leakage in system components, thereby facilitating further assessment of the system's health status, prediction of component service life, and formulation of targeted maintenance and repair plans.
3. The maintained system has not experienced any pressure-holding failures during production operations. This method provides new ideas and approaches for fault diagnosis and maintenance of port hydraulic equipment.

References

1. Yoo, S., Jung, H.J., Lee, K.J., et al.: A convolutional autoencoder based fault diagnosis method for a hydraulic solenoid valve considering unknown faults. Sensors. **23**(16), 7249 (2023)
2. Hari, P., Khiran, S., Manian, S.S., et al.: Vibration-based brake health prediction using statistical features-a machine learning framework. Mater. Today Proc. **46**(P2), 1167–1173 (2021)
3. Yao, X.I.A.O., Jiawei, X.I.A.N.G., Hesheng, T.A.N.G., et al.: Fault diagnosis method for hydraulic waterproof valves based on a multi-sensor information fusion and CNN-BIGRU-attention model. Journal of Mechanical & Electrical Engineering. **41**(09), 1517–1528 (2024)
4. Guanhua, X.: Research on Fault Diagnisis and Prediction of Aircraft Landing Gear Door Based on Health Management. Harbin Engineering University (2023)
5. Chun-hui, Z., Cheng-gang, Y., Hong-xu, W.: Date acquisition and health prediction of dumper hydraulic system. Mining machinery. **51**(3), 9–13 (2023)
6. Zhaoli, Z., Jianqiu, Z., Xin, W., et al.: Health status prediction of hydraulic automatic grab beam based on CNN-LSTM. Machine Tool & Hydraulics. **52**(23), 215–220 (2024)
7. Feng, Y.A.N., Chengjun, X.I.A.O., Yiwei, S.U.N., et al.: Intelligent hydraulic system fault diagnosis method based on MSIDCNN-BOA-SVM. Machine Tool & Hydraulics. **53**(8), 174–181 (2025)

Layered Control Strategy for the Maneuvering Stability of Distributed Electric Drive Semi-trailer Trucks

Jian Zhang[1(✉)], Jianyi Quan[1], Lei Yan[1], Jianyi Lu[1], Jiarong Liang[1], Yaoyu Cui[2], and Yong Shu[3]

[1] School of Vehicle and Traffic Engineering, Taiyuan University of Science and Technology, Taiyuan 030024, Shanxi, China
zhangj@tyust.edu.cn

[2] School of Chemical Engineering and Technology, Taiyuan University of Science and Technology, Taiyuan 030024, Shanxi, China

[3] School of Mechanical Engineering, Taiyuan University of Science and Technology, Taiyuan 030024, Shanxi, China

Abstract. This paper addresses the issues of lateral instability and rollover that often occur when semi-trailer trucks maneuver to avoid obstacles at high speeds. It proposes a distributed electric drive anti-skid and torque coordination control strategy for semi-trailer trucks based on LQR. Using TruckSim and Matlab/Simulink, a high-precision distributed electric drive semi-trailer truck dynamics model was constructed for dual-lane high-adhesion and single-lane low-adhesion road conditions. An upper-layer controller based on a PID speed-following controller and linear quadratic optimal control theory (LQR) was designed for the tractor and trailer, respectively, to apply additional lateral torque. The lower layer employs an extremum optimization method to allocate drive torque among the wheels, and a fuzzy PID anti-slip controller is designed to output corrective torque to prevent excessive wheel slip. The results show that this strategy effectively improves the lateral stability of the semi-trailer truck train. Under dual lane-changing high-adhesion conditions, the vehicle's center of gravity lateral deviation angle is reduced by 27% compared to the uncontrolled condition. Under single lane-changing low-adhesion conditions, the vehicle's center of gravity lateral deviation angle is reduced by 31% compared to the uncontrolled condition.

Keywords: Distributed electric drive semi-trailer truck · LQR control · Fuzzy PID control · Yaw moment control

1 Introduction

In 2024, the number of motor vehicles in the country reached 453 million, with 353 million being automobiles, and 96 cities having over 1 million automobiles. The number of new energy vehicles was 31.4 million, accounting for 8.90% of the total vehicle number, with pure electric vehicles accounting for 70.34% of new energy vehicles. As of

S. Halgamuge et al. (Eds.): ICACMVE 2025, LNME, pp. 329–336, 2026.
https://doi.org/10.1007/978-981-95-6911-3_31

June 2024, there were 7.23 million heavy trucks in China, and 3.632 million heavy truck tractors [1]. Semi-trailer truck trains are important for mainline transportation but have stability issues like cross-swing instability, rollover, and tire skid. Vehicle stability control mainly uses strategies like trailer active steering, wheel braking intervention, and drive torque regulation, with algorithms such as DYC, Fuzzy PID, etc. Domestic scholar Shuo Wang [2] controlled semi-trailer steering to address trajectory deviation. Qinghui Zhou [3] designed an MPC controller to improve high-speed lane-changing stability. Yonghui Jin [4] proposed a sliding-mode-based control strategy to avoid vehicle dumping. Yue Wang [5] designed a hierarchical control strategy to improve vehicle stability. Foreign researchers Chen Y [6] compared the lateral characteristics of multi-trailer trucks, and Jagelčák J [7] studied the impact of semitrailer liftable axles on maneuverability. Previous semi-trailer train model control mainly focused on ideal two-degree-of-freedom values, ignoring the trailer's independent transverse characteristics and lacking research on driving torque control.

Based on the above research background, this paper employs TruckSim dynamics modeling software to establish a complete dynamic model of distributed electric-drive semi-trailer combinations. To validate the established kinematic model, a 3-degree-of-freedom reference model of ideal distributed electric-drive semi-trailer combinations is constructed in the MATLAB/Simulink environment for comparative simulation and validation. In the control architecture design, the upper layer integrates a PID speed-following controller and linear-quadratic optimal control theory to develop additional yaw moment controllers for the tractor and trailer separately. The lower layer adopts an extremum optimization method for driving torque distribution, while a fuzzy PID drive anti-slip controller is designed to output corrective torques, preventing excessive slip of driving wheels. The effectiveness of the proposed control strategy is verified through simulations under high-speed high-adhesion double lane-change and high-speed low-adhesion single lane-change conditions. Results demonstrate that the control strategy can accurately generate the required yaw moments, featuring fast transient response and strong dynamic control capability, and effectively avoids lateral instability phenomena such as sideslip, tail swing, and jackknifing.

2 Distributed Electric Drive Semi-trailer Truck Train Whole Vehicle Dynamics Modeling and Validation

2.1 TruckSim Vehicle Model Creation

In this paper, TruckSim software is used to model the body model, tire model, powertrain model, suspension system model, steering system model, braking system model and other subsystems of the three-axle semi-trailer train dynamics model, and the whole vehicle model is mainly composed of a 4×2 drive two-axle large truck loaded trailer-type hauler, trailer, and towing platform. The front axle of the tractor alone performs the steering function, while the rear axle of the tractor and the rear axle of the trailer reach a distributed drive with the help of permanent magnet synchronous motors. The vehicle model schematic is shown in Fig. 1.

In this paper, the second-order transfer function is used to simplify the calculation of the actual output torque of the motor to meet the highresponse demand, and its specific

Fig. 1. Schematic diagrams of the three body models.

expression is as follows:

$$G(s) = \frac{T_d}{T_d^*} = \frac{1}{1 + 2\zeta s + 2\zeta^2 s^2} \tag{1}$$

In Eq. (1): T_d is the motor output torque; T_d^* is the desired torque of the motor; ζ is a constant of the physical characteristics of the motor.

2.2 Driver Modeling

To achieve tracking control of train speed and stable constant speed operation, this paper does not use the default speed control module of Turcksim software. Instead, a longitudinal driver model based on PID (proportional-integral-derivative) control theory is designed using MATLAB/Simulink software. This model uses the difference between the actual vehicle speed and the desired speed as input and outputs the total driving torque $T_{\_vx}$ required to achieve the target speed. The longitudinal driver model diagram is shown in Fig. 2.

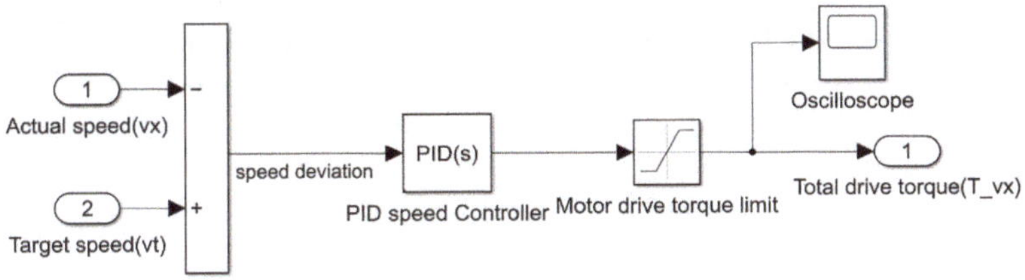

Fig. 2. Longitudinal driver model diagram.

2.3 Linear Three-Degree-of-Freedom Automotive Train Dynamics Model

The three-degree-of-freedom car train model is shown in Fig. 3 [8].

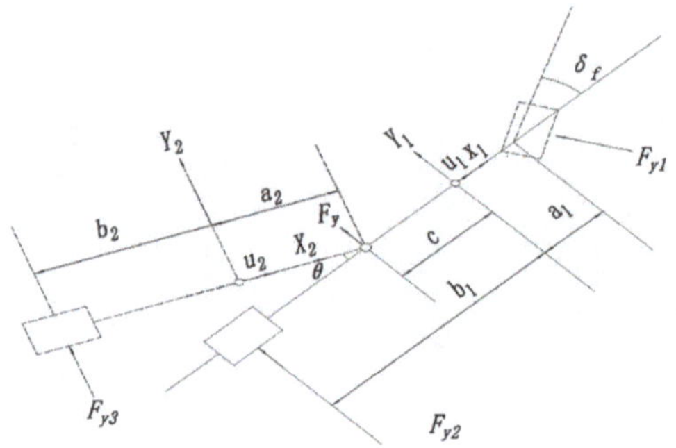

Fig. 3. Three-degree-of-freedom car train model.

The standard expression (2) for the equation of state of this simplified model is given below:

$$M\dot{X} = BX + PU \tag{2}$$

3 Lateral Stability Control Strategy for Distributed Electrically Driven Semi-trailer Vehicle Trains

3.1 General Framework for Hierarchical Control Strategies

The overall control flow is shown in Fig. 4.

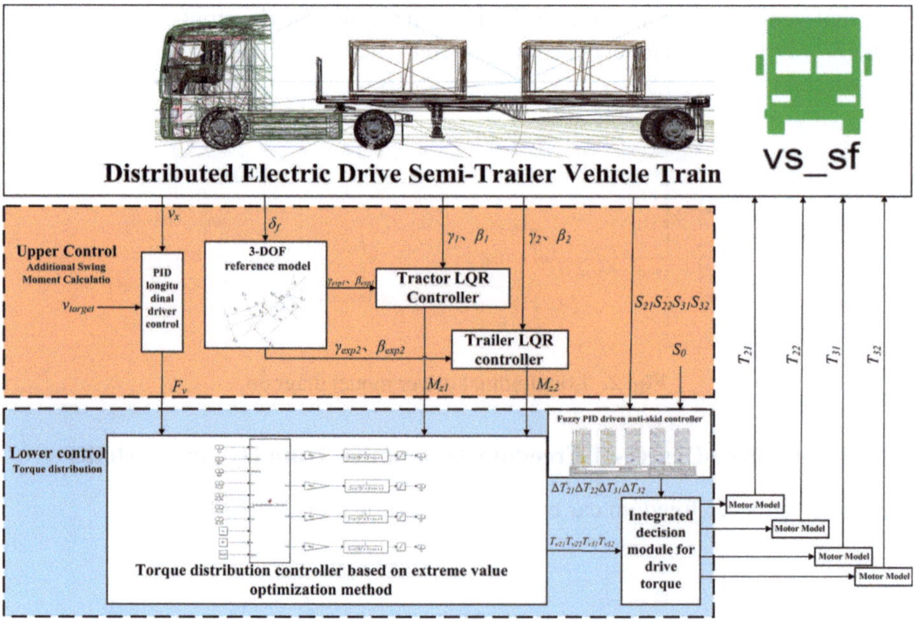

Fig. 4. Overall control process diagram.

In this paper, the dynamics model outputs the front wheel angle δ_f and the desired vehicle speed v_{x_target} to the three-degree-of-freedom reference model through Channels interface via TruckSim software, and calculates and outputs the ideal traversing angular velocities γ_{exp1}, γ_{exp2} and the ideal center-of-mass lateral deflections β_{exp1}, β_{exp2} of the tractor and the trailer, respectively, by the reference model and compares them with the actual traversing angular velocities γ_1,γ_2 and center-of-mass lateral deflections β_1, β_2 as inputs to the LQR-based upper-level controller, respectively. TruckSim outputs of actual traverse angular velocities γ_1,γ_2 and center-of-mass side deflections β_1, β_2 as inputs to the LQR-based upper controller, decides the total active traverse moments M_{z1},M_{z2} required by each vehicle, and uses its total moments and the driving moments T_v required for speed following outputted from the driver speed-following model as inputs to the lower controller, and finally carries out the optimal allocation of traverse moments to each driving wheel The transverse moment of each driving wheel is optimally distributed to each hub motor to achieve the stability control effect.

3.2 Fuzzy PID Drive Anti-skid Controller

The surface diagram of the fuzzy PID controller input variables e, Δe and output variables ΔK_p, ΔK_i, ΔK_d is shown in Fig. 5.

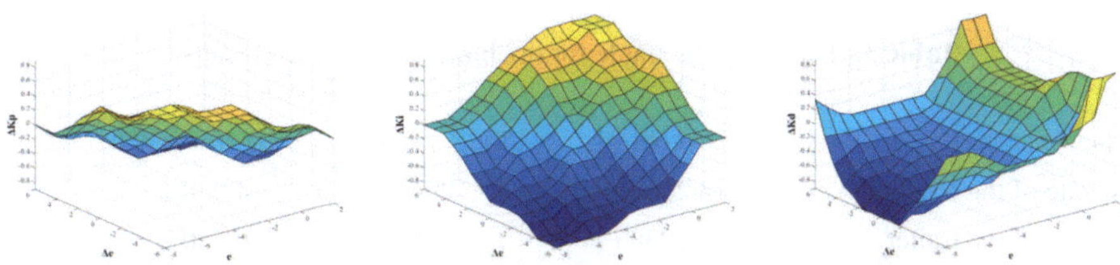

Fig. 5. Surface plots of e, Δe and variables ΔKp, ΔKi, ΔKd.

3.3 Control Strategy Simulation and Verification

The analyzed graphs of the results of the distributed electric drive semi-trailer vehicle train model developed in this paper are shown in Fig. 6 for the ISO3888-2:2002 double-shift line condition (roadway adhesion coefficient of 0.8) without control and LQR control at a speed of 90 km/h.

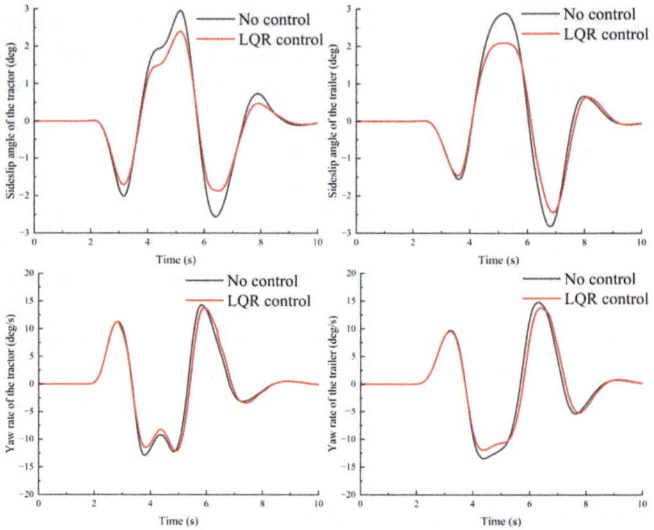

Fig. 6. Double-shift operation condition result analysis curve diagram.

The simulation response results are shown in Table 1.

Table 1. Peak response table for double-shift operation results.

response peak	No control	LQR control	Improvement rate
yaw rate of the tractor	14.28661	13.70968	4.038%
yaw rate of the trailer	14.79383	13.79817	6.730%
sideslip angle of the tractor	2.94999	2.39381	18.854%
sideslip angle of the trailer	2.88764	2.0939	27.487%

The analysis curves for the results of the uncontrolled and LQR-controlled vehicles under single lane change conditions at a speed of 90 km/h (road surface friction coefficient of 0.3) are shown in Fig. 7.

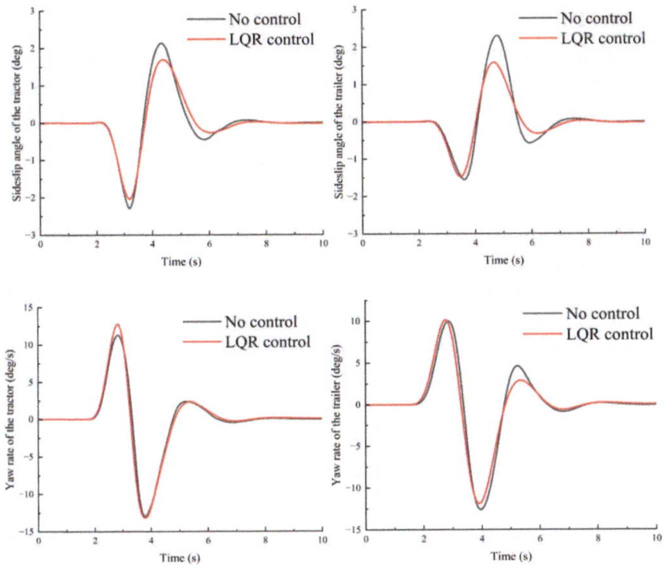

Fig. 7. Single-line transfer condition result analysis curve diagram.

The simulation response results are shown in Table 2.

Table 2. Single-phase line operating condition response peak table.

response peak	No control	LQR control	Improvement rate
yaw rate of the tractor	11.28255	12.73033	−12.832%
yaw rate of the trailer	9.9702	10.13848	−1.688%
sideslip angle of the tractor	2.12982	1.6887	20.712%
sideslip angle of the trailer	2.30054	1.58129	31.264%

4 Conclusion

(1) The established linear three-degree-of-freedom reference model can effectively track the target vehicle speed, with the target trajectory deviation remaining within the ideal range. The state estimation variables (center of gravity lateral deviation angle and yaw rate) exhibit minimal deviation and remain within the ideal range, with consistent trends in the response curves, demonstrating the excellent trajectory tracking capability of the established distributed electric drive semi-trailer truck model.

(2) This paper addresses the issue of lateral instability and rollover that semi-trailer vehicle trains are prone to during high-speed obstacle avoidance. The proposed LQR-based distributed electric drive semi-trailer vehicle train drive anti-slip and torque coordination control strategy effectively improves the lateral stability of semi-trailer

vehicle trains. Under dual-lane high-adhesion conditions, the vehicle's center of gravity lateral deviation is reduced by 27% compared to the uncontrolled condition. Under single lane shift low adhesion conditions, the lateral deviation angle of the vehicle's center of gravity was reduced by 31% compared to the uncontrolled condition.

Acknowledgments. This research is funded by the Science and Technology Innovation Plan of Colleges and Universities of Shanxi Province of China (CN) (Grant No. 2024L232), and the Excellent Doctoral Award Fund for Working in Shanxi (Grant No. 20242037), and the Doctoral Start-up Foundation of Taiyuan University of Science and Technology (Grant No. 20232113), and Shanxi Engineering Research Center of Internal Combustion Engine Power Technology, and Advanced Technology Innovation Center of Zero Carbon Power Special Vehicle.

References

1. Ministry of public security: Data on national motor vehicles and drivers in 2024. Road Traffic Manage. **25**(1), 9 (2025)
2. Wang, S., Wang, W.J., Cheng, B.: Steering control method for semi-trailer tractor-trailer. Journal of Jilin University (Engineering and Technology Edition). Accessed 24 July 2025
3. Zhou, Q.H., Zhang, B.Y.: High-speed lateral stability of double semi-trailer under cooperative control. Sci. Technol. Eng. **25**(17), 7365–7372 (2025)
4. Jin, Y.H.: Research on Drive Anti-Slip and Torque Coordination Control for Distributed Electric Drive Semi-Trailer Tractor-Trailer. Hubei University of Automotive Technology, Shiyan (2023)
5. Wang, Y., Li, C.M., Xiao, L.: Research on yaw moment of distributed electric vehicles based on torque optimal distribution. Veh. Power Technol. **1**, 16–22 (2019)
6. Chen, Y., Peterson, W.A., Zhang, C., et al.: A simulation-based comparative study on lateral characteristics of trucks with double and triple trailers. Int. J. Veh. Saf. **11**(2) (2019)
7. Jagelčák, J., Kiktová, M., Frančák, M.: The analysis of manoeuvrability of semi-trailer vehicle combination. Transport. Res. Procedia, 44176–181 (2020)
8. Nie, Z.G.: Research on Control Strategy of Heavy-Duty Semi-Trailer Based on Integrated Rear Axle Active Steering and Differential Braking. Jilin University, Changchun (2014)

Fatigue Life Prediction of Casting Crane Bridge Structure Based on Digital Twin

Rongpei Zheng[1], Xin Wang[1(✉)], Keqin Ding[2], Li Chen[2], and Jingyu Zhai[1]

[1] Dalian University of Technology, Dalian 116024, China
wangxbd21@163.com
[2] China Special Equipment Inspection and Research Institute, Beijing 100029, China

Abstract. Due to the high operation frequency and intensity, casting cranes are prone to structural fatigue damage. In this paper, a life prediction method is proposed based on digital twin. A prediction model based on LSTM network is established. By combining real-time monitoring data with finite element simulation data and integrating the Miner linear cumulative damage theory, the life prediction of key structures of casting cranes is achieved. The large amount of data generated by the structural digital twin model is used to make up for the shortage of measured data, and to train the LSTM model. Through the analysis of the training and validation loss curves and the R^2 accuracy analysis of the validation data set, it is shown that this proposed model can effectively predict the life of the structure.

Keywords: Casting Crane · Digital Twin · Life Prediction · LSTM Network Model

1 Introduction

With the rapid economic development and the continuous iteration of various technological fields, many traditional mechanical industries are rapidly moving towards informatization and intelligence. Digital twin technology, as a comprehensive engineering technology integrating modeling, simulation, data analysis and intelligent control, is regarded as one of the key technologies supporting the intelligence of future industrial systems. The integration of traditional lifting machinery industry with digital twin technology not only enhances the production efficiency of the industry, but also greatly improves the safety and maintainability of mechanical structures. The realization of lifting machinery's intelligence and informatization has great significance to the development of the lifting machinery industry.

At present, scholars at home and abroad realize the operation state detection, stress state detection and life prediction of cranes by constructing digital twin systems or surrogate models. Qiu Qihang [1] et al. proposed the framework of digital twin monitoring system for container gantry crane, which used the numerical fitting method to realize the real-time monitoring of the metal structure stress of the crane. Chen Yuhao [2] et al. proposed a health prediction method of lifting machinery based on convolutional neural

S. Halgamuge et al. (Eds.): ICACMVE 2025, LNME, pp. 337–344, 2026.
https://doi.org/10.1007/978-981-95-6911-3_32

network and bidirectional encoder-decoder long short-term recurrent neural network. Jia Zhengnan [3] et al. established a tower crane motion model and simplified the model grid, and used artificial neural network technology to realize online prediction. Rama [4] et al. proposed an intelligent system based on support vector machine to detect and locate cracks. Zhidchenko [5] et al. created a simplified dynamic and kinematic model of the crane to achieve fast motion prediction.

However, the above studies only realize the real-time mapping of the stress and displacement state of the solid structure, and lack the analysis and prediction of the fatigue life of the actual structure. Therefore, a certain type of casting crane is as the research object in this paper. Through the fusion of historical monitoring data and finite element simulation data, combined with digital twin technology and long short term memory neural network (LSTM), a crane life prediction model is established. The life prediction of the key structure position of the casting crane is realized, which is convenient for predictive maintenance and ensures the operation safety.

2 Digital Twin System

The digital twin system is to establish a one-to-one correspondence twin model with the physical space entity in the digital space. The data of various working conditions of entities in the real physical space are transferred to the twin model in the digital space. The Siamese model can map the state of the actual object in real time. Therefore, the digital twin system is mainly composed of three parts, as shown in Fig. 1: (1) The physical space, including the crane, the sensing system composed of various sensors, the signal transmission equipment, the industrial computer and other hardware equipment; (2) digital space, including finite element simulation model, stress prediction surrogate model, crane twin model and other virtual models; (3) Data space, including all kinds of data generated and derived from the fusion of physical space and Siamese space [6].

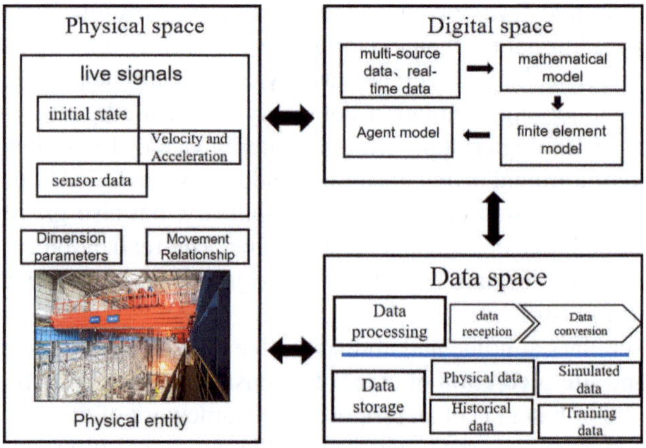

Fig. 1. System Framework Diagram

All kinds of monitoring data in the physical space are transmitted and stored in the data space. Various models in the digital space determine the boundary conditions and inputs according to the data collected in the physical space, and obtain the real-time state of the structure by calculation. The results of the computation in the digital space guide the physical space. Through the digital twin technology, the force situation of the crane is simulated in the virtual environment, so as to obtain the state data of the crane structure. The collected data is preprocessed and used as the data set for training LSTM together with the historical data of the device operation. Through the life prediction of the structure, the position of easy fatigue can be found in time, predictive maintenance can be carried out, and the overall life of the crane can be improved.

3 Stress-Life Prediction Model Based on LSTM

Due to the high frequency and strength of crane operation, it is easy to cause fatigue failure of bridge structure section mutation and other positions. Therefore, the life prediction of bridge structure is very important. The casting crane is as the research object, as shown in Fig. 2. The LSTM network model is established to predict the corresponding lifetime through the Stress time series.

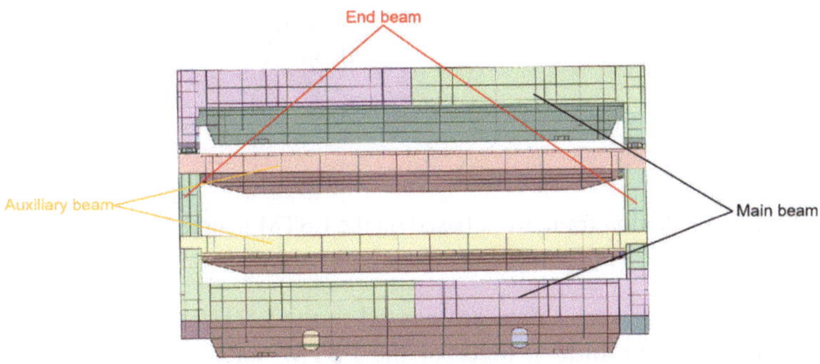

Fig. 2. Casting crane

LSTM (Long Short-Term Memory) is a recurrent neural network (RNN) structure with gating mechanism. LSTM is used to solve the vanishing or exploding gradient problem that occurs in traditional RNNS when dealing with long-term dependencies. By introducing a gating mechanism, LSTM can effectively "remember" important information and "forget" useless information in the learning process, so as to realize the modeling of long-term dependent features in time series.

In this paper, a PyTorch based LSTM recurrent neural network is constructed to extract stress timing features. The network architecture includes a 5-layer structure of input layer, LSTM layer, 2 fully connected layers and output layer, as shown in Fig. 3. The network input is a 200-step stress timing. The network uses a single-layer unidirectional LSTM to extract timing features, where the number of hidden units is 64. The selection of 64 LSTM hidden units in this study is to strike a balance between model complexity and generalization ability. Too few hidden units may make the model unable to learn, and too many hidden units may lead to overfitting. 64 hidden units are sufficient to capture the nonlinear fatigue evolution characteristics in the time series while maintaining good interpretability and visualization capabilities. The output of the LSTM layer connects two fully connected layers, which compress the 64-dimensional feature vector map obtained from the LSTM into 32 dimensions respectively, and then map the 32-dimensional vector to a scalar. Finally, the results are presented by the output layer to realize the life regression prediction [7]. The total number of parameters of the model is 19,009, of which the number of LSTM layer parameters is 16,896, accounting for the main number of parameters.

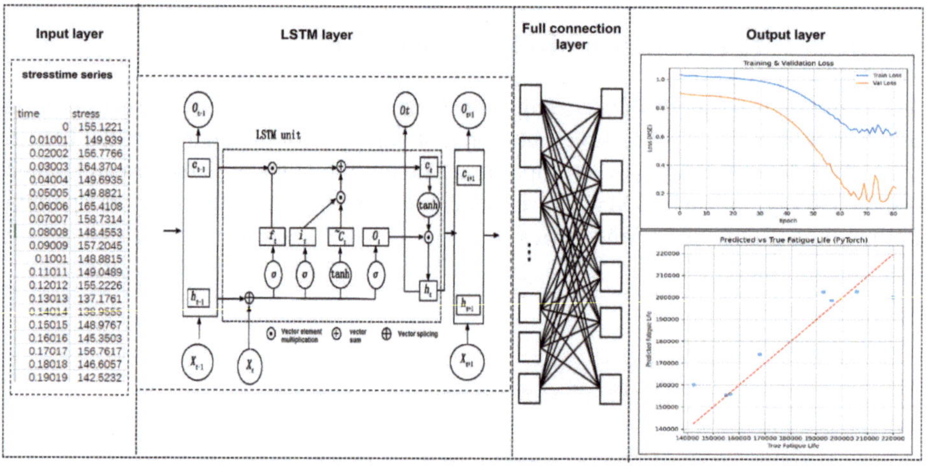

Fig. 3. Prediction framework based on the LSTM neural network model

The input stress time series data are passed to the LSTM layer after normalization and normalization. In the hidden layer of the LSTM, each time step takes an output and the cell state from the previous time step. Gating units are used to control the flow of information and also to adjust the cell state, including input gates, forget gates, and output gates. The gating unit determines what information is added, what information is forgotten and what information is output. The Sigmoid function is usually used as the activation function for each gate, and the value after activation is between 0 and 1, which is used to control the opening and closing of the gate [8]. Each LSTM neuron has two inputs and two outputs. The two input values are the sequence value x_t and the previous memory c_{t-1}. The two output values are the updated memory unit c_{t-1} and the updated value of the neuron h_t. The aforementioned input and output values are controlled by the LSTM's three gate mechanisms and a memory refresh structure, thereby overcoming the gradient explosion and gradient vanishing issues of traditional recurrent neural networks [9].

1) Forgetting gate: The forgetting gate f_t is a function of the previous state h_{t-1} and the new input x_t. The sigmoid activation function between 0 and 1 acts on each element of h_{t-1} and x_t to obtain the forgetting gate f_t. The forgetting gate f_t is multiplied by

the previous memory unit c_{t-1} to determine what percentage of the memory will be used next time. The expression of the forgetting gate is as shown in Eq. (1):

$$f_t = \sigma\left(W_f x_t + U_f h_{t-1} + b_f\right) \tag{1}$$

In the equation, W_f and U_f is the parameter matrix to be estimated, and b_f is the bias.

2) Input gate: The calculation method of the input gate is very similar to that of the forget gate, as shown in (2):

$$i_t = \sigma(W_i x_t + U_i h_{t-1} + b_i) \tag{2}$$

The input gate i_t is also a function of the previous old state vector h_{t-1} and the new input vector x_t, the sigmoid activation function between 0 and 1 acts on each element of the h_{t-1} and x_t to obtain the input gate i_t. [9].

3) Memory unit: The memory unit is updated jointly by the candidate memory unit $\tilde{c}_t$ and the previous memory unit c_{t-1}, and is regulated by the input gate i_t and the forget gate f_t. Its expression is shown in Eq. (3):

$$c_t = f_t * c_{t-1} + i_t * \tilde{c}_t \tag{3}$$

Among them: The expression for the backup memory unit is shown in Eq. (4):

$$\tilde{c}_t = \tanh(W_c x_c + U_c h_{t-1} + b_c) \tag{4}$$

In the formula, tanh is the hyperbolic tangent activation function.

4) Output gate: The output of the LSTM is first updated through the output gate: The expression is shown in (5):

$$o_t = \sigma(W_o x_t + U_o h_{t-1} + b_o) \tag{5}$$

Next, combine the output gate o_t to transmit internal information to the external state h_t:

$$h_t = o_t * \tanh(c_t) \tag{6}$$

In the equation, the tanh function is a hyperbolic tangent function with a range between -1 and 1. This function is applied to the updated memory unit c_t and multiplied by the output gate to obtain the LSTM output.

4 Model Training and Validation

4.1 Data Acquisition

Through the crane digital twin surrogate model, 1000 stress timing points can be obtained by applying a variable amplitude load between 500t and 350t to the main beam in 10s. From the 1000 stress time series points, 200 stress time series points were randomly selected to form a group, resulting in a total of 40 groups. The fatigue life corresponding to each set of stresses was calculated by using the rainstream counting method and Miner's linear cumulative damage theory [10]. Stress values and lifetime data are used for regression training of the LSTM model.

4.2 Model Training

Based on the stress time series obtained above and the calculated fatigue life, the LSTM neural network is used to find the regression relationship. An early stopping mechanism is added to the model training process to prevent overfitting. Dynamic learning rate adjustment was adopted to improve the stability of model training. The loss function used in the model is the mean square error function. The plot of training and validation loss is shown in Fig. 4. Both the training and validation loss curves decrease and plateau, indicating that the model learned well and did not overfit or underfit.

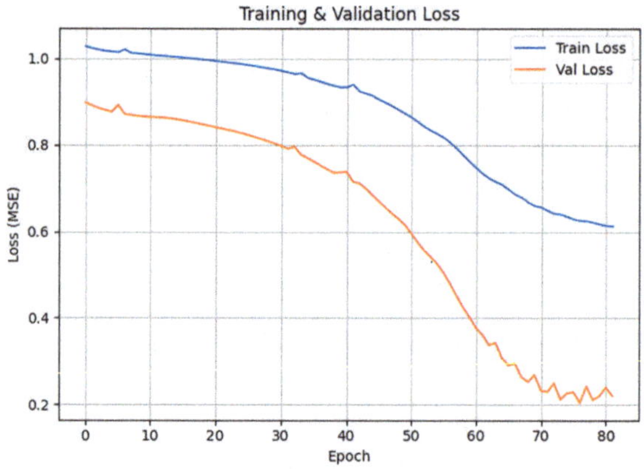

Fig. 4. Training and validation loss curve diagram

4.3 Model Validation

20% of the dataset is used as the experimental data (test set). LSTM network model is used for lifetime prediction. See Fig. 5 for the distribution of the predicted and calculated true values. The red dotted line in the plot represents the ideal case where the predicted value equals the true value. It can be seen from the figure that the predicted value points all fall near the red dashed line, indicating that the predicted value is very close to the true value and the prediction error is very small. The accuracy of the model is evaluated by the correlation coefficient R2, which measures how well the regression curve fits the observed values. The closer it is to 1, the better the fit is. The fitting optimization degree of this training is 0.795, which is close to 1. This shows that the model has a good fitting effect and can realize the prediction of the structure life. Compared with the traditional Miner theory, the LSTM model shows higher prediction accuracy and generalization ability when dealing with complex loading history and nonlinear fatigue response. This provides a more reliable data-driven method for structural residual life prediction.

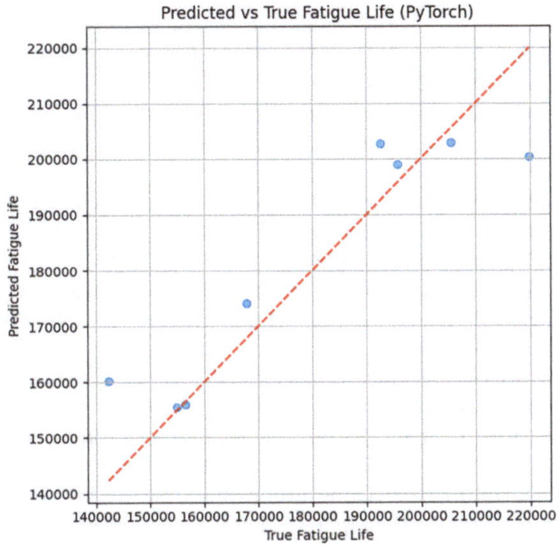

Fig. 5. Scatter plot of true and predicted values

5 Conclusion

In this paper, the digital twin model is established for the main structure of the casting crane. The rainstream counting method and Miner's linear cumulative damage theory are combined to obtain structural stress and life data to create a data set. Then, a life prediction model is established based on LSTM neural network to characterize the mapping relationship between stress time series and fatigue life. A gating mechanism is introduced into the LSTM, which enables the network to selectively forget, store and output information. so as to prevent effectively the vanishing gradient problem. At the same time, the introduced cell state improves the stability of the network, so the network can deal with long time series information more freely. The neural network is trained using the data set generated by the digital twin model. According to the trend of the loss function curve, the effectiveness of the model training is verified. The analysis of R2 accuracy error indicates the fitting effect of the model. As a result, the trained LSTM model can effectively predict the life of the structure.

Acknowledgements. This project is supported by National Key Research and Development Program (Project No.: 2024YFC3014900).

References

1. Qiu, Q., Xiao, H.: Research on stress state monitoring of container gantry cranes based on digital twin technology. Crane Transp. Mach. (19), 28–34 (2024). https://doi.org/10.3969/j.issn.1001-0785.2024.19.010

2. Chen, Y., Yang, Z., Wen, J.: Health prediction method of lifting machinery based on CNN and bidirectional encoder-decoder LSTM fusion. J. Chongqing Univ. (6) (2025). https://doi.org/10.11835/j.issn.1000-582X.2025.06.007
3. Jia, Z., Wang, X., Gao, S., et al.: An online monitoring method for the structural performance of tower cranes based on digital twins. Comput.-Integr. Manuf. Syst. **30**(12), 4468–4476 (2024). https://doi.org/10.13196/j.cims.2023.0I03
4. Krishna, S.R., Sathish, J., Tarun, M., et al.: A support vector machine-based intelligent system for real-time structural health monitoring of port tower cranes. J. Fail. Anal. Prev. **24**(6), 2543–2554 (2024). https://doi.org/10.1007/s11668-024-02049-8
5. Zhidchenko, V., Malysheva, I., Handroos, H., et al.: Faster than real-time simulation of mobile crane dynamics using digital twin concept. J. Phys. Conf. Ser., 1096 (2018). https://doi.org/10.1088/1742-6596/1096/1/012071
6. Dong, Q., Nan, F., Xu, G.: Construction method of digital twin system for real-time monitoring of crane operation state. Chin. J. Constr. Mach. (2) (2025)
7. Wang, Y., Cui, H., Liu, Y., et al.: Residual life prediction of rolling bearing based on digital twin and informer-LSTM. Combination Mach. Tools Autom. Process. Technol. (7), 135–140 (2025). https://doi.org/10.13462/j.carolcarrollnkimmtamt. 2025.07.024
8. Liang, M., Peng, X., Liu, D.: Research on fault diagnosis of gantry crane gearboxes based on the LSTM algorithm. Comput. Meas. Control **29**(12), 67–72 (2021). https://doi.org/10.16526/j.cnki.11-4762/tp.2021.12.013
9. Kapteyn, M., Knezevic, D., Huynh, D., et al.: Data-driven physics-based digital twins via a library of component-based reduced-order models. Int. J. Numer. Meth. Eng. **123**(13), 2986–3003 (2022)
10. Hussain, M., Ye, Z., Chi, H., et al.: Predicting degraded lifting capacity of aging tower cranes: a digital twindriven approach. Adv. Eng. Inform. **59**(1), 102310 (2024)

Research on Multi-fidelity Surrogate Models of Ship to Shore Gantry Crane Based on Digital Twin

Fan Yang[1], Xin Wang[1(✉)], Jingyu Zhai[1], Li Chen[2], and Keqin Ding[2]

[1] Dalian University of Technology, Dalian 116024, China
wangxbd21@163.com
[2] China Special Equipment Inspection and Research Institute, Beijing 100029, China

Abstract. For the poor adaptability of multi-fidelity methods for single-surrogate models, a multi-fidelity optimization method to build a hybrid surrogate model is proposed. This method combines three regular surrogate models, including PRS, Kriging, and RBF. According to the characteristics and prediction accuracy of different surrogate models, a hybrid improved surrogate model is constructed. A ship to shore gantry crane is as the research object. High and low fidelity finite element models of this crane are established. Latin hypercube sampling is used to sample load cases of the crane, and stresses are calculated via the finite element model and extracted into a stress database. On this basis, a low-fidelity surrogate model and a residual surrogate model for the stress points of the crane's key parts are established. Furthermore, these models are combined into a multi-fidelity surrogate model. This model can rapidly predict stress values according to new inputs. The accuracy of the predicted outputs is compared with the actual finite element outputs, which verifies the performance of the proposed model.

Keywords: Ship to Shore Gantry Crane · Digital Twin · Stress Prediction · Hybrid Multi-Fidelity Surrogate Model

1 Introduction

A digital twin is a digital representation of a physical object, with a multidisciplinary, multiscale coupled real-time "co-evolutionary" model built by integrating mechanistic models, real-time sensor data, and expert knowledge. Driven by research advances from NASA and the U.S. Air Force as well as progress in computer technology, digital twin studies have grown explosively in recent years, and its dynamic real-time features have gained wide attention and application in the lifting engineering field. Szpytko J et al. developed a gantry crane digital twin integrated with a maintenance decision model, improving efficiency via particle swarm optimization [1]. He Xiwang et al. used a Gaussian process surrogate model to calculate the stress and strain of the main beam, based on the real-time position of the gantry crane trolley detected by sensors and the stress and strain data from finite element simulation.[2]. Jiang et al. established a tower crane scaled-model digital twin to assess structural stability during unsafe lifting [3]. Lai et al.

S. Halgamuge et al. (Eds.): ICACMVE 2025, LNME, pp. 345–353, 2026.
https://doi.org/10.1007/978-981-95-6911-3_33

proposed a shape-integrated digital twin modeling method for jib cranes, based on surrogate models, reduced-order models, and dynamic data, with stress analysis verification [4]. Song Xueguan et al. put forward a multi-model fusion shape-integrated framework, conducting stress-strain monitoring modeling and verification for boom cranes.[5]. In the multi-fidelity method for surrogate models, the multi-fidelity method based on residual improved functions (or bridge functions), is simple and easy to understand. Because of their relatively simple modeling process and ease of implementation, they are widely applied in engineering design. Tyan et al. used low-fidelity models and scale functions to approximate high-fidelity ones for efficient design optimization [6]. Sun et al. built low-fidelity surrogate models via moving least squares, and improved the models with polynomial response surface scale functions [7]. Liu et al. proposed an improved RBF-based multi-fidelity surrogate modeling method, enhancing the fitting flexibility and capability of multi-fidelity surrogate models [8]. Zhang et al. put forward a method to determine the optimal scaling factor of multi-fidelity surrogate models by minimizing LOOCV errors [9].

However, for approximation, current residual improved models are typically studied based on a single surrogate model. When confronted with complex engineering design problems, the residual function between high- and low-fidelity models also grows complex. Thus, accurate approximation of the differences between high- and low-fidelity models is difficult only based on a single surrogate model. Therefore, considering the correlation between low-fidelity and high-fidelity models, a high-accuracy hybrid surrogate model is researched and constructed.

2 System Framework

According to a five-dimensional digital twin model [10], and the overall framework of a digital twin monitoring system for ship to shore gantry cranes is presented in this paper, as shown in Fig. 1. This system is consisted of physical space (including cranes, sensor systems, industrial control computers, etc.), digital space (including finite element simulation models, stress prediction surrogate models, gantry crane twin models, etc.), and data space. Twin data in data space includes various data from physical and digital spaces, with bidirectional data communication enabling mutual control between virtual and physical domains.

3 Establishment of Stress Database

3.1 Latin Hypercube Sampling Selection of Load Conditions

In crane stress calculations, the finite element simulations or experiments for all load cases are extremely costly. Therefore, experimental design is employed to select samples reasonably within the design space, so that the number of experiments is reduced and more information is obtained. Latin hypercube sampling (LHS), as a stratified Monte Carlo sampling method, ensures that the multidimensional design space is uniformly covered by samples, thereby improving data representativeness. The 50-ton Ship to Shore gantry crane shown in Fig. 2 is taken as the research object, with its basic working

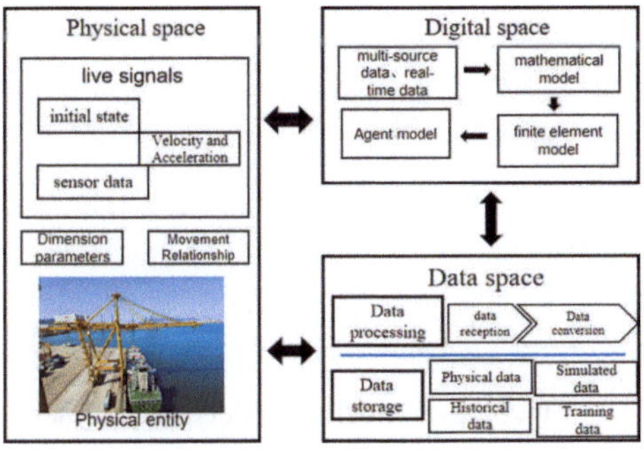

Fig. 1. System framework diagram.

Fig. 2. Physical picture of ship to shore gantry crane.

condition parameters being operating radius and lifting capacity. The extreme backward radius is defined as the 0 point, the operating radius parameter ranging from 0m to 116.2 m, and the crane load parameter from 0t to 50t. These two working condition parameters are normalized using MATLAB. The high-fidelity model randomly selects 20 sample points, while the low-fidelity model selects 40. Combining the actual crane movement, stratified Latin hypercube sampling is applied: the operating radius (0 m to 38.1 m) is assigned to interval 1, and (38.1 m to 116.2 m) to interval 2, with a ratio of 2:3. The sample distributions for both fidelity levels are shown in Fig. 3.

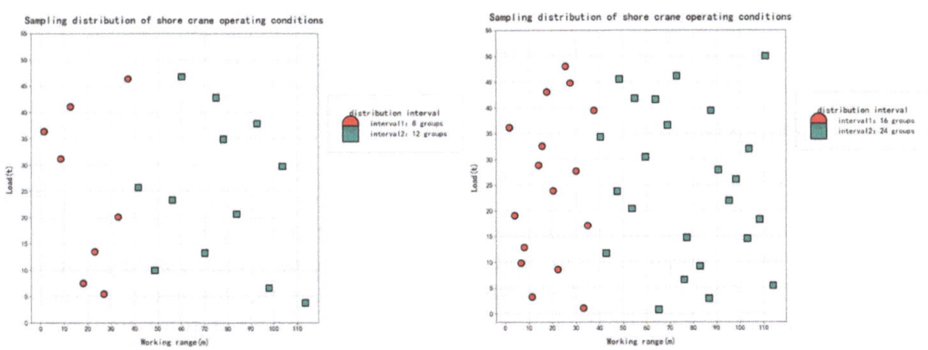

Fig. 3. Two types of fidelity Latin hypercube sampling results

3.2 Stress Extraction from Static Simulation

Based on the sampling results, a static simulation model was established using ANSYS. The high-fidelity model uses plate elements, as shown in Fig. 4. The low-fidelity model uses beam elements, as shown in Fig. 5.

A rotational constraint is applied between the tie rod and main beam; remote displacement is set at the support leg bottom, with z-axis rotational freedom released. A mass point is coupled to the main beam's rear end (electrical control room) to simulate its weight. For loading, a 2 * 4 grid (100 mm sides) is used in the high-fidelity model to

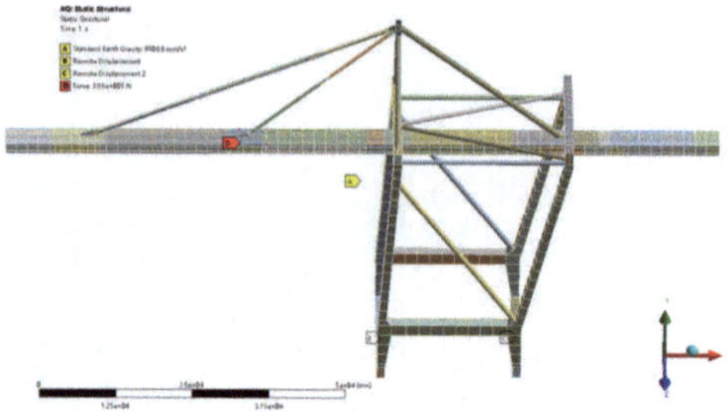

Fig. 4. High-fidelity model establishment.

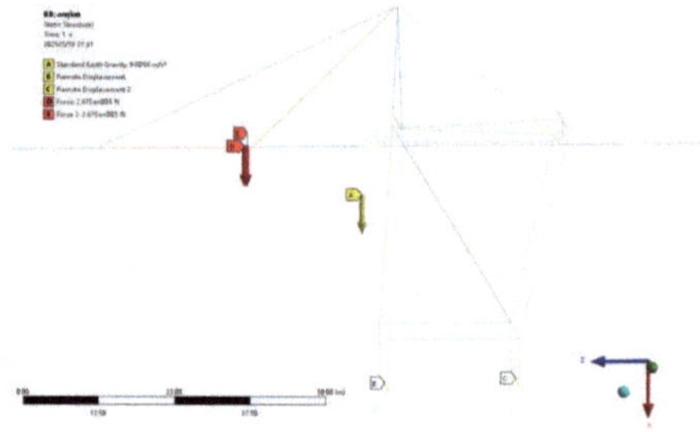

Fig. 5. Low-fidelity model establishment

simulate cart wheel positions, while grid nodes are adopted in the low-fidelity model. Gravitational acceleration is applied to the entire structure. Key parts of the main beam, tie rod, and strut are selected as stress points. Stress point cloud maps of high- and low-fidelity models are obtained through solving, as shown in Fig. 6. A stress database is compiled based on the input calculation conditions of Latin hypercube sampling.

Fig. 6. High- and low-fidelity model stress points cloud map

4 Surrogate Model

Stress prediction is a multi-input multi-output regression problem. Low-fidelity data, though complete, diverse and economical, has low accuracy and suits macro trends; high-fidelity data, accurate but hard and time-consuming to obtain, can improve low-fidelity precision. Correlation of data with different fidelities enables precise prediction, so a multi-fidelity surrogate model is built using small amounts of high-fidelity and large amounts of low-fidelity data.

Research on hybrid surrogate model-based multi-fidelity methods integrates the advantages of models, providing correction models with better accuracy and stability. To ensure capabilities in low-dimensional, nonlinear, and high-dimensional prediction, PRS (polynomial fitting for nonlinear accuracy), Kriging (spatial correlation interpolation with uncertainty estimates, suitable for nonlinear and precise interpolation), and RBF (basis function combination for high-dimensional nonlinear interpolation, precisely matching samples) are selected.

First, a Kriging model is built using low-fidelity data to predict high-fidelity sample values, and errors are obtained by comparison with actual values. Since the three models have advantages for residual and stress points, the construction of surrogate models for difference data can amplify strengths and reduce weaknesses, ensuring prediction accuracy at each stress point. Integration with the low-fidelity model enables more precise prediction of new input-output relationships.

4.1 Establishment of a Low-Fidelity Surrogate Model

The parameters of the Kriging model were optimized using genetic algorithms, and a low-fidelity surrogate model was established. To test the model performance, simple cross-validation was adopted, with 12 sample points reselected for verification—working radii of 10 m, 30 m, 50 m, 70 m, 90 m, and 110 m, and loads of 10 t and 40 t applied respectively. After stress data were obtained, the RMSE and R^2 evaluation indices were calculated against the stress values predicted by the surrogate model. The calculation results are shown in Table 1.

Table 1. Low-fidelity surrogate model accuracy evaluation indicators.

Item	1	2	3	4	5	6	7	8	9	10
Rmse	6.44	0.59	2.13	0.89	2.08	1.68	2.54	1.53	3.95	1.44
R^2	−0.93	0.99	0.76	0.99	0.78	0.80	0.48	0.98	0.79	0.97

Among them, the root mean square error (RMSE) is the arithmetic square root of the mean square error and is used to measure the overall accuracy of the approximation model. The larger the coefficient of determination R2, the better the model fit.

$$RMSE = \sqrt{\frac{1}{m}\sum_{i=1}^{m}\left(y^{(i)} - \hat{y}^{(i)}\right)^2} \tag{1}$$

$$R^2 = 1 - \frac{\sum_{i=1}^{m}\left(\hat{y}^{(i)} - y^{(i)}\right)^2}{\sum_{i=1}^{m}\left(\overline{y} - y^{(i)}\right)^2} \tag{2}$$

where m is the number of sample points, $y^{(i)}$ represents the response value of the i-th sample, $\hat{y}^{(i)}$ represents the predicted value of the surrogate model for the i-th sample, and $\overline{y}$ represents the average response value of the sample.

From Table 1, stress point 1 is $R^2 < 0$, which is relatively distorted. A check of the database shows that the stress at this point undergoes abrupt changes only under a few load cases. This is because the simplified structure of the low-fidelity finite element model cannot fully reflect changes in load cases. The R^2 values at other stress points meet the accuracy requirements, and predictions using the low-fidelity model can be made.

4.2 Establishment of Residual Surrogate Model

Residuals are obtained by predicting the stress at high-fidelity sample locations using a low-fidelity surrogate model. The parameters of three surrogate models are optimized via a genetic algorithm, and respective residual surrogate models are established. Since high-fidelity samples are difficult to obtain and costly, they are all used as the training set without an additional test set. Leave-one-out cross-validation is adopted: one sample is removed in turn from the m samples, and the remaining samples are used to construct the model and predict the response at the removed point. The results obtained are shown in Tables 2 and 3, with corresponding bar charts depicted in Figs. 7 and 8.

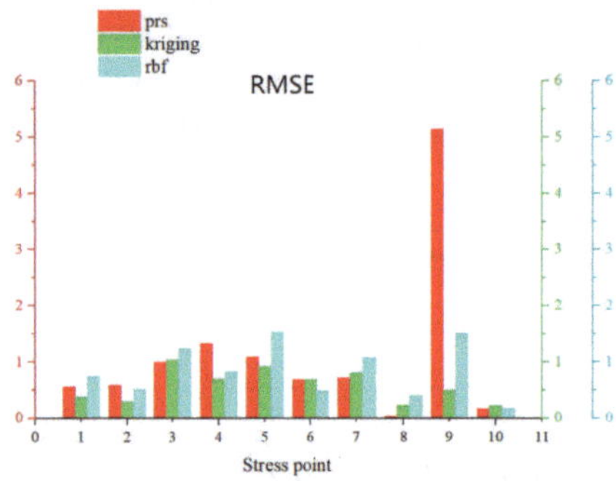

Fig. 7. Three surrogate models Rmse bar chart.

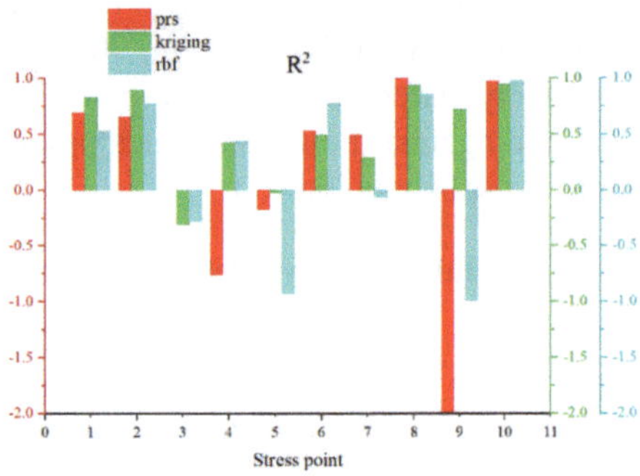

Fig. 8. Three surrogate models R2 histogram.

Table 2. Difference fidelity surrogate model Rmse evaluation index (For each stress point).

Item	1	2	3	4	5	6	7	8	9	10
PRS	0.56	0.59	1.00	1.33	1.08	0.69	0.71	0.04	5.14	0.17
Kri	0.38	0.30	1.04	0.70	0.92	0.69	0.81	0.24	0.50	0.22
RBF	0.75	0.52	1.24	0.82	1.53	0.49	1.08	0.40	1.51	0.17

Table 3. Difference fidelity surrogate model R2 evaluation index (For each stress point).

Item	1	2	3	4	5	6	7	8	9	10
PRS	0.69	0.66	0.00	−0.76	−0.17	0.53	0.49	0.99	−25.4	0.97
Kri	0.83	0.90	−0.31	0.423	−0.02	0.49	0.29	0.94	0.72	0.94
RBF	0.53	0.77	−0.27	0.434	−0.93	0.78	−0.06	0.86	−0.99	0.97

To compare the predictive ability of different models, R^2 can be used as an evaluation index. It is noted that different surrogate models show varying R^2 performances at different stress points: the Kriging model performs better at stress points 1, 2, 5, and 9; the polynomial model at 3, 7, and 8; and the RBF model at 4, 6, and 10. RMSE represents the distance between predicted and true values, and minimizing it is the ultimate goal of the stress surrogate model. Models with larger R^2 are observed to have smaller RMSE at corresponding points, so a combination form can be set to integrate them for maximum advantage.

4.3 Establishment of a Multi-fidelity Surrogate Model and Prediction Testing

Definition:

$$Y_P = \frac{1}{2}\left(\frac{R_{\rm kri} + R_{\rm rbf}}{R_{\rm prs} + R_{\rm kri} + R_{\rm rbf}} Y_{PRS} + \frac{R_{\rm prs} + R_{\rm rbf}}{R_{\rm prs} + R_{\rm kri} + R_{\rm rbf}} Y_{Kri} + \frac{R_{\rm prs} + R_{\rm kri}}{R_{\rm prs} + R_{\rm kri} + R_{\rm rbf}} Y_{RBF}\right) + Y_L \quad (3)$$

Among them, Y_{prs}, Y_{kri}, Y_{rbf}, Y_P, and Y_L represent the outputs of the PRS model, Kriging model, RBFmodel, hybrid prediction model, and low-fidelity model, respectively. R represents the Rmse of the corresponding surrogate model, i.e., the smaller the mean square error in the change weight, the greater the output weight, so that the hybrid model outputs close to the optimal model output in each dimension.

The median value of the sample interval with the most actual load cases was selected for testing, specifically a working range of 77.2 m and a load size of 25 t. Stress was quickly predicted by the surrogate model based on the input. A comparison between the actual high-fidelity model data and the mixed model prediction data is shown in Table 4.

Table 4. Comparison of high-fidelity model and hybrid model data (unit: MPa).

stress point	high-fidelity model	mixed model	accuracy rate
1	203.12	206.62	98.30%
2	60.44	57.35	94.62%
3	12.68	19.41	65.31%
4	28.62	33.38	85.74%
5	8.43	7.42	86.46%
6	35.66	35.02	98.17%
7	12.13	10.74	87.08%
8	125.31	125.63	99.75%
9	73.70	73.72	99.97%
10	210.47	210.27	99.91%

Except for stress point 3, the accuracy of all other stress points exceeded 85%, with an average accuracy of 91.53%. This indicates that the data predicted by the mixed model can be used to identify high-risk crack-prone area.

5 Conclusion

Two finite element models of different fidelities are established for ship-to-shore gantry cranes, with a stress database constructed by extracting and evaluating stress points. A difference surrogate model and a low-fidelity surrogate model are developed, whose mean square error (Rmes) and correlation coefficient(R^2) are assessed via cross-validation before being combined into a multi-fidelity surrogate model. Case predictions demonstrate good accuracy in quickly and precisely forecasting stress under different load cases. However, some correlation coefficients of the proposed model are unsatisfactory due to insufficient samples. Future research will focus on determining the minimum sample size ensuring model accuracy and the high-low fidelity sample ratio.

Acknowledgements. This research was supported by the National Key R&D Program of China (No. 2024YFC3014900).

References

1. Szpytko, J., Duarte, Y.S.: Digital twins model for cranes operating in container terminals. IFAC Papers Online **52**(10) (2019)
2. He, X., Lai, X., Yang, L., et al.: Mechanism and data-driven design of a gantry crane digital twin. In: Proceedings of the 17th China CAE Engineering Analysis Technology Conference, Deqing, Zhejiang, China, pp. 21–28 (2021)
3. Lai, X., Wang, S., Guo, Z., et al.: Designing a shape–performance integrated digital twin based on multiple models and dynamic data: a boom crane example. J. Mech. Des. **143**(7), 71–89 (2021)
4. Zhang, J., Liu, F., Guo, G., et al.: Optimization of tower crane boom structure based on multi-fidelity agent models. Mech. Electr. Eng. (11) (2024). https://doi.org/10.3969/j.issn.1001-4551.2024.11.005
5. Jiang, W., Ding, L., Zhou, C.: Digital twin: Stability analysis for tower crane hoisting safety with a scale model. Autom. Constr. **138**(2), 104–112 (2022)
6. Tyan, M., Nguyen, N.V., Lee, J.W.: Improving variable-fidelity modeling by exploring global design space and radial basis function networks for aerofoil design. Eng. Optim. **47**(7), 885–908 (2015)
7. Sun, G., Li, G., Zhou, S., et al.: Multi-fidelity optimization for sheet metal forming process. Struct. Multidiscip. Optim. **44**(1), 111–124 (2011)
8. Liu, Y., Wang, S., Zhou, Q., et al.: Modified multifidelity surrogate model based on radial basis function with adaptive scale factor. Chin. J. Mech. Eng. **35**(1), 212 (2022)
9. Zhang, S., Liang, P., Pang, Y., et al.: Multi-fidelity surrogate model ensemble based on feasible intervals. Struct. Multidiscip. Optim. **65**(8), 1–13 (2022)
10. Fei, T., He, Z., Qi, Q., et al.: Theory and application of digital twin model construction. Comput.-Integr. Manuf. Syst. **27**(1), 1–15 (2021)

Research on Vehicle Tire Noise Prediction Method Based on Transfer Learning

Mingyu Wu[1(✉)], Yafei Wang[1], Junjie Chen[2], Yichen Zhang[1], and Zexing Li[1]

[1] School of Mechanical Engineering, Shanghai Jiao Tong University, Shanghai, China
wmy_marvin@sjtu.edu.cn

[2] School of Mechanical and Electrical Engineering, Jiangxi University of Science and Technology, Ganzhou 341000, China

Abstract. To address issues in traditional tire noise prediction methods such as simplified assumptions in physical modeling, insufficient generalization ability of data-driven models with small samples, and lack of multi-band noise collaborative analysis, this study proposes a multimodal tire noise prediction method that integrates transfer learning with physical mechanisms. By constructing a transfer learning network to extract geometric topological features of the tread pattern combined with a channel attention mechanism for dynamic weighting, an innovative physically-constrained hybrid loss function and multi-objective regression architecture are designed. Based on 126 sets of semi-anechoic chamber experimental data, a grouped cross-validation strategy is used to validate model performance. The results show that the combination of ResNet101 and LightGBM achieves the best performance in cross-modal feature fusion, with an average absolute prediction error of 0.66 dB, a coefficient of determination R2 of 0.97, and 96.96% of samples having prediction errors below 2 dB. The physical fundamental frequency constraint reduces model error by 15%, effectively balancing the advantages of data-driven and mechanistic models.

Keywords: Transfer Learning · Image Processing · Tire Noise Prediction · Feature Fusion

1 Introduction

Tire noise is a core metric of vehicle NVH (Noise, Vibration, and Harshness) performance, directly influencing user comfort and compliance with environmental regulations. Studies by the WHO indicate that continuous exposure to noise levels exceeding 70 dB(A) can cause hearing damage and adverse health effects. With increasingly stringent regulations such as ECE-R117 and ECE R51.03, and rising consumer demand for quieter vehicles, efficient and accurate tire noise prediction and optimization has become a key industry challenge.

Tire noise primarily arises from external noise, cavity noise, and pattern noise. External noise is road-dependent, affected by pavement conditions and tread design. Research has explored using asphalt texture modeling [1] and groove ventilation (Sandberg) to

S. Halgamuge et al. (Eds.): ICACMVE 2025, LNME, pp. 354–360, 2026.
https://doi.org/10.1007/978-981-95-6911-3_34

reduce noise, though some non-physical methods failed to yield effective reduction. Cavity noise originates from internal structural vibrations and air resonance, modeled by Thompson with verified accuracy. Structural deformation can transmit vibration to the rim and body, amplifying noise. Experimental models have been developed for in-cabin noise prediction. Pattern noise relates to the tire's structural vibration modes and material properties [2]. Data-driven methods, such as CNNs and BP neural networks, have achieved ~ 0.6 dB error in noise prediction [3], while others [4, 5] explored image-based and test correlation approaches.

However, mainstream prediction approaches face three major bottlenecks: 1) Conventional physical models rely on oversimplified assumptions, making it hard to capture nonlinear coupling between micro-geometry and multi-physics fields. 2) Pure data-driven deep models lack interpretability and suffer generalization degradation in small-sample scenarios. 3) Most methods lack adaptive focus on high-contribution noise features, limiting their effectiveness in complex patterns.

To address the above challenges, this paper proposes a transfer learning-based framework for automotive tire noise prediction. By embedding physical fundamental frequency constraints into the neural network training process, the framework enables the co-evolution of data-driven models and physical mechanism models.

2 Research Method

The multi-physics feature coupling in this study encompasses vibration-acoustic fields, aerodynamic noise fields, and friction-induced noise fields. By jointly leveraging multi-dimensional physical features and microscopic tread pattern characteristics, precise tire noise prediction is achieved. As shown in Fig. 1, the proposed technical route adopts a multi-modal feature fusion strategy to construct a noise prediction model based on two dimensions: tire tread images and physical parameters.

First, the tire tread image undergoes systematic processing. A vertical projection segmentation algorithm is then used to divide the preprocessed image into multiple columns of regions of interest (ROIs).

Meanwhile, a feature extractor based on the ResNet152 deep network is constructed. The input image is first processed through a 7×7 convolution layer and a max-pooling layer for low-level feature abstraction, followed by stacked residual blocks to capture deep semantic features. A high-dimensional feature vector is then output through global average pooling, effectively encoding both the macroscopic layout and microscopic groove structure of the tread pattern. This vector is further fed into a channel attention mechanism, where linear layers dynamically compute weights to assign varying importance to different channels. This encourages the model to focus on the tread regions most relevant to noise generation.

Finally, a regression model maps the refined feature vector to the predicted noise level, completing the tire noise prediction process.

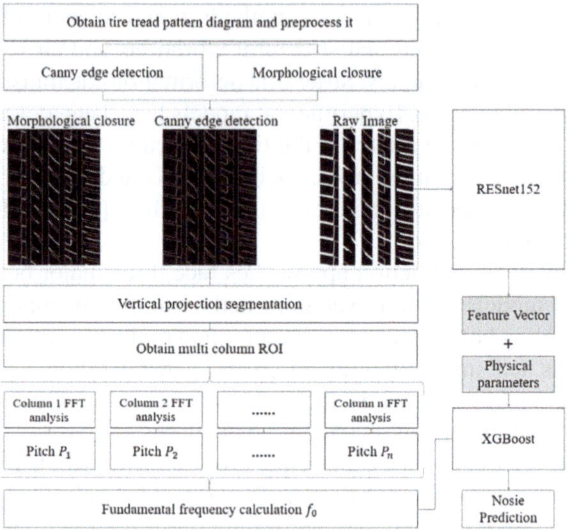

Fig. 1. Technical route for tire noise prediction.

2.1 Data Acquisition and Processing

As shown in Fig. 2, the tire noise measurement test bench is installed in a professional semi-anechoic chamber. It integrates a high-precision drum simulation system and multi-channel acoustic analysis technology, enabling accurate reproduction of the dynamic contact characteristics of tires under varying speeds, loads, and inflation pressures. The test bench consists of an electric-driven drum platform, a tire clamping device, and far-field acoustic microphones. Together with full-frequency wedge-shaped absorbers and a vibration-isolated foundation in the chamber, it effectively isolates external noise and vibration interference.

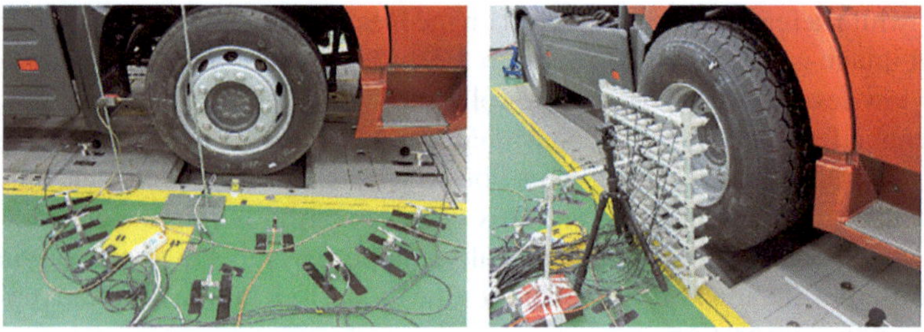

Fig. 2. Technical route for tire noise prediction.

During testing, tire vibration acceleration and radiated sound field data are synchronously acquired. The acceleration sensors and supporting equipment, provided by Anhui Lupital Intelligent Technology Co., Ltd., allow for quantitative analysis of the tire–road noise generation mechanism, offering high-confidence experimental support for low-noise tire design and NVH performance optimization.

The experimental noise acquisition targets the noise level of the tire under steady-state operation, where its temperature and noise values stabilize without significant fluctuation—ensuring valid and reliable measurements. The test speeds are set at 40, 60 80 KPH, 100 KPH, and 120 KPH. For each speed condition, five independent tests are conducted, and the arithmetic mean of the recorded noise values is used as the target value for model fitting. In addition, physical parameters such as temperature and stiffness of the test tire samples are also measured and included as supplementary model inputs.

2.2 Attention Mechanism for Weighted Tread Feature Representation

The attention mechanism enables the model to focus on data features at different hierarchical levels by assigning weights to various feature layers. In our framework, we integrate a channel attention module between the transfer learning model and the regression model. This module computes the average value of all data within each feature vector channel to obtain a single scalar that represents the importance of that channel.

Assuming the input feature map is $X \in R^{C \times H \times W}$, where C is the number of channels, H and W denote height and width respectively, the global average pooling operation can be expressed as:

$$z = \frac{1}{H \times W} \sum_{i=1} HWX_{ci} \tag{1}$$

Subsequently, a fully connected layer is used to process the global average pooling result $z \in R^{C}$. This layer takes the channel-wise importance values obtained from global average pooling as input, and computes the channel importance weights s through neural network weights and activation functions:

$$s = \sigma(W_{fc}^{T} z + b_{fc}) \tag{2}$$

Finally, the SEBlock (Squeeze-and-Excitation Block) applies the computed importance weights to the input feature map to achieve channel-wise feature recalibration. This is done by multiplying the output of the fully connected layer $s \in R^{C}$ with the input feature map $X \in R^{C \times H \times W}$:

$$R_{cij} = s_c \cdot X_{ci} \tag{3}$$

By performing global average pooling and passing the results through a fully connected layer, the importance weight of each channel is computed and applied to the input feature map, enabling channel-wise feature recalibration.

3 Noise Prediction Based on Transfer Learning and Regression Models

As shown in Fig. 3, for each CNN model, we remove its original classification layer and retain the convolutional layers to serve as the feature extraction module. For ResNet50, ResNet101, and ResNet152, the main difference lies in the depth of their residual blocks. Therefore, we modify these models by disconnecting the upper layers from the fully connected classifier and removing the final fully connected layer entirely.

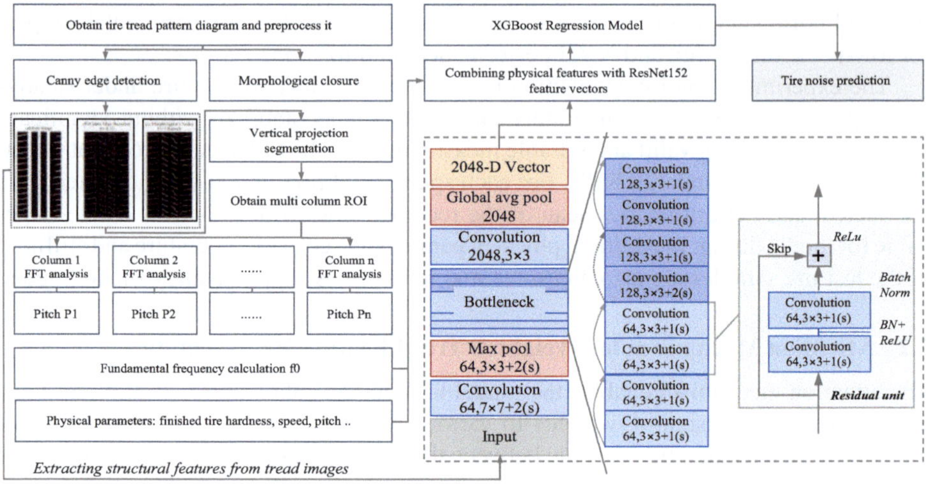

Fig. 3. Tire noise prediction framework based on transfer learning and regression

4 Experiments

In this study, two sets of comparative experiments were designed. On one hand, the performance differences among various CNN architectures based on transfer learning—including the ResNet series and Transformer-based models—were evaluated to investigate the impact of network depth and attention mechanisms on feature extraction effectiveness. On the other hand, the predictive performance of different regression models was assessed by comparing gradient boosting methods (XGBoost, LightGBM, CatBoost), analyzing their accuracy in tire noise parameter prediction under different ensemble strategies, as shown in Table. 1.

Table 1. Comparison of noise prediction performance.

CNN Models	Regression Models	MSE	MAE	R^2	≤1 dB (%)	≤2 dB (%)	≤3 dB (%)	>3 dB (%)
Resnet152	Catboost	1.14	0.79	0.96	69.97	95.77	98.02	1.98
Resnet101	LightGBM	0.78	0.66	0.97	78.97	96.96	99.21	0.79
Resnet101	XGBoost	0.81	0.66	0.97	78.70	96.96	98.81	1.19
Convnext	LightGBM	0.87	0.66	0.97	77.12	96.69	98.55	1.46
Swin Transformer	XGBoost	0.83	0.67	0.97	76.85	96.30	98.94	1.06
Convnext	XGBoost	0.85	0.67	0.97	78.17	96.96	98.68	1.32
Resnet152	XGBoost	0.82	0.68	0.97	76.59	97.35	98.94	1.06

By contrast, ConvNeXt and Swin Transformer models showed generally higher MAE and lower R2 values than ResNet models. This is attributed to the fact that tire tread patterns are dominated by local geometric features (e.g., grooves and pitches), for which convolutional operations are more efficient in capturing translation invariance. The global attention mechanism of Transformer architectures may introduce redundant computations and lacks sufficient generalization ability under limited training samples.

5 Conclusion

This paper proposes a multi-modal tire noise prediction method based on transfer learning, which dynamically weights tread image features by integrating the transfer learning capability of ResNet models, channel attention mechanisms, and the LightGBM regression algorithm to achieve high-accuracy noise prediction. Experimental results demonstrate that the proposed feature representation framework significantly improves model generalization, achieving a mean absolute error (MAE) of 0.66 dB and an R2 of 0.97. The incorporation of physical constraints aligns the prediction more closely with actual noise generation mechanisms, reducing error by approximately 15%. Among the ensemble regression models, LightGBM performed the best, with 96.96% of predicted samples having an error below 2 dB. This method provides an efficient tool for low-noise tire design and NVH performance optimization.

Acknowledgments. This study was funded by: Sichuan Science and Technology Program No. 2025ZNSFSC1320; the Vehicle Measurement, Control and Safety Key Laboratory of Sichuan Province under Project No. QCCK2024-005.

Disclosure of Interests. The authors have no competing interests to declare that are relevant to the content of this article.

References

1. Almirón, J.O., San Segundo, D., García-Pozuelo, D., García, A.: Vehicle road noise prediction using component-based transfer path analysis from tire test-rig measurements on a rolling tire. J. Sound Vib. **523**, 116694 (2022)
2. Lee, S.-K., Kim, J.-H., Park, S.-W., Lee, S.-W.: Prediction of tire pattern noise in early design stage based on convolutional neural network. Appl. Acoust. **172**, 107617 (2021)
3. Hong, S.J., Park, S.-W., Lee, S.W.: Tire-pavement noise prediction using asphalt pavement texture. KSCE J. Civ. Eng. **22**(9), 3358–3362 (2018)
4. Xiang, D., Zhang, C., Wang, H., et al.: Research on tire pattern noise prediction method based on deep learning. Tire Industry **39**(8), 503–506 (2019)
5. Li, Z., Su, Y., Zhang, S., et al.: Tire pattern noise value prediction based on convolutional neural network. Tire Industry **43**(12), 756–761 (2023)

Research on Map Reconstruction and Path Planning of Crawler Cranes Based on Improved RRT

Shibo Liang[1], Xin Wang[1(✉)], Xuyang Cao[1], Weijun Cao[2], Kun Zhang[2], and Dazhi Wang[1]

[1] Dalian University of Technology, Dalian 116024, China
wangxbd21@163.com
[2] Taiyuan Heavy Industry Co., LTD, Taiyuan 030001, China

Abstract. To achieve automated crawler crane path planning, environment construction, modeling, algorithm design, simulation, and real vehicle testing are studied. For map reconstruction, environmental data is gathered using unmanned aerial vehicles (UVA). And a 3D model is built processed by mapping software, and is converted into 2D gridded height information through bounding box algorithms for path planning. The crane's mathematical model is built according to forward and inverse kinematics. Then, the connection between the end pose of the lifted object and the joint angles is established. In path planning, the RRT algorithm is enhanced with motion constraints and heuristic strategies, so as to improve the smoothness and search efficiency of the paths. Simulation comparisons show that the improved algorithm can generate better paths than other algorithms. Real testing confirms that the lifted object is moved quickly and stably to the target node, successfully avoiding obstacles.

Keywords: Crawler crane · Path planning · Map reconstruction · Improved RRT algorithm · Collision detection · Kinematic model

1 Introduction

At present, traditional crane operations rely heavily on operators' experience, which leads to safety and efficiency issues. With increased focus on crane efficiency, functionality, and safety, global industrialization drives cranes toward automation and informatization [1]. As a result, research has been focused on automatic motion, 3D environment reconstruction, and path planning. Outdoor 2D scenes were reconstructed by Liu Wentao using UAV images [2]. UAVs were used by Pileun Kima et al. to assist robots in collecting data in unstructured terrain, solving laser scanning occlusion [3]. For path planning, mathematical models for cranes were established [4], and algorithms like RRT [5, 6], Dijkstra [7], A* [8], and their improvements [9, 10] were proposed for path and motion optimization.

Although theoretical research is mature, practical application still needs improvement. In this paper, multi-degree-of-freedom rigid body and kinematic models of a

S. Halgamuge et al. (Eds.): ICACMVE 2025, LNME, pp. 361–369, 2026.
https://doi.org/10.1007/978-981-95-6911-3_35

180t crawler crane are built. UAVs are used to establish working scenes, so as to avoid the deficiencies due to traditional vision/lidar occlusion and installation. Bounding box algorithms are applied for environmental information extraction. The RRT algorithm is optimized to improve path fitness, smoothness, and search efficiency. Effectiveness is verified through simulation and real testing, supporting crane automated operation.

2 Reconstruction of the Working Environment Map

Crane path planning needs detailed environmental info. In this paper, the Mavic 3 Enterprise drone is used to assist perception, solving occlusion problems of traditional perception systems. Aerial images are stitched and reconstructed into an OBJ-type 3D model using DJI Terra software. The model is then converted into grid-based height information stored in a 2D array, with each element representing the highest node of the corresponding grid. The coordinates correspond to the X and Y axes of the 3D model. The model consists of vertices and triangular faces. Let x_{min} be the minimum value of all x_i vertices and x_{max} the maximum value, with the same approach for other dimensions. The grid division number is set as S (per dimension), and the block_size is calculated as $block\_size = \max(x_{max} - x_{min}, y_{max} - y_{min})/S$.

For a grid $P(i, j)$, its actual coordinates $P(x_b, y_b)$ are calculated via the equation below; conversely, the grid containing $P(x_b, y_b)$ is inversely determined using the equation.

$$P(x_b, y_b) = (x_{min} + i \times block\_size, y_{min} + j \times block\_size) \tag{1}$$

The separating axis theorem is used to determine intersecting grids between the OBJ model's triangles and 2D grids by calculating projection ranges along each normal direction. Heights of intersecting grids are updated using triangle vertices' Z-coordinates. The grid height is expanded to surrounding areas to reduce subsequent collision detection calculations. In this process, e_1 is a triangle edge vector, r_1 and r_2 are rectangle edge vectors, n_1, n_2, n_3 are normal vectors, P_{rojn} denotes node P's projection on normal n, $h_{expanded}(x, y)$ is the expanded grid height, and expand_distance is the height expansion safety margin. The equation is as follows:

$$\begin{cases} e_1 = P_2 - P_1, n_1 = (-e_{1y}, e_{1x}), n_2 = (-r_{1y}, r_{1x}), n_3 = (-r_{2y}, r_{2x}) \\ P_{rojn}(P) = P \cdot n \\ Height_{grid} = max(Height_{grid}, max(Z_{P_1}, Z_{P_2}, Z_{P_3})) \\ h_{expanded}(x, y) = max(h_{x,y}, max(h_{x+1,y}, h_{x-1,y}, h_{x,y+1}, h_{x,y-1}) + expand\_distance) \end{cases} \tag{2}$$

The results of before and after optimization are shown in Fig. 1, which reveal the model to be transformed into a regular geometric entity with expanded surrounding grids.

3 Establishment of the Mathematical Model of the Crane

The TGZC180 crawler crane with a main boom is as the research object (Fig. 2). It consists of the carriage (crawlers and chassis) and the upper structure (superstructure, boom, counterweight, luffing rods). Its mechanisms include lifting, luffing, slewing, and

$$\begin{cases} e_1 = P_2 - P_1,\ n_1 = (-e_{1y}, e_{1x}),\ n_2 = (-r_{1y}, r_{1x}),\ n_3 = (-r_{2y}, r_{2x}) \\ P_{rojn}(P) = P \cdot n \\ Height_{grid} = max(Height_{grid}, max(Z_{P_1}, Z_{P_2}, Z_{P_3})) \\ h_{expanded}(x, y) = max(h_{x,y}, max(h_{x+1,y}, h_{x-1,y}, h_{x,y+1}, h_{x,y-1}) + expand\_distance) \end{cases} \quad (2)$$

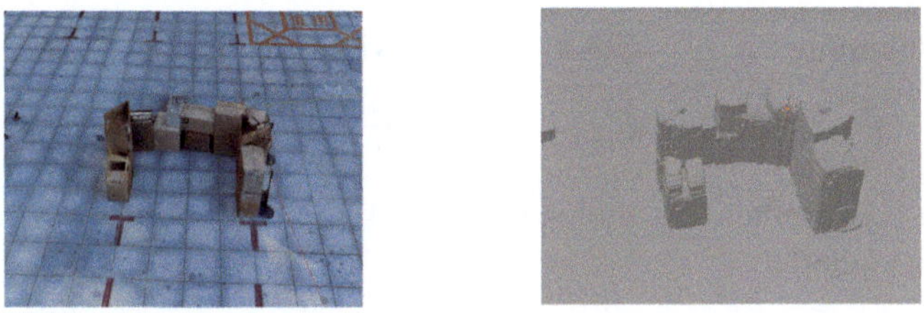

a) Unoptimized model diagram b) Optimized model diagram

Fig. 1. Model diagrams before and after optimization.

travelling. Since the crane rarely travels with a load, a rigid body model with a fixed station is established to simulate the other three movements, as shown in Fig. 3.

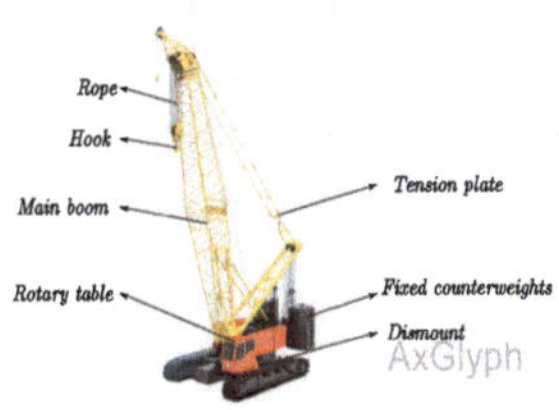

Fig. 2. TGZC180-type crawler crane.

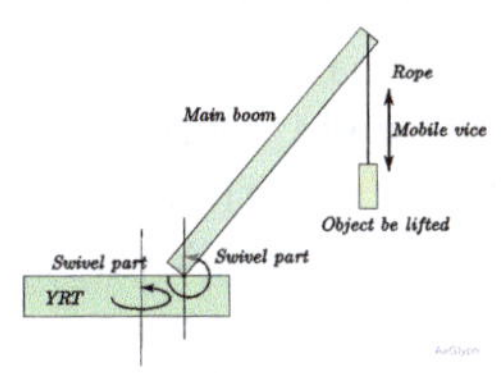

Fig. 3. Rigid Body Model.

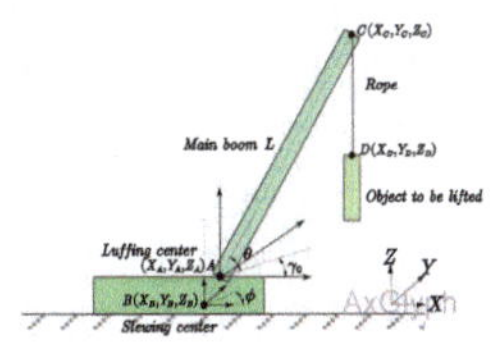

Fig. 4. Kinematic Model.

When determining the crane's posture, sensor limitations typically allow only key parameters: coordinates of the slewing center, luffing center, hook, and boom length. A kinematic model is established, as shown in Fig. 4. B is the slewing center, A is the luffing center, C is the boom top, and D is the hook position. ϕ is the slewing angle, θ is the boom's XY projection angle, γ_0 is the boom's absolute direction angle, and L is the boom length. The initial station information includes B, initial A_0, initial D_0, L, initial slewing angle α_0, initial γ_0, and turning radius r (the XY distance between A_0 and B). These values are used as initial inputs for kinematic calculations.

In forward kinematics, given the slewing angle ϕ, the luffing angle θ, and the hoisting rope length *rope_len*, the position of the hook can be determined using Eq. (3):

$$\begin{cases} A_x = x_B + r \cdot cos(\alpha_0 + \phi), A_y = y_B + r \cdot sin(\alpha_0 + \phi), A_z = A_{0z} \\ C_x = A_x + L \cdot cos(\theta) \cdot cos(\gamma_0 + \phi), C_y = A_y + L \cdot cos(\theta) \cdot sin(\gamma_0 + \phi), C_z = A_z + L \cdot sin(\theta) \\ D_x = C_x, D_y = C_y, D_z = C_z - rope\_len \end{cases} \quad (3)$$

In inverse kinematics, ϕ, θ, and *rope_len* are solved given D. The reverse equation is as follows.

$$\begin{aligned} D_x &= B_x + r \cdot cos(\alpha_0 + \phi) + L \cdot cos(\theta) \cdot cos(\gamma_0 + \phi), \\ D_y &= B_y + r \cdot sin(\alpha_0 + \phi) + L \cdot cos(\theta) \cdot sin(\gamma_0 + \phi) \end{aligned} \tag{4}$$

Squaring and summing Eq. (4), with $u = \cos(\theta)$, yields the following equation:

$$L^2 \cdot u^2 + 2rL \cdot u \cdot cos(\alpha_0 - \gamma_0) - (D_x^2 + D_y^2 - r^2) = 0 \tag{5}$$

θ is obtained using Eq. (5), then substituted into Eq. (4) to find ϕ. *rope_len* is calculated using the following equation:

$$A_z = A_{0z}, C_z = A_z + L \cdot sin(\theta), rope\_len = C_z - D_z \tag{6}$$

4 Path Planning Algorithm Research

4.1 Collision Detection

In hoisting operations, collisions between the main boom, lifted object, and obstacles should be avoided. A grid map with a height field grid is used to check if the coordinates of both are higher than the grid height. During lifting, only the lifted object's height changes, so only its coordinates $P(x, y, z)$ are compared with the grid height $h(x, y)$. The no-collision detection equation is as follows:

$$\begin{cases} true & if\ P_z \le h(x, y) \\ false & if\ P_z \ge h(x, y) \end{cases} \tag{7}$$

The luffing and slewing motions follow a circular path, requiring collision detection for both the lifted object and the boom. The path is approximated as multiple boom segments for obstacle intersection detection. For a segment with endpoints P_1 and P_2, XY direction vector d is calculated. The parameter t represents any node on the segment, and the step size t to reach the next grid is calculated using Eq. (8).

$$\begin{aligned} t_{maxX} &= \frac{x_{bound} - x_1}{d_x}, t_{deltaX} = \frac{block\_size}{|d_x|}, \\ t_{\max Y} &= \frac{y_{bound} - y_1}{d_y}, t_{deltaY} = \frac{block\_size}{|d_y|} \end{aligned} \tag{8}$$

The following Equation calculates the minimum height z_{min} of the segment within the current grid and traverses all grids containing the segment. t_{min} is the t value when entering the current grid, and t_{max} is the t value when leaving the current grid.

$$\begin{cases} t_{min} = min(t_{\max X}, t_{\max Y}), t_{\max} = \min(0, \max(t_{\max X}, t_{\max Y})), z_{min} \\ = min(z_1 + t_{min} \cdot (z_2 - z_1), z_1 + t_{max} \cdot (z_2 - z_1)) \\ \mathrm{t_{maxX}} = \mathrm{t_{maxX}} + \mathrm{t_{deltaX}} \quad if\ \mathrm{t_{maxX}} < \mathrm{t}_{\max Y},\ \mathrm{t}_{\max Y} = \mathrm{t_{maxY}} + \mathrm{t_{deltaY}} \quad else \end{cases} \tag{9}$$

For the arc path, segment count $N = D/block\_size + 1$ (D: planar distance between boom endnodes). Each segment's position is calculated via Eq. (12) (t_i: interpolation parameter for segment i), with collision detection using *collide_line.*

$$t_i = \frac{i}{N}(1 \le i \le N), (C_i, A_i) = (\phi_{ti}, \theta_{ti}, rope\_len_{ti}), collide\_line(C_i, A_i) \quad (10)$$

4.2 Improved RRT Algorithm

The core of RRT involves random sampling for node generation, as shown in Fig. 5. New nodes are added iteratively to expand the search tree from start to target, forming an obstacle-free path. In Fig. 5, q_{nearst} is the tree node closest to random sample q_{rand}, l is the extension step size, and q_{new} is the new node.

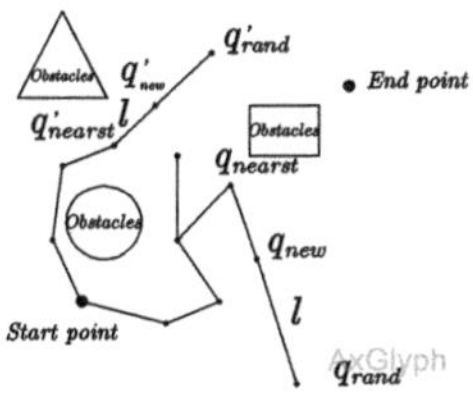

Fig. 5. Principles of RRT node expansion.

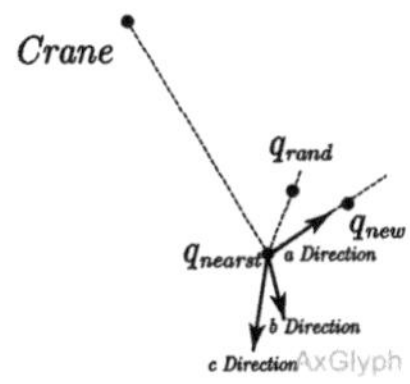

Fig. 6. Direction of crane movement.

Basic RRT generates irregular, unsmooth paths unsuitable for practice. An improved RRT is proposed in this paper, with random node generation enhanced, q_{nearst} selection optimized (to speed up search and reduce detours), and direction generation/selection restricted (to fit crane movement and reduce action switches). For direction handling, crane movements are a (slewing), b (luffing), and c (hoisting) as shown in Fig. 6. According to Eq. (11), the action with the largest $K(m)$ value is selected as the extended action. Where, $V(m)$ is the direction vector of action m, V_p is the vector from q_{nearst} to q_{rand}, and $F(m)$ is the correction parameter.

$$\begin{cases} K(m) = |V(m) \bullet Vp| + F(m) \\ F(m) = \begin{cases} C1 \ \ (m = m_{q_{nearst}}) \\ 0 \ \ (m! = m_{q_{nearst}}) \end{cases} \end{cases} \quad (11)$$

In random node generation, target bias sampling is used, with the target node selected probabilistically. For q_{nearst} selection, a direction/distance correction factor and target proximity reward mechanism are added (*Dir_fac*: distance correction factor, *Dir_fir*: direction correction factor, *weight*: node weight). The node with the smallest *weight*, after evaluating all tree nodes, is chosen as q_{nearst}.

$$\begin{cases} Dir_n = q_{end} - q_{nearst}, Dir_r = q_{rand} - q_{nearst} \\ Dist_n = \sqrt{(q_{end}(x) - q_{nearst}(x))^2 + (q_{end}(y) - q_{nearst}(y))^2 (q_{end}(z) - q_{nearst}(z))^2} \\ Dist_r = \sqrt{(q_{rand}(x) - q_{nearst}(x))^2 + (q_{rand}(y) - q_{nearst}(y))^2 (q_{rand}(z) - q_{nearst}(z))^2} \\ Dir\_fac = k_a + \frac{1}{Dist_n} \\ Dir\_fix = \frac{(q_{end}(x) - q_{nearst}(x))(q_{rand}(x) - q_{nearst}(x)) + (q_{end}(y) - q_{nearst}(y))(q_{rand}(y) - q_{nearst}(y)) + (q_{end}(z) - q_{nearst}(z))(q_{rand}(z) - q_{nearst}(z))}{Dist_n \cdot Dist_r} \\ weight = Dist_r(Dir\_fac - Dir\_fix), weight = weight * k_b \ \ if(Dist_n < k_c) \end{cases} \quad (12)$$

5 Algorithm Simulation and Real Vehicle Test

5.1 Simulation

To verify the practicality of the improved RRT algorithm, it is compared with the traditional RRT algorithm and the RRT algorithm with motion constraints. Simulation results are shown in Fig. 7 and Table 1. The blue, red, and yellow curves represent the paths of the three algorithms respectively. The comparison shows that the improved algorithm produces paths better aligned with crane movement, with stronger goal orientation, smoother motion sequences, and better performance in planning time, path length, and action switches. This proves its higher efficiency and stability.

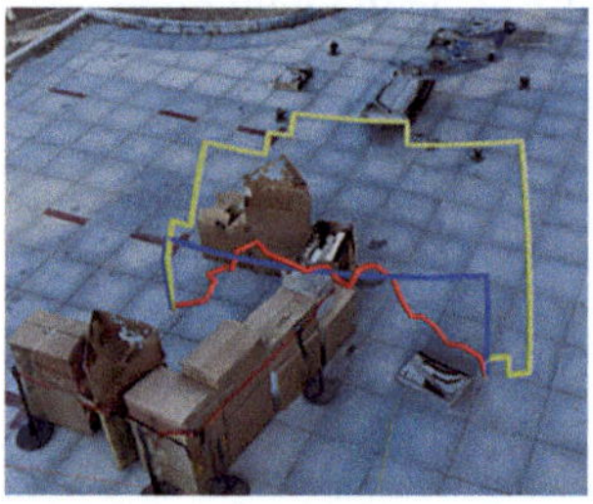

Fig. 7. Simulation results.

Table 1. Simulation results data.

Algorithm type	Planning time (s)	Path length (mm)	Motion Switching Times
RRT	143	11432	29
RRT-MC	33	18334	14
Improved RRT	8	10731	3

5.2 Real Testing

The real test object is the TGZC180 crawler crane (Fig. 3). Two test scenarios are designed. Scene 1 simulates a complex environment with obstacles around the target (not directly observable), as shown in Fig. 8. Using drone scanning and modeling, relevant location information is fed into the path algorithm, generating a motion sequence that is then imported into the industrial computer to complete the 5t load automatic hoisting. Figure 9 shows the simulation path, and Fig. 10 shows the successful test where the load is precisely placed within the target circle, confirming the success of the experiment.

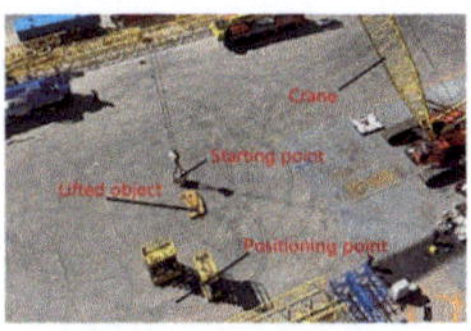

Fig. 8. Scene 1.

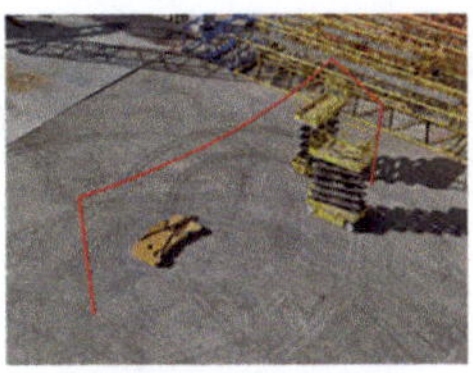

Fig. 9. Scene 1 simulation path.

Fig. 10. Scene 1 result image.

Table 2. Data of 8 Tests in Scenario One (5t Load).

No.	Target Distance (m)	Operation Time (s)	Positioning Accuracy: Eccentric Distance within Circle (cm)
1	15.36	198	7
2	14.80	204	10
3	14.10	183	9
4	14.60	191	5
5	15.20	195	12
6	15.70	203	6
7	22.50	212	7
8 (Manual)	22.50	305	12

From Table 2, it is shown that the distance to the target node has little impact on hoisting duration. It can be inferred that the farther the target node and the heavier the load, the more apparent the advantages of intelligent hoisting. In scenario one, automatic lifting is 30% more efficient and accurate than manual, confirming the improved algorithm enhances path planning efficiency and lifting precision.

Scenario 2 (Fig. 11) simulates a long-distance, complex hoisting scenario with multiple obstacles along the path. The test procedure is the same as Scenario 1. Figure 12 shows the simulation path, and Fig. 13 shows the successful test.

Fig. 11. Scene 2.

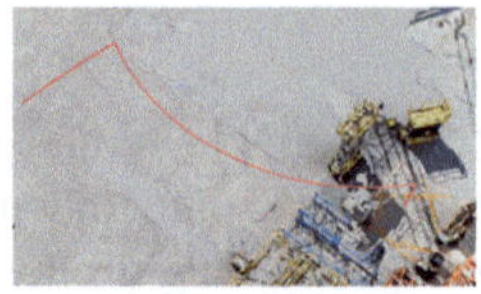

Fig. 12. Scene 2 simulation path.

Fig. 13. Scene 2 result image.

Table 3. Data of 6 Tests in Scenario Two (5t Load).

No.	Target Distance (m)	Operation Time (s)	Positioning Accuracy: Eccentric Distance within Circle (cm)
1	25.50	209	5
2	21.00	220	3
3	22.10	217	6
4	22.30	207	5
5	23.50	210	9
6	21.30	214	6

From Table 3, it is shown that, although the target node in Scenario 2 is farther and the motion sequence is more complex, the work speed and efficiency surpass those of Scenario 1. This indicates that automation in long-distance operations significantly improves both efficiency and accuracy.

6 Conclusion

In this paper, crane hoisting path planning is simplified to a three-degree-of-freedom mathematical model. Kinematic and inverse kinematic models are solved. A 2D bounding box grid with a height field is established using drone imaging. For path planning, the improved RRT algorithm is studied to enhance smoothness and search efficiency by adding motion constraints and guidance, and better fitting crane movement characteristics. Collision detection is based on a "3D grid," considering obstacle height and simplifying the process without needing obstacle shape details. Furthermore, simulations and real testing validate the improved algorithm. It effectively enhances crane hoisting path planning efficiency. The algorithm also smooths motion sequences, shortens paths, and meets hoisting accuracy requirements.

Acknowledgements. This research was supported by the National Key R&D Program of China (No. 2024YFD2400200).

References

1. Xiao, Z.: Research on Intelligent Path Planning for Tower Cranes. Taiyuan University of Science and Technology (2023)

2. Liu, W.: Research on Path Planning for Unmanned Vessels Based on Drone Vision Assistance. Southeast University (2022)
3. Kim, P., Park, J., Yong, K.C., et al.: UAV-assisted autonomous mobile robot navigation for as-is 3D data collection and registration in cluttered environments. Autom. Constr. **106** (2019)
4. Zhu, A., Zhang, Z., Pan, W.: Crane-lift path planning for high-rise modular integrated construction through metaheuristic optimization and virtual prototyping. Autom. Constr. **141**, 10443 (2022)
5. LaValle, S.M., Kuffner, J.K.: Randomized kinodynamic planning. Int. J. Robot. Res. **20**(5), 378–400 (2001)
6. Kuffner, Jr.JJ., La Valle, S.M.: RRT-connect: an efficient approach to single-query path planning. In: IEEE International Conference on Robotics and Automation, San Francisco, CA, USA, pp. 995–1001 (2000)
7. Ahuja, R.K., Magnanti, T.L., Orlin, J.B.: Network Flows: Theory, Algorithms, and Applications, pp.108–112. Prentice Hall (1993)
8. Yershov, D.S., LaValle, S.M.: Simplicial Dijkstra and A* algorithms for optimal feedback planning. In: 2011 IEEE/RSJ International Conference on Intelligent Robots and Systems (IROS), San Francisco, CA, pp. 3862–3867 (2011)
9. Li, C.: Multi-task Lifting Planning for Automotive Cranes. Dalian University of Technology (2022)
10. Guo, W.: Research on Real-time Obstacle Avoidance of Crawler Cranes Based on Deep Reinforcement Learning. Dalian University of Technology (2022)

Evaluation Index System for Identifying High-Risk Fracture-Prone Areas of Ship to Shore Gantry Crane

Qiwei Feng[1], Xin Wang[1(✉)], Li Chen[2], Keqin Ding[2], Junli Pi[3], and Fan Yang[1]

[1] Dalian University of Technology, Dalian 116024, China
wangxbd21@163.com
[2] China Special Equipment Inspection and Research Institute, Beijing 100029, China
[3] Beijing Materials Handling Research Institute Co., Ltd., Beijing 100010, China

Abstract. This study focuses on the compound risks throughout the entire life cycle of ship to shore gantry cranes. Additionally, it constructs a high-risk fracture-prone area evaluation system that covers the four stages of design, manufacturing, service, and maintenance. Guided by the evolution mechanism of structural fatigue cracks, six typical stress concentration areas on the main girder, drag rods and gantry of ship to shore gantry cranes are selected as evaluation points. Quantitative indexes are obtained through the fusion of finite element simulation and measured data. The qualitative indexes are quantified by a five-level language scale combined with triangular fuzzy numbers. The weight distribution introduces the game theory combination mechanism, and the optimal combination coefficient is solved through the Least Square Method. Subjective weighting is carried out through the Analytic Hierarchy Process(AHP), and objective weighting is conducted through the Coefficient of Variation Method(CV) and the Entropy Weight Method(EWM), achieving the integration of subjective and objective weighting. The indexes evaluation adopts the Grey Relational Degree -TOPSIS composite model(GRA-TOPSIS). By comparing the ranking results of the combined weight method and the single EWM proposed in this paper through numerical examples, it can be seen that the combined weight method is superior. And it is verified that the index evaluation system is applicable to the identification of high-risk fracture-prone areas of ship-to-shore gantry cranes.

Keywords: ship to shore gantry crane · index system · weight strategy · game theory

1 Introduction

Ship-to-shore gantry cranes, as large-scale lifting equipment, are used for loading and unloading container ships at port terminals. Their high operational frequency and intensity often lead to a significant number of safety accidents at terminal facilities [1]. Enterprises and special equipment inspection institutions urgently need to conduct safety condition evaluations on cranes that have been in service for a long time and are in poor

S. Halgamuge et al. (Eds.): ICACMVE 2025, LNME, pp. 370–377, 2026.
https://doi.org/10.1007/978-981-95-6911-3_36

condition. The purpose of this evaluation is to extend the service life of the cranes and reduce production costs as much as possible while ensuring safe production. In conclusion, conducting research on the safety assessment methods of ship to shore gantry cranes is of great practical significance for improving the safety production management level of ports and enhancing economic benefits.

At present, there are many theoretical methods for crane safety evaluation, such as Neural Network Evaluation Method, Grey Comprehensive Evaluation Method, Unknown Measurement Theory, Fuzzy Comprehensive Evaluation Method, Expert System Method, etc. Based on data statistics, Ruli Lu et al. [2] analyzed and studied the safety evaluation system of ship to shore gantry cranes in the middle and later stages of service through on-site investigations, expert evaluations and other methods. Wen Zhang [3] determined the safety level of the evaluation object layer by layer from the specific index layer to the ship to shore gantry cranes whole machine through the improved variable fuzzy set method and verified the feasibility of the established ship to shore gantry crane evaluation system. Omidvar M et al. [4] combined the Fuzzy Comprehensive Evaluation Method with the AHP to determine the weights of risk factors for bridge cranes.

To this end, this study closely focuses on the four key stages of ship to shore gantry cranes: design, manufacturing, service, and maintenance. It systematically constructs an evaluation index system for identifying high-risk fracture-prone areas. Furthermore, the practical effectiveness of this index system is proved through case analysis.

2 The Construction of the Evaluation Index System

The high-risks fracture-prone areas of ship to shore gantry cranes result from the accumulation of risks in multiple links including design, manufacturing, inspection and operation. This study constructs a five-level index system covering the entire life cycle, forming a closed-loop logic of "prevention - control - monitoring - feedback". Based on the framework of "risk source - control element - quantifiable representation", it is decomposed into 16 s-level indexes. The classification table of thc index system is shown in Fig. 1:

Fig. 1. Index system classification table.

In this study, differentiated treatment is adopted in the evaluation of indexes: quantitative indexes directly used measured values, while qualitative indexes are evaluated through a five-level scale of "excellent(E), good(G), average(A), fair(F), and poor(P)". Qualitative evaluation results are transformed though triangle fuzzy numbers (corresponding to five group of fuzzy numbers: (0.8, 0.9, 1), (0.6, 0.7, 0.8), (0.4,0.5,0.6), (0.2,0.3,0.4), (0,0.1,0.2)), finally constructing the fusion of quantitative and qualitative evaluation system.

3 Determination of the Weights of Evaluation Indexes

In this study, AHP is adopted to determine the subjective weights. EWM and CV are combined to calculate the objective weights. The optimal integration of subjective and objective weights is achieved through the game theory combination model. GRA - TOPSIS model is used to complete the comprehensive evaluation.

3.1 The Analytic Hierarchy Process (AHP) Determines Subjective Weights

AHP was first proposed by Professor T.L. Saaty, a renowned American operations researcher from the University of Pittsburgh, in the early 1970s. This method combines quantitative analysis with qualitative analysis and integrates mathematical processing with human experience and subjective judgment. It can effectively analyze the non-sequential relationships among the levels of the target criterion system.

In the AHP, the evaluation factors at the same level are pairwise compared based on the evaluation scales [5], and the judgment matrix $A_{ij} = (a_{ij})_{m \times n}$ is constructed. The maximum eigenvalue λ_{max} and the maximum eigenvector W of the judgment matrix A are calculated.

To calculate a reasonable combination weight that meets the consistency requirements of the judgment matrix A, the consistency index $C.I.$ is calculated based on Eq. 1, the consistency ratio $C.R.$ is calculated based on Eq. 2. When $C.R. < 0.1$, the consistency of the judgment matrix A is acceptable. The normalized maximum eigenvector W is the index weight of the judgment matrix A. If $C.R. > 0.1$, the consistency of the judgment matrix A does not meet the requirements, and the judgment matrix needs to be corrected again [6].

$$\mathrm{C.I.} = \frac{\lambda_{\mathrm{max}} - \mathrm{n}}{\mathrm{n} - 1} \tag{1}$$

$$\mathrm{C.R.} = \frac{\mathrm{C.I.}}{\mathrm{R.I.}} \tag{2}$$

3.2 The Determination of the Objective Weights of Evaluation Indexes

This study selects EWM and CV to calculate the objective weights. These two methods have obvious advantages in data adaptability, risk sensitivity and engineering practicality.

They can more scientifically reflect the contribution degree of the indexes to the safety risk of the ship to shore gantry cranes.

Entropy Weight Method

Information entropy was proposed by ClaudeE. Shannon in 1948 and is used to measure the uncertainty or randomness of information. EWM is based on the concept of information entropy. It determines the weight by calculating the information and information utility value of each index, which can reflect the degree of dispersion and importance of index data [7]. When using EWM to calculate the weights, the original data is first normalized through the range method to eliminate the dimensional differences. Subsequently, calculating the feature proportion P_{ij} of each index in the sample set and quantifying the degree of dispersion of the index based on the information entropy theory (Eq. 3). The difference coefficients are derived through information entropy, and the objective weight coefficients of each index are finally determined.

$$e_j = -k\sum\nolimits_{i=1}^{m} P_{ij} \times \mathrm{In}\,\mathrm{P_{ij}}, \mathrm{k} = \frac{1}{\mathrm{In}\,\mathrm{m}} \tag{3}$$

Coefficient of Variation Method

CV is a statistical tool for evaluating the degree of data dispersion, often used to compare the relative volatility or stability of different datasets, especially when the average values of the datasets vary greatly [8]. When using CV to calculate weights, the index data is first standardized, then the mean $\overline{y_j}$ and standard deviation S_j of each index are calculated. The discrete characteristics of the data are quantified based on the coefficient of variation formula (Eq. 4), and finally the weights are determined through the normalization processing of CV.

$$\mathrm{V_j} = \frac{\mathrm{S_j}}{\left|\overline{\mathrm{y_j}}\right|} \tag{4}$$

3.3 Game Theory-Based Combined Weighting Method (GT-CWM)

GT-CWM is to unify and coordinate the weights of each index obtained by the subjective and objective weighting methods, then find the maximum common interest points among each index [9]. This method is based on the index weights of the three methods to establish a linear combined weight vector set. The least square method is adopted to optimize different weight combinations to minimize the deviation between w and w_k. (Eq. 5). The calculated optimized combination coefficients α_k are normalized. Finally, the weights $w^* = (w_1, w_2, \cdots, w_n)$ assigned by the game theory combination are obtained through Eq. 6.

$$\min\|\sum\nolimits_{i=1}^{k}\left(\alpha_k w_k^T - w_k\right)^2\|(k = 1, 2, \cdots, K) \tag{5}$$

In the formula: w^* represents the weight of GT-CWM; The weight coefficient α_k^* obtained through normalization processing α_k. w_k^T is the transpose matrix of the basic weight vector w_k.

$$w^* = \sum\nolimits_{i=1}^{k} \alpha_k^* w_k^T \tag{6}$$

4 Grey Relational Degree -TOPSIS Model

The TOPSIS Method is mainly used to calculate and evaluate the relative superiority and inferiority of the decision-making object. By introducing the Grey Correlation Degree, this method not only takes into account the distance between the scheme and the ideal optimal and worst solutions, but also considers the intrinsic connections and changing trends among the elements of the scheme. Thus, a more comprehensive evaluation is achieved [10]. The calculation steps are as follows: first, standardizing the rows of the initial data matrix to obtain the standardized matrix $\left(z_{ij}\right)_{m\times n}$; then based on the combined weights, a weighted decision matrix is constructed to determine the positive ideal solution z_j^+ and the negative ideal solution z_j^- of each index; Subsequently, calculating the Euclidean distance between the evaluated object and the positive and negative ideal solutions (Eq. 7), and combining the grey relational analysis to calculate the correlation coefficients ζ_j^+ and ζ_j^-; The grey correlation degree is calculated through Eq. 8; The two types of distance indexes are integrated through dimensionless processing; finally, the comprehensive evaluation value is synthesized and the relative alignment progress is calculated.

$$D_i^+ = \sqrt{\sum\nolimits_{j=1}^{n} \left(z_{ij} - z_j^+\right)^2}, D_i^- = \sqrt{\sum\nolimits_{j=1}^{n} \left(z_{ij} - z_j^-\right)^2} \tag{7}$$

$$r_i^+ = \frac{1}{n}\sum\nolimits_{j=1}^{n} \zeta_j^+(i), r_i^- = \frac{1}{n}\sum\nolimits_{j=1}^{n} \zeta_j^-(i) \tag{8}$$

5 Case Analysis

To verify the scientificity and applicability of the constructed index system, this study closely combines the structural characteristics and actual operating conditions of ship to shore gantry cranes, and selects six representative evaluation points (the specific locations are shown in Fig. 2). The quantitative index data are shown in Table 1, and the qualitative index data are shown in Table 2. The selected load condition for stress acquisition: at the frequently used working range, that is, a rated load of 51 tons is applied at the center of the distance between the first rods and the middle rods on the main beam as shown in the figure.

The combined weights of subjective and objective integration obtained through the collaborative calculation of AHP, EWM and CV in this study are shown in Table 3, The weight coefficient calculated by GT-CWM is $\alpha_{AHP} = 0.4085$, $\alpha_{EWM} = 0.345$, $\alpha_{CV} = 0.247$.

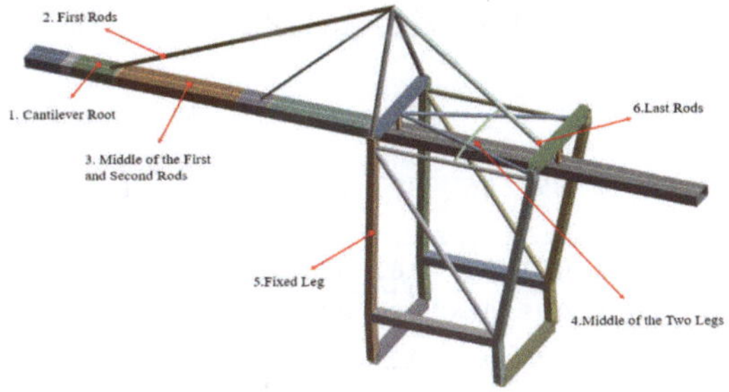

Fig. 2. Evaluation point location.

Table 1. Quantitative Index Data.

Evaluation points	1	2	3	4	5	6
Stress(MPa)	11	226	42	25	34	75
Circulating Number (time × 10^5)	1.55	1.65	1.66	1.46	1.23	1.63
Installation Dimension Error Ratio	0.07	0.15	0.13	0.08	0.06	0.11

Table 2. Qualitative Index Data.

First-level indexes		A1		A2			A3			A4		A5		
Second-level indexes		B1	B2	B4	B5	B6	B7	B8	B9	B10	B11	B14	B15	B16
Language scale	1	F	G	A	G	G	G	G	A	A	G	G	A	A
	2	A	A	A	A	F	A	A	A	A	E	A	A	A
	3	F	A	A	G	A	G	G	A	A	G	A	A	A
	4	F	G	A	A	A	G	G	G	G	G	G	G	G
	5	A	G	G	G	A	G	G	G	A	A	E	G	G
	6	A	A	A	A	A	A	A	A	A	A	G	G	G

The index weights are determined respectively by combined weights and the single entropy weight method. The comprehensive performance evaluation of the evaluation points is conducted based on GRA -TOPSIS. The results are shown in Table 4.

By comparing the output results of the two evaluation methods (Table 4), it can be seen that the core differences are concentrated in the ranking changes of evaluation point 3 (middle of the first and second rods). The combined weight method ranks the risk level of this point as the third (0.283), while the entropy weight method ranks it as the fifth (0.452). Further analysis of the weight distribution of indexes reveals that there are significant differences in the weights of AHP and EWM in five indexes: structural

Table 3. Index Weight of Residual Weighting Methods.

First indexes	A1			A2			A3			A4		A5				
Secondary indexes	B1	B2	B3	B4	B5	B6	B7	B8	B9	B10	B11	B12	B13	B14	B15	B16
AHP	0.63	0.26	0.10	0.62	0.23	0.13	0.62	0.23	0.13	0.16	0.83	0.44	0.15	0.05	0.08	0.26
EWM	0.38	0.38	0.22	0.65	0.25	0.08	0.21	0.21	0.57	0.79	0.20	0.26	0.08	0.09	0.27	0.27
CV	0.35	0.24	0.28	0.58	0.26	0.15	0.25	0.25	0.50	0.73	0.26	0.28	0.11	0.13	0.22	0.22
Combined Weight	0.55	0.20	0.23	0.50	0.35	0.14	0.50	0.34	0.14	0.45	0.54	0.38	0.15	0.08	0.15	0.22

Table 4. Evaluation Results of GRA – TOPSIS.

Evaluation points	2	6	3	1	4	5
Combined method score	0.240	0.267	0.283	0.332	0.442	0.522
Evaluation points	2	6	1	4	3	5
Single EWM score	0.246	0.263	0.337	0.440	0.452	0.519

discontinuity, welding process regulations, welder qualifications, inspector qualifications, and inspection (magnetic particle/ultrasonic/radio graphic). EWM quantifies the discrimination ability of indexes through information entropy, and its weight distribution is completely dependent on the degree of dispersion of data samples. In the case of a small amount of sample data, EWM is prone to cause distortion in weight distribution due to its excessive response to the degree of data dispersion. In contrast, the combined weight method balances subjective experience and objective data by optimizing the weight coefficients. Its evaluation results are more consistent with the risk evolution law of ship to shore gantry crane structures, and thus are more reliable in this scenario.

6 Conclusion

(1) Based on the structural characteristics and working environment of ship to shore gantry cranes, an index evaluation system for high-risk fracture-prone areas is established. According to the evaluation criteria, qualitative indexes are quantified to accurately and objectively analyze and assess the risk level of high-risk fracture-prone areas of cranes.
(2) To integrate the advantages of subjective and objective weighting, the game theory combined weighting method is adopted. The subjective weights obtained by AHP, the objective weights obtained by EWM and CV are combined to determine the final weights.
(3) GRA-TOPSIS is used to rank the risk of evaluation points. The ranking differences between the combined method and the single entropy weight method are compared. The evaluation method with high adaptability and more accurate risk reflection is selected in combination with the characteristics of the index system.

Acknowledgements. This project is supported by National Key Research and Development Program (Project No.: 2024YFC3014900).

References

1. Shenping, H., Jinpeng, Z.: Risk assessment of marine traffic safety at coastal water area. Navigation China **45**, 31–37 (2012)
2. Ruli, L., Hui, Y.: Quayside container crane safety evaluation. Port Load. Unload. (5), 5 (2020)
3. Wen, Z.: Safety Evaluation of Quay Cranes Based on Improved Entropy Weight Method and Variable Fuzzy Set Method. Wuhan University of Technology (2020)
4. Yanqing, W., Song, Q., et al.: Research on fuzzy comprehensive evaluation of influencing factors of airport security personnel work efficiency based on ANP. In: International Academic Symposium on Behavioral Safety and Safety Management (2014)
5. Omidvar, M., Nirumand, F.: Risk assessment using FMEA method and on the basis of MCDM, fuzzy logic and Grey theory - a case study of overhead cranes. J. Health Safe. Work **7**(1), 10 (2017)
6. Dong, H.: Monitoring and reliability analysis of condenser lining in submerged arc furnaces. Adv. Mater. Res. (2012)
7. Zhi-Hong, Z., Yi, Y., Jing-Nan, S.: Entropy method for determination of weight of evaluating indexes in fuzzy synthetic evaluation for water quality assessment. J. Environ. Sci. English Edition **18**(5), 4 (2006)
8. Dongfang, F., Kai, L., Zhihong, J.: Based on the variation coefficient method of social logistics cost aggregation index build. Highway Traf. Sci. Technol. **40**(1), 245–251 (2023)
9. Rong, G., Hao, X., Guodong, L., et al.: Application of matter-element extension model based on game theory comprehensive weights in groundwater quality evaluation. Hydro Ener. Sci. **33**(1), 5 (2015)
10. Yang, X., Jing, C., Huan, X., et al.: Evaluation of water resources carrying capacity in Huai 'an City based on DPSR-improved TOPSIS model. J. Water Resour. Water Eng. **30**(4), 7 (2019)

Enhancing Map Matching Accuracy for Autonomous Rollers via a Sliding Window Polynomial Fitting Approach

Changwei Song[1,2], Haiyuan Yan[1,2], Haiying Cheng[1,2(✉)], and Xuebin Wang[1,2(✉)]

[1] School of Construction Machinery, Chang'an University, Xi'an 710064, Shaanxi, China
chychdktz@163.com, wangxuebin@chd.edu.cn

[2] Engineering Research Center of Ministry of Education for Highway Construction and Maintenance Equipment and Technology, Chang'an University, Xi'an 710064, Shaanxi, China

Abstract. To address the increasing demand for high-precision map modeling in operational environments involving unmanned rollers, this study proposes a robust map matching method that integrates RTK-GNSS positioning data with a sliding window-based polynomial fitting algorithm. Static experiments were conducted to characterize the error properties of RTK-GNSS measurements, upon which a Gaussian error distribution model and corresponding confidence intervals were developed. Furthermore, a composite smoothing strategy, integrating sliding window filtering with polynomial fitting, was developed to reduce noise disturbances along linear and curved road boundaries, enhancing coordinate accuracy and enabling precise boundary estimation. Experimental and simulation results demonstrate that the proposed approach reduces the mean boundary fitting errors to 0.138 cm for straight sections and 0.178 cm for curved sections, effectively suppressing positioning noise and fully satisfying the requirements for high-precision map matching in engineering applications. The findings provide robust technical support for high-precision mapping and trajectory planning in autonomous roller operations.

Keywords: Autonomous Roller · RTK-GNSS · Sliding Window Algorithm · Map Matching

1 Introduction

In recent years, with the rapid development of artificial intelligence, sensor technology, and global navigation satellite systems (GNSS), autonomous driving technologies have been increasingly applied in the field of construction machinery. In particular, autonomous rollers have emerged as essential equipment for improving operational efficiency, precision, and safety in road construction projects [1]. High-precision map matching constitutes the foundation for realizing fully autonomous roller operation. Among various positioning technologies, Real-Time Kinematic (RTK) GNSS has been widely adopted in unmanned construction machinery owing to its centimeter-level positioning accuracy [2, 3].

S. Halgamuge et al. (Eds.): ICACMVE 2025, LNME, pp. 378–389, 2026.
https://doi.org/10.1007/978-981-95-6911-3_37

Distinct from conventional maps for autonomous vehicles, construction maps for rollers focus more on precisely delineating the boundaries of compaction zones and differentiating between processed and unprocessed areas [4]. However, the complexity of construction site environments presents significant challenges. GNSS signals are susceptible to multipath effects and random noise, often resulting in considerable measurement errors when delineating operational boundaries [5]. Following the transformation of RTK-GNSS data into a planar coordinate system, it is necessary to further smooth the boundary coordinates to construct map models that accurately represent the true contours of the work zone. Therefore, optimizing positioning data and achieving high-precision map modeling for work zones have emerged as key challenges in the advancement of autonomous roller technology.

Common approaches for smoothing map boundaries include Kalman filtering, sliding window algorithms, and B-spline curve fitting. The Kalman filter (KF), a recursive estimation algorithm based on mathematical models, is widely applied in navigation and integrated positioning systems. However, its performance is highly dependent on the accuracy of system and measurement noise models. Model inaccuracies may result in filter instability or even divergence [6, 7]. The B-spline curve fitting accomplishes data smoothing via local polynomial approximation or convolution filtering, thereby maintaining higher-order smoothness along boundaries. Nevertheless, this method can become complex when handling highly variable boundaries, as it requires elaborate parameter tuning and boundary management [8]. In contrast, the sliding window algorithm provides an efficient means of processing sequential data with linear computational complexity. By continuously updating a moving sub-sequence window, this method enables real-time smoothing of trajectories or boundaries. This approach is particularly suitable for online processing and real-time response scenarios [9, 10].

Therefore, to overcome the aforementioned challenges, this study presents a map boundary smoothing approach that integrates sliding window algorithms with polynomial fitting, specifically adapted to the characteristics of RTK-GNSS positioning data. The static error distribution of RTK-GNSS measurements is analyzed through experiments to construct a confidence interval model. Moreover, for both linear and curved road boundaries, the proposed method employs a sliding window algorithm to process coordinate points, thus improving the accuracy of map modeling.

2 RTK-GNSS-Based Map Construction Method

An RTK-GNSS system typically comprises a reference station receiver and multiple rover receivers. The reference station determines the three-dimensional coordinates of the rovers in real time by receiving carrier phase differential corrections from navigation satellites [11]. In this study, RTK-GNSS equipment was employed to collect standard NMEA messages, from which latitude, longitude, and heading information were extracted. The obtained three-dimensional WGS-84 coordinates were then projected into two-dimensional planar coordinates using the Gauss-Krüger projection. Subsequently, the boundary coordinate points were smoothed using a sliding window algorithm, ultimately enabling the construction of a precise operational area map.

2.1 Coordinate System Transformation Method

The transformation of the WGS-84 coordinate system primarily involves two approaches: coordinate format conversion and coordinate projection conversion [12]. Coordinate format conversion refers to the transformation of geographic latitude and longitude into a Cartesian coordinate system, following these equations:

$$\begin{aligned} X &= (N + h) \times \cos(\varphi) \times \cos(\lambda) \\ Y &= (N + h) \times \cos(\varphi) \times \sin(\lambda) \\ Z &= [N \times (1 - e^2) + h] \times \sin(\varphi) \end{aligned} \tag{1}$$

where N denotes the radius of curvature of the reference ellipsoid and e is the eccentricity of the ellipsoid.

However, the converted Cartesian coordinates still reside within the WGS-84 geodetic reference system, and the vertical Z component remains as elevation, which requires further processing. Coordinate projection conversion projects the three-dimensional WGS-84 ellipsoidal coordinates onto a two-dimensional Gauss-Krüger plane coordinate system. In this study, the Gauss-Krüger projection method was adopted for coordinate transformation, as illustrated in Fig. 1 [13]. The choice of projection zone typically involves either 3-degree or 6-degree belts; for improved accuracy, the 3-degree projection belt was employed in this research.

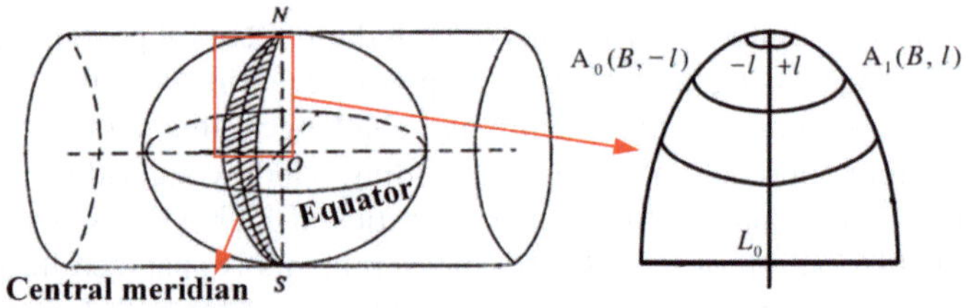

Fig. 1. Principle of Gauss-Krüger Projection.

In Fig. 1, the $A_0(B, -l)$ and $A_1(B, l)$ axes exhibit symmetry, and the projected points $A'_0(x, -y)$ and $A'_1(x, y)$ are likewise symmetric. This geometric relationship can be formulated as follows:

$$\begin{cases} x = F_1(B, \pm l) \\ \pm y = F_2(B, \pm l) \end{cases} \tag{2}$$

where the difference between the geographic longitude L and the central meridian longitude L_0 is denoted as l. Due to the symmetry between variables F_1 and F_2, they serve as even and odd functions of l, respectively.

The reference ellipsoid of the geodetic coordinate system is divided into several equal-width projection zones along the central meridians. Within each projection zone,

l is selected to remain small, enabling the expansion of x and y into the following power series expressions:

$$\begin{cases} x = m_0 + m_2 l^2 + m_4 l^4 + \ldots \\ y = m_1 l + m_3 l^3 + m_5 l^5 + \ldots \end{cases} \tag{3}$$

where $m_i (i = 1, 2, ...)$ is a function of the geodetic latitude B.

The concept of isometric latitude q is introduced to establish the mathematical correspondence between the latitude parameter (B, -l) in the WGS-84 reference frame and the (x, y) coordinates in the Gauss-Krüger plane coordinate system.

$$q = \int_0^B \frac{M}{N \cos B} dB \tag{4}$$

where $M = a(1 - e^2)/(1 - e^2 \sin^2 B)^{3/2}$ represents the radius of curvature of the meridian, $N = a/(1 - e^2 \sin^2 B)^{1/2}$ is the radius of curvature of the prime vertical, $a = 6399698.9018$ denotes the semi-major axis of the ellipsoid, and $e^2 = 0.00669437990013$ is the first eccentricity of the ellipse.

Moreover, the Gauss–Krüger projection must satisfy the Cauchy–Riemann conditions, which are essential for establishing the transformation relationship between (l, q) and (x, y).

$$\begin{cases} \dfrac{\partial x}{\partial q} = \dfrac{\partial y}{\partial l} \\ \dfrac{\partial x}{\partial l} = -\dfrac{\partial y}{\partial q} \end{cases} \tag{5}$$

By differentiating Eq. (3) and substituting into Eq. (5), Since the Gauss projection preserves the length along the central meridian, which leads to the following equality:

$$x = m_0 = X \tag{6}$$

where $X = 6367558.49686 * B/57.29577951308$ represents the meridian arc length from the equator to the geodetic latitude B, and it satisfies $dX = MdB$.

Therefore, by integrating the above equations, we obtain:

$$\begin{aligned} x = X + \frac{Nt}{2}\cos^2 B l^2 + \frac{Nt}{24}\cos^4 B(5 - t^2 + 9\eta^2 + 4\eta^4) l^4 + \\ \frac{Nt}{720}\cos^6 B(61 - 58t^2 + t^4) l^6 \end{aligned} \tag{7}$$

$$\begin{aligned} y = N \cos B l + \frac{N}{6}\cos^3 B(1 - t^2 + \eta^2) l^3 + \\ \frac{N}{120}\cos^5 B(5 - 18t^2 + t^4 + 14\eta^2 - 58\eta^2 t^2) l^5 \end{aligned} \tag{8}$$

where $t = \tan B$, $\eta^2 = e'^2 \cos^2 B$, and $e'^2 = 0.00673949674227$ denotes the second eccentricity of the ellipsoid.

2.2 Two-Dimensional Coordinate Smoothing Based on the Sliding Window Algorithm

A sliding window algorithm is utilized to smooth two-dimensional coordinates, thereby mitigating map boundary matching errors induced by RTK-GNSS positioning noise [10]. By specifying a fixed window width, coordinates within each window are fitted using linear or polynomial models, reconstructing the original coordinates to attenuate the impact of local noise. Deviations between the original values and the fitted results are compared to identify and refine boundary coordinates, while adjacent boundary points are clustered to accurately delineate the boundary position. The fitting error is quantified by the point-to-line distance (Eq. 9), providing a basis for parameter selection and performance evaluation. The strategy for selecting window width is discussed in Sect. 3.

$$d_l = \frac{|Ax + By + C|}{\sqrt{A^2 + B^2}} \tag{9}$$

3 Experiments and Analysis

RTK-GNSS positioning is subject to uncertainties caused by factors such as multipath effects, atmospheric delays, and satellite geometry. The feasibility of the sliding window algorithm was assessed through a two-stage experimental procedure. First, static RTK-GNSS positioning error was evaluated by converting coordinates into the Gauss-Krüger plane and determining the 95% confidence interval to establish an error model. Second, the error model was used to generate noise in a simulation environment; this noise was superimposed onto boundary sampling points, and the effectiveness of the sliding window smoothing algorithm was subsequently validated.

3.1 Hardware Platform for Autonomous Roller

A retrofitted 3-ton double-drum vibratory roller served as the experimental platform, as shown in Fig. 2. The system adopted a modular design, supporting both autonomous control (steering, speed regulation, braking, and vibration activation) and manual driving modes. The hardware architecture comprises an upper controller (Nvidia Jetson TX2 embedded computer) and a lower controller (Rexroth BODAS RC 95202/02.12).

A MGDF-02R dual-antenna GNSS receiver (equipped with a Trimble BD982 chipset) was employed, supporting GPS, Galileo, and BeiDou systems, with a maximum update rate of 50 Hz. The single-baseline RTK positioning accuracy is 8 mm + 1 ppm horizontally and 15 mm + 1 ppm vertically. During testing, the sampling frequency was set to 20 Hz, which meets the spatial resolution requirements for roller operation speeds of 2–5 km/h, resulting in an average distance of approximately 6.9 cm between adjacent positioning points.

Fig. 2. Experimental platform of the autonomous roller.

3.2 Static Positioning Error Test and Error Analysis

Static positioning error tests were conducted at both the starting and ending points of the operational area. While the roller remained stationary, RTK-GNSS was used to continuously collect positioning data, which was then converted into Gauss–Krüger plane coordinates for error analysis (as shown in Fig. 3). The results indicate that the data points at the starting location are densely clustered (Fig. 3(a)), whereas those at the ending location are more widely dispersed (Fig. 3(b)). The positioning errors for both locations are statistically summarized in Table 1. The covariance values for both data sets are close to zero, indicating no correlation between longitude and latitude coordinates. This confirms that the RTK-GNSS equipment functioned properly and that the data are reliable.

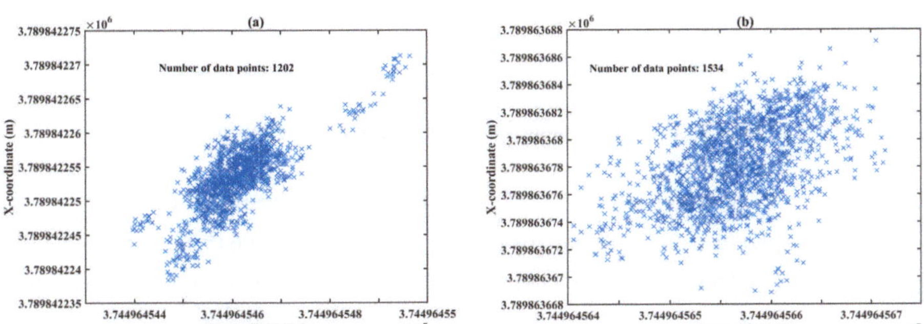

Fig. 3. Distributions of positioning coordinates and deviations in the Gauss–Krüger plane coordinate system. (a) Distribution of starting point positioning coordinates, (b) Distribution of ending point positioning coordinates.

Table 1. Covariance analysis of positioning data in the Gauss–Krüger plane coordinate system.

$Cov(x_1, y_1)$	x_1	y_1	$Cov(x_2, y_2)$	x_2	y_2
x_1	2.34×10^{-5}	2.96×10^{-5}	x_2	9.05×10^{-5}	7.49×10^{-5}
y_1	2.96×10^{-5}	6.22×10^{-5}	y_2	7.49×10^{-5}	2.79×10^{-5}

Further analysis was performed on the positioning accuracy of individual points. The distribution of the transformed X-coordinate values for both datasets is shown in Fig. 4, with the Y-coordinate values exhibit a similar variation pattern. The starting point coordinates are highly concentrated and their errors are approximately normally distributed. In contrast, the endpoint coordinate distribution is more dispersed, with greater variance, but still exhibits general normality. Calculation of the 95% confidence intervals (Table 2) indicates that the errors in both X and Y coordinates remain at the centimeter level, thus meeting the precision requirements for single-point positioning.

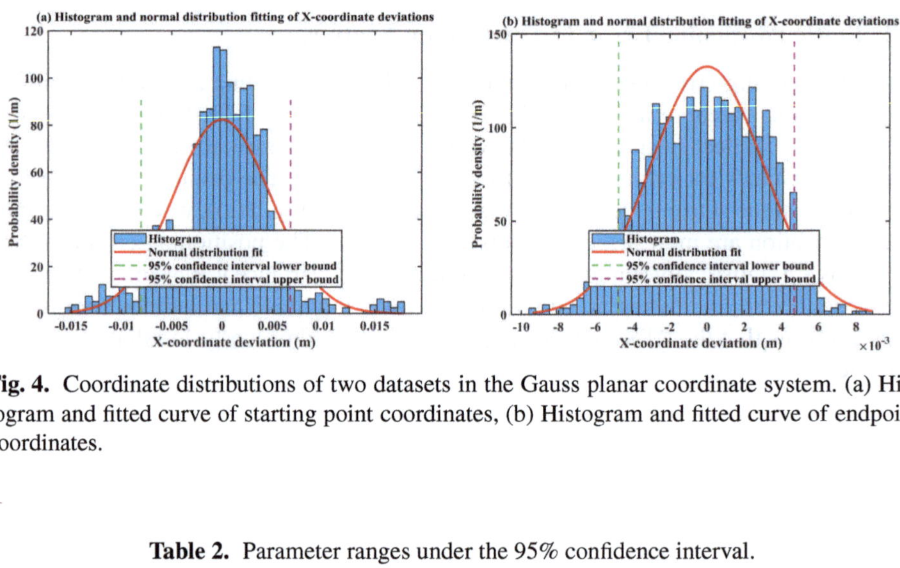

Fig. 4. Coordinate distributions of two datasets in the Gauss planar coordinate system. (a) Histogram and fitted curve of starting point coordinates, (b) Histogram and fitted curve of endpoint coordinates.

Table 2. Parameter ranges under the 95% confidence interval.

Coordinate	Value ranges (m)	Coordinate	Value ranges (m)
x_1	[3789842.24, 3789842.26]	y_1	[37449645.45, 37449645.47]
x_2	[3789863.67, 3789863.68]	y_2	[37449645.64, 37449645.66]

The deviations and deviation distances of the X and Y coordinates for both experimental groups were subjected to the Kolmogorov–Smirnov (K-S) test [14]. After excluding points exhibiting drift greater than 2 mm, the results confirm that the samples are drawn from a normally distributed population, supporting the construction of an error normal distribution model, as shown in Fig. 5. The calculated model mean is 1.36×10^{-4}

meters, and the variance is 1.30×10^{-5} meters, which serve as parameters for subsequent sliding window algorithm evaluation.

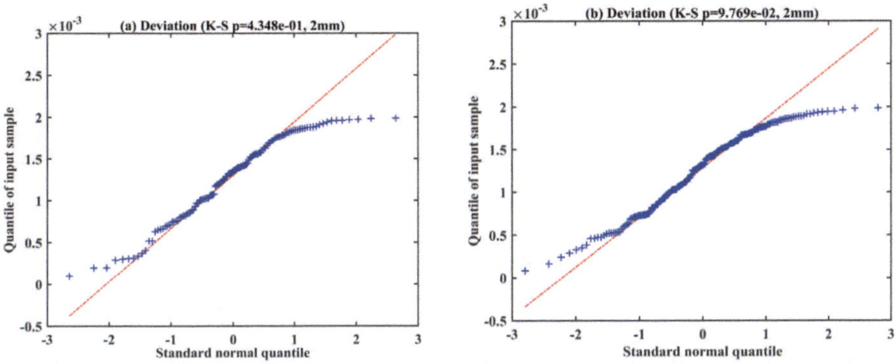

Fig. 5. Kolmogorov–Smirnov test results for positioning errors in both experimental groups. (a) Q-Q plot of deviation distances for the starting point, (b) Q-Q plot of deviation distances for the endpoint.

3.3 Feasibility Simulation and Analysis of the Sliding Window Algorithm

Simulation of Straight Road Boundaries. For straight road construction scenarios, boundaries can be determined either by placing RTK-GNSS receivers at the four vertices and connecting their coordinates, or by moving an RTK-GNSS receiver along the boundary to collect edge data. However, the ground-truth positions of road boundaries cannot be directly acquired by RTK-GNSS, making field performance evaluation challenging. In simulation, two boundary vertices with known positions are defined, and additional boundary coordinates are generated by linear interpolation. Random noise, produced according to the error distribution model from Sect. 3.2, is superimposed on these interpolated coordinates to simulate the process of map modeling as the RTK-GNSS receiver moves along the road boundary.

According to the "Technical specifications for construction of highway asphalt pavements (JTG F40-2004)" [15], the length of a compacted work area typically does not exceed 60 m. In this simulation, the boundary length of the straight road work area is set to approximately 20 m to facilitate clearer observation of the results. The start and end coordinates of the boundary are defined, and 1,000 uniformly distributed boundary points are extracted by linear interpolation. Random noise generated by the error distribution model is then added to each point. The sliding window algorithm described in Sect. 2.2 is applied, with a preset window width of 5, and polynomial fitting is performed on the resulting coordinates. As illustrated in Fig. 6, the sliding window algorithm accurately fits the boundary, achieving a mean fitting error of 1.38×10^{-3} meters, which meets the accuracy requirements for practical construction processes.

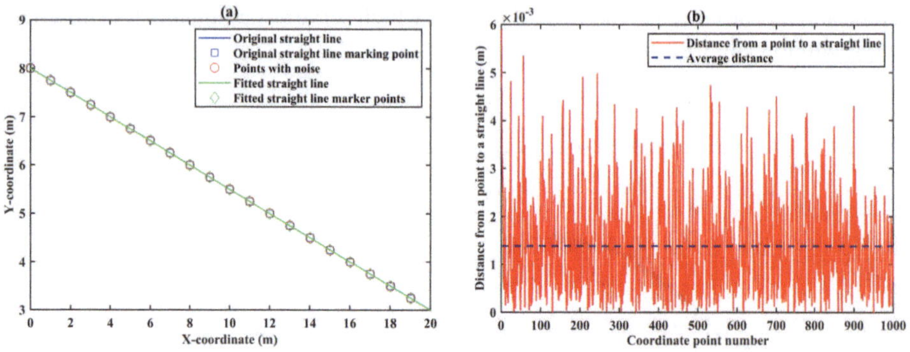

Fig. 6. Algorithm test results for a straight boundary with a window width of 5. (a) Results of the sliding window algorithm, (b) Fitting error between the fitted and original boundary coordinates.

Subsequently, the effect of different sliding window widths on the algorithm's boundary fitting performance was evaluated using three metrics: mean error, maximum error, and standard deviation. Simulations were conducted with sliding window widths set to 2, 3, 4, 5, 6, 7, and 8, and the corresponding error metric trends are presented in Fig. 7. As depicted in the figure, both mean error and standard deviation exhibit a clear decreasing trend with increasing window width, indicating that appropriately enlarging the window effectively mitigates noise interference and improves the overall fitting accuracy of curved boundaries. When the window width exceeds 5, the variations in mean error and standard deviation stabilize, suggesting diminishing returns in accuracy improvements from further enlarging the window size. Notably, the maximum error reaches its minimum at a window width of 5, but slightly increases with larger window sizes, implying that an excessively large window may hinder the algorithm's ability to smooth local outliers effectively.

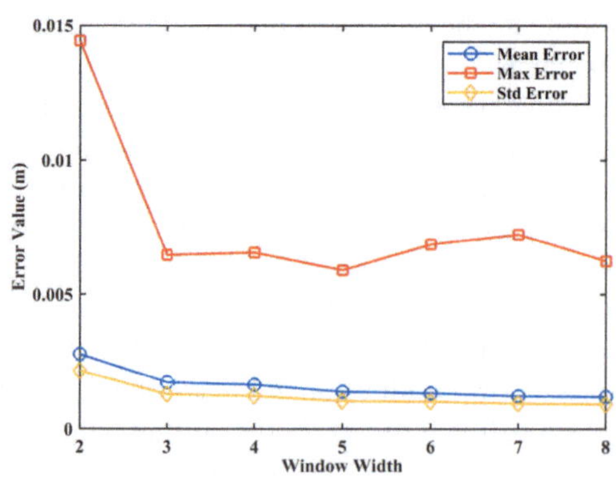

Fig. 7. Statistical results of fitting errors under different sliding window widths for straight road boundaries.

Simulation of Curved Road Boundaries. In practical curved road scenarios, the curvature of highway boundaries is generally predetermined. According to the "Technical standard of highway engineering (JTG B01-2014)" [16], the minimum radius of a circular curve under maximum superelevation for highways with a design speed of 60 km/h is 125 m. In this study, a curvature of 0.08 was specified and the boundary length was set to approximately 10 m to rigorously evaluate the algorithm's performance. The simulation procedure is consistent with that used for the straight road boundary scenario. As depicted in Fig. 8, the sliding window algorithm effectively fits the curved boundary, achieving an average deviation of 1.78×10^{-3} meters. The overall error exhibits a fluctuating upward trend and increases with distance, mainly attributed to cumulative errors in the RTK-GNSS positioning signals, while remaining at the centimeter scale, thereby satisfying engineering precision demands.

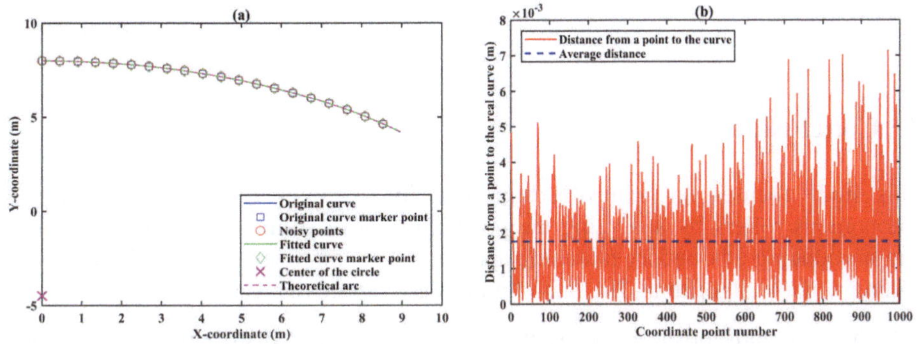

Fig. 8. Algorithm performance for a curved boundary with a window width of 5. (a) Results of the sliding window algorithm, (b) Fitting errors between the fitted and original boundary coordinates.

The boundary fitting performance for curved segments was assessed using the same sliding window widths and evaluation metrics as for the straight road boundaries. The results are presented in Fig. 9. The figure reveals a distinct decreasing trend in both mean error and standard deviation of curved segments with increasing window width. When the window width is small, the fitting errors are larger and exhibit substantial fluctuation. As the window width reaches 5 or above, the variations in mean error and standard deviation stabilize, suggesting diminishing benefits from further widening the window, which aligns with observations from linear boundary analysis. The maximum error reaches its minimum value at a window width of 5 before experiencing a slight increase, indicating that an excessively wide window impairs the effective smoothing of localized outliers in curved boundaries. Overall, considering both linear and curved segments, a window width of 5 provides an optimal balance between global fitting accuracy and local stability, effectively mitigating extreme errors and preserving high fitting precision.

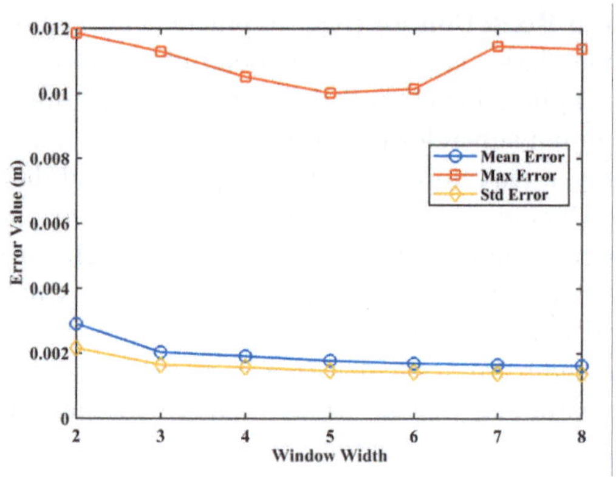

Fig. 9. Statistical results of fitting errors under different sliding window widths for curved road boundaries.

4 Conclusions

This study focuses on achieving high-precision mapping of operational areas for autonomous rollers by analyzing the positioning error characteristics of RTK-GNSS and introducing a boundary matching approach that combines RTK-GNSS data with a sliding window algorithm. The main findings are summarized as follows:

(1) Through static positioning experiments, the statistical distribution and 95% confidence intervals of RTK-GNSS positioning errors were obtained. This facilitated the construction of a normal distribution error model, which serves as a data-driven foundation for subsequent accuracy analysis and algorithm optimization.
(2) The integration of sliding window smoothing and polynomial fitting was utilized to model both linear and curved work area boundaries. The resulting average fitting errors were 0.138 cm for straight boundaries and 0.178 cm for curved boundaries, effectively mitigating positioning noise and meeting engineering accuracy requirements.
(3) The results indicate that a window width of 5 provides an optimal balance between global fitting accuracy and local stability, providing practical guidance for engineering applications.

Future work will focus on validation in real-world construction scenarios, incorporating adaptive parameter optimization and the fusion of multi-source sensor data (e.g., IMU and vision systems), to further enhance the adaptability and robustness of the mapping methodology.

References

1. Huang, H., Hu, Y., Wang, X.: A fusion positioning system based on camera and LiDAR for unmanned rollers in tunnel construction. Sensors **24**(13), 4408 (2024)

2. Chang, G.K., Gilliland, A.L., TaghaviGhalesari, A.: Aggregating high-precision GNSS intelligent construction data for quality asphalt pavements. In: International Conference on Trends on Construction in the Post-Digital Era, pp. 349–368 (2023)
3. Zhang, Q., An, Z., Liu, T., et al.: Intelligent rolling compaction system for earth-rock dams. Autom. Constr. **116**, 103246 (2020)
4. Wang, Z.Q.: Research on location and motion planning of compaction area of autonomous roller. Chang'an University (2023)
5. Nguyen, H.A., Ha, Q.P.: Robotic autonomous systems for earthmoving equipment operating in volatile conditions and teaming capacity: a survey. Robotica **41**(2), 486–510 (2023)
6. Liu, Y.Z., Qu, Q., Cao, D.: Improved multi-sensor information fusion algorithm and its application. Trans. Microsys. Technol. **43**(07), 161–164 (2024)
7. Wang, R., Becker, D., Hobiger, T.: Stochastic modeling with robust Kalman filter for real-time kinematic GPS single-frequency positioning. GPS Sol. **27**(3), 153 (2023)
8. Peng, L.H., Sun, D.H., Cheng, S.S., et al.: Research and application of levelling techniques in airborne gravity data. Prog. Geophys. **38**(05), 2326–2335 (2023)
9. Wang, Z., Liu, J.J., Zhang, W.Z., et al.: Adaptive filtering algorithm for UAV integrated navigation in complex environments. Trans. Microsyst. Technol. **44**(01), 135–140 (2025)
10. Che, L.N., Ren, X.L.: Anomaly detection algorithm based on sliding windows and confidence for WSN. Chin. J. Sens. Actuat. **36**(11), 1801–1807 (2023)
11. Stateczny, A., Specht, C., Specht, M., et al.: Study on the positioning accuracy of GNSS/INS systems supported by DGPS and RTK receivers for hydrographic surveys. Energies **14**(21), 7413 (2021)
12. Omali, T.: Coordinate transformation of GPS measurement results using the Cartesian-to-Ellipsoidal transformation system. Int. J. Sci. Res. Math. Stat. Sci. **10**(4), 9–13 (2023)
13. Li, X., Li, H., Liu, G., et al.: Optimization of complex function expansions for Gauss-Krüger projections. ISPRS Int. J. Geo Inf. **11**(11), 566 (2022)
14. Zhou, W., Lv, Z., Li, G., et al.: Detection of spoofing attacks on global navigation satellite systems using Kolmogorov-Smirnov test-based signal quality monitoring method. IEEE Sens. J. **24**(7), 10474–10490 (2024)
15. Institute of Highway Science, Ministry of Communications of China: Technical specifications for construction of highway asphalt pavements: JTG F40-2004. Industry Standard – Transportation (2004)
16. Department of Highway, Ministry of Communications of China: Technical standard of highway engineering: JTG B01-2014. Industry Standard – Transportation (2014)

Multi-factor Analysis of Optimal Overhanging Height Ratio for Tower Cranes Under Typhoon Conditions

Yongxin Zhang, Hongsheng Zhang, and Yue Wang(✉)

School of Mechatronics Engineering, Harbin Institute of Technology, Harbin 150001, Heilongjiang, China
wang_yue@hit.edu.cn

Abstract. To mitigate tower crane collapse risks during typhoon events, this study investigates the optimal overhanging height ratio (OOHR) for two tower structures: a uniform-section tower with a flexible tie-in structure and a stepped-column tower with a rigid tie-in structure. Employing differential equation methods, the equivalent stiffness provided by the lower tower section to the upper section was derived. By introducing two dimensionless coefficients - the flexible tie-in structure stiffness coefficient and the inertia moment ratio between tower sections, the principle of equal maximum bending moments is applied between the tied-in cantilever tower segment and the independently working maximum-height segment. Functional relationships between these coefficients and the overhanging height ratio were established. The analysis shows that for axial force coefficients ranging from 0.01 to 0.2, variations in flexible tie-in stiffness exert negligible influence on the overhanging height ratio, and significant differences exist between the OOHRs for flexible and rigid tie-in structures within this range. Meanwhile, for a stepped-column tower with a rigid tie-in structure, the inertia moment ratio of the tower's upper part to the lower shows a positive correlation with the OOHR. Under the condition of identical axial force coefficients, the OOHR of a stepped-column tower with a rigid tie-in structure is consistently smaller than that of a uniform-section tower with a flexible tie-in structure. Both configurations yielded OOHRs exceeding the stability-based ratios derived for rigid-tie-in towers under normal working conditions. Therefore, under typhoon conditions, the OOHR of a tower crane with a tie-in structure can be determined based on the stability criterion established for rigidly tied-in towers under normal working loads.

Keywords: Tower with flexible tie-in structure · Optimal overhanging height ratio · Typhoon invasion effects · Differential equation method · Multiple influencing factors

1 Introduction

With the increasing prevalence of high-rise buildings, the application of tower cranes with tie-in structures has become widespread. Typhoons pose a significant risk to these tall structures. Thus, analyzing the safety of tied-in tower cranes during typhoons is essential.

S. Halgamuge et al. (Eds.): ICACMVE 2025, LNME, pp. 390–397, 2026.
https://doi.org/10.1007/978-981-95-6911-3_38

However, most safety calculations assume that the tie-in structure is rigid; meanwhile, the towers were modeled as a uniform-section beam column. These assumptions often lead to calculation errors. Therefore, this study extends the safety analysis of attached tower cranes under typhoon loads by incorporating flexible tie-in structures and stepped column configurations to investigate their effects on optimal overhanging heights. This approach yields more realistic and engineering-practical results for determining the OOHR than the previous approaches.

Current research on the strength and stability of towers with tie-in structures mainly follows the design rules [1]. Luo et al. [2] found the OOHR (0.7419) for a single tied-in supported tower crane using Euler's critical load equivalence principle, with its validity confirmed by Lin et al. [3] and Lou [4] through MATLAB simulations. However, these studies generally assumed the condition with a rigid tie-in structure until Liu [5] introduced a flexible tie-in analysis by deriving internal forces and equivalent stiffness expressions via exact differential equation methods. But these approaches were still limited to normal operating conditions, they did not address tower collapse mechanisms caused by typhoons. Pan et al. [6] and Huang [7] found that typhoons caused the collapse of numerous tower cranes, with the strength failure at the tie-in points of tower cranes being one of the primary reasons for the collapses. These researches prompted advanced research by Wang et al. [8] who applied CFD simulations to quantify wind load coefficients at various heights and distances, while Wang et al. [9] and Huang et al. [10] proposed structural optimizations through height reduction and lightweight designs for typhoon resistance.

The axial force is minimal when determining the OOHR of a tower with a tie-in structure under typhoon conditions. Thus, the equation cannot be established based on Euler's critical buckling load. Instead, our research group employed the principle of an equal maximum bending moment to calculate the OOHR of a rigidly tied-in tower crane under typhoon loads [11]. Building on this foundation, this study further extends the analysis to investigate the multifactorial influences of tied-in tower cranes under typhoon conditions.

2 OOHR for Flexible Uniform-Section Tower Crane with a Tie-In Structure

The tower with a flexible tie-in structure, as shown in Fig. 1(a), is equivalently divided into upper and lower tower sections using the tie-in support point as the boundary. The lower tower section provides an equivalent rotational stiffness to that of the upper section. The upper cantilevered segment is equivalently modeled as a cantilever beam with an elastic rotational support, as illustrated in Fig. 1(c).

The load conditions of the lower tower section are illustrated in Fig. 1(b). With the spring stiffness at the tie-in support point denoted as k_0 and the horizontal displacement at the tie-in support point as δ, the total bending moment includes the applied end moment M, horizontal spring reaction force $k_0\delta$, and second-order bending moment caused by axial force $N(\delta - y)$. By defining $k = \sqrt{N/EI}$, the beam deflection differential equation can be expressed as

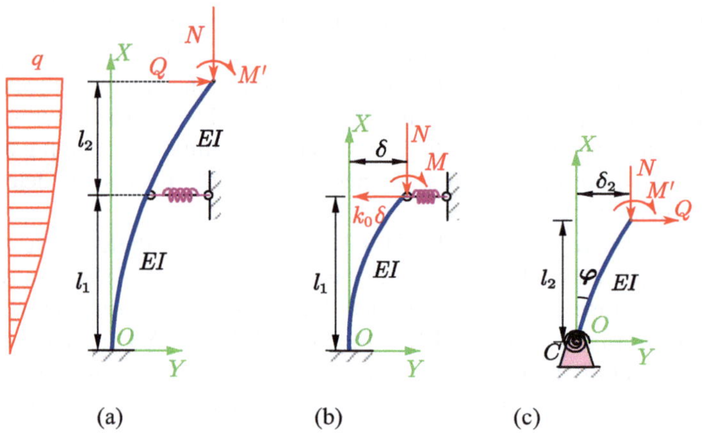

Fig. 1. Schematic diagram of uniform-section tower with a flexible tie-in structure.

$$y'' + k^2 y = [M - k_o \delta(l_1 - x)]/EI + k^2 \delta \tag{1}$$

The solution to the second-order differential equation expressed in Eq. (1) takes the form of a general solution plus a particular solution. Applying boundary condition Eq. (2) eliminates the undetermined coefficients, producing Eq. (3) that establishes the moment-rotation relationship.

$$\begin{cases} y(0) = 0, y'(0) = 0 \\ y(l) = \delta, y'(l) = \varphi \end{cases} \tag{2}$$

$$M = \frac{E^2 I^2 k^4 \cos(kl_1) - EIk^2 k_0 l_1 \cos(kl_L) + EIkk_0 \sin(kl_1)}{2k_0 - 2k_0 \cos(kl_1) + EIk^3 \sin(kl_1) - k_0 kl_1 \sin(kl_1)} \varphi \tag{3}$$

From Eq. (3), it can be seen that the coefficient of φ represents the equivalent stiffness C provided by the lower tower crane section to the upper tower section under flexible tie-in conditions. By introducing dimensionless axial force coefficient $\Omega = N_{cr}/N = \pi^2 EI/(2l_1)^2/N$, dimensionless spring stiffness coefficient $\gamma = EIk^3/k_0$, and defining the symbols, $\varepsilon_1 = l_1\sqrt{N/EI} = kl_1 = \pi\sqrt{\Omega/2}$, $\beta = l_2/l_1$. The expression for the equivalent stiffness can be obtained as

$$C = \frac{EI}{l_1} \cdot \frac{\varepsilon_1 \sin(\varepsilon_1) - \varepsilon_1^2 \cos(\varepsilon_1) + \gamma \varepsilon_1 \cos(\varepsilon_1)}{2[1 - \cos(\varepsilon_1)] - \varepsilon_1 \sin(\varepsilon_1) + \gamma \sin(\varepsilon_1)} \tag{4}$$

Reference [11] formulated the second-order moment (W_2) equation by equating the maximum end moments between the tied-in cantilever tower segment and the free-standing working maximum-height segment. In this equation, δ_1 represents the end displacement of the tower crane's ideal cantilever beam model, and δ_2 denotes the end displacement of the upper tower section under flexible tied-in conditions.

$$0.0363 + 0.00386(1 + \beta)^{2.6} - 0.157\beta(1 + \beta)^{1.6} = N(q_o l_1^{2.6})\delta_1 - \delta_2 = W_2$$

When $\Omega = 0.01$, by substituting C into δ_2 and incorporating the resulting second-order moment difference into the above equation, we obtain a bivariate transcendental equation in terms of β and γ. By moving all terms to one side of the equation, we construct the bivariate function $fun(\beta, \gamma)$ and plot its three-dimensional surface as shown in Fig. 2. The relationship between β and γ can be deduced indirectly by solving the contour line where $fun(\beta, \gamma) = 0$. From Fig. 3, it can be observed that when $\Omega = 0.01$: (1) The variation of γ has a negligible effect on $fun(\beta, \gamma)$ when γ is not in the neighborhood of zero. (2) $fun(\beta, \gamma) = 0$ is satisfied at $\beta = 0.7790$ for non-zero γ values. However, it is noteworthy that: When $\gamma = 0$, $fun(\beta, \gamma) = 0$ is only satisfied at $\beta = 0.7741$.

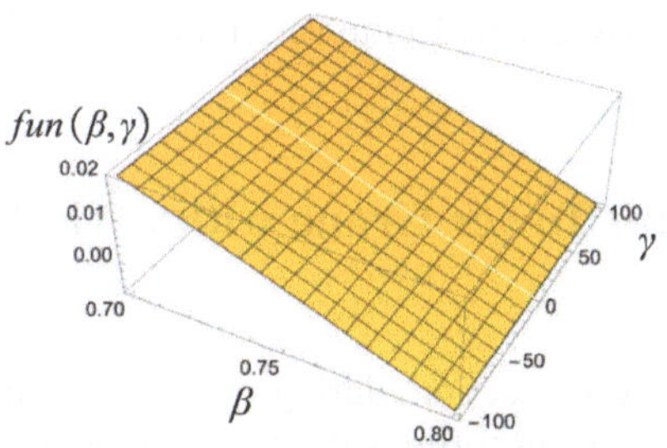

Fig. 2. 3D plot of the function $fun(\beta, \gamma)$

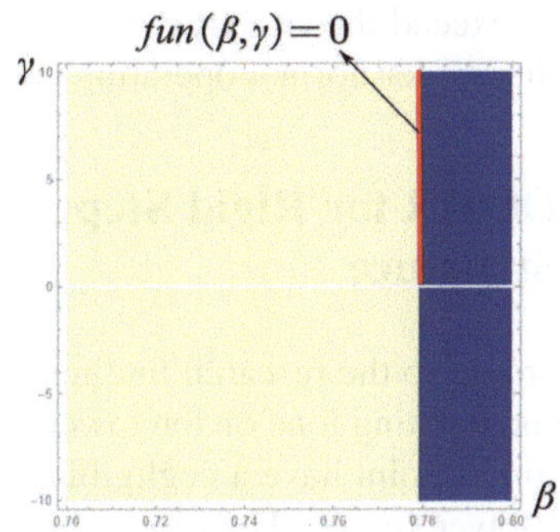

Fig. 3. Contour plot of $fun(\beta, \gamma) = 0$

By analyzing the second-order bending moment expression at $\Omega = 0.01$, it can be observed that a common factor $(1 + 776.4\gamma)$ can be approximately extracted from both the numerator and denominator. This leads to the elimination of γ from the second-order moment expression. This demonstrates that when Ω is minimal (minimal axial force), the variation in spring stiffness at the tie-in support point has a negligible effect on the equivalent stiffness provided by the lower tower section to the upper section. The underlying mechanism can be explained as follows: Under minimal axial forces, the axial end develops minimal displacement δ. The primary contribution of the flexible tie-in spring stiffness lies in generating the support reaction force $k_o\delta$. Consequently, the influence of spring stiffness is insignificant in this regime.

Similarly, the second-order moment difference and equivalent rotational stiffness can be derived for load cases with $\Omega = 0.05, 0.1, 0.15, 0.2$. This enables determination of the overhanging height ratio coefficient β_F under identical maximum bending moment conditions. When $\gamma = 0$, the model reduces to a rigid tie-in structure, yielding the rigid tie-in case's β_R. These results are summarized in Table 1.

As shown in Table 1, a significant numerical discrepancy between β_F and β_R can be observed. This phenomenon can be further verified from Eq. (4). When the axial force approaches an infinitesimal value (i.e., $\varepsilon_1 = 0$), the equivalent stiffness C in the flexible tie-in case reduces to E/l_1, demonstrating that it becomes independent of the spring stiffness k_0 at the tie-in support point. In contrast, the equivalent stiffness C' in the rigid tie-in case approaches $4EI/l_1$.

Table 1. OOHR of uniform-section towers with flexible and rigid tie-in structures under varying loads.

Ω	0.01	0.05	0.1	0.15	0.20
β_F	0.7790	0.7973	0.8219	0.8494	0.8796
β_R	0.7741	0.7720	0.7706	0.7705	0.7715

Therefore, under identical axial loads, the OOHR of a uniform-section tower crane with a flexible tie-in structure is greater than that of a rigid tie-in structure. However, both values exceed the OOHR (0.7419) derived for rigidly tied-in tower cranes, considering stability under normal operating conditions.

3 OOHR for Rigid Stepped-Column Tower Crane with a Tie-In Structure

According to the research findings in the preceding section, during typhoon conditions with no hoisting load on the tower crane, variations in spring stiffness at the flexible tie-in support point have a negligible effect on the OOHR owing to the relatively minimal axial tower forces. Therefore, this section will investigate the OOHR of the stepped-column tower section under rigid tie-in conditions. I_2 denotes the moment of inertia of the upper tower section and I_1 denotes the moment of inertia of the lower tower section. The corresponding mechanical model is shown in Fig. 4.

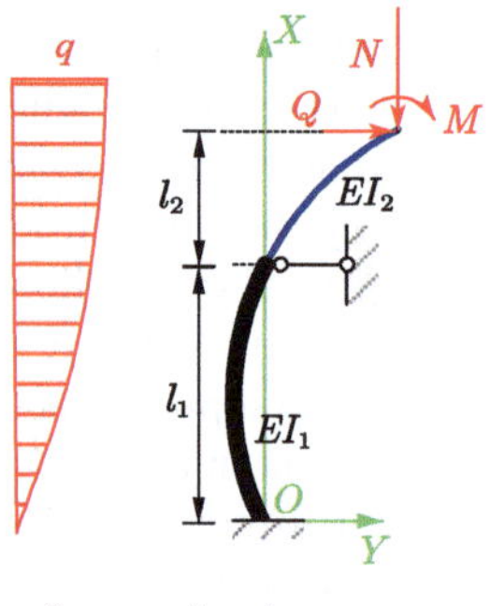

Fig. 4. Schematic diagram of stepped-column tower with a rigid tie-in structure.

The axial force can be expressed as $N = \Omega N_{cr} = \Omega\pi^2 EI_1(2l_1)^2$. Then, the dimensionless coefficient $\alpha = \sqrt{I_2/I_1}$ is defined as the ratio of the moments of inertia between the upper and lower tower section. Thus, $k_1 = \sqrt{N/EI_1} = \pi\sqrt{Q/2l_1}$, $k_2 = \sqrt{N/EI_2} = \pi\sqrt{\Omega/2\alpha l_1}$. According to the principle of equal maximum bending moments, $M_1/W_1 = M_2/W_2$. For the tied-in tower crane studied in this paper, the section flexural modulus ratio can be approximated as $W_1/W_2 = I_1/I_2$. Therefore, the condition of equal maximum bending moments transforms into: $\alpha^2 M_1 = M_2$. Based on this, a new second-order bending moment equation can be derived as

$$\begin{aligned} &\alpha^2(0.09662 - 0.06039 + 0.060386(1+\beta)^{2.6} - 0.157\beta(1+\beta)^{1.6}) \\ &= N(q_o l_1^{2.6})(\delta_2' - \alpha^2\delta_1') = W_2' \end{aligned} \tag{5}$$

By substituting W_2' at $\Omega = 0.01$ into Eq. (5), a transcendental equation relating β and α can be obtained. Applying the corresponding α values for the three cases of $I_1/I_2 = 1.0, 1.1, 1.2$, the respective β values are determined as 0.7741, 0.7733 and 0.7724. Following the same approach, the overhanging height ratio coefficient β under identical maximum bending moment conditions can be derived for axial force coefficients $\Omega = 0.05, 0.1, 0.15, 0.2$, each corresponding to three different ratios of the lower-to-upper tower section moments of inertia, as summarized in Table 2.

Table 2. OOHR of stepped-column tower with a rigid tie-in structure under varying loads and moment of inertia ratios.

Ω	0.01			0.05		
I_1/I_2	1.0	1.1	1.2	1.0	1.1	1.2
β	0.7741	0.7733	0.7724	0.7720	0.7683	0.7643

Table 3. OOHR of stepped-column tower with a rigid tie-in structure under varying loads and moment of inertia ratios.

Ω	0.1			0.15			0.02		
I_1/I_2	1.0	1.1	1.2	1.0	1.1	1.2	1.0	1.1	1.2
β	0.7706	0.7638	0.7566	0.7705	0.7609	0.7511	0.7716	0.7596	0.7475

As shown in Table 3, the overhanging height ratio coefficient β demonstrates a positive correlation with the moment of inertia ratio coefficient α between the upper and lower tower sections. Moreover, under identical axial force conditions, the OOHR of the stepped-column tower crane with a rigid tie-in structure is smaller than that of both the uniform-section tower crane with flexible and rigid tie-in structures.

4 Conclusion

Based on the second-order deflection differential equation and the equivalent elastic support model for towers with tie-in structures, the principle of equal maximum bending moments was applied to analyze freestanding towers and those with tie-in structures. Second-order bending moment equations were subsequently established for two specific configurations: a uniform-section tower with flexible tie-in structures and a stepped-column tower with rigid tie-in structures. The analysis of these equations elucidates the relationships between the OOHR coefficient, the spring stiffness coefficient of the flexible tie-in structure, and the inertia moment ratio of the upper-to-lower tower sections under varying axial force coefficients. The key findings are as follows:

(1) At relatively low axial forces, variations in the spring stiffness of the flexible tie-in structure exert a negligible influence on the equivalent stiffness provided by the lower tower section to the upper section. However, a significant disparity persists between the equivalent stiffness offered by flexible versus rigid tie-in structures. Consequently, for identical axial loads, a flexible tie-in structure results in a larger OOHR compared to a rigid tie-in structure.
(2) Under conditions of identical maximum bending moment, the inertia moment ratio between the upper and lower tower sections exhibits a positive correlation with the OOHR. Furthermore, under identical axial loads, the OOHR of a stepped-column tower with a rigid tie-in structure is smaller than that of uniform-section towers equipped with either flexible or rigid tie-in structures.
(3) Regardless of whether a tower crane employs a flexible tie-in (uniform-section design) or a rigid tie-in (stepped-column design), the calculated OOHR under typhoon invasion conditions exceeds the stability OOHR of rigid-tie-in tower cranes designed for normal working conditions. Therefore, for typhoon conditions, it should be determined based on the stability criterion applied to rigidly tied-in tower cranes.

Acknowledgements. This work was Supported by the Longjiang Science and Technology Talents Spring Goose Support Program (No. CYCX24011) and the Heilongjiang Provincial Key Research and Development Program (No. 2022ZX01A0).

References

1. National Technical Committee for Standardization of Lifting Appliance (SAC/TC 227). Design Rules for Tower Cranes: GB/T 13752-2017. Standards Press of China (2017)
2. Luo, B., Lu, N.L., Xia, Y.J.: Determination of reasonable attachment distance and maximum overhang height of tower crane. Construct. Mach. (10), 12–15+82 (2006)
3. Lin, B., Liu, S.G., Cong, J., et al.: Research on the calculation method of attachment distance and cantilever height of tower crane. Construct. Mech. **40**(02), 18–20 (2019)
4. Lou, K.M.: Analysis of crane stability against overturning. China Equip. Eng. (03):102–103 (2019)
5. Liu, J.H.: Research on analysis method of tower body structure of tower crane with attached flexibility. Harbin Institute of Technology (2019)
6. Pan, L., Zhang, Z.P., Xiao, M.: Analysis of severe typhoon damage to tower cranes and typhoon prevention measures. Construct. Safety **35**(05), 8–11 (2020)
7. Huang, Z.D.: Safety management essentials and practices for tower cranes during typhoon and flood seasons. Construct. Mech. **45**(04), 116–119 (2024)
8. Wang, H., Liu, Y., Xu, Z.D., et al.: Wind-induced buffeting comfort assessment of tower cranes considering the wake effect of super-high bridge towers. J. Wind Eng. Indust. Aerodyn. **240** (2023)

9. Wang, Y.: Analysis and plan of emergency measures of tower crane under typhoon based on Ansys. Lift. Transp. Mach. **23**, 62–67 (2023)
10. Huang, J.T., Gao, C.R., Wang, G.T., et al.: Optimization analysis of boom structure of flat head tower crane based on Ansys Workbench. Lift. Transp. Mach. **16**, 37–42 (2023)
11. Wang, X.X., Zhang, H.S., Wang, Y.: Effect of typhoon load on the attachment height of a rigid attached tower crane. In: ICACMVE (2024)

Study on Voltage Characteristics and Surface Damage Behavior of Bearings in Energy Electric Vehicle Motors Under Different Speed Conditions

Chuanbo Liu[1,2](✉), Shili Cheng[1,2], Leidi Zhao[1,2], Long Kuang[1,2], and Shutian Liu[1,2]

[1] College of Mechanical and Electrical Engineering, Wuhan University of Technology, Hubei, Wuhan 430070, China
lchb72@whut.edu.cn

[2] Shanghai General Motors Wuling (SGMW), Guangxi, Liuzhou 545027, China

Abstract. The discharge damage behavior of drive motor bearings in DHT systems is a key cause of abnormal vibration and noise in new energy vehicles. This issue stems from the shaft voltage generated during motor operation and is closely related to the bearing's operating state. This study focuses on the DHT system of a new energy vehicle, analyzing the relationship between voltage characteristics and surface damage behavior in motor bearings under different conditions. The results show that at low speeds, the bearing voltage amplitude is low, and no abnormal surface damage occurs. At high speeds, the drive motor (P3) bearings exhibit discharge damage in the form of voltage grooves, while the generator motor (P1) bearings show no damage. This is attributed to the significantly higher voltage amplitude in P3 motor bearings compared to P1. Additionally, due to high voltage characteristics, the discharge damage in the front bearings of the P3 motor is more severe than in the rear bearings, confirming that high voltage amplitude is the key factor causing discharge damage.

Keywords: New energy vehicles · DHT system · Bearing voltage characteristics · Rotational speed · Discharge damage

1 Introduction

Discharge damage behavior in new energy vehicle drive motor bearings typically manifests as electric pitting, frosting, and grooves [1]. In recent decades, the field of electric vehicle technology has expanded significantly, with manufacturers striving to meet growing market demands. Particularly in drive motor design, the pursuit of higher performance, efficiency, and user experience has led to trends toward higher speeds and voltages [2, 3]. While higher voltages can reduce resistance losses in electric drive systems, improve efficiency, and lower battery requirements—thereby reducing costs and vehicle weight—these advancements come at a cost. The inevitable consequence is discharge damage in drive motor bearings [4–6]. Reports indicate that over 40% of motor

S. Halgamuge et al. (Eds.): ICACMVE 2025, LNME, pp. 398–410, 2026.
https://doi.org/10.1007/978-981-95-6911-3_39

failures are attributed to bearing failures [6], many of which result from lubrication issues caused by shaft voltage and bearing currents. These problems lead to bearing instability, vibration, noise, and more severe mechanical failures [7], significantly shortening motor lifespan [8].

The discharge damage behavior in new energy vehicle drive motor bearings originates from shaft voltage. Differences in bearing operating states result in varying voltage characteristics, leading to different discharge damage behaviors. Electric pitting is caused by single discharge pulses with high current amplitude and long duration.In terms of distribution, the corroded pits corrosion appears as a random pattern, and they are sparser than frosting [9, 10]. Frosting results from weak but dense discharges, while grooves are caused by periodic discharges.

The voltage characteristics in new energy vehicle motor bearings are generally influenced by rotational speed, lubricant, temperature, and load. At low speeds, shaft voltage and current amplitudes are small. At high speeds, shaft voltage increases, leading to higher current amplitudes but reduced discharge frequency [11, 12]. Research on the impact of lubricating grease on bearing electrical erosion focuses on oil film state, composition, and type. Different base oils in lubricants lead to varying surface damage, with breakdown primarily influenced by the base oil. Temperature alters lubricant properties and oil film state; at the same speed, oil film thickness decreases as temperature rises, affecting discharge activity [13]. Load is a critical operating parameter for motors, affecting oil film thickness. High loads reduce oil film thickness, limiting shaft voltage amplitude and resulting in dense discharge activity. For lower loads, increased oil film thickness allows higher shaft voltage and current amplitudes, leading to more intense discharge activity [14, 15]. However, there is a lack of research on the impact of bearing voltage and discharge damage behavior in new energy vehicle DHT systems.

In this paper, we take the new energy vehicle DHT system as the research object, build a bearing internal voltage acquisition platform, and analyze the influence mechanism of different motor bearing internal voltage characteristics and surface damage behaviors by carrying out the research on the motor bearing voltage characteristics at different rotational speeds. The results obtained can promote the research and development of electric vehicle drive motors.

2 Research Content

2.1 Test Bench and Materials

In this study, we based all experiments on an actual new energy vehicle (EV) (supplied by Shanghai General Motors Wuling (Liuzhou SGMW Co., Ltd., China)). Figure 1 shows the disassembled parts of the DHT motor and the internal structure of the motor and a three-motor EV test rig. The DHT motor system is mainly composed of two motors, the P3 motor as a drive motor to provide a power source for the vehicle, and the P1 motor as a power generator motor whose main job is to recover the power and convert it into electrical energy. The three-motor electric vehicle test bench includes a power dynamometer system (consisting of an input dynamometer and two load motors), cooling system, battery simulator, power analyzer, environmental warehouse and upper computer control system, etc. The input dynamometer can simulate the load of the vehicle driving to measure the torque, speed and output power output of the motor, and the two sides of the load motors to simulate the resistance of the tires, and the cooling system is for the motor, The cooling system provides different temperature flow for the motor, controller and other components to ensure thermal stability during the testing process, the battery simulator replaces the real battery to provide programmable voltage and current inputs to simulate the performance of the motor in different power states, the power analyzer can collect the controller, the motor voltage and current to accurately calculate the efficiency data, etc., the environmental chamber simulates the different temperatures and humidity, etc., and the upper computer control system is an integrated system for the control of various automation programs. The upper computer control system is integrated with test software to realize various automated program control. The test bench can simulate the running process of new energy drive motor from 0 to 20,000rpm. During the entire test bench operation, an oscilloscope is used to detect the voltage signal between the DHT motor shaft and the outer raceway of the bearing. Due to the inherent characteristics of the motor, the voltage detected on the bearing comes from the shaft voltage generated during motor operation.

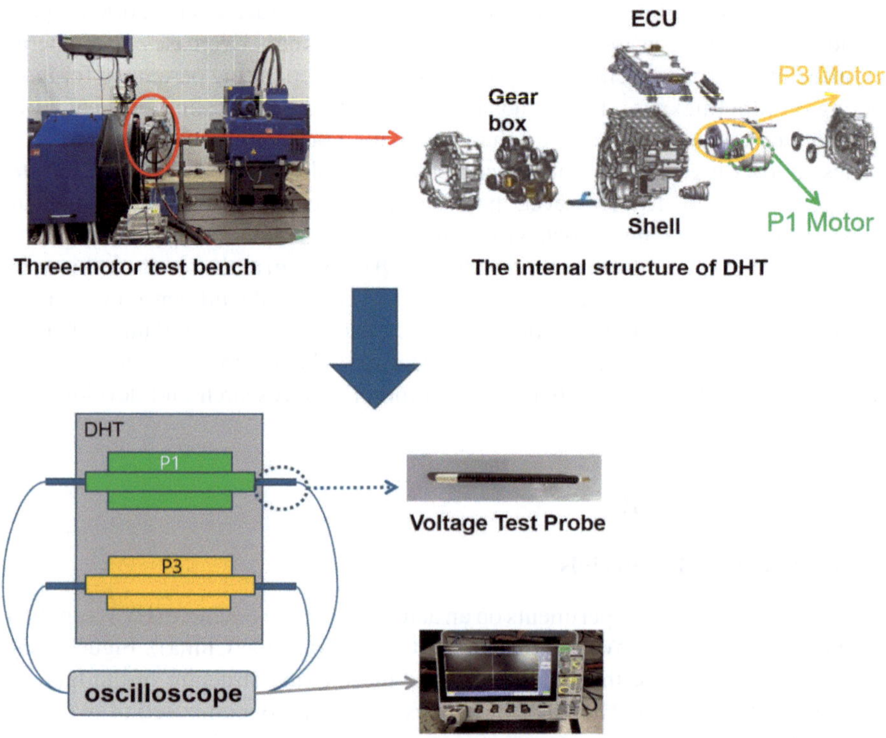

Fig. 1. Shaft voltage testing method and DHT structure.

The output shaft of the motor is fitted with radial ball bearings of specific type 6206 (manufactured by FAG, Nuremberg, Germany) as shown in Fig. 2. The key parameters of the bearing are shown in Table 1.

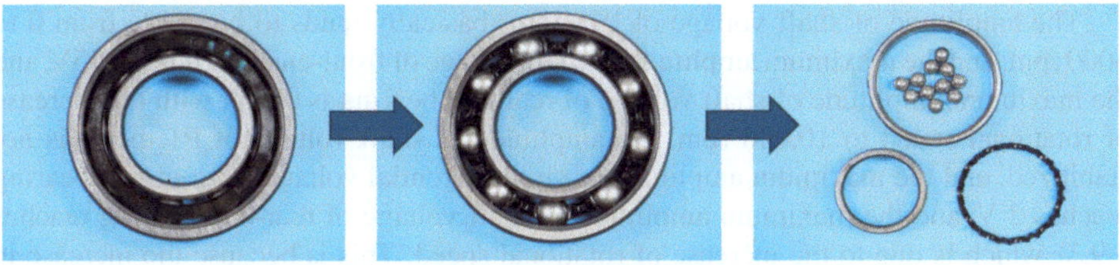

Fig. 2. 6206 bearing.

Table 1. Radial ball bearing 6206 parameters.

Bearing Type	6206	/
Number of Balls	9	Count
Outer Diameter	62	mm
Inner Diameter	30	mm
Bearing Width	16	mm

2.2 Test Conditions

During the test, the motors were operated without external loads at an ambient temperature of 20°C. Different test conditions were established by varying the motor speed, and the purpose of the rotational experiment was to keep the motors P1 and P3 in a constant rotational state for 7 days. The test consisted of seven constant rotational speeds of 2000, 6000, 8000, 10000 and 12000 rpm.

2.3 Measurement Techniques and Equipment

The wear surfaces were analyzed using a scanning electron microscope (SEM) (VEGA3, Tescan, Czech) and a confocal laser scanning microscope (CLSM) (VX-X1000, Keyence, Japan). A VHX-2000c extended depth-of-field microscope from Keyence, Japan was used to observe the discharge damage. Voltage signals were acquired using a high acquisition frequency oscilloscope (MDO3000, Tektroni, USA) and a test probe (TDP1000, Tektronix, USA).

3 Experimental Results

3.1 Voltage Characteristics of P1 and P3 Motor Bearings Under Different Speeds

Figure 3 shows the shaft voltage waveforms of the front-end bearings of the P1 and P3 motors at different constant rotational speeds within 0–0.4 ms. Figure 4 shows the waveforms of the shaft voltage of the rear-end bearings of P1 and P3 motors at different constant rotational speeds. The lubricant film under constant speed conditions is fully formed between the bearing rollers, inner and outer rings.

The amplitude of shaft voltage of P1 motor basically tends to be stable from 0 to 6000 rpm, and the maximum amplitude of shaft voltage of front-end bearing is 0.55V, and the maximum amplitude of shaft voltage of rear-end bearing is 0.5 V; with the increase of rotational speed to 10,000 rpm, the amplitude of shaft voltage of P1 motor is not stabilized, and the maximum amplitude of circumferential voltage of front-end bearing reaches 2 V, and the maximum amplitude of shaft voltage of rear-end bearing reaches 1.9 V, which is due to the increase of rotational speed. This is because the increase in rotational speed exacerbates the accumulation of charge at both ends of the bearing oil film, when the charge accumulates enough to break through the capacitance composed of the bearing outer ring, roller and inner ring will be when it will be an instantaneous discharge, the voltage amplitude instantly returns to zero, as shown in Fig. 3 (3–1); when the rotational speed is increased to 12,000 rpm, the shaft voltage maintains the characteristics of the waveform of 10,000 rpm, but the amplitude of the voltage and the frequency of the discharge increases, the maximum shaft voltage amplitude of the front-end bearing at 12000rpm reaches 2.4 V, and the maximum shaft voltage amplitude of the rear-end bearing reaches 3 V.

P3 motor in 0-2000 rpm when the shaft voltage amplitude basically tends to stabilize, the front end in the bearing maximum shaft voltage amplitude of 0.5V,the rear end in the bearing maximum shaft voltage amplitude of 1V; with the increase in speed to 6000rpm,as shown in Fig. 3 (3–2),the capacitance of the bearing state on the emergence of the breakdown characteristics of the beginning of the discharge, the front end of the maximum shaft voltage amplitude in the bearing 2.5V, the rear end of the maximum shaft voltage amplitude in the bearing 2.2 V; with the increase of speed to 10000 rpm, as shown in Fig. 3 (3–3),the discharge characteristics of the front-end bearing becomes more intense, the maximum voltage amplitude of 5.5 V, while the rear-end bearing discharge behavior began to slow down and the maximum voltage amplitude of 2 V; when the speed reaches 12,000 rpm, the voltage amplitude of the front-end bearing and the discharge behavior compared to the 10,000 rpm has also weakened the maximum voltage amplitude of 3V, the rear-end bearing voltage amplitude is reduced to 3 V. When the rotational speed reaches 12000 rpm, the voltage magnitude and discharge behavior of the front bearing are also weakened compared to 10000 rpm, and the maximum voltage magnitude is 3 V.

The reason for this phenomenon is that when the rotational speed is low, the charge generated does not accumulate in large quantities, resulting in oil film breakage when the accumulated charge is released, resulting in a smaller shaft voltage; when the rotational speed is high, the charge will be rapidly and a large number of accumulation, the accumulated charge can reach the maximum value that can be withstood in the capacitive state, and the shaft voltage also reaches the maximum, more than the maximum value of the oil film will be broken, in the oil film After the oil film is damaged, the oil film damage to repair this process during the bearing is on state, during the charge generated can flow to form the shaft current can be released, the voltage is instantly zero; when the speed is too high, the oil film damage to repair this process during the process will consume more charge and thus can not be a large number of accumulated voltage, so the voltage amplitude will be reduced in more than 10,000 rpm.

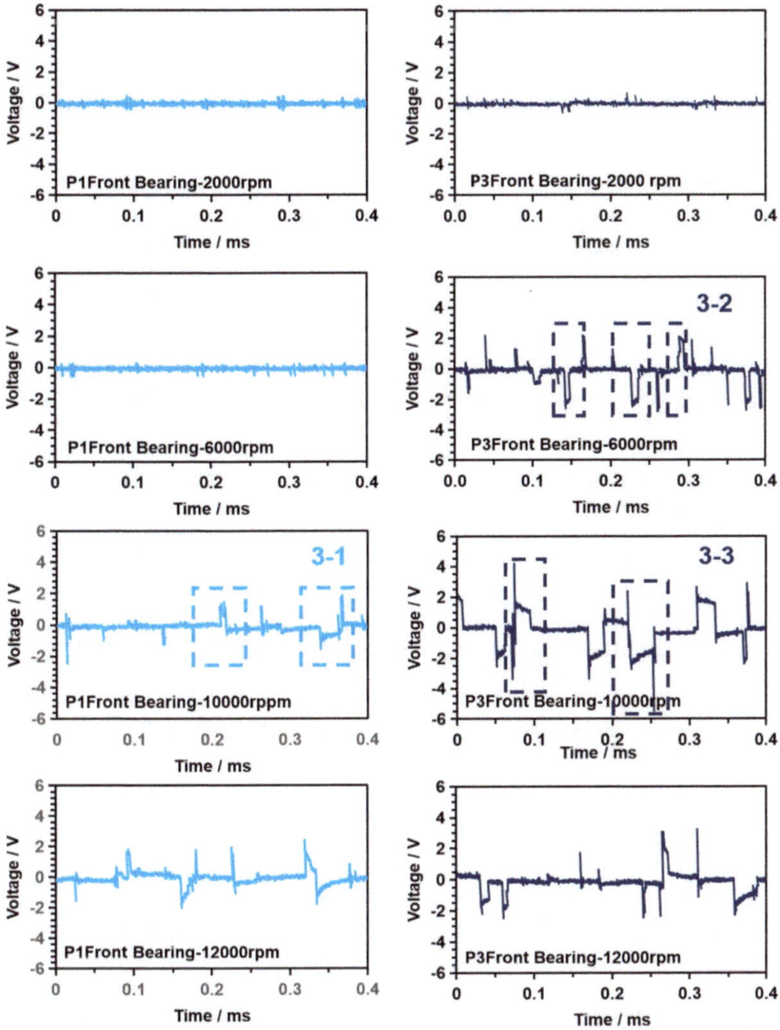

Fig. 3. P1 and P3 motor front-end voltage characteristics of the change rule.

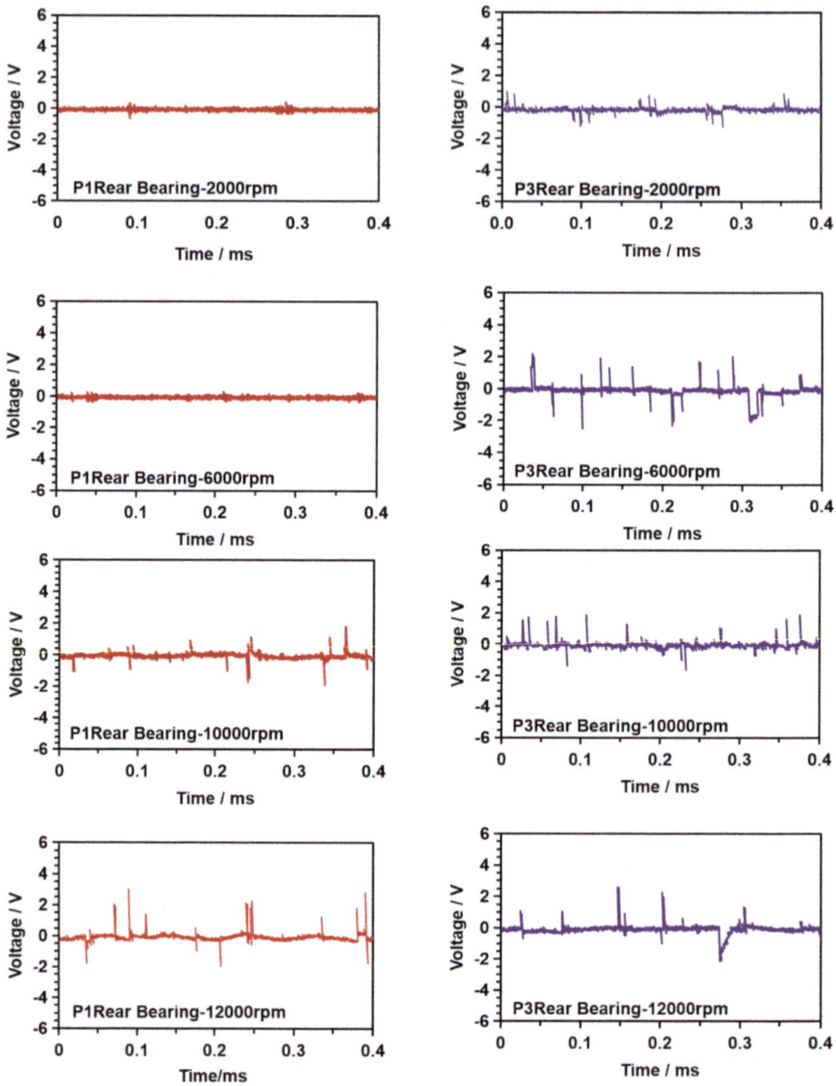

Fig. 4. P1 and P3 motor back-end voltage characteristic change rule.

Figure 6 shows the front and rear end bearing shaft voltage characteristics of P1 and P3 motors at constant speeds of 2000 rpm and 8000 rpm within 0–0.1 ms. It can be found that in the low speed (2000 rpm), the front and rear end bearing shaft voltage tends to stabilize and voltage amplitude is very small, there is no obvious discharge phenomenon. When the motor is at high speed (8000 rpm), the front and rear bearings of P1 motor start to produce resistive discharge as shown in Fig. 5 (5–1, 5–2), during this period, the shaft current flows through the low impedance paths such as the bearing metal contact and the grounded carbon brushes, etc., which is characterized by the stable amplitude and frequency of the shaft voltage, and the discharge of the front end bearings is more serious than that of the rear bearings, with a higher frequency of discharge; the front and rear bearings of P3 motor produce capacitive discharge, with a higher frequency of discharge; and the front and rear bearings of P3 motor produce capacitive discharge, with a higher frequency of discharge. P3 motor front and rear bearings produce capacitive discharge as shown in Fig. 5 (5–3), the shaft voltage amplitude of the front bearing reaches the upper limit of the capacitance of the breakdown of the oil film phenomenon; the shaft voltage amplitude of the rear bearing is lower than the voltage amplitude of the front bearing. The amplitude of the shaft voltage of the rear bearing is lower than that of the front bearing. It is characterized by high amplitude but very short duration.

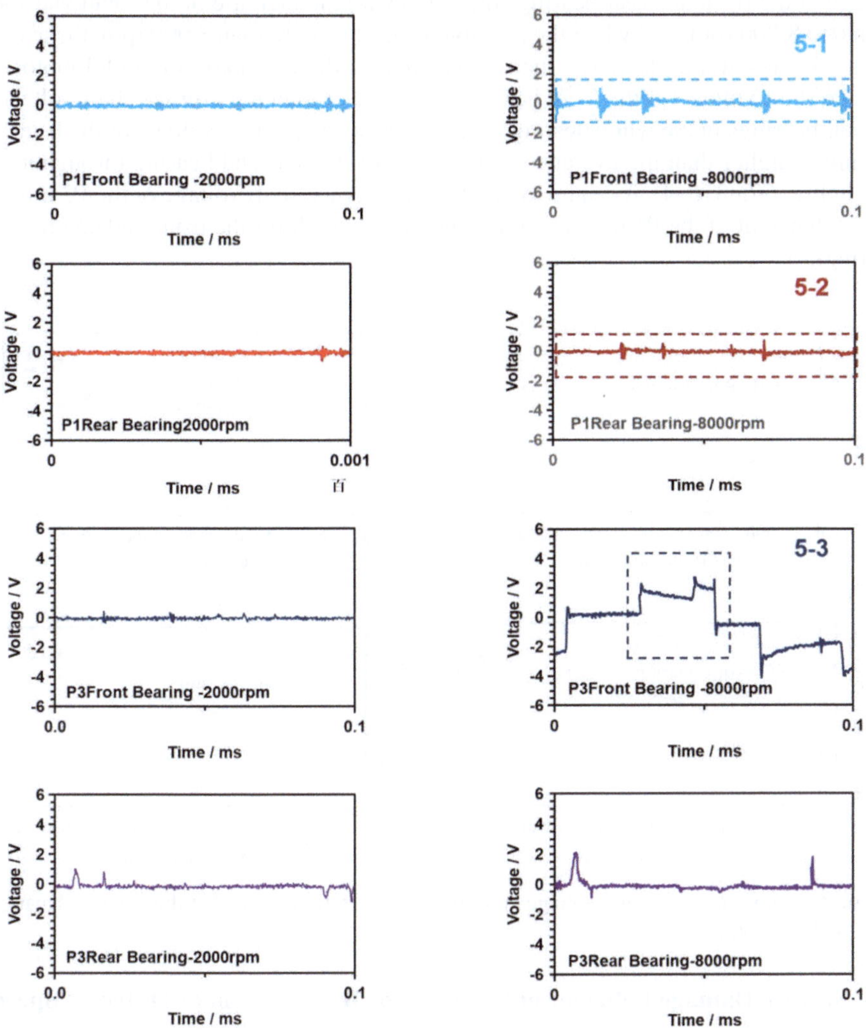

Fig. 5. Detailed analysis of voltage characteristics before and after P1 and P3 motors.

In order to have a clearer understanding of the relationship between the speed of the drive motor and the shaft voltage, the average shaft voltage is obtained from the shaft voltage generated in 0–0.4ms at our different speeds as shown in Fig. 6. Figure 6 (6–1) shows the average circumferential voltage of the front and rear end bearings of the P1 motor. It is easy to see that the average shaft voltage has been in an upward trend with the increase of rotational speed, and the growth rate of the average shaft voltage increases when it is at a high rotational speed (8,000 rpm–12,000 rpm), and the average shaft voltage of the front-end bearings is higher than that of the rear end bearings throughout the entire testing process. Figure 6 (6–2) shows the average circumferential voltage of the front and rear bearings of the P3 motor, and it can be observed that the average shaft voltage of the front and rear bearings increases with the increase of the rotational speed from 0 to 8,000 rpm, and when the rotational speed is more than 8,000 rpm, the average shaft voltage starts to decrease; the same as that of the P1 motor, in the P3 motor, the average shaft voltage of the front bearings was higher than the average shaft voltage of the rear bearings in the whole testing process. The average shaft voltage of the front end bearing is higher than the average shaft voltage of the rear end bearing throughout the test. In Fig. 6(6–3, 6–4), it can be found that the average shaft voltage of the P3 motor is higher than that of the P1 motor throughout the test, both for the front-end and rear-end bearings.

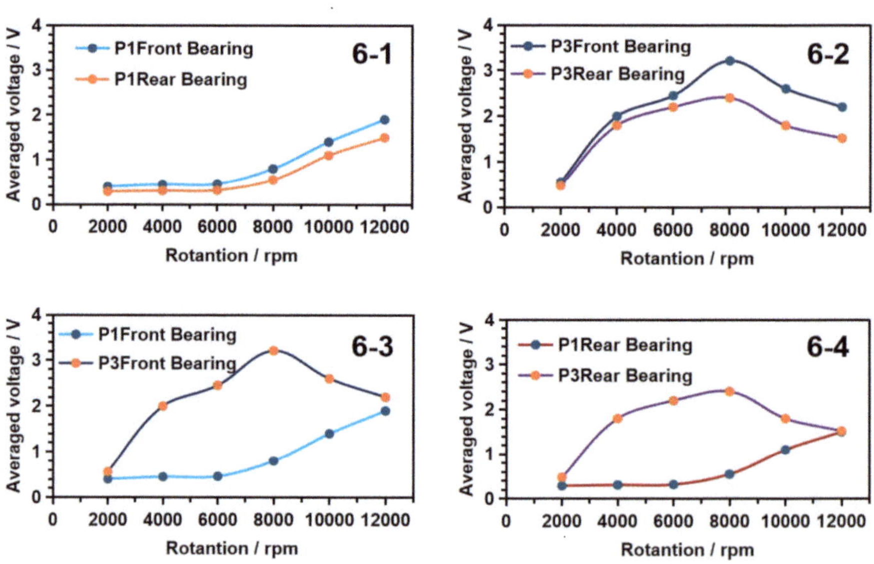

Fig. 6. Change rule of average voltage amplitude with rotational speed at the front and rear ends of P1 and P3 motors.

3.2 Surface Damage Behavior of Drive Motor Bearings Under Different Speeds

After completing the entire test at a constant speed, the surface damage caused by electrical discharge on the bearing surface was examined. Figure 7 shows the surface damage on the outer rings of the front and rear bearings of the P1 motor after the entire test. It can be seen that there was no damage to the outer rings of the front and rear bearings of the P1 motor. Discharge damage to the bearing surface after completing the entire test at constant rotational speed, the damage to the outer raceway of the front-end bearing of the P3 motor is shown in Fig. 8, where the difference in damage between the outer track of the front and rear-end bearings is clearly observed; the outer track of the front-end bearing is characterized by prominent bright streaks and dark streaks shown in Fig. 8 (8–1). With the appearance of pitting damage, the bright streaks exhibit significant electrical pitting as shown in Fig. 8 (8–2), which indicates severe discharge damage. In contrast, the rear end bearing outer track has no visible grooves nor electrical pitting as shown in Fig. 8 (8–3, 8–4).

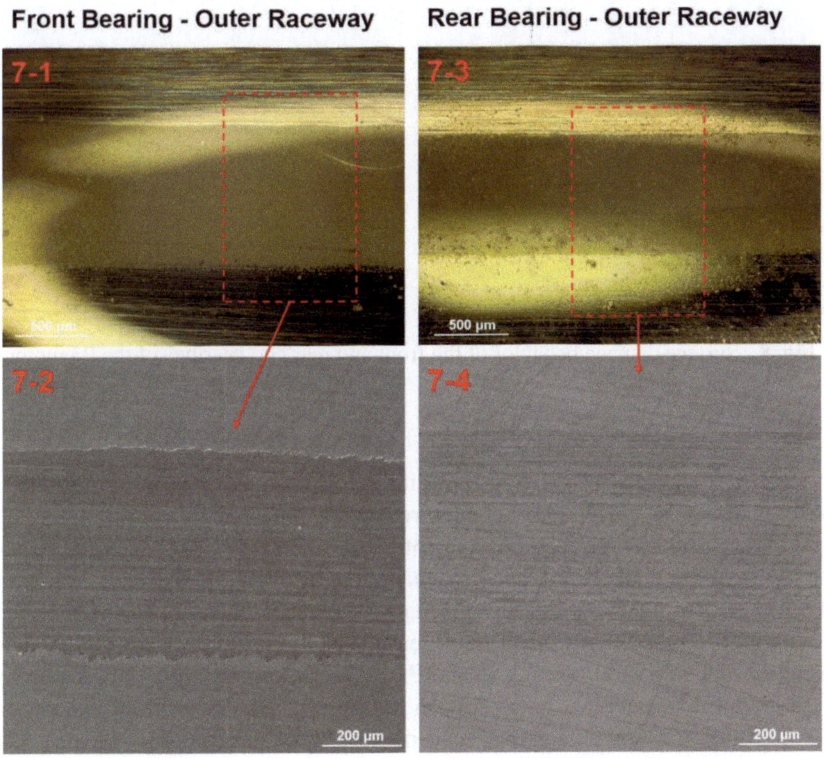

Fig. 7. Surface damage behavior of the front and rear end bearings of the P1 motor.

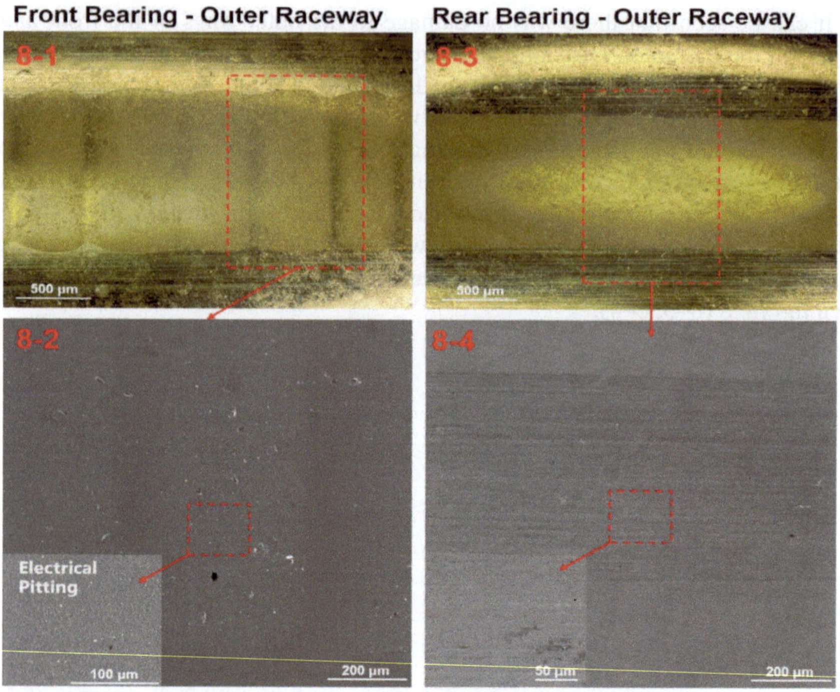

Fig. 8. Surface damage behavior of the front and rear end bearings of the P3 motor.

4 Conclusion

In this work, we performed discharge experiments on an actual electric vehicle (EV) DHT motor to examine the correlation between rotational speed, shaft voltage and discharge damage in the front and rear end bearings of P1 and P3 motors. The results showed that the P1 and P3 motors produced fewer discharges at low rotational speeds (<6000 rpm.) The P3 motor exhibited more severe discharge behavior during operation, the average shaft voltage produced by the front-end bearing of the P3 motor was higher than that produced by the front-end bearing of the P1 motor at any rotational speed throughout the experiment, and galvanic corrosion on the outer track of the front-end bearing of the P3 motor was more severe than that of the P1 motor front-end bearing is more serious. Moreover, both P1 and P3 motors showed more severe discharge behavior of the front-end bearing than the rear-end bearing throughout the experiment. The results demonstrate that the high amplitude voltage inside the bearing is the key reason for the discharge damage behavior. These results provide certain scientific basis and data reference for the reduction of bearing damage and prolongation of drive motor life in practical engineering applications of new energy vehicle drive motors.

References

1. He, F., Xie, G., Luo, J.: Electrical bearing failures in electric vehicles, Revised 26 November 2019, Accepted 17 December 2019

2. Ibrahim, M., Järg, O., Seppago, R., Rassõlkin, A.: Performance optimization of a high-speed permanent magnet synchronous motor drive system for formula electric vehicle application. Sensors **25**(10), 3156 (2025)
3. Zheng, X., Lin, H., Li, Y., Wang, J., Wen, Q.: Analysis and optimization design of a brushless power feedback PM adjustable speed drive with bilayer wound rotor. Actuators **14**(5), 241 (2025)
4. He, F., Xie, G.X., Luo, J.B.: Electrical bearing falures in lectric vehile, riction 8, 4–28 (2020)
5. Teshale, A., Biru, G.: Overview of shaft voltage and bearing current mitigation methods applied on the victim machine. Electr. Eng. **104**, 1727–1739 (2024)
6. Xiao, S., Zhu, T., Wu, G., et al.: Electrical damage mechanism analysis of traction motor bearings on high-speed trains considering overvoltage impulse. Int. J. Rail Transp. **13**(4), 839–863 (2025)
7. Jie, H., Songhua, L., Zhongxian, X., Yonghua, W., Weinan, W., Guangda, S.: Experimental study on electric corrosion damage of bearing and solution. Proc. Inst. Mech. Eng. Part C J. Mech. Eng. Sci. **236**(19), 10349–10358 (2022)
8. Êvo, M.T.A., Alzamora, A.M., Zaparoli, I.O., et al.: Inverter-induced bearing currents: a thorough study of the cause-and-effect chains. IEEE Ind. Appl. Mag. **29**(3), 57–66 (2023)
9. Sun, Y.M., Song, C.F., Sun, Y.M., et al.: Effect of rolling speed on electrical damage of GCr15 during tribology with current. J. Henan Univ. Sci. Technol. (Natural Science) **40**(3), 6–10 (2019)
10. Suzumura, J.: Prevention of electrical pitting on rolling bearings by electrically conductive grease. Quart Rep RTRI **57**(1), 42–47 (2016)
11. Khan, K., Gyllensten, F.: Experimental investigation of bearing currents in low voltage motors. In: IEEE XIII International Conference on Electrical Machines (ICEMS), Alexandroupoli, Greece, pp. 218–224 (2018)
12. Schuster, M., Springer, J., Binder, A.: Comparison of a 1.1 kW-induction machine and a 1.5 kW-PMSM regarding common-mode bearing currents. In: 2018 International Symposium on Power Electronics, Electrical Drives, Automation and Motion (SPEEDAM), Amalfi, Italy, pp. 1–6 (2018)
13. Tu, W., Wang, H., Liu, C., Hu, D., Yu, W.: Influence of temperature on dynamic contact characteristics of oil-jet lubricated rolling bearings. Proc. Inst. Mech. Eng. Part K J. Multi-body Dyn. **238**(2), 167–181 (2024)
14. Grujičić, R., Tica, M., Stojanović, B., Ivanović, L., Mitrović, R., Tomović, R.: Analysis of impact of shaft speed and external load on the radial ball bearing lubrication regimes. FME Trans. **50**(1) (2022)
15. Chetti, B., Crosby, W.A.: Preload effects on the static characteristics of three-lobe journal bearings lubricated with a couple stress fluid. Indust. Lubrication Tribol. **71**(10), 1136–1143 (2019)

Research on Synchronous Lifting Cloud Control System Based on Wireless Local Area Network

Yulin Pu[1], Meng Yang[1,2](✉), Zhicheng Yuan[1,2], Yongming Bian[1,2], and Guangjun Liu[1,2]

[1] School of Mechanical Engineering, Tongji University, Shanghai 201804, China
yangmeng@tongji.edu.cn
[2] National Engineering Technology Research Center for Prefabrication Construction in Civil Engineering, Tongji University, Shanghai 200092, China

Abstract. With the continuous expansion of the application field of hydraulic synchronous lifting technology, the traditional wired control method has been difficult to adapt to some complex construction environments, and the development of information technology has also made the synchronous lifting technology begin to change in the direction of digitalization and intelligence. In order to improve the existing synchronous lifting control network structure, a cloud-based synchronous lifting control system based on wireless local area network is proposed. The synchronous lifting wireless communication system is built by AC + AP networking architecture, and the synchronous lifting pump station system is connected to the LAN through WiFi-CAN devices; the pump station control host computer is designed and developed by using B/S architecture and front-end/back-end separation technology, which realizes the remote control and monitoring of the synchronous lifting pump station. Finally, the effectiveness and feasibility of the synchronous hoisting cloud control system are verified through the experiment of connecting the pump stations in the field. Finally, the effectiveness and feasibility of synchronized lifting cloud control system is verified through the experiment of actual pumping station.

Keywords: WLAN · Synchronous Lifting Control System · Host Computer

1 Introduction

Synchronous lifting technology is a technology with good application prospects, which is widely used in construction engineering, bridge engineering and the installation of large-scale electromechanical equipment [1]. Compared with the increasingly mature application of synchronous lifting technology, the development of construction process monitoring and control methods is slow, and the traditional monitoring methods failing to meet the increasingly complex construction scenarios [2]. In the construction process of synchronous lifting, PLC or dedicated industrial control machine is often used to monitor the operation of the construction equipment, and this kind of monitoring method has a low level of informatization, which is difficult to adapt to the development trend of synchronous lifting technology clustering, networking and digitization.

S. Halgamuge et al. (Eds.): ICACMVE 2025, LNME, pp. 411–418, 2026.
https://doi.org/10.1007/978-981-95-6911-3_40

The use of advanced measurement equipment and emerging technologies to achieve the level of informationization and intelligence of the simultaneous lifting construction and monitoring process is an important research direction at present. Wei et al. adopted the measurement mode of combining BIM software and total station for the lifting process of shaped metal roofing mesh frame, and obtained the deformation amount of metal roofing mesh frame and uneven lifting height value during the lifting process of shaped metal roofing mesh frame through the prism-free mode tracking and monitoring of the total station to understand the actual deformation of the mesh frame [3]. Liang et al. utilized the 3D laser scanner to Non-contact automatic multi-target displacement monitoring in the main beam jacking project was realized [4]. Zhang et al. proposed a synchronized lifting construction cloud platform system, which collects construction process data and sends them to the remote end based on the side cloud collaborative architecture to realize the remote monitoring of the construction site [5].

This paper proposes a synchronous lifting cloud control system based on a wireless local area network (WLAN). With AC + AP networking as the core, a synchronous lifting wireless communication system is built. Synchronous lifting cloud control host computer is designed and developed by using front-end and back-end separation technology to realize synchronous lifting remote control and monitoring. Through the experiment of connecting the pump station, it verifies the effectiveness and feasibility of remote control of the pump station through WLAN between the pump station control host computer and the back-end server.

2 Overall Design Scheme

The synchronized lifting cloud control system architecture consists of three parts: the cloud, human-machine interface (HMI) and hydraulic actuation system.

The cloud includes WLAN and server, which mainly realizes wireless communication, equipment interaction and data processing. The HMI includes control and monitoring functions, allowing operators to remotely control the hydraulic pump station and monitor the cylinder operation status through the host computer equipment such as smartphones, tablets and computers. The hydraulic actuating system includes WiFi-CAN devices, lifting cylinders, cylinder sensors and pump stations, to realize command reception and control cylinder movement.

3 Wireless Communication System Architecture

The network topology of the wireless communication system is shown in Fig. 1, which adopts the AC + AP networking method to build a WLAN, and manages multiple access points (APs) through a access controller (AC) to realize the unified management and control of the whole network, which is characterized by high scalability and fast response speed [6].

The server is the data center of the synchronous lifting cloud control system, which is used to realize the data processing and information interaction between the host computer and pump station. The WiFi-CAN devices realize the mutual conversion of WiFi signals and CAN signals, which is used to send sensor data and receive the control commands of pump station. The host computer connects to the WLAN through WiFi, controlling cylinder movements through the control platform while displaying cylinder operational status through the monitoring platform.

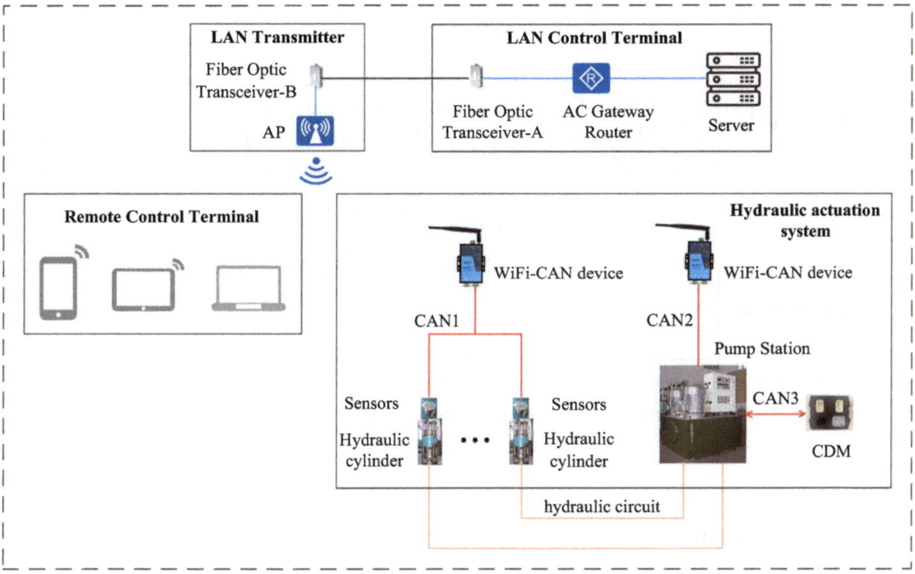

Fig. 1. Network topology diagram of wireless communication system.

4 Remote Control System Architecture

Figure 2 shows the software architecture of the synchronous lifting remote control system, which consists of three parts: the execution side, cloud side and the control side.

The execution side collects the status information of the cylinder, anchorage and pumping station through sensors, and receives control commands through the control drive module (CDM) of pumping station. The communication between the execution side and cloud side is established through TCP socket protocol.

The cloud side realizes data processing, control frame definition and data storage functions. Sensor data is processed and then forwarded to monitoring platform. Generate CAN message control command set according to the data flow definition of WiFi-CAN device and pump station protocol. Store the data and operation records during the operation of the cylinder and pump station through the database.

The control side mainly consists of control interface and monitoring interface. The control interface realizes remote control of the pumping station system, and the monitoring interface displays the running status of cylinder and pumping station system in real time. The control side and cloud side establish communication through HTTP and WebSocket protocols.

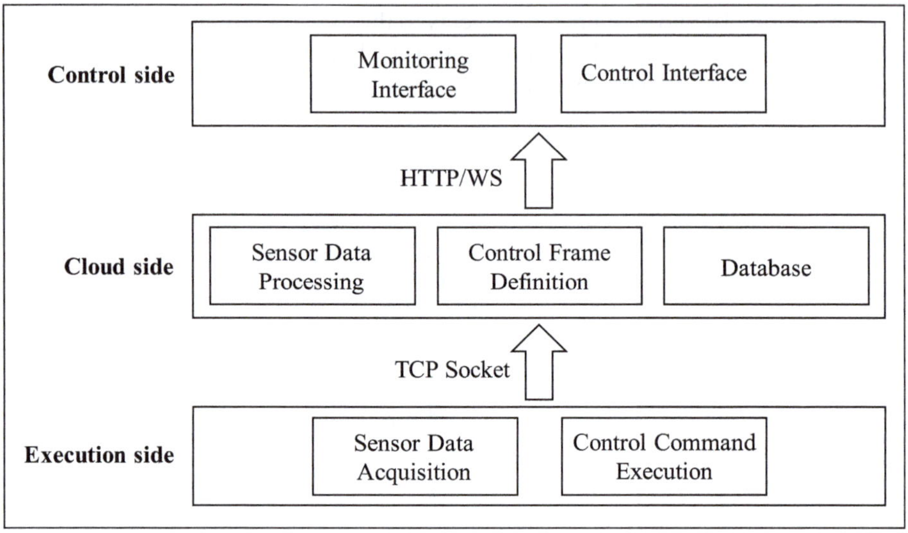

Fig. 2. Software architecture of synchronous lifting remote control system.

4.1 Control System

Based on front-end and back-end separation technology development pumping station remote control system. The front-end is based on uni-app framework, which is developed using Vue.js with excellent compatibility. The back-end is based on Spring Boot framework, which is developed using annotations and is the latest Java Web back-end framework developed from Spring framework [7]. For database design, the system utilizes the MySQL database management system along with the Navicat database management tool.

The pump station remote control program is designed to realize the mapping relationship between the control components and CAN messages. When the control interface performs a control operation, the server responds by sending the corresponding CAN messages to the WiFi-CAN device, i.e., sending control commands to the pump station. Figure 3 shows the data flow definition of WiFi-CAN module, a data frame contains 20 bytes. According to the CAN protocol, only the frame ID and frame data are specified, so the remaining byte values are defined by the program.

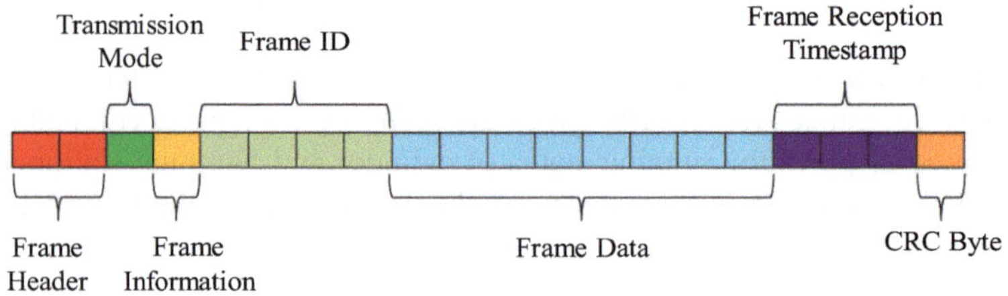

Fig. 3. WiFi-CAN device data flow definition.

After completing the development of the pump station control system, package and deploy it and carry out the control simulation test. Press the control button in the control interface, the server responds and sends CAN messages to the WiFi-CAN device, which is connected to the test computer through the USBCAN module, and the CAN messages received by the WiFi-CAN device are displayed through the CANTest software. The results of the test are shown in Fig. 4, which indicates that the WiFi-CAN device receives data successfully.

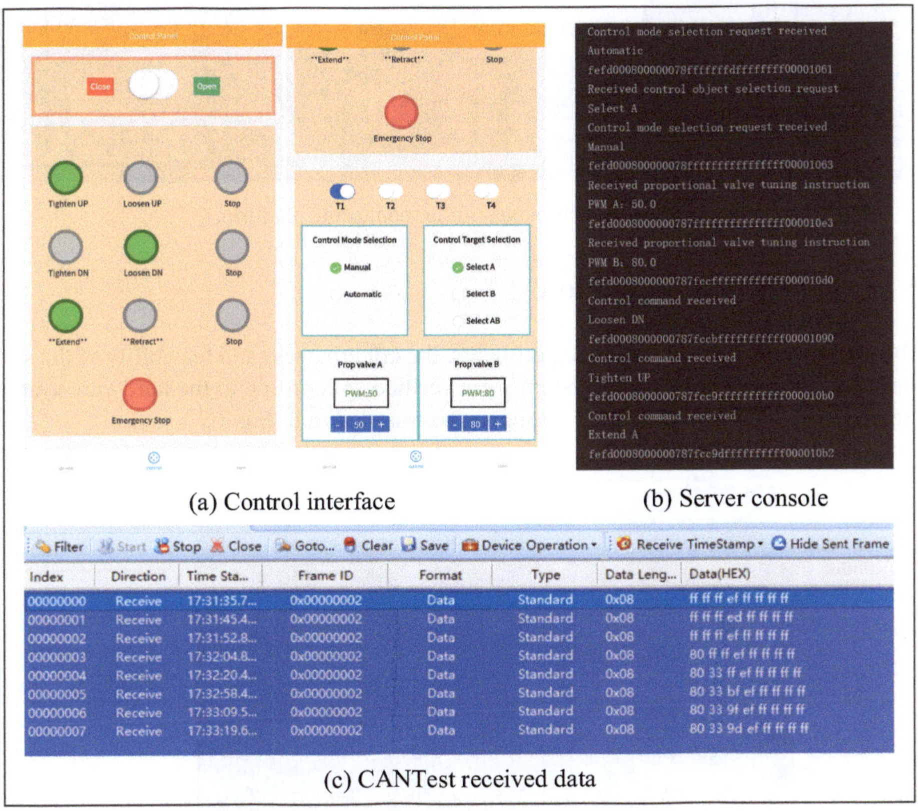

(a) Control interface (b) Server console

(c) CANTest received data

Fig. 4. Control system communication test.

4.2 Monitoring System

Synchronous lifting monitoring system is developed based on B/S architecture, and the cloud sends the analyzed and processed sensor data in JSON format to the front-end, using real-time data to combine the actual working conditions of the cylinder with the virtual model, and at the same time, key data such as anchorage status, cylinder expansion and contraction status, cylinder stroke, cylinder pressure data, and number of times of cylinder expansion and contraction are visualized, as shown in Fig. 5.

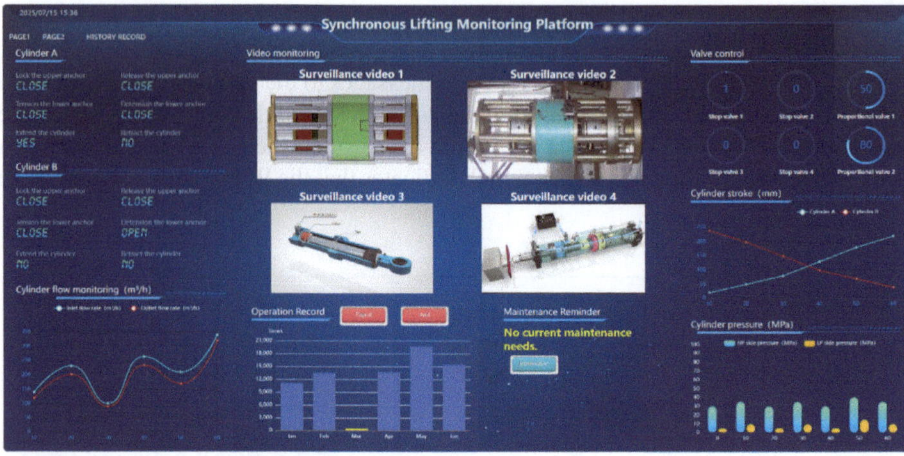

Fig. 5. Synchronous lifting monitoring platform.

5 Experimental Validation

Through practical pump station experiments, the effectiveness and feasibility of the synchronous lifting cloud control system were verified. According to the hardware system architecture, the field equipment arrangement was shown in Fig. 6.

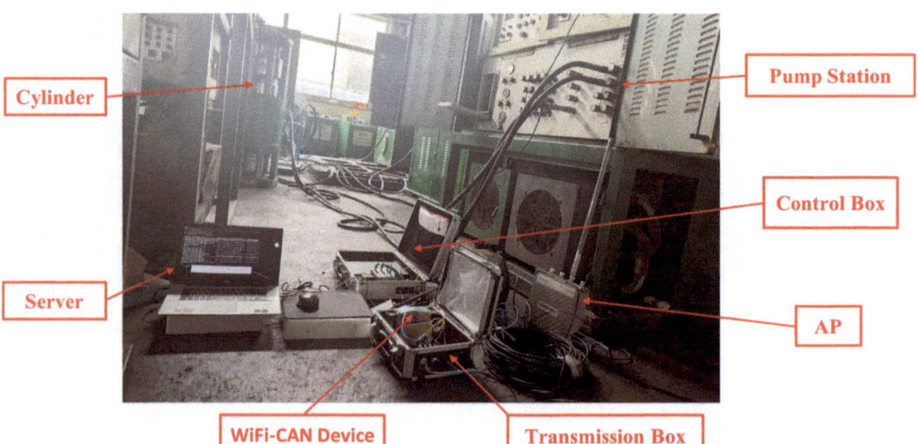

Fig. 6. On-site equipment arrangement.

Turn on the pump station, carry out the corresponding "Extend", "Retract" and "Stop" operation in the control interface, observe the cylinder movement state, and use the cylinder stroke sensor to read the cylinder's elongation. The control effect is shown in Fig. 7.

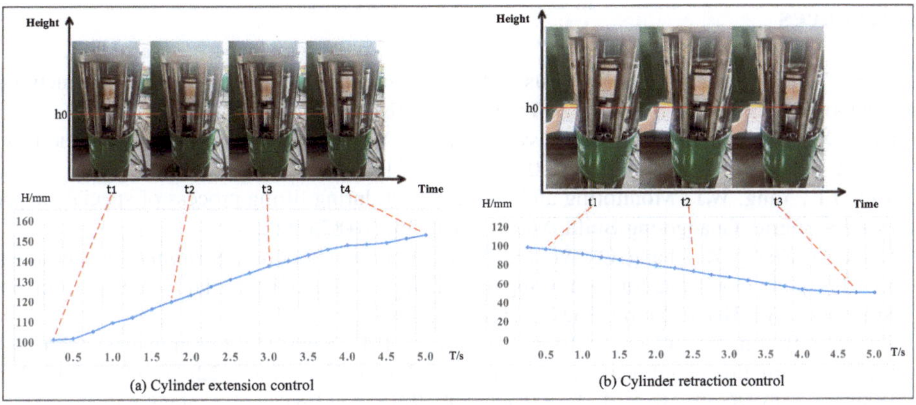

Fig. 7. Cylinder movement control.

The experimental results show that when the control interface presses the control button, the cylinder can respond to the real-time action, which proves that the frame ID and frame data in the CAN message sent by the server are in line with the control protocol of the pumping station, and that it is able to realize the remote control of pump station through WLAN, which verifies the feasibility of the system.

6 Conclusions

- This paper establishes a synchronous lifting wireless communication network with AC + AP as the core networking architecture.
- Based on the front-end and back-end separation technology and B/S architecture, the pump station control host computer was developed to realize the remote control and real-time monitoring of the synchronous lifting system.
- Conducted experiments on the actual pumping station and successfully realized the remote action control of cylinders by applying the host computer via WLAN, which verified the effectiveness of the synchronous lifting cloud control system.

Acknowledgments. We express our profound gratitude to the anonymous reviewers for their insightful comments and constructive suggestions, which have significantly contributed to the enhancement of this manuscript. This research was supported by the National Natural Science Foundation of China (52205279), the National Key Research and Development Program of China (2023YFF0613200 & 2023YFC3806603) and the Top Discipline Plan of Shanghai Universities-Class I.

Disclosure of Interests. The authors have no competing interests to declare that are relevant to the content of this article.

References

1. Bian, Y.M.: Hydraulic Synchronous Lifting Technology for Large-scale Constructional Components. Shanghai Scientific & Technical Publishers, Shanghai (2015)
2. Chen, X.J.: Application of hydraulic synchronous lifting technology in gas holder demolition. Indust. Safety Environ. Protect. **47**(10), 71–74 (2021)
3. Wei, Y.F., Yang, W.J.: Monitoring and measurement during lifting process of special-shaped metal roof grid. Guangdong Build. Mater. 38(6), 36–38+82 (2022)
4. Liang, D., Zhang, Z., Zhang, Q., Wu, E., Huang, H.: Fast extraction algorithm of planar targets based on point cloud data for monitoring the synchronization of bridge jacking displacements. Struct. Control. Health Monit. **2024**(1), 9687805 (2024)
5. Zhang, X.R., Lu, Y.C., Chen, L., Bian, Y.M., Wang, R.F., Yang, M.: Research on synchronous lifting construction cloud platform system. Chin. J. Construct. Mach. **22**(3), 363–368 (2024)
6. Iacoboaiea, O., Krolikowski, J., Ben Houidi, Z., Rossi, D.: Real-time channel management in Wlans: deep reinforcement learning versus heuristics. In: 2021 IFIP Networking Conference (IFIP Networking), pp. 1–9. Espoo and Helsinki, Finland (2021)
7. Mythily, M., Arun Raj, A.S., Joseph, I.T.: An analysis of the significance of spring boot in the market. In: 2022 International Conference on Inventive Computation Technologies (ICICT), Nepal, pp. 1277–1281 (2022)

Research Status and Development Trends of Automotive Electronically Controlled Suspension Systems

Rui Mao[1], Qile Li[1], Kaize Wu[1], Xingxing Xu[2], and Xiaomin Cheng[2(✉)]

[1] Ningbo Tuopu Group Co., Ltd., Ningbo, Z.J. 315806, China
[2] Hangzhou City University, Hangzhou, Z.J. 310015, China
chengxiaomin@hzcu.edu.cn

Abstract. To systematically clarify the research context and future directions of automotive electronically controlled suspensions, this paper reviews the latest progress in three categories of active/semi-active suspensions—hydraulic, pneumatic, and electromagnetic. It categorically describes the composition, principles, and application status of three electronically controlled suspension systems: electro-hydraulic active suspension, electro-pneumatic active suspension, and electromagnetic active suspension. It analyzes classic control methods currently applied in these systems, including PID control, state-feedback control, and neural-network control, as well as modern control methods such as robust control, sliding-mode control, and adaptive control. A brief overview of domestic and international research achievements related to these control methods is provided. Finally, the paper summarizes future research directions and development trends in active suspension control strategies.

Keywords: Active suspension system · Control method · Fusion control

1 Introduction

The primary role of an automotive suspension is to bridge the force/torque transmission path between the wheels and the chassis while attenuating stochastic road excitations transmitted to the vehicle body. Three key metrics for evaluating active suspensions are: vehicle-body vertical acceleration, tire dynamic load, and suspension working travel. Reducing body vertical acceleration is critical for improving ride comfort, whereas effectively suppressing tire dynamic load and ensuring suspension travel remains within allowable limits are essential for maintaining excellent handling safety [1].

Passive suspensions, typically composed of a spring-damper combination, feature simple mechanical structures and low cost, providing adequate vibration isolation on well-paved roads. However, they possess fixed spring stiffness and damping coefficients, unable to adapt dynamically to real-time road conditions. Under stochastic road surfaces, their damping performance deteriorates significantly. Softening the spring improves ride comfort yet undermines handling stability, whereas stiffening it enhances stability at the expense of ride comfort.

S. Halgamuge et al. (Eds.): ICACMVE 2025, LNME, pp. 419–426, 2026.
https://doi.org/10.1007/978-981-95-6911-3_41

Electronically controlled suspensions are classified into semi-active and active types. Semi-active suspensions adjust either spring stiffness or damper force as required, while active suspensions can vary both stiffness and damping parameters in real time, enabling dynamic responses to external excitations. This paper reviews the current research status of mainstream automotive electronically controlled suspensions and their classic control methods, followed by a summary and outlook on different suspension systems and control strategies.

2 Research Progress in Automotive Electronically Controlled Suspension Systems

An automotive electronically controlled active suspension system primarily consists of sensors and control switches, an electronic control unit (ECU), and actuators, forming a closed-loop control system. Sensors continuously monitor vehicle motion states and road excitation inputs, transmitting signals to the ECU. The ECU performs real-time computations and issues control commands to damper controllers. Based on these commands, actuators actively adjust suspension stiffness, damping coefficients, and ride height, dynamically optimizing system parameters. This closed-loop mechanism aims to keep the suspension system near optimal damping continuously while suppressing body attitude motions (pitch, roll, heave). Currently, mainstream electronically controlled active suspensions achieve parameter adjustability via pneumatic, hydraulic, or magnetorheological (MR) actuators. Thus, the core feature of active suspensions lies in their ability to respond dynamically to external inputs (e.g., road conditions) and vehicle states [2].

2.1 Electro-Hydraulic Active Suspension

The principle of the hydraulic control system is shown in Fig. 1. Dou Wenbo [3] established a half-car active suspension mathematical model using a hydraulic servo system as the actuator and designed two control strategies. Co-simulation results showed that the PPC controller reduced vertical acceleration and pitch angular acceleration by over 90% compared to PID, significantly improving comfort. Zeng Xi [4] proposed an intelligent hydraulic suspension architecture integrating sensors, ECU, and hydraulic actuators, enabling real-time adjustment of suspension stiffness/damping to markedly enhance handling and comfort. Nie Jiamei et al. [5] introduced a hydraulic inerter suspension combining gas-oil springs and inerters, integrating spring, damper, and inerter functions into a single hydraulic structure to address structural complexity. Vehicle model simulations revealed that compared to dual-chamber hydraulic suspensions, the hydraulic inerter suspension reduced body acceleration by 9.2%–17%, suspension travel by 5.1%–12.5%, and tire dynamic load by 10.8%–18.5%. Zhang Xiang [6] developed a "model-in-the-loop (MHIL) test platform" based on a four-poster road simulator for a vehicle-specific hydraulic active suspension. The platform's test results under four conditions—height adaptation with speed, steady-state cornering, slalom, and double-lane-change—closely matched real-vehicle data, validating its authenticity, accuracy, and engineering applicability.

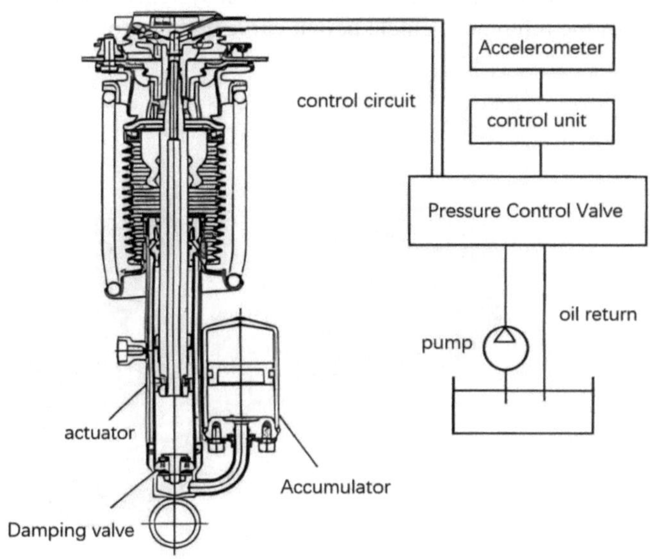

Fig. 1. Structural principle of pressure actuator in hydraulic suspension system.

2.2 Electro-Pneumatic Active Suspension

The structural composition of the electronically controlled air suspension system is shown in Fig. 2. Yi et al. [7] proposed an air spring power system for wheel hub motor electric vehicles. By establishing an air spring wheel hub motor electric vehicle dynamic model that considers the electromechanical coupling effect, a quasi infinite horizon nonlinear model predictive control is designed to improve the longitudinal and longitudinal dynamic performance. The results indicate that the proposed control algorithm can improve ride comfort, reduce motor vibration, and enhance longitudinal and vertical dynamic performance. Huang et al. [8] used adaptive differential evolution algorithm to dynamically control the air suspension of automobiles. The system adopts adaptive differential evolution algorithm as the core control strategy, dynamically adjusting the parameters and control strategies of the suspension system to adapt to different road conditions and driving conditions. Under medium speed conditions, the root mean square value of vertical acceleration at the center of mass of the vehicle decreased by 15.2%. The root mean square value of the dynamic deflection of the four-wheel suspension decreased by 8.46%. The root mean square value of the dynamic load on the four-wheel tires only slightly increased by 0.47%. Yin et al. [9] designed a small parameter continuous finite time feedback stable controller for a quarter car model, and the results showed that the height adjustment system can achieve finite time stability. Finally, the simulation results under sinusoidal and random disturbances show that compared with traditional partial discharge control, the designed finite time controller not only ensures faster response speed and better control accuracy of the system, but also achieves better anti-interference performance. Zhang et al. [10] proposed an adaptive control algorithm to address the uncertainty issues of nonlinear active air suspension systems, which solves the problems of vehicle height adjustment and ride comfort in such systems. The algorithm achieves precise tracking of the reference vehicle height curve and improves ride comfort.

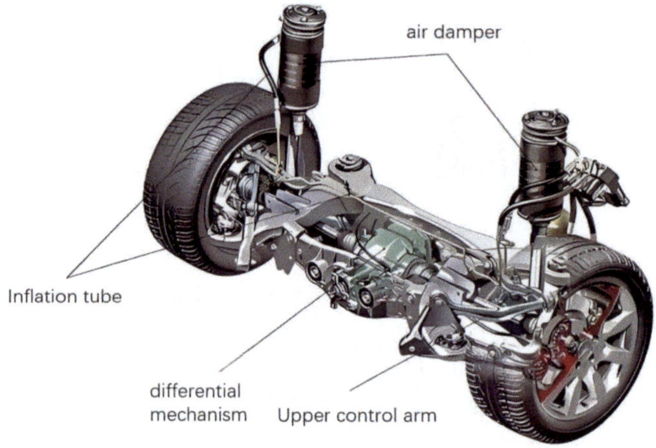

Fig. 2. Structural diagram of air suspension.

2.3 Electromagnetic Active Suspension

The structural composition of the electromagnetic suspension system is shown in Fig. 3. Feng Guizhen et al. [11] constructed a quarter-car semi-active suspension model incorporating electromagnetic and road excitations, designing a robust H∞ state-feedback controller to handle parametric uncertainties. The H∞ algorithm outperformed passive and skyhook controls under stochastic and impulse roads: body acceleration RMS reduced by 27.4%, suspension deflection by 44.7%, and tire dynamic load by 19.3%. Pei Qianqian et al. [12] proposed a magnetorheological (MR) semi-active suspension based on fuzzy-PID control. Using a quarter-car semi-active model and fuzzy rules to optimize PID parameters, simulations showed reductions in body vertical acceleration, suspension travel, and tire dynamic load, outperforming conventional PID. Xie Haifeng [13] developed an improved hyperbolic-tangent MR damper model and air-spring model, designing an adaptive backstepping controller based on radial-basis neural networks. Simulations confirmed the strategy stabilized ride-height tracking error within 1 mm under mass variations and time delays, enhancing stability and comfort. Zhou Min et al. [14] built a quarter-car semi-active suspension model, designing a PID controller to adjust damping force. Compared to passive suspension on C-class roads, the PID-controlled semi-active suspension reduced body acceleration RMS by 55.78%, suspension deflection by 12.8%, and tire dynamic load by 24.64%, significantly improving ride comfort within effective travel.

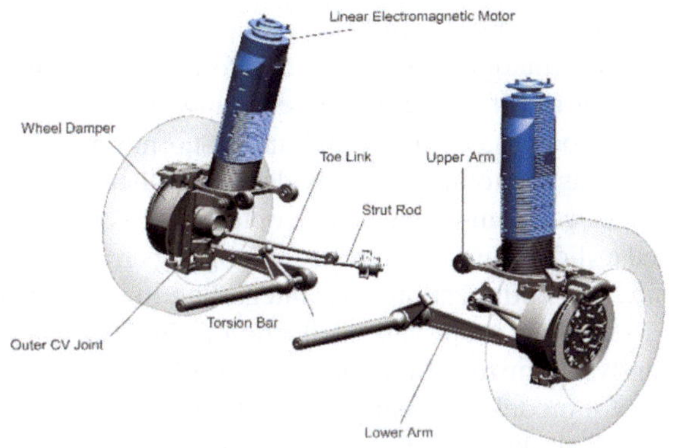

Fig. 3. Electromagnetic active suspension developed by Bose Corporation.

3 Typical Control Methods for Electronically Controlled Suspension Systems

In automotive design, advanced control strategies in electronically controlled suspensions are widely regarded as key to enhancing vehicle dynamics. Applied algorithms are highly diverse, broadly categorized into classic methods—PID, state-feedback, and neural-network control—and modern methods—robust, sliding-mode, and adaptive control.

3.1 Classic Control Methods

Nie et al. [15] proposed a hierarchical control strategy for a full-vehicle model to suppress multi-DOF motions, employing a double-skyhook-based fuzzy control method. Compared to double-skyhook suspension, ride comfort was maintained while handling safety improved, proving the necessity and superiority of hierarchical control. Sun Yudun et al. [16] designed an active-suspension fuzzy-PID controller, integrating PID and fuzzy algorithms for a half-car four-DOF model. Simulations under measured ditch-road excitation showed the fuzzy-PID controller reduced body vertical acceleration by 49.46% (C-class road) and pitch angular acceleration by 41.31% (ditch road), outperforming standalone algorithms. Uddin et al. [17] proposed an optimal control design for a half-car active suspension, achieving a 94.81% ride-comfort improvement over passive systems. Du et al. [18] developed a nonlinear extended-state-observer-based output-feedback stabilizing controller for a half-car active suspension, demonstrating robustness and effectiveness. Li Jiapeng [19] built a quarter-car two-DOF active-suspension model, introducing a fuzzy-neural-network structure to adapt PID parameters in real time. Results showed body acceleration reduced by 37.4%, suspension deflection by 14.6%, and tire displacement by 27.2% on stochastic roads, balancing comfort, safety, and real-time performance.

3.2 Modern Control Methods

Jin Xianjian et al. [20] proposed a μ-synthesis robust control strategy for an active suspension system with directional hub-mounted vibration absorbers. Simulations on stochastic and bump roads showed significant reductions in body acceleration, suspension travel, and tire deformation, improving comfort. Mona et al. [21] introduced a robust cost controller for a half-car active suspension with gain stochastic fluctuations, using Markov-process-coordinated gains and an H2-norm-constrained optimal multi-modal design to ensure asymptotic stability under given disturbance attenuation. Wang et al. [22] built a quarter-car semi-active model, optimizing a cabin cover model as a reference, and designed a sliding-mode controller combining error tracking and saturation-function convergence rate optimization, later integrating fuzzy control. Results showed body acceleration (−26.9%), suspension deflection (−36.7%), and tire dynamic load (−28.4%) significantly optimized. Kou Farong et al. [23] coupled a quarter-car active suspension with a motor model, proposing a dual-sliding-surface adaptive sliding-mode strategy referencing ideal skyhook. Compared to traditional skyhook, body acceleration dropped 16.1% and suspension deflection improved 4.2%. Wang Xianming [24] proposed three model-free adaptive control (MFAC) variants—improved MFAC, MFAC predictive control, and full-vehicle pseudo-inverse decoupling control. Improved MFAC enhanced body acceleration and pitch angular acceleration by 5%–7% over MFAC predictive control, which itself achieved > 30% comfort improvement under stochastic/impulse roads, outperforming improved MFAC. All three methods are viable for full-vehicle suspension control.

4 Conclusion

As a key component connecting the wheels and frame, the performance of the car suspension directly affects the ride comfort and handling safety of the vehicle. A passive suspension with fixed parameters is difficult to cope with complex road conditions; Electronic control suspension can dynamically adjust parameters and is divided into semi-active and active types. In electronic control suspension systems, active suspension achieves adjustable parameters through pneumatic, hydraulic, or magnetorheological actuators, forming closed-loop control. In terms of control methods, fuzzy control, PID and their fusion algorithms can effectively optimize indicators such as vertical acceleration of the vehicle body; Robust control, sliding mode control, adaptive control, etc. have shown excellent performance in improving system robustness and comfort. The current active suspension system is facing problems such as high energy consumption and low energy recovery rate. The control strategies of active suspension are mostly focused on the simulation test stage, and the variable price conditions set in simulation analysis tend to be idealized, lacking analysis and research on practical application situations. With the increasing number of active suspension control methods, the future should not be limited to simulation experiments alone, but should match the working conditions provided by experimental research equipment with actual working conditions, and the low energy consumption and energy recovery rate of active suspension systems are also research topics worth further exploration.

Acknowledgments. This study was funded by Major Scientific and Technological Special Projects of Zhejiang Province and Ningbo City (grant number 2024Z149, 2025Z012, 2024C01039, 2023Z032 and 2022Z071).

References

1. Shuo, M., Yongming, L., Shudong, Y.: A survey of control methods of automobile active suspension systems. Control Eng. China **31**(04), 695–702 (2024)
2. Hui, Y.: Present research and developing trends of automobile electronic control suspension. Equip. Manufact. Technol. **04**, 204–206 (2013)
3. Wenbo, D.: Research on Active Suspension Control Strategy Based on Hydraulic Servo System. Yanshan University (2024)
4. Xi, Z.: Design and implementation of intelligent hydraulic suspension system for automobile based on intelligent control. Auto Mainten. Repair **08**, 44–45 (2025)
5. Jiamei, N., Hao, W., Xiaoliang, Z., et al.: Simulation analysis and experiment research on hydraulic inerter suspension. Mach. Des. Manufact. Eng. **52**(02), 20–25 (2023)
6. Xiang, Z.: Design and Testing of Hydraulic Active Suspension System in Loop Test Bench Based on Four-Column. Shandong University (2024)
7. Yu, Y., Li, Z., Zhou, Y., et al.: A nonlinear model predictive control for air suspension in hub motor electric vehicle. Proc. Inst. Mech. Eng. Part D J. Automob. Eng. **239**(2–3), 564–585 (2025)
8. Huang, X.: Design of A-DE dynamic control system for automotive air suspension. Int. J. Veh. Struct. Syst. **16**(2), 280–286 (2024)
9. Chunfang, Y., Dean, Z., Jun, Z., et al.: Body height robust control of automotive air suspension system using finite-time approach. Asian J. Control **24**(2), 859–871 (2021)
10. Zhang, J., Yang, Y., Hu, C.: An adaptive controller design for nonlinear active air suspension systems with uncertainties. Mathematics **11**(12) (2023)
11. Guizhen, F., Zhengchuan, L., Xiangyu, H., et al.: Modeling and control of magnetorheological semi-active suspension for electric vehicle. J. Vibr. Measur. Diag. **45**(03), 438–445+617 (2025)
12. Qianqian, P., Yujie, H.: Research on fuzzy-PID control of semi-active suspension based on magnetorheological damper. New Technol. New Prod. China (11), 15–17 (2025)
13. Haifeng, X.: Research on Leveling and Posture Control Strategy Design of Automotive Active Air Suspension Eqiupped with Magnetorheological Damper. Guizhou Normal University (2024)
14. Min, Z., Yanlong, Lü.: Research on control of automotive magnetorheological semi-active suspension. J. Suihua Univ. **45**(05), 150–153 (2025)
15. Nie, J., Huang, J., Zhang, X., et al.: Hierarchical control for car attitude with double-skyhook semi-active ISD suspension. Proc. Inst. Mech. Eng. Part I J. Syst. Control Eng. **239**(7), 1180–1197 (2025)
16. Yudun, S., Xiaojia, W., Gen, L., et al.: Research on fuzzy PID control of vehicle active suspension for ditch pavement. Mod. Manufact. Eng. (09), 84–92 (2023)
17. Uddin, N., Manurung, A., Wijaya, A.N.R.: Optimal state-feedback control design of half-car active suspension system. IAENG Int. J. Appl. Math. **51**(3), 1–9 (2021)
18. Du, M., Zhao, D., Yang, M., et al.: Nonlinear extended state observer-based output-feedback stabilization control for uncertain nonlinear half-car active suspension systems. Nonlinear Dyn. **100**, 1–21 (2020)
19. Jiapeng, L.: Research on Ride Comfort Analysis and Control Strategy of Active Suspension Based on Fuzzy Neural Network. Hainan University (2023)
20. Xianjian, J., Jiadong, W., Liwei, X., et al.: μ-Synthesis robust control for active suspension of in-wheel-motor-driven electric vehicles. J. Mech. Eng. **60**(16), 259–269 (2024)
21. Mona, N.F., Fatemeh, K.: Robust guaranteed-cost control for half-vehicle active suspension systems subject to Markovian controller uncertainties. IETE J. Res. **69**(5), 2701–2709 (2023)

22. Wang, W., Xu, J., Zheng, J., et al.: Fuzzy sliding-mode control of automotive semi-active suspension based on optimization convergence law. Mech. Based Des. Struct. Mach. **52**(12), 9909–9922 (2024)
23. Farong, K., Jiahuan, W., Qiangqiang, L., et al.: Adaptive sliding mode control of active suspension for hub motor-driven vehicle. J. Vibr. Shock **44**(09), 188–194 (2025)
24. Xianming, W., Yuan, G., Wen, L., et al.: Research on model free adaptive control method for automotive active suspension system. J. Guangxi Univ. Sci. Technol. **34**(02), 76–83 (2023)

Positioning Control of Water Landing Point for Fire-Fighting Water Cannon Based on Fuzzy Algorithm

Zhangsheng Wang[1], Meng Yang[1](✉), Xinran Zhang[1], Yongming Bian[1], Zeyu Duan[2], and Chenchen Jiang[3]

[1] School of Mechanical Engineering, Tongji University, Shanghai 201804, China
yangmeng@tongji.edu.cn
[2] Department of Civil and Environmental Engineering, The Hong Kong University of Science and Technology, Hong Kong 999077, China
[3] Leon Robotics Co., Ltd, Shenzhen 518048, China

Abstract. The widespread use of large buildings has greatly promoted the development of fire-fighting water cannon, especially the positioning control of water landing point is an urgent problem to be solved. In this paper, the positioning error model of fire water cannon is established by analyzing the water jet trajectory and its dynamic equation. Then, an intelligent positioning algorithm of water landing point based on fuzzy theory is designed. To evaluate the performance of the proposed method, simulation studies under the MATLAB/Simulink are performed with initial conditions. Finally, simulation results show that the fire-fighting water cannon can control the position of water landing point accurately, and make it coincide with the position of flame burning point, proving its effectiveness and feasibility.

Keywords: Positioning Control · Water Jet Trajectory · Fire-Fighting Water Cannon · Fuzzy Algorithm

1 Introduction

Fire is a major threat to human life, causing not only property losses, but also casualties. According to statistics from the National Fire and Rescue Administration in China, the fire caused a total of 2001 deaths and 2665 injuries, resulting in a direct property loss of 7.74 billion yuan in 2024. Meanwhile, the United States, Australia and other countries have also witnessed major fire accidents. With global warming and various extreme weather appearing, it is noted that the fire will further threaten human lives and property. Consequently, there is no doubt that a reliable and effective method for the fire extinction is needed.

As is known to all, water is often used as an excellent fire inhibitor in most cases of non-electric fire. In terms of intelligent fire extinguishing technology, there have been numerous studies published recently. McNeil et al. detected fire through ultraviolet and

S. Halgamuge et al. (Eds.): ICACMVE 2025, LNME, pp. 427–437, 2026.
https://doi.org/10.1007/978-981-95-6911-3_42

infrared radiation sensors, and suppressed fire by visual positioning and feedback control [1, 2]. The positions of fire and water were determined by the IR vision system, based on the segmentation process and the real-time probabilistic classification method. The decision tree model is applied to develop feedback control law for the movements of nozzle, so that the water jet could reach the target position of fire. Chen et al. searched fire by computer vision theory via a CCD camera, which is fixed at the end of a fire monitor chamber [3]. The position of fire was calculated by the height and the pitch angle of the fire monitor. Then, the fire was extinguished by the continuous swing water jet in about 75°. In [4], the position of fire was obtained with the flame and smoke detection [4]. Afterwards, the fire was extinguished by moving the mobile device along a fixed path quickly and starting the water cannon to spray water. However, in the actual application of extinguishment, these methods described above cannot satisfy the requirements of response speed and control accuracy.

What's more, several studies also have developed various firefighting methods and designed systems, such as water mist system [5], automatic sprinkler system [6], and fire extinguishing system composed of fire-fighting water cannons [7]. In the past years, with the development of super-large cities all over the world, the use of large buildings has become more and more widespread. As a result, the severe effects caused by fire have been highlighted, and fire-fighting water cannon used for intelligent firefighting have also become research hotspots [8–10]. Xin carried out a series of experiments, and found that the water cannon responds to fire in large-open spaces more quickly than sprinklers [11]. Therefore, to further enhance the operational efficiency and practical applicability of fire-fighting water cannons, while minimizing casualties and property losses caused by fires to the greatest extent, it is of great significance to study the control technology of water jet landing point of fire-fighting water cannon.

This paper aims to realize the positioning control of water landing point for fire-fighting water cannon. Firstly, the mathematical model of water jet trajectory is established, which is used to predict its motion trajectory. Then, an intelligent positioning controller based on fuzzy algorithm of water landing point is designed. To evaluate the control performance of the proposed method, some simulation studies under the MATLAB/Simulink environment are performed. Finally, the effectiveness and feasibility of the proposed method are confirmed by the simulation results.

2 Mathematical Model of Water Jet Trajectory

Modeling the water jet trajectory of fire-fighting water cannon is to study the motion law of the jetted water fluid. The actual water jet trajectory research should be described as a three-dimensional problem. However, it is assumed that the horizontal rotation angle is determined in this paper, which can transform the water jet trajectory from a three-dimensional problem to a two-dimensional problem, and simplify the computation of the research problem greatly.

For the convenience of mathematical modeling, the effects of environmental factors such as temperature, humidity and wind speed on the trajectory of the water jet are all ignored. The water jet is divided into micro-element bodies of unit volume and unit flow as the research object. The movement trajectory of the water jet is composed of the positions which include the centroid of the micro-element body.

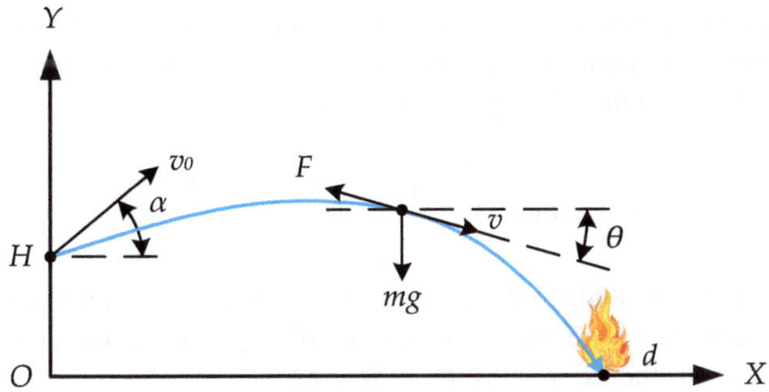

Fig. 1. Trajectory model of the water jet.

As shown in Fig. 1, in the rotation plane of the water cannon, a rectangular coordinate system is established to analyze the force condition of the water jet micro-element bodies, which regards the horizontal ground and its vertical lines as the coordinate axes. The air resistance is $F = -kv$, where k is the air resistance coefficient, v is the velocity of the water jet micro-element bodies. The mass of the water jet micro-element bodies is m, and the acceleration of gravity is g. The angle between the velocity v and the positive direction of the X-axis is θ, and it is a positive value when rotating clockwise. According to the horizontal and vertical force decomposition of the water jet micro-element bodies moving in the air, the linear differential equations can be described as:

$$\begin{cases} m\frac{dv_x}{dt} + F\cos\theta = m\frac{dv_x}{dt} + kv_x = 0 \\ m\frac{dv_y}{dt} + F\sin\theta + mg = m\frac{dv_y}{dt} + kv_y + mg = 0 \end{cases} \tag{1}$$

In the plane corresponding to the rotation angle of the water cannon, the height of the water cannon muzzle from the ground is H, the distance of the water landing point is d, the initial velocity of water jet is v_0, and the angle between the muzzle and the positive direction of the X-axis is the pitch angle α, $\alpha \in (-90°, 90°)$ $\alpha \in (-90°, 90°)$. The pitch angle α is a positive value when rotating counterclockwise. According to the initial condition, the linear differential equations can be solved, and the equation of the water jet trajectory can be derived as:

$$y = \left(\frac{mg}{kv_0\cos\alpha} + \tan\alpha\right)x + \frac{m^2 g}{k^2}\ln\left(1 - \frac{kx}{mv_0\cos\alpha}\right) + H \tag{2}$$

To simplify the computation, the logarithm term in Eq. (2) expands in the second-order Taylor series. Then, the expression of the final water jet trajectory model can be represented as:

$$y = -\frac{g\left(\tan^2\alpha + 1\right)}{2v_0^2}x^2 + \tan\alpha \cdot x + H \tag{3}$$

The expected control result of water cannon is to make the landing point of the water jet trajectory coincide with the burning flame point. Therefore, in order to obtain the required pitch angle of the water cannon based on the target landing point, which is also the position of the flame burning point, it is necessary to construct a functional analytical equation with the pitch angle α as the dependent variable. By solving the Eq. (3), the functional analytical equation is expressed as follows:

$$\alpha = \arctan \frac{v_0^2 - \sqrt{v_0^4 - g^2 d^2 + 2ghv_0^2}}{gd} \tag{4}$$

Similarly, in order to analyze the characteristics of the water jet trajectory, the functional analytical equation with the distance of landing point d as the dependent variable can be computed from the Eq. (4), as the following:

$$d = \frac{v_0^2 \tan \alpha + \sqrt{v_0^4 \tan^2 \alpha + 2ghv_0^2 \left(\tan^2 \alpha + 1\right)}}{g\left(\tan^2 \alpha + 1\right)} \tag{5}$$

3 Positioning Control of Water Landing Point Based on Fuzzy Algorithm

Fuzzy theory formulating rules based on expert experience, which can avoid the impact of inaccuracy of the model and nonlinear factors. Considering the complexity and accuracy of the water jet trajectory model, as well as the influence by the actual environmental factors such as wind speed, temperature and humidity, this paper selects the fuzzy theory to design the algorithm for positioning control of water landing point. The control target of this algorithm is to make the water landing point coincide with the flame burning point by controlling the pitch angle and the rotation angle of the water cannon, and realize intelligence and automation.

As depicted in Fig. 2, in the world coordinate system $O_W - X_W Y_W Z_W$, I represents the muzzle position of the water cannon, whose coordinate is (h, 0, 0), ϕ indicates the rotation angle, C is the current landing point of the water jet, whose coordinate can be described as (0, Y_C, Z_C) by projecting the distance of the water landing point d, onto the Y-axis and Z-axis, T represents the target landing point, whose coordinate is (0, Y_T, Z_T).

Through using the geometric relationship shown in Fig. 3, the deviation model of positioning control between the current landing point of the water jet and the target landing point can be expressed as:

$$\begin{pmatrix} e_L \\ e_\phi \end{pmatrix} = \begin{pmatrix} \sqrt{Y_C^2 + Z_C^2} - \sqrt{Y_T^2 + Z_T^2} \\ \phi_C - \phi_T \end{pmatrix} \tag{6}$$
$$\phi_C = \arctan \frac{Z_C}{Y_C} \quad \phi_T = \arctan \frac{Z_T}{Y_T}$$

where e_L is the deviation of the landing point, e_ϕ is the deviation of the rotation angle, ϕ_C is the rotation angle corresponding to the current landing point, ϕ_T is the rotation angle corresponding to the target landing point.

From the Eq. (6), it can be found that the target of the positioning control is to make the deviations converge to zero gradually, by selecting the appropriate control input

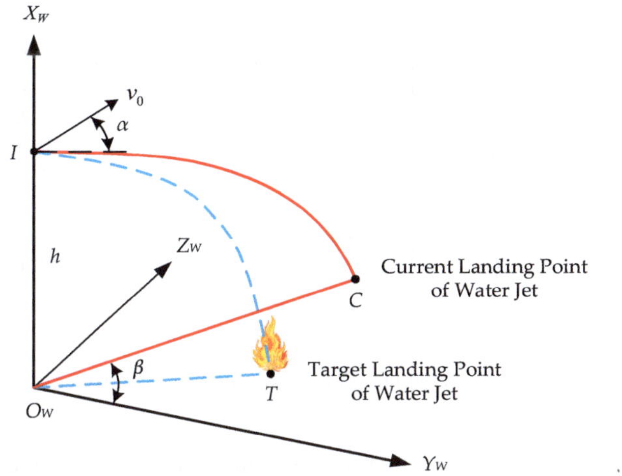

Fig. 2. Schematic diagram for the water landing point's positioning control.

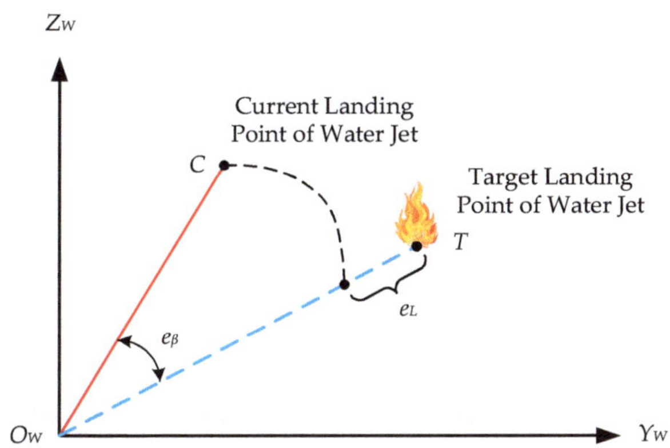

Fig. 3. Schematic diagram for the deviation model of positioning control.

based on the deviation model. As the position of the landing point is adjusted by the fuzzy controller, it finally will reach the target landing point.

According to the Fig. 4, the necessary parts in designing a fuzzy controller include input variable, output variable, fuzzification, defuzzification and fuzzy rules. The first step is to determine the input variables and output variables of the controlled object. Here, the input variables are composed of the deviation of the landing point and its rate of change, the deviation of the rotation angle and its rate of change, which are respectively expressed as e_L, $\dot{e}_L$, e_ϕ and $\dot{e}_\phi$. And, the output variables include the increment of the pitch angle and the increment of the rotation angle, which are respectively expressed as $\Delta\alpha$ and $\Delta\phi$.

Moreover, the fuzzy subsets are defined as seven levels, called {PB, PM, PS, ZE, NS, NM, NB}. In order to make the basic domain of all input variables correspond to the domain of fuzzy sets, the quantization factors of e_L, $\dot{e}_L$, e_ϕ and $\dot{e}_\phi$ are set to 0.1, 0.05,

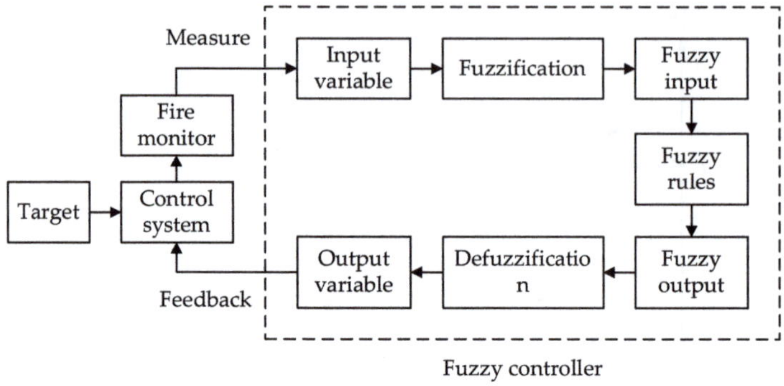

Fig. 4. Structure diagram of the fuzzy controller.

0.35 and 0.2, respectively. A symmetrical triangular membership function is adopted to input and output variables, as shown in Fig. 5.

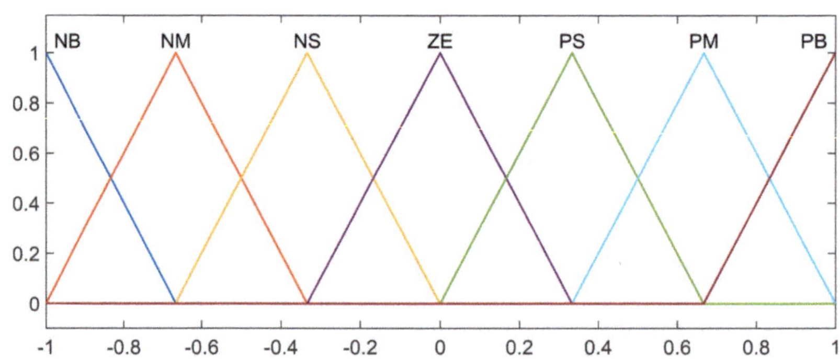

Fig. 5. Membership function of fuzzy controller for the input and output variables.

It is worth pointing out that the fuzzy control rules are designed in the form of "if-then" structure as follows:

$$R_i : \textit{if} \;\; e_L \text{ is NB and } \dot{e}_L \text{ is NB } \textit{then} \;\; \Delta\alpha \text{ is PB}$$

where the left-hand side of a rule is called the antecedent, and the right-hand side of a rule is called the consequent.

The formulation of the fuzzy control rules in this paper is described as following. If the deviations have the same sign with their rate of change, the output in the opposite direction needs to be taken. In order to reduce the deviation as soon as possible, the greater the deviation and its rate of change, the larger the output amount. On the contrary, if the sign between the deviations and their rate of change is different, the output in the same direction as the deviation change rate needs to be taken. In order to prevent the occurrence of overshoot, a smaller amount needs to be output when the sign between the deviations and their rate of change is different. The increment of the pitch angle is determined by the deviation of landing position and its rate of change, and the increment of the rotation angle is determined by the deviation of rotation angle and its rate of change. The fuzzy control rules are listed in Table 1.

Table 1. The fuzzy control rules for the increments of pitch and rotation angles.

Deviation of e_L and e_θ	Change rate of $\dot{e}_L$ and $\dot{e}_\theta$						
	NB	NM	NS	ZE	PS	PM	PB
NB	PB	PB	PM	PM	PS	PS	ZE
NM	PB	PM	PM	PS	PS	ZE	ZE
NS	PM	PM	PS	PS	ZE	ZE	ZE
ZE	PS	PS	ZE	ZE	ZE	NS	NS
PS	ZE	ZE	ZE	NS	NS	NM	NM
PM	ZE	ZE	NS	NS	NM	NM	NB
PB	ZE	NS	NS	NM	NM	NB	NB

Besides, the Mamdani method based on "max-min" is used for fuzzy inference, where the "and" operation uses the minimum, and the "or" operation uses the maximum.

Note that the output results of fuzzy inference are fuzzy sets, which need to be deblurred before being output from the fuzzy controller. The defuzzification method used in this paper is the regional centroid method. Similarly, in order to make the basic domains of all output variables correspond to the domains of fuzzy sets, the quantization factor of the increment of the pitch angle is set to 0.01, and the quantization factor of the increment of the rotation angle is set to 0.02.

4 Analysis of Simulation Results

To evaluate the performance of the positioning control algorithm proposed in this paper, another simulation study of the fire-fighting water cannon was also carried out under the MATLAB/Simulink 2016b programming environment. The structure block diagram of the simulation model is presented in Fig. 6.

As illustrated in Fig. 6, the initial condition module mainly provides the initial velocity, gravitational acceleration, muzzle position, initial pitch angle and initial rotation angle used for the water jet trajectory model. And, the water jet trajectory model module mainly calculates the real-time data of water jet trajectory, which includes the real-time water jet trajectory curve and the current water landing point, according to the given input parameters, the initial pitch angle and the initial rotation angle. The positioning deviation model mainly calculates the deviation of the landing point and the deviation of the rotation angle in the world coordinate system, based on the current landing point and the target landing point. The landing point's positioning controller module mainly uses the fuzzy theory to output the increment of the pitch angle and the increment of rotation angle, relying on the deviation of landing point and the deviation of rotation angle given by the positioning deviation model module.

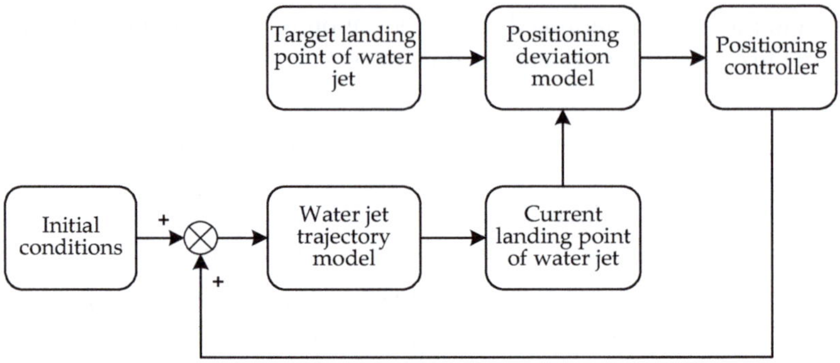

Fig. 6. Diagram of the simulation model for water landing point's position control.

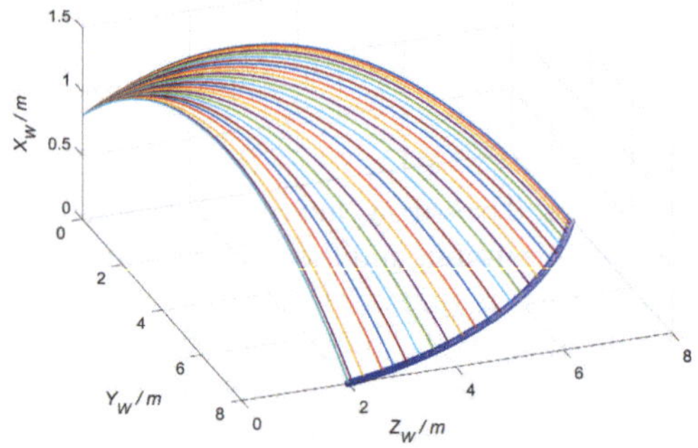

Fig. 7. 3D model of real-time trajectory of the water jet under positioning control of water landing point.

In the world coordinate system, the initial velocity of the water cannon is set to 10.88 m/s, the acceleration of gravity is set to 9.8 m/s^2, the coordinate of muzzle is set to (0.815, 0, 0), the initial pitch angle is set to 0.301 rad, the initial rotation angle is set to 1.212 rad. According to the Eq. (5) and the geometric relationship, the coordinate of initial landing point and target landing point can be calculated as (0, 3, 8) and (0, 8, 2), respectively. Time of the simulation study is set to 5 s. The real-time trajectory of the water jet under the positioning control of the water landing point is shown in Fig. 7, and the results curve of the simulation study is depicted in Fig. 8.

From Fig. 7, it can be observed that the movement trajectory curve of the water jet forms a surface, and the movement trajectory of the landing point forms a circular curve in the $Y_W Z_W$ plane. Fig. 8(a) and 8(b) depict that the deviation changes drastically at the beginning of the simulation. The deviation of the landing point changes rapidly, and it converges to zero after about 1 s. The deviation of the rotation angle changes relatively stable, and it converges to zero after about 3 s. It also can be seen from Figs. 8(c) and 8(d) that the output of the fuzzy controller is relatively large in the initial stage of the simulation, which contributes to adjusting the position of the landing point quickly. In the final stage of the simulation, as the deviation gradually decreases, the output of

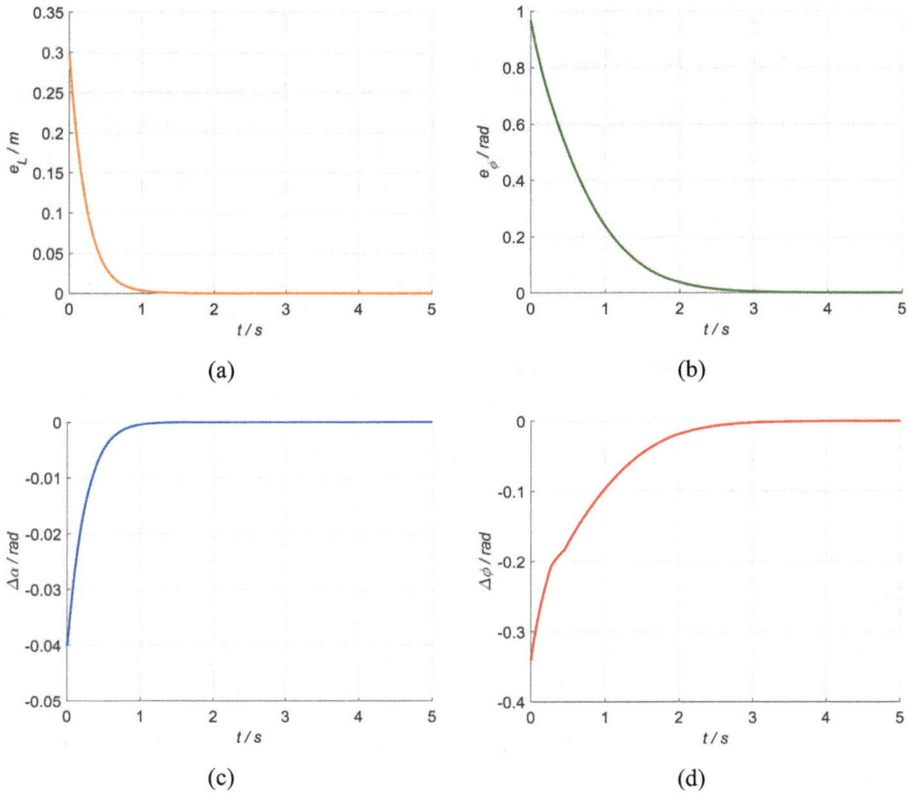

Fig. 8. Simulation results of the positioning control of the landing point: (a) the deviation curve of landing point; (b) the deviation curve of rotation angle; (c) the increment curve of pitch angle; (d) the increment curve of rotation angle.

controller also slightly decreases, so that the trajectory of the water jet can reach the target landing point smoothly.

Finally, the simulation results show that the algorithm for the position control of water landing point based on fuzzy control proposed in this paper can achieve accurate control, and make the landing point of the water cannon move to the flame burning point in the world coordinate system. The proposed algorithm was proved to have great effectiveness and feasibility. Furthermore, intelligent positioning control and extinguishment of the fire-fighting robot can be realized by combining with the aforementioned technologies.

5 Conclusion

This paper presents an intelligent positioning and extinguishing technology of fire-fighting water cannon based on fuzzy control. Through the force analysis and the dynamic equation of the jet trajectory of the fire-fighting water cannon, a water jet trajectory model is established. Aiming at the problem of extinguishing the fire by positioning and controlling the water landing point, a positioning control algorithm based on fuzzy control theory is designed according to the positioning deviation model of the water jet motion trajectory, so that the fire-fighting water monitor can control the position of the water landing point to reach the flame burning point accurately. The performance of the control algorithm is evaluated by MATLAB/Simulink. In conclusion, the effectiveness and feasibility of the proposed control algorithm are confirmed by the simulation results, which may have implications and wide application in the intelligent positioning control and extinguishment of the fire-fighting robot.

Acknowledgments. We express our profound gratitude to the anonymous reviewers for their insightful comments and constructive suggestions, which have significantly contributed to the enhancement of this manuscript. This research was supported by the National Natural Science Foundation of China (52205279), the National Key Research and Development Program of China (2023YFC3806603 & 2023YFF0613200) and the Top Discipline Plan of Shanghai Universities-Class I.

Disclosure of Interests. The authors have no competing interests to declare that are relevant to the content of this article.

References

1. McNeil, J.G., Lattimer, B.Y.: Autonomous fire suppression system for use in high and low visibility environments by visual Servoing. Fire. Technol. **52**(5), 1343–1368 (2016). https://doi.org/10.1007/s10694-016-0564-8
2. McNeil, J.G., Lattimer, B.Y.: Real-time classification of water spray and leaks for robotic firefighting. Int. J. Comput. Vis. Image Process. **5**(1), 1–26 (2015)
3. Chen, T., Yuan, H.Y., Su, G.F., Fan, W.C.: An automatic fire searching and suppression system for large spaces. Fire Safety J. **39**(4), 297–307 (2004). https://doi.org/10.1016/j.firesaf.2003.11.007
4. Yuan, F.N.: An integrated fire detection and suppression system based on widely available video surveillance. Mach. Vis. Appl. **21**(6), 941–948 (2010)
5. McBride, W.E.: Fine water mist fire protection system. In: Record of Conference Papers. IEEE Incorporated Industry Applications Society. Forty-Eighth Annual Conference. 2001 Petroleum and Chemical Industry Technical Conference (Cat. No.01CH37265)., Toronto, ON, Canada, pp. 245–252 (2001). https://doi.org/10.1109/PCICON.2001.960564
6. Ye, X., Ma, J., Shen, Y.X., Lin, L.Y.: Suppression effect of sprinkler system on fire spread in large commercial buildings. Procedia Eng. **135**, 455–462 (2016)
7. Yuan, W., Yuan, J., Li, Y.D., Wang, X.W., Zhang, Y.M.: Video-based integrated fire detection, positioning and water cannon suppression intelligent system. Appl. Mech. Mater. **391**, 482–487 (2013)
8. Zhang, Y., Shen, L.Y., Ren, Y.T., Wang, J.H., Liu, Z., Yan, H.: How fire safety management attended during the urbanization process in China? J. Clean. Prod. **236**, 117686 (2019)
9. Egodage, N., Abdeen, F.N., Sridarran, P.: Fire emergency evacuation procedures for differently-abled community in high-rise buildings. J. Facil. Manag. **18**(5), 505–519 (2020). https://doi.org/10.1108/Jfm-07-2020-0043

10. Luo, J., Qiu, J., Han, J.: Research status and Prospect of fire rescue equipment in high-rise building. In: 2018 International Conference on Sensing, Diagnostics, Prognostics, and Control (SDPC) (2018)
11. Xin, Y.B., et al.: An experimental study of automatic water Cannon Systems for Fire Protection of large open spaces. Fire. Technol. **50**(2), 233–248 (2014). https://doi.org/10.1007/s10694-013-0363-4

Path Tracking Control of Robotic Rollers on Unstructured Road Surfaces

Meng Yang[1(✉)], Yuyang Cai[1], Siyu Wu[1], Zeyu Duan[2], and Jian Zhang[3]

[1] School of Mechanical Engineering, Tongji University, Shanghai 201804, China
yangmeng@tongji.edu.cn

[2] Department of Civil and Environmental Engineering, The Hong Kong University of Science and Technology, Hong Kong 999077, China

[3] Leon Robotics Co., Ltd, Shenzhen 518048, China

Abstract. With the continuous advancement of construction technologies toward intelligence, robotic rollers equipped with intelligent compaction techniques have gradually become a research focus in the field of infrastructure engineering. During operation, the path-tracking accuracy of robotic rollers plays a vital role in ensuring compaction quality and improving efficiency. This study addresses the path-tracking control problem of robotic rollers operating on unstructured road surfaces. A motion model and a multi-body dynamic model are developed, based on which a Model Predictive Control (MPC) strategy is introduced. To validate the effectiveness and robustness of the proposed control method, a co-simulation platform integrating RecurDyn-MATLAB-EDEM is established to simulate the robotic roller's behavior under complex conditions. Comparative analyses are conducted between the MPC approach and traditional methods, including Proportional-Integral-Derivative (PID) control and Pure Pursuit Control (PPC). Simulation results demonstrate that under initial tracking deviations, the MPC algorithm can reduce and stabilize errors within ±5 cm in 15 s. Compared to PID and PPC methods, the MPC strategy reduces both the maximum and average tracking errors by over 60%, highlighting its superior control stability and adaptability to varying environments. These findings provide theoretical and methodological support for intelligent compaction in complex construction scenarios.

Keywords: Robotic Roller · RecurDyn-MATLAB-EDEM Co-Simulation · Path Tracking · Model Predictive Control

1 Introduction

In recent years, with rapid advances in satellite positioning, sensing, and control technologies, robotic rollers equipped with autonomous operational capabilities have been increasingly developed to adapt to extreme environments [1]. However, existing research mainly focuses on traditional rollers and their application in intelligent compaction during road construction [2]' [3], while studies on robotic rollers with autonomous decision-making capabilities remain limited. Path tracking is a fundamental component of intelligent compaction for robotic rollers and holds significant research value.

S. Halgamuge et al. (Eds.): ICACMVE 2025, LNME, pp. 438–448, 2026.
https://doi.org/10.1007/978-981-95-6911-3_43

In recent years, path tracking control algorithms for autonomous vehicles have been extensively studied [4]. Among various control approaches, Proportional-Integral-Derivative (PID) controllers remain dominant due to their simple structure and ease of implementation. However, PID control typically relies on significant tracking errors to initiate correction, and is prone to instability under external disturbances—manifesting as oscillations or "snaking motion", which can severely degrade tracking accuracy and compaction quality in robotic rollers. To overcome this, Model Predictive Control (MPC) has been increasingly explored due to its predictive nature and optimization capability [5]. Although MPC demonstrates superior performance in terms of robustness [6], its strong reliance on predictive models for future states limits its applicability. Due to the articulated structure of robotic rollers and their significantly different kinematic characteristics from traditional vehicles, control strategies based on conventional vehicle kinematic models are difficult to apply directly to path tracking tasks for robotic rollers. Moreover, while MPC can achieve favorable control results under relatively ideal road conditions [7, 8], robotic rollers often operate in vibrating conditions on unpaved, unstructured, and uneven surfaces. Under such circumstances, infeasibility may occur during the MPC solving process.

To address the aforementioned issues, this study establishes a dynamic model tailored to the motion characteristics of robotic rollers, which is incorporated as the prediction model within a Model Predictive Control (MPC) framework. In addition, relaxation factors are introduced into the traditional MPC structure to mitigate potential infeasibility during execution, thereby enhancing the robustness and adaptability of the system under complex and unstructured road conditions. Subsequently, a RecurDyn–MATLAB–EDEM co-simulation platform is employed to validate the proposed control method, and comparative analyses are conducted with other algorithms to evaluate its control accuracy and robustness.

2 System Modeling of the Robotic Roller

2.1 Nonlinear Kinematic Model of the Robotic Roller

Due to the articulated structure unique to robotic rollers, conventional vehicle models cannot be directly applied to their path tracking control. Therefore, a dedicated mathematical model must be established. This study constructs a kinematic model to describe the motion process of the robotic roller. Compared to dynamic models, the kinematic model features lower computational complexity, making it suitable for applications with high real-time requirements. As illustrated in Fig. 1, the front vibrating drum and rear drive unit are connected via the articulated frame structure.

In Fig. 1, OXY denotes the global Cartesian coordinate system. $P(x_1, y_1)$ represents the center of the front vibrating wheel, and $P(x_2, y_2)$ denotes the center of the rear frame. θ_1 and θ_2 are the heading angles of the vibrating wheel and the rear frame relative to the positive X-axis, respectively. V_1 and V_2 denote the forward velocities of the vibrating wheel and the rear frame, and α is the articulation angle between the vibrating wheel and the rear frame. l_1 and l_2 represent the distances from the articulation point G to the centers of the vibrating wheel and the rear frame, respectively. Based on the schematic

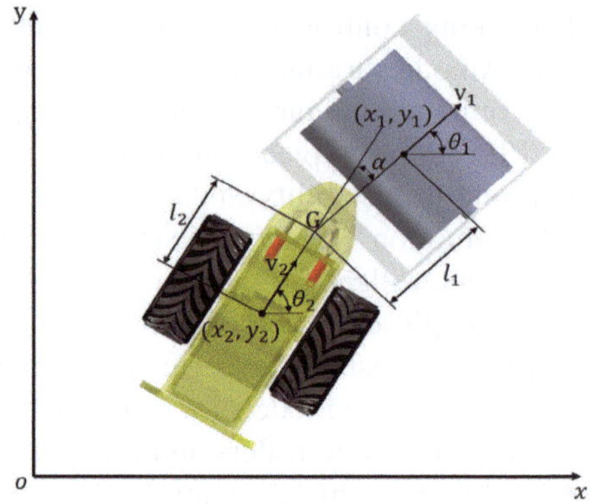

Fig. 1. Kinematic model of the robotic roller.

diagram of the robotic roller and the proposed modeling method, the kinematic model can be derived as shown in Eq. (1).

$$\begin{cases} \dot{x} = vcos\theta \\ \dot{y} = vsin\theta \\ \dot{\theta} = (-vsin\alpha - \dot{\alpha}l_2)/(l_2 + l_1cos\alpha) \end{cases} \tag{1}$$

2.2 Multi-Body Dynamics Model of the Robotic Roller

To validate the performance of the proposed adaptive MPC path tracking controller, a complete vehicle model was constructed in RecurDyn by importing the robotic roller along with an eccentric block, as shown in Fig. 2. The model parameters are listed in Table 1.

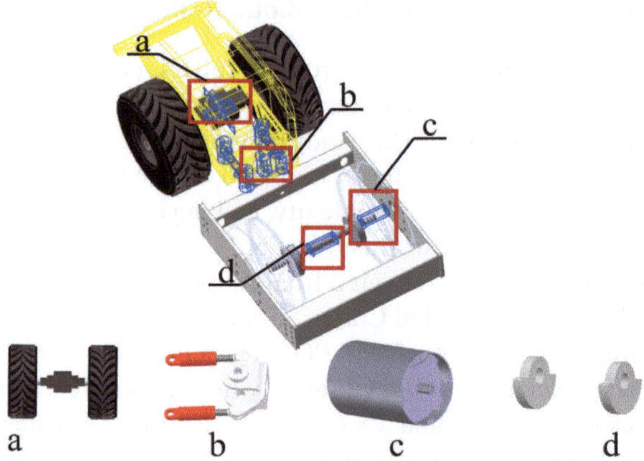

Fig. 2. Multi-body dynamics model of the robotic roller.

Table 1. Model parameters of the robotic roller.

Parameters	Unit	Value	Parameters	Unit	Value
Overall mass	(Kg)	12100	Vibrational frequency	(Hz)	30
steel wheel mass	(Kg)	7100	Theoretical amplitude	(mm)	1.7/0.85
steering angle	(°)	30°	Excitation force	(KN)	247/127

In the RecurDyn environment, appropriate constraints are applied at a, b, c, and d, corresponding to the drive module, articulation module, vibrating wheel module, and eccentric wheel module, respectively. The eccentric wheel module functions as the working component of the vibrating wheel. By adjusting the rotation angle of the eccentric block, the vibration amplitude can be varied. The rotation of the eccentric block generates an excitation force.[9] This excitation force is the vertical component produced as the eccentric block rotates about the vibration axis of the wheel, and it follows the principles of simple harmonic motion. Its expression is given by:

$$F = mA_0\omega^2 sin\omega t = em_d\omega^2 sin\omega t \tag{2}$$

Where $A_0 = \frac{em_d}{m}$ represents the nominal amplitude, with m being the mass of the vibrating wheel, e the eccentric distance, and m_d the mass of the eccentric block.

To ensure correct operation of all components, motion functions were applied to the corresponding constraints in RecurDyn, and the motion subcomponents and drive function configurations are listed in Table 2.

Table 2. Settings of the kinematic pair and driving function for each component.

Constraint member	Constraint type	Driving function
Rear frame—Wheel axle	Rotating pair	1.25 rad/s
Rear frame—Articulation	Rotating pair	PIN
Front frame—Vibrating wheel	Rotating pair	/
Vibrating wheel—Eccentric block	Rotating pair	$em_d\omega^2 sin\omega t$

PIN serves as the interface between RecurDyn and MATLAB/Simulink, and is used for inputting the subsequent control algorithm.

3 Path Tracking Controller Design

3.1 MPC Path Tracking Algorithm Based on Kinematic Model

From Eq. (1), it can be seen that when the vibrating wheel roller turns, the system can be regarded as a control system with input $u(v, \alpha)$ and state variables $X(x, y, \theta)$. The following provides an analytical calculation of the turning process. The state equation of the vibrating wheel roller is given by Eq. (3):

$$\begin{pmatrix} \dot{x} \\ \dot{y} \\ \dot{\theta} \end{pmatrix} = \begin{pmatrix} cos\theta \\ sin\theta \\ -sin\alpha/(l_2+l_1 cos\alpha) \end{pmatrix} v + \begin{pmatrix} 0 \\ 0 \\ -l_2/(l_2+l_1 cos\alpha) \end{pmatrix} \dot{\varnothing} \tag{3}$$

Eq. (3) represents the state equation in the form $\dot{X} = f(x, y, u)$. For a planned reference trajectory, all reference points satisfy this equation. Let r denote the reference parameter, then the target state is given by:

$$\dot{X}_r = f(x_r, y_r, u_r) \tag{4}$$

Here, $X_r = [x_r \; y_r \; \theta_r]$ and $u_r = \varnothing_r$. By performing a Taylor expansion of the state equation at the reference point, Eq. (5) is obtained:

$$\dot{X} = f(x_r, u_r) + \frac{\partial f(x,u)}{\partial x}\bigg|_{\substack{x=x_r \\ y=y_r \\ u=u_r}} (X - X_r) + \frac{\partial f(x,u)}{\partial u}\bigg|_{\substack{x=x_r \\ y=y_r \\ u=u_r}} (u - u_r) \tag{5}$$

By subtracting Eq. (4) from Eq. (5), the linearized error model of the vibrating wheel roller can be obtained, discretized and reformulated into a linear error model of the robotic roller.

$$\widetilde{X}(k+1) = A_k \widetilde{X}(k) + B_k \widetilde{u}(k) \tag{6}$$

To predict future outputs based on the linearized error model of the robotic roller, a new state vector is constructed accordingly, and then reformulated into a new state-space representation.

$$\begin{gathered} \xi(k+1|k) = \begin{bmatrix} \widetilde{X}(k+1|k) \\ \widetilde{u}(k|k) \end{bmatrix} = \\ \begin{bmatrix} A_k & B_k \\ 0 & I_m \end{bmatrix} \xi(k|k) + \begin{bmatrix} B_k \\ I_m \end{bmatrix} \Delta\widetilde{u}(k|k) = \widetilde{A}_k \xi(k|k) + \widetilde{B}_k \Delta\widetilde{u}(k|k) \end{gathered} \tag{7}$$

Output equation:

$$\eta(k|k) = [I_n \; 0] \begin{bmatrix} \widetilde{X}(k|k) \\ \widetilde{u}(k-1|k) \end{bmatrix} = C\xi(k|k) \tag{8}$$

Here, I_m, I_n are identity matrices. Thus, the system's predicted output within the prediction horizon N_p is expressed as Eq. (9):

$$Y(t) = \Psi\xi(k|k) + \Theta\Delta U(k) \tag{9}$$

Where:

$$Y(t) = \begin{bmatrix} \eta(k+1|k) \\ \cdots \\ \eta(k+N_c|k) \\ \cdots \\ \eta(k+N_p|k) \end{bmatrix}, \Psi = \begin{bmatrix} C\tilde{A}_k \\ \cdots \\ C\tilde{A}_k^{N_c} \\ \cdots \\ C\tilde{A}_k^{N_p} \end{bmatrix}, \Theta = \begin{bmatrix} C\tilde{B}_k & 0 & \cdots & 0 \\ C\tilde{A}_k\tilde{B}_k & C\tilde{B}_k & \cdots & 0 \\ \cdots & \cdots & \ddots & 0 \\ C\tilde{A}_k^{N_p-1}\tilde{B}_k & C\tilde{A}_k^{N_p-2}\tilde{B}_k & \cdots & C\tilde{A}_k^{N_p-N_c-1}\tilde{B}_k \end{bmatrix}, \Delta U = \begin{bmatrix} \Delta\tilde{u}(k|k) \\ \cdots \\ \Delta\tilde{u}(k+N_c|k) \end{bmatrix}$$

N_p is the prediction horizon, and N_c is the control horizon.

To enable the robotic roller to rapidly and stably track the reference trajectory, the articulation angle increment is selected as the control variable in the cost function. The optimization objective function for path tracking is formulated as follows:

$$J(k) = \min \sum_{i=1}^{N_p} \|\eta(k+i|k) - \eta_r(k+i|k)\|^2 Q + \sum_{i=1}^{N_c-1} \|\Delta u(k+i|k)\|^2 R + \rho\varepsilon^2 \tag{10}$$

Let $\eta(k+i|k) - \eta_r(k+i|k)$ denote the deviation between the predicted and reference outputs, and let Q , R , ρ be the weighting matrices, while ε denotes the relaxation factor. The relaxation factor not only directly constrains the control increment, but also prevents infeasibility during optimization execution. Considering the physical constraints of actual robotic rollers, bounds are applied to both the articulation angle and its increment as follows:

$$-20^\circ \leq \varnothing \leq 20^\circ$$

$$-0.57^\circ \leq \Delta\varnothing \leq 0.57^\circ$$

The objective function is then transformed into a standard quadratic programming form, and combined with constraint conditions to yield:

$$J = \left[\Delta U(t)^T, \varepsilon\right]^T H_t \left[\Delta U(t)^T, \varepsilon\right] + G_t \left[\Delta U(t)^T, \varepsilon\right] \tag{11}$$

Where $H_t = \begin{bmatrix} \Theta^T Q \Theta + \mathrm{R} & 0 \\ \mathbf{0} & \rho \end{bmatrix}$, $G_t = [2e^T Q\Theta \quad 0]$, with e denoting the tracking error over the prediction horizon.

3.2 Stability Analysis

To further verify the feasibility of the proposed MPC control strategy in practical applications, a theoretical stability analysis is conducted. According to Eq. (11), the constructed objective function is a positive semi-definite function. When the lower bound $J = 0$, both the tracking error e and control increment Ø converge to zero. Therefore, the objective function $J(k)$ is considered as a Lyapunov function candidate and is simplified into a general form as shown in Eq. (12) for analysis:

$$J(x_0) = \sum_{i=0}^{N-1} q(x_i, u_i) + px_N \tag{12}$$

Solving Eq. (12) yields the optimal solution $J^*(x_0)$. Assume that the state progresses from x_0 to x_1 in the next step. Since this is an optimization problem, the objective value corresponding to any feasible solution must be no less than the optimal value. That is:

$$J^*(x_1) \le \sum_{i=1}^{N} q\left(x_i, u_i^*\right) + p(Ax_N + Bu_N)$$

$$= J^*(x_0) - q\left(x_0, u_0^*\right) + p(Ax_N + Bu_N) - p(x_N) + q(x_N, u_N)$$

Where $p(x) = p(Ax_N + Bu_N) - p(x_N) + q(x_N, u_N) \le 0$, which implies:

$$J^*(x_1) - J^*(x_0) \le -q(x_0, u_0^*) < 0 \tag{13}$$

According to Eq. (13), since Eq. (11) is a valid Lyapunov function that is monotonically decreasing, the cost function J will converge to 0 over time. This means both the tracking error e and control increment Ø will converge to 0. Therefore, under the MPC control law designed in Sect. 3.1, the Robotic Roller is guaranteed to asymptotically converge to the equilibrium point from any given initial deviation, thus achieving desired trajectory tracking.

4 Control Strategy Validation and Performance Analysis

4.1 Construction of the co-Simulation System

To capture the dynamic response characteristics of the Robotic Roller during operation and to validate the effectiveness of the proposed trajectory tracking control algorithm, this study employs the EDEM software to generate a granular road surface model that replicates real working conditions. Combined with the controller design in the MATLAB/Simulink platform, and the high-fidelity dynamic model established in RecurDyn, a precise control of the robotic system is achieved. This leads to the construction of a co-simulation framework integrating discrete element modeling, control strategy implementation, and multi-body dynamics simulation. The input–output structure of the co-simulation system model is illustrated in Fig. 3.

4.2 Simulation Analysis

To verify the control performance of the MPC algorithm during straight driving and negotiating turns on a flat road surface, comparative experiments were conducted with the Pure Pursuit Control (PPC) algorithm and the PID algorithm. The system sampling rate was set to 0.1 s, and the driving speed of the road roller robot was set to 1 *m/s*.

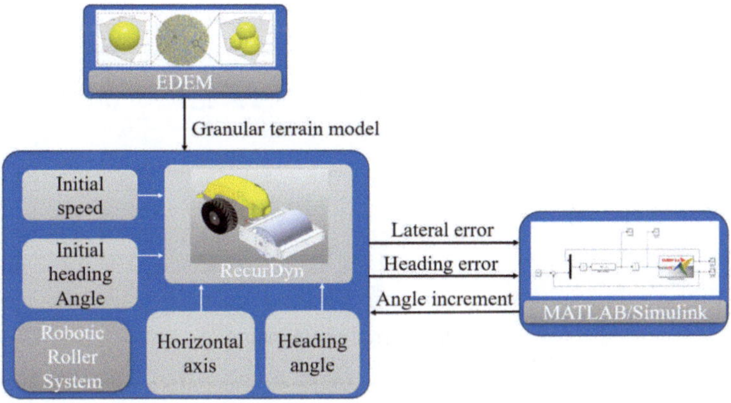

Fig. 3. Structure of the Co-simulation System Model.

The initial lateral deviation was 2 *m*, and the initial heading angle was 0°. Under the same initial deviation conditions, three control strategies PID, PPC, and MPC were respectively adopted for comparative experiments on compaction control. Fig. 4a-b show the lateral deviation and heading angle deviation curves of the road roller under the same initial conditions for PID, PPC, and MPC control strategies. Table 3 presents the comparison of straight-line compaction errors of the robotic roller under three different control algorithms.

Simulation results demonstrate that the MPC controller outperforms PID and PPC in straight-line compaction tracking of the robotic roller. Compared to PID (55.43 cm) and PPC (37.86 cm), MPC achieves significantly smaller lateral errors, with a maximum of 5.28 cm, and exhibits substantially lower heading errors. While MPC, PID, and PPC all drive the errors toward zero over time, MPC ensures faster convergence, better overshoot suppression, and more precise response regulation. The superior performance of MPC is attributed to its real-time optimization and prediction capabilities, which enhance its adaptability to complex and dynamic working conditions—especially on unstructured roads and under vibration disturbances—unlike PID and PPC, which suffer from post-compaction oscillations. Quantitatively, MPC reduces the average and standard deviation of lateral errors by 78.4% and 73.7% over PID, and by 62.4% and 60.3% over PPC; heading errors are reduced by 74.1% and 57.3% over PID. These results confirm that MPC offers better control precision, robustness, and adaptability, establishing its advantage in straight-line tracking tasks for robotic rollers.

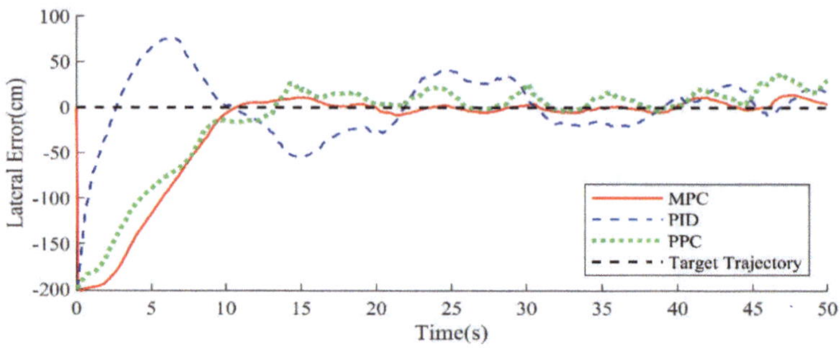

a Lateral error comparison.

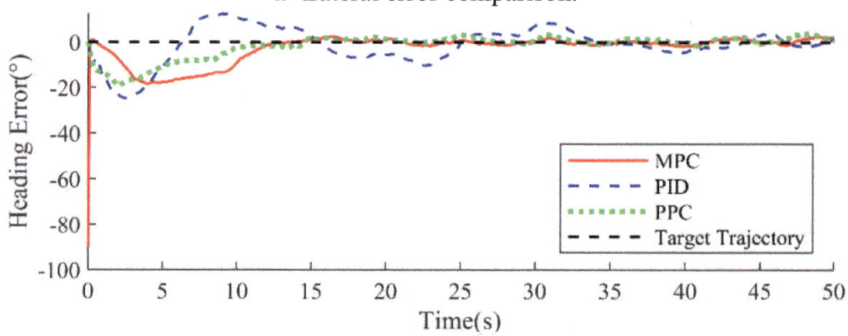

b Heading error comparison.

Fig. 4. Comparison of control errors under PID, PPC, and MPC.

Table 3. Comparison of straight-line tracking errors under MPC, PID, and PPC control for the robotic roller.

Controller	Category	Lateral error (*cm*)	Heading error (°)
MPC	Average	4.48	1.06
	Standard Deviation	3.41	1.17
	Maximum	5.28	9.21
PID	Average	20.81	4.11
	Standard Deviation	12.97	2.74
	Maximum	55.43	11.02
PPC	Average	11.91	1.26
	Standard Deviation	8.58	0.91
	Maximum	37.86	4.22

5 Conclusions

This paper first establishes a dynamic model and a multibody dynamics model of the road roller robot, and then applies the constructed dynamic model to the MPC control algorithm, avoiding the infeasibility issues that may arise due to actuator saturation or other constraints during the control execution process. To further evaluate the effectiveness of the proposed control strategy, a co-simulation platform based on RecurDyn-MATLAB-EDEM is employed to simulate the working conditions of the road roller robot traveling on an uneven surface while transitioning to a flat road. The performance of the proposed MPC algorithm is compared against two typical control strategies: PID and PPC. Simulation results show that, under conditions of relatively large initial lateral deviation, the MPC algorithm keeps the lateral error stabilized within ±5 cm and outperforms the other two algorithms in both convergence speed and robustness. Compared to the PID algorithm, the maximum and average lateral errors of the MPC algorithm are reduced by 85.2% and 78.4%, respectively. Compared to the PPC algorithm, the maximum and average lateral errors are reduced by 78.1% and 62.4%, respectively. These findings indicate that the MPC control strategy demonstrates superior performance in road trajectory tracking of the road roller robot, showing strong potential for practical applications.

Acknowledgments. We express our profound gratitude to the anonymous reviewers for their insightful comments and constructive suggestions, which have significantly contributed to the enhancement of this manuscript. This research was supported by the National Natural Science Foundation of China (52205279), the National Key Research and Development Program of China (2023YFC3806603 & 2023YFF0613200).

Disclosure of Interests. The authors have no competing interests to declare that are relevant to the content of this article.

References

1. Yang, M., Bian, Y., Liu, G., Zhang, H.: Path tracking control of an articulated road roller with sideslip compensation. IEEE Access. **8**, 127981–127992 (2020)
2. Tang, C., et al.: Coupled vibratory roller and layered unsaturated subgrade model for intelligent compaction. Comput. Geotech. **177**, 106827 (2025)
3. Yao, Y., Song, E.: Intelligent compaction methods and quality control. Smart Constr. Sustain. Cities. **1**(1), 2 (2023)
4. Yao, Q., Tian, Y., Wang, Q., Wang, S.: Control strategies on path tracking for autonomous vehicle: state of the art and future challenges. IEEE Access. **8**, 161211–161222 (2020)
5. Vivek, K., Sheta, M.A., Gumtapure, V.: A comparative study of Stanley, LQR and MPC controllers for path tracking application (ADAS/AD). In: In 2019 IEEE International Conference on Intelligent Systems and Green Technology (ICISGT), pp. 67–674. IEEE (2019, June)
6. Amer, N.H., Zamzuri, H., Hudha, K., Kadir, Z.A.: Modelling and control strategies in path tracking control for autonomous ground vehicles: a review of state of the art and challenges. J. Intell. Robot. Syst. **86**(2), 225–254 (2017)
7. Goli, M., Eskandarian, A.: MPC-based lateral controller with look-ahead design for autonomous multi-vehicle merging into platoon. In: In 2019 American Control Conference (ACC), pp. 5284–5291. IEEE (2019, July)
8. Wang, H., Liu, B., Ping, X., An, Q.: Path tracking control for autonomous vehicles based on an improved MPC. IEEE access. **7**, 161064–161073 (2019)
9. Peng, H., Jiazhen, C., Lejin, Z., Kun, W., Shuping, W., Lianyang, C.: Discrete element simulation of vibration compaction of slag subgrade. Sci. Rep. **14**(1), 5039 (2024)

Deep Reinforcement Learning Based Speed Control of Pure Methanol Engine

Yafeng Ren(✉), Jiaxin Qin, and Hong Jia

Taiyuan University of Science and Technology, Shanxi Taiyuan 030000, China
renyafeng@tyust.edu.cn

Abstract. In order to improve the speed stability and dynamic responsiveness of methanol engine, this study proposes an intelligent control strategy based on deep reinforcement learning (DRL). By applying the flexible actor-critic (SAC) algorithm and deep deterministic policy gradient (DDPG) algorithm, a DRL controller was constructed that can replace the traditional PID speed control loop. Based on the MATLAB/Simulink platform, a high-precision crankcase power output model are established by combining the engine experimental data. Simulation results show that the SAC algorithm exhibits faster response time, lower error level and smaller overshoot rate in speed control.

Keywords: deep reinforcement learning · methanol · SAC · DDPG

1 Introduction

With the increasing global demand for sustainable energy and environmentally friendly technologies, methanol engines are gradually becoming an important alternative to conventional fuel engines due to their low carbon emissions and excellent combustion characteristics [1], The physical and chemical properties of methanol with high latent heat of vaporization and low air-fuel ratio make it more sensitive to control deviations than conventional gasoline engines. In this context, the control accuracy and performance stability of methanol engines are particularly important, especially the speed control as the core of engine control, whose performance directly affects the engine power output, fuel economy and emission level. Therefore, this paper focuses on reinforcement learning algorithms represented by DDPG and SAC to construct a speed control agent and systematically compares it with traditional PID control methods [2, 3].

2 Methanol Engine Test Research Platform and Engine Digital Modeling

2.1 Establishment of Test Platform

The test platform is mainly composed of six-cylinder four-stroke test machine, fuel system, coolant thermostat system, upper computer measurement and control system, electric power measurement and control machine and other systems.

S. Halgamuge et al. (Eds.): ICACMVE 2025, LNME, pp. 449–454, 2026.
https://doi.org/10.1007/978-981-95-6911-3_44

2.2 Modeling of Methanol Engine

The simulation model of this article is a four-stroke water-cooled inline six-cylinder ignition engine. The Matlab\Simulink model of the methanol engine is divided into the engine crankcase power output model.

Crankcase Power Output Model. The power generated by the crankshaft combustion process is converted into the crankshaft mechanical physical output through the simscape module library in Matlab\Simulink, and the engine drive train is established through the physical port. The power output from the crankshaft is specified as follows.

The piston force is determined by the cylinder pressure and cross sectional area. The model obtains the combustion cylinder pressure parameters through a parametric look-up table, which includes the crankshaft angle, crank speed and engine throttle opening.

The nonlinear function of gas pressure $p(\theta)$ in the combustion chamber corresponding to the crankshaft turning angle θ, throttle opening α and rotational speed N is established as shown in (1) below.

$$p(\theta) = f(\theta, \alpha, N) \tag{1}$$

This mapping relationship is collected by the established experimental platform and a database is created.

The gas pressure is converted to a linear force F_g, expressed by Eq. (2), through the piston area A_P.

$$F_g = \mathrm{p}(\theta) \cdot A_\mathrm{p} \tag{2}$$

The crankshaft tangential force F_t, is obtained by the derivation of geometric relationship of crank-linkage mechanism and expressed by Eq. (3, 4).

$$\mathrm{F_t}(\theta) = \mathrm{F_g}(\theta) \cdot \sin(\phi) \tag{3}$$

$$\phi = \arcsin\left(\frac{r \cdot \sin\theta}{l + r \ \cdot \cos(\theta)}\right) \tag{4}$$

where r is the crank radius; $_l$ is the length of connecting rod; ϕ is the angle between crank and connecting rod.

Single-cylinder torque T_{cyl} is determined by the product of tangential force F_t and force arm r, as shown in Eq. (5).

$$\mathrm{T_{cyl}} = \mathrm{F_t}(\theta) \cdot \mathrm{r} \tag{5}$$

The whole engine torque output of multi-cylinder engine needs to solve the phase synchronization problem, firstly, the ignition phase is modeled, for a six-cylinder four-stroke engine, the working interval of each cylinder is $\Delta\theta = 720°/6$, and the instantaneous torque of the whole engine, T_e (θ), is shown in Eq. (6) as follows.

$$\mathrm{T_e}(\theta) = \sum_{\mathrm{i}=6}^{6} \mathrm{T_{cyl}}\ (\theta - \Delta\theta_\mathrm{i}) \tag{6}$$

The crankshaft system motion follows the law of rotation as shown in Eq. (7).

$$J\frac{d\omega}{dt} = T_e(\theta) - T_{load} - T_f(\omega) \tag{7}$$

where J is the rotational inertia, T_{load} is the load torque, and the friction torque T_f is a complex function and is therefore modeled with experimental data.

Model Validation. Figure 1 shows the experimental situation of the engine's speed control compared with the simulation data.

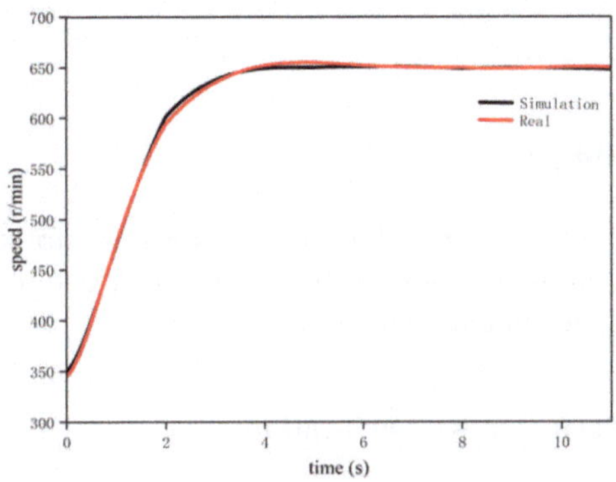

Fig. 1. Comparison of experimental and simulation speed control.

Based on the data presented in Fig. 1, the methanol engine model developed in this study demonstrates an average speed deviation of 5.23 r/min, confirming that it meets the requirements for simulation validation.

3 Control System Design Based on Deep Reinforcement Learning

Create and configure SAC and DDPG agents, specifying hyperparameter configurations such as network architecture, exploration strategy, learning rate, and discount factor, as shown in Tables 1. This reinforcement learning intelligent body interacts with the Simulink model of the engine during training and evaluation [4, 5]. Observations are continuous and are defined as 3 observations: the engine speed, the difference between the speed and the target speed, and the integral of the difference. The action space is defined as a continuous action with the dimension of the action space [1 1]. The absolute difference |e|, between the actual speed and the desired speed, is used as an input to the reward function. The RL-Agent will act as a controller to decide the desired throttle angle [6] .

Table 1. SAC Algorithm Hyperparameter Settings.

Item	Parameter
Value Network Learning Rate	0.0001
Policy Network Learning Rate	0.0002
Discount Factor	0.95
Entropy Weight	0.1
Adaptive Temperature Parameter	0.15
Target Entropy	-1

4 Results and Analysis

This section aims to compare the performance of deep reinforcement learning (SAC, DDPG) with proportional-integral-derivative (PID) as a controller for speed control performance in the engine simulation model.

4.1 Anti-Jamming Performance Verification

This paper compares and analyzes the control performance of three control strategies—traditional PID control, SAC algorithm, and DDPG algorithm—under two typical disturbance conditions: sudden load addition and sudden load removal. In order to evaluate the anti-disturbance performance of the controller, the load inertia step test is designed, the system speed is stabilized at the set value of 1500 r/min, and the load step disturbance is applied at 22 s to last until 26 s to recover, comparing the dynamic regulation capability of the controller. The simulation results are shown in Fig. 2.

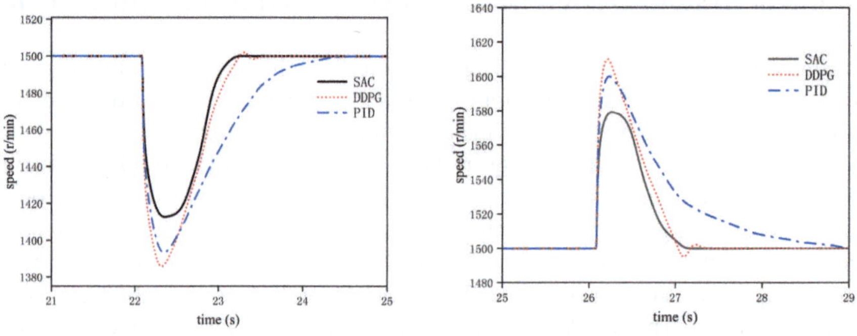

(a) Load sudden addition. (b) Load sudden removal.

Fig. 2. Comparison graph of load inertia sudden change.

Comprehensively, the SAC algorithm performs optimally in the core performance indexes such as stabilization time and error control overshoot rate under two typical disturbing conditions, namely, sudden load increase and sudden load decrease. The algorithm combines fast dynamic response, high steady state accuracy and strong anti-disturbance ability, and shows remarkable robustness and adaptability, which can effectively guarantee the system to realize smooth transition and precise control when the load is suddenly changed.

4.2 Application Analysis of Traditional PID and DRL Control Algorithms

As shown in Table 2, in embedded control system deployment, the PID controller demonstrates significant lightweight and real-time advantages due to its extremely low resource consumption and sub-millisecond inference latency, making it suitable for simple control scenarios. In contrast, SAC and DDPG reinforcement learning controllers exhibit substantially increased resource overhead owing to their neural network architectures. Despite higher resource consumption and latency, SAC and DDPG deliver superior control precision, adaptability, and robustness in complex dynamic operating conditions, with their latency remaining acceptable in non-extreme real-time scenarios. Consequently, traditional PID controllers remain the ideal choice for resource-constrained, high-realtime applications, while SAC/DDPG represent a cutting-edge approach to intelligent control on mid-to-high-performance hardware platforms, offering viable deployment solutions for high-precision control tasks.

Table 2. Comparison of PID and intelligent agent embedded implementations.

	PID	SAC	DDPG
ROM Usage/KB	8.11	215	132
RAM Usage/KB	0.84	78	55
Inference Latency/ms	<1	4	3

5 Conclusion

(1). SAC and DDPG significantly outperform traditional PID control in methanol engine speed control regarding response speed, error reduction, and overshoot suppression. SAC delivers the best performance, shortening startup stabilization time by over 30% while exhibiting strong robustness and rapid recovery under sudden load disturbances. This substantially enhances dynamic regulation performance and control accuracy.

(2). PID controllers require minimal resources and enable ultra-fast responses. SAC/DDPG demand higher memory resources due to their neural network architectures. Despite increased resource consumption, reinforcement learning algorithms offer superior precision and robustness in complex dynamic conditions, demonstrating practical deployment value.

References

1. Liu, F., Guo, H., Zheng, X., Li, H., Wang, X.: Mechanism of methanol and formaldehyde emissions from methanol-fueled engines. Fuel Process. Technol. **268**, 108177 (2025)
2. Han, R., Lian, R., He, H., Han, X.: Continuous reinforcement learning-based energy management strategy for hybrid electric-tracked vehicles. IEEE J. Emerg. Sel. Topics Power Electron. **11**(1), 19–31 (2021)
3. Egan, D., Zhu, Q., Prucka, R.: A review of reinforcement learning-based powertrain controllers: effects of agent selection for mixed-continuity control and reward formulation. Energies. **16**(8), 3450 (2023)
4. Huang, R., He, H.: Naturalistic data-driven and semission reduction-conscious energy management for hybrid electric vehicle based on improved soft actor-critic algorithm. J. Power Sources. **559**, 232648 (2023)
5. Gao, W., Zhou, X., Pan, M., Zhou, W., Lu, F., Huang, J.: Acceleration control strategy for aero-engines based on model-free deep reinforcement learning method. Aerosp. Sci. Technol. **120**, 107248 (2022)
6. Zhu, Z., Lin, K., Jain, A.K., Zhou, J.: Transfer learning in deep reinforcement learning: a survey. IEEE Trans. Pattern Anal. Mach. Intell. **45**(11), 13344–13362 (2023)

Research on Distributed Data Interaction Technology Based on Dual-Mode WiFi-CAN Node Control Modules

Wenyuan Jiang[1], Meng Yang[1,2(✉)], Zhicheng Yuan[1,2], Yongming Bian[1,2], and Guangjun Liu[1,2]

[1] School of Mechanical Engineering, Tongji University, Shanghai 201804, China
yangmeng@tongji.edu.cn

[2] National Engineering Technology Research Center for Prefabrication Construction in Civil Engineering, Tongji University, Shanghai 200092, China

Abstract. The CAN protocol has been widely used in automotive control, sensor signal transmission, and industrial control due to its high reliability and strong real-time performance. However, in scenarios such as intelligent building retrofitting, mine equipment monitoring, large-scale production line engineering control, and outdoor synchronized construction operations—where large-scale distributed deployment, long-distance transmission, and real-time monitoring of CAN messages are required—traditional CAN communication solutions often fall short due to limited wiring space and the restricted transmission distance of CAN signals. To address this challenge, this paper designs and tests a WiFi-CAN node control module with an AP + STA dual-mode capability. By introducing WiFi signals as relays, a dedicated wireless local area network environment for control nodes is constructed over a large area. In this environment, device deployment is more flexible and convenient, no longer constrained by factors such as cabling and power supply. Through practical testing, the proposed solution meets the requirements for control accuracy and temporal precision.

Keywords: CAN · WiFi-CAN Module · Latency

1 Introduction

Traditional schemes are typically unidirectional, either merely reading CAN device messages for data monitoring or sending limited control messages to CAN devices. When applied to large-scale distributed layouts of CAN devices, these traditional methods face challenges such as difficult wiring and uncontrollable delays.[1] With the advent of the intelligent wave and the emergence of concepts like digital twins and smart construction sites, traditional data communication methods can no longer meet the demands of big data interaction and flexible addition or removal of control nodes under intelligent control and management scenarios.[2]

This paper designs a distributed data interaction scheme using a dual-mode WiFi-CAN node control module. In this scheme, the AP (Access Point) mode module serves

S. Halgamuge et al. (Eds.): ICACMVE 2025, LNME, pp. 455–463, 2026.
https://doi.org/10.1007/978-981-95-6911-3_45

as a relay to expand the wireless control local area network (WLAN). The STA (Station) mode module interacts with CAN devices by sending corresponding control signals from the WLAN to the CAN devices and feeding back CAN device messages in real-time to the control system connected to the WLAN. This data interaction scheme is highly scalable, flexible in layout, and supports remote control, offering significant advantages over traditional schemes in intelligent transformation scenarios with numerous and widely distributed CAN devices.[3]

Focusing on the critical issue of delay in CAN device control, this paper sets up a test environment in an actual factory setting. The tests measured the delay in data interaction between CAN device nodes and the control system at distances of 20 m, 50 m, and 100 m, with CAN devices distributed over a range exceeding 100 m and controlled remotely by the data interaction control system. The average delay was kept within 100 ms, meeting the requirements of the vast majority of application scenarios.

2 Overall Design of Data Interaction Scheme

The distributed data interaction scheme designed in this paper is shown in Fig. 1, with an overall architecture in a "star" topology. Each WiFi-CAN node control module is equipped with three CAN signal interfaces, capable of managing three CAN devices simultaneously. During use, users can group up to three CAN devices into a device group, which is then controlled by a single WiFi-CAN node control module. Several such nodes are connected via an inherent wireless local area network (WLAN) or a WLAN generated by a module in AP (Access Point) mode, thus forming a data interaction WLAN. The data interaction control system is established on the main control server side, and it can interact with the corresponding WiFi-CAN node control modules connected within different data interaction networks through device IDs.

3 Functional Design of WiFi-CAN Node Control Module

The dual-mode WiFi-CAN node control module designed in this paper is shown in Fig. 2. It is designed around three CAN interfaces, a communication serial port, and an ESP01-F Soc chip with WiFi functionality. To meet the needs of diverse application scenarios, the module uses a switchable Type-C and 5 V lithium battery dual-mode power supply, which is compatible with most low-voltage power supply methods, maximizing its applicability. To ensure real-time data transmission, the module uses the UDP protocol and has expanded SRAM to temporarily store transmitted data, preventing data accumulation from causing message loss.[4]

Considering that actual application scenarios are often harsh environments such as factory workshops and outdoor construction sites, the module has been specially designed in terms of signal bus, power supply, and antenna layout to minimize external environmental interference on the signal.[5]

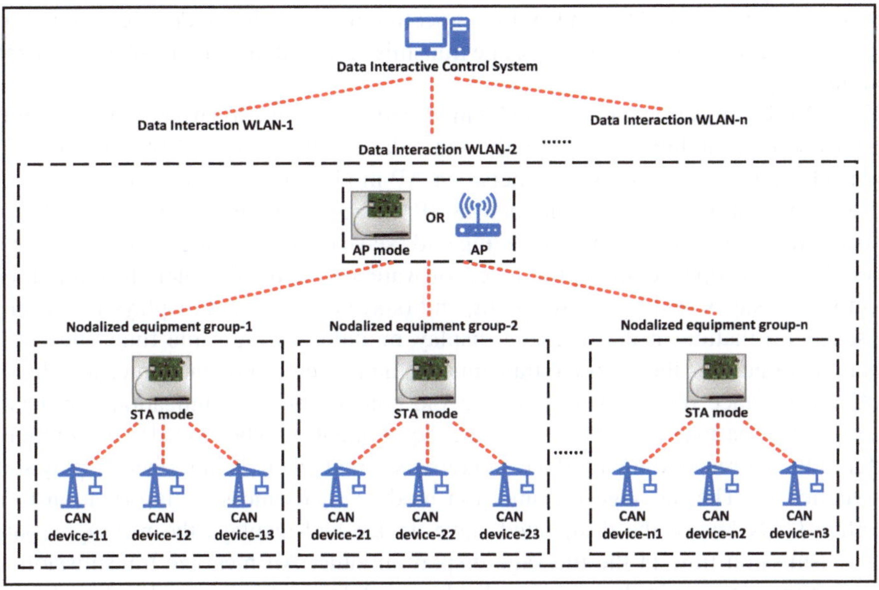

Fig. 1. Overall Data Interaction Scheme Design.

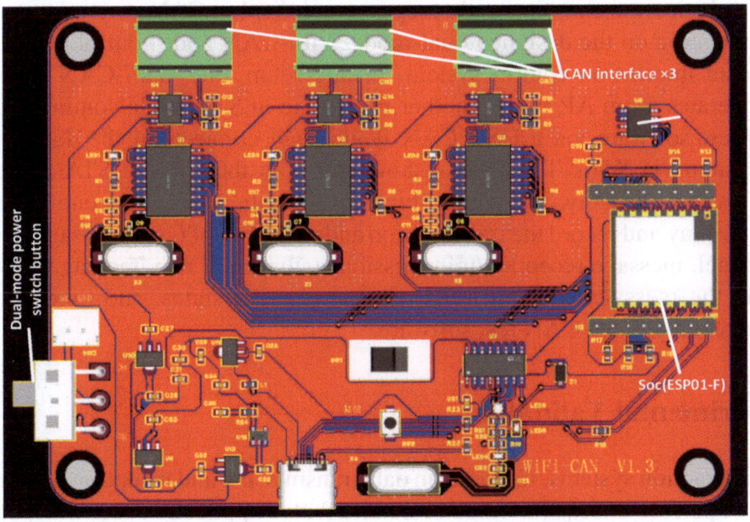

Fig. 2. The Modular PCB Hardware Design.

4 Design of Data Interaction Control System

As shown in Fig. 3(a), the host computer software of the data interaction control system is developed based on the Windows desktop application environment. It primarily consists of two main components: the main window and the prompt dialog box. The main window includes five major modules: the WiFi connection module, the mode selection module, the port connection module, the message transmission module, and the message display module.

The WiFi connection module can be modified at any time to ensure that the control-end host and all modules are within the same data interaction local area network. The mode selection module can switch between AP mode and STA mode for the modules and display required information such as IP addresses within the module. The port connection module specifies the receiving end (slave device) for message transmission. When the port is opened, the upper-level software automatically enters listening mode, monitors messages sent by the receiving end (slave device), and displays the received message information in the message display module. At the same time, the control system distinguishes the message transmission and reception of each interface through the CAN interface ID and baud rate to avoid concurrent communication conflicts.[6] Message information that does not meet the requirements will be moved to the temporary buffer of the corresponding interface and retransmitted when the interface display status is switched to. The message transmission module sends message information to the specified CAN bus of the designated slave device and displays the message content and transmission time in the message display module. All message transmission and reception information in the message display module can be exported in Excel format for storage and analysis.

As shown in Fig. 3, this paper uses the CANTest software, which can simulate real-time CAN device transmission and reception and display CAN messages, to perform functional testing on the designed dual-mode WiFi-CAN node control module and its adapted data interaction control system. In this setup, one WiFi-CAN node control module operates as an AP, while another m module in STA mode communicates with the CANTest software via a USB CAN device and connects to the data interaction control system via WiFi. Test results show that both the AP and STA modes of the module function normally. The host software of the data interaction control system can normally display and record the transmission and reception of CAN messages (including CAN channel, message reception/transmission status, baud rate, message content, and time), and the message content, transmission/reception status, and time information match the records in the CANTest software.

5 Experimental Validation

For data interaction systems, the delay in data transmission and reception is an important indicator of performance.[7] In large-scale distributed application scenarios, delay control is even more critical. This paper conducts theoretical analysis and practical testing of the internal delay of the designed WiFi-CAN node control module and its usage delay in actual deployment environments based on actual application requirements.

5.1 Delay Analysis and Testing Within Module Nodes

As shown in Fig. 4, within each CAN node in the module, messages primarily interact between three components: the SoC, the CAN controller, and the CAN transceiver.[8]

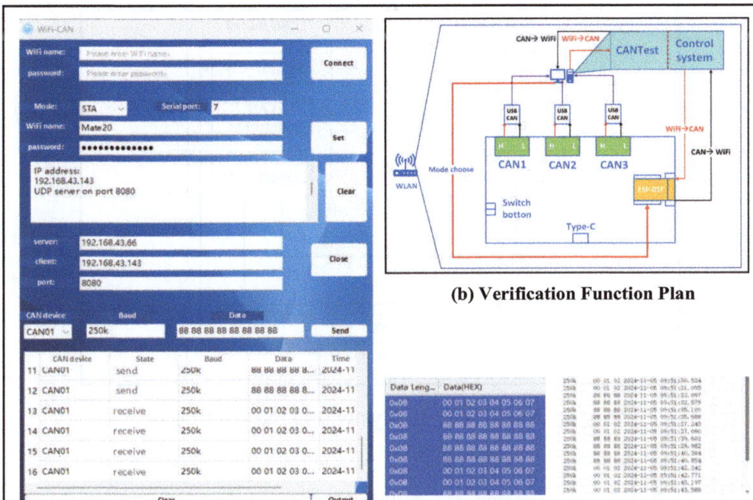

(a) Control System Interface (b) Verification Function Plan (c) CANTest data (d) saved data

Fig. 3. The Control System Interface and Functional Testing.

Subsequently, the node communicates with the CAN bus via the CAN transceiver to exchange data. As shown in Fig. 6, the delay can be divided into the delay t3 on the CAN bus, the delay t2 = t4 within the CAN node, and the delay t1 = t5. Among these, the delays t1 and t5 between the SoC and the CAN controller are generated by the execution of the program and primarily depend on the performance of the system network, such as the CPU processing speed. These time delays are generally considered fixed and have a delay magnitude in the microsecond range, which can be neglected relative to the total delay in the communication process. Setting timing interrupt points at different stages of communication enables the measurement of delays in different sections. Considering a complete CAN communication transmission and reception process, the total delay t of CAN communication can be expressed as:

$$t = t2 + t3 + t4 \tag{1}$$

For the WiFi-CAN node control module, the data transmission process is "CAN bus - CAN node 1 - WiFi signal - CAN node 2 - CAN bus." The t3 portion of the above model does not occur on the bus and can be approximated as the delay t3' caused by the transmission of the WiFi signal. Therefore, the delay performance is primarily reflected in the delays t1 and t2 within the CAN node and the delay t3' during the WiFi signal transmission process. Since t3' is influenced by external factors such as transmission distance, the testing focuses on the delay generated during the process from receiving the UDP data packet to processing and packaging it into a CAN message data packet. The delay in this process and its reverse process is measured by the module's built-in program.

Test results show that the average delay in data processing by the module is 2.00 ms, with the maximum delay not exceeding 2.10 ms. Within the 50-m transmission range of

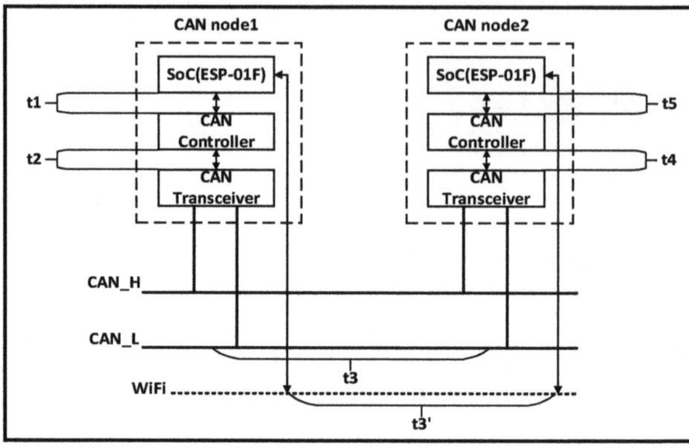

Fig. 4. Analysis of Delays within the CAN Node.

the core SoC (ESP-01F), the transmission delay can be maintained within 5.00 ms, which is comparable to the delay performance of similar modules available on the market.

5.2 Delay Testing of Modules in Actual Scenarios

In actual use, the delay t3' mentioned in Sect. 5.1 cannot be ignored, and the total communication delay is ΔT.

$$\Delta T = (t1 + t5) + (t2 + t4) + t3' \tag{2}$$

This paper conducted on-site testing of the latency of the entire signal transmission process from WiFi to CAN modules within a typical factory building. For practical application scenarios, the specific test setup is shown in Fig. 5, using two computers connected to WiFi-CAN node control modules as test platforms 1 and 2. Test Platform 1 is located in the monitoring area of an office building adjacent to the factory area. The main AP is deployed in the monitoring area, and WiFi signals are covered throughout the monitoring area and factory area via fiber optic cables and extended APs. Test Platform 2 is deployed within the factory area near extended AP-2, and the system latency is tested at straight-line distances of 20 m, 50 m, and 100 m from Test Platform 1.

In the experiment, the three CAN interfaces of the test platform are all connected to the computer to simulate the situation where multiple nodes are simultaneously connected to the distributed control system. Test platform 1 and test platform 2 use the data interaction system's upper computer software to take turns as the sender and receiver of CAN messages respectively. The average time of each node of test platform 1 is recorded as T1, and the average time of each node of test platform 2 is recorded as T2.The timestamp information obtained by the host computer software depends on the computer platform itself and is influenced by factors such as the system clock not being fully synchronized and differences in the hardware time accuracy of the two computers. As a result, the timestamps T1 and T2 obtained have errors compared to the actual time and cannot be directly subtracted to obtain the delay time. During the experiment, it was

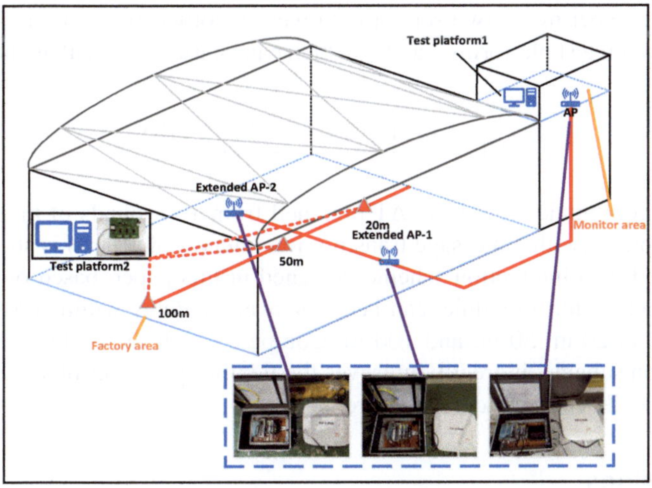

Fig. 5. Schematic Diagram of the Delay Test Site.

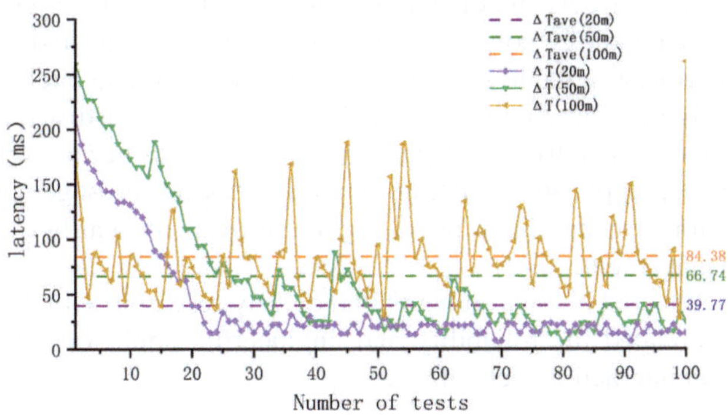

Fig. 6. The Delay Test Results under Different Module Distribution Distances.

assumed that the system clocks of the two computers did not drift within a short period of time,[9] so the difference in their timestamps was relatively fixed and denoted as Δt.

$$\Delta t = T2 - T1 \tag{3}$$

When Test Platform 1 sends a message, the time recorded is T1s; when Test Platform 2 receives the message, the time recorded is T2r. When Test Platform 2 sends a message, the time recorded is T2s; when Test Platform 1 receives the message, the time recorded is T1r. Where:

$$T2r = T1s + \Delta t + \Delta T \tag{4}$$

$$T1r = T2s - \Delta t + \Delta T \tag{5}$$

During the experiment, two sets of data were recorded from the transmission and reception between test platforms 1 and 2, which can be used to calculate the total system delay ΔT.

$$\Delta T = \frac{(T2r - T1s) + (T1r - t2S)}{2} \quad (6)$$

The total communication delay ΔT obtained from 100 rounds of mutual transmission and reception (with a message transmission interval of 10 ms) is shown in Fig. 6. The distributed data interaction scheme designed in this paper, based on a dual-mode WiFi-CAN node control module, can keep the average delay within 100 ms when the node distance is 20 m, 50 m, and 100 m. Compared with the traditional CAN-based data interaction scheme, it greatly improves the flexibility of control system layout and modification while ensuring delay requirements.

6 Conclusions

- This paper proposes a distributed data interaction technology based on an AP + STA dual-mode WiFi-CAN module node control module. Compared with traditional solutions, this technology solution is flexible in deployment and highly scalable, making it particularly suitable for engineering applications and intelligent renovation scenarios involving a large number of devices that are widely distributed.
- A control system host software adapted to the module has been developed based on Windows desktop applications. It can monitor and send messages in real time and has message storage capabilities, providing the software and hardware foundation for intelligent management.
- Delay tests conducted in environments that closely resemble actual application scenarios have validated the reliability and engineering application potential of this distributed data interaction technology.

Acknowledgments. We express our profound gratitude to the anonymous reviewers for their insightful comments and constructive suggestions, which have significantly contributed to the enhancement of this manuscript. This research was supported by the National Natural Science Foundation of China (52205279), the National Key Research and Development Program of China (2023YFF0613200 & 2023YFC3806603) and the Top Discipline Plan of Shanghai Universities-Class I.

Disclosure of Interests. The authors have no competing interests to declare that are relevant to the content of this article.

References

1. Wei, B., Liu, H.J., Zhang, T.: Research and application of construction machinery controller combined with WiFi and CAN. Build. Mechanization. **39**(06), 28–30 (2018)
2. Teh, K.H., Ng, W.L., Ng, C.K., Noordin, N.K.: Home appliances management system using controller area network (CAN). In: 17th Asia-Pacific Conference on Communications (APCC), pp. 899–904, Kota Kinabalu, Malaysia (2011)

3. Hong, Y.Q., Li, Q.D., Mou, H.M., et al.: A distributed EtherCAT-to-CAN high-speed control module. Electron. Sci. Technol. **38**(08), 49–56 (2025)
4. Wang, Q.H., Wu, W.J., Fan, B.T., et al.: Development of data transmission system for UDP to CAN protocol. Sci. Technol. Innov. **21**, 98–99 (2019)
5. Li, X.J., Liu, S.: Application research of CAN bus communication relay in stratospheric unmanned aerial vehicles. Power Technol. **48**(01), 174–180 (2024)
6. Wei, C.S., Du, X.G., Tan, K.W.: A multi-Tire pressure monitoring system based on distributed CAN bus network. Mech. Electr. Eng. Technol. **54**(14), 154–158 (2025)
7. Li, W.X.: Design of Distributed Pump Station Frequency-Conversion Synchronous Control System Based on CAN bus protocol. Autom. Appl. **66**(14), 247–249 (2025)
8. Chin, J.-C., Thapliyal, H., Cultice, T.: CAN bus: the future of additive manufacturing (3D printing). IEEE Consum. Electron. Mag. **13**(5), 23–29 (2024)
9. Xie, Y.K., Wang, J.D., Zhu, C., et al.: Research and implementation of high-precision timestamps in network measurement. J. Comput. Res. Dev. **47**(12), 2049–2058 (2010)

Design of a Virtual LiDAR System Simulation Training Platform for Port Automation Equipment Based on Digital Twin

Yujie Zhang[1], Xintai Man[2], Mengjie He[3], Chao Mi[2], Octavian Adrian Postolache[4], and Yang Shen[5](✉)

[1] School of Technology and Architecture, ISCTE-Instituto Universitário de Lisboa, Lisbon 1649-026, Portugal
yzgeu@iscte-iul.pt
[2] Logistics Engineering College, Shanghai Maritime University, Shanghai 201306, China
202430210134@stu.shmtu.edu.cn, chaomi@shmtu.edu.cn
[3] Information Engineering College, Shanghai Maritime University, Shanghai 201306, China
[4] Instituto de Telecomunicações and also with ISCTE-Instituto Universitário de Lisboa, Lisbon 1649-026, Portugal
aopostolache@lx.it.pt
[5] Higher Technology College, Shanghai Maritime University, Shanghai 201306, China
yangshen@shmtu.edu.cn

Abstract. As the core sensing method for port automation equipment, LiDAR is widely used in spatial distance measurement, three-dimensional positioning, and environmental perception. However, its ranging mechanism, parameter configuration, and role in complex port operations are relatively abstract, making traditional training difficult to achieve intuitive understanding and efficient application. To address this issue, this paper proposes a virtual LiDAR simulation training platform based on digital twin technology. The platform constructs a virtual environment highly consistent with real port operation processes, designs a virtual LiDAR module with adjustable horizontal/vertical field of view, resolution, channel count, and ranging distance, and achieves bidirectional synchronization of data and control signals between physical and twin spaces. Using a bulk cargo handling machine as the verification object, consistency comparison experiments were conducted on point cloud density, ranging accuracy, and geometric similarity. The results show that the virtual point clouds generated by the platform are highly consistent with real point clouds in key indicators, featuring low cost, high flexibility, and scalability, and can provide reliable support for operation training, algorithm testing, equipment deployment assessment, and operational safety analysis of port automation equipment.

Keywords: Port Automation Equipment · Digital Twin · LiDAR · Virtual Sensor · Simulation Training Platforms

S. Halgamuge et al. (Eds.): ICACMVE 2025, LNME, pp. 464–472, 2026.
https://doi.org/10.1007/978-981-95-6911-3_46

1 Introduction

Automated port equipment serves as the core component for achieving intelligent and unmanned port operations, significantly enhancing efficiency in loading/unloading, storage, and transportation.[1] As a critical non-contact sensing device, LiDAR provides precise spatial perception data for automated equipment such as gantry cranes.[2] Taking bulk terminal gantry cranes as an example, LiDAR is required to measure the 3D pose of grab buckets, ship hatches, and discharge hopper positions in real-time.[3] However, its working principles (ranging mechanisms, parameter configuration) remain too abstract for engineers to intuitively comprehend the mapping relationships between point cloud data and physical targets through traditional training methods. On-site sensor teaching experiments prove prohibitively expensive and pose safety risks,[4] potentially leading to operational errors.

To address abstract cognition challenges of LiDAR, semi-physical simulation technology has been introduced into training methodologies. Mi et al.[5] proposed an electrical engineering experimental teaching approach based on semi-physical simulation, establishing a new paradigm for complex system training. For port applications, Zhang et al.[6] developed a vision measurement simulation platform utilizing digital twin technology, while He et al.[7] employed virtual reality to evaluate risk behaviors of quay crane operators.

Significant advancements have been achieved in LiDAR simulation: Wang et al.[8] revealed ranging principles through synthetic point cloud generation, Yang et al.[9] enhanced data authenticity by optimizing scenarios, and Abu et al.[10] generated highly realistic data by integrating virtual sensors with real backgrounds. Digital twin technology provides theoretical support: Tao et al.[11] established its fundamental framework, with Shen et al.[12] detecting truck lifting accidents using improved PointNet++ for container terminals.

To address the aforementioned issues, this paper makes the following key contributions:

1) To resolve the abstraction challenges of LiDAR, this paper proposes a virtual LiDAR digital twin platform architecture tailored for port automation equipment, achieving organic integration of virtual scenarios, physical spaces, and engineering validation modules.

2) To address the insufficient consideration of LiDAR parameters in existing methods, this paper designs a virtual LiDAR modeling approach that allows adjustable horizontal/vertical field of view, resolution, channel count, and ranging capabilities.

3) Using bulk cargo portal cranes as the application object, a comprehensive training simulation platform was constructed, covering the interaction between physical space and digital twin space, communication modules, and engineering validation modules. Consistency verification experiments were conducted in real port scenarios.

2 Simulation Training Platform Architecture

The virtual LiDAR simulation training platform for port automation equipment consists of physical space, digital twin space, communication modules, virtual LiDAR parameter configuration, and engineering validation modules, as shown in Fig. 1.

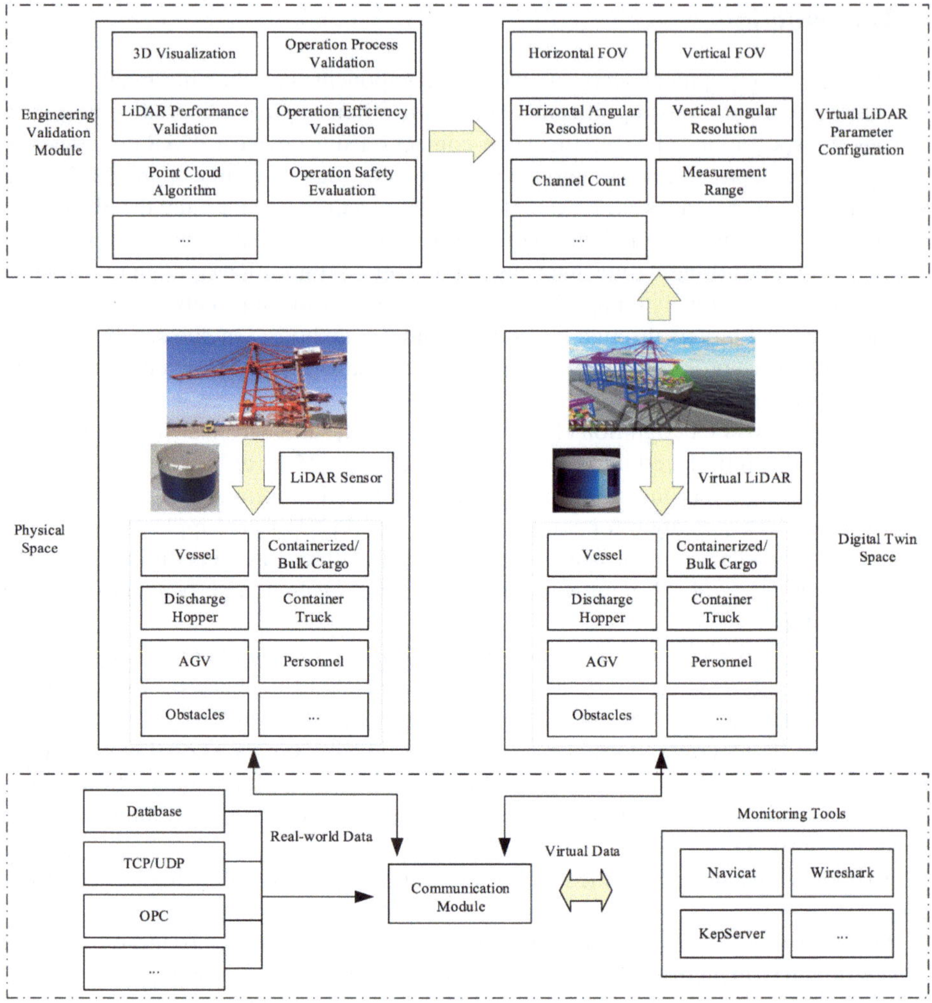

Fig. 1. Architecture Diagram of the Digital Twin Simulation Training Platform.

The physical space includes real operational objects such as port machinery, ships, operational targets (containers or bulk materials), discharge hoppers, AGVs, personnel, and obstacles. It is equipped with multi-line LiDAR to collect point cloud data of the operational environment, which is transmitted in real-time through communication modules to provide dynamic data support for the digital twin space.

The digital twin space performs high-precision modeling of the physical operating environment, deploys configurable virtual LiDAR, and enables flexible adjustment of parameters such as horizontal/vertical field of view, resolution, number of channels, and ranging distance, while generating point cloud data consistent with the physical scene. The communication module supports various methods including databases, TCP/UDP, and OPC, ensuring bidirectional synchronization of data and control signals between the physical and virtual environments.

The engineering validation module integrates functionalities such as 3D visualization, LiDAR performance verification, and automated operation efficiency and safety assessment, etc. It supports diverse training and validation scenarios based on data generated by virtual LiDAR. Through the coordinated operation of physical and twin spaces, the platform provides an intuitive, flexible, and scalable simulation environment for training and validating LiDAR applications in port automation equipment

3 Virtual LiDAR Parametric Modeling

The virtual LiDAR module developed in this study aims to accurately replicate the working mechanism and data characteristics of multi-line LiDAR in a digital twin environment, ensuring its authenticity and controllability in the training and verification of port automation equipment. This module features adjustable parameters, configurable precision, and controllable noise, supporting simulations of various LiDAR models and application scenarios.

3.1 Coordinate System Definition and Angle Sampling

The virtual LiDAR uses the sensor center O as the coordinate origin and adopts a right-handed Cartesian coordinate system, where x, y, and z represent the spatial axes for forward, left, and vertically upward directions, respectively. The emission direction of each laser beam is determined by horizontal α and vertical β angles, with the calculation process as shown in Eq. (1):

$$\begin{aligned} \alpha_i &= \alpha_{\min} + i \cdot \Delta\alpha, \quad i \in [0, N_h - 1] \\ \beta_j &= \beta_{\min} + j \cdot \Delta\beta, \quad j \in [0, N_v - 1] \end{aligned} \tag{1}$$

where i represents the vertical direction's i-th laser beam, j represents the horizontal direction's j-th sampling point, and N_h, N_v are the number of sampling points in the horizontal and vertical directions respectively, and $\Delta\alpha$, $\Delta\beta$ are the horizontal and vertical angular resolutions respectively, and $\alpha_{\min}$, $\beta_{\min}$ are the minimum scanning angles in the horizontal and vertical directions.

32Distance Measurement and Noise Modeling

To more closely approximate the distance measurement behavior of real sensors, the module incorporates distance measurement accuracy parameters and a random noise model in the calculation for each laser beam. The ideal distance measurement value is, and the actual measured value can be expressed as:

$$d_{i,j} = d_{i,j}^{\text{deal}} + \epsilon_r \tag{2}$$

where $d_{i,j}$ is the actual measured value of the j-th vertical scan line at the i-th horizontal sampling point, $\epsilon_r \sim \mathcal{N}(0, \sigma^2)$ follows a zero-mean Gaussian distribution, σ is the ranging accuracy parameter, which can be configured according to the device model or on-site environmental conditions. This model can reflect the ranging fluctuations caused by hardware characteristics and environmental conditions.

3.2 Point Cloud Coordinate Calculation

Based on the coordinate system and angle sampling method defined in the previous subsection, as well as the ranging and noise model, each laser beam's horizontal angle α_j, vertical angle β_i, and actual ranging value $r_{i,j}$ can be converted into 3D point cloud coordinates. The calculation method is shown in Eq. (3):

$$\begin{cases} x_{i,j} = d_{i,j} \cdot \cos\beta_i \cdot \cos\alpha_j \\ y_{i,j} = d_{i,j} \cdot \cos\beta_i \cdot \sin\alpha_j \\ z_{i,j} = d_{i,j} \cdot \sin\beta_i \end{cases} \tag{3}$$

where $(x_{i,j}, y_{i,j}, z_{i,j})$ represents the spatial coordinates of the i-th vertical scan line at the j-th horizontal sampling point; $r_{i,j}$ is obtained by combining the ideal ranging value with the ranging accuracy parameter and random noise correction, ensuring that the coordinate calculation results are closer to the actual sensor output. By performing the above calculations for all scan lines and sampling points, a complete 3D point cloud frame can be obtained, providing foundational data for subsequent visualization, recognition algorithm verification, and training tasks.

4 Application Validation

In preliminary work, we conducted tests on real port LiDAR and accumulated experience. To validate the effectiveness of the constructed digital twin virtual LiDAR simulation training platform in the training of port automation equipment and engineering verification, this study focuses on the dry bulk cargo gantry crane operation scenario as the application object, conducting experiments on point cloud generation consistency and performance evaluation.

4.1 Experimental Environment and Platform Configuration

The experimental platform consists of two parts: the physical space and the digital twin space. The physical space covers the actual scenario of the dry bulk cargo gantry crane operation area, including the gantry crane body, ships, hoppers, transport vehicles, and operators. Multi-line LiDAR are deployed at key locations to collect real-time 3D point cloud data of the operating environment, as shown in Fig. 2(a). The digital twin space is constructed in the Unity3D platform based on measured data and equipment parameters, meticulously replicating the geometric structure, equipment layout, and dynamic operation processes of the physical space. It integrates a virtual LiDAR module with adjustable horizontal/vertical field of view, resolution, line count, and ranging range, achieving bidirectional synchronization of data and control signals with the physical space, as illustrated in Fig. 2(b).

4.2 Scene and Equipment Deployment

In a real port, select typical working conditions in the dry bulk cargo gantry crane operation area, deploy a 16-line LiDAR, and collect point cloud data for a complete

(a) (b)

Fig. 2. Physical Space and Digital Twin Space of Bulk Cargo Gantry Crane Operation.

operation cycle. The virtual scene strictly replicates the geometric features and equipment layout of the real working environment, with parameters consistent with the real sensor, including field of view, resolution, number of lines, and ranging range, as detailed in Table 1.

To validate the geometric authenticity and ranging accuracy of the virtual LiDAR modeling, the platform generated 3D point clouds in the virtual scene with the same configuration as the real sensor, as shown in Figure 3(a). This virtual point cloud not only fully reflects the spatial distribution of equipment, operational objects, and the surrounding environment but also exhibits high consistency with measured data in geometric details. Additionally, to simulate potential ranging fluctuations and sampling disturbances that the sensor may encounter in complex environments, this study introduced a random noise model, with the resulting point cloud effect shown in Figure 3(b).

Table 1. Comparison of Real and Virtual Scene Configurations.

Parameter	Physical Configuration	Virtual Configuration
LiDAR Model	16-line	Virtual 16-line
Vertical FOV	+11.6° ~ -12.8°	+11.6° ~ -12.8°
Vertical Resolution	1.6°	1.6°
Horizontal FOV	0° ~ 359.8°	0° ~ 180°
Horizontal Resolution	0.2°	0.2°
Points per Line	1800	1800

4.3 Consistency Comparison and Performance Evaluation

The experiment collected 570 sets of real point cloud data and virtual point cloud data, with comparison metrics including:

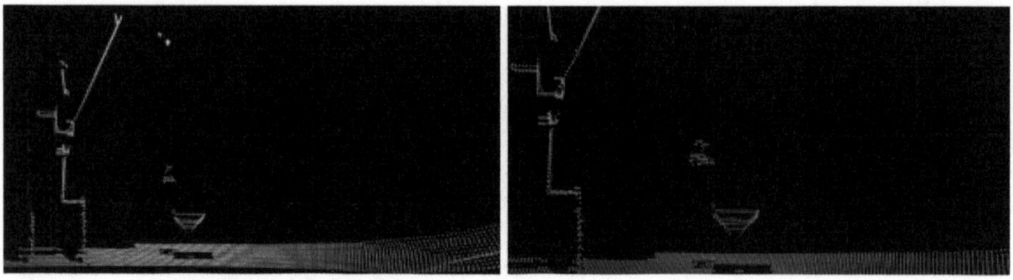

(a) (b)

Fig. 3. Original and Random noise virtual point cloud rendering.

1. Point cloud density consistency:

$$C_{\text{density}} = \frac{\left|N^{virtual} - N^{real}\right|}{N^{real}} \times 100\% \tag{4}$$

where $N^{virtual}$ and N^{real} represent the number of points per unit line for the real and virtual point clouds, respectively.

2. Range Measurement Error (RMSE):

$$E_{RMSE} = \sqrt{\sum_{i=1}^{N_v} \sum_{j=1}^{N_h} \left(d_{i,j}^{virtual} - d_{i,j}^{real}\right)^2} \tag{5}$$

where $d_{i,j}^{\text{virtual}}$ and $d_{i,j}^{\text{real}}$ denote the range measurement values of the j-th horizontal sampling point on the i-th vertical scan line in the virtual and real point clouds, respectively; N_v and N_h are the sampling counts in the vertical and horizontal directions, respectively, and $N = N_v \times N_h$ is the total number of aligned points. A smaller E_{RMSE} value indicates higher consistency in range measurements between the two.

3. Geometric Similarity (IoU-based):

$$S_{geo} = \frac{\left|P^{\text{virtual}} \cap P^{real}\right|}{\left|P^{\text{virtual}} \cup P^{real}\right|} \times 100\% \tag{6}$$

where P^{virtual} and P^{real} represent the virtual and real point cloud sets, respectively, calculating the intersection-over-union ratio based on spatially overlapping voxels.

Based on the above metric definitions and algorithms, Table 2 presents a comparative analysis of the consistency between virtual and real point clouds in terms of density, ranging accuracy, and geometric similarity.

As shown in Table 2, the virtual point cloud is completely consistent with real data in terms of density, with a ranging error of 98.3% and a geometric similarity of 96.5%. The experimental results demonstrate that the virtual point cloud generated by this platform exhibits high consistency with real point clouds in key metrics, proving the effectiveness of the digital twin modeling method in terms of geometric authenticity

Table 2. Comparison of Consistency Between Virtual and Real Point Clouds.

Metric	Physical	Virtual	Consistency (%)
LiDAR Model	1800/line	1800/line	100
Horizontal Resolution	0.05 m	0.04 m	98.3
Geometric Similarity	——	——	96.5

and measurement accuracy. The platform not only provides intuitive demonstrations of sensor working principles for the operational training of port automation equipment but also serves as a low-cost, highly flexible engineering validation tool, offering reliable data support for equipment deployment evaluation, and operational safety analysis.

5 Conclusion

This paper addresses the issues of high abstraction, elevated training costs, and the lack of specialized platforms in LiDAR training for port automation equipment by designing and implementing a virtual LiDAR simulation training platform based on digital twin technology. The platform integrates high-precision virtual port operation scenarios, parameterizable virtual LiDAR modeling methods, bidirectional communication between physical and twin spaces, and an engineering validation module, enabling comprehensive support from perception mechanism demonstrations to performance evaluation. Experimental validation shows that the platform achieves high consistency with real sensor outputs in terms of point cloud density, ranging accuracy, and geometric similarity, effectively enhancing the intuitiveness and specificity of training while demonstrating strong scalability and adaptability to multiple scenarios.

Acknowledgments. This research was supported by 2023 Education Science Research Project of Shanghai Municipality (No. B2023003)

References

1. Kon, W.K., Abdul Rahman, N.S.F., Md Hanafiah, R., Abdul Hamid, S.: The global trends of automated container terminal: a systematic literature review. Marit. Bus. Rev. **6**(3), 206–233 (2021)
2. Singh, D., Bhamu, J., Kumar, P.: Machine vision in industry 4.0: applications, challenges and future directions. In: Proceedings of the International Conference on Industry 4.0, pp. 1–12. Springer, Cham (2021)
3. Huang, Q., Huang, Y., Zhang, Z., Zhang, Y., Mi, W., Mi, C.: Truck-lifting prevention system based on vision tracking for container-lifting operation. J. Adv. Transp. **2021**, 9612480 (2021)
4. Raut, R., Krit, S., Chatterjee, P.: Machine Vision for Industry 4.0: Applications and Case Studies. CRC Press, Boca Raton (2022)
5. Mi, C., Huang, Y., Liu, Z., Mi, W.: A novel experimental teaching approach for electrical engineering based on semi-physical simulation. World Trans. Eng. Technol. Educ. **12**(4), 778–783 (2014)

6. Zhang, Y., Luo, K., Postolache, O., Mi, C.: Design of a simulation platform for visual measurement of port automation equipment based on digital twin technology. In: In: 2025 IEEE International Instrumentation and Measurement Technology Conference (I2MTC), pp. 1–6. IEEE, Piscataway (2025)
7. He, M., Zhang, Y., Liu, Y., Shen, Y., Mi, C.: Automated evaluation method for risk behaviors of quay crane operators at ports using virtual reality. Algorithms. **17**(11), 508 (2024)
8. Wang, F., Zhuang, Y., Gu, H., Hu, H.: Automatic generation of synthetic LiDAR point clouds for 3-D data analysis. IEEE Trans. Instrum. Meas. **68**(7), 2671–2673 (2019)
9. Yang, Y., Wang, J., Guo, X., Yang, X., Qin, W.: Methods for improving point cloud authenticity in LiDAR simulation for autonomous driving: a review. IEEE Access. **13**, 4562–4580 (2025)
10. Abu Alhaija, H., Mustikovela, S.K., Mescheder, L., Geiger, A., Rother, C.: Augmented reality meets computer vision: efficient data generation for urban driving scenes. Int. J. Comput. Vis. **126**(9), 961–972 (2018)
11. Tao, F., Qi, Q., Wang, L., Nee, A.Y.C.: Digital twins and cyber-physical systems toward smart manufacturing and industry 4.0: correlation and comparison. Engineering. **5**(4), 653–661 (2019)
12. Shen, Y., Man, X., Wang, J., Zhang, Y., Mi, C.: Truck lifting accident detection method based on improved PointNet++ for container terminals. J. Mar. Sci. Eng. **13**(2), 256 (2025)

Application Research of Digital Twin System in Operation Risk Assessment of Container Gantry Crane Drivers

Mengjie He[1], Xintai Man[2], Chao Mi[2], and Yang Shen[3](✉)

[1] Information Engineering College, Shanghai Maritime University, Shanghai 201306, China
mjie1780@163.com

[2] Logistics Engineering College, Shanghai Maritime University, Shanghai 201306, China
202430210134@stu.shmtu.edu.cn, chaomi@shmtu.edu.cn

[3] Higher Technology College, Shanghai Maritime University, Shanghai 201306, China
yangshen@shmtu.edu.cn

Abstract. With the continuous improvement of port automation, remote control of container gantry cranes has become mainstream. However, real machine training poses significant safety risks, where novice drivers (Trainee) are prone to equipment collisions and violent container swings during operations. To address this issue, this paper designs and implements a digital twin system integrated with an operation risk assessment module, used for safety reproduction and risk quantification of gantry crane operations. The system adopts a time series prediction method based on attention mechanism, establishing a safety benchmark by modeling skilled drivers' (Trainer) operational data, and predicts and compares novice drivers' operational behaviors to quantify risk deviations. Experimental results demonstrate that the system can effectively distinguish operation risks across different driver experience levels, showing excellent application value in avoiding real machine training hazards, improving training efficiency, and achieving personalized risk assessment.

Keywords: Digital Twin · Rail-Mounted Gantry Crane · Operation Risk Assessment · Time Series Prediction · Attention Mechanism

1 Introduction

Ports, as core nodes of the water transportation system, are accelerating their transformation towards intelligence. Rail-Mounted Gantry Cranes (RMGs) are critical equipment in port operations, responsible for container loading and unloading tasks between the yard and horizontal transport equipment. The application of automated remote control technology has significantly improved the operational efficiency of gantry cranes. However, during real machine training, novice drivers are prone to causing accidents such as equipment collisions, violent container swinging, and even personnel injuries due to lack of experience. To mitigate the risks of real machine training, digital twin technology has gradually become an important means in port training.

S. Halgamuge et al. (Eds.): ICACMVE 2025, LNME, pp. 473–482, 2026.
https://doi.org/10.1007/978-981-95-6911-3_47

Existing research on container lifting equipment driver operation behavior and safety primarily focuses on swing control, and accident detection. For example, Chen Bin et al. and Wang Tiexin proposed risk control measures from human factor design and evaluation system perspectives [1, 2]. In accident detection research, Huang et al. and Shen et al. used visual tracking and PointNet++ to detect truck lifting accidents [3, 4]; Wang et al. and Zhang et al. achieved high-precision pose measurement through YOLO algorithms [5, 6]. While these achievements have laid the foundation for operational safety improvement, most focus on single data dimensions, lacking multi-source data fusion and digital twin support, which limits real-time risk simulation in complex scenarios.

Digital twin technology is widely applied in crane training, utilizing immersive virtual environments to improve operational skills and safety awareness while reducing training costs and risks. In the port domain, Song et al. and Guo Xiaoqiang et al. verified the effectiveness of VR training [7, 8]; Zhang Liandi et al. developed an immersive virtual driving system with automatic evaluation support [9]; Zhang et al. designed a digital twin visual measurement platform to improve detection accuracy [10]. However, these systems mostly focus on skill training, lacking systematic evaluation of safety protocol compliance and risk prediction capabilities.

In the automated evaluation of operational behaviors, data-driven and machine learning methods have been widely applied. Huang Lirong et al. employed deep belief networks and multi-indicator evaluation to quantify driving behaviors [11]; He et al. constructed a quay crane operational risk evaluation model using deep Q-networks [12]. While these methods perform well in vehicle driving and port scheduling domains, research on automated risk assessment for rail-mounted gantry crane operators remains scarce, particularly lacking real-time evaluation frameworks integrating digital twins.

Addressing these issues, this paper designs an integrated digital twin system with an operational risk assessment module, achieving safety reproduction through high-fidelity simulation of rail-mounted gantry crane work scenarios. We propose a time series prediction algorithm based on attention mechanisms, modeling safety benchmarks using expert operator data, then inputting novice operator data into the model for prediction and comparison to quantify risk deviation. The innovations of this paper include:

(1) Innovatively introduce a Trainer-Trainee comparative evaluation framework based on digital twin technology, achieving risk-free reproduction of real scenarios and automated quantitative risk assessment in container crane operator training;
(2) Propose a time series prediction model incorporating multi-head self-attention mechanism to enhance the ability to capture long-sequence dependencies and critical risk moment features in complex container crane operations (especially during stabilization processes);
(3) Focusing on high-risk work scenarios such as stabilization operations, provide quantifiable risk assessment criteria for novice drivers through predictive deviation quantification and curve comparison.

2 Digital Twin System Design

The digital twin system constructed in this study consists of three parts: the physical entity, the virtual twin, and the evaluation module, as shown in Fig. 1.

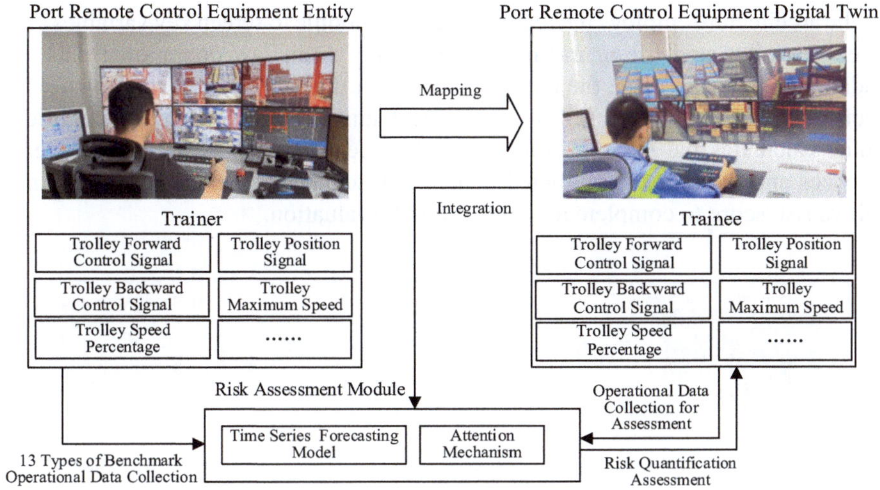

Fig. 1. Digital Twin System Architecture.

At the physical layer, multi-source data is collected during operations performed by experienced operators, including trolley forward/backward control signals, position, speed percentage, and maximum speed, serving as the safety benchmark. Simultaneously, control signals and state data from novice operators working within the twin environment are collected in real-time. At the virtual layer, the system builds a rail-mounted gantry (RMG) crane model based on the Unity3D physics engine, replicating dynamic characteristics such as gantry travel, trolley movement, spreader hoisting/lowering, and container sway. This ensures novice operators gain a control experience consistent with a real port environment under risk-free conditions. At the evaluation layer, the system integrates a risk assessment module that employs a time series forecasting model based on a multi-head self-attention mechanism. This model is trained using 13 types of benchmark operational data from experienced operators to generate an ideal operational sequence. Subsequently, the operational sequences of novice operators are input into the model and compared against the benchmark sequence. Metrics including speed control error, operation delay, and trajectory deviation are calculated and converted into risk scores using the Mean Absolute Error (MAE), thereby completing the quantitative assessment of the novice operators' performance.

3 Rail Crane Driver Operation Risk Assessment Model

3.1 Operation Risk Assessment Model Architecture

The core source of rail-mounted gantry crane operation risks stems from critical control link errors, especially during anti-swing operations where sudden changes in trolley speed can easily trigger violent container swinging, significantly increasing collision accident potential. To quantitatively assess such risks, this model architecture is constructed from three core modules: feature selection, time series prediction, and risk quantification, as shown in Fig. 2.

The architecture employs the feature selection module to screen key variables closely related to trolley and lifting mechanism dynamics from raw operational data. After standardized preprocessing, the data is input into a time series prediction model based on an improved multi-head self-attention mechanism. Finally, the risk quantification module calculates the mean absolute error between novice drivers' actual operation sequences and the model's predicted benchmark sequences, transforming this into an intuitive risk score to complete a systematic risk evaluation.

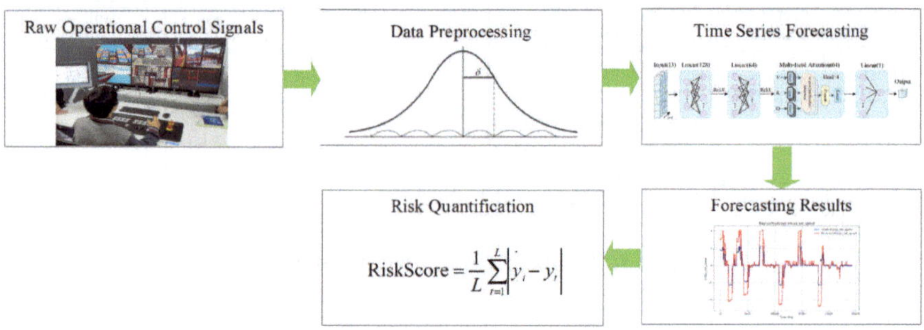

Fig. 2. Rail-Mounted Gantry Crane Driver Operation Risk Assessment Model Architecture.

3.2 Feature Selection and Data Preprocessing

To accurately reflect the driver's risk characteristics during anti-swing operations, 13 key variables related to the trolley and spreader were selected as inputs, including control signals, speed percentages, position coordinates, etc. All original feature data x underwent standardization processing, with the standardized features x' expressed as:

$$x' = \frac{x - \mu}{\sigma} \tag{1}$$

where μ and σ represent the feature mean and standard deviation, respectively. Subsequently, a sliding window method was employed to construct input sequences suitable for time series prediction from continuous time-step data.

3.3 Improved Time Series Prediction Model

To characterize the temporal features of a container crane driver during anti-swing operations, this paper proposes an improved time series prediction model based on multi-head self-attention mechanism. The model consists of a complete process through input representation, feature mapping, multi-head self-attention, and output layer, as shown in Fig. 3.

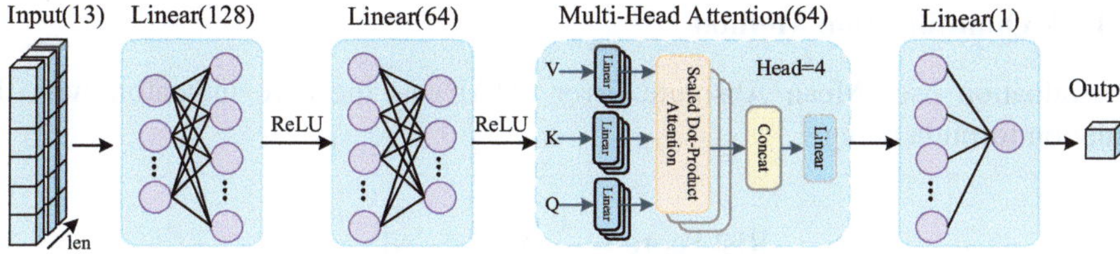

Fig. 3. Improved Time Series Prediction Model Network Structure.

(1) Input Representation

The operational data consists of d vector features, including trolley speed percentage, crane hook position coordinates, and control signals. Using the sliding window method, consecutive time segments of length L are constructed into the input sequence X.

$$X = \{x_{t-L+1}, \ldots, x_t\} \tag{2}$$

(2) Feature Mapping Layer

The input sequence first passes through two fully connected network layers to achieve dimensionality enhancement and compression, obtaining the hidden representation H:

$$H = ReLU(XW + b) \tag{3}$$

where W, b represents weights and bias terms, and ReLU is the linear rectification function.

(3) Multi-Head Self-Attention Mechanism

Based on the implicit representation H, multi-head self-attention mechanism is introduced to enhance the model's ability to capture long-sequence dependencies and critical moment features. Its calculation formula is:

$$\text{MultiHead}(Q, K, V) = \text{Concat}(\text{head}_1, \ldots, \text{head}_h)W^O \tag{4}$$

where $\text{head}_i = \text{Attention}\left(QW_i^Q, KW_i^K, VW_i^V\right)$ represents the number of attention heads.

(4) Output Layer

The multi-head attention layer output is mapped linearly to obtain a single-dimensional prediction value:

$$\hat{y}_{t+1} = HW + b \tag{5}$$

where $\hat{y}_{t+1}$ represents the trolley's set speed for the next moment. Through a recursive method, a complete operation prediction sequence can be generated, providing a benchmark for risk quantification.

3.4 Risk Quantification Method

Risk evaluation uses Mean Absolute Error (MAE) as the core indicator, with the calculation formula being:

$$\text{RiskScore} = \frac{1}{L}\sum_{t=1}^{L}|\hat{y}_t - y_t| \tag{6}$$

Set the risk threshold τ, and determine it as high risk when RiskScore $> \tau$. By combining curve comparison and deviation analysis, it is possible to intuitively identify risk points for novice drivers during the operation process.

4 Experimental Verification

4.1 Dataset Construction

The dataset for this study consists of operational data from two categories of operators: Trainer and Trainee. Among them, Trainers are 10 skilled drivers with over 5 years of experience, collecting a total of 500 sequence data sets, with each set containing 20 time steps; Trainees are 20 novice drivers with no prior experience, collecting a total of 100 sequence data sets, which are ultimately divided into training and validation sets according to the proportion of 8:2.

4.2 Model Training Process and Convergence Analysis

During the model training process, this study employs an improved time series prediction model, using driver operation data sequences as input and risk indicators as prediction targets. The training process enhances the model's generalization capability through cross-validation. To ensure comparability across different training scales, Epochs were set at 300, 500, 1000, and 2000, and the loss function's change trend was analyzed, with training results shown in Fig. 4.

The experimental results show that the loss function rapidly decreases within the first 10 Epochs, indicating the model's ability to quickly learn key features. At Epoch 300, the loss curve basically converges and remains at a low level, demonstrating high stability; at Epoch 500, the loss is further reduced, though the improvement is limited compared to 300 rounds; at Epochs 1000 and 2000, while the loss curve remains stable, the convergence speed and fluctuation amplitude are essentially consistent with 500 rounds, with significantly increased training costs. Comprehensively considering convergence speed, prediction accuracy, and computational cost, Epoch $= 500$ is ultimately selected as the optimal number of training rounds.

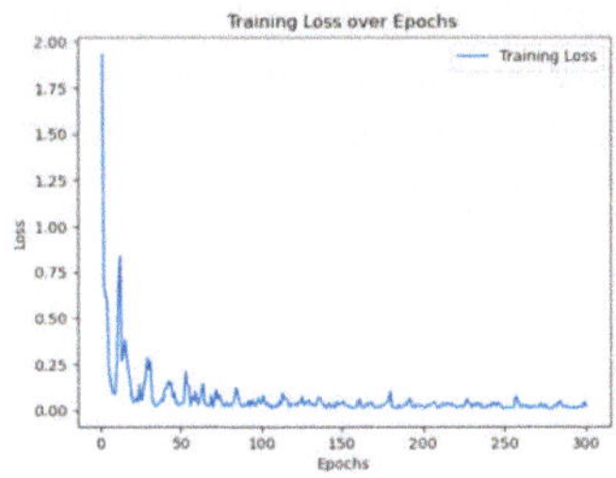

(a) epoch=300

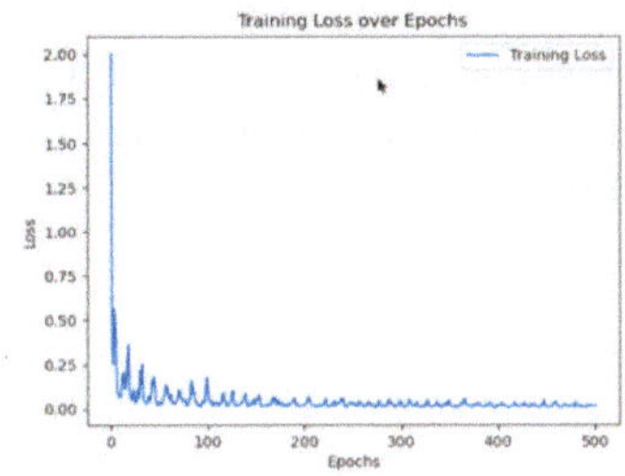

(b) epoch=500

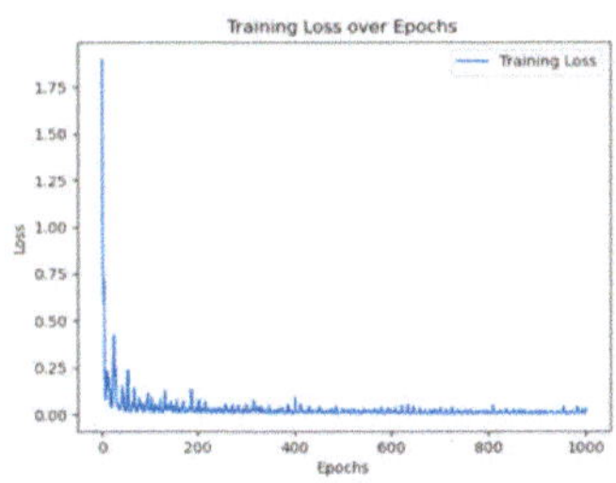

(c) epoch=1000

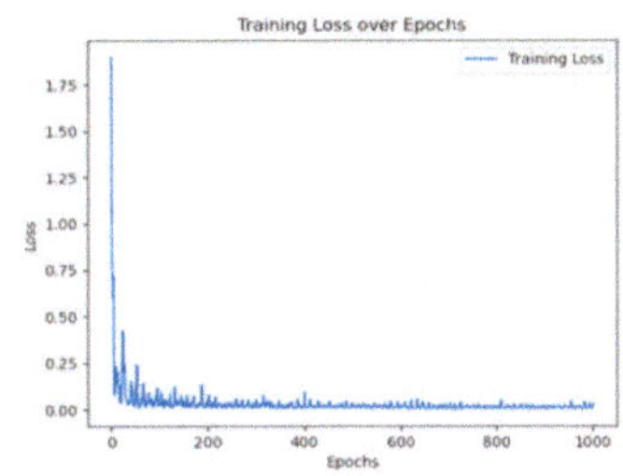

(d) epoch=2000

Fig. 4. Loss Convergence Curve.

4.3 Experimental Comparative Analysis

To verify the effectiveness of the proposed method in practical applications, under the anti-sway scenario, operation data from Trainer and new drivers with different work durations were input, and prediction curves were generated using the improved time series prediction model, followed by risk score calculations. The experimental results are shown in Table 1.

Table 1. Quantitative Comparison of Operational Risk for Different Driver Groups.

Operational Object	Average MAE	Risk Score (Mean)
Trainer (Experienced Driver)	0.021	0.08
Trainee (1 Month on the Job)	0.162	0.52
Trainee (3 Months on the Job)	0.141	0.41
Trainee (6 Months on the Job)	0.112	0.28
Trainee (12 months on the job)	0.087	0.19

As shown in Table 1, the risk score of novice drivers gradually decreases with job tenure, indicating that their operational level progressively approaches that of experienced drivers. Specifically, the risk score is highest (0.52) at 1 month on the job, characterized by significant operational lag and reverse errors. By 12 months, the risk score drops to 0.19, nearly reaching the Trainer level, which validates the effectiveness of this system in long-term training performance evaluation.

To further examine the model's performance under different time-dependent conditions, sequence lengths of 20, 50, 100 and 500 were set, and predictions were made on the same set of Trainee operational data, with results compared to the actual curve as shown in Fig. 5.

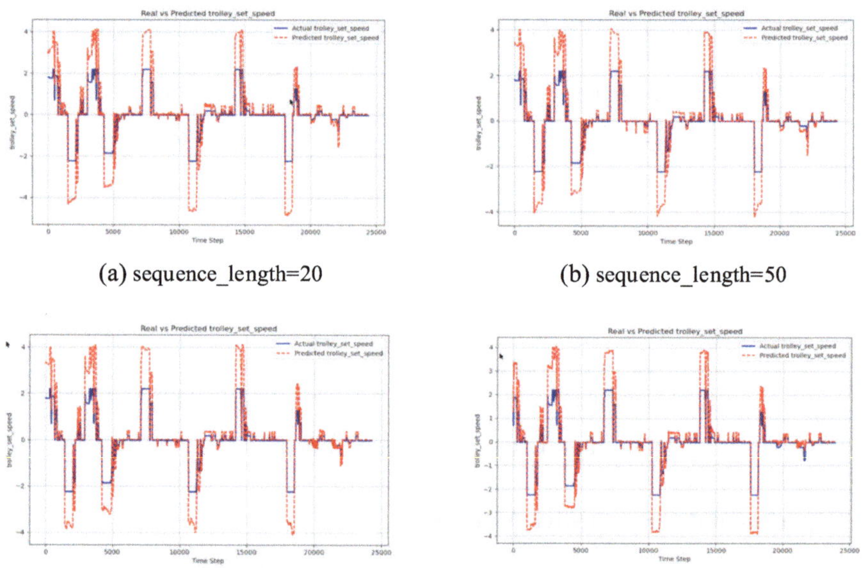

(a) sequence_length=20

(b) sequence_length=50

(c) sequence_length=100

(d) sequence_length=500

Fig. 5. Comparative Analysis of Trolley Velocity Curves under Different Sequence Lengths.

Overall, the prediction curves under four different sequence lengths can well reproduce the dynamic variation characteristics of the trolley's speed, maintaining high consistency with the actual curve. This demonstrates the model's good temporal fitting capability. There are slight differences in curve details across different sequence lengths, especially in the prediction accuracy at positive and negative peak values. However, the overall trend remains consistent, accurately reflecting the operational rhythm and speed change patterns.

5 Conclusion

The digital twin system with an integrated operational risk assessment module designed in this paper can effectively address safety hazards in container gantry crane real-machine training. The system precisely reproduces on-site operations through a high-fidelity virtual environment and employs a time series prediction model based on multi-head attention mechanism, constructing a safety evaluation system benchmarked against experienced drivers' operations. Experiments show that the system can effectively quantify novice drivers' operational deviations in high-risk scenarios such as swing reduction.

This not only significantly reduces training accident risks but also provides data support for personalized teaching through visualization feedback, achieving a risk-free and efficient intelligent training approach.

Acknowledgments. This research was supported by 2023 Education Science Research Project of Shanghai Municipality (No. B2023003).

References

1. Chen, B., Pan, Y., Xia, Z., Zhang, R., Zhao, H.: Analysis of human factors in remote control of shore cranes under semi-automated mode. Port Technol. **9**, 1–3 (2019). https://doi.org/10.3969/j.issn.1673-6826.2019.09.001
2. Wang, T.: Research on safety evaluation method of container terminal shore cranes based on human factors. Master's thesis, Ningbo University, Ningbo (2021). https://doi.org/10.27256/d.cnki.gnbou.2021.000488
3. Huang, Q., Huang, Y., Zhang, Z., Zhang, Y.: Truck-lifting prevention system based on vision tracking for container-lifting operation. J. Adv. Transp. **2021**(1), 9612480 (2021). https://doi.org/10.1155/2021/9612480
4. Shen, Y., Man, X., Wang, J., Zhang, Y., Mi, C.: Truck lifting accident detection method based on improved PointNet++ for container terminals. J. Mar. Sci. Eng. **13**(2), 256 (2025). https://doi.org/10.3390/jmse13020256
5. Wang, J., He, M., Zhang, Y., Zhang, Z., Postolache, O., Mi, C.: High-precision pose measurement of containers on the transfer platform of the dual-trolley quayside container crane based on machine vision. Sensors **25**(9), 2760 (2025). https://doi.org/10.3390/s25092760
6. Zhang, Y., Song, Y., Zheng, L., Postolache, O., Mi, C., Shen, Y.: Improved YOLOv5 network for high-precision three-dimensional positioning and attitude measurement of container spreaders in automated quayside cranes. Sensors **24**(17), 5476 (2024). https://doi.org/10.3390/s24175476
7. Song, H., Kim, T., Kim, J., Ahn, D., Kang, Y.: Effectiveness of VR crane training with head-mounted display: double mediation of presence and perceived usefulness. Autom. Constr. **122**, 103506 (2021). https://doi.org/10.1016/j.autcon.2020.103506
8. Guo, X., Liu, X., Huang, D., Feng, W.: Application of virtual reality technology in port crane training. Mech. Electr. Inf. **33**, 80–81 (2020)
9. Zhang, L., Chai, J., Xing, S., Li, L., Meng, F.: Virtual driving training system for cranes based on virtual reality technology. Eng. Mach. Maintenance **S1**, 115–118 (2020)
10. Zhang, Y., Luo, K., Postolache, O., Mi, C.: Design of a simulation platform for visual measurement of port automation equipment based on digital twin technology. In: Proceedings of the 2025 IEEE International Instrumentation and Measurement Technology Conference (I2MTC), pp. 1–6. IEEE, Chemnitz (2025). https://doi.org/10.1109/I2MTC62753.2025.11079158
11. Huang, L.: Analysis and application of driving behavior habits based on DBN. Master's thesis, Jiangsu University, Zhenjiang (2019)
12. He, M., Zhang, Y., Liu, Y., Shen, Y., Mi, C.: Automated evaluation method for risk behaviors of quay crane operators at ports using virtual reality. Algorithms **17**(11), 508 (2024). https://doi.org/10.3390/a17110508

Study on the Characteristics of Digital Hydraulic Pumps and LUDV Multi-way Valves for Excavators

Yi Zhang, Zhiqiang Zhang(✉), Bin Bian, Dongliang Chen, and Zhiqi Liu

School of Mechanical Engineering, Taiyuan University of Science and Technology, Taiyuan 030024, Shanxi, China
zhiqiangzhang1@tyust.edu.cn

Abstract. In response to the development needs of energy conservation and intellectualization in construction machinery, a design is proposed where the displacement of a digital hydraulic pump is regulated in groups by eighteen solenoid on-off valves to achieve multi-stage flow output. A Load-Independent Flow Distribution (LUDV) multi-way valve is used as the main valve, and pressure compensating valves are employed to maintain stable pressure differences between the inlets and outlets of the main valve's throttling orifices. Based on the AMESim-Simulink co-simulation platform, a hydraulic system model and a control model of the digital hydraulic pump combined with the LUDV multi-way valve are established, and simulation analyses of the actuator motion under different working conditions are conducted. Theoretical analysis and simulation results show that the digital pump can accurately match the flow demand of the actuator with a response time of 50 ms; under the condition of unsaturated flow, the LUDV multi-way valve maintains the pressure difference fluctuation of the main valve within ±0.2 MPa, and the flow of each actuator circuit only depends on the opening degree of the main valve's throttling port.

Keywords: Digital hydraulic pump · LUDV multi · way valve · AMESim · Simulink co · simulation

1 Introduction

As the core power unit of construction machinery, the energy efficiency and controllability of hydraulic systems significantly influence the overall performance of the equipment [1]. Digital pumps output precise flow using discrete signals, having significant advantages, including rapid response, discretized flow control, and strong anti-interference capability [2]. LUDV which is the load-independent flow distribution multi-way valve with post-valve compensation, solves flow unsaturated problem during flow distribution. For high coordination requirement of several actutors, the hydraulic system with post-valve compensation demonstrate distinct advantages over the system with traditional pre-valve compensation [3].

S. Halgamuge et al. (Eds.): ICACMVE 2025, LNME, pp. 483–490, 2026.
https://doi.org/10.1007/978-981-95-6911-3_48

Zhiguang Wu et al. [4] developed the digital hydraulic pump using several high-speed on-off valves controlled by a dual fuzzy PID strategy, validated through simulations and experiments. The pump demonstrated a 350ms displacement step response, $\leq 0.5\%$ displacement hysteresis, and achieved high-precision multi-mode closed-loop control (with pressure errors within 1.2MPa and power errors within 1.4kW). To solve the lack of guidelines for selection and configuration digital hydraulic component, Dimitri Oliveira e Silva et al. [5] proposed a design method for variable-speed digital hydraulic pump. The method employs optimization algorithms to determine the number of pump units and their displacement allocation using pre-defined patterns, facilitating the screening of optimal designs. Experimental validation confirmed that the method effectively identifies optimal pump unit combinations.

This study investigates the synergistic control characteristics between digital pumps and LUDV multi-way valves. The hydraulic system was modelled and simulated using AMESim and Simulink, and the rapid flow response of digital pumps and the precise flow distribution of LUDV multi-way were analyzed under different operating conditions.

2 Working Principles of Digital Hydraulic Pumps and LUDV Hydraulic Systems

Figure 1 shows the system including a digital pump, LUDV multi-way valve, pressure sensor, and three hydraulic cylinders. The digital hydraulic pump comprises eighteen pistons, those are divided into six working groups with three identical pistons per group. The ratio of piston cross-sectional areas of the six groups are 1:2:2:2:4:4. Each piston has an electromagnetic on-off valve and a check valve. The pump regulates its output displacement by controlling the states of its solenoid valves. The pump achieves 16 different displacements and automatically selects optimal flow output for actual working conditions [6].

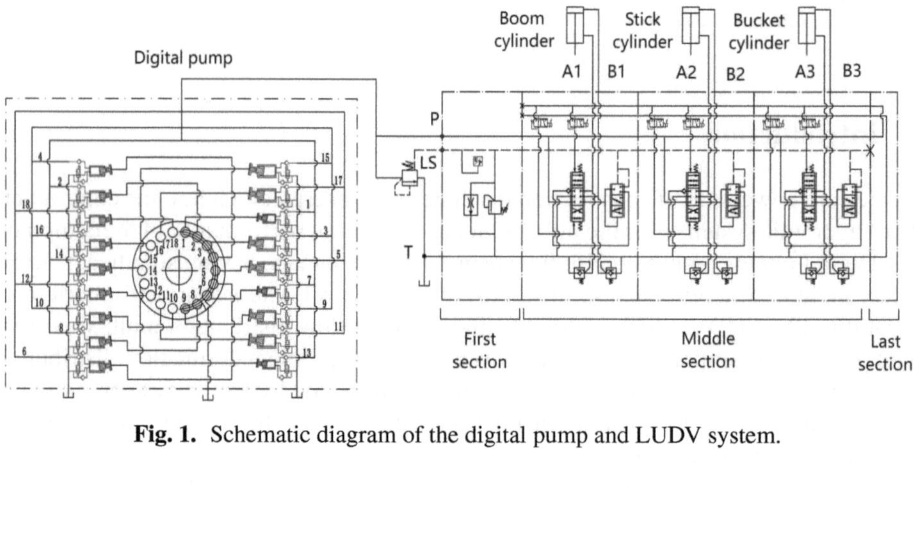

Fig. 1. Schematic diagram of the digital pump and LUDV system.

The LUDV multi-way valve utilizes post-valve pressure compensation technology. The pressure compensator is located downstream of the throttle orifice and regulates flow by maintaining a constant pressure drop across the throttle orifice. For multi-actuator operation, it provides load-independent flow distribution according to the spool opening ratio. Its saturation-tolerant flow distribution ensures proportional flow allocation without prioritizing the low-load circuit during pump flow insufficiency, thereby maintaining actuator speed synchronization. Load pressures are fed back to the pressure compensator and system relief valve via shuttle valves, with the compensator set pressure being $P_{max} + \Delta P_P$. Port relief valves are installed on work ports A and B. As shown in Fig. 1, the valve assembly comprises main spool, pressure compensator valve, shuttle valves, and electrohydraulic proportional relief valve. This system significantly enhances the smoothness and coordination of construction machinery [7].

3 Simulation Analysis of Digital Pump with LUDV Load Sensing System

3.1 Digital Pump Paired with LUDV Load-Sensing System Simulation Modelling

The physical model of the hydraulic system was implemented in AMESim [8], as shown in Fig. 2. The AMESim-Simulink co-simulation was performed for the digital pump-LUDV multi-way valve system. The key simulation parameters are provided in Table 1.

Table 1. Simulation parameters of digital pump, LUDV multi-way valve, and hydraulic cylinder system.

parameters	numerical value	parameters	numerical value
Piston bore A/B/C diameter/mm	10.5/14.85/21	Motor speed/r/min	2200
Swash plate angle/°	8	Main spool land diameter/mm	15
Main spool stem diameter/mm	7.8	Piston pitch circle diameter / mm	90
Boom cylinder bore diameter / mm	100	Boom cylinder rod diameter / mm	85
Stick cylinder bore diameter / mm	85	Stick cylinder rod diameter / mm	55
Bucket cylinder bore diameter/mm	80	Bucket cylinder rod diameter / mm	50

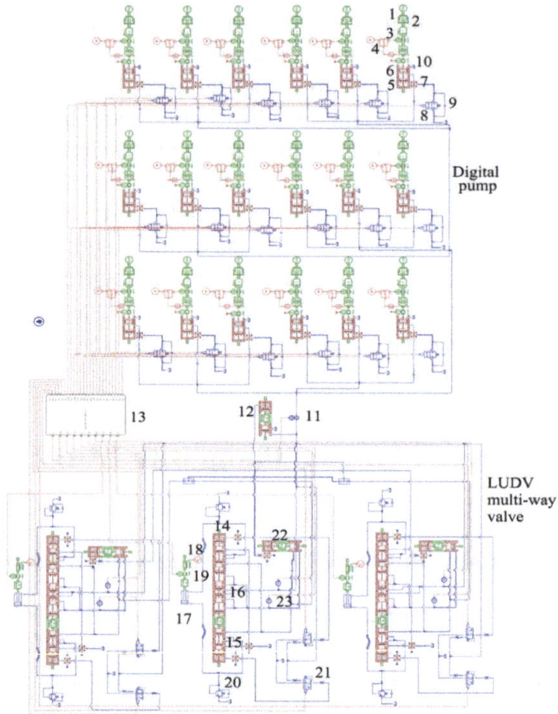

Fig. 2. Hydraulic system physical model of digital pump and LUDV multi-way valve (1-Motor; 2-Coupling; 3-Rotational inertia; 4-Swash plate angle; 5-Piston; 6-Leakage; 7-Dead volume; 8-Solenoid valve; 9-Check valve; 10-Oil tank; 11-Flow sensor; 12-System relief valve; 13-AMESim-Simulink co-simulation interface; 14-Main spool of multi-way valve; 15-Directional control valve; 16-Metering valve; 17-Hydraulic cylinder; 18-Load; 19-Displacement sensor; 20-Port relief valve; 21-Pilot proportional pressure reducing valve; 22-Post compensator; 23-Pressure sensor).

3.2 Performance Simulation of Digital Pump-LUDV System Under Multi-Condition Operations

To analyze the system's dynamic characteristics under accelerated operating conditions, a time compression coefficient k_t is defined: $k_t = 1.0$ indicates normal operating conditions (reference duration T_0); $k_t = 0.5$ indicates high-speed operating conditions (operating duration $0.5T_0$). The load remains constant across conditions, while time compression is implemented by accelerating the hydraulic cylinders.

To verify the dynamic characteristics of the digital pump-LUDV system during composite actions, the coordinated movement timing sequence for the three hydraulic cylinders is established as follows [9]. The initial positions of the excavator's hydraulic cylinders are as follows: boom cylinder at 0.40 m, stick cylinder at 0.55 m, and bucket cylinder at 0.60 m. All operating conditions remain stationary for 0–1 s to eliminate disturbances. Under normal speed conditions, from 1–5 s, the stick cylinder and bucket cylinder retract synchronously by 0.40 m; from 5 to 7 s, the boom cylinder retracts independently by 0.30 m; from 7 to 11 s, the bucket cylinder extends by 0.50 m to perform bucket closing; from 11 to 15 s, the boom cylinder extends by 0.30 m while the stick cylinder extends by 0.20 m. In high-speed conditions, with the displacement of hydraulic cylinders unchanged, the action time is compressed according to k_t, and the speed is correspondingly increased to twice the original value. The peak flow demand increases to 200% of the baseline condition ($Q_{max} = 109.55$ L/min). This test design creates three evaluation scenarios for the digital pump-LUDV valve system-pressure impact, flow distribution, and coordinated control-through staggered cylinder start-stop sequences and directional switching [10].

The digital pump delivers discretized flow according to system input signals, providing 16 discrete displacement levels from zero to maximum flow. It achieves a switching transient time of 50 ms with a peak flow of 108.4 L/min. These characteristics demonstrates rapid response, high flow capacity, excellent energy efficiency, and precise flow control [11].

Figure 3 shows the displacement tracking errors of the boom cylinder during retraction under normal-peed and high-speed working conditions are 2.51% and 4.27% respectively, and during extension are 1.53% and 5.89% respectively; the tracking errors of the stick cylinder during retraction are 1.13% and 0.6% respectively, and during extension are 0.15% and 1.31% respectively; the tracking error of the bucket cylinder during retraction are 1.28% and 0.99% respectively, and during extension are 0.16% and 4.23% respectively. The high-speed condition significantly increases tracking error in the boom cylinder (peaking at 5.89% during extension), indicating greater velocity-dependent dynamic effects in the large-inertia actuators. In contrast, the stick cylinder maintains errors below 1.31%, highlighting actuator-specific dynamic responses.

Figure 4(a) shows that under normal speed working conditions, the compensator pressure drop maintains approximately 20 bar throughout all stages. During compound actions, load pressures (downstream of metering orifices) remain consistent despite motion compounding, directional switching, or velocity change. This verifies both the effectiveness of post-valve compensation and load-independent flow distribution. Figure 4(b) reveals that during sole boom actuation (3–4 s), the metering orifice of the boom valve spool reaches maximum opening, reducing the compensator pressure drop to 6.2 bar. During boom-stick compound motion (6–8 s), peak flow reaches 200% (109.5 L/min), inducing flow saturation. Here, the compensator pressure drop decreases to 19 bar, This significant pressure reduction critically impacts the LUDV system's core operation. LUDV relies on maintaining a constant 20 bar pressure differential across compensating valves for proportional flow distribution per spool commands. When system flow demand exceeds pump capacity (saturation), the compensator pressure differential deviates from the design value, impairing precise operation of the compensating valves. While load pressure consistency persists, this shows compensating valves still attempt to coordinate post-valve pressure to maintain flow sharing. Yet insufficient pressure differential may cause actual actuator flow proportions to deviate from spool opening commands (flow distribution distortion), exposing the system's flow supply limitation under extreme conditions. In other phases, the compensator pressure drop stabilizes at 20 bar with maintained load pressure consistency. Pressure fluctuations during actuation switching stem from response lag of the digital pump's solenoid valves and dynamic coupling of LUDV compensators, yet these do not compromise multi-actuator coordination. However, during flow saturation, insufficient pump supply and compensation pressure reduction-induced flow distortion may disrupt actuator synchronization. For example, boom-stick speed matching in composite excavation may deviate from expectations, impairing operational coordination and accuracy.

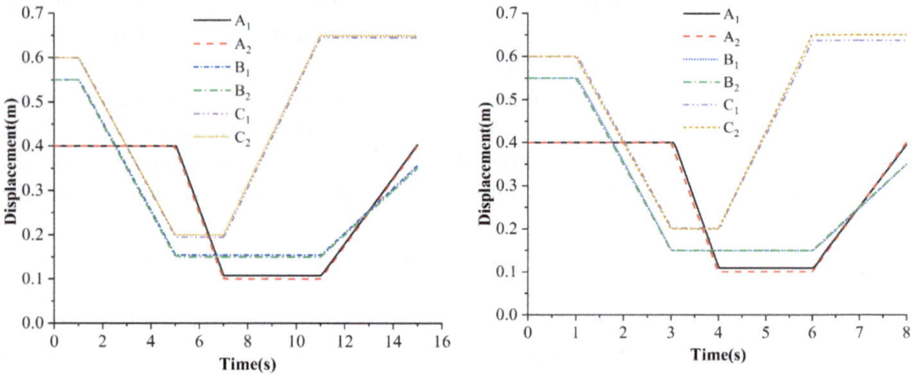

(a) (b)

Fig. 3. Displacement results of hydraulic cylinders under normal speed (a) and high speed (b) conditions (A_1- Actual displacement of the boom cylinder; A_2-Ideal displacement of the boom cylinder; B_1-Actual displacement of the stick cylinder; B_2-Ideal displacement of the stick cylinder; C_1-Actual displacement of the bucket cylinder; C_2-Ideal displacement of the bucket cylinder).

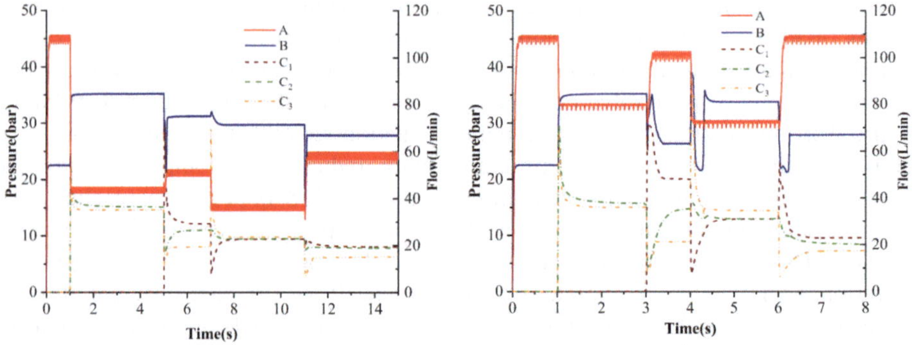

(a) (b)

Fig. 4. Pressure-flow characteristics of pump-valve system normal speed (a) and high speed (b) operating conditions (A-Pump outlet flow; B-Pump outlet pressure; C_1-Pressure after the throttle valve of the main spool of the boom valve; C_2-Pressure after the throttle valve of the main spool of the stick valve; C_3-Pressure after the throttle valve of the main spool of the bucket valve).

4 Conclusion

This study aims to develop a high-performance hydraulic system integrating a digital hydraulic pump, an LUDV multi-way valve, and an actuator. In the design of the digital pump, the displacement combination of 6 groups of pistons is controlled by electromagnetic on-off valves to achieve 16 discrete displacement levels. The LUDV multi-way valve utilizes a pressure compensator to stabilize the pressure difference across the main valve, enabling on-demand flow distribution. Through co-simulation analysis and verification using AMESim and Simulink, the results show that:

1. The digital pump has a rapid response with a response time of up to 50 ms and a maximum output flow of 108.4 L/min.
2. The system output flow is only related to the opening degree of the throttle valve of the main valve and is not affected by load fluctuations. Under both normal-speed and high-speed (non-saturated) working conditions, the system can maintain a constant pressure difference across the main valve and achieve precise flow control, confirming its high efficiency and stability.

References

1. Xu, Y.Z.: Trend of hydraulic technology in industry 4.0 and its intelligence hydraulics. Hydraulics Pneumatics Seals **42**(02), 1–7 (2022)
2. Wang, C.L., Ye, F., He, T., et al.: Analysis of flow response characteristics of PWM valve-distributed digital piston pump. Mach. Tool Hydraulics **53**(08), 134–141 (2025)
3. Bai, C.H., Du, K., Mi, X.M., et al.: Simulation and test of flow control and micro-motion characteristics of LUDV multi-way valve. Chin. Hydraulics Pneumatics **48**(07), 86–92 (2024)
4. Wu, Z.G., Sun, H.D.: Research on multi-mode control of digital hydraulic pump based on high speed on-off valve. Mach. Tool Hydraulics **53**(09), 187–194 (2025)
5. e Silva, D.O., Nostrani, M.P., Waltrich, G., De Negri, V.J.: A method for designing digital hydraulic pumps with variable angular speed. Meccanica **60**(4), 1–15 (2025)
6. Li, C.S.: Research on control characteristics and application of digital pump based on discrete variable technology. Doctoral dissertation, Jilin University, CNKI (2022). https://doi.org/10.27162/d.cnki.gjlin.2022.007768
7. Xu, Z.G.: Flow control of load sensing multiple unit valve with electro–hydraulic double control. Chin. Hydraulics Pneumatics (06), 135–140 (2019)
8. Fu, S., Wang, L., Lin, T.: Control of electric drive powertrain based on variable speed control in construction machinery. Autom. Constr. **119** (2020)
9. Shi, J.P., Zhang, X.G., Ge, L., et al.: A combined control strategy of velocity and position for excavator arm. Mach. Tool Hydraulics **45**(24), 101–105+159 (2017)
10. Lee, M., Choi, H., Kim, C., et al.: Precision motion control of robotized industrial hydraulic excavators via data-driven model inversion. J. IEEE Robot Autom. Lett. **99**, 1 (2022)
11. Kadu, C.B., Kothule, P.S., Vikhe, P.S., et al.: Design, analysis and motion control of hydraulic excavator using discrete sliding mode control with inertial delay. In: International Conference on Advanced Computing and Communication Systems (2017)

Advanced Manufacturing and Materials

Study on Cylinder Pressure Identification and Improvement Measure

Shuhai Wang, Biao Jin(✉), Nanfeng Zhang, and Qiang Fei

College of Mechanical and Electrical Motoring, Guangdong University of Science and Technology, Dongguan 523083, Guangdong, China
53340450@qq.com

Abstract. Identification of motor cylinder pressure mainly refers to the identification of dynamic loading of structure through assessing the response of structures (e.g., movements, speed, acceleration and response) and according to dynamic characteristics as well. This is an approach to inverse problems in dynamic structure and is called identification of structural loads, aiming at achieving the model of its system information as to an unchanging system of single input and output. Its inverse $H^{-1}(f)$ can be obtained once after achieving FRF (frequency response function). Based on the response, loading can be identified after distinguishing the loading spectrum and conducting a Fourier converse transformation. The under problems will be researched: Identification of Cylinder Pressure of Motor, Measures on Improving Accuracy Identification of Cylinder Pressure, Coherent amendment of the frequency response function, Inverse Spectrum Level and Smoothing of FRF. These problems can be solved by inverse spectrum windowing with nonlinear inverse filtering. However, the types and width of windows are difficult to solve and there is a great amount of relative calculation. As a matter of fact, the key reason why the inverse method of frequency function affects the results of loading identification is that there are a number of sharp low points in the amplitude of frequency function and these low points will be amplified, resulting in serious distortion of inversion signals. As long as the amplitude of the transfer function is smoothened, identification results can be improved.

Keywords: Motor · Cylinder Pressure · Improvement measure

1 Introduction

Conducting the fault diagnosis of the diesel motor, no matter which method and means are used, is always inseparable from the feature extraction of the fault signal. Without effective feature extraction, it is impossible to make in-depth and accurate diagnosis decisions. Traditional vibration signal analysis has shortcomings, mainly in two aspects. First, the traditional spectral analysis is time-independent average spectral density. In the stochastic process spectrum composition, assuming the process is stable, the statistical characteristics do not change with time. However, in fact, most mechanical vibration is smooth. For example, combustion shock and injector needle valve seat impact shock

S. Halgamuge et al. (Eds.): ICACMVE 2025, LNME, pp. 493–501, 2026.
https://doi.org/10.1007/978-981-95-6911-3_49

are smooth. When using the traditional spectral analysis method to analyze these non-stationary signal in the total duration after average frequency composition, it often difficult to reflect the non-stationary, time-varying characteristics. The second deficiency in traditional vibration signal analysis is the importance of ignoring signal time domain analysis. For the recognition of combustion cylinder pressure in single-cylinder or fewer-cylinder diesel motors, the general frequency sound inverse method [6] or a reverse filter can achieve accuracy. However, for multi-cylinder diesel motors, the traditional method obviously can not solve the problem of accuracy identification, wavelet noise reduction of input and output signal, using the coherent.

Based on the above problems, this paper focuses on four aspects. The first one is structural load identification. Pressure wave recovery of the cylinder of a motor is mainly assayed by structural response, such as movement, speed, acceleration and reaction. These responses are easily assayed, facilitating the identification of structural dynamic loading, according to the structural mobile features that have been known. The second is the identification of the structural loads, specifically the cylinder pressure identification of the motor. It involves the cylinder pressure recognition of the motor in an automobile chassis, and the characteristic parameters of the identified cylinder pressure are extracted [3]. The third is the extraction of fuel pressure signal based on time domain analysis, containing Wavelet noise reduction, Coherent amendment of frequency response function, Inverse Spectrum Level and Smoothing of FRF. The fourth is an improvement measure on the Identification of Cylinder Pressure, containing Wavelet noise reduction, Coherent amendment of frequency response function, Inverse Spectrum Level and Smoothing of FRF [5].

2 Structural Loads Identification

This is an approach to inverse problems in dynamic structure and is called identification of structural loads, aiming at achieving the model of its system information as to an unchanging system of single input and output. Its inverse $H^{-1}(f)$ can be obtained once after achieving FRF (frequency response function). Based on the response, loading can be identified after distinguishing the loading spectrum and conducting a Fourier converse transformation. Loading spectrum and Fourier converse transformation are shown in Fig. 1.

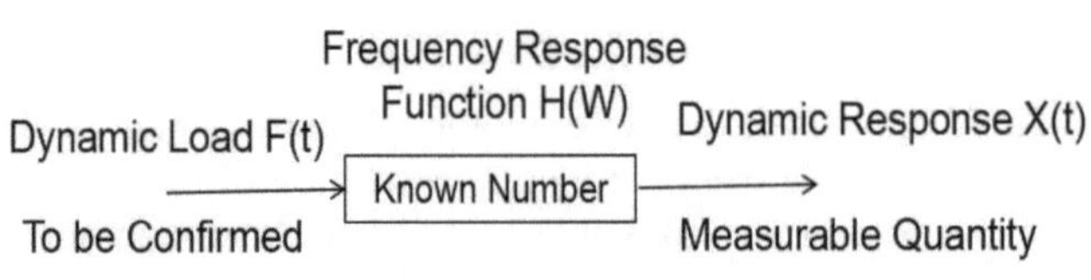

Fig. 1. Loading spectrum and Fourier converse transformation.

Cylinder pressure waveform recovery mainly refers to the determination of the structural response (such as displacement, velocity, acceleration, strain, etc., which should be easy to measure) based on the known dynamic characteristics of the structure [2].

This is a treatment of the inverse problem in structural dynamics, called structural load identification, as shown in Fig. 1. For a single input single output linear time-invariant system, its system information model is shown in Formula (1):

$$Y(\mathrm{f}) = \mathrm{H(f)X(f)} \tag{1}$$

where *Y(f)* is the spectrum of the cylinder head vibration signal, *H(f)* is the frequency sound function of the cylinder head system, and *X(f)* is the spectrum of the cylinder pressure. Through Formula (1), the frequency sound function $H^{-1}(f)$ can be obtained. With the known response, the load spectrum can be identified, and then the load can be identified by doing the Fourier inverse transformation, as in Formula (2) and Formula (3).

$$\mathrm{H(f)} = \frac{\mathrm{Y(F)}}{\mathrm{X(F)}} \tag{2}$$

$$\mathrm{X(F)} = \mathrm{H^{-1}(f)Y(f)} \tag{3}$$

The general method of finding $H^{-1}(f)$ is to design a reverse phase filter whose growth rate is the gain of the negative function, with the phase of the negative transfer function as its phase component, as in Formula (4). Figure 2 shows the identified cylinder pressure.

$$\mathrm{H^{-1}(f)} = 1/H(\mathrm{f}) \tag{4}$$

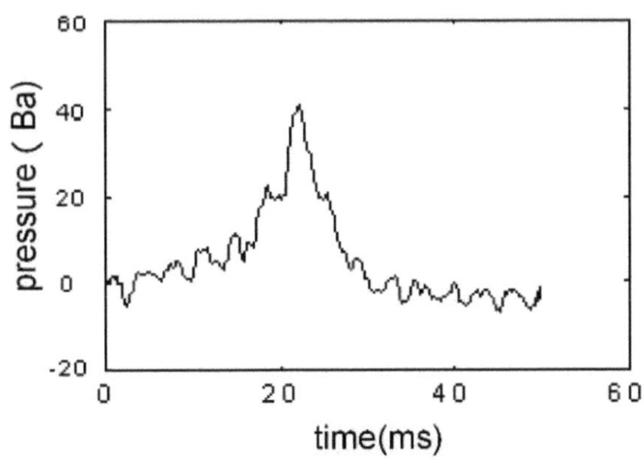

Fig. 2. Identified cylinder pressure.

3 Cylinder Pressure Identification of the Motor

A converse filter is designed with its growth rate as a gain of negative function and the phrase of negative transmission function as its phrase component. Then, a Fourier converse transformation is conducted so as to identify cylinder pressure. Compared with cylinder pressure assessed by actual measurement, either amplitude or phrase has some errors [1]. The main reasons are that there are disturbing signals from the frequency domain in vibratory response and meanwhile there is no amending and smoothening with the function of frequency response [7].

3.1 The Cylinder Pressure Recognition of the Motor for the Automobile Chassis

The car chassis uses the cylinder pressure recognition of the motor. Figure 2 shows the cylinder pressure identified by Formula (3) and then by the Fourier inverse Fourier transform. Compared with the measured cylinder pressure, both the amplitude and the waveform have certain errors. The main reason is that there is still a frequency domain interference signal in the vibration response, and the frequency sound function is not corrected and smoothed.

3.2 Identified Cylinder Pressure Characteristic Parameters Were Extracted

Two characteristic parameters are selected, namely, the highest combustion pressure [4] and the maximum combustion pressure angle (according to the crankshaft rotation angle relative to the upper stop point) as shown in Fig. 3. Table 1 shows normal (the oil supply advance point is 34 deg CA [9], and the opening pressure of the injector needle valve is 210 kg/cm^2), ten kinds of working process faults and four kinds of injector faults (the poor atomization in the working process fault is simulated by the spring fracture in the injector fault). Data for two characteristic parameters were collected, and each data was averaged over ten times.

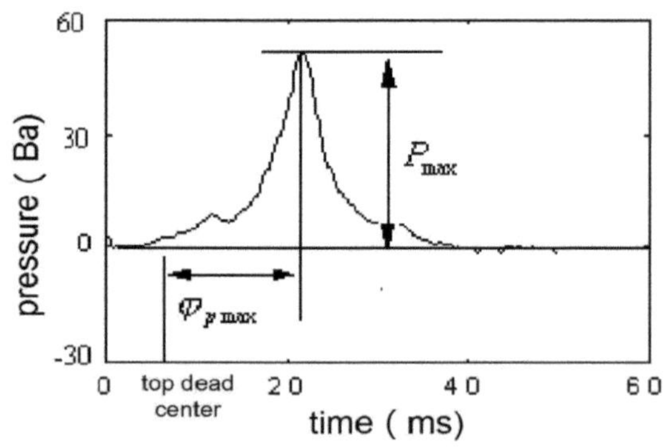

Fig. 3. Representation of the identified cylinder pressure characteristic parameters.

Table 1. Ten states for P_{max} and $\varphi_{p\,max}$.

Order Number	Status	$P_{y\,max}$	α_p	I_f
1	Normal	2.52	34.06	7.60
2	The early Angle of oil supply is 25 deg CA	2.38	26.15	8.01
3	The early Angle of oil supply is 40 deg CA	2.29	40.52	5.47
4	Spring fracture (simulated poor atomization)	1.09	35.40	3.10
5	Insufficient intake	2.62	34.92	6.50
6	Air cylinder leakage	2.66	36.26	3.90

(*continued*)

Table 1. (*continued*)

Order Number	Status	$P_{y\max}$	α_p	I_f
7	Oil injector 4 holes are blocked	2.73	33.12	3.71
8	The pin valve is stuck	2.75	31.98	3.33
9	Turn on pressure for 150 kg/cm^2	2.11	36.33	5.42
10	Turn on pressure for 260 kg/cm^2	2.88	32.88	5.23

4 Extraction of Fuel Pressure Signal Based on Time Domain Analysis

Normal turn-on pressure is 210 kg/cm^2[15]. The signal of the exhaust gate cam is taken as the approximate stop point (location point). In order to improve the temporal resolution, the sampling frequency is 40 K, and the signal of the window for the characteristic segment signal is 256 points, about 6.4 mm [8], 40° crankshaft Angle. In normal conditions, the fuel pressure signal measured by the gripped pressure sensor produced by Austria company AVL is shown in Fig. 3.

In the time domain, according to the timing relationship [6], three characteristic parameters are selected, the maximum oil pressure, the oil injection advance Angle (the crankshaft corner of the starting oil injection point to the upper stop point), and the pulse factor I_f:

$$I_f = \frac{\max(z) - \min(z)}{\frac{1}{N}\sum_{n=1}^{N}|z|} \tag{5}$$

where z is the oil pressure signal. Table 2 is the specific data of the above three characteristic parameters obtained by 10 states after 10 averaging. Among them, α_p is the actual static oil supply advance corner after correction.

Table 2. Ten state fuel pressure.

Order Number	Status	$P_{\max}(Ba)$	$\varphi_{p\max}$
1	Normal	58.95	8.05
2	The early Angle of oil supply is 25 deg CA	55.18	9.21
3	The early Angle of oil supply is 40 deg CA	50.78	6.92
4	Spring fracture (simulated poor atomization)	54.25	7.81
5	Insufficient intake	55.04	7.35
6	Air cylinder leakage	50.12	8.24
7	Oil injector 4 holes are blocked	44.72	7.14

(*continued*)

Table 2. (*continued*)

Order Number	Status	$P_{max}(Ba)$	$\varphi_{p\,max}$
8	The pin valve is stuck	42.77	7.05
9	Turn on pressure for 150 kg/cm^2	57.12	7.33
10	Turn on pressure for 260 kg/cm^2	52.62	7.27

5 Improvement Measure on Identification of Cylinder Pressure

5.1 Wavelet Noise Reduction

Based on the theories of wavelet analysis, a function system, formed by extension and translation of a generating function $\psi(x)$ [14], can make up the orthonormal basis of an integrable space $L^2(R)$ of square function. Any function in $L^2(R)$ space can be expressed with this base line [10]. It is also the form of function generation into wavelet progression.

As to restructural vibrating signals which are mutually removed passing the time scope, we adopt Daubechis6 (D6) to disassemble its three layers of waves, getting its small wave coefficients d1, d2, and d3 and scale coefficient a3 and restructural signals of vibrating response d1, d2, d3 and a3, which is shown in Figs. 4 and 5. Figure 6 obviously shows the effect of reducing noise.

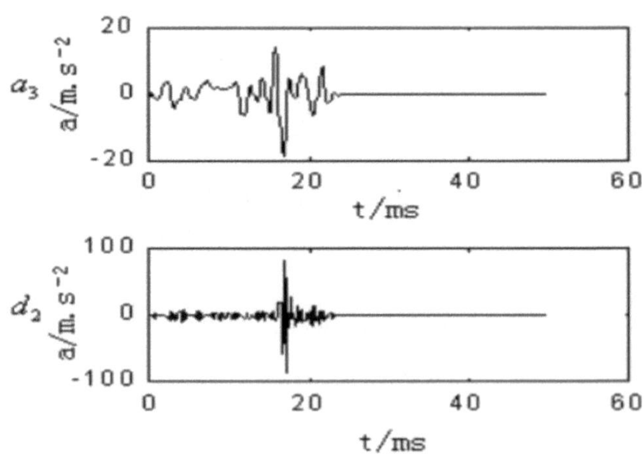

Fig. 4. d2 and a3.

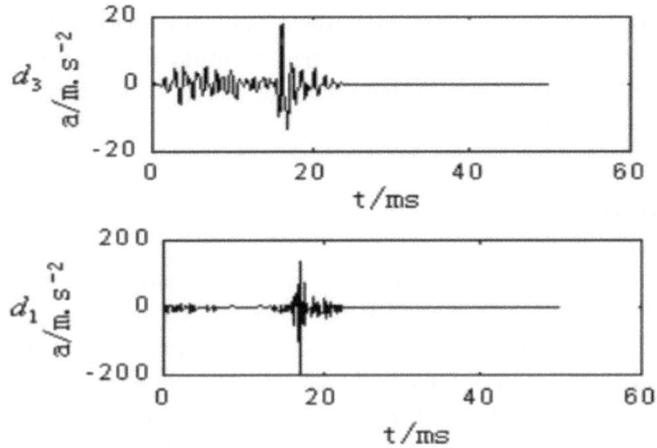

Fig. 5. d1 and d3.

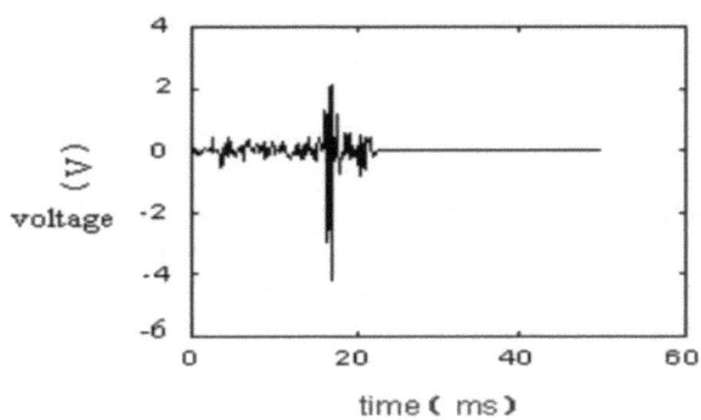

Fig. 6. d2, d3 and a3.

5.2 Coherent Amendment of Frequency Response Function

Identification accuracy of force can be accelerated by employing coherent amendment. Coherent function is used in this measure to carry through frequency domain amendment on identification results. Coherent function proceeds amendment operation on frequency function [11]. It can also be interpreted as an amendment operation on the frequency spectrum of vibratory response. No matter what interpretation it is, it accords with the physical significance of coherent function: coherent function $\gamma_{xy}(f)$ indicates the amount of output of some frequency caused by input [13].

5.3 Inverse Spectrum Level and Smoothing of FRF

These problems can be solved by inverse spectrum windowing with nonlinear inverse filtering. However, the types and width of windows are difficult to solve and there is a great amount of relative calculation. As a matter of fact, the key reason why the inverse method of frequency function affects the results of loading identification is that there are a number of sharp low points in the amplitude of frequency function and these low points will be amplified, resulting in serious distortion of inversion signals. As long as the amplitude of the transfer function is smoothened, identification results can be improved. Figure 1 shows loading spectrum [12] and Fourier converse transformation, and Figs. 2 and 3 disassemble its three layers of waves.

6 Conclusions

The cylinder pressure of the motor in the car chassis is recognized through the Fourier inverse Fourier transform. Compared with the measured cylinder pressure, both the amplitude and the waveform have certain errors. The main reason is that there is still a frequency domain interference signal in the vibration response, and the frequency sound function is not corrected and smoothed. In this paper, a converse filter with its growth rate as a gain of negative function and the phrase of negative transmission function as its phrase component is designed. Then, a Fourier converse transformation is conducted to identify cylinder pressure.

Identification of motor pressure mainly refers to the identification of dynamic loading of structures through assessing the response of structures. As a matter of fact, the key reason why the inverse method of frequency function affects the results of loading identification is that there are a number of sharp low points in the amplitude of frequency function and these low points will be amplified, resulting in serious distortion of inversion signals. Based on the response, loading can be identified after distinguishing the loading spectrum and conducting a Fourier converse transformation. The above-mentioned difficult problems are solved through a great amount of relative calculation. In addition, the amplitude of the transfer function is smoothened. The improvement measures were proposed in the paper.

References

1. Wang, S.: Identification of motor pressure of motors and measures on its accuracy improvements, pp. 1–131. ISMT, Suzhou, China (2011)
2. Yang, Z.: Design of autonomous fire hose docking system for intelligent fire-fighting robots. RICAI, pp. 1–6 (2023)
3. Neslen, R.B.: An Introduction to Copulas. Springer, New York (2006)
4. Du, X.: Research on the wall-climbing of TOFD inspection for welding seam of spherical tank, pp. 35–38 (2023)
5. Hu, C.: Application of a modified IEKF algorithm in mobile robot localization. In: RICAI 2022, Dongguan, China, 16–18 December 2022
6. Wang, S.: Diesel engine used for gun chassis failure Diagnostic method study, pp. 51–70. China Ship Research Institute (2020)
7. Chen, Z.: Research on AGVS dynamic path planning method based on UGNL. In: RICAI 2020, Shanghai, 978–1–4503–8830–6, 25 December 2023
8. Neslen, R.B.: An Introduction to Copulas. Springer, New York (2021)
9. Wu, Y.: Design and research of a highly adaptive autonomous wall-climbing robot. In: RICAI 2020, Shanghai, pp. 12–16, December 2020
10. Yang, Z.: Design of autonomous fire hose docking system for intelligent fire-fighting robots. In: RICAI 2023, pp. 1–6 (2023)

11. Fernando, T., Murray, N., Tan, K., Wimalaratne, P.: Software architecture for a constraint-based virtual environment. In: Proceedings of the ACM International Symposium on Virtual Reality Software and Technology (VRST 2022), London, UK, pp. 147–153 (2022)
12. Washington: School of mechanical and materials engineering, Washington State University, USA (2024). http://www.mme.wsu.edu/people/jay.htm
13. Gan, M.: Maintainability Design and Verify. Deference Industry Publishing House (1995)
14. Tao, F.: The study of maintainability model in mechanical system and equipment. Ordnance Engineering College Master dissertation, March 2023
15. Krause, G.S.: Generic automated model for early MTTR prediction. In: Proceedings of Annual Reliability and Maintainability Symposium 2021, pp. 280–285 (2021)

Analysis on the Impact of Electrically Assisted Turbocharging System on the Performance of High-Power Methanol Engines

Lin Sun, Fanhong Fan, Xianyu Jia, Shaopeng Gong, Songyue Chai, and Yongfeng Liu(✉)

Taiyuan University of Science and Technology, Taiyuan 030024, China
2025033@tyust.edu.cn

Abstract. Methanol has currently become an excellent alternative fuel for reducing emissions and improving engine performance due to its wide range of production sources and excellent combustion characteristics. To address issues such as insufficient power at low speeds, excessive supercharging at high speeds, and poor economy of methanol engines, a one-dimensional model of a high-power six-cylinder methanol engine was established based on GT-Power. A one-dimensional simulation model was built, calibrated, and validated. On the basis of the original engine's exhaust turbine, an electrically assisted turbocharging system was established by coupling a motor. Specifically, electric supercharging is adopted at low speeds, and turbine energy is recovered for power generation at high speeds to achieve energy conservation and emission reduction of the methanol engine. The influence law of motor/generator power on the comprehensive thermal efficiency of the methanol engine was studied. The results show that the efficiency of the compressor and turbine is the key factor affecting the comprehensive efficiency. The electrically assisted turbocharging system can improve the overall efficiency of the methanol engine by 0.7% and enhance the low-speed power performance, with a power increase of 6.9%.

Keyword: Methanol Engine · Electrically Assisted Turbocharging · GT-Power

1 Introduction

Amid global energy transition and worsening environmental issues, the unsustainability of traditional fossil fuels has become prominent, making the development of alternative energies an urgent need. Methanol stands out among clean energy options due to its low carbon emissions, high octane number, and excellent combustion performance; optimizing methanol engine technology is crucial for emission reduction and energy diversification [1–4].

Studies have shown that electromechanical compound supercharging systems, through the synergistic coupling of motors and turbines, can break through the energy constraints of traditional pneumatic connections, significantly improve the low-speed

S. Halgamuge et al. (Eds.): ICACMVE 2025, LNME, pp. 502–513, 2026.
https://doi.org/10.1007/978-981-95-6911-3_50

economy and transient response of methanol engines, and represent a key future development direction of supercharging technology [5–8]. The core of electromechanical compound supercharging technology is integrating a motor/generator into the intermediate shaft system of the turbocharger or at the compressor end, enabling the engine and turbocharger to operate independently or semi-independently [9–11]. Li Yang [12] found that electrically assisted turbocharging technology achieves partial decoupling between diesel engines and turbochargers by integrating a motor into traditional turbochargers; after optimization, it can enhance recovery power, comprehensive thermal efficiency, and low-load torque. Hu Chao from Shanghai Jiao Tong University [13] took marine diesel engines as the research object, analyzed the influence laws of key electromechanical compound supercharging factors on diesel engine performance via simulation, explored optimized turbine-compressor matching schemes, and constructed efficient control strategies. Su Dengke [14] used GT-Power simulations to compare front-mounted and mid-mounted motor schemes for six-cylinder diesel engines, finding the mid-mounted scheme superior in improving low-speed performance and high-speed energy recovery (maximum recovery power 13 kW, recovery efficiency 19%).Yang Pantao [15] demonstrated that electric compound supercharging expands the low-speed high-torque region of diesel engines by 18% and reduces fuel consumption by 16.7%.Frigo S et al. [16, 17] showed that at 4000 r/min full load, the electrically assisted turbocharging system reduces fuel consumption by 5.5% under low operating conditions compared to the split turbocharging system. Pasini G et al. [18], however, noted that split turbocharging shows more prominent energy-saving advantages under medium and high loads, lowering diesel engine fuel consumption by 8%.

Existing studies on electromechanical compound supercharging systems have mostly focused on diesel engines, confirming their value in improving diesel engine performance. However, methanol and diesel exhibit fundamental differences in fuel properties and combustion mechanisms, resulting in distinctly different requirements for supercharging matching and energy utilization. Therefore, studying the impact of electrically assisted turbocharging on the external characteristics of methanol engines is of great significance. This paper takes high-power methanol engines as the research object, studies the influence of motor power in the electrically assisted turbocharging system on its comprehensive thermal efficiency, and analyzes the influence laws of key factors, providing a theoretical basis for the application of electrically assisted turbocharging systems in methanol engines.

2 Numerical Modeling and Experimental Verification

2.1 Numerical Model

In this paper, a one-dimensional numerical simulation calculation method is adopted to conduct research on the electrically assisted turbocharging system. Firstly, a steady-state model of the engine is built and calibrated based on experimental data. Then, on the basis of the steady-state model of the original engine, relevant structures are modified to complete the construction of the steady-state model of the engine with the electrically assisted turbocharging system. The experimental engine is a spark-ignition high-power methanol engine modified from the Weichai WP13 series natural gas engine, and its performance parameters are shown in Table 1.

Table 1. Main Parameters of the Experimental Methanol Engine.

Project	Parameter
Type	Inline 6-cylinder engine
Firing Order	1-5-3-6-2-4
Cylinder diameter * stroke/mm	127*165
Compression ratio	12.5
Peak torque/N·m	2000
Rated power/kW	338

In this paper, the GT-Power software is used to establish the simulation model of the original engine, which is calibrated based on bench test experimental data. The establishment of this model includes sub-models such as the exhaust turbocharging system, intake and exhaust systems, fuel injection system, cylinders, and crankshaft, and adopts the Weber combustion model suitable for spark-ignition engines. The established one-dimensional thermodynamic model of the methanol engine is shown in Fig. 1.

The target engine originally used exhaust turbocharging, and this paper focuses on studying the impact of the electrically assisted turbocharging system on its steady-state performance. This system breaks the original power balance between the compressor and turbine by supplementing and recovering energy transmitted from the turbine shaft to the compressor. A power module (PowerRot) is installed on the turbocharger shafting to replace the motor/generator model (without considering the influence of motor speed, power, or voltage): positive values are input at low speeds to simulate the motor supplying energy to drive the shafting, while negative values are input at high speeds to simulate the generator absorbing energy from the shafting for exhaust energy recovery. The specific model is shown in Fig. 2.

2.2 Empirical Test

Figure 3 is a schematic diagram of the methanol internal combustion engine test bench. On this bench, tests on relevant parameters of the internal combustion engine regarding power performance, economy, and emission performance under different operating conditions were conducted. Based on the test experimental data, parameter calibration and model validation of the steady-state model of the original engine were completed.

After the calibration of the original engine test was completed, based on the external characteristic experimental data, seven calibrated operating points at rotational speeds of 700 rpm, 900 rpm, 1100 rpm, 1300 rpm, 1500 rpm, 1700 rpm, and 1900 rpm, which are consistent with the external characteristic calibration speeds, were selected for simulation. Figure 4 shows the comparison between the simulation results and experimental data of engine output power, fuel consumption rate respectively. It can be seen from the

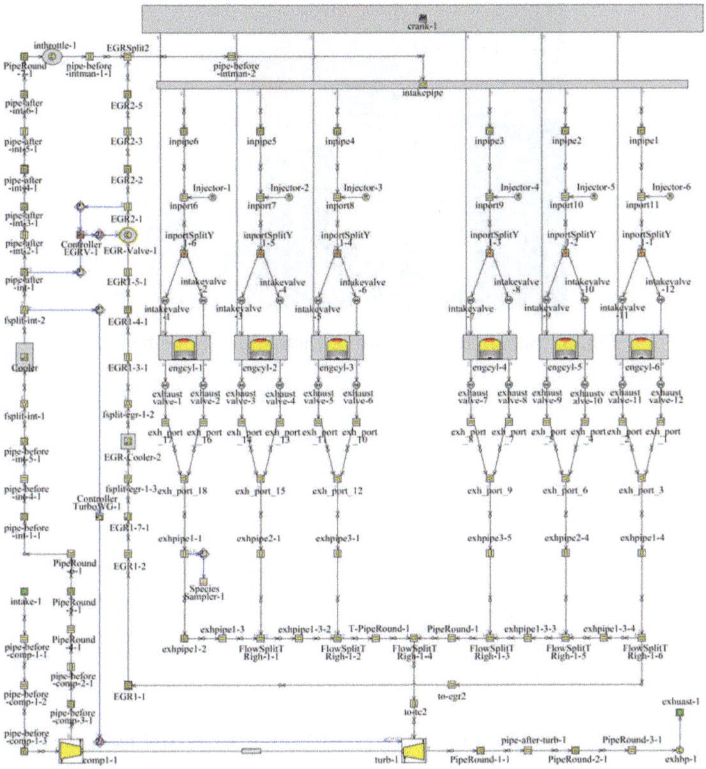

Fig. 1. One dimensional thermodynamic simulation model of methanol engine.

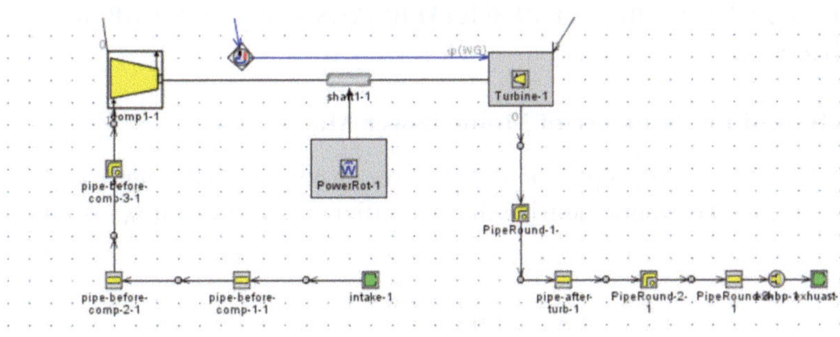

Fig. 2. Model of Electric Auxiliary System in Motor.

figure that the fuel consumption rate is higher than the experimental data at medium and high rotational speeds, but the maximum error does not exceed 5%, which allows the continuation of subsequent work.

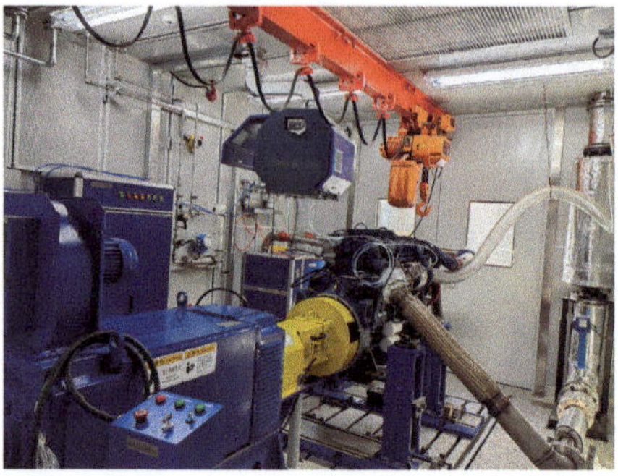

Fig. 3. Schematic diagram of methanol engine bench.

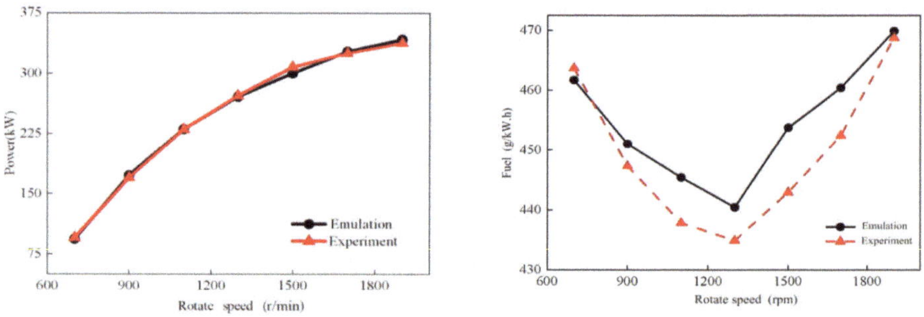

(a) Power Comparison Diagram (b) Comparison of Fuel Consumption Rates

Fig. 4. Comparison between simulation results and experimental.

3 Analysis of the Impact of Electric Auxiliary Systems on Engine Performance

3.1 Analysis of the Influence of Motor Power Matching

To effectively evaluate the impact of motor power on engine performance under different rotational speeds, the comprehensive thermal efficiency formula η_{EAT} is introduced as follows:

$$\eta_{\mathrm{H}} = \frac{E_{\mathrm{crank}}}{E_{\mathrm{fuel}}} \tag{1}$$

$$\eta_{\mathrm{EAT}} = \frac{E_{\mathrm{crank}} - E_{\mathrm{motor}}}{E_{\mathrm{fuel}}} \tag{2}$$

Among them, E_{crank} represents the effective power output by the engine through the crankshaft; E_{fuel} represents the total energy consumed by fuel per second; E_{motor} represents the power consumed by the motor in electric mode and recovered by the generator in power generation mode.

Figure 5 shows the trend of comprehensive thermal efficiency varying with motor recovery/supplementary power under high/low rotational speeds. Analysis of each curve in the figure reveals that, theoretically, there exists an optimal motor power at each rotational speed to maximize the engine's comprehensive thermal efficiency. At 900 rpm, the optimal electric power is 4 kW; At 1900 rpm, the optimal electric power is −9 kW; By connecting the optimal electric power at each operating point, it is found that there is an optimal electric power switching point, i.e., when the motor power is 0. As shown in Fig. 6, 1500 rpm is the switching point between electric mode and generator mode.

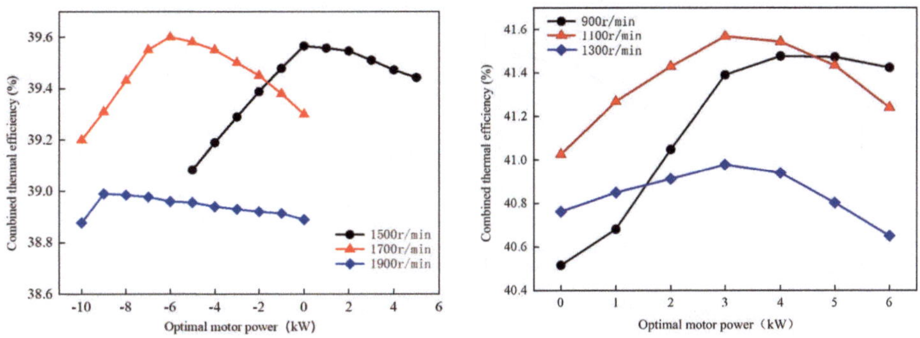

(a) Comprehensive Efficiency at Low Speeds (b) Comprehensive Efficiency at High Speeds

Fig. 5. Changes in Comprehensive Efficiency with Motor Efficiency.

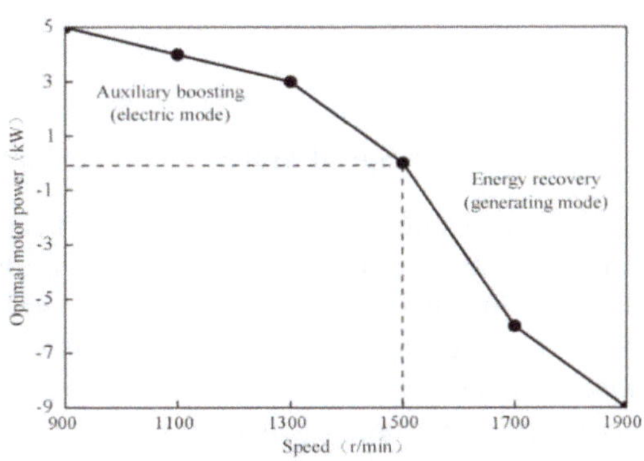

Fig. 6. Optimal Motor Power Variation.

3.2 Comparison of External Characteristics of Electric Assisted Boosting

Figure 7 shows the comparison of power and fuel consumption rate between the original engine and the methanol engine equipped with an electrically assisted turbocharger, respectively. The simulation results in the figure indicate that after adopting the electrically assisted turbocharging system, the engine power increases significantly at medium and low rotational speeds, with a maximum increase of 6.9%, and the fuel consumption rate decreases slightly at the same time, showing good economy and power performance. However, as the rotational speed increases, the power decreases slightly compared with the original engine, and the fuel consumption rate also increases. It is worth noting that the power reduction at high rotational speeds is offset by the recovered electric power, and the overall efficiency still improves. This indicates that the electrically assisted turbocharging system has significant advantages at medium and low rotational speeds, while further optimization is required at high rotational speeds to improve performance.

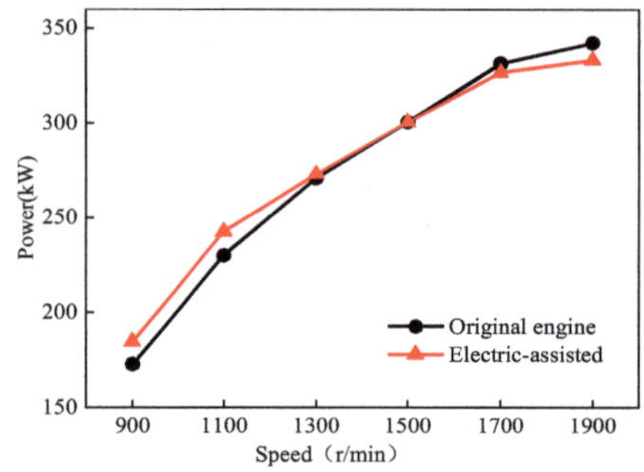

(a) Comparison of original engine and electric assisted boost power

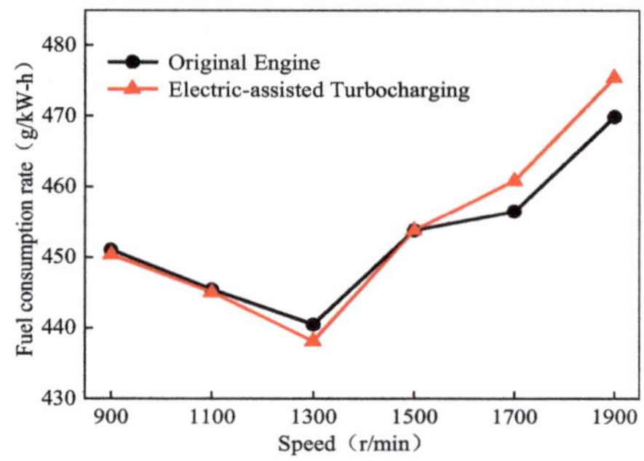

(b) Comparison of original engine and electric assisted boost fuel consumption rate

Fig. 7. Performance comparison between original engine and electric assisted turbocharged engine.

3.3 Analysis of the Impact of Low Speed Motor Mode Performance

Next, we delve into the key factors through which the motor affects the comprehensive thermal efficiency of the engine. The main factors influencing the engine's comprehensive thermal efficiency include the following five: engine thermal efficiency, turbine efficiency, compressor efficiency, exhaust gas energy at the turbine inlet, and compressed air energy at the compressor outlet.

Since the compressor is prone to surge when the engine idles at 700 rpm, 900 rpm is selected as the low-speed operating point for analysis. At an engine speed of 900 r/min, with the maximum cylinder average effective pressure corresponding to the maximum torque as the boundary condition, the energy provided by the motor to the turbocharger is adjusted, and simulations are conducted to obtain changes in compressor efficiency, turbine efficiency, and thermal efficiency. In the following analysis, the compressor and turbine efficiencies of the original engine and other performance parameters are uniformly used as reference indices, corresponding to the origin in the figures. The performance parameters corresponding to different electric powers are normalized to analyze their variation trends.

To further analyze the influencing factors of different optimal motor powers, analysis of Fig. 8b reveals that pre-turbine temperature rises and pre-turbine pressure drops significantly, indicating that motor power alters the energy distribution of the turbine, enabling more sufficient conversion of exhaust gas energy into compressor work. This also indirectly explains the phenomenon where compressor efficiency increases while turbine efficiency decreases. Comprehensive research shows that the engine's comprehensive efficiency is jointly determined by the traditional thermal efficiency of the engine, as well as turbine and compressor efficiencies, with the impact of turbine efficiency being more prominent.

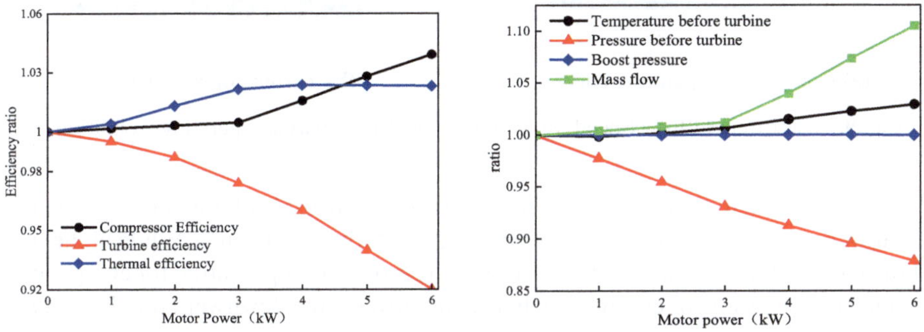

(a) The Effect of Low Speed Power on Efficiency

(b) Changes in High Speed Engine Parameters

Fig. 8. Trend of Main Parameter Changes in Low Speed Motor Mode.

To investigate the impact of motor power on the efficiency of the compressor and turbine at this rotational speed, a further analysis of the operating points of the turbine and compressor was conducted, as shown in Map Diagram 9. As motor power increases, the operating point of the compressor gradually moves toward the high-efficiency region with high pressure ratio and high flow rate, which also explains why the compressor efficiency improves with increasing motor power. Meanwhile, when the motor power is 0 kW, the operating point (B) of the original engine's turbine is already located within the optimal efficiency region. As motor power increases, the turbine operating trend shows a decreasing pressure ratio, reducing rotational speed, and moves through the high-efficiency region to the medium and low efficiency regions (Fig. 9).

3.4 Analysis of the Impact of High Speed Generator Mode Performance

At an engine speed of 1900 r/min, the maximum cylinder average effective pressure corresponding to peak torque serves as the boundary condition. As shown in Fig. 10, exhaust gas temperature at the turbine inlet rises continuously with increasing power generation output. When motor power reaches 9 kW, intake flow rate and compressor boost pressure decrease—this is because part of the turbine power to the compressor is diverted for power generation, reducing the compressor boosting capacity. Though this lowers boosting capacity, it reduces the compressor's excessive consumption of exhaust energy, allowing more energy to be recovered as electricity and thereby improving comprehensive efficiency.

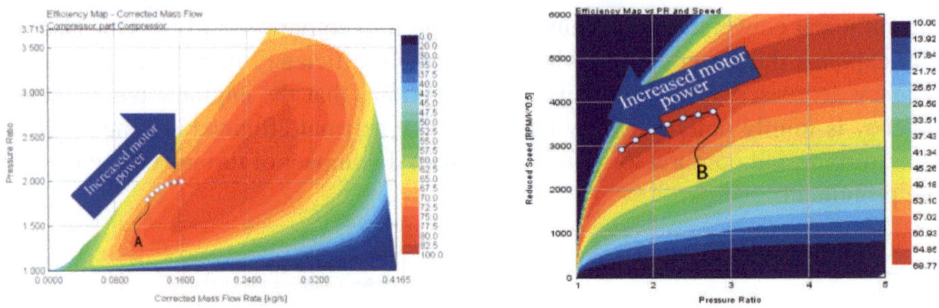

(a) Compressor operating curve

(b) Turbine operating curve

Fig. 9. The Effect of Low Speed Motor Power on Turbocharger Operating Conditions.

Meanwhile, increased pre-turbine pressure raises exhaust backpressure, and reduced mass flow rate cuts exhaust energy, further lowering turbine efficiency. While higher pre-turbine temperature increases exhaust enthalpy (partially compensating for energy loss), flow rate and pressure effects alleviate but cannot reverse the downward trend of exhaust energy.

As motor recovery power increases, turbine efficiency drops continuously (maximum 7.5%), while compressor efficiency rises slightly. This confirms turbine efficiency as the main factor affecting the engine's comprehensive thermal efficiency at high speeds. This trend reveals the principle behind peak comprehensive thermal efficiency in high-speed power generation mode, with the corresponding compressor efficiency aligning basically with the generation power.

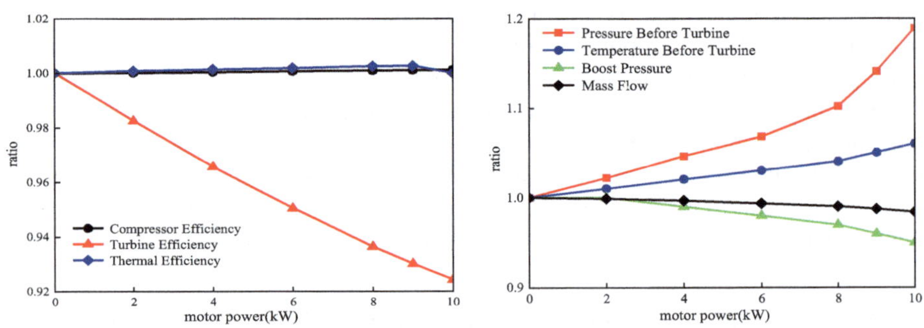

(a) The Effect of High Speed Motor Power on Efficiency

(b) Changes in High Speed Engine Parameters

Fig. 10. Trend of Main Parameter Changes in High Speed Generator Mode.

To investigate the variation law of compressor efficiency and turbine efficiency with the increase of motor recovery power under high-speed operating conditions, an analysis of the operating conditions of the compressor and turbine was conducted based on the characteristic curves (Map diagrams) in Fig. 11. The operating point (Point C) of the original engine's compressor is located at the edge of the high-efficiency region. As the recovery power increases, it gradually moves toward the high-efficiency region in the area with high flow rate and high pressure ratio, but the trend is relatively gentle and not obvious. In the turbine Map diagram, the operating point (Point D) of the original engine is already at the edge of the optimal efficiency region. As the motor power increases, the turbine operating point gradually moves toward the direction of low pressure ratio and low rotational speed. This is the reason for the continuous decrease in turbine efficiency shown in the diagram. Moreover, compared with compressor efficiency, the change in turbine efficiency is greater, and its impact on the engine's comprehensive thermal efficiency is more prominent.

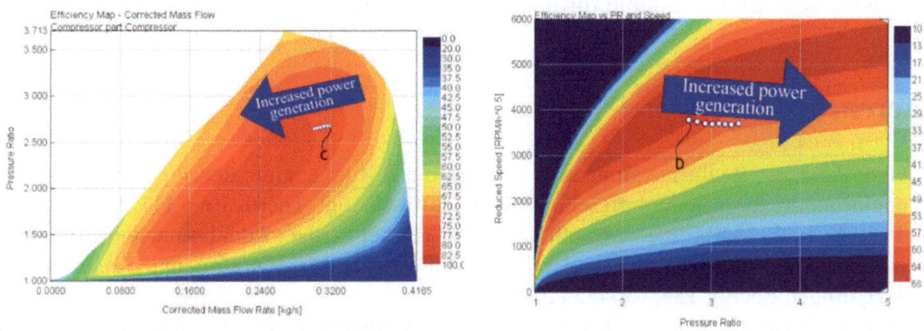

(a) High speed compressor operating curve (b) High speed turbine operating curve

Fig. 11. Influence of High Speed Motor Power on Turbocharger Operating Conditions.

4 Conclusion

To analyze the impact of the electrically assisted turbocharging system on methanol engines, this paper established a one-dimensional thermodynamic simulation model of a methanol engine with an electrically assisted turbocharging system using GT-Power. The influence of motor power on the performance of methanol engines under different operating conditions was explored, and the impact of compressor speed on the external characteristics of methanol engines was analyzed. The following conclusions are drawn:

(1) There exists a definite optimal motor power at each rotational speed to optimize the engine's comprehensive thermal efficiency. The mechanism behind this phenomenon is that motor power changes the efficiency of the compressor and turbine, driving the operating points of the turbine and compressor to move toward high-efficiency regions, thereby improving the engine's comprehensive thermal efficiency. By optimizing the motor power of the electrically assisted turbocharging system, the comprehensive thermal efficiency of the engine at low speeds can be increased to a maximum of 41.4%, with an improvement of 0.7%; at high speeds, it can be increased by a maximum of 0.4%, reaching 39.6%.

(2) After adopting the electrically assisted turbocharging system, the engine power at medium and low speeds is significantly increased by 6.9% with a reduced fuel consumption rate, showing good economy and power performance. At high speeds, however, the power decreases slightly and the fuel consumption rate increases, but part of the increase in fuel consumption is offset by the recovered electrical power, resulting in a certain improvement in overall efficiency.
(3) The original engine's small exhaust turbine with poor flow capacity leads to low exhaust energy utilization, hindering energy recovery, thus requiring rematching of turbine and compressor. Additionally, comprehensive thermal efficiency is more sensitive to compressor efficiency at low speeds and to turbine efficiency at high speeds; selecting a suitable turbine can further enhance system efficiency.

References

1. Guo, Z., et al.: Real-time simulation of marine dual-fuel engine: development of 0-D full-engine model for methanol-diesel mixed combustion. Appl. Therm. Eng. **112**, 127526 (2025)
2. Tang, Y., et al.: Multi-objective optimization of a diesel-methanol dual-direct injection engine integrated with on-board methanol reforming based on RSM-MOPSO coupled algorithm. Int. J. Hydrogen Energy **156**, 150432 (2025)
3. Rao, X., Yuan, C., Guo, Z., Xu, Y., Sheng, C.: Methanol as an alternative fuel for marine engines: a comprehensive review of current state, opportunities, and challenges. Renewable Energy **201**, 123562 (2025)
4. Li, S., Li, Y., Chen, Z., Zhou, A.: Optimization of combustion characteristics in methanol/gasoline medium-scale internal combustion engines based on homogeneous pre-treatment. Int. J. Hydrogen Energy **157**, 150203 (2025)
5. Li, R., Liu, Q., Li, X., Yang, D., Yue, H., Meng, Y.: The influence of methanol mixture formation mode on diesel engine combustion process. Appl. Therm. Eng. **111**, 127279 (2025)
6. Xu, Z., Wang, L., Jia, M., Chang, Y., Ding, R., Du, L.: A novel method for high-efficiency methanol compression-ignition engine based on onboard fuel reforming strategy: experimental study. Energy **324**, 137126 (2025)
7. Tian, J., Wang, Y., Long, Y., Cheng, Y.: Experimental investigation of a pure methanol reforming gas engine: performance, emission, and combustion characteristics. Int. J. Hydrogen Energy **135**, 20–30 (2025)
8. Spinner, G., Dahinten, F., Dauscher, S., et al.: Electrified boosting systems in today's and future automotive applications, pp. 1–5. SAE International, Monte Carlo (2018)
9. Li, H., Chen, X.X., Li, X.N., et al.: Current status of turbocharging technology development for marine diesel engines. J. Propul. Technol. **41**(11), 2438–2449 (2020)
10. Yang, L.: Pneumatic Design of Electric-Assisted Turbocharging System and Research on Matching Performance with Diesel Engines, p. 98. Harbin Engineering University, Harbin (2021)
11. Hu, C., Yang, M.Y., Deng, K.Y., et al.: Review of research progress on electromechanical hybrid boosting technology. Veh. Engine **1**, 1–6 (2019)
12. Yu, H.P., Song, K., Xie, H.: Research on optimal control of air system for electric-assisted turbocharged diesel engine based on model prediction. Engine Eng. **39**(4), 39–46 (2018)
13. Frigo, S., Lutzemberger, G., Martini, F., et al.: Performance evaluation of a medium size diesel vehicle equipped with different electric-turbo compound layouts. IET Electr. Syst. Transp. **8**(2), 71–79 (2018)

14. Su, D.K.: Design and Matching of Electric-Assisted Turbocharging System. Beijing Institute of Technology, Beijing (2017)
15. Yang, P.T., Cui, T., Zhao, Y.K., et al.: Influence of electric supercharger on low-speed steady-state performance of diesel engine. Acta Armamentarii **42**(9), 1829–1837 (2021)
16. Mazanec, J.M., Vang, N.S., Kokjohn, S.L.: Enabling off-highway diesel engine downsizing and performance improvement using electrically assisted turbocharging. Int. J. Engine Res. **24**(9), 4104–4126 (2023)
17. Pasini, G., Frigo, S., Antonelli, M.: Electric turbo compounding applied to a CI engine: a numerical evaluation of different layouts. In: Proceedings of Internal Combustion Engine Division Fall Technical Conference. American Society of Mechanical Engineers, New York, vol. 50503, pp. 115–147 (2016)
18. Vitek, O., Macek, J.: Thermodynamic potential of electrical turbocharging for the case of small passenger car ICE under steady operation. SAE International, Warrendale, vol. (1), pp. 512–526 (2017)

Research on the Working Characteristics of Electro-proportional Multi-way Valves in Excavators

Zhengjun Li, Zhiqiang Zhang(✉), Bin Bian, Dongliang Chen, and Zhiqi Liu

School of Mechanical Engineering, Taiyuan University of Science and Technology, Taiyuan 030024, Shanxi, China
zhiqiangzhang1@tyust.edu.cn

Abstract. In order to meet the requirements of accurate flow and stable control of excavator hydraulic system, this paper systematically studies the core performance of electric proportional multi-way valve based on the flow curve of mainstream multi-way valve in the market. Firstly, the physical models of proportional pressure reducing valve and electric proportional multi-way valve are built in AMESim software. On this basis, the influence of three key structural parameters of spring stiffness, spring preload and spool cover on the outlet flow performance curve of multi-way valve is mainly explored. Then, the orthogonal test method is used to carry out multiple sets of simulation tests and optimization analysis on the above variable parameters. Taking the flow curve of the mainstream multi-way valve as the target reference, the optimal parameter configuration is finally determined by comparing and screening the simulation results under different parameter combinations. The optimized simulation results show that the flow error between the flow characteristics of the electric proportional multi-way valve and the mainstream multi-way valve curve is only about 3.14%, and the system outlet flow curve is more gentle, which effectively improves the stability of the flow output. It provides appropriate research theoretical guidance for the later research and design of electric proportional multi-way valve.

Keywords: Electric proportional multi-way valve · AMESim · Flow

1 Introduction

Hydraulic technology came out at the beginning of the 20th century. At the beginning, it was only a transmission method. However, with the progress of modern science and technology, hydraulic technology has become a very important technical means of modern science and technology. At present, 95% of construction machinery mainly adopts hydraulic transmission system [1]. Nowadays, the development of hydraulic technology tends to be more and more integrated with electronic technology. Hydraulic technology and electronic technology are two different technical means. When components are combined, the system can meet the requirements of high precision, fast response, and intelligent control, and has a wide range of applications in industrial and military fields

S. Halgamuge et al. (Eds.): ICACMVE 2025, LNME, pp. 514–521, 2026.
https://doi.org/10.1007/978-981-95-6911-3_51

[2]. In the 1980 s, excavators adopted mechatronics technology, and the energy saving effect was remarkable [3]. Among them, the multi-way directional valve is the core control component of the excavator control system. The multi-way directional valve is referred to as the multi-way valve [4]. It is a multi-functional composite valve that integrates directional valve, one-way valve, pressure compensation valve, and safety relief valve [5].Therefore, it can realize the simultaneous control and operation of multiple actuators of the excavator. The high pressure and large flow of the hydraulic system is becoming more and more prominent [6], and the control performance of the hydraulic system is valued;

2 The Working Principle of the Electric Proportional Multi-way Valve

The structure of the electro-hydraulic proportional multi-way valve is shown in Fig. 1. Since the influencing factors of the proportional multi-way valve in this study are mainly related to the pressure characteristics of the electro-hydraulic proportional pressure reducing valve and the displacement of the main valve core, the electro-hydraulic proportional pressure reducing valve, as the pilot control element of the multi-way directional valve, directly exerts its output pilot pressure on the spring cavity of the main valve core. The pilot pressure enables the main valve spool to overcome the friction force and the spring preload pressure and move proportionally, ultimately achieving the flow regulation entering the actuator [7].

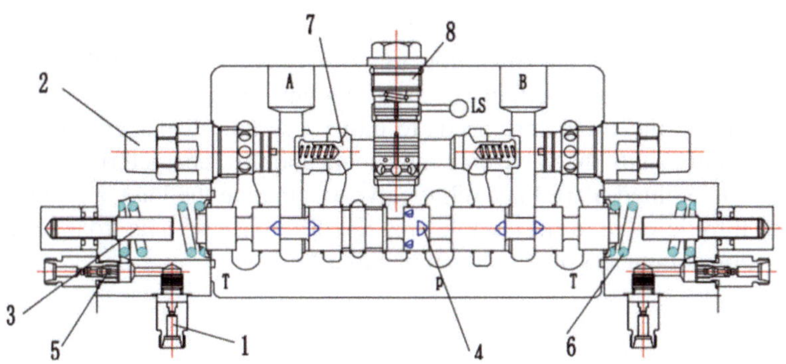

Fig. 1. Structural diagram of the electric multi-path valve for mining machinery (Part1-pilot pressure port; Part2-secondary relief valve; Part3-stroke limiter; Part4-throttle groove; Part5-pilot pressure shuttle valve; Part6-spring; Part7-load holding valve; Part8-pressure compensator).

2.1 Modeling of the Proportional Pressure Reducing Valve System

As shown in Fig. 2 below, in fact, the use of PWM technology enables the proportional pressure reducing valve system to achieve electrical proportional control [8]. It is the structure diagram of the proportional pressure reducing valve. The electric proportional pressure reducing valve proportionally controls the pressure at the oil port A according to the current of the input electromagnet.

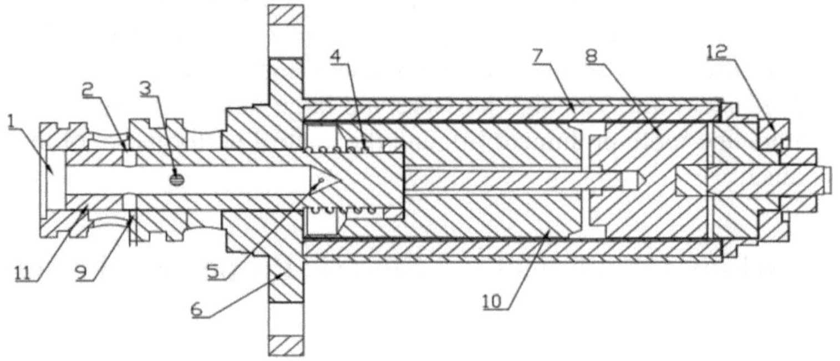

Fig. 2. Structural Schematic Diagram of Proportional Pressure Relief Valve (Part1-Port A; Part 2-Port P; Part3-Port T; Part4-Reset spring; Part5-Damping orifice; Part6-Shell structure; Part7-Coil; Part8-Armature; Part9-Overlap; Part10-Soft magnet guide sleeve; Part11-Spool; Part12-valve pocket).

3 Modeling of the Main Valve Core System of the Proportional Multi-way Valve

As shown in Fig. 3, a multi-valve model is built in the AMESim software. Part 1 is the encapsulated module of the optimized proportional pressure reducing valve model. Part 2 is the valve core throttling groove of the multi-valve. At the same time, to prevent low-frequency jitter when the small flow rate is applied, when the valve core is in the middle position, there is a certain covering amount between the valve core and the valve body. Then, the valve core needs to move a certain displacement before it can open [9]. To better study the valve core movement and flow characteristics, the pressure compensator is omitted. Part 3 is a relief valve to simulate the load, a fixed pressure is set, and the throttle valve forms a stable pressure difference with its front and rear parts. The simulation parameters are as shown in Table 1 below.

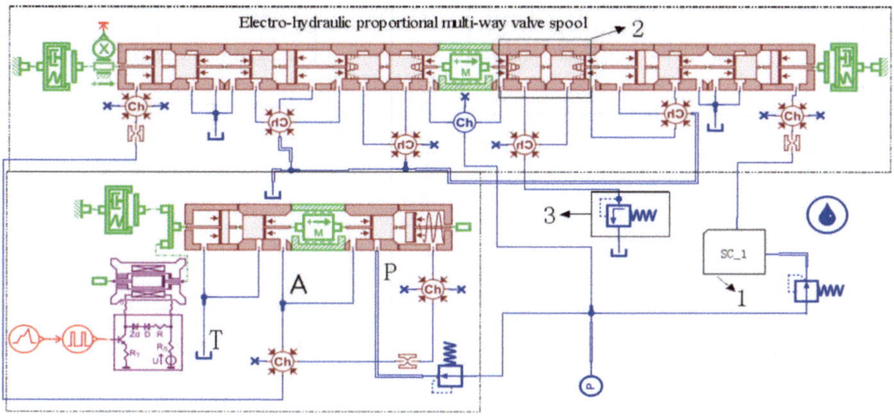

Fig. 3. Electro-pneumatic proportional multi-channel valve simulation model.

Table 1. Electrical proportional multi-way valve simulation parameters.

parameters	numerical value	parameters	numerical value
Valve core mass / Kg	0.3	Input pressure at port P / MPa	10
Reset spring stiffness / N/mm	115	Valve core diameter / mm	20
Spring preload force / N	120	Relief valve pressure / MPa	8.0
Rod diameter / mm	12	Valve core coverage / mm	1.0
The diameter of the first U-shaped / mm	3.0	The diameter of the second U-shaped / mm	7.0
Depth of the first U-shaped groove / mm	2.0	Depth of the second U-shaped groove / mm	4.0

3.1 Analysis of the Dynamic Characteristics Curve of the Outlet Flow of the Proportional Multi-path Valve

According to the basic principle of the hydraulic valve, in the case of stable pressure difference, the flow required by each actuator in the system is independent of the load, and only depends on the opening size of the corresponding multi-way valve [10]. In order to further analyze the dynamic characteristics, when the input is a transient signal, how the variable parameters affect the dynamic response of the spool displacement, such as the current signal of the proportional pressure reducing valve. This providing a more comprehensive theoretical basis for the precise control and optimization design of the system.

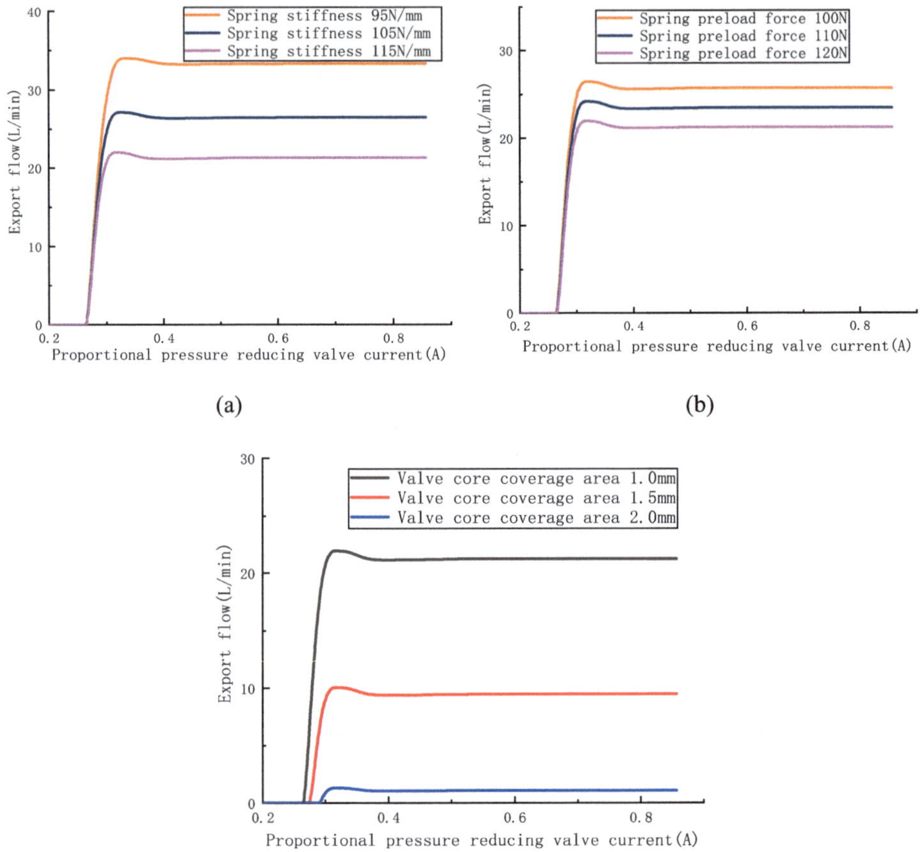

Fig. 4. The influence of different spring stiffness on the outlet flow rate (a). The influence of different spring preload forces on the outlet flow rate (b). The influence of different valve core coverage amounts on the outlet flow rate(c).

These three sets of curves respectively reveal the influence laws of spring stiffness, preload force, and valve core coverage on the dynamic response of the electric proportional multi-way valve's flow. At an electric current of approximately 280 mA, as shown in Fig. 4 (a), when the proportional pressure reducing valve current jumps to 0.28 A and triggers the flow response, the stiffness curve of 115 N/mm shows that due to the large valve core resistance, the steady-state flow is 14 L/min lower than the 95 N/mm curve. As shown in Fig. 4(b), the blue curve with a preload force of 120 N requires overcoming a greater initial resistance, resulting in a delayed flow start-up and a steady-state flow that is 4 L/min lower than that of the 100 N preload force. As shown in Fig. 4(c), the curve with a 2.0 mm coverage shows that due to the restricted flow passage of the valve port, the flow peak is only 1.2 L/min, and the steady-state flow is nearly 1 L/min, which is 20 L/min lower than that of the curve with 1.0 mm coverage. All three show significant differences in the delay time of the flow dynamic response, steady-state flow, and error rate by changing the valve core motion resistance and the initial valve port opening. The greater the stiffness, preload force, and coverage, the slower the flow response, the smaller the steady-state flow, and the higher the error rate. This provides a quantitative design basis for the optimization of the dynamic characteristics of the electric proportional multi-way valve (such as matching spring parameters and controlling the valve core coverage), and improves the pressure-flow control accuracy of the system.

3.2 Orthogonal Experiment Optimization and Comparative Analysis

The single factor analysis cannot reflect the working characteristics of the multi-way valve flow, and there may be a large error compared with the actual product characteristic curve. Therefore, the orthogonal test method is adopted for the above three sets of variable parameters. The main index is the flow characteristic curve of the mainstream multi-way valve on the market. Through the statistical analysis of multiple sets of test data, the parameter combination that minimizes the maximum error rate is finally selected, that is, the optimal parameter. This process not only excludes the one-sidedness of single factor analysis, but also fully considers the synergistic effect between parameters to ensure that the optimized flow characteristics of the multi-way valve are closer to the working performance of the actual product, which provides a more reliable basis for subsequent structural design and performance improvement.

Based on a comprehensive analysis of the orthogonal experiment results, the optimal combination of parameters was determined [11],as shown in Table 2 below.

Table 2. Orthogonal experiment result table.

test number	spring stiffness (A)/N/m	spring preload force (B)/N	valve core coverage area (C)/mm	error rate /%
1	95	100	1.0	4.05
2	95	110	1.5	4.61
3	95	120	2.0	4.85
4	105	100	1.5	3.72
5	105	110	2.0	4.01
6	105	120	1.0	3.62
7	115	100	2.0	3.55
8	115	110	1.0	3.14
9	115	120	1.5	3.35
mean value (K_1)	4.5	3.67	3.6	
mean value (K_2)	3.78	3.95	3.89	
mean value (K_3)	3.24	3.94	4.13	
range	1.26	0.28	0.53	

It can be known that when the spring stiffness is 115N/mm, the spring preload force is 110N, and the valve core coverage is 1mm, the maximum error rate is relatively small.

And according to the size of the range, the influence order of each factor on the error rate is A > C > B. That is, the spring stiffness has the greatest impact on the flow error rate, followed by the coverage, and finally the valve core coverage. At the same time, as shown in Fig. 5 curve, the optimized simulation flow curve basically coincides with the flow characteristic curve of a mainstream multi-way valve. The maximum flow rate is 120L/min and the curve is more gentle, and the maximum error rate is only about 3.14%. This result indicates that the parameter combination selected through orthogonal testing can accurately reproduce the flow characteristics of the actual multi-way valve. At the same time, such a small error rate also indicates that this simulation model can effectively reflect the influence mechanism of different parameters on the performance of the multi-way valve, providing a reliable reference basis for subsequent research on valve core structure optimization and control strategy design.

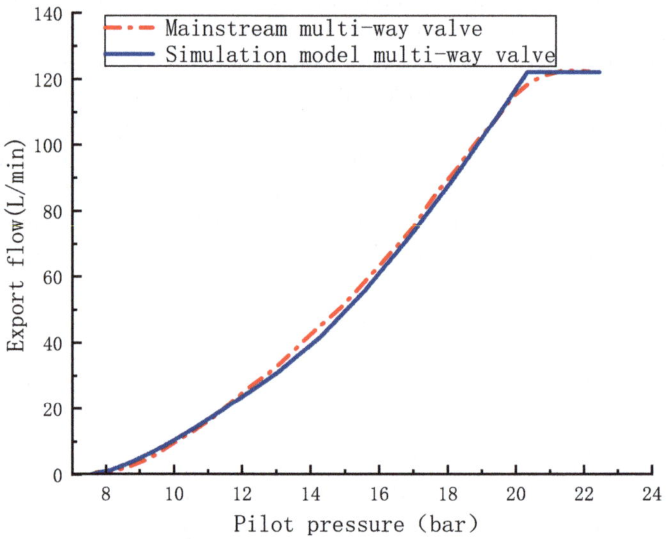

Fig. 5. Optimized simulation model and flow characteristic curve of the M4–12 multi-way valve

4 Conclusion

This paper established a simulation model of the electro-hydraulic proportional multi-way valve using the AMESim software, and conducted simulation studies on different parameters that affect its working characteristics. Firstly, a proportional pressure reducing valve model was built, Then, a proportional multi-way valve model was built, and the orthogonal experimental method was used to conduct reasonable optimization design of the variable parameters of the multi-way valve. Finally, it was compared with the mainstream multi-way valve, and the characteristic results showed an error of 3.14% after optimization, and the flow rate of the optimized system was more stable. This provides certain theoretical support for the design, control, construction of the hydraulic system of domestic load-sensitive proportional multi-way valves, as well as the research and development of new products.

References

1. Zhang, T.: Energy saving, emission reduction and consumption reduction-Hydraulic technology how to adapt to the requirements of sustainable development. Hydraul. Pneumatic Seal **31**(01), 74–77 (2011)
2. Hu, Q.Y.: Research on the valve core position control algorithm of proportional multi-way valve based on model predictive control (Master's Thesis, Yanshan University). Master's Degree (2022). https://link.cnki.net/doi/https://doi.org/10.27440/d.cnki.gysdu.2022.002 36610.27440/d.cnki.gysdu.2022.002366
3. Zhang, B.: Analysis of energy-saving technologies and development strategies for hydraulic systems in construction machinery. Intern. Combust. Engine Accessories **14**, 196–197 (2021). https://doi.org/10.19475/j.cnki.issn1674-957x.2021.14.088
4. Yang, H.Y., et al.: The development history and research prospects of multi-way directional valves. J. Mech. Eng. **10**, 1–5 (2005)
5. Cao, W.B., et al.: Simulation study on the influence of combined throttle slot on the flow field characteristics within the multi-way valve. Hydraul. Pneum. **46**(11), 17–25 (2022)
6. Yang, Y.L., et al.: Research on energy saving of hydraulic excavators based on negative flow control. Mach. Tool Hydraul. **41**(23), 125–127+131 (2013)
7. Xiao, N.X., et al.: Stability analysis of electro-hydraulic proportional multi-channel valve system under flutter signals. Mach. Tool Hydraul. **50**(10), 26–30 (2022)
8. Topçu, E.E., Bali, E.: Modeling and analysis of an electro-pneumatic brake valve with on-off type solenoid driven by PWM technique. Open Access J. **36**(3), 1417–1430 (2021)
9. Bai, Z., Du, K., Mi, X.: Simulation and experiment on flow control and micro-motion characteristics of LUDV multi-way valve. Hydraul. Pneum. **48**(07), 86–92 (2024)
10. Zhou, S.Y., et al.: Simulation analysis of the working circuit performance of electrically controlled load-sensitive mountain harvesting machine. Mach. Tool Hydraul. **53**(02), 207–212 (2025)
11. Xie, N., Wei, W., Ba, J., Yang, T.: Operation parameters study on the performance of PEMFC based orthogonal test method. Case Stud. Therm. Eng. **61**, 105035 (2024)

Study on Unsteady Pumping Characteristics of Concrete Pump Truck Based on CFD-DEM Three-Phase Flow Coupling Model

Jinning Zhi[1](✉), Shuxuan Yang[1], Yabo Wei[2], and Lingying Zhao[1]

[1] School of Mechanical Engineering, Taiyuan University of Science and Technology, Taiyuan 030024, Shanxi, China
jinning_zhi@tyust.edu.cn

[2] College of Vehicle and Traffic Engineering, Taiyuan University of Science and Technology, Taiyuan 030024, Shanxi, China

Abstract. Gas suction during concrete pumping induces a transition from steady-state to unsteady flow, significantly compromising boom stability. This study investigates unsteady flow characteristics in different pipelines of a pump truck. First, a geometric fluid domain model of C40 concrete pipelines was established, and grid independence was verified. A three-phase flow mathematical model was subsequently developed, comprising gas-liquid control equations, solid particle motion equations, and a laminar flow model. The gas and liquid phases were simulated using Fluent, while solid particle interactions were modeled with Rocky DEM. Flow characteristics in straight pipes, elbows, and combined pipelines were analyzed via coupled CFD-DEM simulations. Finally, a comparative analysis evaluated the effects of pipeline structures and pumping parameters on concrete flow. Results indicate that increasing gas volume fraction reduces pressure in straight pipes and elbows while decreasing excitation forces in straight pipes. Excitation forces in elbows remain unaffected. Increasing the straight pipe attitude angle elevates pressure but reduces excitation forces. Larger elbow curvature radii increase pressure with negligible impact on excitation forces.

Keywords: Concrete pump truck · Gas-liquid-solid three-phase flow · CFD-DEM · Unsteady pumping · Numerical simulation

1 Introduction

A concrete pump truck is a machine that continuously transports concrete through pipelines using pressure. Inadequate suction induces unsteady flow, generating pipe wall excitation forces and boom vibration [1], thereby impairing operational stability. Concrete flow involves complex multiphase interactions (e.g., sand, gravel, air, water, cement). Traditional methods are inadequate for analyzing this complex fluid dynamics problem. Current numerical approaches for fresh concrete include: continuous fluid models (Bingham, Herschell-Bulkley) [2], discrete particle models [3], and suspended particle two-phase models [4]. CFD-DEM coupling is widely adopted for fluid-solid pipeline flow.

S. Halgamuge et al. (Eds.): ICACMVE 2025, LNME, pp. 522–533, 2026.
https://doi.org/10.1007/978-981-95-6911-3_52

Extensive research has examined concrete rheology and multiphase flow effects on pumping pressure. Yuan S [5] differentiated concrete flow behaviors, leading to pumpability models and pressure prediction methods. Martin A. Haustein [6] used PIV to show how material segregation reduced pressure loss by 60%, reflecting actual pumping rheology including particle migration. Fumin Li [7] proposed a new pressure loss equation for specific concrete mixtures and pipe lengths. To address complex gas-liquid-solid flow, CFD-DEM coupling has been widely applied: Tong et al. [8] simulated slurry transport with pebbles in curved shield pipelines via CFD-DEM coupling, studying migration and deposition effects on flow fields. Li Qi [9] established a CFD-DEM model analyzing fresh concrete pumping flow characteristics and elbow wear. Xiao-Fang Lv [10] confirmed the feasibility of hydrate slurry multiphase simulations.

Current studies primarily model concrete pumping using fluid-solid coupled models, neglecting gas effects and struggling to accurately simulate actual flow. While focused on predicting pressure and flow rate, they largely ignore pipeline excitation forces. This study addresses these gaps by developing a gas-liquid-solid three-phase flow model based on the CFD-DEM coupling algorithm, controlling gas column volume fraction within the pipe. It numerically simulates the influence of different structural parameters and gas volume fractions on unsteady concrete pumping. The research specifically analyzes the variation laws of pipeline pressure and excitation forces during this unsteady flow process.

2 Three-Phase Coupling Equations for Concrete Flow

2.1 Liquid Phase Control Equation

Concrete mortar is modeled as a uniform fluid, following mass and momentum conservation, governed by the continuity equation and Navier-Stokes equation:

$$\frac{\partial \alpha_f \rho_f}{\partial t} + \nabla \cdot (\alpha_f \rho_f U_f) = 0 \tag{1}$$

$$\frac{\partial}{\partial t}(\alpha_f \rho_f U_f) + \nabla \cdot (\alpha_f \rho_f U_f) = -\nabla p_f + \nabla \cdot \alpha_f \tau + \alpha_f \rho_f g + F_f \tag{2}$$

where α_f, ρ_f, U_f, P_f are the volume fraction, density, velocity vector and pressure component of the fluid phase, respectively; τ is a continuous stress tensor; g is the acceleration due to gravity, taken as 9.81 m/s^2; F_f is the force exerted on the mortar.

Concrete in the pipe is a non-Newtonian Bingham fluid, with flow characteristics determined by yield stress and plastic viscosity. Its rheological model is:

$$\begin{cases} \tau = \tau_0 + \eta\dot{\gamma}(\tau \geq \tau_0) \\ \dot{\gamma} = 0(\tau < \tau_0) \end{cases} \tag{3}$$

where τ is the shear stress, τ_0 is the yield stress, γ is the shear rate, and η is the plastic viscosity of the mortar.

2.2 Gas Phase Control Equation

In CFD-DEM, the gas phase is continuous and follows Navier-Stokes equations, with energy equations ignored (no exothermic reactions). It is governed by mass and momentum conservation:

$$\frac{\partial}{\partial t}(\varepsilon_g \rho_g) + \nabla \cdot (\varepsilon_g \rho_g \mu_g) = 0 \tag{4}$$

$$\frac{\partial}{\partial t}(\varepsilon_g \rho_g \mu_g) + \nabla \cdot (\varepsilon_g \rho_g \mu_g u_g) = -\varepsilon_g \nabla p + \nabla \cdot (\varepsilon_g \tau_g) + \varepsilon_g \rho_g g - S^A \tag{5}$$

where p, ρ_g, ε_g, τ_g and μ_g represent the pressure, density, volume fraction, viscosity and velocity of the air flow field, respectively, g is the acceleration due to gravity, taken as 9.81 m/s^2, S^A represents the momentum exchange between the gas-liquid phase and the particle phase as:

$$S^A = \frac{1}{V_{cell}} \sum_{i=1}^{N_P} F_{d,i} \tag{6}$$

2.3 Discrete Phase Control Equation

Larger stones in the mud are discrete phases, analyzed via DEM and Newton's laws. Forces on stones include gravity, contact stress from pipe wall collisions, and drag from mud shear flow. The Hertz-Mindlin soft sphere contact model is used. The governing equation for each particle at any given time t can be expressed as follows:

$$m_i \frac{dU_{P,i}}{dt} = m_i g + \sum_{j=1}^{n} (F_{c,ij} + F_{d,ij}) + (F_{c,iw} + F_{d,iw}) + F_{f-g,i} \tag{7}$$

$$I_i \frac{dw_i}{dt} = \sum_{j=1}^{n} T_{ij} + T_{iw} \tag{8}$$

where m_i, $U_{P,i}$, I_i and w_i are the mass, velocity, inertial motion and rotational velocity of particle i, respectively; $F_{c,ij}$, $F_{d,ij}$ and T_{ij} are the contact force between particles, the viscous damping force and the torque between particles i and j respectively; $F_{c,iw}$, $F_{d,iw}$, and T_{iw} are the contact force, viscous damping force and torque between particle i and pipe wall; $F_{f-g,i}$ is the interaction force between continuous phase and discrete phase.

2.4 Gas-Liquid-Solid Three-Phase Coupling

Gas, liquid, and solid phases are uniformly mixed (no velocity difference, mass transfer, adiabatic). Interaction forces include drag, buoyancy, pressure, and gradient forces; this study considers only buoyancy and drag. The coupling interaction force formula:

$$F_{f-g,i} = F_{B,i} + F_{d,i} \tag{9}$$

In the formula:

$$F_{B,i} = -\rho_g g V_i \tag{10}$$

where V_i, $F_{B,i}$ and $F_{d,i}$ represent the volume, buoyancy, and drag of particle i, respectively. The coupling between the gas-liquid-solid three phases is realized by momentum exchange, the momentum exchange amount is:

$$S = \frac{1}{\Delta V}\sum_{i=1}^{n} F_{f-g,i} \tag{11}$$

where ΔV is the volume of each phase, and n is the number of particles.

3 Three-Phase Flow Coupling Simulation Model for Concrete

3.1 Coupling Process

This study used CFD-DEM coupling: Fluent solved the continuous phase (gas/liquid) flow, while Rocky DEM calculated discrete particle motion and stress. The interaction between them is realized by two-way coupling. The CFD-DEM coupling process is shown in Fig. 1.

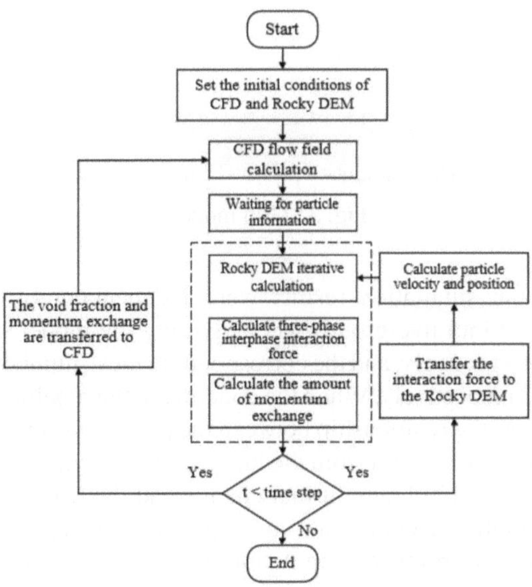

Fig. 1. Coupling flow chart.

3.2 Calculation Model

The pump truck pipeline and particle model are shown in Fig. 2. A 125 mm diameter, 1100 mm length straight pipe with horizontal posture is selected. The particle model simulates irregular spherical stones. A pressure monitoring surface is set up at a distance of 0.1 m from the inlet of the pipeline to monitor the inlet pressure, and the latter one-meter pipeline is used to extract the excitation force on the pipe wall.

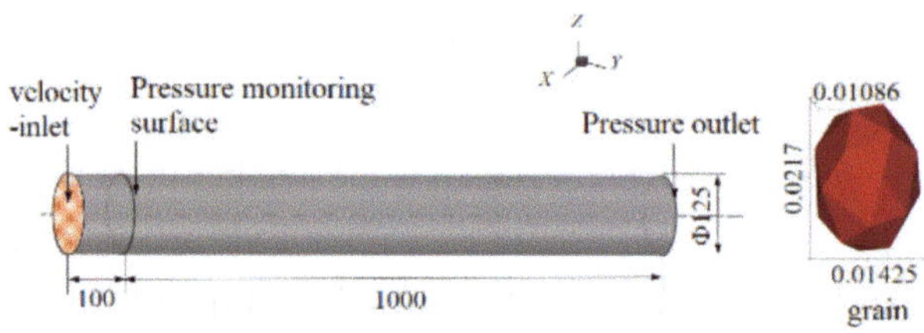

Fig. 2. Pipe and particle model.

3.3 Grid Division and Independence Verification

The fluid domain is divided into boundary and core layers via tetrahedral mesh (finer near walls, coarser internally) as shown in Fig. 3.

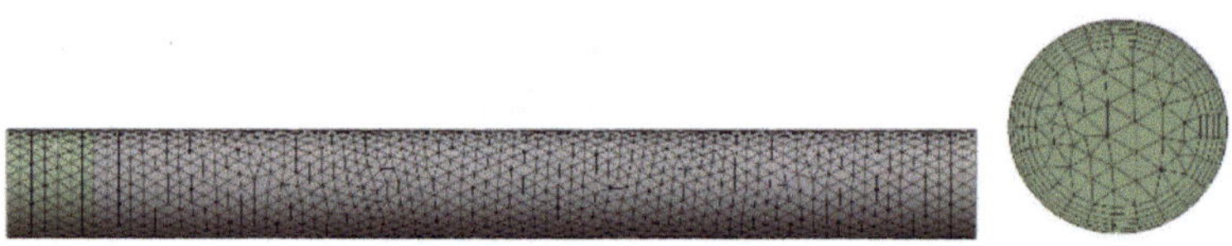

Fig. 3. Grid model.

To ensure that the simulation results were independent of the mesh size, the mesh size was divided into five groups when dividing the model structure, which and 20,18,16,15,14 mm respectively, and the corresponding mesh numbers are 22194, 27297, 35329, 41214, 47942 respectively. When the concrete in the pipeline flowed periodically at a flow rate of 2.5 m/s, the relationship between the pressure at the stable stage of 1.93 s and the number of grids at 100 mm from the inlet was obtained, as shown in Fig. 4, when the number of grids was 41214, any further change had the least effect on the pressure. Considering the calculation accuracy and convergence speed, approximately 41214 grid elements were finally set up for the simulation grid of all working conditions, and the grid size was 15 mm.

3.4 Calculation Parameters and Boundary Conditions

To ensure convergence stability and speed, the SIMPLE algorithm solved pressure-velocity coupling. Since the Reynolds number of concretes in the pipeline (Re $\approx$1,200)

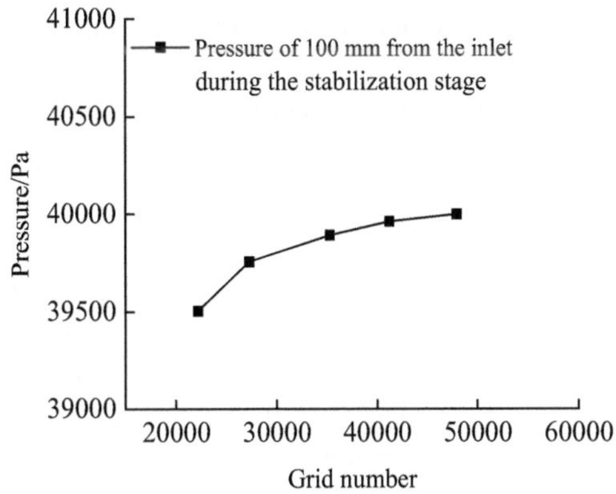

Fig. 4. Grid independence verification.

is lower than the critical value for turbulent flow (Re < 2,000), and the behavior of the lubricating layer conforms to the characteristics of non-Newtonian fluids. Therefore, this paper adopts the laminar flow model, which can better simulate the flow trajectory of concrete in the pipeline at low speeds. The density of C40 concrete is defined as 2400 kg/m^3, and under the laminar flow model, the Herschel-Bulkley model is used to define its viscosity (with parameters shown in Table 1). A time step of 0.001 s with 100 iterations per step ensured calculation stability and accuracy.

Rocky DEM utilizes the Hertz-Mindlin contact model for particle and wall forces, with parameters detailed in Table 2.

Table 1. CFD numerical calculation parameters.

Material properties	Core concrete	Lubricating layer	Gas
Consistency index / (kg/(m·s))	30	2	0.001
Yield stress / Pa	70	5	/
Critical shear rate /s^{-1}	1.52	2	/
Power-law index n	1	1	/

The boundary conditions of the model are velocity inlet, pressure outlet, and no-slip wall. The outlet pressure is atmospheric pressure, whose impact on the flow characteristics of concrete inside the pipeline can be neglected.

To match the periodic flow from concrete pump trucks' dual-cylinder system, an ideal inlet speed was calculated based on pumping pressure signals. A User-Defined Function (UDF) was then developed to control the pipeline inlet flow rate accordingly.

Table 2. DEM numerical calculation parameters.

Particle property		Interparticle contact model		Particle-wall contact model	
Name	Numerical value	Name	Numerical value	Name	Numerical value
Particle Poisson's ratio	0.25	Restitution coefficient	0.6	Restitution coefficient	0.5
Particle density /(kg/m^3)	1400	Coefficient of static friction	0.2	Coefficient of static friction	0.3
Particle shear modulus / Pa	3×10^8	Coefficient of kinetic friction	0.02	Coefficient of kinetic friction	0.03

The velocity function at the entrance of the pipeline is:

$$V(t) = \begin{cases} \frac{v}{2}\cos[\frac{\pi}{(t_2-t_1)}(t-t_2)] + \frac{v}{2}\,(t_2,t_1) \\ v(t_3,t_2) \\ \frac{v}{2}\cos[\frac{\pi}{(t_4-t_3)}(t-t_3)] + \frac{v}{2}\,(t_4,t_3) \end{cases} \tag{12}$$

where v is the maximum (concrete) flow during constant-speed piston motion, $t_1 - t_2$ is the initial acceleration interval, $t_2 - t_3$ is the uniform motion interval, and $t_3 - t_4$ is the deceleration interval.

The concrete inlet velocity is shown in Fig. 5.

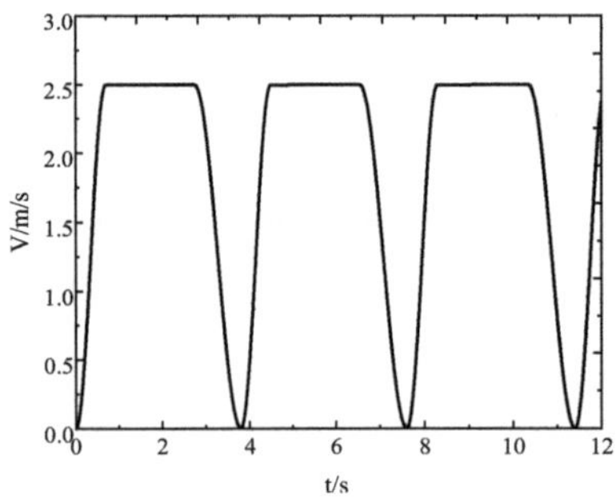

Fig. 5. Concrete inlet velocity.

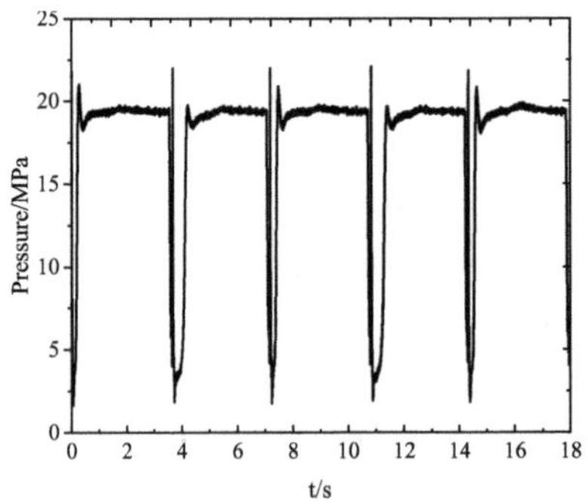

Fig. 6. Time-domain diagram of pumping pressure in a real vehicle.

4 Simulation Results and Discussion

4.1 Model Verification

The horizontal posture of the pump truck is selected for experimental research. Pressure monitors tracked the conveying pipe inlet pressure during concrete pumping. Operating at a 3.8 s period, 58% displacement, 1485 r/min, and 129° boom angle, the measured cylinder pressure matched the simulated pumping pressure trend (Fig. 6), validating the model.

To verify model accuracy, this study compared its gas-liquid-solid three-phase simulation results for unsteady concrete flow against Choi's [11] experimental data from a 170 m horizontal coil at flow rates of 0.68, 0.92, and 1.13 m/s. Introducing 10% gas, other model parameters are the same. As shown in Fig. 7, the results exhibit strong agreement with experiments and Choi's simulations, validating the model. Slightly higher-pressure losses occurred, primarily due to the deviation between the straight pipe model used for verification and the actual coiled pipe geometry simplified in the simulation.

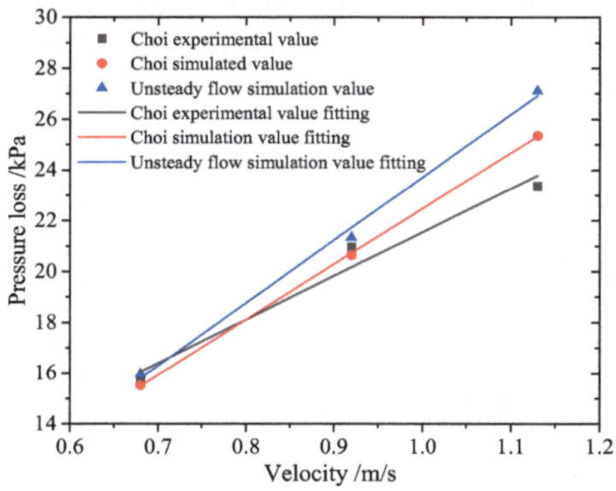

Fig. 7. Three-phase coupled simulations versus Choi's experimental and simulated values.

4.2 Flow Properties of Straight Pipeline and Elbow Pipe

Researching concrete pumping is complicated by non-uniform pipeline gas distribution. This study quantitatively analyzes effects of gas volume fractions (10–40%) on pipeline pressure and excitation force. With gas distribution frequency at 3.75 s and concrete flow velocity at 2.5 m/s, a three-phase flow coupling simulation is conducted.

Figure 8 shows the variation of pressure and excitation force in straight and elbow pipes with time under different gas volume fractions. Periodic piston pumping causes cyclic pipeline pressure changes: peaking during concrete filling, dipping slightly with gas entry but recovering rapidly, then falling as concrete discharges. When the gas volume fraction increases, pressure decreases in both straight pipes and elbows. The straight pipe wall's excitation force amplitude remains below 150 N and reduces with higher gas volume fraction. For the elbow, excitation force fluctuates below 100 N and is largely unaffected by gas volume fraction.

When a pipeline is rotated counter clockwise from horizontal (0° to 90°), greater inclination increases pressure due to the elongated flow path and higher resistance, requiring more pressure to overcome it. Conversely, during pumping, the pipe wall excitation force fluctuates periodically below 250 N and decreases as the inclination angle increases. As shown in Fig. 9.

In order to analyze the influence of the curvature radius of the elbow on the flow characteristics of concrete during the pumping process, four common curvature radiuses of the elbow (350 mm, 500 mm, 1000 mm, 1400 mm) were numerically simulated. The gas column distribution time interval was set to 3.75 s, the entry speed was 2.5 m/s, and the gas volume fraction entering different elbows was controlled at 30%. Figure 10 shows pipe pressure increases with larger elbow curvature radii, this phenomenon seems to be in contradiction with the intuitive cognition of ' resistance decreases with the increase of radius '. This occurs because larger radii reduce the R/d ratio, significantly lowering the local resistance coefficient (ζ). Reduced ζ diminishes turbulent friction losses, conserving kinetic energy and improving flow uniformity. Per Bernoulli's principle, this enhanced uniformity enables more complete dynamic-to-static pressure conversion at the elbow outlet, elevating downstream pressure. Additionally, larger radii mitigate the centrifugal force-induced low-pressure zone on the outer bend, raising the overall pressure level. Excitation forces remain consistently below 120 N with cyclic fluctuations, showing minimal variation across radii.

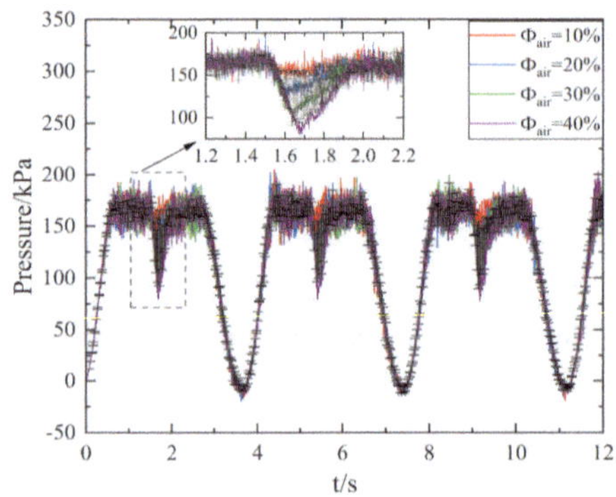

(a) Straight pipe pressure.

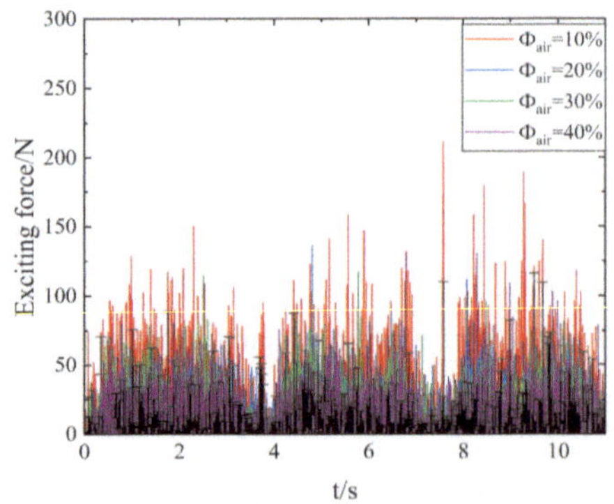

(b) Straight pipe excitation force.

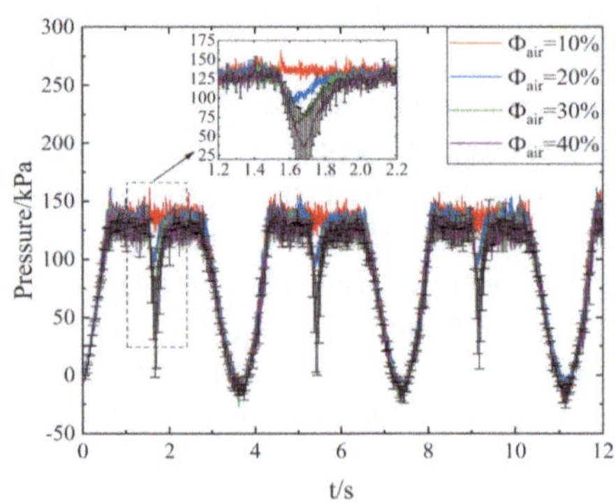

(c) Elbow pipe pressure.

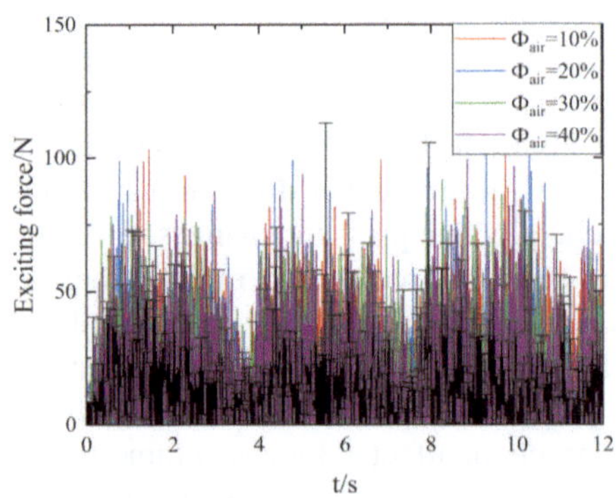

(d) Excitation force of elbow pipe.

Fig. 8. The changes of pressure and excitation force in straight and elbow pipes under different gas volume fractions during unsteady pumping.

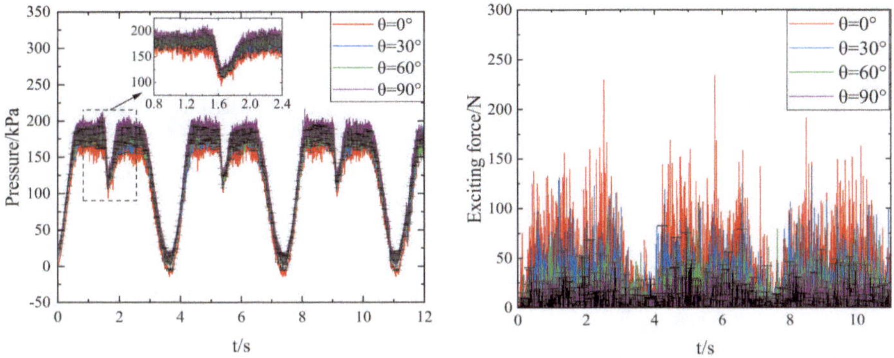

Fig. 9. Pressure and excitation force changes in straight pipes at various attitude angles during unsteady pumping.

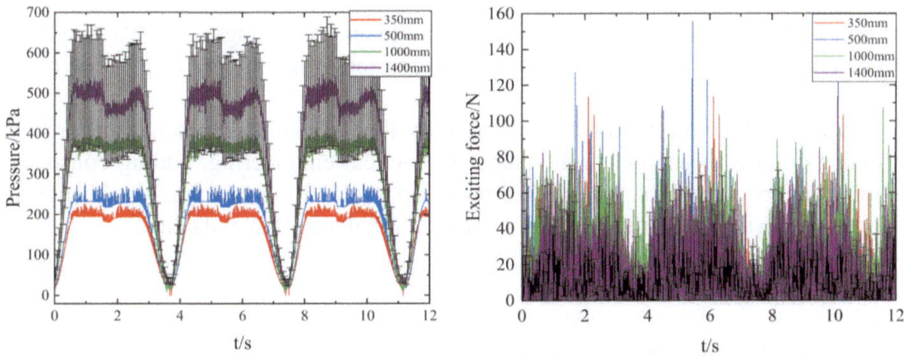

Fig. 10. Elbow curvature radius effects on pressure and excitation force variations during unsteady pumping.

4.3 Flow Characteristics of Combined Pipe

Using three 125-mm-diameter, 300-mm-long straight pipes and two 140-mm-radius, 30° elbows, a combined pipeline was modeled in a horizontal configuration. A gas-liquid-solid coupling approach simulated unsteady concrete flow at 85 m^3/h. The pressure change is shown in Fig. 11, periodic pressure fluctuations occur at the inlet.

The excitation force of the combined pipe wall after 0.1 m from the inlet of the horizontal combined pipe was monitored, as shown in Fig. 12. The pipe wall's excitation force rises to 20 kN with concrete inflow, then fluctuates; drops to 13 kN when gas enters, rising again as gas exits. During concrete outflow, it decreases until the next hydraulic cylinder activates.

4.4 The Influence of Excitation Force on Structural Fatigue

Periodic excitation force fluctuations during concrete pumping critically impact pump truck pipe safety and longevity. Structurally, fatigue life relates to cyclic stress amplitude.

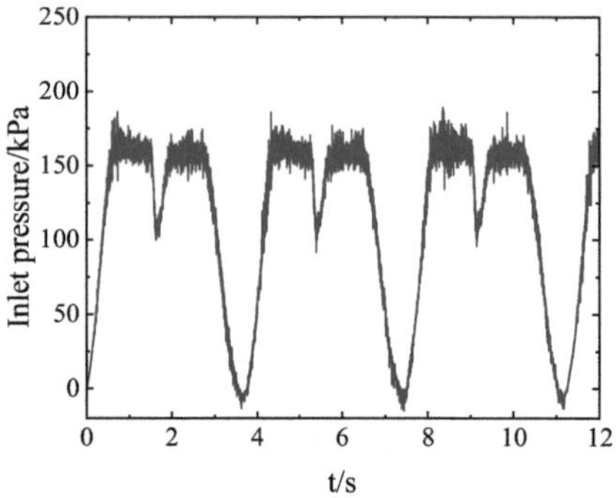

Fig. 11. Pressure change of horizontal attitude combined pipeline under unsteady flow.

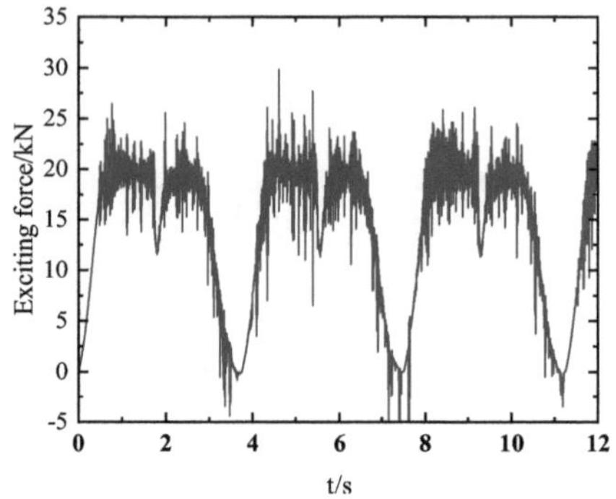

Fig. 12. The change of excitation force of horizontal attitude combined pipeline under unsteady flow.

Per GB/T 1591–2018 S-N curves, stress amplitudes exceeding 100 MPa may cause fatigue cracks after 10^6 cycles. Uncontrolled long-term operation risks pipeline fatigue failure.

In view of the vibration problem caused by the fluctuation of excitation force, the following control measures can be taken in practical engineering: adding constrained damping layers at elbows, optimizing pumping cycles to reduce gas-liquid interface impacts, and using real-time pressure feedback to balance flow pulsations.

5 Conclusions

In this study, the unsteady flow characteristics of gas-liquid-solid three-phase concrete in the pipeline were systematically studied by using the coupling simulation method of CFD and DEM. The unsteady characteristics of pipeline pressure and excitation force under three key configurations of straight pipe, elbow pipe and combined pipe are analyzed. Based on the simulation results and analysis, The following conclusions are drawn regarding the influence of different structural parameters and pumping parameters on the unsteady flow characteristics of concrete are obtained:

1) CFD-DEM coupling enables simulating gas-liquid-solid three-phase concrete flow.
2) When the gas volume fraction entering the pipeline increases, the pipeline pressures of straight and bent pipes decrease; The excitation force of the straight pipe decreases (with a fluctuation amplitude less than 150 N) while that of the elbow pipe does not change significantly (with a fluctuation amplitude less than 100 N). When the attitude angular of the straight pipe increases, the pipeline pressure increases but the excitation force decreases; when the curvature radius of the elbow increases, the pipeline pressure increases while the excitation force does not change significantly.
3) The pressure and excitation force of the horizontal attitude combined pipe show periodic changes.

Acknowledgments. This work was supported by the Fundamental Research Program of Shanxi Province (202403021222209).

Disclosure of Interests. The authors declare no conflict of interest.

References

1. Ma, J., Tu, Y., Xie, Y., et al.: The performance change and mechanism of concrete before and after 800 m ultra-long coil pumping. J. Build. Mater. **25**(12), 1293–1299 (2022)
2. Li, Y., Tian, X., Jin, C., et al.: Numerical simulation of rheological properties of pumped concrete based on CFD. Silic. Bull. (2020)
3. Tan, Y., Song, J., Zhang, H., et al.: Discrete element simulation of the wear process of the pipe wall in concrete pumping. China Mech. Eng. **25**(15), 2091–2097 (2014)
4. Jiang, S.Q., Zhang, W., Chen, X.D., et al.: CFD-DEM simulation research on optimization of spatial attitude of concrete pumping boom based on evaluation of minimum pressure loss. Powder Technol. **403**, 117365 (2022). https://doi.org/10.1016/j.powtec.2022.117365
5. Yuan, S., Xu, Z., Liu, J.: Insights and challenges of predicting concrete pumpability: a state-of-art review. J. Build. Eng. **95** (2024). https://doi.org/10.1016/j.jobe.2024.110265
6. Haustein, M.A., Kluwe, M.N., Rüdiger, S.: Experimental Investigation of the pumping of a model-concrete through pipes. Materials **13**(5), 1161 (2020). https://doi.org/10.3390/ma13051161
7. Fumin, L., Shen, W., Ji, Y., et al.: Prediction of nonlinear pumping pressure loss of concrete induced by rheological property changes. Cem. Concr. Res. **169**, 107182 (2023). https://doi.org/10.1016/j.cemconres.2023.107182
8. Li, T., Li, D., Li, S., et al.: Study on the pebble stagnation of slurry discharge pipe of slurry balance shield based on CFD-DEM coupling. J. Shandong Univ. (Eng. Ed.), 1–15, 20 February 2025
9. Li, Q.: Study on the flow characteristics of concrete in pumping pipelines under fluid-solid coupling simulation. Taiyuan Science and Technology, Taiyuan (2023)
10. Lv, X.F., Xu, K.Y., Liu, Y., et al.: Numerical simulation analysis of stable flow of hydrate slurry in gas-liquid-solid multiphase flow. Ocean Eng. **300** (2024). https://doi.org/10.1016/j.oceaneng.2024.117336
11. Choi, M., Roussel, N., Kim, Y., et al.: Lubrication layer properties during concrete pumping. Cem. Concr. Res. **45**, 69–78 (2013). https://doi.org/10.1016/j.cemconres.2012.11.001

Research on Segmented Micro Sliding Friction Control Strategy Based on Front-Installed Wet Clutch Torque Collapse Gradual Decline

Zegui Wang(✉), Hong Yin, Cuirong Huang, and Jiuying Zhu

Research Center of Mechanical Reliability & Intelligent Diagnosis of Frontier Equipment, Shanghai Technical Institute of Electronics and Information, College of Mechanical and Energy Engineering, Shanghai, China
wangzegui@stiei.edu.cn

Abstract. To enhance the driving dynamics of automatic transmissions equipped with front-mounted clutches, this study presents an innovative segmented micro-slip control methodology that utilizes the clutch torque pre-collapse gradient characteristic. Through extensive real-vehicle calibration and validation, the proposed approach effectively regulates target speed differentials, leading to a significant improvement in vehicle propulsion efficiency. This research not only advances overall driving performance but also establishes a foundational framework for future adaptive control of clutch torque transfer characteristics.

Keywords: Clutch · Micro sliding friction · Front collapse and gradual decline · Control Strategy

1 Introduction

Automobile automatic transmissions are a vital component of modern vehicles, significantly impacting driving performance, fuel efficiency, and user experience. In both domestic and international markets, these systems can be broadly categorized into four types: traditional hydraulic mechanical automatic transmissions (AT), continuously variable transmissions (CVT), electronically controlled mechanical automatic transmissions (AMT), and dual-clutch automatic transmissions (DCT) (Gerlach et al., 2020). Among these categories, AMT and DCT have garnered increasing attention due to their potential for high efficiency, low cost, and compatibility with electric vehicle integration. However, both architectures depend on a torque input front clutch, whose precise control is crucial for ensuring smooth operation—particularly during gear shifts and transient states (Liu & Zhang, 2021). This paper aims to enhance the driving performance of a specific vehicle by proposing a micro-slip control strategy for the front clutch in fixed gear positions while utilizing the front collapse gradient as an innovative control parameter.

S. Halgamuge et al. (Eds.): ICACMVE 2025, LNME, pp. 534–546, 2026.
https://doi.org/10.1007/978-981-95-6911-3_53

1.1 Transmission Types and Clutch Functionality

Traditional automatic transmissions (ATs) dominate the market due to their established technology and reliable operation; however, they are plagued by significant energy losses attributed to liquid friction and intricate hydraulic systems (Wong & Raman, 2019). Continuously variable transmissions (CVTs) provide an infinite range of gear ratios, thereby enhancing fuel efficiency. Nonetheless, their dependence on a metal belt or chain constrains both torque capacity and durability (Kandagal et al., 2022). In contrast, automated manual transmissions (AMTs) and dual-clutch transmissions (DCTs) utilize either dry or wet clutches for torque transmission, rendering them more efficient and cost-effective alternatives. However, the operational dynamics of these systems necessitate frequent clutch engagements and disengagements, which pose challenges related to managing slip, wear, and overall driver comfort (Zhang et al., 2023).

The front clutch in Automated Manual Transmission (AMT) and Dual Clutch Transmission (DCT) systems fulfills several essential functions: it transmits torque during locked states, modulates power flow during slipping conditions, and isolates the engine during open states. Micro-slip control—characterized by minimal relative slip between the clutch plates under partial locking—is particularly crucial. This mechanism strikes a balance between minimizing wear and ensuring precise torque delivery, which directly influences shift quality and vehicle dynamics (Li et al., 2022). Existing research often conceptualizes clutch control in binary terms (locked/open), overlooking the nuanced advantages of micro-slip regulation (Smith & Kumar, 2021).

1.2 Research Gaps and Motivation

Despite advancements in clutch control, several challenges continue to persist. First, conventional strategies primarily aim to eliminate slip entirely, which can lead to abrupt torque transfer and compromise NVH (noise, vibration, harshness) performance (Muller et al., 2020). Second, fixed control parameters are inadequate for adapting to varying operating conditions such as temperature fluctuations, clutch wear, or driver behavior (Kim et al., 2023). Third, the real-time estimation of the micro-slip state remains computationally intensive, thereby limiting its implementation in resource-constrained electronic control units (ECUs) (Lee et al., 2022).

To address these gaps, this study presents a micro-slip control strategy based on the front collapse gradient—a metric derived from the rate of slip reduction during clutch engagement. By monitoring the rate at which slip diminishes when transitioning from a slipping to a locked state, the system dynamically adjusts both clamping force and lockup timing. This approach offers two significant advantages: (1) it mitigates excessive heat and wear induced by slip while preserving torque capacity, and (2) it enhances adaptive control by correlating collapse gradients with physical properties of the clutch (e.g., friction coefficient, thermal degradation).

1.3 Novel Contributions and Structure

The proposed strategy builds upon previous research by integrating three key innovations:

1. **Front Collapse Gradient Analytics**: A mathematical model establishes a connection between the gradient of slip velocity and clutch lockup readiness, thereby facilitating predictive adjustments.
2. **Adaptive Torque Control**: Real-time feedback derived from the collapse gradient informs actuator commands, which diminishes reliance on pre-calibrated maps.
3. **Hybrid Stability Criteria**: This approach combines slip speed thresholds with gradient-based metrics to enhance both responsiveness and durability.

The paper begins with a review of clutch dynamics in Automated Manual Transmission (AMT) and Dual Clutch Transmission (DCT) systems in Sect. 2. This is followed by the derivation of the front collapse gradient model presented in Sect. 3. In Sect. 4, we introduce the micro-slip control algorithm along with its experimental validation. Section 5 discusses the implications for vehicle performance. Finally, the conclusions outline potential future directions for scaled implementation.

1.4 Literature Review and Theoretical Foundations

Prior research on clutch control has investigated various methodologies, including PID tuning (Wang et al., 2019), fuzzy logic (Chen & Li, 2020), and model predictive control (MPC) (Patel & Singh, 2022). However, these approaches often prioritize either speed or stability, neglecting the potential benefits of the intermediate micro-slip regime. Recent studies underscore the significance of clutch thermodynamics; for instance, Zhang et al. (2023) demonstrated that temperatures induced by slip account for 60% of clutch wear in heavy-duty applications. Building upon this foundation, our approach utilizes the front collapse gradient as a proxy for frictional consistency, which is consistent with theories related to transient elastohydrodynamic lubrication (EHL) in clutch interfaces (Evans & Johnson, 2018).

In summary, this study addresses a significant gap in adaptive clutch control by formalizing the micro-slip behavior through the front collapse gradient. By integrating theoretical models with practical control requirements, the proposed strategy seeks to enhance shifting smoothness, prolong clutch lifespan, and simplify calibration processes for next-generation AMT/DCT systems.

In both domestic and international markets, automobile automatic transmissions can generally be categorized into four types: traditional hydraulic mechanical automatic transmissions (AT), continuously variable transmissions (CVT), electronically controlled mechanical automatic transmissions (AMT), and dual-clutch automatic transmissions (DCT). Notably, both AMT and DCT incorporate a torque input front clutch, which is crucial for the control and operation of the transmission system. The functioning of the front-installed clutch can be broadly classified into three states: locked, slipping, and open.

This article primarily focuses on the micro-slip control of the clutch in a specific vehicle operating in a fixed gear position. It proposes a micro-slip control strategy based on the front collapse gradient to enhance vehicle driving performance. Furthermore, the physical characteristics of the clutch during its micro-slip state are leveraged to establish a foundation for adaptive control of clutch torque transmission [1–12].

2 Analysis of the Principle of Micro Sliding Friction in Clutch

When a vehicle operates in a fixed gear, abrupt acceleration or deceleration can lead to substantial variations in engine torque, necessitating corresponding adjustments in clutch torque. As illustrated in Fig. 1, if the clutch master and slave plates are excessively compressed, a pronounced braking sensation will be felt throughout the vehicle. Conversely, as depicted in Fig. 2, if the engagement between the clutch master and slave plates is too loose, there will be an increase in engine speed that results in inadequate overall power for the vehicle. Therefore, it is imperative that when driving in a fixed gear, the clutch is engaged at an optimal position to enhance driving performance; this can be effectively achieved through precise control of clutch micro-slip.

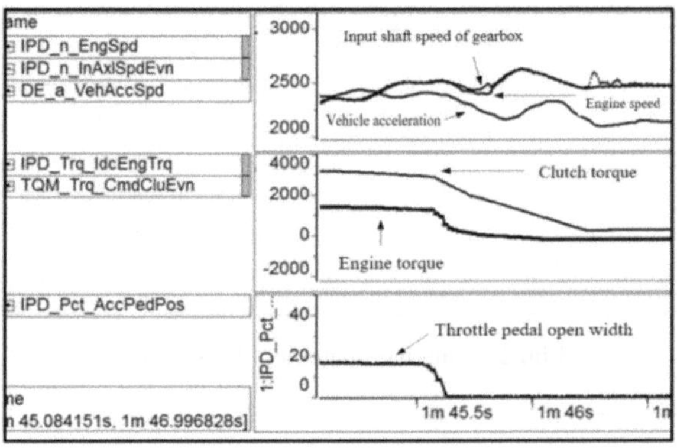

Fig. 1. Clutch engagement tightness curve

Clutch micro-slip control modulates the transmission torque of the clutch to maintain a minimal speed differential between the master and slave plates. When there is a sudden change in engine torque, the resulting oscillations are effectively dampened by the clutch, thereby preventing them from propagating to the vehicle's wheels and enhancing overall smoothness. Due to this small and stable speed difference, as described by formula (1), the angular acceleration ω' during micro-slip friction can be approximated as zero. This condition signifies that both ends of the clutch—master and slave—are in a state of torque equilibrium, where the output torque from the engine is equivalent to that transmitted through the clutch. In instances of abrupt changes in engine torque, the clutch transmission torque exhibits high responsiveness for immediate adjustment, thus averting excessive increases in engine speed and ensuring an optimal power response for the vehicle as a whole.

$$T_{eng} = J\omega + T_{cl} \tag{1}$$

where T_{eng} is the engine output torque, T_{cl} is the clutch load torque, J is the clutch load moment of inertia, and ω is the clutch angular acceleration.

The torque control of the clutch is significantly influenced by its transmission torque characteristics. At varying temperatures or after different durations of usage, the torque transmitted by the clutch at a given engagement position may differ. Moreover, during vehicle operation, accurately measuring the actual torque transmitted by the clutch at each engagement position using sensors presents considerable challenges. However, based on the principle of torque balance in micro-slip friction, it is possible to derive the true torque transmitted by the clutch from the engine output torque [7–9]. Consequently, micro-slip control of the clutch plays an essential role in adaptive control of clutch torque transmission.

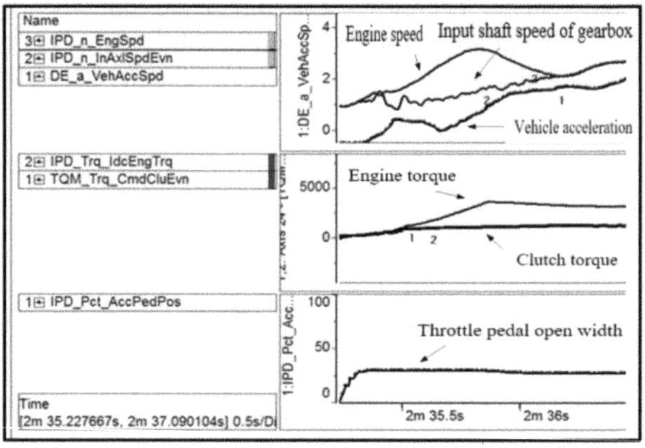

Fig. 2. Clutch engagement loose curve

3 Segmented Control Strategy for Micro Sliding Friction Based on Front Collapse Gradual Descent

When a vehicle is operating steadily in a specific gear, the clutch initially remains in a locked state, which may result in the clutch torque exceeding the engine output torque. During this phase, disengaging the clutch can require considerable time. This article proposes a segmented control strategy for managing micro-slid friction of the clutch by gradually reducing the front failure term illustrated in Fig. 3. The primary objective is to swiftly decrease the clutch torque to a level that approximates the engine torque, utilizing this as the front failure term for clutch control. Subsequently, PID control is employed to maintain a stable speed differential between the clutch and engine. The entire control process is categorized into three distinct stages: rapid reduction of clutch torque, gradual reduction of clutch torque, and implementation of PID control.

3.1 Strategy for Rapid Torque Reduction Stage of Clutch

When the clutch is in a locked state, the torque generated by the clutch exceeds that of the engine output. To facilitate rapid disengagement under these conditions, the clutch torque is reduced directly to a predetermined fixed value at an accelerated rate. This fixed value is theoretically established to be slightly above the instantaneous engine torque, as it must

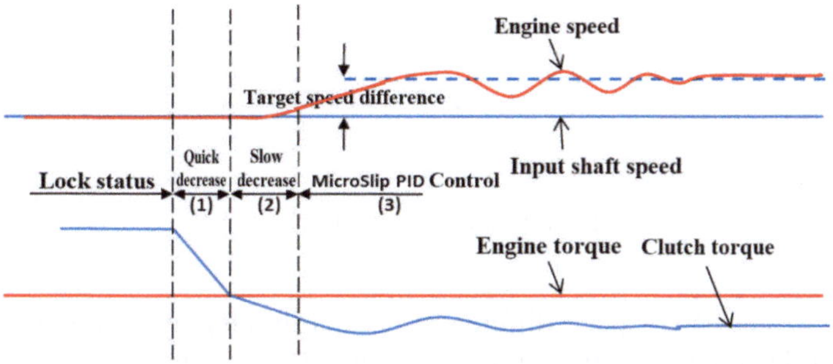

Fig. 3. Segmented control strategy for clutch micro sliding friction based on front collapse gradual reduction

account for system response lag and transient dynamics during the process of torque reduction. The dynamic equilibrium condition—wherein engine torque equals clutch torque—serves as the theoretical foundation for determining this threshold. However, practical calibration must take into consideration nonlinear factors such as wear on the clutch disc, temperature-dependent friction coefficients, and hysteresis effects within the actuation system.

3.2 Strategy for the Slow Reduction Stage of Clutch Torque

Once the clutch torque reaches a predetermined threshold, any subsequent rapid reduction may result in an abrupt disengagement of the clutch, thereby compromising the driving experience. To mitigate this issue, a controlled and gradual reduction of clutch torque is implemented to decrease the rate at which torque is diminished. The ultimate target for clutch torque reduction is aligned with the current engine torque, thereby facilitating subsequent PID control of speed differentials to ensure a smooth disengagement of the clutch.

3.3 PID Control Phase Strategy

As illustrated in Fig. 4, PID control is activated when the clutch torque matches the engine torque. The control inputs consist of both the difference and the rate of change between the actual and target speed differentials of the engine and clutch, while the output corresponds to the clutch compensation torque. The engagement torque of the clutch is dynamically adjusted to sustain a stable and minimal speed differential between the engine and clutch. Based on both the characteristics of various clutches and practical vehicle control experience, it is common practice to set the target speed differential at 30 rpm for wet clutches and 10 rpm for dry clutches.

During the process of clutch micro-slip control, various factors such as the accuracy of engine output torque, clutch transmission torque, and driving conditions can lead to deviations in the speed difference between the engine and the clutch. These deviations may be excessively large or insufficiently small, thereby impacting the effectiveness of micro-slip control. The micro-slip friction control strategy, which is predicated

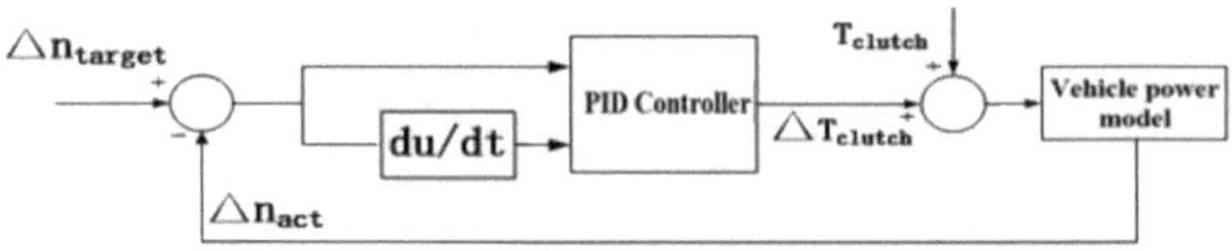

Fig. 4. PID control of clutch torque

on dynamic detection of speed differences, primarily employs Proportional-Integral-Derivative (PID) control. This approach dynamically monitors the speed difference between the engine and clutch, classifies it into three distinct levels, and applies corresponding control parameters for each level to ensure a stable speed difference between these two components.

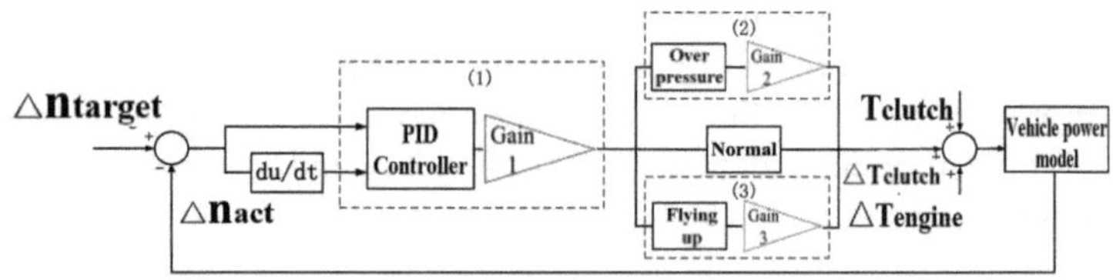

Fig. 5. Micro sliding friction pid control based on dynamic detection of speed difference

As illustrated in Fig. 5, the micro-slide friction control strategy, which is predicated on the dynamic detection of speed differences, comprises three distinct modules: the PID basic control module, the clutch overpressure module, and the engine speed rise module. The method for calculating clutch torque is delineated in formula (2), wherein the engine torque (T_eng) functions as the primary failure term. The coefficient Gain1 denotes the fundamental coefficient of the PID control module; Gain2 represents the amplification coefficient associated with the clutch overpressure module; and Gain3 indicates the amplification coefficient pertinent to the engine speed rise module.

$$T_{cl} = T_{eng} + T_{PID} * Factor(Gain1; Gain2; Gain3) \tag{2}$$

1) PID Basic Control Module: This module utilizes the discrepancy between the actual and target speed, as well as the rate of change of this discrepancy, to derive control inputs related to both the engine and clutch. The output generated is the clutch compensation torque, which dynamically adjusts the engagement torque of the clutch. Given that the complexity of controlling speed discrepancies varies under different engine torque conditions, a PID control base coefficient, referred to as Gain1, has been introduced. As illustrated in Table 1, for a specific vehicle model, the calibration value of Gain1 is appropriately increased in accordance with rising engine torque levels, thereby expanding the PID output range as required.
2) Clutch Overpressure Module: Taking the fixed-gear acceleration condition as an example, it is essential that the target speed difference for micro-slip control remains positive. Should the actual speed difference between the engine and clutch be detected

Table 1. PID correction coefficient Gain1 based on engine torque

TEng/Nm	-30	0	30	70	100	130	160	200
Gain1	1.3	1	1.5	1.7	2	2.3	2.8	3

as negative and persist over a specified duration, this indicates that the clutch is transmitting excessive torque and has entered an overpressure state. Consequently, the clutch overvoltage module is activated. The compensation torque for the clutch is computed by multiplying the output of the PID base module by Gain2, thereby facilitating a more rapid reduction in clutch torque and restoring the speed difference to its normal range.

3) Engine Speed Boost Module: The engine speed boost module is activated when the actual speed difference between the engine and clutch exceeds a specified target speed difference within a defined range and persists for a certain duration. This condition indicates that the current torque transmitted by the clutch is insufficient. When this occurs, it becomes crucial to manage the engagement of clutch torque effectively; engaging too rapidly can result in a pronounced braking sensation throughout the vehicle, while engaging too slowly may hinder the recovery of the micro-slip target speed difference. To address these challenges, Gain3 has been introduced and expanded based on both the actual speed difference and its rate of change. Through real-vehicle calibration, an effective control mechanism for fast and smooth clutch micro-slip has been achieved.

4 Analysis and Comparison of Real Vehicle Test Results

Using the Matlab/Simulink software platform, a model for the clutch front failure gradient micro-slip control strategy was developed and subsequently calibrated and validated on an actual vehicle. As illustrated in Fig. 6, the test results indicate that the target speed difference was calibrated to 30 rpm. The transition from micro-slip to a stable speed difference took approximately 1.02 s, with no significant fluctuations observed in vehicle acceleration, thereby demonstrating an effective micro-slip control performance.

As illustrated in Fig. 7, the actual vehicle test results for clutch overpressure control, based on dynamic detection of speed difference, are presented. The target speed difference is calibrated at 10 rpm, and it takes approximately 1.4 s to stabilize this speed difference following the activation of the clutch overpressure control. No significant fluctuations in overall vehicle acceleration were observed, thereby confirming the effectiveness of micro-slip control.

As illustrated in Fig. 8, the actual vehicle test results for engine speed rise control, based on dynamic detection of speed difference, are presented. The target speed difference is calibrated at 10 rpm, and it takes approximately 0.4 s to stabilize this speed difference following the activation of the engine speed rise control. Notably, there is no significant fluctuation in vehicle acceleration, which indicates a satisfactory micro-slip control effect.

Further analysis of the calibration and testing results indicates that the clutch feedforward gradient micro-slip control method can effectively and smoothly regulate the speed

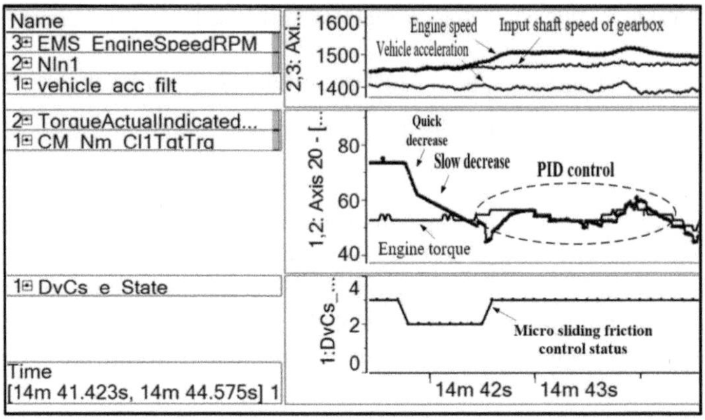

Fig. 6. Real vehicle test results of segmented control of clutch micro sliding friction based on front collapse gradient

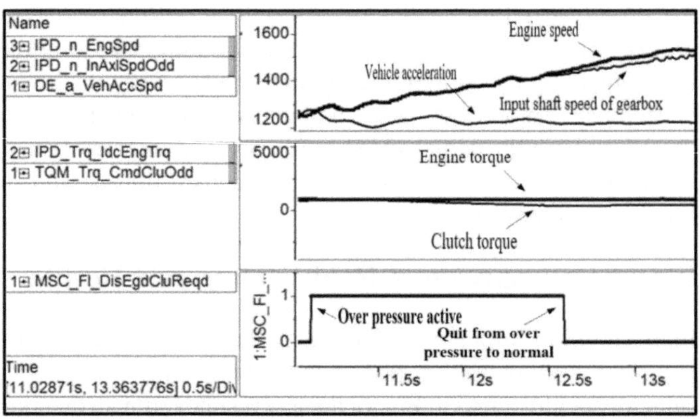

Fig. 7. Test results of clutch overpressure control based on dynamic detection of speed difference

differential between the engine and clutch within the designated target range, exhibiting a relatively rapid response time. However, its performance is significantly influenced by the accuracy of both engine output torque and clutch transmission torque. Enhanced accuracy leads to improved control outcomes [20–22].

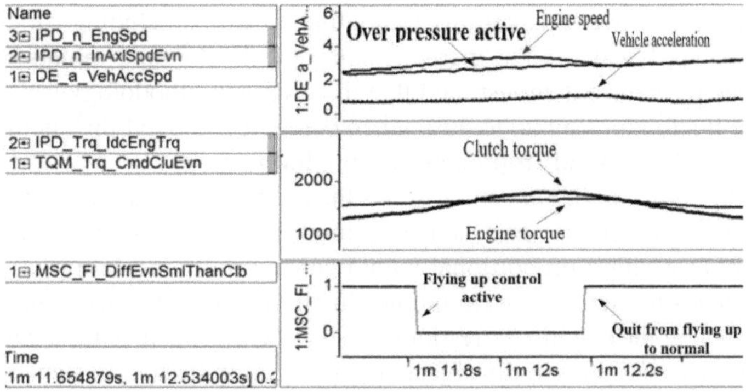

Fig. 8. Test results of engine speed rise control based on dynamic detection of speed difference

5 Conclusions

To improve the driving performance of automatic transmissions equipped with a front clutch and to facilitate the adaptive torque transmission function of the clutch, this paper proposes a segmented control micro-slip friction strategy that is based on the gradual reduction of clutch torque during front failure. Comprehensive calibration and analysis were conducted using real vehicles, leading to the following conclusions [13–24]:

1. **Rapid Regulation and Performance Improvement**
   The gradual reduction micro-slip control strategy for front collapse enables rapid and smooth regulation of the speed differential between the engine and clutch within a specified target range. This approach not only enhances the overall driving performance of the vehicle but also establishes a stable foundation for subsequent adaptive control of clutch torque transmission. Under various operating conditions, such as startup and gear shifting, this strategy exhibits significant robustness and adaptability, effectively mitigating torque fluctuations and minimizing driveline vibrations.
2. **Segmented Control Accuracy and Dependence on Torque Estimation**
   The segmented control method for clutch micro-slip friction, which is based on forward collapse descent, demonstrates rapid response and high control precision. However, its effectiveness is critically dependent on the accurate estimation of engine output torque and clutch transmission torque. High-fidelity torque models (such as MAP-based or real-time estimation algorithms) along with precise sensor feedback (including crankshaft and clutch shaft speed sensors) are essential for minimizing control errors. Experimental results indicate that torque estimation errors exceeding 5% significantly impair micro-slip control performance, underscoring the necessity for advanced torque monitoring techniques.
3. **Dynamic Speed Difference Detection for Comprehensive Optimization**
   To enhance the effectiveness of micro-slip control and accommodate complex vehicle operating conditions, it is advisable to combine a segmented control strategy with dynamic speed difference detection. By integrating real-time speed difference analysis techniques, such as sliding mode control or model predictive control, this approach allows for the dynamic adjustment of clutch torque compensation. This

enhancement improves adaptability to nonlinear factors including variations in road gradient, driver behavior, and thermal fading of the clutch. Simulation studies and real-vehicle tests have demonstrated that this hybrid methodology can reduce speed difference fluctuations by up to 40% when compared to fixed-threshold methods.

4. **Tradeoffs and Practical Implementation Challenges**

   Experimental validation demonstrates the existence of trade-offs among response speed, control precision, and actuator durability. Employing aggressive control parameters (e.g., high proportional gains) enhances the regulation of speed differences but leads to increased wear on the clutch. In contrast, conservative parameters mitigate wear at the expense of responsiveness. Achieving an optimal balance among these factors necessitates multi-objective optimization strategies, such as minimizing the integral of speed difference while constraining fluctuations in clutch pressure. Furthermore, practical implementation must consider computational latency associated with real-time torque estimation and processing of control signals.

5. Future Research Directions

   To further advance clutch torque adaptive transmission, future work should focus on:

- **Intelligent Torque Estimation:** Utilizing machine learning techniques, such as neural networks, to enhance the accuracy of real-time torque predictions under transient conditions.
- **Hybrid Control Architectures:** Combining rule-based segmented control with data-driven algorithms, including reinforcement learning, to effectively manage unmodeled dynamics.
- **Durability-Oriented Optimization:** Formulating wear-aware control strategies that dynamically adjust micro-slip thresholds in response to clutch aging and operational history.

These findings offer both theoretical and experimental validation for the implementation of segmented micro-slip control in automated transmissions. They provide valuable insights into achieving a balance among performance, accuracy, and durability in practical applications.

Acknowledgements. Thanks a lot for funding support from Shanghai Technical Institute of Electronics and Information (GCC2024007).

References

1. Gerlach, J., Liu, Y., Petri, D.: Automotive Transmissions: Fundamentals and Evaluation. Springer (2020).
2. Liu, X., Zhang, Q.: Clutch control optimization for dual-clutch transmissions using adaptive sliding mode. IEEE Trans. Vehic. Technol. **70**(2), 1435–1444 (2021)
3. Wong, J.H., Raman, K.: Comparative analysis of AT, CVT, and DCT efficiency. Int. J. Autom. Technol. **21**(3), 513–528 (2019)
4. Kandagal, P., Bhandari, R., Somanathan, S.: Life prediction of CVT steel belts under variable loading. Proc. IMechE Part D: J. Autom. Eng. **236**(5), 1243–1257 (2022)

5. Smith, T.L., Kumar, P.: Micro-slip detection in automotive clutches via machine learning. SAE Int. J. Engines **13**(2), 2021–2035 (2021)
6. Muller, M., Schmidt, F., Woida, M.: NVH optimization in DCT systems through advanced clutch control. J. Sound Vibr. **475**, 115468 (2020)
7. Kim, D., Lee, S., Choi, S.: Temperature-dependent friction compensation in AMT clutches. Tribol. Int. **180**, 110547 (2023)
8. Lee, H., Park, G., Rhee, C.: Real-time slip estimation using recurrent neural networks for automotive clutches. Mech. Syst. Sign. Process. **172**, 109409 (2022)
9. Wang, Y., Li, Z., Huang, H.: PID parameter self-tuning for clutch control in AMT systems. Chin. J. Mech. Eng. **32**(4), 89–98 (2019)
10. Chen, X., Li, W.: Fuzzy logic-based clutch engagement control for improved start moving comfort. Proc. Inst. Mech. Eng. Part D **234**(7), 2345–2356 (2020)
11. Patel, R., Singh, S.: Model predictive control for dual-clutch transmissions: a review. Vehicles **3**(4), 678–705 (2022)
12. Evans, H.T., Johnson, K.L.: Friction modeling in automotive clutches: from macro to nanoscale perspectives. Wear **416**, 14–28 (2018)
13. Yang, Q., Wu, G., Zhang, S.: Smart control of DCT proportional solenoid valve based on data mining. J. Int. J. Autom. Technol. **25**(3), 673–687 (2024)
14. Canale, M., Cerone, V., Corigliano, E., Osella, G.: Driveline oscillation attenuation through clutch micro-slip and model predictive control. Control. Eng. Pract. **140**, 1016–1025 (2023)
15. Shin, S., Choi, S.B.: Observer-based iterative learning control for a high relative degree nonlinear system and its application to a vehicular wet-clutch system. Mech. Mach. Theory **180**, 105158 (2023)
16. Canale, M., Cerone, V., Corigliano, E., Osella, G.: Driveline oscillation attenuation through Clutch Micro-Slip and Model Predictive Control]. Mech. Mach. Theory **180**, 105672 (2023)
17. Wang, X., Lu, T.: Offline model predictive control approach to micro-slip control in gearshifts of dual clutch transmission. Proceed. Inst. Mech. Engineers, Part K: J. Multi-body Dyn. **236**(1), 84–98 (2022)
18. Deng, T., Hu, F.B., He, Z.Y., Yin, Y.L.: Simulation, experimental testing and optimization of starting and shifting control strategies of DCT wet dual clutches with respect to sliding friction. Iranian J. Sci. Technol. Trans. Mech. Eng. **43**(4), 693–705 (2019)
19. Bao, W., Ge, J., Xiao, X.: General cao research on key issues of wet dual clutch micro sliding friction control. Autom. Eng. **40**(10), 1192–1199 (2018)
20. Tall and towering Research on the Sliding Friction Control Characteristics of DSG Clutch Engagement Master's thesis from Beijing Institute of Technology (2016)
21. Xue, D., Chang, N., Guo, X.: Research on start sliding friction control of hydraulic torque converter lock up clutch. Mech. Trans. **40**(05), 38–42 (2016)
22. Wang, X., Chen, J., Hao, W.: Research on sliding friction control strategy of hydraulic torque converter. Shanghai Autom. **04**, 10–15 (2023)
23. Zhang, X., Liu, Z., Chen, G., Yang, Y., Li, S.: Optimization strategy for powertrain control to improve low-speed vibration of passenger cars. J. Autom. Safety Energy Effi. **12**(02), 193–200 (2021)
24. Jiang, J., Cui, Y., Lu, C.: Research on AMT automatic clutch control based on model predictive control. J. Changchun Inst. Technol. (Nat. Sci. Ed.) **25**(02), 33–38 (2024)

Simulation of Delamination Damage in Composite Laminates Under Low-Velocity Impact

Zhi Huang(✉)

Aircraft Design and Research Institute, COMAC, Shanghai 200000, China
huangzhi@comac.cc

Abstract. To improve the accuracy of delamination damage prediction in composite laminates under low-velocity impact, three distinct delamination failure criteria were selected. The low-velocity impact process was numerically simulated using ANSYS finite element analysis software, and the impact damage morphology of the laminates was obtained. The simulation results were compared with experimental data. The findings demonstrate that the Yeh delamination failure criterion exhibits significant prediction errors (exceeding 30%), whereas the Weng delamination criterion and the modified Hou delamination criterion provide accurate predictions, with errors remaining within 7%. Notably, the Weng criterion overestimates damage areas relative to experimental values, while the modified Hou criterion yields conservative predictions. Additionally, this study validates the capability of the developed numerical simulation model to reliably predict low-velocity impact damage in composite laminates.

Keywords: Composite laminate · Low-velocity impact · Failure criteria

1 Introduction

The proportion of composite materials in civil aircraft structures has become a key indicator of technological advancement. In the Airbus A350 and Boeing 787, composites have replaced a substantial amount of metallic structures, with composite weight proportion exceeding 50%. China's domestic large aircraft industry is developing rapidly, with composite material proportion increasing from 2% in regional jet ARJ21 to 12% in narrow-body mainline jet C919. The newly approved C929 wide-body aircraft project is expected to achieve 50% composite content [1]. Composite structures have transitioned from secondary load-bearing components such as nacelles and slats to primary structural components, including fuselages and wings. However, during daily aircraft operations and maintenance, these structures are highly susceptible to impacts from foreign objects like gravel, hail, and maintenance tools. While such low-velocity impacts typically cause minimal visible surface damage to the laminate, they may induce matrix cracking on the bottom of the laminate and extensive interlaminar delamination. These barely visible impact damage (BVID) significantly compromises the strength and service life of composite components [2]. Therefore, accurate simulation of low-velocity impact damage in composite structures is imperative.

S. Halgamuge et al. (Eds.): ICACMVE 2025, LNME, pp. 547–556, 2026.
https://doi.org/10.1007/978-981-95-6911-3_54

Currently, the combination of experimental investigations and numerical simulations serves as the primary methodology for studying impact damage in composite materials. Zhang et al. [3–5] have conducted extensive experimental studies on impact damage in laminates. Their findings reveal that under low-energy impacts, matrix cracking and interlaminar delamination between plies with different orientations constitute the predominant failure modes. Xu et al. [6] explored correlations between impact energy and damage characteristics (length, width, depth, and area), identifying damage area as the parameter most strongly correlated with impact energy. Regarding numerical simulation of low-velocity impacts, scholars [7] have developed various failure criteria and implemented finite element analysis to evaluate low-velocity impact damage in composite structures.

This paper conducts low-energy impact tests on composite laminates under two distinct energy levels and conducts numerical simulations to analyze the low-velocity impact process on composite laminates by adopting three distinct delamination failure criteria. By comparing the damage areas derived from experimental results and simulation outputs, an error analysis is performed to determine the delamination failure criteria best suitable for predicting low-velocity impact damage.

2 Failure Criteria for Laminates

2.1 In-Plane Failure Criteria

Regarding fiber and matrix in-plane damage, this paper adopts the following failure criteria, which can effectively predict fiber and matrix damage under impact loads [8]:

(a) Fiber Failure:

$$\left(\frac{\sigma_{11}}{X_T}\right)^2 + \left(\frac{\sigma_{12}}{S_f}\right)^2 + \left(\frac{\sigma_{13}}{S_f}\right)^2 \geq 1 \quad (\sigma_{11} \geq 0) \tag{1}$$

$$\left(\frac{\sigma_{11}}{X_C}\right)^2 + \left(\frac{\sigma_{12}}{S_f}\right)^2 + \left(\frac{\sigma_{13}}{S_f}\right)^2 \geq 1 \quad (\sigma_{11} < 0) \tag{2}$$

(b) Matrix Failure:

$$\left(\frac{\sigma_{22}}{Y_T}\right)^2 + \left(\frac{\sigma_{12}}{S_{12}}\right)^2 + \left(\frac{\sigma_{13}}{S_{13}}\right)^2 \geq 1 \quad (\sigma_{22} \geq 0) \tag{3}$$

$$\frac{1}{4}\left(\frac{\sigma_{22}}{S_{12}}\right)^2 + \left(\frac{\sigma_{12}}{S_{12}}\right)^2 + \frac{Y_C^2\sigma_{22}}{4S_{12}^2Y_C} - \frac{\sigma_{22}}{Y_C} \geq 1 \quad (\sigma_{22} < 0) \tag{4}$$

In the equations:

σ_{ij}–the stress components in the material directions of each element;

X_T, X_C, Y_T, Y_C–the longitudinal and transverse tensile and compressive strengths;

S_{ij}, S_f –the in-plane shear strength and shear strength considering fiber failure.

When the stress in any element satisfies the above criteria, it is considered to have experienced the corresponding failure mode. The appropriate reduction coefficients are then applied to degrade the material properties of that element.

2.2 Delamination Failure Criteria

The selection of delamination failure criteria significantly influences the prediction of the final damage area. To compare the effects of different delamination failure criteria on the prediction of low-velocity impact damage in laminates, this study adopts three distinct delamination failure criteria.

Wen et al. [8] proposed a new criterion based on the original Hou delamination failure criterion. The criterion comprehensively considers the influence of interlaminar stress and matrix cracking damage on delamination, allowing for the prediction of whether delamination occurs at the upper or lower interface of the laminate. The model's validity has been verified through experimental tests.

$$\left(\frac{n_{\sigma 23}}{S_{23}}\right)^2 + \left(\frac{n+1_{\sigma 13}}{S_{13}}\right)^2 + \left(\frac{n+1_{\sigma 22}}{Y_T}\right)^2 + \left(\frac{n+1_{\sigma 33}}{Z_T}\right)^2 \geq 1 \quad (\sigma_{33} \geq 0) \tag{5}$$

Yeh's delamination failure criterion was adopted as a three-dimensional complement to the Hashin failure criteria, which exactly predict the impact damage mode and degree [9]. The failure criterion is as follows:

$$\left(\frac{\sigma_{33}}{X_{Tm}}\right)^2 + \left(\frac{\sigma_{13}}{S_{13}}\right)^2 + \left(\frac{\sigma_{23}}{S_{23}}\right)^2 \geq 1 \quad (\sigma_{33} \geq 0) \tag{6}$$

$$\left(\frac{\sigma_{13}}{S_{13}}\right)^2 + \left(\frac{\sigma_{23}}{S_{23}}\right)^2 \geq 1 (\sigma_{33} < 0) \tag{7}$$

Gong [10] adopted a Weng delamination failure criterion in stitched laminates' impact-induced damage simulation and demonstrated good agreement with experimental results. The criterion adopted the first stress invariant as the reference parameter and predicted the damage mainly based on the matrix strength. This paper extends its application to unidirectional tap laminates:

$$\left(\frac{\sigma_{11}+\sigma_{22}+\sigma_{33}}{X_{Tm}}\right)^2 + \left(\frac{\sigma_{12}}{S_m}\right)^2 + \left(\frac{\sigma_{13}}{S_m}\right)^2 + \left(\frac{\sigma_{23}}{S_m}\right)^2 \geq 1 \quad (\sigma_{11}+\sigma_{22}+\sigma_{33} \geq 0) \tag{8}$$

$$\left(\frac{\sigma_{11}+\sigma_{22}+\sigma_{33}}{X_{Cm}}\right)^2 + \left(\frac{\sigma_{12}}{S_m}\right)^2 + \left(\frac{\sigma_{13}}{S_m}\right)^2 + \left(\frac{\sigma_{23}}{S_m}\right)^2 \geq 1 \quad (\sigma_{11}+\sigma_{22}+\sigma_{33} < 0) \tag{9}$$

In the equations:
X_{Tm}, X_{Cm}, S_m–the tensile, compressive, shear strength of matrix.

3 Case Study Analysis

3.1 Carbon Fiber/Epoxy Composite Laminate

The analysis focuses on a carbon fiber/epoxy composite laminate fabricated using unidirectional prepreg tape. The specimen consists of 20 plies with a symmetric layup sequence: $[45/-45/90/90/-45/0/45/90/90/0]$ s. The specimen dimensions are 200 mm $\times$ 40 mm $\times$ 2.5 mm. The ply material properties are shown in Table 1.

Table 1. Parameters of unidirectional composite

E_1 (GPa)	E_2 (GPa)	G_{12} (GPa)	X_T (MPa)	X_C (MPa)	Y_T (MPa)	Y_C (MPa)	S (MPa)	ν
128.8	8.3	4.1	1446.5	1171.4	69.6	187.3	95.9	0.355

3.2 Low-Velocity Impact Test

The tests were conducted using a drop-weight impact testing machine. The test equipment includes an impactor, counterweights, guide rails, and an anti-secondary impact device. The test specimen was installed into a specially designed fixture with a tight fit, then mounted onto the test bench via clamps. The equipment and fixtures are illustrated in Fig. 1. The effective impact area is 30 mm $\times$ 30 mm. Impact energies are set at 6 J and 8.5 J according to the specimen size.

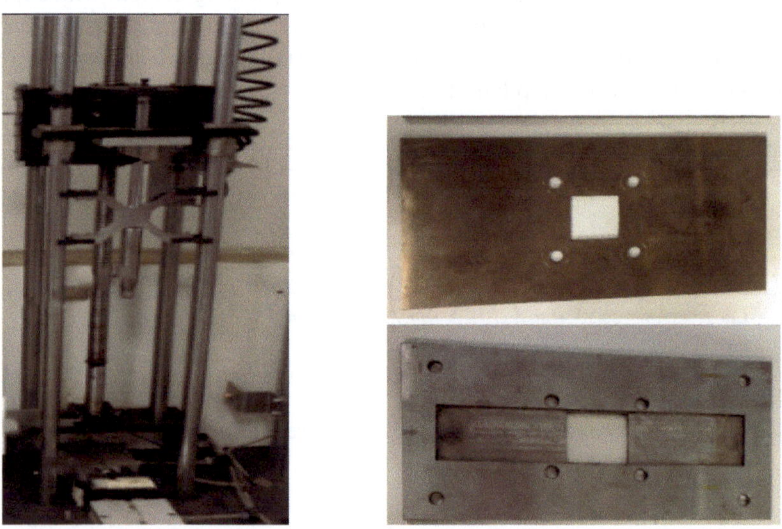

(a) Equipment (b)fixtures

Fig. 1. Test Equipment & Fixtures

3.3 Finite Element Modeling

This study employed the ANSYS APDL language for secondary development. For stress analysis, the LS-DYNA solver was invoked to perform explicit dynamic analysis, while damage assessment was implemented based on predefined damage criteria. A progressively refined mesh strategy was adopted, with grid density gradually increasing from peripheral regions to the central impact zone. The SOLID164, a three-dimensional solid element type specifically designed for nonlinear dynamic analysis, was selected, supporting nearly all material types. Automatic surface-to-surface (ASTS) contact was defined between the impactor and the specimen. To optimize computational efficiency, a minimal initial gap was set between components, with residual energy converted into the impactor's initial kinetic energy by assigning an appropriate initial velocity. Fixed boundary conditions were applied to the top and bottom surfaces except for the central 30 mm × 30 mm impact area. The finite element model is illustrated in Fig. 2.

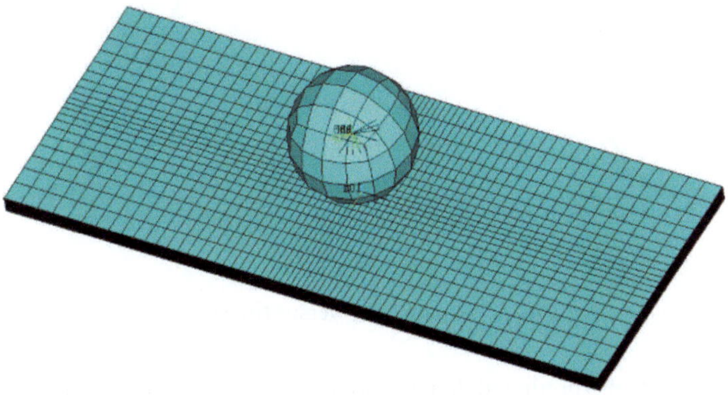

Fig. 2. Finite element model

Figure 3 shows the time-history curves of impact force. Under the same impact energy, the curves of impact force over time for different delamination failure criteria largely overlap. As the impact energy increases, the duration of the impact process shortens, the maximum impact force rises, and the peak impact force occurs earlier.

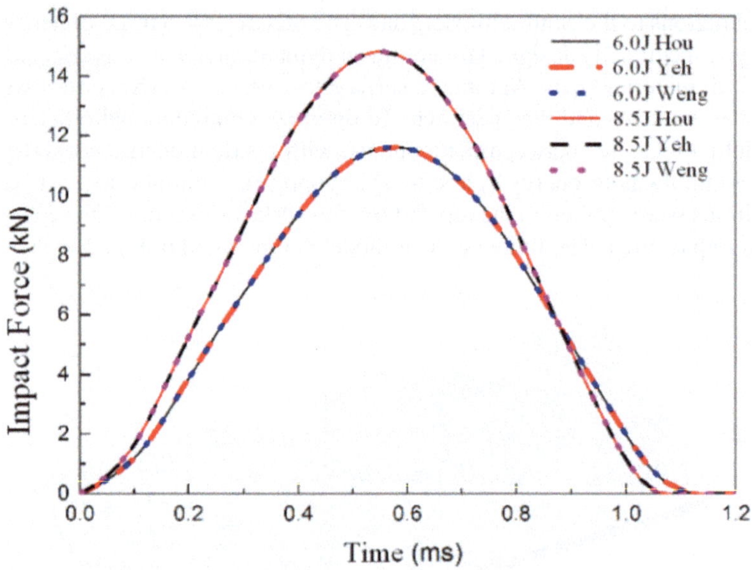

Fig. 3. Impact force versus time curve

Table 2 presents the predicted damage areas under different delamination failure criteria. With increasing impact energy, fiber damage intensifies, and the projected damage area expands accordingly.

Table 2. Total damage area under different failure criteria

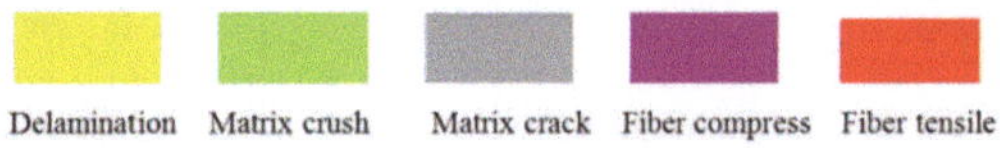

Impact Energy	Yeh criterion	Hou criterion	Weng criterion
6 J	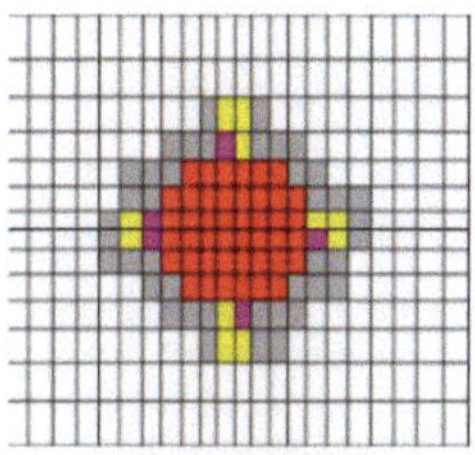	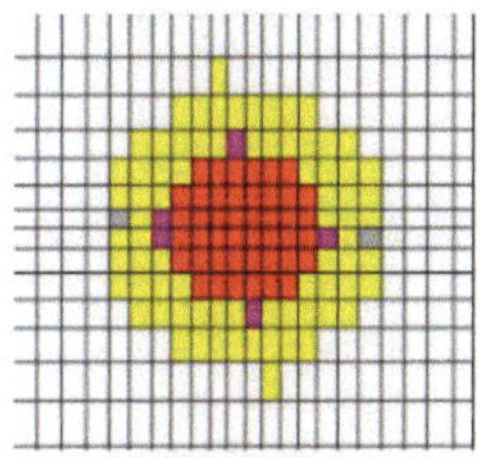	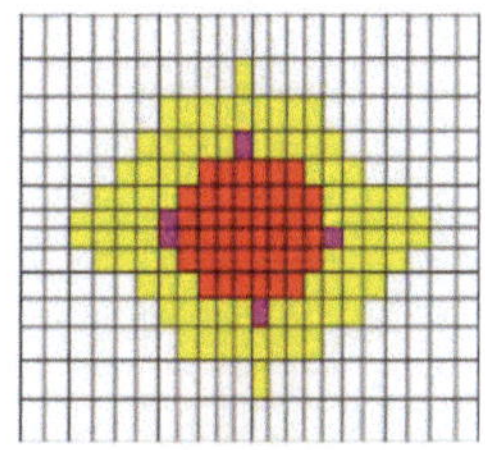
8.5 J	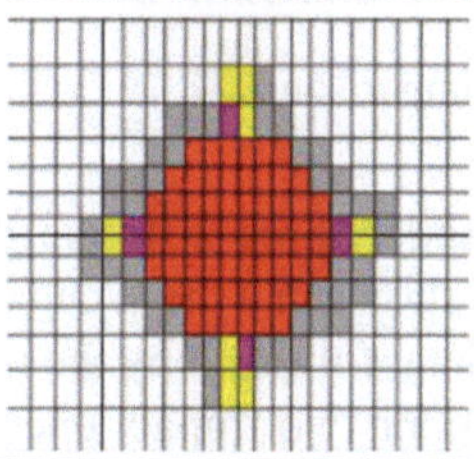	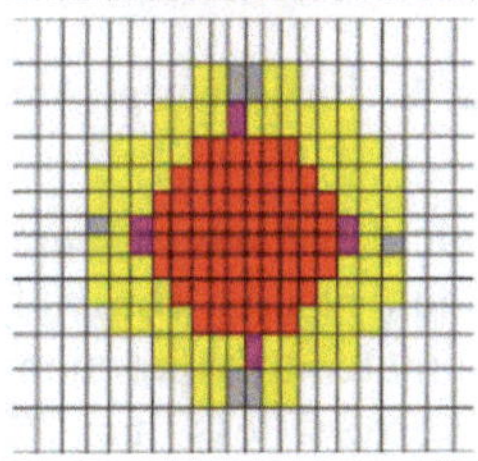	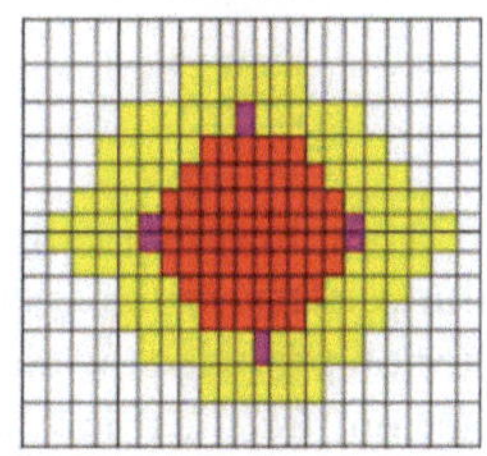

3.4 Result Comparison

Due to the low impact energy, only small dents were observed on the top surface of the laminates. Notably, visible matrix cracking appeared on the bottom surface. The projected damage areas, measured via ultrasonic C-scan, yielded average values of 169.56 mm^2 for 6 J impacts and 235.50 mm^2 for 8.5 J impacts.

As shown in Fig. 4, the Weng delamination criterion yields the largest damage area, followed by the modified Hou criterion, while the Yeh delamination criterion predicts the smallest damage. Current experimental studies by numerous scholars demonstrate that under low-velocity impact, matrix cracking occurs first, which then induces interlaminar delamination. Additionally, interlaminar shear stresses further contribute to delamination, making delamination damage the predominant contributor to the overall damage area in impacted laminates. The Yeh criterion inaccurately predicts delamination damage areas smaller than matrix damage areas.

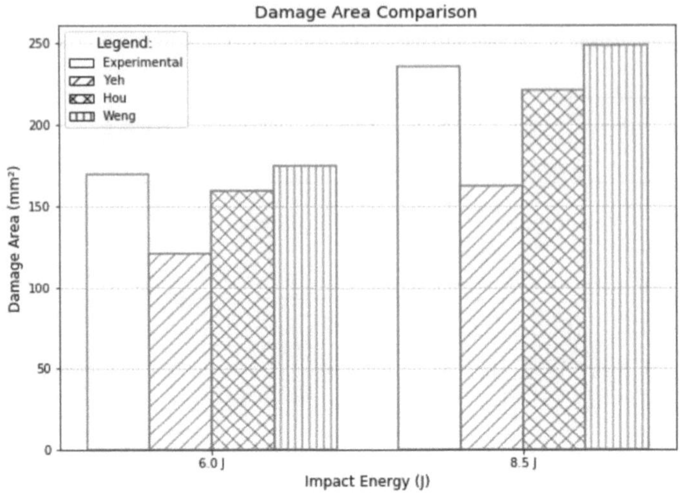

Fig. 4. Total Damage Area Comparison

Table 3 compares the calculated damage areas under three different delamination failure criteria with experimental results. The analysis shows that the Weng delamination criterion provides the closest match to experimental data, followed by the modified Hou criterion. Both exhibit small errors (less than 7%) under two impact energy levels. In contrast, predictions using the Yeh criterion deviate significantly from experimental results, with errors reaching 30%.

Table 3. Experimental vs. Simulated Impact Damage Areas

Impact Energy (J)	Experimental Damage Areas (mm^2)	Delamination Failure Criteria	Simulated Damage Areas (mm^2)	Error (%)
6.0	169.56	Yeh	120.57	-28.9
		Hou	159.51	-5.9
		Weng	174.35	2.8
8.5	235.50	Yeh	162.55	-31.0
		Hou	220.76	-6.3
		Weng	248.77	5.6

4 Conclusions

Based on the results and discussions presented above, the conclusions are as below:

(1) Three distinct delamination failure criteria were combined with the modified Hou in-plane failure criteria to simulate impact damage in composite laminates using finite element analysis. Results demonstrate that the Weng delamination criterion and the modified Hou delamination criterion provide accurate predictions, with errors below 7%.
(2) Under low-velocity impact, composite laminates exhibit small dents on the top surface, significant matrix cracking on the bottom surface, and extensive interlaminar delamination. The total damage area increases with higher impact energy.
(3) Greater impact energy leads to an increase in maximum contact force and an earlier occurrence of peak force during the impact event.

References

1. Zhang, C., Liu, J.Q., Li, J., et al.: Application of composite materials in commercial large aircraft. Fiber Compos. **4**, 77–84 (2024)
2. Zhao, T., Li, Y., Zhang, C., et al.: Fundamental mechanical problems in high-performance aerospace composite structures: state-of-the-art review. Acta Aeronaut. et Astronaut. Sinica **43**(6), 526851 (2022)
3. Zhang, Q., Huang, G.Q., Song, G.B., Xiong, H.F.: Experimental study on impact threshold of composite laminates. In: The 6th China Aeronautical Science and Technology Conference. Zhejiang, pp. 696–700 (2023)
4. Zhang, M.T., Liu, F., Li, X.J.: Experimental research on the static strength of carbon fiber laminated panels after low-energy impact. New Technol. & New Process **1**, 62–66 (2022)
5. Huang, S.F., Chen, X.Z., Liu, D., et al.: Research on mechanical property degradation and failure process of carbon fiber reinforced polymer. Aerospace Mater. & Technol. **5**, 14–20 (2022)
6. Xu, Z., Li, X.Y., Zhou, E., Wu, G.H.: Correlation analysis on the impact damage shape and impact energy for carbon fiber laminate. Non-destr. Examination **4**(38), 6–14 (2016)
7. Yu, H.Y., He, H.W., Xing, P.: Development of carbon fiber reinforced plastic failure models considering different stiffness degradation modes (2024)
8. Wen, W.D., Xu, Y., Cui, H.P.: Damage analysis of laminated composites under low-velocity impact loading. material engineering, 7: 6–11. J. Mech. Eng. **60**(2): 197–208 (2007)
9. Yi, T.Y., Yu, Z.F., Wang, H.: Stiffness degradation methodology for low-velocity impact simulation in composite laminate. Chin. Quart. Mech. **33**(3), 469–475 (2012)
10. Gong, B.B.: Investigation of damage model of stitched laminates subjected to low-velocity impact. Nanjing University of Aeronautics & Astronautics (2017)

Research on Modeling and Simulation Parameterization Methods for Helical Gear

Linhao Ji[1(✉)], Xiaohui Duan[1], Junfeng Li[1], Jinghua Jiang[1], Pengchao Wang[1], and Mingming Yao[2]

[1] Henan Pinggao Electric Co., Ltd, Pingdingshan, Henan, China
jilinhao868@163.com
[2] Pinggao Group Co., LTD., ZhengZhou, Henan, China

Abstract. To solve the difficulties in modeling, designing, and simulation analysis of helical gear in high-pressure equipment, a parametric method for modeling and simulation of helical gear is proposed, and a user-oriented CAD-FEM coupled modeling and simulation analysis system is designed and developed. The geometry equation and data structure are used to describe the state of the modified helical gear model, the point, and line, and surface feature structures are established by the parametric design method, the integrated software platform for modeling, and grid division. Simulation analysis of the modified helical gear is established, the model data and its affiliated structures are shared, and the CAD-FEM module coupling is realized. The platform has the advantages of strong interactivity, good stability, high modeling and simulation analysis efficiency, small data volume, etc. Direct modeling and simulation analysis without data reconstruction and conversion provides a parametric method and software platform for modeling, design, and simulation analysis of modified helical gear.

Keywords: Modified helical gear · Structural modeling · FEM · Parameterization

1 Introduction

Power equipment is one of the important components of the power grid system, among which high-voltage equipment is an important protective execution device, including mechanical, electrical, thermal, and other disciplines. The transmission system serves as the action execution unit of high-voltage equipment [1]. It directly determines the working ability of the equipment. Gear transmission is also one of the important transmission methods. Accurately calculating the static and dynamic characteristics of gear systems and designing them reasonably is crucial for reducing working power, improving the smoothness of transmission pairs, and extending the service life of gears [2]. Therefore, it is necessary to evaluate the motion characteristics, gear meshing, and dynamic characteristics of the system through simulation analysis. Among them, the structure of helical gears is complex, and the establishment and simulation analysis of numerical model structures have become major challenges faced by researchers.

S. Halgamuge et al. (Eds.): ICACMVE 2025, LNME, pp. 557–568, 2026.
https://doi.org/10.1007/978-981-95-6911-3_55

The general method for simulation analysis is to use external software (such as CATIA, UGNX, and SOLID) to establish models. CATIA has few standard databases, only equipped with simple standard accessories such as bolts, nuts, and gaskets; UGNX generates a large amount of data after modeling and reduces the software running speed, which brings many inconveniences to user operation and storage [3]. Although SOLID gives a gear part in the design library, it still needs a lot of modification and the modeling efficiency is low. In addition, the above software not only lacks the pertinence to FEM pre-processing, main processing, and post-processing but also needs to import the model data into the engineering simulation software after the model is established. Then we gradually carry out the simulation process (such as the network division method and boundary condition establishment, analysis, and solution setting).

This method has three shortcomings: 1. Model establishment requires repeated data processing, and any modification of geometric parameters requires repeated modeling, which increases the design and analysis time and reduces repeatability; 2. The process of correctly dividing the grid is cumbersome, prone to generating singular elements, and has low accuracy in unfolding line contours; 3. Due to data compatibility and software communication, the transmission model is distorted, resulting in repeated transmission and modeling. So, Liu et al. proposed a parametric integration method based on the NX platform to address the problems of loss of conversion accuracy and repetitive modeling in traditional gear models; This provides an efficient and convenient simulation approach for the integrated design and analysis of gear CAD/CAE [4]. Zhang et al. proposed a NURBS-based parameterized modeling and single tooth meshing contact simulation method for involute cylindrical gears to address issues such as design simulation fragmentation and the effectiveness of the proposed method was verified [5]. Hu et al. have derived the machining process of non-standard cylindrical gear-cutting tools, developed parameterized models, and achieved real-time and accurate changes in the tooth profile of complex non-standard cylindrical gears, as well as efficient analysis and simulation [6]. The above-mentioned scholars have only conducted relevant research on relatively simple cylindrical spur gears, and have not yet touched on the parameterized modeling and simulation methods of helical gears and modified helical gears.

Software menu and interface design shall be carried out based on the database requirements of the Pyprogram development platform, and the geometric model data synchronization method [7] shall be adopted to provide metadata for simulation analysis, to greatly reduce the data information of the entity model of the modified helical gear structure and greatly improve the modeling efficiency. At the same time, ANSYS self-programming language is used to develop CAD-FEM coupling system [8], build parametric programs such as displacement bevel gear modeling and grid generation, fully include the whole process of simulation analysis into the software platform, reduce repeated design [9], and realize data sharing.

2 CAD-FEM Coupled System Architecture

2.1 Relationship Between Modules in a Coupled System

The user-oriented gear CAD-FEM coupling system mainly includes four parts: CAD modeling of modified helical gear, FEM pre-processing, FEM main processing, and result display post-processing. The relationship between the four parts is shown in

Fig. 1. FEM pre-processing module is mainly used for finite element mesh division and attribute transfer, and FEM main processing module is mainly used for boundary condition setting, degree of freedom control, load application, and mechanical property calculation. The results show that the post-processing module is mainly used for calculation output and contour evaluation, such as mode identification, stress evaluation, and displacement deformation evaluation [10]. The CAD modeling module plays an important role in structural design and data provision parameterization in this system [11], and automatically converts gear entity into an FEM model through CAE pretreatment.

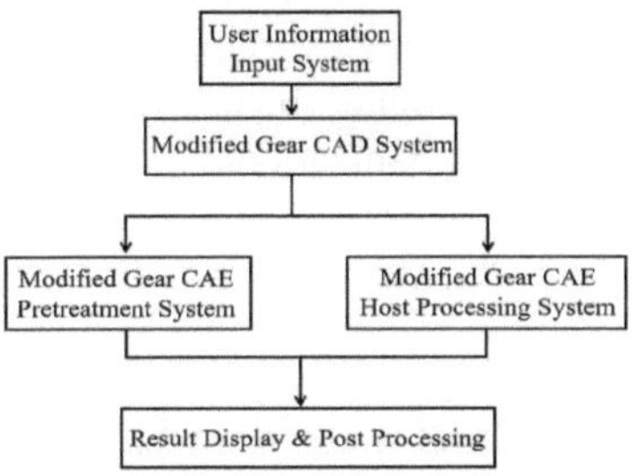

Fig. 1. Relationship between module.

2.2 Software Structure

Based on the basic progressive relationship between modules, the system adopts a design principle that combines partitioning and layering and is divided into the user interaction layer, data processing layer, and graphic display layer.

The user interaction layer mainly consists of three parts: modeling, simulation, and solving, responsible for processing user requests, completing modeling simulation parameter settings [2] and data modification. Parameters include unit, material, load, etc. [12].

The data processing layer is mainly responsible for data integration and transmission to relevant program variables for storage and extraction, handling the start-stop, and calling various functions.

The graphic display layer is mainly responsible for displaying graphics, and selecting display modes through parameter input, option settings, etc. [13], such as entities, cells, and perspective.

3 Organizational Relationship of the Modified Helical Gear System Model

The CAD and FEM models of the modified helical gears consist of two parts: 1. Geometry, including involutes, gear faces, gear teeth solids, bores, etc.; 2. The model data mainly includes the correlation and collection among the structural relationship, material properties, element node properties, and other data of the geometric model.

3.1 Organizational Management Relationship of Variable Position Helical Gears

The geometrical modeling of the modified helical gear follows its own rotational symmetry, and the direct modeling technology is used. First, any tooth profile involute is generated by the involute function, and the second involute is established by the mirror symmetry; Then, the part intersecting with the tooth tip circle is cut to obtain the tooth profile end face. Secondly, an auxiliary spiral line based on the reference circle is established according to the spiral angle, and the involute surface of the gear profile and the gear tooth entity are generated by end surface stretching. Finally, the first tooth is rotated and copied using a cyclic symmetry operation to produce a complete gear. At this time, start data cleaning, such as merging overlapping surfaces, deleting auxiliary lines, auxiliary surfaces and other intermediate objects, reducing data storage and avoiding affecting modeling efficiency. The organizational relationship of the model is shown in Fig. 2.

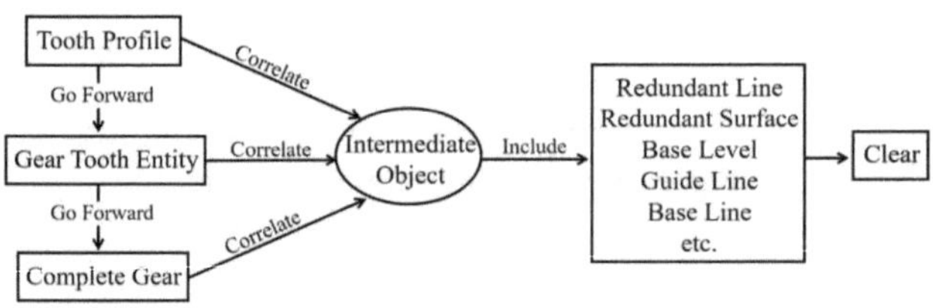

Fig. 2. Model organizational relationship.

4 Direct Modeling and Parameterization Method for Modified Helical Gears

4.1 Direct Modeling of Primary and Secondary Parameters and Algorithms

Based on the structural characteristics, modeling practice, and involute function equation of the modified helical gear, the driving parameters for structural modeling are mainly determined [14]. After analysis, the number of teeth, normal modulus, and other main variables are determined, as shown in Table 1. We can see Table 2 for the determination of secondary variables such as reference circle diameter and tip circle diameter. The basic algorithm and underlying logic for the direct modeling of the modified helical gear are established through the relationship between the primary and secondary variables. Based on the algorithm relationship of primary and secondary variables and the involute function equation, as shown in Formula (1), the algorithm relationship of any point i($x(j)$ and $y(j)$) of the modified helical gear tooth profile in the Cartesian coordinate system is derived, as shown in Formula (2).

$$inv\alpha_i=\theta_i=tan\alpha_i-\alpha_i \tag{1}$$

$$\begin{cases} x(j) = -\frac{(\tan\frac{\pi}{9}\cos\varphi - \sin\varphi) * (m_n z/2)}{1+\tan^2(\pi/9)} \\ \quad + \frac{(\tan\frac{\pi}{9} sin\varphi + \cos\varphi) * (\frac{\pi m_n}{4} - \frac{m_n z\varphi}{2} + d\tan\frac{\pi}{9})}{1+\tan^2(\pi/9)} \\ y(j) = \frac{(\tan\frac{\pi}{9}\cos\varphi - \sin\varphi) * (\frac{\pi m_n}{4} - \frac{m_n z\varphi}{2} + d\tan\frac{\pi}{9})}{1+\tan^2(\pi/9)} \\ \quad + \frac{(\tan\frac{\pi}{9} sin\varphi + \cos\varphi) * \frac{m_n z}{2}}{1+\tan^2(\pi/9)} \end{cases} \tag{2}$$

The meanings of the symbols in the formula are shown in Table 2.

During modeling and grid generation, we accurately calculate the interpolated node coordinates of the modified helical gear tooth profile through the relationship between the primary and secondary variables and their algorithm and realize the direct modeling parameterization of the modified helical gear.

Table 1. Direct modeling of main variables for a pair of modified helical gears.

Name	Symbol
Number of teeth	z_1
Normal plane modulus	m_n
Normal pressure angle	α_n
Angle of helix	$\beta\beta$
Normal crest clearance coefficient	c_n
Normal crest height coefficient	h_n
Normal displacement coefficient	x_{n1}
Width of tooth	b_1

Table 2. Algorithm relationship of primary and secondary variables.

Name	Symbol	Algorithm relationship
Reference circle diameter	d_1	$d_1 = z_1 m_n$
Diameter of gear tip circle	d_{a1}	$d_{a1} = d_1 + 2h_{an}m_n$
Base circle diameter	d_{b1}	$d_{b1} = d_1 cos(\alpha_n)$
Root circle diameter	d_{f1}	$d_{f1} = d_1 - m_n * (h_{an} + c)$
Height of any round tooth	h_i	$h_i = m_n * (3h_{an} + c)/(N - 1)$
The diameter of any circle	d_i	$d_i = d_1 - m_n * (h_{an} + c) + i * h_i$
The center angle of any tooth thickness	φ	φ =2*(s/d − $inva_f$ + $inva$)
Normal pressure angle of any circle	α_{ni}	$\alpha_{ni} = acos\,(r * cos(\alpha_n)/r_i)$

4.2 Modeling Method

Due to the high complexity of the structure of the modified helical gear, it is very difficult to model the 3D solid of the gear. First, a plurality of interpolation nodes is generated in a Cartesian coordinate system; Secondly, the higher order B-spline curve function equation is used to connect the nodes successively to the smooth and complete modified helical gear tooth profile. Then, based on the transition curve equation, the tooth profile at the tooth root is transited to establish a complete tooth profile curve. Finally, a guideline is established, and a gear tooth structure is established by extending the lead wire on the end face of the gear tooth, and the whole structure of the modified helical gear is generated by rotating and copying. The basic method and implementation process are shown in Fig. 3.

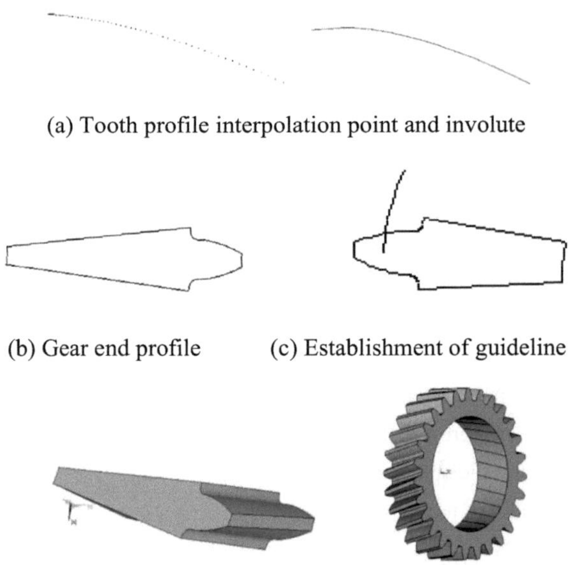

(a) Tooth profile interpolation point and involute

(b) Gear end profile (c) Establishment of guideline

(d) Stretch to create a tooth (e) Rotate copy

Fig. 3. Direct modeling of helical gears with variable displacement.

5 Composition of a Parameter Simulation Platform

The program adopts a high-level, interpretable, compiled, interactive, and object-oriented efficient high-level language Python. It uses direct modeling technology and parametric control methods to parameterize the design of the variable-modified helical gear and establishes a window application program to achieve user interaction and develop a user-friendly interface. This facilitates the setting and modification of boundary conditions such as loads and constraints in the modeling and finite element model generation of the helical gear structure, as well as simulation analysis. The parameterization of the entire process of modeling, analysis, and simulation result display lays the foundation for the development of a CAD-FEM coupling system for a gear and a pair of gears.

5.1 Structural Entity Modeling Platform and Visualization

The parameterized modeling platform for variable helical gear structure entities is designed based on the main and auxiliary parameters of variable helical gears and their related algorithm programming to create a user interaction interface. According to actual engineering, material attribute setting options are added to achieve parameterization of modeling, material attribute definition, data communication, and storage for one variable helical gear. The entity modeling and visualization are shown in Fig. 4.

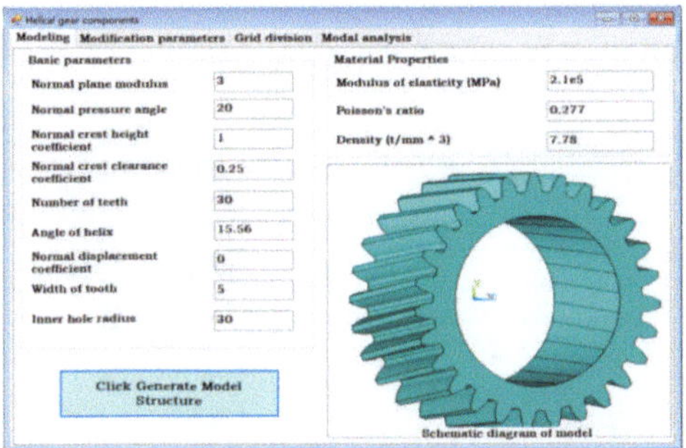

Fig. 4. Parameterized modeling platform.

According to the requirements of gear design, to reduce the meshing impact or local stress, it is necessary to carry out gear tooth modification design or gear modification design according to specific working conditions. In the development process of the CAD-FEM coupling system for modified helical gear, based on the basic principles of modification and gear modification, a CAD modeling visualization platform for modification and gear modification of modified helical gear is built, as shown in Fig. 5.

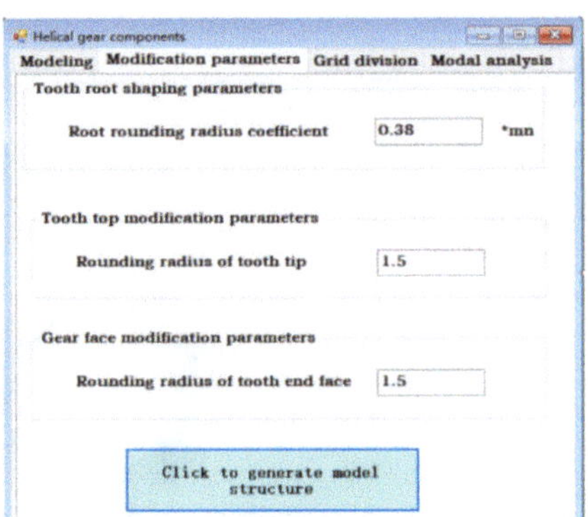

Fig. 5. Helical gear modification platform.

5.2 FEM Platform and Visualization

After the model is established, grid partitioning is required accordingly. Based on the finite element method, we establish a mesh division module for modified helical gear, design mesh division methods and options for setting element types and assigning element sizes, transfer relevant variable values through Python programming, establish a window application, develop a user-friendly interface, and achieve user interaction to facilitate the generation of finite element models, setting and modification of parameters such as loads and constraints, and parameterize the entire process of modeling, analysis, and simulation result display post-processing. The composition of the user-oriented finite element modeling and parameterization software platform for helical gears is shown in Fig. 6.

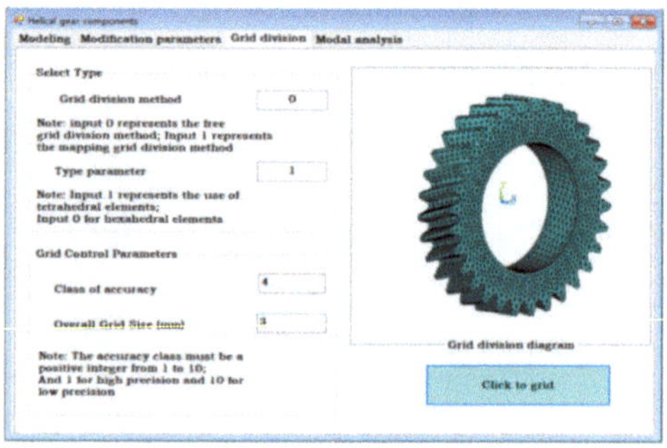

Fig. 6. Grid generation platform.

5.3 Analysis Platform and Visualization

According to the requirements of gear design and finite element analysis, based on the finite element analysis process and analysis option settings, we design parameter input or modification tabs, program, and implement a user-oriented finite element simulation parameterization platform and user interface visualization for variable helical gears, as shown in Fig. 7.

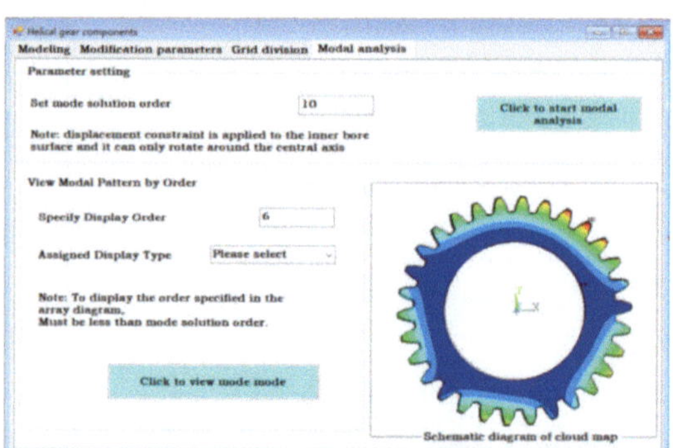

Fig. 7. Finite element simulation platform.

Through the modal analysis module in this parameterized simulation platform, the natural frequency of the variable displacement helical gear is analyzed, and the natural frequency, vibration form, and deformation trend of the variable displacement helical gear are calculated, as shown in Fig. 8. This provides basic support for further research on system-level dynamic response, reducing meshing impact, and achieving optimized design.

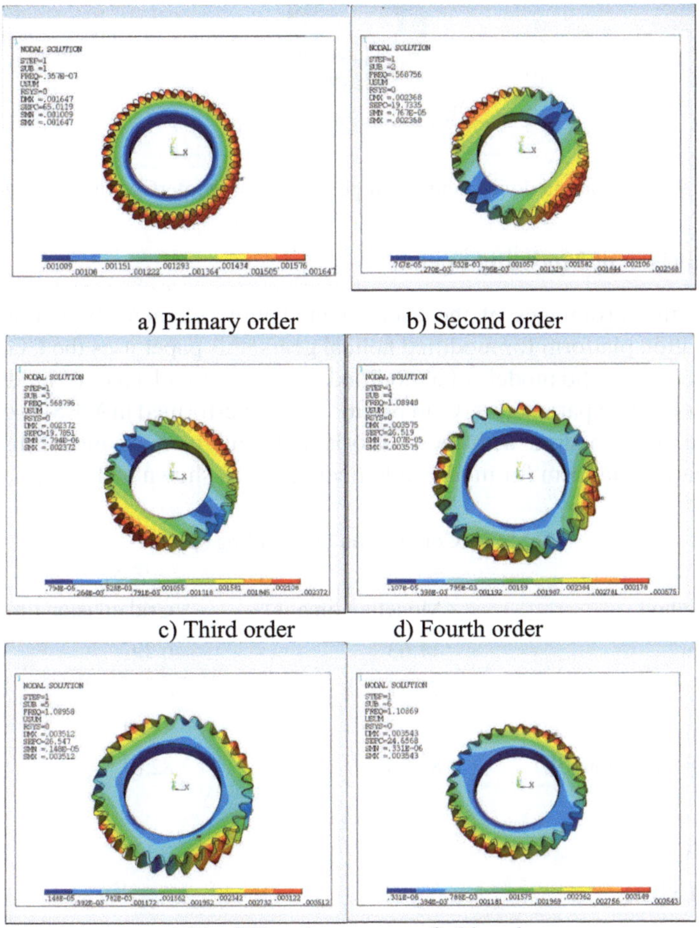

a) Primary order b) Second order

c) Third order d) Fourth order

e) The fifth order f) 6th order

Fig. 8. Modal analysis vibration mode diagram.

5.4 Model and Result Visualization

After the simulation solution is completed, it is necessary to analyze the software calculation results. Therefore, a result display module based on ANSYS post-processing is designed to directly read the finite element modeling and analysis results through the background, and we establish a user-oriented model and result visualization interface, as shown in Fig. 9.

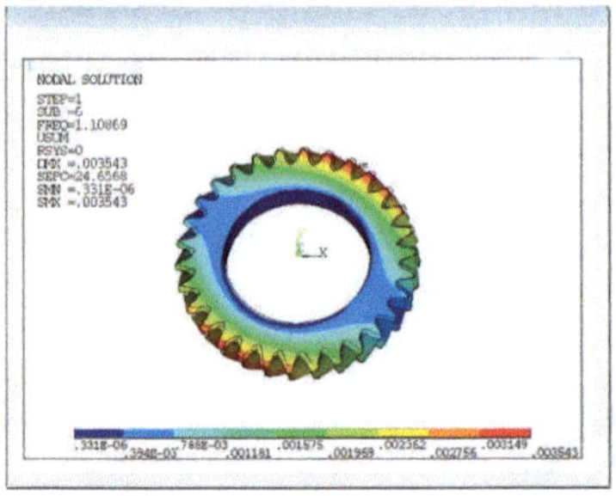

Fig. 9. User-oriented view and result post-processing interface of helical gear.

5.5 Comparison of Modeling Speed

To illustrate the advantages of the user-oriented finite element modeling simulation parameterization platform for modified helical gears, this paper uses the UG platform to establish structural solid models of a modified helical gear and a pair of modified helical gears with consistent parameters. Grid partitioning is performed in MESH software, and the time used is compared with the user-oriented finite element modeling simulation parameterization platform for modified helical gears, as shown in Table 3.

Table 3. Comparison of modeling speed.

Modeling method	Modeling time (s)	Grid division time (s)
This platform	11.63	6.29
UG	72.51	/
MESH	/	34.91
Speed improvement rate (%)	83.96	82.81

6 Conclusion

The platform adopts the direct modeling technology, through the establishment of the underlying algorithm of the primary and secondary variable parameter modeling, and puts forward an accurate modeling method of the modified helical gear tooth profile, programming to realize the modified helical gear modeling simulation parameter platform and improve the complexity of traditional modeling analysis.

Based on the basic principle and meshing characteristics of involute modified helical gear modification and modification, the modified helical gear modification and modification parameterization are realized.

We use this parametric software platform for modeling and state response of helical gears. The solid modeling time is 11.63 s, the grid partitioning time is 6.29 s, and the modeling efficiency has increased by more than 77.04%, with a grid partitioning efficiency increase of more than 74.32%.

Acknowledgment. The paper is funded by the Research on Key Technologies and Equipment Development of 252 kV 50 kA GIS Based on C4F7N Mixed Gas (5191BG240002).

References

1. Zhai, G., Liang, Z., Fu, Z.: A mathematical model for parametric tooth profile of spur gears. Math. Probl. Eng. **2020**(1), 7869315 (2020)
2. Motahar, H., Samani, F.S., et al.: Nonlinear vibration of the bevel gear with teeth profile modification. Nonlinear Dyn. **83**(4), 1875–1884 (2016)
3. Liang, X., Haichun, D., Lei, L., et al.: Parametric modeling and finite element simulation research on locomotive traction gears. Mech. Trans. **42**(10), 91–95 (2018)
4. Liu, M., Zhang, R., Du, X., et al. Parametric modeling and geometric analysis of involute cylindrical gears. J. Comput.-Aided Design Graph. 1–15 (2025)
5. Zhang, R., Yu, J., Zhao, Y.: Parametric CAD/CAE integrated contact stress analysis of spur gears. Mechanical Design and Manufacturing, 1–6
6. Hu, S., Wang, S., Shen, G., et al.: Construction of precision parameterized model for nonstandard cylindrical gears. Mechanical Design, 1–6
7. Hu, Z., Liu, S., Chang, J., Zhang, J.: Application of response surface method for contact fatigue reliability analysis of spur gear with consideration of EHL. J. Central South Univ. **22**(07), 2549–2556 (2015)
8. Xuewen, Z., Yaxu, C., Hui, X.: Parametric modeling and finite element analysis of helical gear anti-skid differential gears. Mech. Trans. **40**(03), 146–149 (2016)
9. Jiang, P., Qi, L., Wu, H., et al.: Feature preserving structured mesh generation method based on shard parameterization. J. Comput.-Aided Design Graph. **36**(08), 1286–1297 (2024)
10. Samani, S.F., Molaie, M., Pellicano, F.: Nonlinear vibration of the spiral bevel gear with a novel tooth surface modification method. Meccanica **54**(7), 1071–1081 (2019)
11. Jiandong, W., Jiayuan, Z., Jing, L., et al.: A parameterized modeling method for planning boats. J. Huazhong Univ. Sci. Technol. (Nat. Sci. Ed.) **48**(12), 83–88 (2020)
12. Farshidianfar, A., Saghafi, A.: Global bifurcation and chaos analysis in nonlinear vibration of spur gear systems. Nonlinear Dyn. **75**(4), 783–806 (2014)
13. Tao, Z., Qiang, L., Xi, W., et al.: In situ modal identification of high-speed train gearbox body. J. Mech. Eng. **54**(12), 31–41 (2018)
14. Lin, J., Zhao, Y., Fu, X., et al.: Parametric modeling and contact stress analysis of asymmetric helical gears. Mech. Trans. **48**(06), 117–123 (2024)

Multi-objective Genetic Algorithm Optimization of Energy Efficiency Index for Plate Heat Exchanger in Commercial Electric Vehicle

Yunjia Liang[1], Naijiang Liu[1(✉)], Haidong Wang[1], Zhiao Wang[1], Jiaxin Liu[1(✉)], and Martin Krechel[2]

[1] College of Mechanical Engineering, North China University of Science and Technology, Tangshan 063210, China
{liunaijiang,liujiaxin}@ncst.edu.cn

[2] School of Architecture and Engineering, SRH University of Applied Sciences Heidelberg, Heidelberg 69123, Germany

Abstract. Plate heat exchangers (PHEs) play a vital role in the processes of heat transfer and energy conservation. To find the best combined optimization result using the criteria of PHE heat transfer and energy conservation, a multi-objective genetic algorithm optimization (MOGAO) targeting the minimum -*JF* factor and entropy generation was carried out on the PHE, and 100 optimization results were comprehensively assessed by a quantitative energy efficiency index (EEI). Finally, the comparisons of *JF* factor, entropy generation per mass flow rate (EGPMFR), and EEI between improved scheme 6 (IS6) and original scheme were numerically discussed under the same configuration. It is found that the deviation between experimental data and simulation results could enable further analyses within the specific range. Under the inlet velocity of 0.6 m/s, the *JF* factor, EGPMFR, and EEI of IS6 increase by 12.20%, 19.42%, and 8.82%, respectively. This work may somehow offer a reference for the PHE comprehensive optimization.

Keywords: Multi-objective genetic algorithm · Plate heat exchanger · Energy efficiency index

1 Introduction

Multi-objective Genetic Algorithm Optimization (MOGAO) effectively handles conflicting objectives, generating non-inferior solution sets while avoiding local optima. Sun et al. [1] optimizes Y-fin geometry in PCM units via NSGA-II/TOPSIS, boosting melting power 4.2× with only 14.9% energy storage trade-off. Li et al. [2] used the NSGA-II MOGAO to optimize GaN MMIC PA heat sinks, minimizing friction factor (f) or maximizing Nusselt number (*Nu*) via Pareto solutions. Zimparov and Vulchanov [3] proposed the Nusselt-related performance evaluation criteria (NPEC) as an alternative to JF. Fang et al. [4]'s interrupted-rotation finned tube disrupts thermal boundary layers, boosting heat transfer 9.32% and cutting pressure drop 69.2% via CFD-optimized design. Yu et al. [5] found sinusoidal wavy tapes in HEs yielded higher Nu, f, and NPEC

S. Halgamuge et al. (Eds.): ICACMVE 2025, LNME, pp. 569–576, 2026.
https://doi.org/10.1007/978-981-95-6911-3_56

than triangular wire coils. Niu et al. [6] create an LES-based predictive framework for plate heat exchangers with NSGA-II optimization.

This work introduces an Energy Efficiency Index (EEI) as a unified criterion for PHE heat transfer and energy conservation. EEI aided optimal scheme selection from MOGAO results. Its feasibility was confirmed by comparing JF factor, entropy generation per mass flow rate (EGPMFR), and EEI against original and IS6 schemes. Objectives are: (a) Validate EEI for MOGAO result selection; (b) Establish EEI as a comprehensive criterion for structural optimization.

2 Multi-objective Genetic Algorithm Optimization

2.1 Numerical Investigation and Energy Efficiency Index

An elementary unit of the PHE for commercial electric vehicle was confirmed as the analysis object. The detailed structural parameters and EEI were set from Reference [7].

2.2 Optimization Flow and Setting

The trial of EEI would be initialized by a traditional MOGAO targeting *-JF* and ΔS. The optimized results should firstly meet the initial expectation, and then be assessed with EEI to confirm its effectiveness. Thus, a classic MOGAO was performed for a set of the optimum structure parameters. The entropy generation ΔS and *-JF* from Reference [7] were set as objective functions. The expression of the MOGAO is written as Eq. (1):

$$
\begin{gathered}
f(x) = \min(\Delta S, -JF) \\
s.t.\begin{cases}
3.5 \le x_1 \le 4.5 \\
0.8 \le x_2 \le 1.2 \\
0.3 \le x_3 \le 0.6 \\
B_\text{p} \times L_\text{p} = 58 \times 40 \\
\beta = 60^\circ
\end{cases}
\end{gathered} \tag{1}
$$

The design variables, λ_p (x_1), h_p (x_2) and h_m (x_3), and their corresponding boundaries are listed in Table 1. The MOGAO toolbox in MATLAB R2018a was applied for the optimization with the settings in Table 2. Following the flow diagram of the optimization in Fig. 1, to form the initial population, a number of 500 individuals were randomly generated. Working conditions and design variable ranges were set. Crossover fraction was set to 0.8, and the Pareto front population fraction was set to 0.2. Design variables were set according to Table 1. According to the objective function, the initial population ΔS and *-JF* were minimized as the results of the first iteration, and the population was ranked, selected, cross mutated. Among them, the mutation function was set as constraint dependent, the selection function was set as tournament with the size of 2, the frequency division function was set as intermediate with the ratio of 0.8. The migration function was set as direction forward with the fraction and interval of 0.2 and 20, and minimized to produce ΔS and *-JF* as the results of the second iteration. The results from two generations were compared, if the functional tolerance of the comparison result was greater than 1×10^{-4}, the second iteration result would be recalculated. If not, then the iteration would be considered convergent.

Table 1. Design variables and corresponding boundaries

Parameters (Unit)	Variable	Vector element	Variation rang
Plate width (mm)	λ_p	x_1	$3.5 \leq x_1 \leq 4.5$
Plate height (mm)	h_p	x_2	$0.8 \leq x_2 \leq 1.2$
Plate spacing (mm)	h_m	x_3	$0.3 \leq x_3 \leq 0.6$

Table 2. Main settings for the optimization

Option	Setting	Option	Setting
Size of population	500	Selection function	Tournament with the size of 2
Crossover fraction	0.8	Migration	Direction forward with the fraction and interval of 0.2 and 20
Pareto front population fraction	0.2		
Mutation function	Constraint dependent	Crossover function	Intermediate with the ratio of 0.8
		Stopping criteria	Function tolerance of 10–4

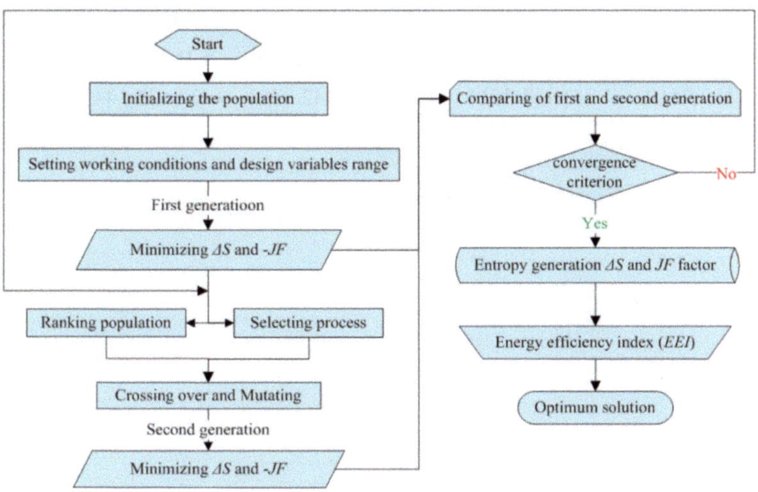

Fig. 1. Flow diagram of optimization

2.3 Optimization Results

After 109 iterations, Pareto front, its average spread, and applicable solutions were obtained and shown in Fig. 2. As for the solution convergence, the Pareto front average spread was 0.113421 with a convergent tendency in Fig. 2(a). In Fig. 2(b), Point A stood for the acceptable dynamic equilibrium between two individual optimization objectives, hence, any Pareto front nearing it would be considered as feasible solution. Once beyond those applicable solutions, *JF* factor or entropy generation would be preferred by the practical working conditions and the requirements of engineering under the main perspectives of heat transfer or energy conservation.

Under the pure consideration of energy conservation in tolerant *JF* factors, 6 improved schemes with minimum EEIs were selected and are listed in Table 3. Among all the improved schemes, Improved Scheme 6 (IS6) was adopted for further analysis because of its *JF* factor approximate to the original one.

Table 3. Optimization solutions of PHE

Variable	Original scheme	Improved scheme					
		1	2	3	4	5	6
λ_p	4	4.493	4.476	4.479	4.466	4.497	4.491
h_p	1	0.803	0.801	0.802	0.802	0.813	0.801
h_m	0.3	0.565	0.563	0.548	0.59	0.582	0.472
JF	0.03470	0.03438	0.03442	0.03448	0.03429	0.03423	0.03485
ΔS	16.7466	15.2126	15.2379	15.2699	15.1849	15.1667	15.4410
EEI	482.6110	442.4716	442.7152	442.8321	442.8422	443.0639	443.0813

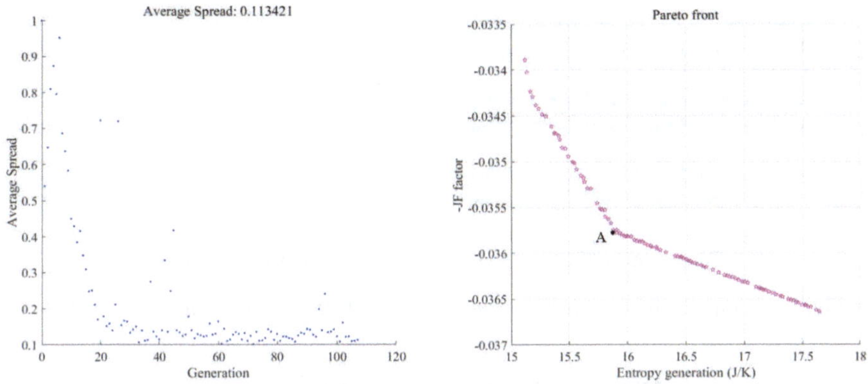

a Pareto front average spread b Pareto front

Fig. 2. Pareto front of the MOGAO

3 Comparison of Improved and Original Schemes

3.1 Numerical Simulation Results of Improved and Original Schemes

The original scheme and IS6 were numerically compared under Reference [8]'s settings. At 0.6 m/s inlet velocity, Fig. 3 shows their temperature and velocity distributions. Figure 3a indicates heat exchange occurred between cold and hot fluids via metal plates. The original scheme's outlet temperatures were approx. 27.90 °C (cold) and 72.16 °C (hot), while IS6's were approx. 26.30 °C and 74.03 °C. Figure 3b shows IS6 had a maximum velocity around 0.72 m/s and smoother flow overall, due to increased plate spacing and decreased peak height. Intense velocity between adjacent cavities was reduced to 1.32 m/s.

For fair comparison of f factors, both schemes were evaluated at equal pump power (Fig. 3). IS6 showed a leading f factor with a 19% higher average mass flow rate compared to the original scheme. This compensates for the slightly reduced j factor and significantly decreases flow resistance. The comprehensive discussion on IS6 aligns with findings in Salman et al. [8].

Under Reference [8]'s settings, the original scheme and IS6 were numerically compared at 0.6 m/s inlet velocity. Figure 3a shows heat exchange via plates. Outlet temperatures were ~ 27.90 °C (cold)/72.16 °C (hot) for the original scheme vs. ~26.30 °C/74.03 °C for IS6. Figure 3b shows IS6 had smoother flow (max velocity ~ 0.72 m/s) and reduced intense inter-cavity velocity (1.32 m/s), attributed to increased plate spacing and decreased peak height.

For equal pump power (P) comparison, which shows in Fig. 3, IS6 exhibited a leading *f* factor with 19% higher average mass flow rate. This compensates for its slightly reduced *j* factor and significantly decreases flow resistance. These IS6 findings align with Salman et al. [8].

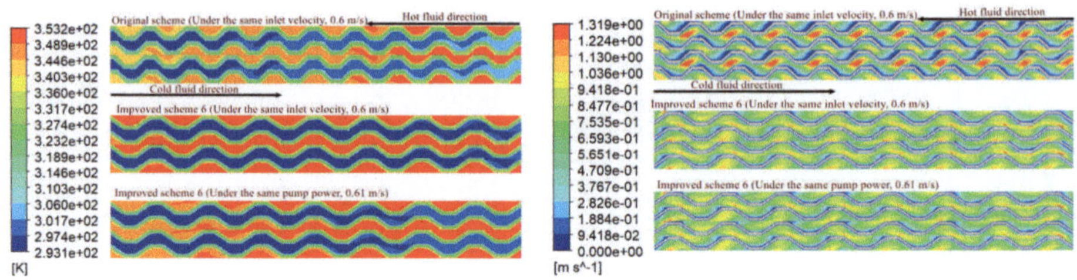

a Temperature (°C) b Velocity (m/s)

Fig. 3. Comparison of temperature and Velocity Distributions from both Schemes

3.2 Comparisons of *JF* Factor, EGPMFR and EEI

More details on those two schemes were further compared in this section. Nusselt number *Nu* and pressure loss Δp at different inlet velocities are listed in Table 4, where IS6 presented a lower Δp and higher *Nu*. *Nu* and Δp of both schemes tended to rise with the increasing inlet velocity, which could be explained by more intense flow resistance and heat transfer brought by the increasing the flow velocity.

Table 4. Comparisons of *Nu* and Δp of those schemes

Velocity	Nu		Δp	
	IS6	Original scheme	IS6	Original scheme
0.6 m/s	28.79	28.43	6126	7657
0.7 m/s	31.73	31.34	7875	9714
0.8 m/s	34.52	34.09	9824	11965
0.9 m/s	37.18	36.72	11980	14407
1.0 m/s	39.74	39.25	14338	17038
1.1 m/s	42.21	41.68	16902	19854

In Fig. 4(a), IS6's larger *JF* factor yielded better performance. As inlet velocity increased from 0.6 to 1.1 m/s, *JF* factors decreased. This occurred because the Colburn factor j decreased more than the friction factor $f^{1/3}$ with increasing velocity. Although both j and f decreased for IS6, its $f^{1/3}$ reduction was greater than the original scheme's. At 0.6 m/s, IS6's JF factor increased by 12.20%. Optimized parameters altered the inlet area, so EGPMFR from Reference [9] assessed entropy generation. Fig. 4(b) shows EGPMFRs decreased over the velocity range. At 0.6 m/s, the improved scheme's energy conservation was 19.42% better. Fig. 4(c) confirmed IS6's smaller EEI throughout the range, decreasing by 8.82% at 0.6 m/s. IS6 was superior to the original scheme in both heat transfer and energy conservation, and the proposed EEI supports performance and economy perspectives simultaneously.

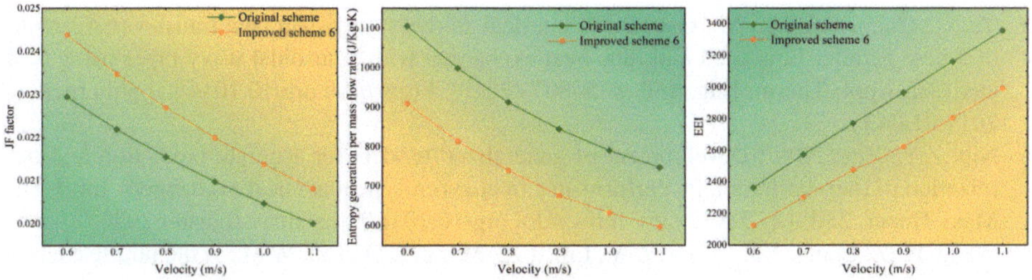

a JF factor b EGPMFR c EEI

Fig. 4. Comparisons of JF factor, EGPMFR and EEI

4 Conclusion

To set up an EEI under the consideration of heat transfer and energy conservation, a numerical analysis of the PHE was validated by the corresponding experiment. Based on entropy generation, together with published Nusselt number and friction factor correlations, the EEI = $\Delta S/\alpha_{jf}$ was derived and expressed in terms of *j* and *f* factors. After setting the minimum *-JF* and entropy generation as objective functions, a MOGAO regarding structural parameters was implemented with the assistance of EEI to the result screening. Under the same configuration, IS6 and original scheme were numerically analyzed, and the obtained Nusselt number *Nu*, pressure loss Δp, *JF* factors, EGPMFR, and EEI were all compared to ensure the effectiveness of the proposal. The major conclusions are made as follows:

1. EEI = $\Delta S/\alpha_{jf}$ could be derived in terms of *j* and *f* factors for convenience and assist the result screening of the MOGAO or other structural optimizations.
2. Compared with the original model, the *JF* factor, EGPMFR, and EEI of the IS6 respectively improved by 12.20%, 19.42%, and 8.82% at 0.6 m/s, which directly could sustain EEI prediction in performance practicing.

Acknowledgments. This work is supported by National Science Foundation (No.52175063).

References

1. Sun, P.B., Wang, X.W., Kong, J.C., et al.: Multi-objective optimization of Y-type fin structure in rectangular phase change energy storage units for thermal energy storage. Appl. Therm. Eng. **276**, 126899 (2025)
2. YH. Li: Numerical investigation of GaN MMIC PA thermal management system and multi-objective genetic algorithm optimization of heat sink parameters, Appl. Therm. Eng., vol. 272, 126446 doi:https://doi.org/10.1016/j.applthermaleng (2025)
3. Zimparov, V.R., Vulchanov, N.L.: Performance evaluation criteria for enhanced heat transfer surfaces. Int. J. Heat Mass Transfer. **37**, 1807–1816 (1994)
4. Fang, Z.J., Deng, S.F., Du, Y.F., et al.: Multi-objective optimization of internally finned tubes with interruption-and-rotation strategy based on RSM and NSGA-II. Int. J. Therm. Sci. **217**, 110051 (2025)

5. Yu, C., Chen, J., Zeng, M., Gao, B.: Numerical study on turbulent heat transfer performance of a new parallel-flow shell and tube heat exchanger with sinusoidal wavy tapes using RSM analysis. Appl. Therm. Eng. **150**, 875–887 (2019). https://doi.org/10.1016/j.applthermaleng.2019.01.043
6. Niu, Z.Y., Zhu, X.W.: Development of generalized models for accurate prediction and optimization of thermal-hydraulic performance in chevron-type plate heat exchangers. Int. J. Heat Mass Transf. **250**, 127298 (2025). https://doi.org/10.1016/j.ijheatmasstransfer.2025
7. Wang, H.D., Liang, Y.J., Wang, Z.A., Liu, J.X., Meng, L.Q., Liu, W.M.: A quantitative energy efficiency index for heat transfer and energy conservation of plate heat exchanger in commercial electric vehicle. In: The 9th International Conference on Advances in Construction Machinery and Vehicle Engineering. in publication (2025)
8. Liu, J.X., Jiang, Y.K., Wang, B.Z., He, S.M.: Assessment and optimization assistance of entropy generation to air-side comprehensive performance of fin-and-flat tube heat exchanger, (in English). Int. J. Therm. Sci. **138**, 61–74 (2019). https://doi.org/10.1016/j.ijthermalsci.2018
9. Al zahrani, S., Islam, M.S., Xu, F., Saha, S.C.: Thermal performance investigation in a novel corrugated plate heat exchanger. Int. J. Heat Mass Transf. **148**, 119095 (2020). https://doi.org/10.1016/j.ijheatmasstransfer.2019

Analysis and Improvement on Cooling System of IGBT for New Energy Wheel Loader

Jiaxin Liu[1(✉)], Xiang Li[1], Wanjun Zhang[1], Naijiang Liu[1], Jingtao Zhao[1(✉)], Peng Liu[2], and Martin Kreschel[3]

[1] College of Mechanical Engineering, North China University of Science and Technology, Tangshan 063210, China
{liujiaxin,zhaojingtao}@ncst.edu.cn
[2] Xushui No.1 Middle school, Baoding 072550, China
[3] School of Architecture and Engineering, SRH University of Applied Sciences Heidelberg, Heidelberg 69123, Germany

Abstract. To address the cooling issue of IGBT (Insulated Gate Bipolar Transistor) chips in a 5-ton pure electric loader, this study experimentally measured the temperature characteristics of the existing forced convection cooling system and proposed an improved liquid cooling solution with performance testing of the cold plate. A system model was built in AMESim, converting experimental data into boundary conditions for simulation validation. Results showed a simulation error of only 2.47%, confirming model reliability. The improved solution reduced chip temperature from 72.46 °C to 66.79 °C, an 8.49% decrease. The study demonstrates that cold plate performance must be pre-measured to ensure simulation accuracy, providing a reference for optimizing IGBT cooling systems in new energy loaders. To enhance the cooling performance of IGBT chips in a 5-ton electric loader, this study tested the thermal characteristics of the original air-cooling system and proposed an improved liquid-cooling solution. The performance of the cooling plate was experimentally evaluated. Based on measured data, an AMESim model was established, with a simulation error of only 2.47%, validating the model's reliability. The improved system reduced the chip temperature from 72.46 °C to 66.79 °C, providing a foundation for further IGBT cooling optimization.

Keywords: New energy · Wheel loader · IGBT · Cooling system · Analysis · Improvement

1 Introduction

With the widespread adoption of new energy vehicles in the construction machinery sector, electric engineering vehicles typically employ an integrated thermal management system architecture combining ECU, AC, BMS, and EMS [1]. The cooling performance of IGBTs directly impacts the stability and reliability of the power system, necessitating in-depth research and optimization [2].

S. Halgamuge et al. (Eds.): ICACMVE 2025, LNME, pp. 577–582, 2026.
https://doi.org/10.1007/978-981-95-6911-3_57

Recent domestic studies have extensively analyzed and compared IGBT cooling solutions [3]. Zhai et al. [4] established a thermal management model for IGBT power modules, Ma et al. [5] conducted experiments on a three-phase half-bridge IGBT to study overload failure characteristics and proposed methods to enhance operational reliability. He et al. [6] Established a liquid-cooled heat sink simulation model for IGBT modules and compared the performance of three series flow channel structures via CFD.

This study focuses on a domestic 5-ton electric loader, collecting original vehicle data and evaluating the cooling performance of a newly designed cooling plate. AMESim was used to predict the performance of the improved system, with experimental and simulation results compared to confirm its effectiveness.

2 Original IGBT Scheme and Temperature Characteristics

2.1 Research Vehicle

The study targets a domestic 5-ton pure electric wheel loader with a rated load capacity of 5,000 kg, an operating weight of 19,000 kg, a maximum unloading height of 3,600 mm, a battery capacity of 292 kWh, and a rated power of 180 kW.

2.2 IGBT Cooling Scheme

The target vehicle uses Fuji 2MBI600VN-120-50-C chips, with a maximum allowable operating temperature of 85 °C. Fig. 1 shows that the IGBT employs straight-fin heat sinks with uniform c ross-sections, cooled by multiple electronic fans for forced convection.

2.3 Test Conditions and Data Collection

The ambient temperature was 32 °C, with an average speed of 6 km/h, primarily performing "V"-shaped operations. IGBT temperatures were measured using temperature sensors installed inside the enclosure, with a sampling frequency of 1 s. Data were recorded via the Dolphin Expert Logger200, thermocouple temperature sensors, and monitoring software as shown in Fig. 2. After one hour of continuous operation, the system stabilized, and the average temperature over a 10-min period was calculated as 72.46 °C.

Fig. 1. Original cooling scheme

Fig. 2. Vehicle testing and data acquisition

3 Water Cooling System

3.1 Proposed Water Cooling System

The new solution proposes a combined system of "electronic water pump + cooling fan + heat exchanger + water-cooling plate." The water-cooling plate is made of aluminum alloy; the cooling fan and heat exchanger specifications are detailed in the full text.

3.2 Water-Cooling Plate Performance Calibration

Figure3(a) shows the experimental setup: an IGBT-sized ceramic heater operating at 33 V with 360.69 W power attached to the water-cooling plate, with FD-2 coolant circulated at 1,000–5,000 ml/min. Tests ran at 14 °C/32%, 20 °C/30%, 26 °C/30%, and 32 °C/28% ambient conditions, triggering at ≥85 °C to record thermal and pressure data.

3.3 Results and Discussion

Using the MATLAB fitting toolbox, the heat transfer and pressure drop data were fitted as functions of ambient temperature and flow rate, and the surface plots are presented as shown in Figs. 3(b) and (c). When the ambient temperature reaches 35 °C and the flow rate is 20 L/min, the heat transfer capacity approximates 185 W, providing crucial reference data for subsequent IGBT cooling simulations.

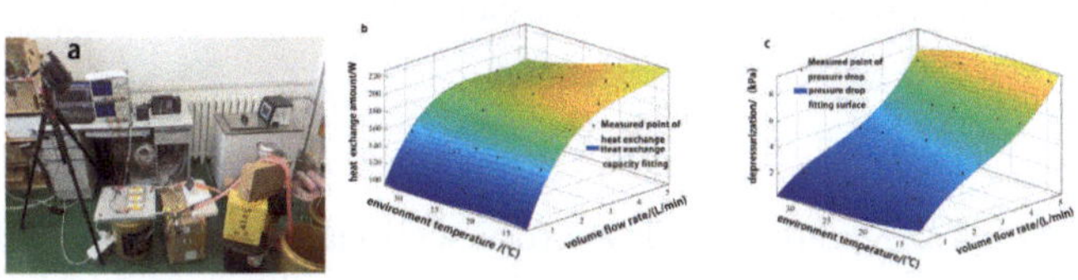

Fig. 3. Experimental design and simulation results.

4 AMESim Simulation of the Improved Scheme

4.1 Electronic Water Pump Model

The AMESim model divided the cooling system into hot and cold sides. The hot side included an equivalent IGBT heating unit, water pump, and radiator. The cold-side coolant circulation achieves heat dissipation through fan-forced convection cooling.

A standard pump model from the component library is adopted as shown in Fig. 4(a). Ports 1 and 2 connected to the piping, while Port 3 controlled startup, shutdown, and speed adjustment. Pump performance curves were provided by the manufacturer, with the flow rate set to 20 L/min.

4.2 Integrated fan and Radiator Model

The cooling fan was modeled as follows:

$$P_s = \left(\frac{\rho \bar{r} v_s}{30k_z R^2(1-\chi_b^2)} - \frac{\rho v_s^2}{2\pi^2 R^4(1-\chi_b^2)}\right)Q^2 \tag{1}$$

where ρ and $\bar{r}$ are the air density and the blade section relative radius, respectively. v_s and R represent the Swirl coefficient behind the fan and the Fan blade radius. χ_b and k_z represent the fan hub ratio and the ratio of rotational speed to volumetric flow rate. Q and P_S represent the air volumetric flow rate through the fan and fan static pressure.

The radiator was modeled using the(ε-Ntu)method [6]

$$\begin{cases} \varepsilon = 1 - e^{\frac{Ntu^{0.22}(1-e^{-C_r Ntu^{0.78}})}{C_r}} \\ Q_r = \varepsilon C_{\min}(T_{thermal,inlet} - T_{could,inlet}) \\ C_r = \dfrac{\min[(C_{p,thermal}\dot{m}_{thermal})_{thermal},(C_{p,cold}\dot{m}_{cold})_{cold}]}{\max[(C_{p,thermal}\dot{m}_{thermal})_{thermal},(C_{p,cold}\dot{m}_{cold})_{cold}]} \end{cases} \tag{2}$$

where ε and Q_r are the Radiator efficiency and the Radiator heat transfer , respectively. Ntu and $T_{thermal,inle}$ represent the Number of transfer units and the Hot fluid inlet temperature. $T_{cold,inlet}$ and $C_{p,\ thermal}$, $C_{p,\ cold}$ represent the Cold fluid inlet temperature and the Specific heat capacities of hot and cold fluids.

References [7] Express the Radiator Pressure Drop as

$$\Delta p = \frac{m_i^2}{2A_i^2\rho_i}(K_{ci} + K_{ei} + \frac{4f_i \cdot Lt}{h_{di}}) \tag{3}$$

where A_i and K_c, K_e are the Flow cross-sectional area and the Inlet and outlet resistance coefficients, respectively. f and Lt represent the Fanning friction factor and the Flow length. h_{di} and represent the Hydraulic diameter.

This study employs the component library's thermal module for system modeling, integrating both cooling fan and radiator. Port 1 serves as the air parameter input, Port 2 connects to the water reservoir, while Ports 3 and 5 function as coolant inlet/outlet respectively. Port 4 controls fan activation/deactivation. The model's computational accuracy is ensured through experimentally validated parameters.

4.3 Boundary Conditions and Model

The final IGBT thermal management system model is shown in Fig. 4(b), with key components include IGBT heating model, cooling plate, and radiator all constructed using aluminum. The coolant selected was a 50% ethylene glycol aqueous solution, with its thermophysical properties properly configured. The simulation environmental parameters were set as: air temperature 35 °C and density 1.12 kg/m^3. The IGBT heating model was set to 82 °C and a power of 360 W, with the cooling plate at 185 W heat exchange. The radiator fan activates between 50–80 °C, remaining off below 50 °C.

Figure 4(c) shows the simulated temperature curves, demonstrating the liquid-cooling system reduced IGBT temperature from 82 °C to 65 °C, meeting the <65 °C operational requirement. The improved solution achieved a temperature of 65.14 °C, representing a 7.32 °C reduction compared to the air-cooling system's 72.46 °C. validating its effectiveness.

Since the coolant flow in the system exhibits high Reynolds numbers, the k-ε turbulence model was applied. The cold plate pressure drop used experimental data from Sect. 3.2 as shown in Fig. 3(c), while the radiator employed Eq. 3 and manufacturer-provided pressure drop curves.

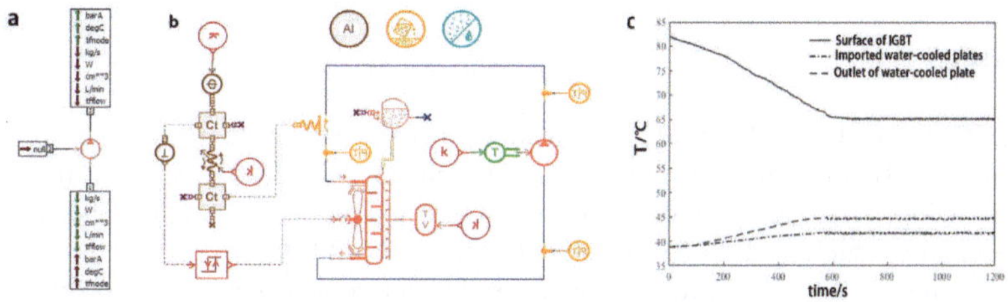

Fig. 4. Simulation process

5 Experimental Validation

Experimental validation of the improved solution was conducted under similar operating conditions. A comparison between simulated and experimentally measured data is presented in Table 1. The simulated values were slightly lower, with an approximate deviation of 2.47%, attributed to measurement and model simplifications. The improved system achieved a chip temperature of 66.79 °C, an 8.49% reduction, confirming its effectiveness.

Table 1. Comparison of results

Original Scheme		Improved Scheme			
Experimental	Percentage	Experimental	Percentage	Simulation	Percentage
72.46 °C	–	66.79 °C	↓8.49%	65.14 °C	↓2.47%

6 Conclusion

This study focuses on the IGBT module of a 5-ton pure electric loader. First, the performance of the original forced air-cooling system was tested and evaluated. Subsequently, experimental calibration was conducted on a new liquid cold plate, and an AMESim simulation model was established and validated based on the measured data.

For IGBT liquid cooling systems in new energy loaders, cold plate performance must be calibrated under varying ambient temperatures before AMESim simulation, as it critically impacts accuracy. With11 proper boundary conditions, the simulation achieved a 2.47% error versus experimental data, demonstrating reliability. The optimized liquid cooling system outperformed the original air-cooling solution, providing valuable insights for future IGBT cooling research.

Declaration of Competing Interest. The authors declare that they have no known competing financial interests or personal relationships that could have appeared to influence the work reported in this paper.

References

1. Su, Z., Zhan, N., Deng, Y., et al.: Application of micro-heat pipe arrays in natural cooling systems for IGBTs in light rail vehicles. J. Jilin Jianzhu Univ. **38**(05), 59–65 (2021)
2. Tong, L., Ding, Y., Li, X.: Heat dissipation structure and thermal simulation analysis of motor controllers for vehicles. Micromotors. **49**(11), 30-33+37 (2021)
3. Shi, J., Zhang, Y., Wu, W.: Simulation analysis of water-cooled heat dissipation characteristics for IGBT modules based on Icepak. Agric. Equipment Veh. Eng. **59**(07), 146–148 (2021)
4. Zhai, X., Li, Y., Li, S., et al.: Thermal management research for IGBT power modules. J. Appl. Technol. **20**(03), 211–219 (2020)
5. Yaqing, M., Jianfeng, L., Fangfang, D., et al.: Power cycling failure of double side cooled IGBT modules for automotive applications. Microelectron. Reliab. **124**(05), 114282 (2021)
6. He, L., Hu, X., Zhang, L., Xing, T., et al.: Performance evaluation and optimization of series Flow Channel water-cooled plate for IGBT modules. Energies. **16**(13), 11–13 (2023)
7. Liu, J., Qin, S., Xu, Z., et al.: Numerical simulation of heat dissipation performance for radiator modules in engineering vehicles. J. Southwest Jiaotong Univ. **47**(04), 623–628 (2012)

3D Modeling and Analysis of Piezoelectric Transducer

Zongyu Bai[1], Jiaxin Liu[1(✉)], Jingtao Zhao[1(✉)], Xiaolei Dong[1], and Martin Kreschel[2]

[1] College of Mechanical Engineering, North China University of Science and Technology, Tangshan 063210, China
{liujiaxin,zhaojingtao}@ncst.edu.cn

[2] School of Architecture and Engineering, SRH University of Applied Sciences Heidelberg, Heidelberg 69123, Germany

Abstract. This study comprehensively investigates the design principles and performance optimization methodologies for piezoelectric transducers within ultrasonic vibration systems. A primary focus is achieving precise modeling and analysis of Langevin-type piezoelectric transducers. The research team developed an advanced three-dimensional vibration model to recognize the limitations of traditional one-dimensional vibration theory, particularly in analyzing complex vibration coupling phenomena within the transducer structure. This novel model employs the apparent elasticity method for a detailed multidimensional analysis of the critical longitudinal-radial coupling vibration characteristics. Mechanical coupling coefficients were systematically incorporated as fundamental parameters throughout the design process. To establish a robust numerical analysis framework, SOLIDWORKS for three-dimensional geometric modeling was seamlessly integrated with COMSOL Multiphysics for multiphysics finite element numerical. This integrated approach facilitated effective vibration modeling, comprehensive numerical analysis of the transducer design. Consequently, the methodology successfully revealed essential transducer characteristics, including nodal plane position 71 mm, resonance frequency 19.393 kHz, head displacement amplitude 36.1 μm and complex vibration mode shapes.

Keywords: Apparent elasticity method · Langevin ultrasonic transducer · Three-dimensional vibration model

1 Introduction

Piezoelectric transducers have gained widespread application in ultrasonic cleaning, grain refinement, and non-destructive testing owing to their superior efficiency, operational stability, and cost-effectiveness [1, 2], while agricultural adaptations feature acoustic breeding enhancement and energy-efficient dehydration [3, 4]. Nevertheless, conventional one-dimensional analytical models exhibit significant limitations in frequency prediction accuracy when the transducer's transverse dimension exceeds quarter-wavelength specifications, primarily due to the neglect of transverse vibration coupling effects [5].

S. Halgamuge et al. (Eds.): ICACMVE 2025, LNME, pp. 583–588, 2026.
https://doi.org/10.1007/978-981-95-6911-3_58

While two-dimensional models improve performance by partially addressing coupling effects, they remain inadequate for fully characterizing three-dimensional energy conversion. Recent three-dimensional vibration analyses using the apparent elasticity method have achieved breakthroughs. Lin [6] achieved a remarkable reduction in frequency prediction error from 9.7% to 0.95% through composite vibration mode decomposition, while Karafi's research team [7] refined electromechanical modeling via stress-based parameterization, which reduced design deviations from 9.7% to 0.95% for large-scale transducers.

Langevin transducer design employs the apparent elasticity method with longitudinal-to-transverse stress ratio coupling parameters. Analytical formulations determine resonant lengths, while integrated SOLIDWORKS and COMSOL numerical precisely identify nodal positions, quantify resonant frequencies, and characterize vibration amplitudes.

2 Theoretical Calculations of Piezoelectric Transducers

Under neglect of shear effects, the axisymmetric cylindrical coupled vibration equation assumes the form

$$\begin{cases} \rho \frac{\partial^2 \delta_r}{\partial t^2} = \frac{\partial \zeta_r}{\partial r} + \frac{\zeta_r - \zeta_\theta}{r} \\ \rho \frac{\partial^2 \delta_z}{\partial t^2} = \frac{\partial \zeta_z}{\partial z} \end{cases} \tag{1}$$

By Poisson's effect, the interdependent longitudinal and transverse stresses define the mechanical coupling coefficient as

$$n = \frac{\zeta_z}{\zeta_r + \zeta_\theta} \tag{2}$$

where, ζ_{r}, ζ_{z}, and ζ_θ are the stresses in radial, axial, and tangential directions, respectively.

2.1 Equivalent Radial Vibration Equation for a Hollow Cylinder

The differential equation for the radial vibration of a hollow cylinder is as follows:

$$\frac{\partial^2 \delta_r}{\partial r^2} + \frac{1}{r}\frac{\partial \delta_r}{\partial r} - \frac{\delta_r}{r^2} + k_r^2 \delta_r = 0 \tag{3}$$

Where, the equivalent radial wave number $k_{\mathrm{r}} = w/\upsilon_r$, the wave speed of radial vibration of the shaker $\upsilon_r = (E_r/\rho)^{1/2}$, respectively. E_r is the equivalent elastic constant for radial vibration and is given by

$$\begin{cases} k_r = \frac{w}{\sqrt{\frac{E(2-\nu n)}{\rho(1+\nu)(1-\nu-2\nu n)}}} \\ E_r = \frac{E(1-\nu n)}{(1+\nu)(1-\nu-2\nu n)} \end{cases} \tag{4}$$

The differential Eq. (3) for the radial vibration of a hollow cylinder is solved as

$$\delta_r(r) = [D_1 J_1(k_r r) + D_2 Y_1(k_r r)] \quad (5)$$

The radial frequency equation for a hollow cylinder is expressed as

$$\frac{k_r a J_0(k_r a) - \frac{(1-\nu-2\nu n)}{(1-\nu n)} J_1(k_r a)}{k_r b J_0(k_r b) - \frac{(1-\nu-2\nu n)}{(1-\nu n)} J_1(k_r b)} = \frac{k_r a Y_0(k_r a) - \frac{(1-\nu-2\nu n)}{(1-\nu n)} Y_1(k_r a)}{k_r b Y_0(k_r b) - \frac{(1-\nu-2\nu n)}{(1-\nu n)} Y_1(k_r b)} \quad (6)$$

2.2 Equivalent Longitudinal Vibration Equations for a Hollow Cylinder

$$\begin{cases} E_z = \frac{E}{1-\frac{\nu}{n}} \\ \mathrm{k_z} = \frac{\mathrm{w}}{\nu_\mathrm{z}} \end{cases} \quad (7)$$

where, k_z is the longitudinal vibration equivalent radial wave number. E_z is the equivalent modulus of elasticity for longitudinal vibration.

2.3 Equivalent Radial Vibration Equation for a Solid Cylinder

$$ak_r J_0(k_r a) - \left[\frac{(1-\nu-2\nu n)}{(1-\nu n)}\right] J_1(k_r a) = 0 \quad (8)$$

2.4 Equivalent Radial Vibration Equations for Piezoelectric Ceramic Rings

Under axisymmetric conditions, the equations of motion can be introduced in the following form

$$\begin{cases} \rho \frac{\partial^2 \delta_r}{\partial t^2} = \frac{\partial T_r}{\partial r} + \frac{1}{r}\frac{\partial T_{r\theta}}{\partial \theta} + \frac{\partial T_{rz}}{\partial z} + \frac{T_r - T_\theta}{r} \\ \rho \frac{\partial^2 \delta_\theta}{\partial t^2} = \frac{\partial T_{r\theta}}{\partial r} + \frac{1}{r}\frac{\partial T_\theta}{\partial \theta} + \frac{\partial T_{\theta z}}{\partial z} + \frac{2T_{r\theta}}{r} \\ \rho \frac{\partial^2 \delta_z}{\partial t^2} = \frac{\partial T_{rz}}{\partial r} + \frac{1}{r}\frac{\partial T_{\theta z}}{\partial \theta} + \frac{\partial T_z}{\partial z} + \frac{T_{rz}}{r} \end{cases} \quad (9)$$

In a piezoelectric body undergoing radial vibration, the boundary condition specifies that the stresses at the inner and outer radii are zero. Based on this condition, the piezoelectric frequency equation for radial vibration can be derived

$$\frac{k_{zu} a J_0(k_{zu} a) - \left[\frac{(s_{11}^E + s_{12}^E + 2ns_{13}^E)}{(s_{11}^E + ns_{13}^E)}\right] J_1(k_{zu} a) k_{zu} b J_0(k_{zu} b) - \left[\frac{(s_{11}^E + s_{12}^E + 2ns_{13}^E)}{(s_{11}^E + ns_{13}^E)}\right] J_1(k_{zu} b) =}{k_{zu} a Y_0(k_{zu} a) - \left[\frac{(s_{11}^E + s_{12}^E + 2ns_{13}^E)}{(s_{11}^E + ns_{13}^E)}\right] Y_1(k_{zu} a) k_{zu} b Y_0(k_{zu} b) - \left[\frac{(s_{11}^E + s_{12}^E + 2ns_{13}^E)}{(s_{11}^E + ns_{13}^E)}\right] Y_1(k_{zu} b)} \quad (10)$$

In the equation, $k_{zu} = w/\upsilon_{zu}$ and $\upsilon_{zu} = (E_{zu}/\rho)^{1/2}$ denote the equivalent radial wave number and wave speed of the radial vibration of the piezoelectric ceramic ring, respectively.

3 Electromechanical Modeling of Piezoelectric Transducers

The analysis assumed purely longitudinal vibration modes in the transducer, perfectly frictionless contact surfaces, and negligible stress from air loading. Table 1 contains the basic data of the transducer. Key operational parameters are summarized in Table 2.

Table 1. Basic parameters of piezoelectric transducer.

Parameters	Values
Resonant frequency	20 kHz
piezoelectric material	PZT-4
backing	St-304 L
Material	Titanium
Power	2 kW

The dimensional structure of the piezoelectric transducer would be derived from the equation.

Table 2. Piezoelectric transducer size distribution.

	Materials	Inner radius (mm)	Outer radius (mm)	Lengths (mm)
1	Titanium	Solid	20	25
2	Titanium	7	22.5	5
3	Titanium	7	25	9
4	Titanium	7	27	4
5	Titanium	7	25	9
6	PZT-4	8.5	25	28
7	St-304 L	10	25	20
8	St-304 L	12	25	10
9	AISI1020	Solid	7	75

4 Finite Element Analysis of Piezoelectric Transducers

Building upon prior theoretical dimensioning foundations, this study conducts piezoelectric transducer analysis through a SOLIDWORKS developed parametric model. Finite element-based modal analysis was implemented via COMSOL Multiphysics computational platform, with auxiliary components including preload bolts, copper electrodes and viscoelastic interface layers excluded to enhance computational efficiency.

Numerical demonstrated a third longitudinal resonance frequency of 19.393 kHz with 3.04% deviation, accompanied by a vibrational amplitude of 36.1 μm and a nodal plane at 74 mm axial position, as documented in Figs. 1 and 2. These computational results confirm the predictive accuracy of the established transducer mode.

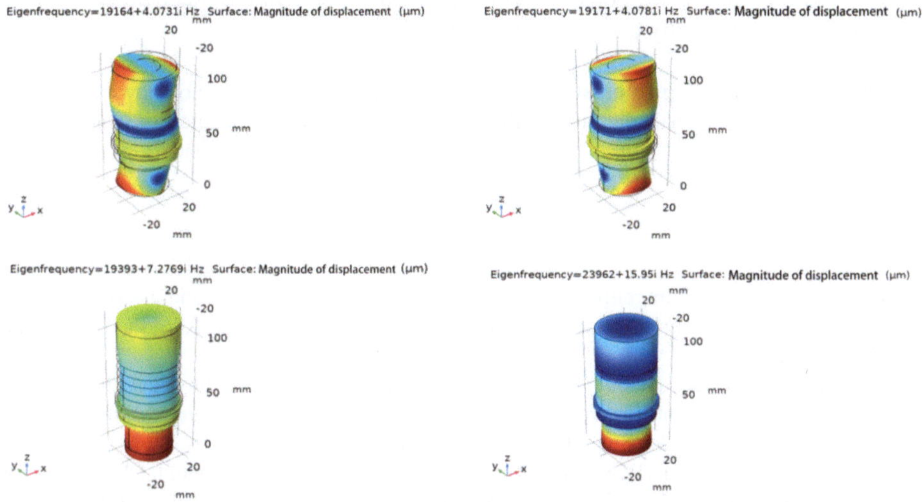

Fig. 1. Piezoelectric transducer modes

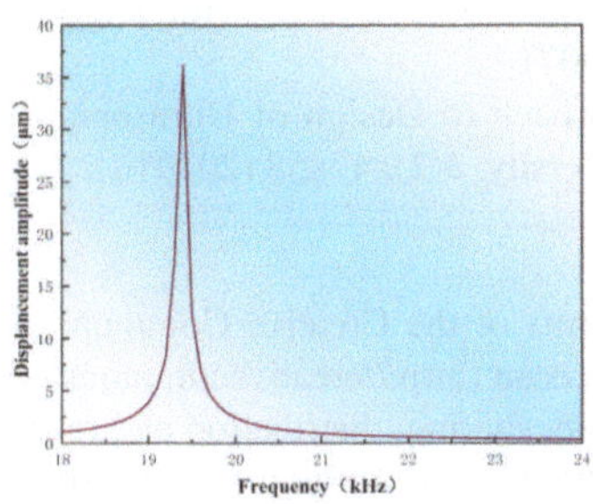

(a) Piezo transducer head amplitude

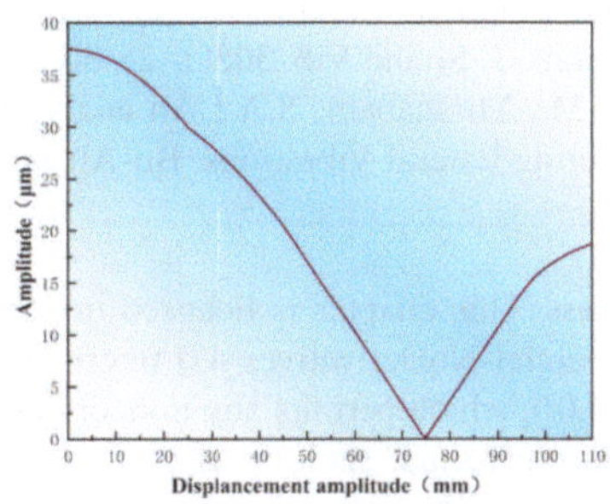

(b) Piezo transducer node position

Fig. 2. Piezoelectric transducer numerical data plot

5 Conclusion

A three-dimensional dynamic model of the Langevin piezoelectric transducer was developed by integrating apparent elasticity theory with finite element numerical. The hybrid methodology demonstrated a 3.04% frequency deviation, 36.1 μm operational amplitude, and a 71.6 mm axial nodal position. This approach established a practical framework for optimizing structurally complex piezoelectric transducers.

Declaration of Competing Interest. The authors declare that they have no known competing financial interests or personal relationships that could have appeared to influence the work reported in this paper.

References

1. Vorontsov, A., et al.: Grain refinement and element partitioning in ultrasound-assisted wire and electron beam additively manufactured Udimet 500 nickel superalloy. J. Alloys Compd. **1020**, 179474 (2025)
2. Lima, H.M., Fernandes Neto, J.A.D., Haach, V.G., Munaiar Neto, J.: Physical and mechanical properties of cement-lime mortar for masonry at elevated temperatures: destructive and ultrasonic testing. J. Build. Eng. **104**, 112291 (2025)
3. Pszczółkowski, P., Barbaś, P., Sawicka, B.: Toward enhanced seed potato yield: Ultrasonication techniques for sustainable agricultural development. Sustainability. **17**(3), 1225 (2025)
4. Namjoo, M., Moradi, M., Niakousari, M.: Evaluation of the effect of high-power ultrasound waves on conventional air drying of cumin seeds. Sustain Energy Technol Assess. **52**, 102262 (2022)
5. Arnold, F.J.: Análise unidimensional de transdutores piezelétricos baseada no teorema de Thévenin. Revista Brasileira De Ensino De Física. **31**(4), 4305–4314 (2009)
6. Lin, S.: Coupled vibration of isotropic metal hollow cylinders with large geometrical dimensions. J. Sound Vib. **305**(1–2), 308–316 (2007)
7. Karafi, M., Mirshabani, S.A.: An analytical approach to Design of Ultrasonic Transducers Considering Lateral Vibrations. Bu-Ali Sina University. **3**(2), 47–58 (2019)

The Influence of Asynchronous Rolling on Magnesium Alloy AZ31 Sheet

Baishun Liu[1], Jiaxin Liu[1(✉)], Li Li[1(✉)], Baozhong Wang[1], and Martin Kreschel[2]

[1] College of Mechanical Engineering, North China University of Science and Technology, Tangshan 063210, China
{liujiaxin,jxlili}@ncst.edu.cn

[2] School of Architecture and Engineering, SRH University of Applied Sciences Heidelberg, 69123 Heidelberg, Germany

Abstract. Finite element analysis was systematically conducted to investigate the plastic deformation behavior of 6mm-thick AZ31 magnesium alloy sheets processed by asynchronous rolling. This advanced rolling technology has significant advantages over traditional rolling methods and the co-directional dynamic rolling method. Particularly, it performs exceptionally well in reducing the rolling force. In this study, the influence of various depositional parameters on three key performance indicators of rolling mill, equivalent stress and strain distribution was comprehensively reviewed. Thanks to parametric optimization, the research has identified that the increase in the allometric ratio and the amount of dislocation leads to a corresponding decrease in the rolling force. The optimal rolling parameters were determined to be an allometric ratio of 1.3 combined with a dislocation amount of 28 mm. Under these conditions, the rolling process achieves superior performance characteristics, exhibiting ideal rolling force levels, favorable stress distribution patterns, and optimal strain uniformity, all of which contribute to enhanced rolling efficiency and product quality.

Keywords: Asynchronous rolling · AZ31 magnesium alloy · Rolling force · Stress-strain

1 Introduction

Magnesium alloys are commonly produced via casting [1], semi-solid molding, and extrusion, but these methods often introduce microstructural defects, weakening mechanical properties. In contrast, plastic deformation enhances strength and ductility, making them suitable for lightweight automotive and household applications [2]. However, their rolling process is temperature-limited [3], typically requiring 200–420 °C.

S. Halgamuge et al. (Eds.): ICACMVE 2025, LNME, pp. 589–595, 2026.
https://doi.org/10.1007/978-981-95-6911-3_59

2 Plate Parameters, Geometric Model Establishment and Method

The material performance is shown in Table 1. Figure 1 illustrates the concept: the left panel shows standard asynchronous rolling, while the right panel highlights the two-dimensional schematic diagram of asynchronous rolling for this article.

Table 1. Material parameters of AZ31 magnesium alloy

Temperature	Young's modulus	Specific heat	Thermal conductivity	CTE	Poisson ratio	Density
20	45.0	1.0150	78.6	2.61 × 10–5	0.35	1.77 × 10–9
93.9	43.7	1.0628	86.6			
149	41.3	1.1011	92.4			
205	35.2	1.1567	98.1			
260	29.4	1.198	102.3			
315.6	23.3	1.236	105.4			

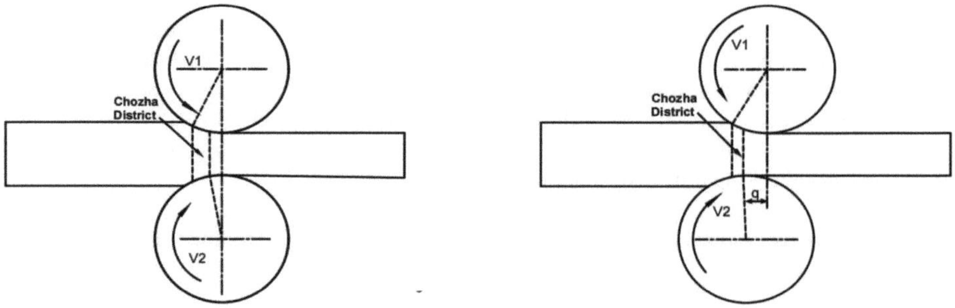

Fig. 1. Comparison of asynchronous rolling and asynchronous rolling in this paper

Figure 2, the roll radius r of the asynchronous rolling is 150 mm. The fixed speed of the upper working roller is 0.5 rad/s. The speed of the lower working roller is set differently according to the different speed ratio, and the single-pass finite element analysis is carried out.

Due to the numerous factors affecting the rolling of the plate, finite element analysis of AZ31 magnesium alloy plate is conducted with a friction coefficient of 0.3, speed ratios of 0.9, 1.1, 1.2, 1.3, 1.4 and dislocations of 20, 22, 25, 28, 30 [4].

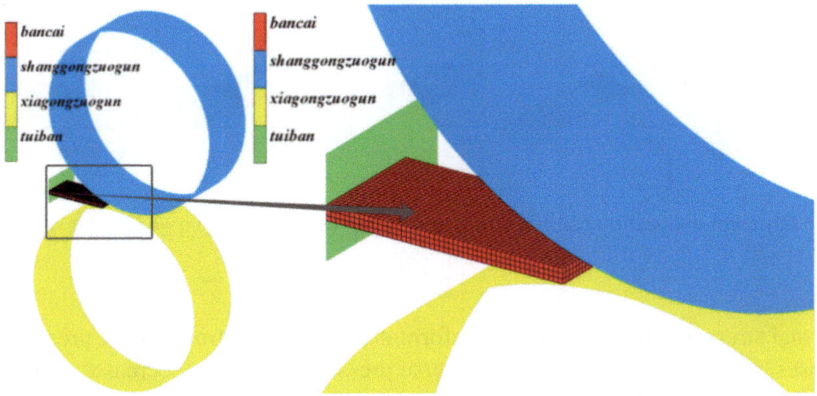

Fig. 2. Three-dimensional geometric model of rolling

3 Results and Analysis

3.1 Effect of Misalignment on Asynchronous Rolling

Figure 3 shows that the rolling force is lower than conventional asynchronous rolling (without dislocation) and initially decreases with increasing dislocation. This occurs because the larger deformation zone reduces unit pressure distribution along the contact surface, thereby decreasing total rolling force. However, as dislocation further increases, the rolling duration extends, requiring higher unit pressure and ultimately increasing total rolling force. Additionally, the force distribution becomes more uneven, with asymmetric rolling forces between the upper and lower work rolls.

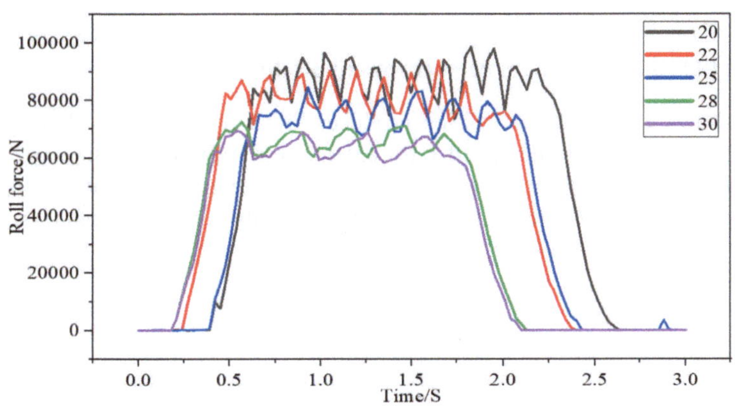

Fig. 3. Rolling force curves of different dislocations

Figure 4 shows that at q = 30 mm, the upper surface exhibits low stress. For q = 20, 22, 25 mm, stress distribution is uneven. At q = 28 mm, stress concentrates in a small but uniform zone, showing distinct properties. This occurs because shear stress (in addition

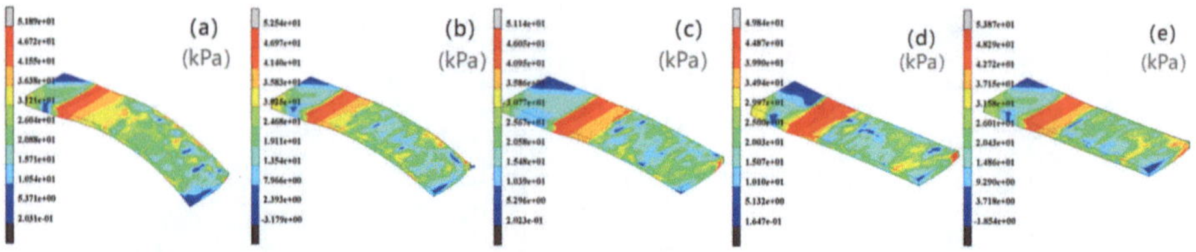

Fig. 4. Different dislocation equivalent stress diagram (a) q = 20, (b) q = 22 (c) q = 25 (d) q = 28 (e) q = 30

to normal stress and friction) reduce deformation resistance from lateral pressure. As q increases, shear stress rises while transverse pressure resistance decreases, making q = 28 mm optimal for rolling.

Figure 5 shows varying equivalent plastic strain distributions under different dislocation conditions. At other dislocation values, strain distribution is notably non-uniform with limited magnitude. The offset rolling forces generate torque, which increases with dislocation distance, inducing strong shear strain in the plate. However, this results in only marginal reduction of shear deformation overall.

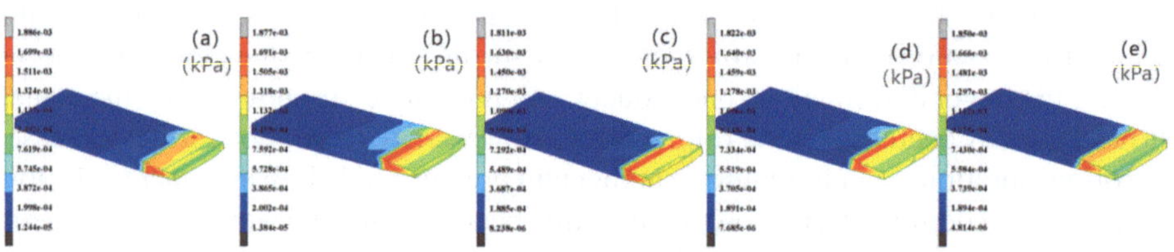

Fig. 5. Different dislocation equivalent strain cloud diagram (a) q = 20 (b) q = 22 (c) q = 25 (d) q = 28 (e) q = 30

3.2 Effect of Speed Ratio on Asynchronous Rolling

Figure 6 shows, asynchronous rolling produces lower rolling forces than differential-speed rolling. Increasing the speed ratio extends force duration while reducing both unit pressure and total rolling force. This occurs because asynchronous rolling creates a larger 'rubbing zone', distributing pressure more evenly across the contact surface.

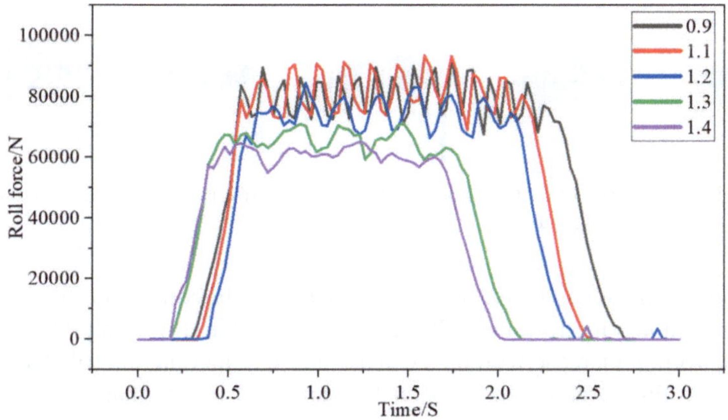

Fig. 6. Rolling force curves of different speed ratios

Figure 7 shows distinct stress distributions for different allometric ratios. At n = 1.3, stress concentrates uniformly in a small deformation zone, indicating minimal equivalent stress disturbance. Smaller friction zones from varying neutral point distributions reduce shear moments, leading to less concentrated stress patterns.

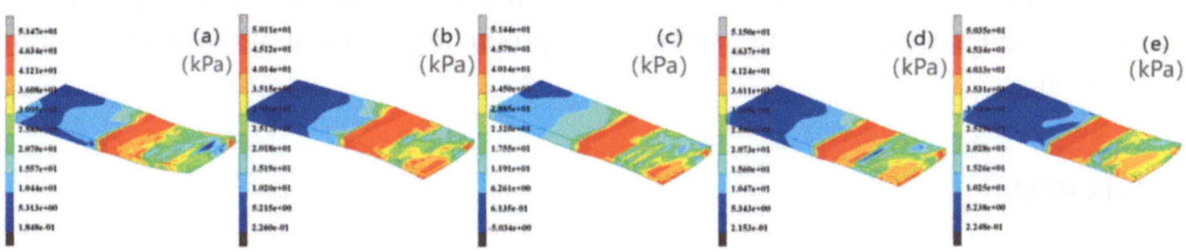

Fig. 7. Equivalent stress cloud diagram of different allometric ratios (a) n = 0.9 (b) n = 1.1 (c) n = 1.2 (d) n = 1.3 (e) n = 1.4

Figure 8 the uniformity of stress distribution at different allometric ratios is different, and the metal flow on the side with faster rotation speed is faster, and the equivalent effect becomes larger, while the equivalent effect on the slower side is relatively smaller. However, the area of the friction zone on the surface of the deformation zone is getting larger and larger. In the friction zone, the shear forces on the sheet are high. As a result, the larger the friction zone, the shear strain of the plate is greater. However, when other parameters are constant, the curvature of the plate increases with the increase of the allovelocity ratio, and the larger curvature will affect the subsequent rolling process. Therefore, it is necessary to choose an appropriate allometric ratio and obtain a large shear strain. When the allovelocity ratio is 1.3, the equivalent effect in the deformation zone is the largest, the deformation ability is better, and the larger deformation can be obtained, and the maximum equivalent effect occurs on the surface of the plate with a larger velocity, the allokinetic ratio of 1.3 is selected.

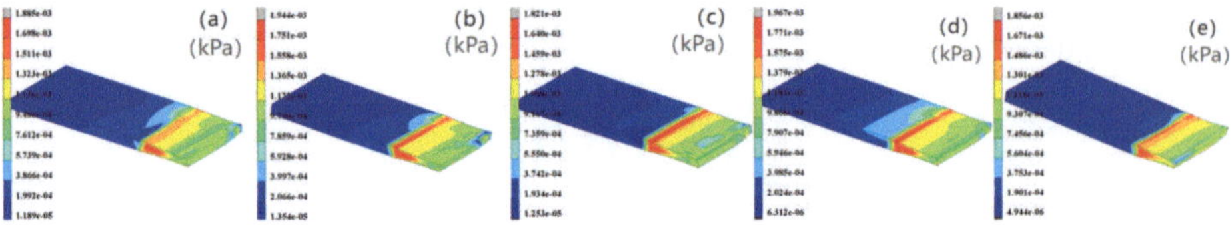

Fig. 8. Equivalent strain cloud diagram of different allometric ratios (a) n = 0.9 (b) n = 1.1 (c) n = 1.2 (d) n = 1.3 (e) n = 1.4

The above analysis shows that at n = 1.3 and q = 28, the equivalent stress distribution is more uniform and the forming is better. The bending and warping degree [5] of the plate during rolling increase with the decrease of the reduction rate [6]. Therefore, it is considered that this parameter is a reasonable parameter for the shaped rolling of AZ31 magnesium alloy sheet.

4 Conclusion

The proposed asynchronous rolling method reduces rolling force by 30% compared to conventional methods, effectively suppressing plate bending and yielding flatter AZ31 magnesium alloy plates. Rolling quality improves with higher dislocation (20–30 mm) and speed ratio (0.9–1.4), though other factors must be considered. Optimal parameters are: 41.6% reduction, q = 28 mm, and n = 1.3, producing the most uniform stress-strain distribution.

References

1. Song, J., Pan, F.: Some thoughts on accelerating the development of magnesium industry in China. Mod. Transport. Metall. Mater. **2**(6), 1–5 (2022)
2. Niu, W., Wang, D.: Recrystallization and anisotropy of AZ31 magnesium alloy by asynchronous rolling. Metals **9**(9), 2031–2038 (2023)
3. Pan, J.: Development and application of plastic processing technology for magnesium alloys. J, Metals **57**(11), 1362–1379 (2021)
4. Li, Q., Deng, B., Wu, Y., et al.: Effect of rolling temperature on microstructure and wear resistance of AZ31 magnesium alloy. Metal Heat Treat. **48**(6), 58–62 (2023)
5. Wu, L., He, B., Qin, M.: Summarization of research on bending control and its influence on properties of dragon-shaped rolled plates. Hot Process. Technol. **53**(9), 8–12 (2023)
6. Jiang, L., Xue, Z.: Modeling and analysis of deformation characteristics for the two-layered metal clad plate produced by asymmetric rolling. J. Mater. Res. Technol. **1**(27), 2031–2051 (2023)

Effect of the Number of Injection Holes on the In-Cylinder Mixing and Performance of Low-Load Methanol Direct Injection Free Piston Engines

Xiaoxu Hu, Huihua Feng(✉), Zihan Wang, and Boru Jia

School of Mechanical Engineering, Beijing Institute of Technology, Beijing, China
{3220230471,fenghh,boru.jia}@bit.edu.cn

Abstract. In this study, the effects of the number of injection holes on the in-cylinder mixing characteristics and combustion performance of a low-load methanol direct-injection free-piston engine generator (FPEG) were systematically investigated through a combination of numerical simulation and experiment. The results show that the increase in the number of injection holes significantly optimizes the in-cylinder equivalence ratio distribution, and the percentage of the extreme equivalence ratio region is reduced dramatically; at the same time, the increase in the number of holes shortens the spraying, the rise in the number of nozzles shortens the spraying time, prolongs the premixing time of oil and gas, and improves the mixing uniformity. In terms of combustion performance, the three-hole nozzle exhibits the best overall performance, with a 19% increase in peak cylinder pressure compared to the dual-hole nozzle, an indicated thermal efficiency above 40%, and the shortest combustion duration. In contrast, the six-hole nozzle produces an excessively dispersed spray, resulting in a dilute mixture in the vicinity of the spark plug, prolonged ignition delay, and lag in the cylinder pressure phase.

Keywords: injection parameters · methanol direct injection · free piston engine generator · mixing equivalence ratio · combustion performance

1 Introduction

The global energy crisis and escalating environmental pollution have heightened the focus on developing efficient, low-emission internal combustion engines for power systems [1, 2]. As an emerging power unit, the FPEG eliminates crankshaft and connecting rod mechanisms, offering structural simplicity alongside critical advantages, including adjustable compression ratio and multi-fuel adaptability [3–5]. Methanol presents a viable alternative to fossil fuels due to its high octane rating, knock resistance, and low emissions [6, 7]. Furthermore, its rapid combustion rate accommodates FPEG's slow-compression, fast-expansion cycle, particularly the piston's brief residence time at the top dead center (TDC). However, the high latent heat of vaporization and low

S. Halgamuge et al. (Eds.): ICACMVE 2025, LNME, pp. 596–606, 2026.
https://doi.org/10.1007/978-981-95-6911-3_60

fuel viscosity of methanol pose a challenge to enhancing its mixing with air for efficient combustion, which is crucial for optimizing the performance of methanol direct injection engines.

The current regulation of in-cylinder mixture distribution primarily depends on fuel injection strategies in direct injection systems [8–10]. Nevertheless, existing research on such strategies mainly investigates parameters such as multiple injections, injection timing, and injector layout while largely overlooking the direct influence of injector structural parameters, notably orifice count, on in-cylinder fuel distribution. Therefore, this paper establishes an FPEG combustion simulation model based on the Converge software platform to simulate the in-cylinder combustion process and its influencing factors numerically. Experimental data verify the reliability of the model. The fundamental influence mechanism of the number of injection holes on the in-cylinder mixture formation process and combustion performance is systematically isolated and quantified under the conditions of fixed orifice diameter, injection volume, and injection timing, which provides a key theoretical basis for the optimization of the injector structure for the application of methanol fuel in FPEG. This study aims to explore the potential of FPEG in enhancing thermal efficiency and facilitating rapid in-cylinder combustion, thereby providing new research ideas for further improving the performance of the prototype.

2 Model Construction and Validation

2.1 Experimental Setup

Figure 1 illustrates the two-cylinder spark-ignition FPEG experimental rig developed in this study. The system features a centrally mounted linear motor for actuation and starting, with two free-piston engines symmetrically positioned on either side of the motor. The linear motor's mover and connecting rods functionally replace conventional crankshaft linkages. Each engine operates on a two-stroke cycle with gasoline direct injection (GDI). The main technical parameters of the prototype are detailed in Table 1.

Injector configurations with 1, 2, 3, and 6 orifices were tested. Figure 2 correlates the injected fuel mass with the injection duration for each configuration.

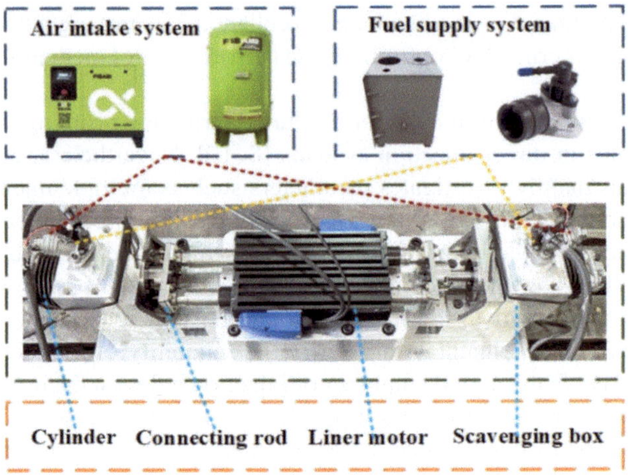

Fig. 1. Twin-cylinder spark ignition FPEG experimental bench diagram.

Table 1. The main parameters of the FPEG prototype for the experiment

Parameters	Unit	Value
Cylinder bore	mm	52.5
Design stroke	mm	65
Engine displacement	cm^3	135
Compress ratio	–	12
Fuel type	–	Methanol
Ignition mode	–	Spark Ignition
Fuel supply system	–	GDI

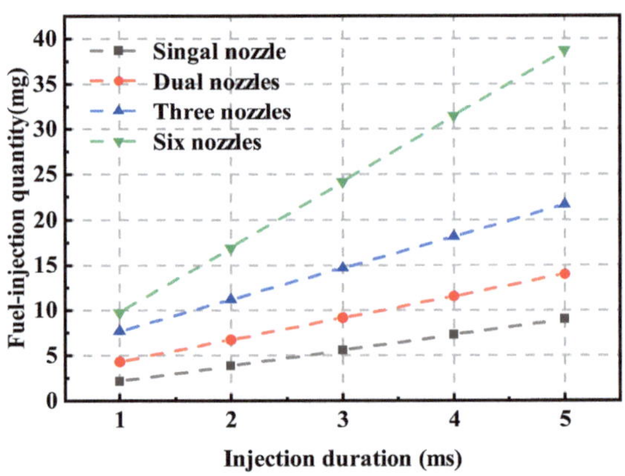

Fig. 2. Effect of injection duration on fuel injection quality.

2.2 Numerical Model Construction

Boundary conditions and key computational parameters are documented in Table 2. The 3D solid model was constructed using SolidWorks software and imported into CONVERGE software. A dynamic computational grid representing the full engine cycle was generated, incorporating critical components, including intake/exhaust tracts, piston, and cylinder walls (Fig. 3). Adaptive mesh refinement was implemented in regions vital to combustion physics: the combustion chamber, fuel spray zones, and spark plug vicinity to resolve flame kernel development.

Table 2. Boundary conditions for a free-piston engine

Boundary conditions	Unit	Value
Inlet mass flow	Kg/s	0.00645
Inlet temperature	K	323
Outlet pressure	bar	1
Outlet temperature	K	600
Cylinder wall temperature	K	500
Piston head temperature	K	500

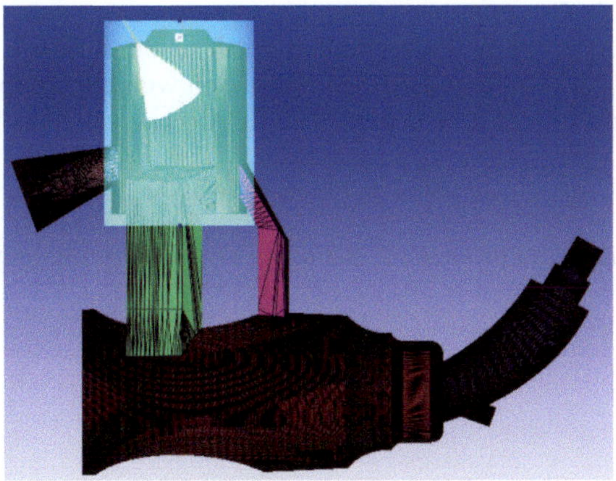

Fig. 3. Computational model and meshing.

2.3 Model Validation

A comparison of the experimental and simulated in-cylinder pressure curves is shown in Fig. 4. As can be seen from the figure, the peak cylinder pressure, the phase corresponding to the peak pressure, and the overall trend of the in-cylinder pressure measured by the simulation and the experiment all show a high degree of consistency. Only in the expansion stroke, due to the difference between the actual sealing conditions and the simulation assumptions, the local deviation is acceptable, and the maximum relative error is controlled within 5%. The comparison results above demonstrate that the simulation model established in this study can accurately characterize the in-cylinder mixture formation and combustion process, and its accuracy meets the requirements of subsequent simulation studies.

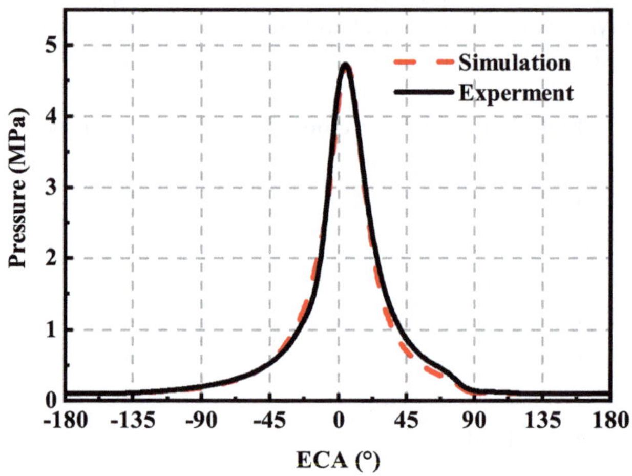

Fig. 4. Model validation of simulated and experimental cylinder pressure.

The ECA in the figure is the equivalent crank angle, which is the recorded time signal converted to ECA to facilitate the comparison with the combustion process in a conventional internal combustion engine. The ECA is calculated as follows:

$$ECA = t - t_0 * f * 360 \tag{1}$$

where t_0 is the time at the moment of the upper stopping point, and f is the operating frequency.

3 Results and Discussion

3.1 Effect of the Number of Spray Holes on the Distribution of Equivalence Ratios

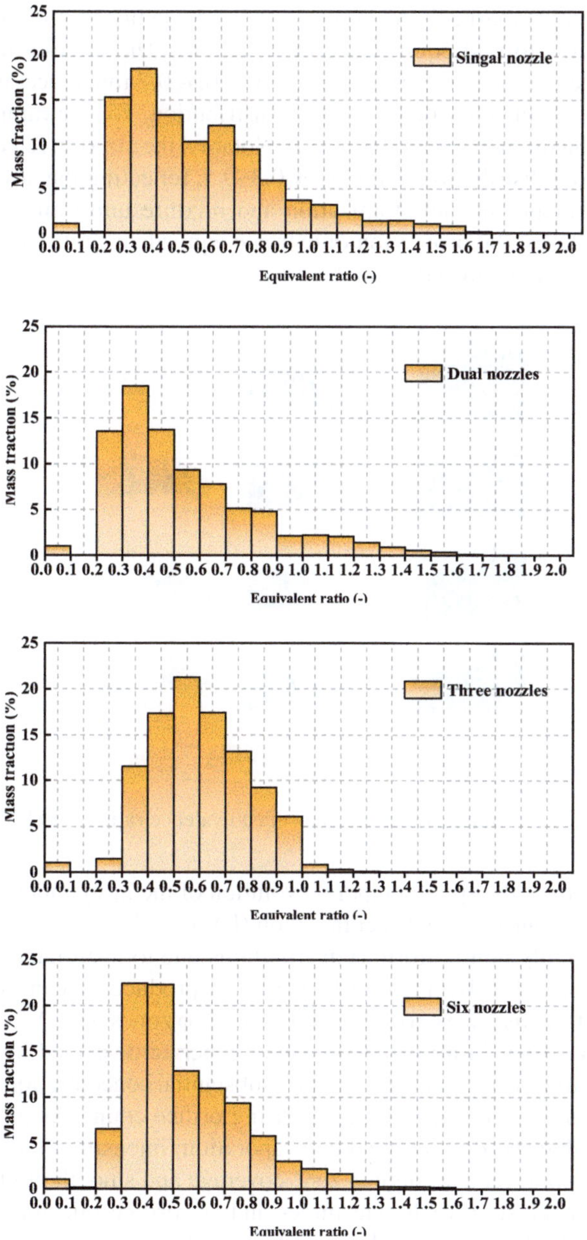

Fig. 5. Percentage of the distribution of equivalence ratios at the moment of ignition

Figure 5 compares equivalence ratio (φ) distributions at ignition timing across injectors with different orifices under matched injection parameters. For the single-orifice nozzle, the equivalence ratio distribution displays significant inhomogeneity: the peak occurs in the $\varphi = 0.3$–0.4 range, but the overall distribution is broad, with the high-probability region (>10%) spanning $\varphi = 0.2$–0.7. It indicates uneven fuel mixing and substantial variations in the local equivalence ratio. With the two-orifice nozzle, the distribution converges markedly toward $\varphi = 0.3$–0.5, demonstrating improved mixing homogeneity in low-to-mid equivalence ratio regions. The three-orifice nozzle shifts the distribution peak to $\varphi = 0.5$–0.6, and the $\varphi = 0.4$–0.7 range becomes dominant. It signifies further optimization of fuel spatial distribution, concentrating mixtures near the theoretical stoichiometric ratio ($\varphi = 1.0$). Notably, the six-orifice nozzle produces a highly concentrated distribution within $\varphi = 0.3$–0.5, reflecting the advantage of multi-hole designs in enhancing fuel atomization and mixture uniformity. Additionally, as orifice count increases, the proportion of extreme equivalence ratio regions ($\varphi < 0.3$ or $\varphi > 1.0$) shows a clear decreasing trend.

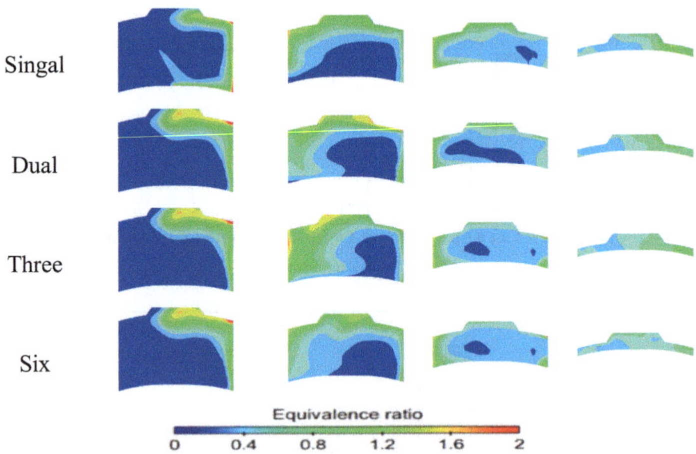

Fig. 6. Comparison of changes in equivalence ratios with EVA

Figure 6 illustrates the spatiotemporal evolution of the in-cylinder equivalence ratio preceding ignition. During initial fuel injection (EVA = -110° CA), all injector configurations exhibit highly non-uniform φ distributions due to insufficient droplet breakup and evaporation. Specifically, single-orifice injectors produce concentrated fuel jets with extended penetration distances, forming narrow high-φ zones ($\varphi > 1.2$) accompanied by significant wall wetting. Multi-orifice injectors generate dispersed high-concentration regions through multiple fuel jets yet maintain substantial concentration gradients. These gradients progressively attenuate with increasing orifice count. As piston motion continues upward, enhanced droplet breakup/evaporation, intensified fuel vapor diffusion, and turbulent transport reduces high-φ zone coverage and smooth gradient transitions.

Comparative analysis reveals that three- and six-orifice injectors exhibit superior resistance to in-cylinder flow perturbations than single- and two-orifice configurations, demonstrating enhanced flow adaptability. This robustness stems from the increased orifice count, which expands the fuel-air interfacial contact area, thereby accelerating droplet fragmentation, enhancing atomization quality, and intensifying fuel-air mixing, ultimately improving mixture homogeneity. Furthermore, maintaining constant injection volume and orifice diameter while increasing orifice count necessarily reduces injection pulse width. It advances fuel delivery timing, extending the mixing lead time before ignition. Therefore, the additional residence time enhances fuel diffusion and air mixture, which constitutes a crucial factor in improving in-cylinder mixture homogeneity.

3.2 Effect of the Number of Holes on Engine Performance

Figure 7 illustrates the influence of injector orifice count (1–6 holes) on in-cylinder pressure and cumulative heat release. As orifice count increases from one to six, peak cylinder pressure and cumulative heat release initially rise and then decline, while combustion efficiency consistently changes. The 3-orifice configuration achieves maximum peak pressure (5.6 MPa at 6.5°CA). However, transitioning to six orifices reduces peak pressure by 8.3% with a 1.5° CA retardation. It is noteworthy that, despite the decrease in peak pressure for the 6-hole nozzle, the cumulative heat release does not decrease significantly, indicating that the combustion process remains relatively complete. This behavior originates from competing mechanisms: increasing orifices reduce the Sauter Mean Diameter (SMD) and mixing duration, thereby enhancing fuel-air homogeneity and accelerating combustion, which in turn increases pressure. Conversely, six orifices diminish per-orifice momentum, compromising spray penetration and hindering mixture diffusion to the combustion chamber peripheries. This spatial limitation weakens flame kernel propagation, ultimately reducing pressure rise rate and delaying phasing, while superior homogeneity preserves heat release.

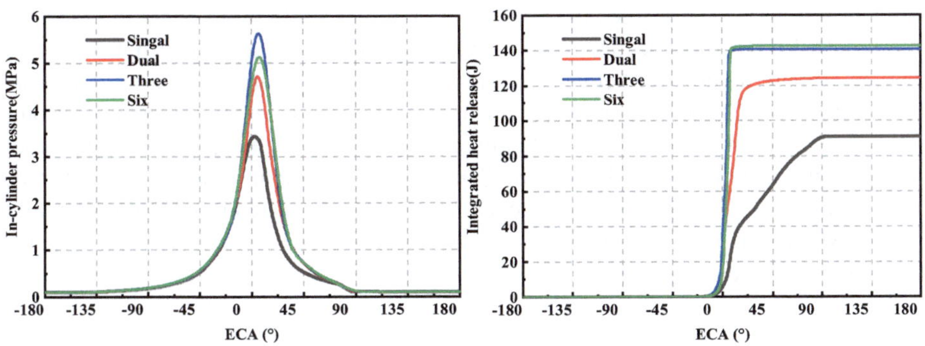

(a) In-cylinder pressure (b) Cumulative heat release

Fig. 7. Changes in in-cylinder pressure and cumulative heat release.

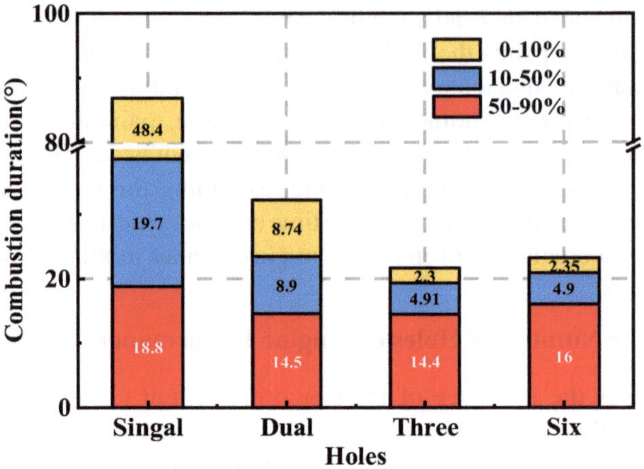

Fig. 8. Phase diagram of combustion duration.

Figure 8 depicts the influence of orifice count on combustion phasing, expressed through crank angles corresponding to 10%, 50%, and 90% mass fraction burned (MFB). Combustion duration is defined as the crank angle interval between MFB10% and MFB90%, where MFB represents the mass ratio of burned fuel to total fuel. These benchmarks characterize distinct combustion phases: MFB10% marks ignition delay completion, MFB50% indicates flame development progression, and MFB90% signifies primary combustion conclusion. Analysis reveals a progressive reduction in total combustion duration (MFB10–90%) with increasing orifice count ($1 \rightarrow 6$). This acceleration is particularly pronounced during the MFB50–90% phase, the diffusion-controlled combustion period - demonstrating that increased orifice count preferentially enhances diffusion combustion kinetics. This combustion acceleration trend aligns with cumulative heat release rate profiles in Fig. 7(b). Notably, the minimum overall combustion duration (MFB10–90%) occurs under 3-orifice operation. Comparative analysis reveals that the primary divergence between 3- and 6-orifice configurations emerges during initial combustion. The underlying mechanism involves excessive fuel dispersion through six orifices, which reduces the local equivalence ratio near the spark plug gap below the optimal ignition threshold($\varphi < 0.8$). This lean mixture compromises ignition reliability and hinders early flame kernel development, thereby extending the ignition delay period (MFB0–10%) and ultimately leading to a prolonged total combustion duration for 6-orifice injectors.

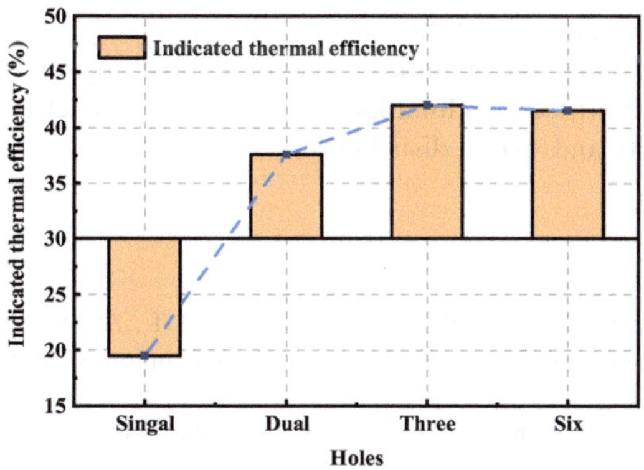

Fig. 9. The variation of indicating thermal efficiency.

Consistent with the preceding analysis, single- and two-hole injectors yield substantially lower work output than three- and six-hole configurations, as evidenced by indicated thermal efficiency in Fig. 9. Using a baseline efficiency of 30%, the single-hole configuration achieves less than 20% thermal efficiency due to incomplete combustion, which results from poor mixture homogeneity. Thermal efficiency improves progressively as the number of holes increases, peaking at three holes before experiencing a slight decrease at six holes. This progression in efficiency corresponds directly with the trends in combustion duration shown in Fig. 8, where shorter combustion intervals concentrate heat release closer to the TDC. This variation enhances the work capacity of the piston, which explains why the three-hole configuration, with its minimal combustion duration, maximizes thermal efficiency, even though the six-hole setup offers better mixture homogeneity.

4 Conclusion

This study examines the impact of injector orifice count on mixture formation and combustion characteristics in methanol direct-injection engines, utilizing an integrated numerical and experimental approach. The main findings are summarized as follows:

1. Increasing orifice count from 1 to 6 narrows the equivalence ratio (φ) distribution at ignition timing from $\varphi = 0.2$–0.8 to $\varphi = 0.3$–0.5. Concurrently, extreme mixture regions ($\varphi < 0.3$ or $\varphi > 1.0$) demonstrate progressive reduction (>40% decrease at six orifices vs. single-orifice), significantly mitigating combustion instability risks.
2. Three- and six-orifice configurations exhibit superior resistance to in-cylinder flow perturbations than single- and two-orifice configurations, demonstrating enhanced flow adaptability. The inherent reduction in injection pulse width extends the pre-ignition mixing duration, enhancing fuel-air homogeneity before spark discharge.
3. The three-orifice configuration achieves the maximum peak pressure and cumulative heat release. MFB10–90% duration decreases with hole count. Indicated thermal efficiency peaks at three orifices vs. < 20% (single-orifice), correlating with TDC-proximal heat release concentration.

4. While six-orifice injectors maximize mixture homogeneity, their reduced spray penetration limits combustion chamber utilization, resulting in lower peak pressure compared to three-orifice benchmarks. It demonstrates the critical balance between atomization quality and spatial distribution control.

References

1. Raheem, A.T., Aziz, A.R.A., Zulkifli, S.A., Rahem, A.T., Ayandotun, W.B.: A review of free piston engine control literature—Taxonomy and techniques. Alexandria Eng. J. **61**(10), 7877–7916 (2022)
2. Reitz, R.D., et al.: IJER editorial: the future of the internal combustion engine. Int. J. Engine Res. **21**(1), 3–10 (2020)
3. Zhang, Z., et al.: Identification and analysis on the variation sources of a dual-cylinder free piston engine generator and their influence on system operating characteristics. Energy **242**, 123001 (2022)
4. Jia, B., Mikalsen, R., Smallbone, A., Roskilly, A.P.: A study and comparison of frictional losses in free-piston engine and crankshaft engines. Appl. Thermal Eng. **140**, 217–224 (2018)
5. Wang, J., Feng, H., Zhang, Z., Wu, L., Yang, F.: Development of a coupling model and parametric analysis of an opposed free-piston engine linear generator. Appl. Thermal Eng. **219**, 119205 (2023)
6. Gilbert, P., Walsh, C., Traut, M., Kesieme, U., Pazouki, K., Murphy, A.: Assessment of full life-cycle air emissions of alternative shipping fuels. J. Clean. Prod. **172**, 855–866 (2018)
7. Vargün, M., Yılmaz, I.T., Sayın, C.: Investigation of performance, combustion and emission characteristics in a diesel engine fueled with methanol/ethanol/nHeptane/diesel blends. Energy **257**, 124740 (2022)
8. Fayad, M.A., et al.: Incorporating of TiO2 with oxygenated fuel and post-injection strategy in CRDI diesel engine equipped with EGR: a step towards lower NOX, PM and enhance soot oxidation reactivity. Case Stud. Thermal Eng. **53**, 103894 (2024)
9. Singh, V., Kumar, N.: Numerical investigation of the effect of nozzle hole diameter on the combustion, emission, and spray characteristics in a diesel engine. Energy Sources Part A: Rec. Utilizat. Environ. Effects **47**(1), 11537–11554 (2025)
10. Jiang, M., et al.: Numerical optimization of injector hole arrangement for marine methanol/diesel direct dual fuel stratification engines. Appl. Thermal Eng. **257**, 124456 (2024)

Simulation Analysis of Mechanical Behavior of Aluminum Alloy Plate Under Multi Working Conditions: Stretching, Stamping and TNT Blasting Impact

Penghui Fan[1], Zongyu Bai[1], Jiaxin Liu[1(✉)], Jingtao Zhao[1(✉)], and Martin Kreschel[2]

[1] College of Mechanical Engineering, North China University of Science and Technology, Tangshan 063210, China
{liujiaxin,zhaojingtao}@ncst.edu.cn

[2] School of Architecture and Engineering, SRH University of Applied Sciences Heidelberg, 69123 Heidelberg, Germany

Abstract. Aluminum alloy plates, characterized by their low density, high strength, and superior processing capabilities, are extensively employed in industries including aerospace, automotive manufacturing, and shipbuilding, making them an indispensable material for industrial development. This study centers on the failure mechanisms of aluminum alloy plates subjected to tensile, stamping, bending, and TNT blasting impact loads, which are prevalent in practical engineering and exert a substantial influence on their performance. The research leverages the ANSYS platform to conduct multi - condition explicit dynamics numerical simulations using the finite element method, offering technical backing for the investigation. The mesoscale model can capture the changes in material microstructure characteristics and properties, and the multiphysics coupling approach ensures the realism of the results. By comparing the simulation results with classical theoretical solutions and engineering empirical data, it is demonstrated that the multi - condition joint simulation method exhibits high accuracy and can forecast the mechanical behavior of aluminum alloy plates under complex loads. This provides theoretical underpinnings for the design and optimization of engineering structures capable of withstanding impacts, assisting engineers in making informed decisions and enhancing the safety and reliability of engineering structures.

Keywords: Aluminum alloy plate · finite element analysis · Stretching · Stamping and bending · TNT blasting impact

1 Introduction

Aluminum alloys, such as 6061 - T6 and 7075 - T6, are exemplary metallic structural materials that are widely used in various industries. The mechanical properties of these aluminum alloys are not constant; instead, they change under different loading conditions. For example, in the aerospace sector, components made from these alloys must be

S. Halgamuge et al. (Eds.): ICACMVE 2025, LNME, pp. 607–613, 2026.
https://doi.org/10.1007/978-981-95-6911-3_61

able to withstand both static tensile loads, which are applied gradually and consistently, and sudden impact loads, which are abrupt and intense forces. In contrast, within the automotive industry, when it comes to body panel forming, there are dynamic plastic deformation processes involved, such as stamping, where a tool is pressed into the metal to create a specific shape, and bending, which involves deforming the metal around a straight axis to form an angle or curve. These different requirements in various industries highlight the versatility and adaptability of aluminum alloys like 6061 - T6 and 7075 - T6 [1, 2]. However, most current research focuses on single - loading conditions and lacks a systematic analysis of failure mechanisms under multiple loading conditions such as tensile, stamping, and blasting impacts. TNT blasting impact, an extreme loading form, can cause complex failure modes in aluminum alloy plates [3, 4], and traditional quasi - static simulation methods struggle to depict dynamic response characteristics accurately. Therefore, it is significant to introduce multiphysics coupling simulation technology to construct a full - condition analysis model considering strain rate effects and material damage evolution for improving the impact resistance design of aluminum alloy structures.

2 Simulation Model and Method

2.1 Analyze Tools and Material Parameters

Firstly, establish the LS - Dyna and Explicit Dynamics modules of the Ansys analysis system. These were explicit dynamics modules with multi - physics solving capabilities, capable of simulating compressible and incompressible fluids, airbag particle methods, metal forming, and Arbitrary Lagrange - Euler (ALE) fluid dynamics [5]. The LS - Dyna module was served for tensile and TNT blasting impact simulation, and the Explicit Dynamics module was served for stamping simulations. The parameters of the aluminum alloy plate were characterized according to the EPS constitutive model (detailed in Table 1). The specific parameters (25°C) are as follows:

Table 1. Aluminum alloy basic parameters.

Materials	Young modulus	Poisson ratio	Yield strength	Tangent modulus	Density	Equivalent plastic strain
Aluminum alloy	7000 MPa	0.33	120 MPa	250 MPa	2700 kg/m^3	0.15

2.2 Geometric Model and Solution Setting

In the DesignModeler module of ANSYS, create aluminum alloy plates. To meet analysis requirements, create three plates of the same material and thickness but different sizes: a 40 mm × 160 mm × 2 mm plate for the tensile model, a 44 mm × 8 mm × 2 mm plate for the stamping model, and an 800 mm × 800 mm × 2 mm plate for the TNT blasting model (shown in Figs. 1a, 1b, and 1c).

Boundary Conditions and Meshing: In the stretching model, one end was fixed, and the other end got a 15 mm displacement load (0.1 s) with a 15 mm clamping part. The grid was divided by adaptive technology at intervals of 3 mm (Fig. 1d). For the stamping model, the structural steel punch has a 10 mm remote displacement (0.1 s), the V - shaped groove die of structural steel was fixed (both are rigid), and the aluminum alloy plate was flexible. Considering contact friction (dynamic friction coefficient 0.1, static 0.15). The contact position between the plate and the punch had a 1 - mm grid density, and the rest was divided at 3 - mm intervals (Fig. 1e). In the TNT blasting impact model, a 1000 mm × 1000 mm × 1200 mm closed area was established. TNT (80 mm × 80 mm × 80 mm) explodes 800mm from the plate in a non - leaking air - filled environment. The air density was 1.1845 kg/m^3, the pressure cut - off (pc) was −10 Pa, and the dynamic viscosity (μ) is 1.844 × 10^{-5} Pa·s. EOS_LINEAR_POLYNOMIAL coefficients c4 and c5 were 0.4, others were 0, and the grid was divided into 3 mm (Fig. 1c, 1f). A novel Fluid-Structure Interaction (FSI) method, which is structured based on the Arbitrary Lagrangian-Eulerian (ALE) framework and incorporates Euler/Lagrange coupling within the fluid-solid domain, is employed in this study. This particular method has the capability to automatically determine and set the coupling points, which significantly enhances computational efficiency. It is worth noting that the coupling direction in this method is strictly normal, and only compressive forces are applied during the interaction process. Furthermore, the analysis time for each simulation step is set to be an extremely short duration of 0.0001 s, ensuring high precision in capturing the transient behavior of the fluid-solid system. The load was applied through ALE-mediated coupling of the air and structural domains, and the TNT detonation products' state equation (JWL model [6]) was defined.

$$\mathrm{p} = \mathrm{A}\left(1 - \frac{\omega}{\mathrm{R_1V}}\right)\mathrm{e}^{-\mathrm{R_1V}} + \mathrm{B}\left(1 - \frac{\omega}{\mathrm{R_2V}}\right)\mathrm{e}^{-\mathrm{R_2V}} + \frac{\omega E}{\mathrm{V}} \tag{1}$$

Here, A = 3.72 × 105 MPa, B = 3.23 × 103 MPa, $R_1 = 4.15$, $R_2 = 0.95$, ω = 0.35.

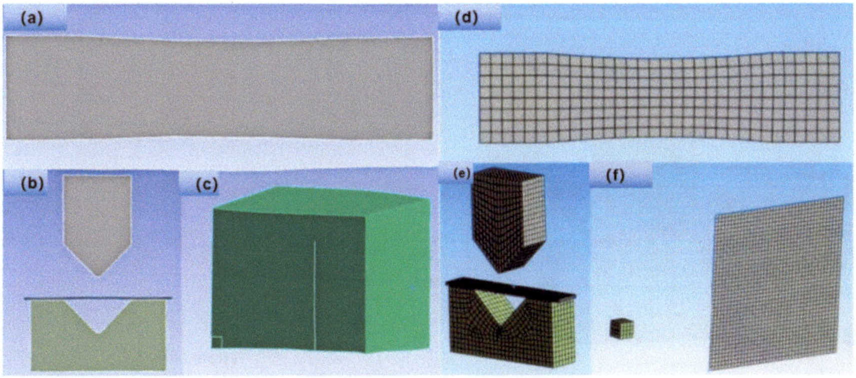

Fig. 1. (a) Stretching model. (b) Stamping model. (c) TNT blasting impact model. (d) Extrude model meshing. (e) Stamping model meshing. (f) TNT blasting model meshing.

3 Results and Discussion

3.1 Simulation Results of Stretching Condition

Through tensile numerical analysis, the stress - strain curve of the aluminum alloy plate was obtained, the equivalent yield stress was 118.38 MPa, conforming to the typical plastic deformation characteristics of aluminum alloy. Failure mode: The maximum pressure area occurs near the clamping part (Fig. 2), ultimately resulting in a shear fracture surface, which is consistent with the theoretical analysis.

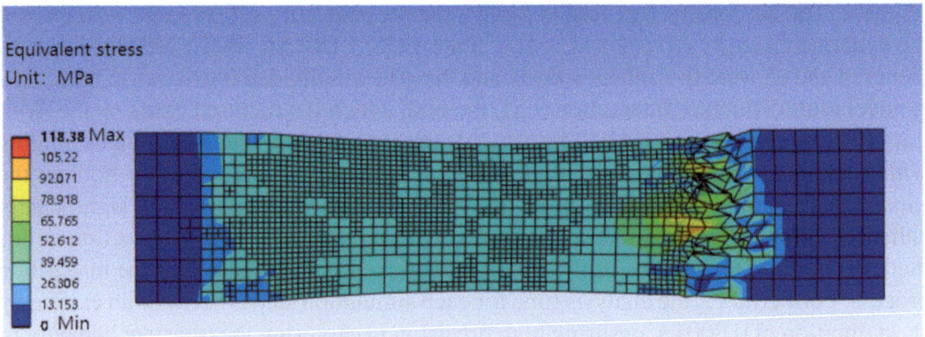

Fig. 2. Equivalent stress diagram during the stretching process of aluminum alloy sheet.

3.2 Simulation Results of Stamping and Bending Conditions

Transformation Process: As shown in Figs. 3a and 3b, upon the stamping displacement attaining 10 mm, the bending angle reaches 90°. Simultaneously, the maximum equivalent plastic strain (11.507 mm) emerges at the rounded corner of the punch, with rupture taking place.

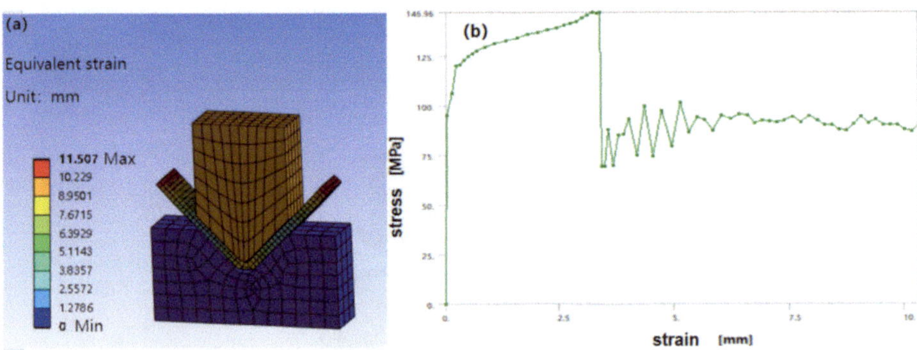

Fig. 3. (a) Equivalent stress diagram. (b) Stress-strain curve of aluminum alloy plate during stamping process.

3.3 Simulation Outcomes Under the Condition of TNT Blasting Impact

Propagation of shock waves: At a time point precisely 100 microseconds after the detonation of TNT, an exceptionally powerful shock wave is produced. This shock wave is a result of the rapid release of energy during the explosion. The maximum pressure experienced on the surface of the aluminum alloy plate reaches 157.29 MPa. This immense pressure has a significant impact on the material. Additionally, the maximum plastic deformation achieved is measured at 379.35 mm. As can be observed in Fig. 4a, within the region near the blasting point, a cross - shaped plastic deformation zone is formed due to the intense stress and strain exerted by the shock wave. This zone is a clear indication of the material's response to the extreme conditions, and fracture failure ultimately occurs as the material's structural integrity is compromised under these circumstances.

A layer-cracking phenomenon was observed on the back side of the plate during the inspection process. This particular type of crack was found to propagate along the direction of the blast wave, which is in perfect alignment with and corroborated by the dynamic fracture mode that was meticulously measured and documented in the Hopkinson rod experiment, thus providing a clear indication of the underlying failure mechanism under such loading conditions. The propagation pattern of the crack suggests a strong correlation with the stress distribution and intensity caused by the blast wave, further validating the experimental findings obtained from the Hopkinson rod test setup [3].

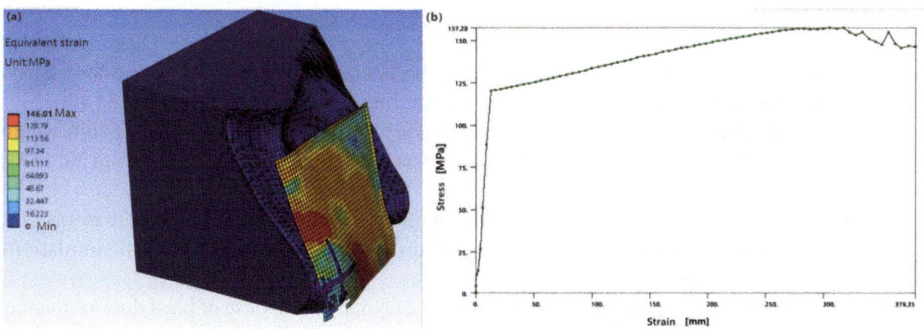

Fig. 4. (a) The equivalent stress of the aluminum alloy sheet during TNT explosion. (b) The stress-strain curve of the aluminum alloy plate during TNT explosion.

4 Conclusion

This research work employs the ANSYS software to carry out simulations on aluminum alloy plates under a variety of different conditions. The key findings obtained from this study are as follows: The EPS (Extended Plasticity Simulation) model has the capability to predict both the yield point and the occurrence of fracture during tensile simulation processes. When it comes to stamping and bending operations, the deformation that occurs is primarily concentrated around the area of the punch fillet. It is worth noting that the accuracy achieved in the simulation proves to be sufficient enough to meet the requirements and demands of engineering applications. In addition, under the extreme conditions created by TNT blasting impact, there exists a nonlinear relationship among the hole diameter, the range of layer - cracking, and the pressure distribution within the material. Furthermore, the ALE (Arbitrary Lagrangian - Eulerian) algorithm demonstrates its effectiveness in capturing and describing the failure process of the aluminum alloy plates under such intense loading conditions.

In order to effectively prevent the occurrence of shock wave oscillation errors when simulating aluminum alloy plates under dynamic impact conditions, it becomes absolutely necessary to carry out fine grid partitioning. This process involves dividing the simulation space into smaller, more detailed sections, which allows for a much higher resolution in the computational model. Such fine-grained division of the grid plays an extremely crucial role in ensuring that the simulation results are both highly accurate and reliable. When this meticulous approach is properly implemented, it greatly enhances the quality of the simulation, as it can significantly minimize—or even completely eliminate—the potential for errors associated with shock wave oscillations that might otherwise compromise the integrity of the results. Therefore, careful attention to grid refinement is essential for achieving optimal outcomes in simulations involving complex dynamic forces.

Declaration of Competing Interest. The authors declare that they have no known competing financial interests or personal relationships that could have appeared to influence the work reported in this paper.

References

1. Cai, D., et al.: Test piece design, mechanical properties and fracture strain in two-step tensile testing of lightweight materials at high strain rates. Int. J. Impact Eng. **206**, 105431 (2025)
2. Liu, X., et al.: Fast light alloy stamping technology (FAST) for manufacturing lightweight pressings from dissimilar aluminium alloy-tailor welded blanks. J. Manuf. Process. **120**, 1141–1156 (2024)
3. Zhang, L., et al.: Effect of polyurea coating with different mechanical properties on blast resistance of aluminum alloy circular tube structures: experiments vs numerical simulations. Thin-Walled Struct. **183**, 110361 (2023)
4. Wu, C., Zhang, G., He, Y., Liu, L., Liu, J., Yang, X.: Dynamic response of blast doors enhanced by enclosed-space TNT explosions: experimental and numerical study. Def. Technol. **48**, 173–186 (2025)
5. Qiu, Y., Lan, X., Liu, J., Wang, G., Huang, Z.: Optimization design of nozzle parameters under the condition of submerged water jet breaking soil based on response surface method. Appl. Ocean Res. **154**, 104369 (2025)
6. Liu, M., Liu, Y., Bai, F., Wang, H., Chu, S., Huang, F.: Determining the parameters and chemical behaviour of the overdriven detonation reaction zone of CL-20-based aluminized explosives. Def. Technol. **47**, 46–66 (2025)

Numerical Analysis of Coolant Temperature Effects on Battery Thermal Management Performance at Low Flow Velocity

Mengchen Bai[1], Naijiang Liu[1], Jiaxin Liu[1](✉), Li Li[1](✉), and Martin Kreschel[2]

[1] College of Mechanical Engineering, North China University of Science and Technology, Tangshan 063210, China
{liujiaxin,jxlili}@ncst.edu.cn

[2] School of Engineering and Architecture, SRH Hochschule Heidelberg, 69123 Heidelberg, Germany

Abstract. Liquid cooling is a crucial method for mitigating runaway battery thermal management systems (BTMS). Among the various factors that influence the effectiveness of liquid cooling, the coolant inlet temperature (CIT) of the coolant plays a crucial role. This study investigates the impact of CIT on BTMS under low flow velocity conditions (0.1 m/s). Under the condition of constant flow rate and a fixed battery heat generation model, the three CIT levels (280 K, 290 K, and 300 K) were analyzed. At CIT levels of 280 K, 290 K, and 300 K, the peak battery temperatures were 303 K, 309 K, and 314 K, respectively, while the minimum temperatures (at the cold plate interface) were 280 K, 290 K, and 300 K, respectively. Results indicate that in the designed cold-plate structure, lowering CIT enhances local heat extraction near the cooling plate but exacerbates thermal inhomogeneity within the battery module. We suggest selecting a CIT of 290 K to achieve the dual objectives of mitigating thermal stress risk and ensuring effective heat dissipation.

Keywords: battery thermal management systems · liquid cooling · electric vehicle

1 Introduction

Lithium-ion batteries, because of their high energy density, long cycle life, and low self-discharge rate, have become the dominant power source for electric vehicles (EVs) and grid-scale energy storage systems [1]. However, their performance, safety, and service life exhibit a strong dependence on operating temperature. Low temperatures significantly reduce ionic conductivity and reaction kinetics, leading to a sharp decrease in usable capacity, diminished charge acceptance capability, and even the risk of lithium plating [2]. Conversely, elevated temperatures accelerate parasitic side reactions and electrolyte decomposition, causing irreversible capacity fade. In extreme cases, this can

M. Bai and N. Liu—These authors contributed equally to this work.

S. Halgamuge et al. (Eds.): ICACMVE 2025, LNME, pp. 614–620, 2026.
https://doi.org/10.1007/978-981-95-6911-3_62

trigger thermal runaway cascades, potentially resulting in catastrophic safety incidents [3, 4]. Therefore, developing efficient and reliable battery thermal management systems (BTMS) is a critical technological for ensuring the safe and stable operation of LIBs and maximizing their performance potential.

Among various BTMS approaches, liquid cooling systems have emerged as the mainstream solution for medium-to-high-power applications, particularly EVs, due to their superior heat capacity, thermal conductivity, and precise temperature control capability [5]. Saraireh et al. [6] found that increased flow rate improved temperature uniformity, but this was accompanied by diminishing returns and increased energy consumption. Some researches [7, 8] have conducted numerous experiments in the performance studies of nanofluids and water under various emission rates. As the fundamental driving force and key boundary condition for system heat exchange, the coolant inlet temperature (CIT) warrants thorough investigation.

This study employs a controlled variable approach. Computational fluid dynamics (CFD) simulations were conducted using ANSYS Fluent® to investigate the heat dissipation effectiveness under different CITs (280 K, 290 K, 300 K) while maintaining a constant coolant flow velocity of 0.1 m/s.

2 Basic Heat Exchange Formula

2.1 Heat Exchange Between the Coolant and the Surface of the Battery

Newton's Law of Cooling is used to calculate the heat dissipation between the battery surface and the coolant.

$$Q = h \cdot A_s \cdot (T_{batt} - T_{cool}) \tag{1}$$

In the formula, Q denotes the heat flow rate, h stands for the convective heat transfer coefficient, A_s represents the effective heat transfer area of the battery, T_{batt} is the local surface temperature of the battery, and T_{cool} refers to the local temperature of the coolant.

2.2 Solve for the Heat Transfer Coefficient of Laminar Flow State

The calculation equation for the Reynolds number is

$$Re = \frac{\rho v D_h}{\mu} \tag{2}$$

In the formula, ρ represents the fluid density, v denotes the flow velocity, D_h stands for the hydraulic diameter, and μ represents the dynamic viscosity.

The calculation equation for the Prandtl number is

$$Pr = \frac{\mu C_p}{k} \tag{3}$$

In the formula, k represents the thermal conductivity coefficient.

Finally, the heat transfer coefficient is calculated using formula (4).

$$Nu = \frac{h \cdot D_h}{k} = 0.023 \cdot Re^{0.8} \cdot Pr^{0.4} \tag{4}$$

3 The Physical Model of the Battery Thermal Management System

Figure 1 illustrates the overall framework of the physical model, where the upper section constitutes the battery module and the lower section accommodates the cooling plate. The battery module comprises three prismatic cells, each equipped with positive and negative electrodes, with adjacent cells electrically connected via busbars.

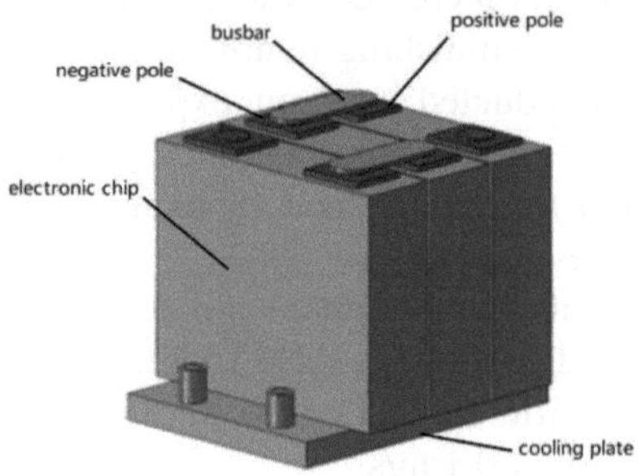

Fig. 1. The overall physical model of the battery thermal management system

Figure 2 depicts the three-dimensional structure of the bottom cooling plate, where the transparent outer portion represents the housing structure, and the internal solid portion constitutes the coolant flow channels.

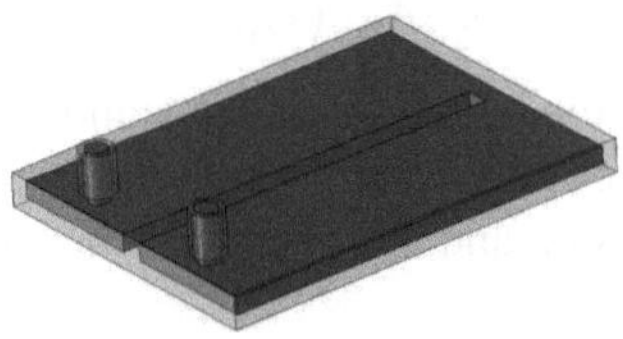

Fig. 2. Schematic diagram of the cooling plate structure

Table 1 summarizes the material properties of constituent components in the physical model.

Table 1. Material parameters for each part.

Parameters	ρ/(kg·m^{-3})	C/[J·(kg·m) $^{-1}$]	K/ [W·(m^2·k)$^{-1}$]	μ/(Pa·s)
electronic chip	2350	980	12.8/12.8/2.6	
cooling liquid	1071.11	3300	0.384	0.00339
cooling plate	2719	871	202.4	
positive pole	2719	871	202.4	
negative pole	8978	381	387.6	

4 Simulation Analysis of Different Coolant Temperature

This study investigates the thermal characteristics of batteries operating under high-power conditions. To reduce computational resource consumption, a fixed heat generation battery model was employed in the simulations. A constant heat generation rate of 10,000 W/m^3 was specified, and based on the battery's physical volume of 0.063 m^3, the heat generation rate per battery unit was calculated to be 630 W.

CFD simulations performed using ANSYS Fluent, with a constant coolant flow velocity of 0.1 m/s and varying CIT, revealed a consistent low-temperature zone proximate to the cooling plate across all cases, as shown in Fig. 3. When the CIT is 300K, the region near the cooling plate reached 300 K, while the maximum temperature of battery was 314 K, resulting in a temperature difference (ΔT) of 14 K. At a CIT of 290 K, the plate-proximal region cooled to 290 K, while the maximum temperature in the distal battery region increased to 309 K, yielding a resultant ΔT of 19 K. Lowering the CIT to 280 K resulted in a corresponding decrease in the plate-proximal temperature to 280 K. This condition led to a peak temperature of 303 K in the battery region, establishing a substantially increased ΔT of 23 K.

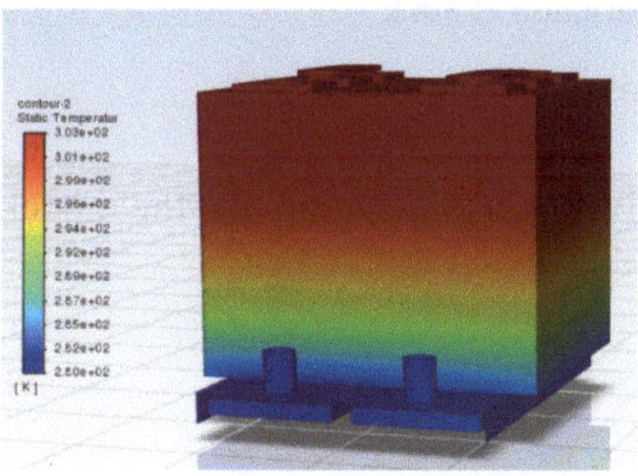

(a) The results at a temperature of 280k

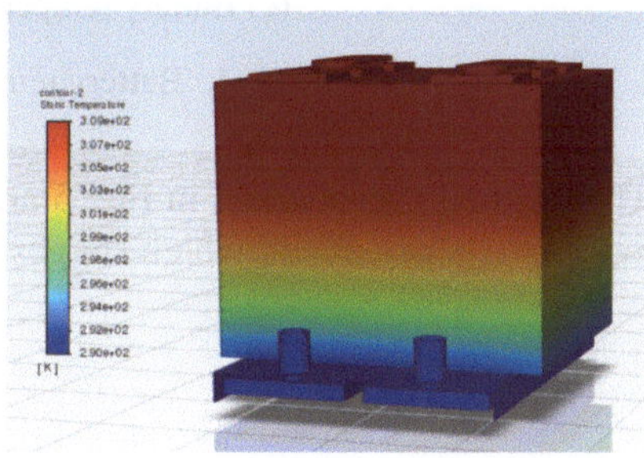

(b) The results at a temperature of 290k

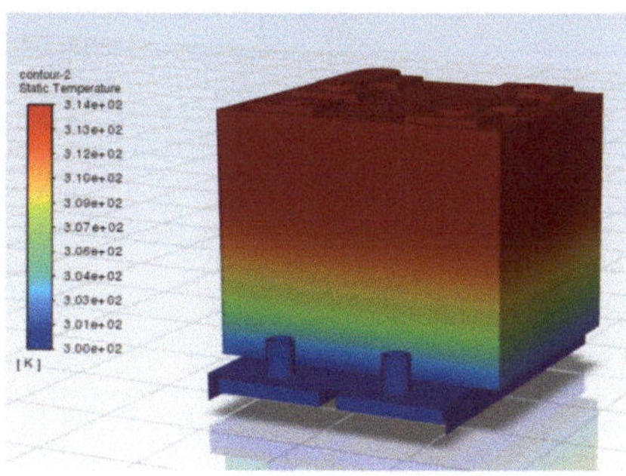

(c) The results at a temperature under 300k

Fig. 3. Schematic diagram of thermal simulation at different temperatures

As shown in Fig. 4, the average temperature of the battery gradually increased at first and then gradually stabilized.

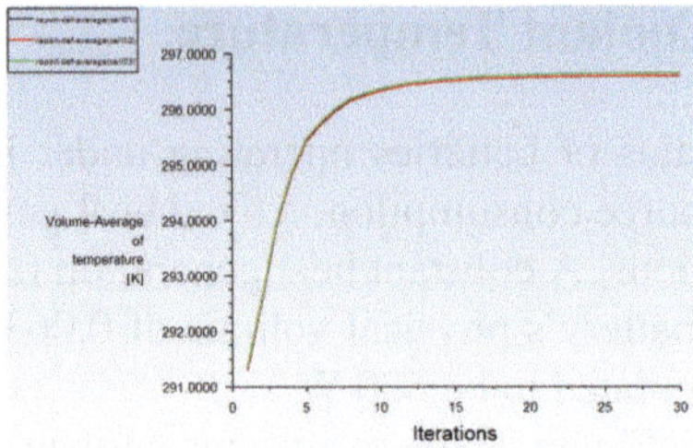

(a) Battery temperature variation at 280k

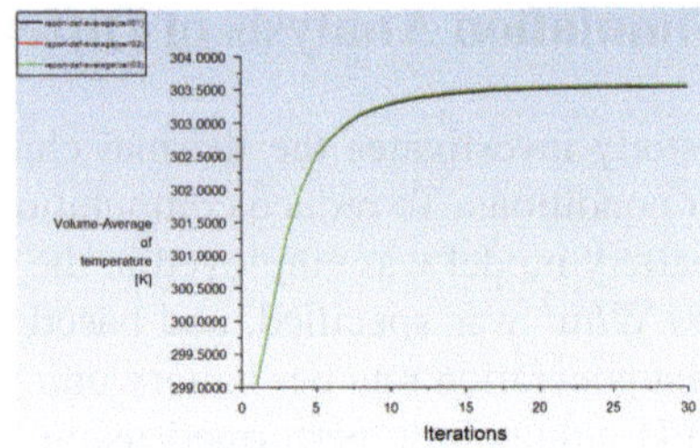

(b) Battery temperature variation at 290k

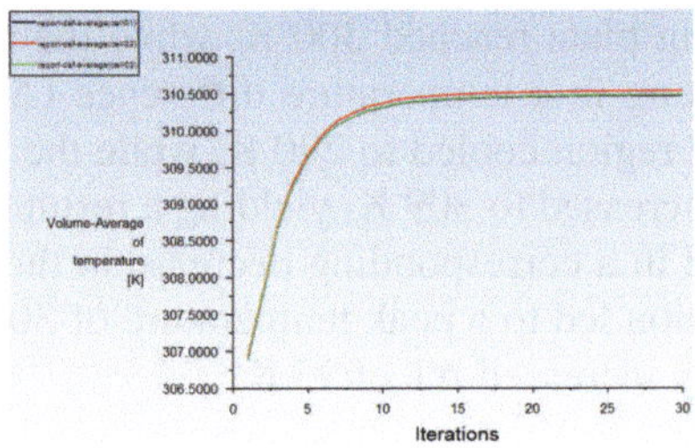

(c) Battery temperature variation at 300k

Fig. 4. Battery temperature variation curve

The residual curve values in Fig. 5 are relatively low, indicating that the model has a good convergence effect, which suggests that this model is effective.

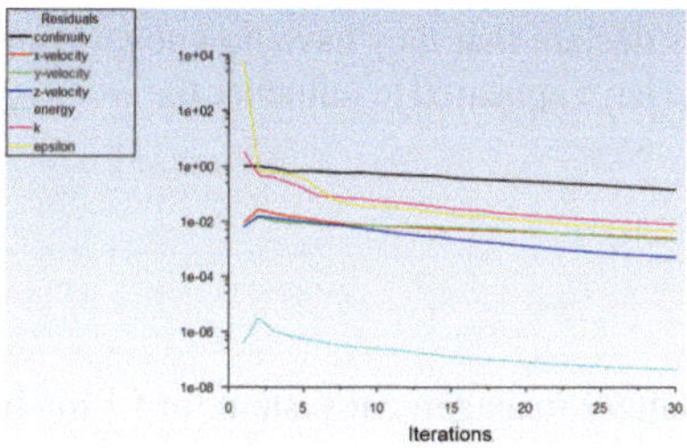

(a) The residual change curve at 280k

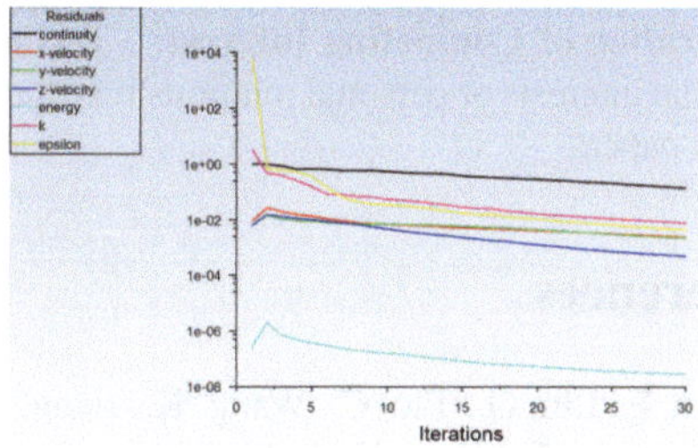

(b) Battery temperature variation at 290k

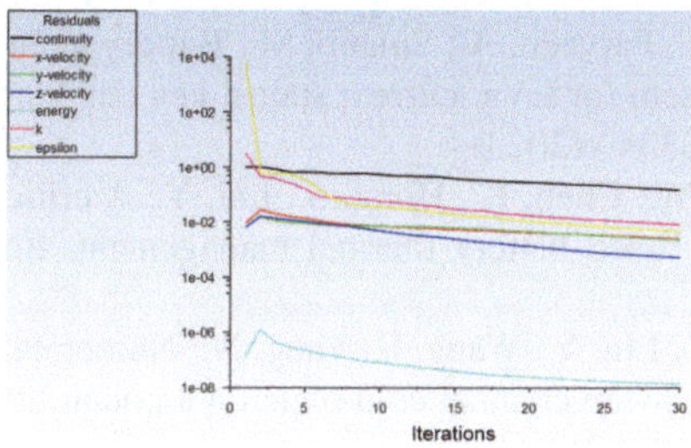

(c) Battery temperature variation at 300k

Fig. 5. Residual curve diagram

While reducing the coolant temperature (e.g., to 280 K) enhances the local heat extraction rate near the cooling plate, it concurrently exacerbates the overall thermal inhomogeneity within the system, manifested by a substantial increase in ΔT. Given that excessive temperature differentials induce elevated thermal stress, potentially accelerating battery degradation, an optimal compromise for the 0.1 m/s flow condition was identified by setting the coolant inlet temperature to 290 K. This configuration effectively limits the battery's maximum operating temperature (309 K) while mitigating the system-wide ΔT (19 K). It ensures adequate thermal management performance while minimizing the potential long-term detrimental impact of thermal inhomogeneity on battery lifespan.

5 Conclusion

Simulation analysis reveals that within the cold plate structure designed in this study, lowering the CIT enhances local heat dissipation near the cold plate. However, it concurrently exacerbates temperature non-uniformity within the battery module (e.g., ΔT reaches as high as 23 K at CIT = 280 K). Excessive temperature differences significantly shorten battery's lifespan. Comparative assessment of performance across different CIT levels demonstrates that a CIT of 290 K achieves the optimal balance between cooling capacity and temperature uniformity. The above conclusions were all derived under the conditions of constant flow rate and a fixed battery heating model.

Declaration of Competing Interest. The authors declare that they have no known competing financial interests or personal relationships that could have appeared to influence the work reported in this paper.

References

1. Zhu, F., Liu, G., Tao, C., Wang, K., Jiang, K.: Battery management system for Li-ion battery. J. Eng. **2017**(13), 1437–1440 (2017)
2. Jaguemont, J., Van Mierlo, J.: A comprehensive review of future thermal management systems for battery-electrified vehicles. J. Energy Storage **31**, 101551 (2020)
3. Waseem, M., Ahmad, M., Parveen, A., Suhaib, M.: Battery technologies and functionality of battery management system for EVs: current status, key challenges, and future prospectives. J. Power Sources **580**, 233349 (2023)
4. Wu, W., Wang, S., Wu, W., Chen, K., Hong, S., Lai, Y.: A critical review of battery thermal performance and liquid based battery thermal management. Energy Convers. Manag. **182**, 262–281 (2019)
5. Wang, J., Liu, X., Liu, F., Liu, Y., Wang, F., Yang, N.: Numerical optimization of the cooling effect of the bionic spider-web channel cold plate on a pouch lithium-ion battery. Case Stud. Thermal Eng. **26**, 101124 (2021)
6. Saraireh, M.: A novel method for heat exchange evaluation in EV. Intell. Autom. Soft Comput. **36**(1), 57–70 (2023)
7. Zhu, C.: Optimizing and using AI to study of the cross-section of finned tubes for nanofluid-conveying in solar panel cooling with phase change materials. Eng. Anal. Bound. Elem. **157**, 71–81 (2023)
8. Smaisim, G.F., Al-Madhhachi, H., Abed, A.M.: Study the thermal management of Li-ion batteries using looped heat pipes with different nanofluids. Case Stud. Thermal Eng. **37**, 102227 (2022)

Application of Dimensional Analysis Method in the Design of Side-Open Crimping Machines

Yangyupu Zhang[1], Junpeng Liu[1], Bole Wang[1](✉), Yunjia Liang[2], Jiaxin Liu[2], and Martin Krechel[3]

[1] Collage of German Engineering, Chongqing College of Mobile Communication, Chongqing 401520, China
bole.wang@qq.com

[2] College of Mechanical Engineering, North China University of Science and Technology, Tangshan 063210, China

[3] School of Architecture and Engineering, SRH University of Applied Sciences Heidelberg, 69123 Heidelberg, Germany

Abstract. The thesis undertakes a rigorous dimensional analysis of the serial design data using the JKK120 crimp machine. The material mechanics approach involves the construction of a mathematical model to simulate the data. This is then used to verify the dimensional analysis design by calculating displacement, strength, and rigidity.

Keywords: Crimp machine · Dimensional analysis · Hierarchical modeling · Scaling · Finite element analyze

1 Introduction

In engineering practice, achieving product serialization or modification often necessitates extensive repetitive design and calculations. The methods of dimensional analysis include Rayleigh's method [1] and Buckingham π method [2], with the latter being the currently widely adopted approach.

Oppenheimer et al. [3] enabled reliable cross-scale predictions for physical systems through neural networks operating in dimensionless space by converting physical quantities into dimensionless parameter groups via the Buckingham-π theorem, achieving prediction errors below 3% for unknown-scale systems. Oliveira et al. [4] applied the Buckingham πmethod to calculate the general equation for slurry filtration with an incompressible filter cake under constant pressure. Dumka et al. [5] implemented the Buckingham π method function using Python, which can be used to evaluate the Pi term for any fluid flow problem. Ayubali et al. [6] used Buckingham-π theorem to predict the quantitative model of the life of industrial components subjected to high temperatures and cyclic stresses. Sellami et al. [7] constructed pipeline property parameters and incorporated the defect apex radius into a data-driven burst pressure model with Buckingham-π theorem.

This paper introduces a hierarchical simulation-based dimensional analysis design method to accomplish these objectives rapidly, simply, and effectively.

S. Halgamuge et al. (Eds.): ICACMVE 2025, LNME, pp. 621–630, 2026.
https://doi.org/10.1007/978-981-95-6911-3_63

2 Analysis and the Buckingham-Π Theorem

2.1 Optimization Flow and Setting

If a physical process involves n dimensional variables x_1, x_2,..., x_n, their functional relationship can be expressed as:

$$Q(x_1, x_2, ..., x_n) = 0 \tag{1}$$

where r fundamental dimensions are involved, yielding n-r independent dimensionless π-groups $(\pi_1, \pi_2, \ldots, \pi_{n-r})$; the functional relationship could then be expressed as:

$$\phi(\pi_1, \pi_2, ..., \pi_{n-r}) = 0 \tag{2}$$

where $n \geq r$, with r being the rank of the dimensional matrix formed by the n variables. The dimensionless π-groups can be expressed as:

$$\pi_j = x_j \prod_{i=1}^{r} x_i^{-\alpha_{ji}} \tag{3}$$

where $j = 1, ..., m \in N^+$, $n \geq m$, $\alpha_{ji} \in R$.

The dimensional matrix associated with variable set x_1, x_2,..., x_n are shown in Fig. 1(a), where exponents of variables x_i are denoted by e_{ij}, with n rows and k columns corresponding to fundamental dimensions s_k.

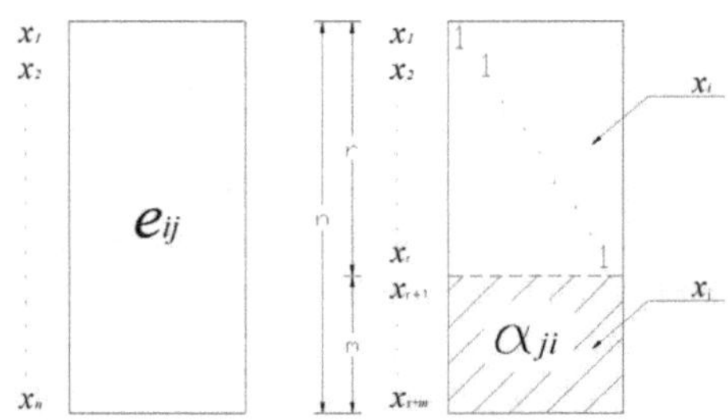

Fig. 1. The definition of the quantity matrix

To compute the dimensionless product π_j using Formula (3), the dimensionless matrix for $x_1, x_2, \ldots, x_n$ shown in Fig. 1(a) must first be constructed according to Reference [8, 9]. Through rank-preserving elementary operations, this matrix is transformed into the upper triangular form displayed in Fig. 1(b). This transformation permits linear combinations of columns (scalar multiples) and row exchange operations. Consequently, the exponents α_{ji} for the dimensionless product in Formula (3) can be directly obtained from the shaded pivot columns of the upper triangular matrix (Fig. 1b).

For Formula (2), it can be expressed in explicit functional form as:

$$x_j = Q\left(x_1, x_2, ..., x_{j-1}, x_{j+1}, ...x_n\right) \tag{4}$$

Multiplying both sides of Formula (4) by $\prod_{i=1}^{r} x_i^{-\alpha_{mi}}$ and applying Formula (4) yields:

$$x_j \prod_{i=1}^{r} x_i^{-\alpha_{mi}} = \prod_{i=1}^{r} x_i^{-\alpha_{mi}} Q(\pi_1, \ldots, \pi_{j-1}, \pi_{j+1}, \ldots, \pi_m) \tag{5}$$

$$\pi_j = \prod_{i=1}^{r} x_i^{-\alpha_{mi}} Q\left(x_1 \prod_{i=1}^{r} x_i^{-\alpha_{1i}}, \ldots x_{j-1} \prod_{i}^{r} x_i^{-\alpha_{(j-1)i}}, x_{j+1} \prod_{i}^{r} x_i^{-\alpha_{(j+1)i}}, \ldots, x_m \prod_{i}^{r} x_i^{-\alpha_{(m-1)i}} \right) \tag{6}$$

2.2 Aspects of Dimensional Simulation

Conventional dimensional analysis typically focuses on single systems; however, complex system models often necessitate decomposition into independent simplified submodels – achieved either by partitioning the system into physical subsystems or through independent simulation of distinct physical quantities – which are subsequently integrated into a complete system model, as exemplified by the hose assembly configuration illustrated in Fig. 2.

In both aforementioned scenarios, dimensional analysis provides a means to determine the coupling parameters required for integrating subsystems into a complete system. The methodology proceeds as follows:

Figure 3 demonstrates the integration of two subsystems with identical rank and dimensions in a simple case [8, 9]. The dimensional matrix merger is accomplished by stacking one subsystem's dimensional matrix vertically above the other, concurrently, dimensionless groups π_j identical to those independently derived for the two subsystems are obtained, while coupling numbers are introduced based on the base variables of System 2.

Fig. 2. Layered simulation

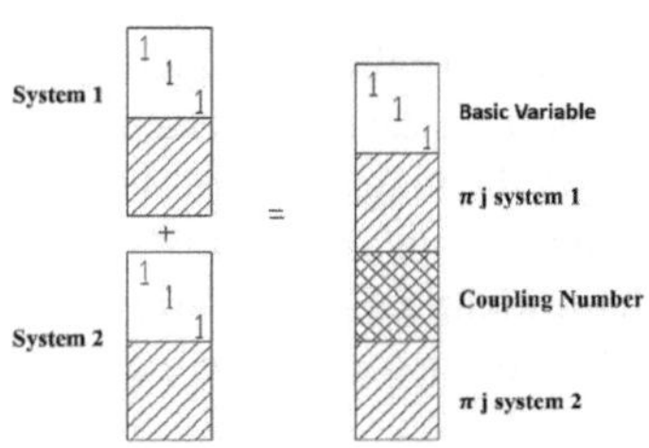

Fig. 3. Integration of two subsystems with identical rank and dimensions

2.3 Hierarchical Simulation

Hierarchical simulation is generally regarded as a functional analog of sub-functions or higher-order sub-functions, where more abstract simulations employ simplified parameter sets, as exemplified by the crimped hose assembly in Fig. 4. In such hierarchical simulations, additional parameters required by subsystems but absent in isolated models – such as flexural rigidity related to other components *EI*, part length *L*, or height *H* – are incorporated into the integrated system. Subsystems are represented by explicit equations and function exclusively as dependent variables (i.e., outputs of subsystem equations) within the integrated system formulation.

Consider k subsystems composed of n_i interrelated physical variables, each with rank r_i yielding m_i dimensionless groups. Their integration results in a combined system of rank r and m dimensionless groups. According to Buckingham-π theorem, the combined system's dimensionless groups m satisfy:

$$
\begin{aligned}
m &= \sum_{i=1}^{k} n_i - r \\
&= \sum_{i=1}^{k} (n_i - r_i) - r + \sum_{i=1}^{k} r_i
\end{aligned}
\tag{7}
$$

$$
m = \sum_{i=1}^{k} m_i + \left(\sum_{i=1}^{k} r_i - r \right) \tag{8}
$$

$$
m_c = \sum_{i=1}^{k} r_i - r \tag{9}
$$

Substituting (9) into (8) yields:

$$
m = \sum_{i=1}^{k} m_i + m_c \tag{10}
$$

Among them, the dimensionless group $\sum_{i=1}^{k} m_i$ is independently determined by each subsystem, whereas the dimensionless group m_c cannot be determined by any single subsystem alone. Instead, it is determined by physical quantities occurring simultaneously in different subsystems, and it was named as the 'coupling number'. Therefore, the number of dimensionless groups for the combined system equals the sum of the total number of dimensionless groups from each subsystem and the coupling number arising from the combination. Moreover, the magnitude of the coupling number depends solely on the rank of the individual subsystems and that of the combined system.

When the subsystems share identical dimensions and rank, as shown by Formula (7), the number of dimensionless groups could be expressed as:

$$
m = \sum_{i=1}^{k} n_i - r
$$

$$= \sum_{i=1}^{k} (n_i - r) - r + k \cdot r$$
$$= \sum_{i=1}^{k} m_i + (k - 1) \cdot r \quad (11)$$

Because of $r = r_i$, and the coupling number from (9) could be expressed as:

$$m_c = \sum_{i=1}^{k} r_i - r$$
$$= (k - 1) \cdot r \quad (12)$$

Combing (12) into (11) yields $m = \sum_{i=1}^{k} m_i + m_c$. Therefore, the dimensionless group number of the combined system remains the sum of the total dimensionless groups of each subsystem and the coupling number generated by the combined system.

3 Application on the C-Frame of a Clamping Machine

3.1 Calculation of Π Coefficient

The design of the JKK240 model uses the JKK120 hose crimper C-frame as its prototype. This demonstrates the application of hierarchical simulation based on dimensional analysis in product series design, as shown in Fig. 4. The C-frame's bending deformation directly influences the crimping quality of hose assemblies. The deformation magnitude has a direct relationship with the crimping force F and the frame's structure.

The beam cantilever underwent division into two subsystems using the hierarchical simulation method, while the column formed the third subsystem. Specifically, the beam section of length L_2 became Subsystem 1, and the beam section of length L_1 became Subsystem 2. Following this sequence, the dimensional matrices yielded their respective coefficients, producing the following results:

$$\begin{aligned}
&\pi_{G1} = \frac{EI}{FX^2}, \pi_{G2} = \frac{L_1}{X}, \pi_{G3} = \frac{Y_{01}}{X}, \pi_{G4}\frac{F}{F} = 1, \pi_{G5} = \frac{L_2}{X}, \pi_{G6} = \frac{EI_1}{FX^2} \\
&\pi_{G7} = \frac{L_2}{X}, \pi_{G8} = \frac{Y_{02}}{X}, \pi_{G9} = \frac{EI_1}{FX^2}, \pi_{G10} = \tfrac{L_2}{X}, \pi_{G11} = \frac{Y_{02}}{X}, \pi_{G12} = \frac{F}{F} = 1 \\
&\pi_{G13} = \frac{X}{X} = 1, \pi_{G14} = \frac{EI_2}{FX^2}, \pi_{G15} = \frac{H}{X}, \pi_{G16} = \frac{Y_1}{X}, \pi_{G17} = \frac{EI}{FX^2}, \pi_{G18} = \frac{L_1}{X} \\
&\pi_{G19} = \frac{Y_{01}}{X}, \pi_{G20} = \frac{F}{F} = 1, \pi_{G21} = \frac{X}{X} = 1, \pi_{G22} = \frac{EI_2}{FX^2}, \pi_{G23} = \frac{H}{X}, \pi_{G24} = \frac{Y_1}{X}
\end{aligned} \quad (13)$$

The coupling conditions are shown as follow:

$$\begin{aligned}
&\hat{\pi}_1 = \pi_{G4} = \frac{F}{F} = 1, \hat{\pi}_2 = \pi_{G5} = \frac{X}{X} = 1, \hat{\pi}_3 = \pi_{G12} = \frac{F}{F} = 1 \\
&\hat{\pi}_4 = \pi_{G13} = \frac{X}{X} = 1, \hat{\pi}_5 = \pi_{G20} = \frac{F}{F} = 1, \hat{\pi}_6 = \pi_{G21} = \frac{X}{X} = 1
\end{aligned} \quad (14)$$

Since the coupling condition has a coupling number $\hat{\pi}_1 = \hat{\pi}_2 = \hat{\pi}_3 = \hat{\pi}_4 = \hat{\pi}_5 = \hat{\pi}_6 = 1$, calibration was not necessary. The resulting coupling condition reduces the number of actual numbers that need to be guaranteed from 24 to 18, and there are duplicate coefficients:

$$\pi_{G6} = \frac{EI_1}{FX^2}, \pi_{G7} = \frac{L_2}{X}, \pi_{G8} = \frac{Y_{02}}{X}, \pi_{G9} = \frac{EI_1}{FX^2}, \pi_{G10} = \frac{L_2}{X}, \pi_{G11} = \frac{Y_{02}}{X} \quad (15)$$

Duplicated coefficients π only require calibration for one in each π according to Reference [10]. Therefore, the coefficients needing calibration after consolidation are:

$$\begin{gathered}\pi_{G1} = \frac{EI}{FX^2}, \pi_{G2} = \frac{L_1}{X}, \pi_{G3} = \frac{Y_{01}}{X}, \pi_{G6} = \frac{EI_1}{FX^2} \\ \pi_{G7} = \frac{L_2}{X}, \pi_{G8} = \frac{Y_{02}}{X}, \pi_{G8} = \frac{Y_{02}}{X}, \pi_{G15} = \frac{H}{X}, \pi_{G16} = \frac{Y_1}{X}\end{gathered} \quad (16)$$

3.2 Calibration of Π Coefficient

The ratio π_p between the prototype parameter π_m and the model parameter λ_i is termed the calibration coefficient (scale coefficient). The following conditions are satisfied:

$$\{\pi_i\}_m = \{\pi_i\}_p, \mathrm{i} = 1, 2, \ldots, \mathrm{k} \text{ or } \lambda_i = \frac{\{\pi_i\}_p}{\{\pi_i\}_m} = 1 \quad (17)$$

The two equations above indicate that complete similarity does not necessarily require geometric similarity.

Based on the JKK120 prototype, series-based calibration is performed. Given the crimping force N and unchanged deformation, Formulas (16) demonstrate that when the crimping force increases by a factor of k while material properties and displacement affecting crimping accuracy remain constant, it suffices to scale the relevant dimensions of the prototype machine by a factor of $\sqrt{k}$. According to the physical experiments in Reference [10], JKK120 scheme used a side-opening hose crimper matching the simulation model, which obtained the π coefficient, further calibrated to K-series results are similarly presented in Table 1.

Table 1. Calibration results

Item	Series Model (JKKXXX)	Calibration Coefficient of Series Model	Item	Series Model (JKKXXX)	Calibration Coefficient of Series Model
Clamping Force F	N	K	Column Height H	$420\sqrt{k}$	$\sqrt{k}$
Loading Position X	$2.15\sqrt{k}$	$\sqrt{k}$	Column Width W	$325\sqrt{k}$	$\sqrt{k}$
Beam Length L	$750\sqrt{k}$	$\sqrt{k}$	Column Thickness B	$75\sqrt{k}$	$\sqrt{k}$

(*continued*)

Table 1. (*continued*)

Item	Series Model (JKKXXX)	Calibration Coefficient of Series Model	Item	Series Model (JKKXXX)	Calibration Coefficient of Series Model
Beam Height h	$360\sqrt{k}$	$\sqrt{k}$	Beam Deformation Y_1	1	1
Beam Width B	$75\sqrt{k}$	$\sqrt{k}$	Column Deformation Y_2	1	1
Total Deformation Y_3	$Y_3 = Y_1 + Y_2$	1			

4 Finite Element Analysis Verification

4.1 Finite Element Analysis

The dimensional analysis method was adopted for the design above. Verification calculations must now be performed on the results to determine whether the design approach is justified.

During simulation, a model of JKK240 was first established based on dimensionally derived data. The operating pressure was set to P = 10 MPa. The base plate was fully fixed (rigidly constrained), with material properties defined by an elastic modulus of 206 GPa and Poisson's ratio of 0.28-node SOLID-BRICK45 tetrahedral elements were employed, with a total of 9,324 elements meshed and 14,951 nodes generated. Loads and constraints were then applied to the model (Fig. 5). System analysis results are presented in Fig. 6.

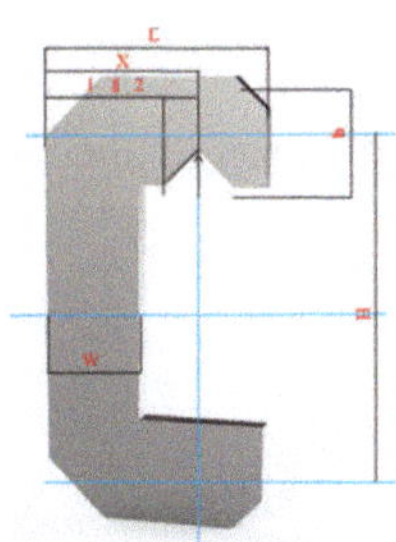

Fig. 4. C-frame crimping machine

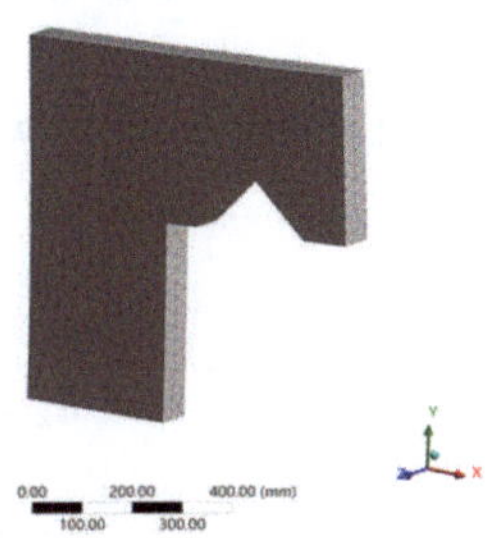

Fig. 5. Finite element model and meshing

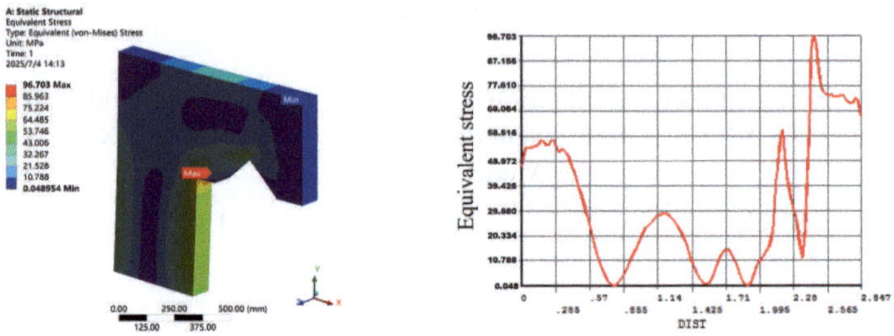

a Equivalent stress

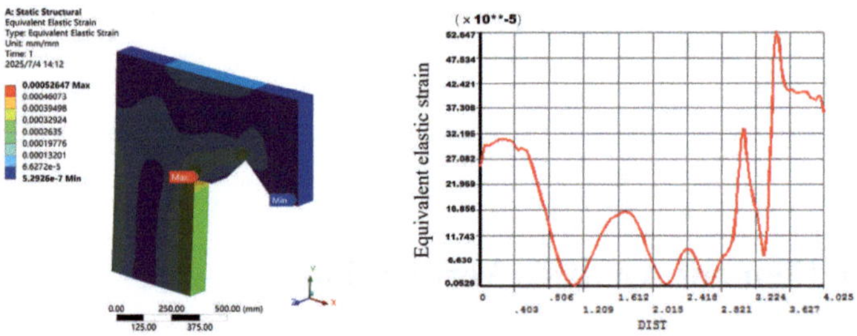

b Equivalent elastic strain

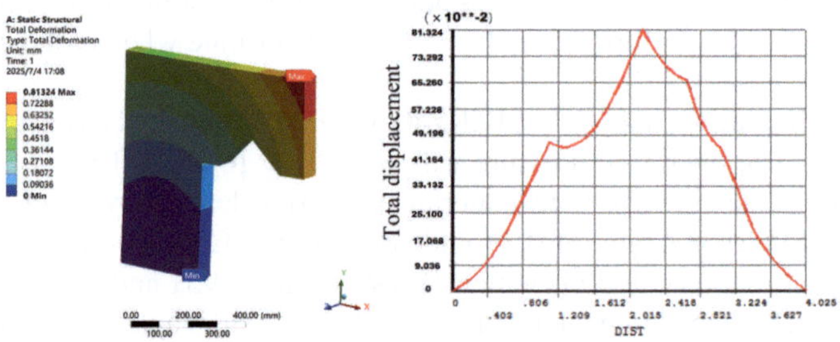

c Total displacement

Fig. 6. Finite element analysis results

The C-type stiffness analysis was verified using the fourth strength theory as follows:

$$
\begin{aligned}
\varepsilon_1 &= \frac{1}{E}[\sigma_1 - v(\sigma_2 + \sigma_3)] \\
\varepsilon_2 &= \frac{1}{E}[\sigma_2 - v(\sigma_3 + \sigma_1)] \\
\varepsilon_3 &= \frac{1}{E}[\sigma_3 - v(\sigma_1 + \sigma_2)]
\end{aligned}
\tag{18}
$$

Thus, it was confirmed that $10.432MPa \leq \sigma_{\min} - \sigma_{\max}(MPa) \leq 368MPa$, the frame operates within safe parameters under the pump pressure of 10 MPa. The displacements measure $0.107 * 10^{-4}$ and $0.537 * 10^{-5}$, respectively, falling within acceptable error margins. These results align with the dimensional design predictions, confirming the validity of the dimensional design methodology.

4.2 Result Comparison

Traditional Finite Element Method utilized a workflow of geometric modeling → trial mesh generation → parameter adjustment for computation. The dimensional hierarchical method proposed directly scaled dimensions based on the JKK prototype machine. Table 2 compared the design accuracy results between the methods, while Table 3 demonstrated the efficiency gains achieved by the dimensional analysis method. While maintaining accuracy (error $< 3\%$), this approach achieved an efficiency improvement exceeding 85%, eliminating unnecessary iterations.

Table 2. The comparison design accuracy

Parameters	FEA simulation	Dimensional method	Relative error
Maximum stress	98.57 MPa	96.70 MPa	1.93%
Key displacement	0.83 mm	0.81 mm	2.42%

Table 3. The comparison design efficiency

Name	FEA simulation	Dimensional method	Improvement
Workload	72 h	4 h	↘ 94.4%
Computational resources	120 cores	40 cores	↘ 66.7%
Iterations	5	1	↘ 80%

5 Conclusion

This study employed dimensional analysis with hierarchical simulation for the critical C-frame component, while validation was conducted using ANSYS finite element software. The results confirm the feasibility of applying dimensional analysis and hierarchical simulation to product series design. For the JKK-series C-frame design, an empirical formula was derived: by determining the crimping force ratio K between the target and prototype machines, and scaling all dimensioned parameters in the prototype drawings by a factor of $\sqrt{k}$, the target machine design can be directly obtained.

Acknowledgment. This work is supported by S&T Program of Hebei Province (No. 22282203Z), Natural Science Foundation of Hebei Province (No. E2022209082, E2024203065), Basic Scientific Program of North China University of Science and Technology (No. JQN2022003), National Science Foundation Doctoral Start-up Project (No. 28406999).

References

1. Chen, Z.N., Yang, W.Z., et al.: Engineering Simulation Experiments, pp. 1–85. Central South University of Technology Press, Changsha (1989)
2. Bhattacharya, M., Hanna, M.A.: Mathematical modeling of food extruder. Lebensmittel-Wissenschaft Technol. **19**(1), 34–38 (1986)
3. Oppenheimer, M.W., Doman, D.B., et al.: Multi-scale physics-informed machine learning using the Buckingham Pi theorem. J. Comput. Phys. **474**, 111810 (2023)
4. Oliveira, M.D.B., Ferreira, B.E.P., et al.: Scale up validation tests for Buckingham Pi theorem applied to leaf test experiments for iron ore slurry. Drying Technol. **42**(6), 997–1008 (2024)
5. Dumka, P., Chauhan, R., Singh, A., et al.: Implementation of Buckingham's Pi theorem using python. Adv. Eng. Softw. **173** (2022)
6. Ayubali, A., Prabu Shanmugavel, B., et al.: On material-agnostic fatigue life prediction using Buckingham Pi theorem. Eng. Fracture Mech. **280**, 109021 (2023)
7. Sellami, A., Franchek, M.A., et al.: Strain-based modeling of burst pressure in pipelines with selective seam weld corrosion. Int. J. Press. Vessels Pip. **217**, 105558 (2025)
8. Rudolph, S.: On topology, size and generalization of non-linear feed-forward neural networks. Neurocomputing **16**(1), 1–22 (1997)
9. International Workshop on Similarity Methods, Hierarchical Modeling using Dimensional Analysis. Universität Stuttgart (2000)
10. Quan, J.L.: Mechanical modeling and finite element analysis of the main vertical plate in side-opening hose crimper. Central South University (2006)

Research of Semi-active Control Strategy for Secondary Lateral Suspension of High-Speed Vehicles

Jinghan Wen[1], Qinghua Dong[2], and Xingyu Liu[3](✉)

[1] Shenyang Vehicle Section of China Railway Shenyang Bureau Group Co., Shenyang, China
[2] Wuhan Optics Valley Traffic Operation Co., Wuhan, China
dqh00@foxmail.com
[3] State Key Laboratory of Traction Power Southwest, Jiaotong University, Chengdu, China

Abstract. With the development of high-speed railways, passive suspension systems struggle to adapt to complex working conditions. This paper focuses on the research of semi-active control strategies for secondary lateral suspension to improve vehicle ride comfort and operational stability. A co-simulation platform combining SIMPACK and MATLAB/SIMULINK is established, and traditional damping control strategies (skyhook, acceleration damping) and fuzzy control strategies are proposed and verified. The results show that the fuzzy skyhook hybrid control strategy effectively balances the vibration of the car body and bogie frame. Its control effect is close to that of continuous skyhook damping, while offering lower costs.

Keywords: High-speed vehicle · Semi-active control Skyhook damping · Fuzzy control · Hunting motion

1 Introduction

With the development of modern industry, the backward means of transportation can not meet the demand for transportation of major commodities, in 1825, the world's first railroad was built in England. China's first operating railroad is the Wusong Railway completed in 1876 [1], Mr. Sun Yat-sen said in Shanghai in 1912: "Today's world, non-railway no country". China's ancient agricultural country to develop into an industrial country, the development of railroads is the national economy of the "artery", is indispensable. After the reform and opening up, China in addition to more railroads, railroads in the "speed up" aspect also put enough effort, in 2007 for the sixth railroad speed up, China formally stepped into the "high-speed railroad" era.2010 CHR380In 2010, the CHR380A electric locomotive set was officially put into use, and in 2017, the Fuxing locomotive set realized 350km commercial operation on the Beijing-Shanghai Railway, which also represents that China's high-speed railways have real independent intellectual property rights, and set a new benchmark for the commercial operation of high-speed railways in the world [2].

S. Halgamuge et al. (Eds.): ICACMVE 2025, LNME, pp. 631–640, 2026.
https://doi.org/10.1007/978-981-95-6911-3_64

Vehicle body as the main body to carry passengers, train body vibration directly affects the comfort of passengers, so the study of vehicle body vibration has always been the focus of the railroad industry [3]. One of the methods usually used to improve the abnormal vibration of the vehicle system is to improve the level of the train track line conditions, and most of the railroad line environment in China is at the top of the world, it is difficult to improve the smoothness of the vehicle's operation through this method [4]. It can be seen that the dynamics of passive suspension system can meet the requirements to a certain extent, and has a simple structure, low cost and no stability problems caused by feedback. However, the system itself does not have the self-correction ability, can not fundamentally improve the dynamic performance of the train, foreign scholars in the 1960s for the shortcomings of the passive suspension system proposed active suspension system [5].

2 Vehicle Suspension System

Vehicle suspension system is an important part of vehicle vibration reduction, which directly transmits all coupling forces between wheelset and bogie frame and body. At present, suspension system is mainly divided into the following three categories, as shown in Fig. 1.

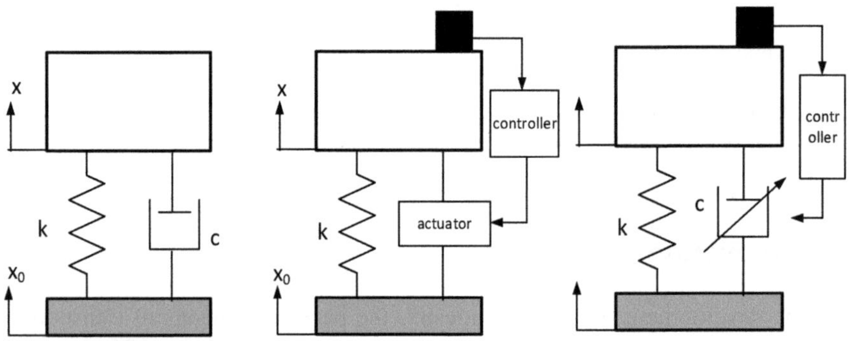

a) Passive suspension system b) Active suspension system c) Semi-active suspension system

Fig. 1. Vehicle Suspension System

1) Passive suspension

Passive Suspension is a traditional vehicle suspension system that does not require external energy. Its elastic components and damping components are only temporary energy storage and dissipation, and all parameters are constant during train operation, which also makes it difficult for the system to cope with more complex train operation routes and meet the requirements of vehicle dynamic performance [6]. In general, the optimization of the parameters of the two-system transverse damper results in the most obvious improvement of the vehicle body vibration. Yao Yuan et al. [7] obtained the suspension parameters through two methods, multi-objective optimization and parameter screening, which enabled the vehicle to have a better system robustness index and improved the linear adaptability of the train vehicle in a wide range of speeds. Teng Wanxiu et al. [8] and Luo et al. [9] found that low temperature increases the damping coefficient of the damper and the stiffness of the rubber parts, leading to the deterioration of the smoothness index of the vehicle.

2) Active suspension system

Active Suspensionis to replace the shock absorber with an actuator, which requires a large amount of energy from the outside world. Based on active control technology, the real-time operating environment and state information of the vehicle can be detected by sensors, and the vehicle suspension system parameters can be changed to adapt the vehicle system to different train routes and different operating environments.

3) Semi-active suspension

Semi-active Suspension (SAS) is a system that uses a damped variable damper instead of a very energy-intensive actuator in order to overcome the high control energy and cost required by active suspension systems. Semi-active suspension systems consume a small amount of internal energy during operation and require no external energy input [10]. The control performance of semi-active suspension system is close to that of active suspension system, and semi-active suspension system has good safety performance [11].With reference to the canopy damping control strategy, many domestic scholars [12] conducted further research on the semi-active suspension system of rail vehicles and developed many deformations Wang Yueming [13] analyzed the control law and system amplitude-frequency characteristics of the semi-active suspension system under the velocity feedback switching control mode. The control performance of the active suspension system is close to the active suspension system, the structure cost is close to the passive suspension system, and can ensure the safety of the train in the control system failure, combined with the actual situation of China's railroads, the semi-active suspension system is currently the best comprehensive situation of the vehicle suspension system [14].

3 Control Strategy of Vehicle Semi-active Suspension System

With the further increase of train speed, the ride comfort and running stability of vehicles will be further deteriorated, so in practical application, in order to improve the running stability index of vehicles, more suitable semi-active control strategy is needed.

3.1 Control Method of Sky-Hook Damping

Among the control strategies of vehicle semi-active suspension system published publicly, Karnopp et al. [15] proposed the sky-hook damping control strategy, which has good simplicity and practicality, and is also one of the most widely applied semi-active control strategies. The schematic diagram of the sky-hook control strategy is shown in Fig. 2. In simple terms, when the damping force output from the damper is in the same direction as the vibration direction of the vehicle body, the damper tries to reduce its damping coefficient as much as possible, and when the damping force output from the damper is in the opposite direction of the vibration direction of the vehicle body, the damper tries to increase its damping coefficient as much as possible, in order to arrive at the optimization of the control of the vehicle body's transverse vibration. Through the different nature of the adjustable damping damper, the canopy damping control strategy can be classified into switching-type and continuous type [16].

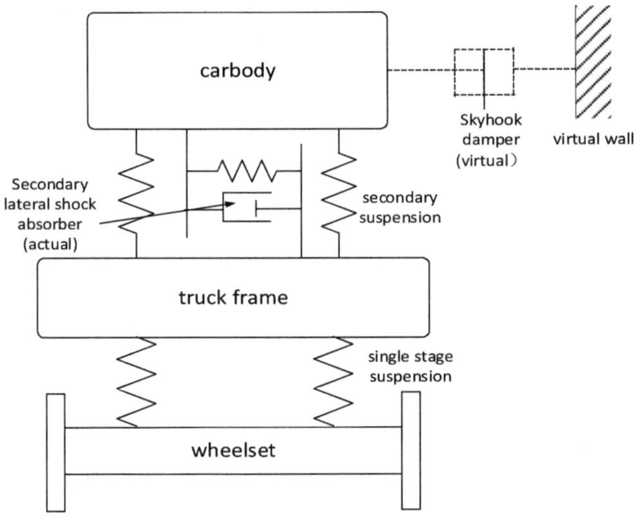

Fig. 2. Schematic Diagram of Ceiling Control Strategy

3.2 Switch Type Ceiling Damping Control

When a suspension system is composed of a switch-type controllable damping damper and a spring in parallel, the damping coefficient provided by the controllable damper is:

$$C = \begin{cases} C_{max}, v_c \bullet (v_c - v_b) \geq 0 \\ C_{min}, v_c \bullet (v_c - v_b) < 0 \end{cases} \quad (1)$$

The maximum and minimum damping coefficients provided by the controllable shock absorber are indicated respectively. At that time, when the vehicle body speed is greater than or equal to the bogie frame speed, the vibration of the vehicle body will be reduced regardless of the extension or contraction stroke of the shock absorber. In order to minimize the lateral vibration of the vehicle body, the damping coefficient of the shock absorber shall be the maximum value; similarly, when the vehicle body speed is less than the bogie frame speed, the damping coefficient of the shock absorber shall be the minimum value [17].

This control strategy is simple and practical, and the controller consists of state judgment element and damping switching element, wherein the setting is 60kN·s/m and 3kN·s/m.

3.3 Continuous Sky-Hook Damping Control

The damping coefficient of continuous controllable damping damper can be changed continuously in a certain range, which is close to the ideal ceiling damping coefficient to the greatest extent relative to the switch type. The damping coefficient formula is:

$$C = \begin{cases} C_{max}, C_{SH}\frac{v_c}{v_c - v_b} \geq C_{max} \\ C_{SH}\frac{v_c}{v_c - v_b}, C_{min} < C_{SH}\frac{v_c}{v_c - v_b} < C_{max} \\ C_{min}, C_{SH}\frac{v_c}{v_c - v_b} \leq C_{min} \end{cases} \quad (2)$$

C_{SH} is the ideal ceiling damping coefficient, the other parameter have that same meaning as switch type. in the above control method, the skyhook damping coefficient is constantly chan and approaches ideal skyhook damping force to the maximum extent, and the control method can reduce high-frequency flutter of vehicle body and effectively increase the service life and reliability of semi-active shock absorber.

The controller consists of a state judging element and a damping adjusting element, wherein the state judging element is the same as the switch type ceiling damping controller, and the damping adjusting element is different. C_{max} is setting as 60 kN·s/m, C_{min} is setting as 3 kN·s/m, C_{SH} is setting as 40 kN·s/m.

3.4 Classical PID Control

In the use of PID control method can not need to know the mathematical model of the controlled object, only need to adjust the controller parameters in real time, you can get a more satisfactory performance of the target, but when the parameters of the controlled object changes, the control effect of the PID control will have a large change [18]. PID control by the deviation of the proportional (P), the integral (I) and the differential (D) for the linear combination of thePID control is the most classical and widely used control method in the field of industrial automation [19], as shown in Fig. 3.

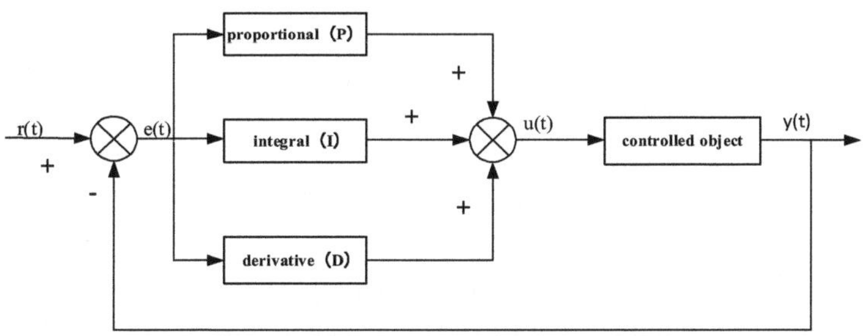

Fig. 3. PID control principle diagram

According to different controlled objects, adjust the different parameters of PID controller properly until satisfactory results are obtained. Briefly, the function of each coefficient of PID control is as follows [20]:

Proportionality coefficient: speeds up the response speed of the system and improves the regulation accuracy of the system.

Integral coefficients: elimination of steady state errors in the system.
Differential coefficients: improving the dynamic properties of the system.

3.5 Fuzzy Control

In practical work, vehicle dynamics system has large complexity, nonlinearity and coupling, traditional control is difficult to establish accurate mathematical model, its accurate model is difficult to express in the form of mathematical analytical formula, but the system has higher requirements for real-time performance, in this case fuzzy control is more suitable [21].

4 Vehicle System Dynamics Modeling

In order to simulate the motion state of the vehicle more realistically, this paper takes a train in China as the research object and establishes the corresponding vehicle dynamics model. The vehicle system considers 50 degrees of freedom of 15 inertia bodies such as a car body, two bogie frames, eight axle boxes and four wheels, among which the axle box only considers the nodding degree of freedom, and the other inertia bodies consider the 6 degrees of freedom. All rigid bodies are connected together by suspension system. The body and frame are connected by a secondary suspension system, and the frame and wheelsets are connected by a primary suspension system. The overall dynamics model of the vehicle system is shown in Fig. 4.

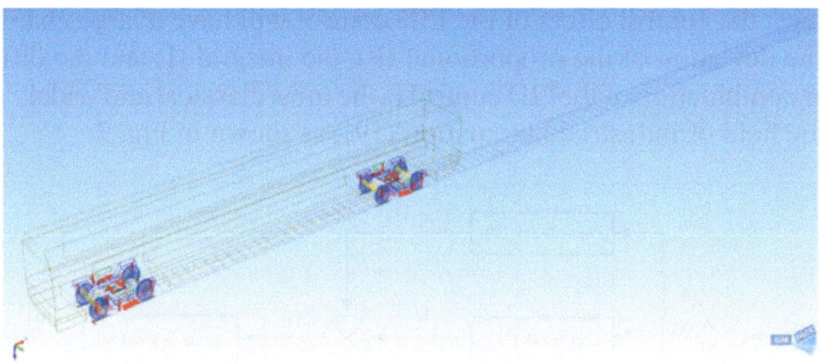

Fig. 4. Vehicle Dynamics Model

Vehicle semi-active suspension system is mainly composed of sensor device, controller and adjustable damping damper. The sensor is responsible for collecting the real-time data of vehicle required in the semi-active control strategy. The semi-active control module is established in MATLAB/SIMULINK. The joint simulation calculation is carried out with the vehicle dynamics model established in SIMPACK. The advantages of each are exerted, which makes the calculation accuracy higher and the calculation speed faster. The data communication between the two is carried out through simat interface. The control signal is transmitted to the vehicle dynamics model established in SIMPACK, so as to suppress the lateral movement of the vehicle.

In order to verify the accuracy of the joint simulation platform, the damping coefficient output in the joint simulation platform is the same as that of the passive damper output in the single simulation, and then simulation calculation is carried out. It can be seen that the two simulation results are basically consistent, indicating the accuracy of the data results of the joint simulation platform. Relevant research on semi-active control algorithm can be carried out on this platform.

5 Semi-Active Control Strategy for Vehicle Lateral Running Stability Based on Fuzzy Control Theory

5.1 Fuzzy Control Strategy Based on Skyhook Damping Control

In the switching type skyhook damping control, the damping coefficient of the secondary lateral shock absorber switches back and forth between the maximum value and the minimum value, and the shock absorber is always in the frequent switching state. In the continuous type skyhook damping control, due to the continuous change of the output damping of the shock absorber, the high frequency flutter of the vehicle body can be reduced, and the life and reliability of the semi-active shock absorber can be effectively increased. However, problems such as complex structure, difficult maintenance and increased cost of the semi-active shock absorber will be encountered [22].The fuzzy control strategy used in this section can meet the requirements of grading the adjustable shock absorber, adding several gears between the damping coefficient "soft" and "hard" can achieve the control effect close to the continuous ceiling damping control, and cannot increase the damping gear of the shock absorber too much to avoid excessive increase in cost [23].

5.2 Establishment of Fuzzy Ceiling Controller

According to the working principle of fuzzy controller [24], this controller adopts Mamdani type two-dimensional structure, and min-max gravity method is adopted for defuzzification. Similar to sky-hook damping control, it takes absolute speed of vehicle body and relative speed of vehicle body and frame as input, and damping coefficient of semi-active adjustable shock absorber as output. The speed range of input signal vehicle body is limited to $[-0.2, 0.2]$, the relative speed range of vehicle body and frame is limited to $[-0.15, 0.15]$, the unit is, 7 language variables are defined, which are [NB, NM, NS, ZE, PS, PM, PB], which respectively represent negative large, negative medium, negative small, zero, positive small, middle and positive, and the input membership function adopts trigonometric function. The damping coefficient of the output signal damper is limited to [3,60] in kN·s/m, also defined as 5 linguistic variables, the same as the input signal [25]. The fuzzy control structure is shown in Fig. 5.

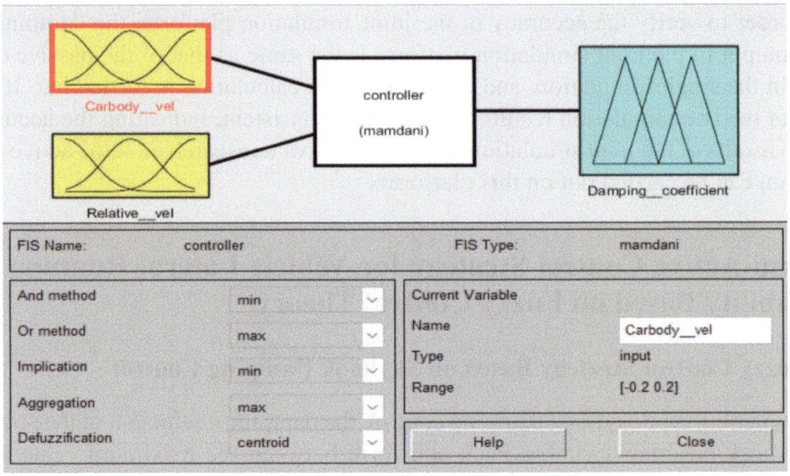

Fig. 5. Fuzzy control structure

5.3 Simulation Result

The control effect after vehicle simulation is shown in Table 1. The control effect of fuzzy ceiling control strategy is better than passive control strategy. Compared with passive control strategy, acceleration amplitude is reduced by 36.1%, acceleration root mean square is reduced by 28.6%, stability index is optimized by 12.1%, and main frequency amplitude of vehicle body is reduced by 39.7%. It can be seen that the control effect of fuzzy ceiling control is close to continuous ceiling damping control, but the cost is close to switching ceiling damping control. It can obtain superior control performance with low cost and has good practical engineering significance.

Table 1. Comparison of effects under three control algorithms

Control strategy	Amplitude of acceleration m/s2	Root mean square of acceleration m/s2	Stationarity index W
Passive	0.361	0.108	2.087
Fuzzy ceiling	0.231	0.0771	1.836
switching mode	0.249	0.0786	1.871
continuity	0.218	0.0711	1.817

In this chapter, fuzzy control principle is introduced, and fuzzy control system is established based on this principle. Then simulation experiments are carried out under two kinds of wheel wear conditions. From the experimental results, it can be seen that compared with traditional damping control strategy, this fuzzy control strategy can avoid the disadvantages of excessive optimization of car body and deterioration of bogie frame vibration, and at the same time, it can take into account the vibration control of car body and frame, and has better line adaptability.

6 Conclusion

Based on the merits and demerits of switch type and continuous type skyhook damping control strategy and fuzzy control principle, fuzzy skyhook hybrid control strategy is proposed. Simulation results show that the control effect of fuzzy skyhook control strategy is close to continuous type skyhook damping control, and the cost of semi-active damper used by fuzzy skyhook control strategy can be lower than continuous type skyhook control in practical application. Therefore, fuzzy skyhook control method has lower cost advantage and control effect close to continuous type skyhook damping control. Has greater engineering significance.

References

1. Fu, Z., Liu, Z., Li, Z.: Hundred years of development and innovation of China's railroads. China Railway **7**, 1–7 (2021)
2. Shi, H., Luo, R., Zeng, J.: A review of domestic and international high-speed train dynamics evaluation criteria[J]. J. Transp. Eng. **21**(1), 36–58 (2021)
3. Ni, J.: Research on semi-active suspension control of high-speed train based on canopy damping. Southwest Jiaotong University, Chengdu (2020)
4. Yang, L.: Research on loop-forming vibration control strategy for transverse semi-active suspension system of high-speed train. Southwest Jiaotong University, Chengdu (2021)
5. Meng, S.-Y.: Research on semi-active control of high-speed train serpentine motion. Southwest Jiaotong University, Chengdu (2018)
6. Chen, C.-J.: Active and semi-active control of high-speed trains. Southwest Jiaotong University Press, Chengdu (2015)
7. Yao, Y., Li, G., Liang, S., et al.: Optimized matching method of high-speed train suspension parameters based on robust stability[J]. Journal of Railway **43**(8), 35–44 (2021)
8. Teng, W., Luo, R., Shi, H., et al.: Dynamic performance of alpine locomotives in -40°C environment. J. Mech. Eng. **55**(4), 148–153 (2019)
9. Luo, R., Teng, W., Wu, X., et al.: Dynamics simulation of a high-speed railway car operating in low-temperature environments with stochastic parameters. Veh. Syst. Dyn. **58**(12), 1914–1934 (2020)
10. Wang, Q., Zeng, J., Zhu, B., et al.: Semi-active suspension of high-speed train undercar suspension system based on optimal control theory. J. Mech. Eng. **56**(4), 160–167 (2020)
11. Han, Z.: Research on control strategy of semi-active suspension system of train. Lanzhou Jiaotong University, Lanzhou (2020)
12. Xu, W., Liang, S., Chi, M., et al.: Study on fuzzy canopy control for transverse semi-active suspension of moving trains. Railway Rolling Stock **40**(5), 23–28 (2020)
13. Yueming, W.A.N.G.: Research on control characteristics of vehicle semi-active suspension switch. J. Mech. Eng. **38**(6), 148–151 (2002)
14. Qu S, Wang L, Wang Z, et al. Simulation and analysis of semi-active control of bogie suspension system of rolling stock based on MATLAB software[J]. Urban Rail Transportation Research, 2021, 24(2): 88–93+96
15. Zhao, Y.W., Liu, Y.Q., Yang, S.P., et al.: Analysis on new semi-active control strategies to reduce lateral vibrations of high-speed trains by simulation and hardware-in-the-loop testing. Proceedings of the Institution of Mechanical Engineers, Part F: Journal of Rail and Rapid Transit **236**(8), 960–972 (2022)
16. Wang, Q.: Study on vibration behavior of undercar suspension system of high-speed train. Southwest Jiaotong University, Chengdu (2018)

17. Yan, H., Zhang, Z.: Research on semi-active control strategy based on canopy damped two-system suspension system. Sci. Technol. Innov. Appl. **10**, 103–104 (2019)
18. Aljarbouh, A., Fayaz, M., Qureshi, M.S., et al.: Hybrid sliding mode control of full-car semi-active suspension systems. Symmetry **13**(12), 2442 (2021)
19. Yu, J.: Research on flexible control method for robot-environment interaction with electro-magnetic variable stiffness. Shanghai Jiaotong University, Shanghai (2019)
20. Wang, Z.: Research on PMSM control based on advanced control theory. Jiangnan University, Wuxi (2009)
21. Gao, R.: Snaking motion control of high-speed trains based on magnetorheological anti-snaking damper. Southwest Jiaotong University, Chengdu (2021)
22. He, F.: Research on semi-active control of high-speed train dynamic performance. Southwest Jiaotong University, Chengdu (2015)
23. Sui, J.: Modeling of high-speed train dynamics and simulation study of its transverse semi-active control. Hunan University, Changsha (2012)
24. Li, C.: Transverse vibration control of high-speed vehicle body based on semi-active damper. Southwest Jiaotong University, Chengdu (2020)
25. Han, C.-C.: Characterization and suppression method of primary serpentine motion of vehicle system. Southwest Jiaotong University, Chengdu (2020)

Comparative Study of Methods for Friction-Induced Vibration in Aircraft Disc Brakes

Qingshan He, Jing Li(✉), Jiancong Yang, and Qianjun Guo

School of Mechanical Engineering, Tongji University, Shanghai 201804, China
cynthia_li@tongji.edu.cn

Abstract. Friction-induced vibration in aircraft disc brakes is particularly complex, and different analysis methods often produce inconsistent results. In the paper, the performance of complex mode analysis (CMA) and explicit transient dynamic analysis (TDA) based on the finite element (FE) method is compared to investigate such vibration. First, a simplified FE model of an aircraft disc brake is developed. Then, under a constant friction coefficient ($\mu = 0.2$), the unstable modes predicted by CMA and TDA show good agreement, and the validity of the results is confirmed. CMA offers higher efficiency, and TDA provides greater accuracy. Next, CMA is applied under constant friction conditions varying from 0 to 0.5 at an interval of 0.1, revealing that increased friction leads to greater instability, while sensitivity analysis reveals that rotational speed has a negligible influence in this method. Furthermore, exponential friction can induce stronger system nonlinearity, which is investigated using TDA. Results show that exponential friction induces higher vibration intensity, and larger initial braking speeds result in greater vibration amplitudes. Overall, CMA consumes fewer computational resources and is better suited for simple scenarios and parametric studies, whereas TDA yields more accurate and intuitive results under complex conditions. Method selection should depend on practical engineering needs and the complexity of the system.

Keywords: Friction-induced vibration · Aircraft disc brake · Finite element method · Complex mode analysis · Explicit transient dynamic analysis

1 Introduction

Friction is a key factor in braking systems and plays a vital role during aircraft taxiing and deceleration. However, friction may induce self-excited vibration during the braking process, potentially leading to structural resonance and noise radiation [1, 2]. Friction-induced vibration can adversely affect passenger comfort and, in severe cases, pose safety risks [3, 4]. Therefore, friction-induced vibration prediction is an important research objective in the aviation field.

Due to its low cost and suitability for parametric studies and structural optimization, numerical simulation has been extensively employed in the analysis of friction-induced vibration [5]. Complex mode analysis (CMA) in the frequency domain and transient

S. Halgamuge et al. (Eds.): ICACMVE 2025, LNME, pp. 641–652, 2026.
https://doi.org/10.1007/978-981-95-6911-3_65

dynamic analysis (TDA) in the time domain are the most commonly used finite element (FE) methods. Wu et al. [6] developed a FE model of a wheel-rail-disc braking system, using CMA and TDA to investigate the impact of friction-induced instability on wheel polygonalization in high-speed trains. Zhang et al. [7] established a simplified FE model of an aircraft disc brake, conducted CMA to analyze key parameters affecting brake squeal, and used TDA to validate the squeal response, proposing optimization strategies. Zhu et al. [8] constructed a FE model of a disc brake, validated CMA results with field tests, and studied the effects of material and geometry on squeal.

Different friction models significantly influence the dynamic response of braking systems. The constant friction model defines a fixed ratio between the friction and normal force, offering a simple form but lacking the ability to capture velocity dependence and other realistic frictional behaviors [9]. In contrast, the exponential friction model more effectively captures velocity dependence and the nonlinear behavior of friction, aligning better with real-world conditions [10]. To further investigate friction-induced vibration under different conditions, combined with the contributions of previous researchers, this paper adopts the FE method to construct an aircraft disc brake model, and based on ABAQUS, a comparative study is conducted to evaluate the applicability of CMA and TDA in the analysis of friction-induced vibration.

2 Aircraft Disc Brake Modeling

2.1 Model Introduction

The simplified aircraft disc brake consists of three rotors and four stators, as shown in Fig. 1.

The geometrical dimensions of the disc brake are shown in Table 1 [11]. The material properties of rotors and stators are identical with a density of 1800 kg/m^3, a Young's modulus of 7.1 GPa, and a Poisson's ratio of 0.33.

The boundary and contact settings of the model are defined as follows: the bottom stator is completely fixed, while only axial movement is allowed for the others. And rotors retain axial and circumferential degrees of freedom. The normal interaction between rotors and stators is hard contact, and the tangential interaction is friction contact. Small sliding algorithm is employed in CMA, and finite sliding algorithm in TDA.

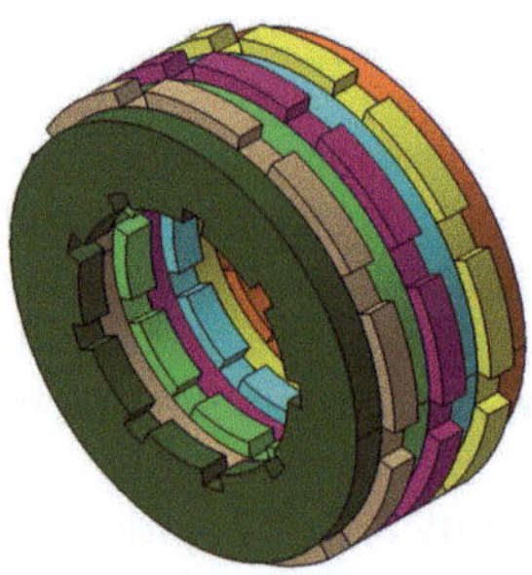

Fig. 1. Simplified aircraft disc brake

Table 1. Geometrical dimensions of the disc brake

Part	Outer radius(mm)	Middle radius(mm)	Inner radius(mm)	Thickness(mm)
Stator	156.0	106.0	94.0	22.0
Rotor	168.0	156.0	106.0	20.5

2.2 Mesh Optimization

Mesh optimization is applied in the contact zone between rotors and stators, namely, the mesh is refined within a 5mm distance from the contact interface. The medial axis algorithm is adopted for higher mesh quality. The rotor consists of 16051 elements, and the stator consists of 21203 elements. The disc brake is meshed using C3D8R elements, as illustrated in Fig. 2.

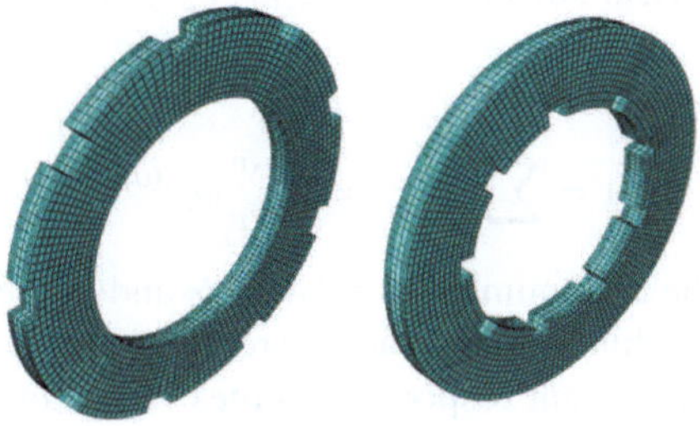

Fig. 2. Meshing of disc brake

3 Methodology and Validation

3.1 CMA

The governing differential equation of the disc brake is as follows:

$$\mathrm{M}\,\ddot{\mathrm{x}} + \mathrm{C}\,\dot{\mathrm{x}} + \mathrm{K}\,\mathrm{x} = \mathrm{F} \tag{1}$$

where M, C and K are the system mass matrix, damping matrix, and stiffness matrix, F is the external force; $\ddot{x}$, $\dot{x}$ and x are the system acceleration, velocity and displacement vector. Additionally, the damping matrix and stiffness matrices need to incorporate frictional effects.

The eigenvalue equation corresponding to Eq. (1) is given by

$$(\lambda^2 M \; + \; \lambda C \; + \; K)\phi = 0 \tag{2}$$

where ϕ is the eigenvector and λ is the eigenvalue.

In ABAQUS/Standard, the solution is performed using the subspace projection method. First, natural modes are computed according to Eq. (3) without consideration of the damping matrix and the asymmetric term of the stiffness matrix. Second, Eqs. (4), (5) and (6) are projected onto the subspace formed by eigenmodes. In the end, eigenvalue analysis is performed based on Eq. (7).

$$\left(-\omega^2 M + K_s\right) z = 0 \tag{3}$$

$$M^* = [z_1, z_2, \ldots, z_N]^T M \, [z_1, z_2, \ldots, z_N] \tag{4}$$

$$C^* = [z_1, z_2, \ldots, z_N]^T C \, [z_1, z_2, \ldots, z_N] \tag{5}$$

$$K^* = [z_1, z_2, \ldots, z_N]^T K \, [z_1, z_2, \ldots, z_N] \tag{6}$$

$$(\lambda^2 M^* + \lambda C^* + K^*)\phi^* = 0 \tag{7}$$

where K_s is the symmetrical stiffness matrix, $z_1, z_2, \ldots z_N$ are eigenvectors obtained by solving Eq. (3).

The final solution is as follows:

$$x(t) = \sum\nolimits_i^n \phi_i e^{\lambda_i t} = \sum\nolimits_i^n \phi_i e^{(\alpha_i + j\omega_i)t} \tag{8}$$

where $n = 30$ denotes the total number of modes; ϕ_i and λ_i are the i-th eigenvector and eigenvalue, respectively, while α_i and ω_i are the real part and imaginary part of λ_i. When $\alpha_i > 0$, the system displacement response diverges indicating system instability. And the larger its value, the more unstable the system is.

3.2 Explicit TDA

Explicit TDA employing the central difference method is a commonly used time integration method, which is particularly suitable for high-speed dynamic problems.

In ABAQUS/Explicit, the FE motion equation is:

$$M \ddot{x}^{(t)} = f_{ex}^{(t)} - f_{in}^{(t)} \tag{9}$$

where $\ddot{x}$ is the acceleration vector, M is the diagonal lumped mass matrix, f_{ex} and f_{in} represent the external load vector and the internal force vector.

The velocity and displacement can be derived:

$$\dot{x}^{(t+0.5\Delta t)} = \dot{x}^{(t-0.5\Delta t)} + \frac{\Delta t^{(t+\Delta t)} + \Delta t^{(t)}}{2} \ddot{x}^{(t)} \tag{10}$$

$$x^{(t+\Delta t)} = x^{(t)} + \Delta t^{(t+\Delta t)} \dot{x}^{(t+0.5\Delta t)} \tag{11}$$

where t is the time increment, Δt is the step length, and $t \pm 0.5\Delta t$ represents the middle increment step.

3.3 Analytical Procedure

CMA in ABAQUS is divided into four analysis steps.

The first step is nonlinear static analysis: a fully fixed constraint is applied to the bottom surface of the discs, while a uniformly distributed brake pressure of 4 MPa is applied to the upper surface.

The second step is nonlinear static analysis: the rotational speed is applied to stators via keyword *Motion, as shown in Fig. 3. Rotor is the node set of rotors, 2 refers to the rotational speed in radians, and the following six values represent the start and the end coordinates of the rotation axis.

```
*Motion,Rotation
Rotor,2,0,0,0,0.1,0,0
```

Fig. 3. Application of rotational speed to rotors

The third step is modal analysis to extract the natural frequency of the undamped system.

The last step is complex eigenvalue analysis: incorporating the effects of friction and calculating complex eigenvalues.

As shown in Fig. 4, TDA is divided into three stages. The first stage is to gradually apply the pressure to 4 MPa within 0.02s. The rotational speed is ramped up to the target value over another 0.02s in the second stage. The rotation is applied by coupling the outer surfaces of rotors to a reference point on the rotation axis, where a boundary condition is imposed. In the last stage, a simulation of 0.06s is performed under constant rotational speed and pressure.

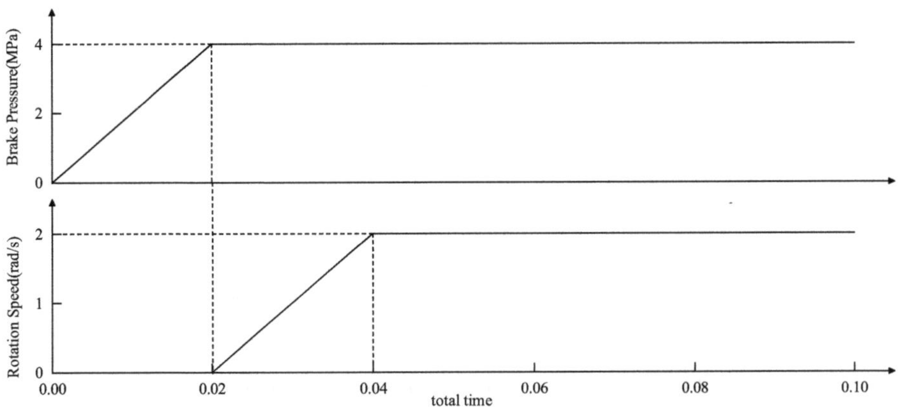

Fig. 4. TDA process

3.4 Method Validation

The results obtained by the two methods are analyzed under the friction coefficient $\mu = 0.2$, which is defined using the penalty method in friction properties. The results obtained from CMA are shown in Fig. 5, and the system is probably unstable at 3400 Hz, 3700 Hz, 3900 Hz, 4100 Hz, 4500 Hz and 5200 Hz. Characteristics of unstable modes corresponding to the instability frequencies can be clearly observed through the displacement contour plot in Fig. 6. In the TDA, Fig. 7 shows the axial displacement of a reference point on stators in the last 0.06s period, with data points sampled at 0.1ms intervals. As shown in Fig. 8, the Fourier transformation denotes that acute vibration takes place around 50 Hz, 1400 Hz, 2700 Hz, 3400 Hz, 4100 Hz and 4500 Hz.

In terms of accuracy, both methods predict the unstable modes at 3400 Hz, 4100 Hz and 4500 Hz, indicating the consistency of two methods in identifying main unstable modes. According to literature [7], aircraft disc brakes are prone to squeal vibrations above 3 kHz under the same material properties, which is consistent with the frequency range predicted in the study. In addition, Liu et al. [11] observed unstable modes below 3 kHz when different boundary conditions were applied to the same model, with axial vibration amplitudes on the order of micrometers, which agrees well with the TDA results. Further analysis shows that CMA also predicts several high-frequency unstable modes and fails to capture certain low-frequency instabilities, most notably the 50 Hz mode. Although discrepancies exist between the two methods, the overall differences are considered to be within acceptable limits. In terms of computational efficiency, TDA takes 1.5 h on a 10-core, 16-thread CPU, significantly longer than the 40 min needed for CMA. Fundamentally, CMA, being a linear method, tends to either over-predict or under-predict unstable modes in strongly nonlinear systems; in contrast, TDA is capable of capturing the vibration evolution process and nonlinear responses, which has a higher accuracy at the cost of greater computational resources [12].

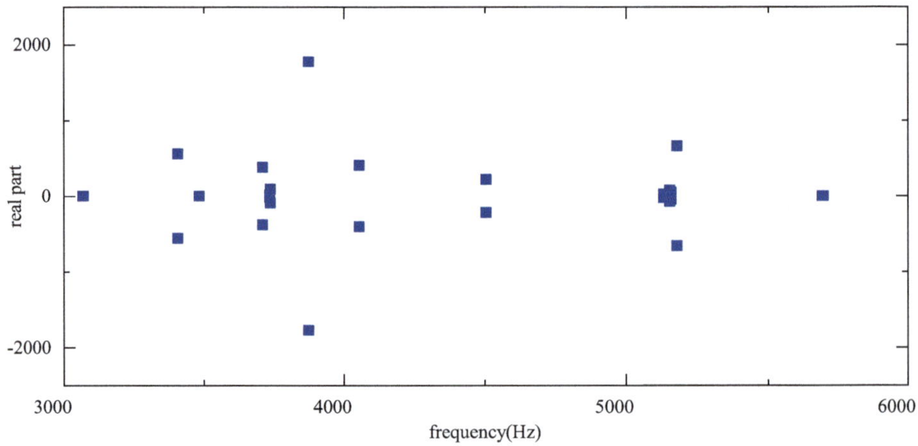

Fig. 5. Complex eigenvalue scattergram at $\mu = 0.2$

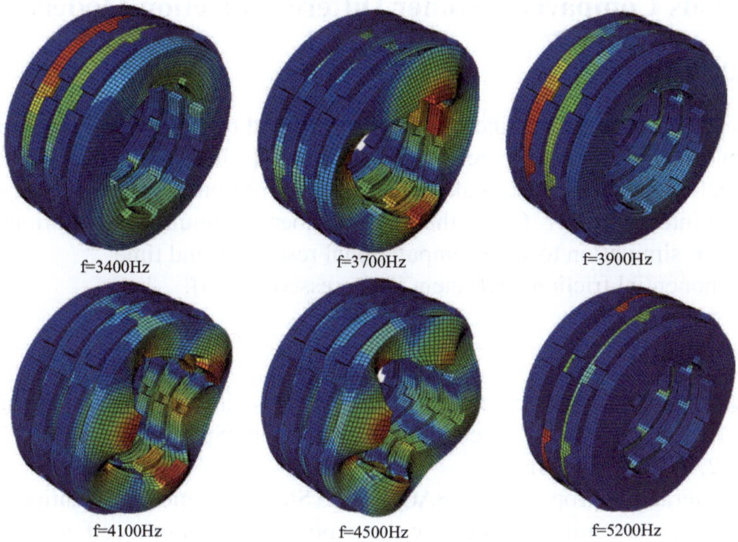

Fig. 6. Unstable vibration modes

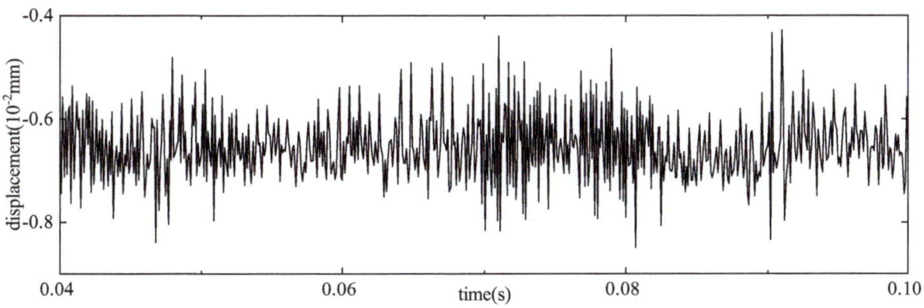

Fig. 7. Time domain axial displacement

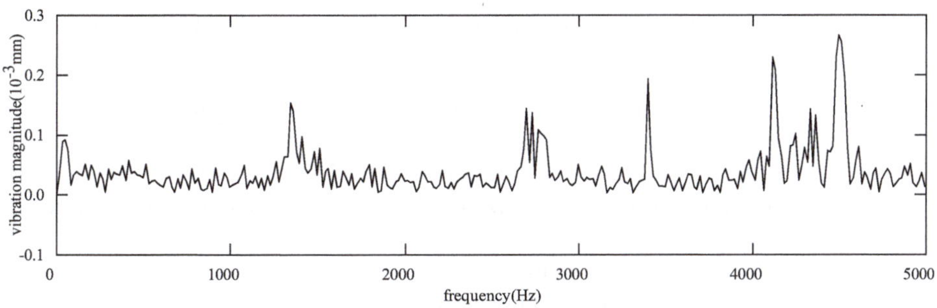

Fig. 8. Frequency domain axial displacement

4 Methods Comparison Under Different Friction Models

4.1 Method Selection

A constant friction coefficient indicates that the ratio of the frictional force to the normal force between contact surfaces remains fixed. To fully investigate the influence of the friction coefficient on the disc brake vibration, the friction coefficient is varied from 0 to 0.5 at an interval of 0.1. Given the large number of simulation conditions, CMA is employed for simulation to save computational resources and time.

The exponential friction coefficient is expressed as [13]:

$$\mu = \mu_k + (\mu_s - \mu_k)e^{-\alpha|v_{rel}|} \tag{12}$$

where μ_s is the static friction coefficient, μ_k is the kinetic friction coefficient, α is the decay coefficient, and v_{rel} is the relative sliding velocity. Specifically, $\mu_s = 0.3$, $\mu_k = 0.2$ and $\alpha = 0.2$, as shown in Fig. 9.

In the interaction property of ABAQUS, the Static-Kinetic Exponential Decay friction formulation is employed to define the exponential friction coefficient. Due to the stronger nonlinearity of exponential friction systems, TDA is used for more accurate predictions.

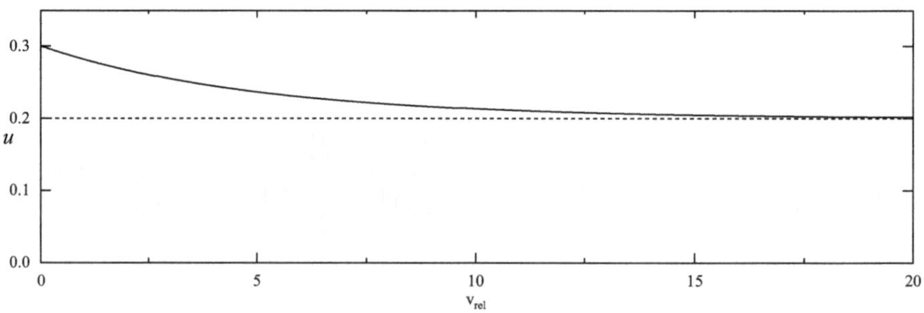

Fig. 9. Exponential friction coefficient

4.2 Constant Friction-Induced Vibration

The results of the CMA are shown in Fig. 10 and Fig. 11, with detailed analysis as follows:

(1) In the absence of friction, the real parts of complex eigenvalues of the system remain almost zero, indicating that the system is consistently stable.
(2) The introduction of friction leads to significantly positive real parts of complex eigenvalues at 3400 Hz, 3700 Hz, 3900 Hz, 4100 Hz, 4500 Hz and 5200 Hz, suggesting a high risk of instability.
(3) The real parts of complex eigenvalues at 3700 Hz, 3900 Hz, 4100 Hz, 4500 Hz and 5200 Hz rise as the coefficient of friction increases, which means that stronger friction exacerbates the system's instability.

(4) As shown in Fig. 11, the real and imaginary parts of the unstable modes at 3400 Hz, 4100 Hz and 4500 Hz are obtained by CMA at 2 rad/s for the system with $\mu = 0.2$, and their relative vibrations are less than 1% compared to the results at 10, 20, 30, 40, 50, 60 rad/s. The sensitivity analysis indicates that the rotational speed has a negligible effect on the overall modal characteristics in the CMA. Consequently, CMA has difficulty capturing the differences in system response under varying rotational speeds.

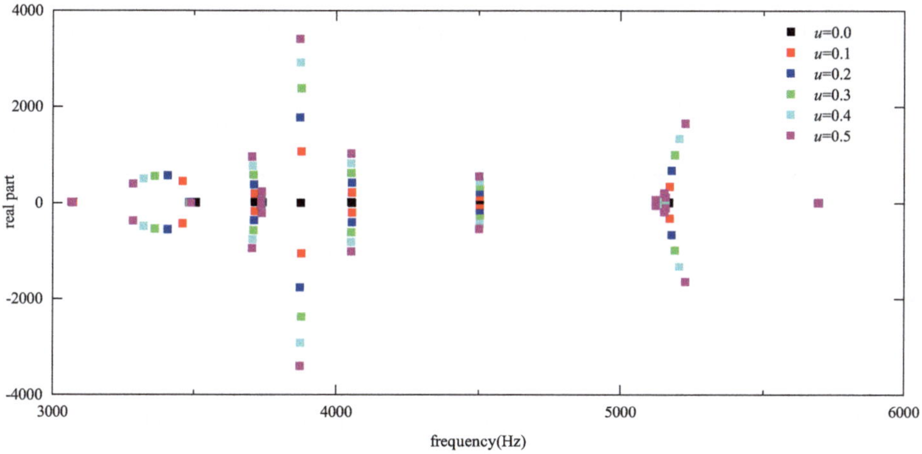

Fig. 10. Complex eigenvalue scattergram of different friction coefficients at 2rad/s

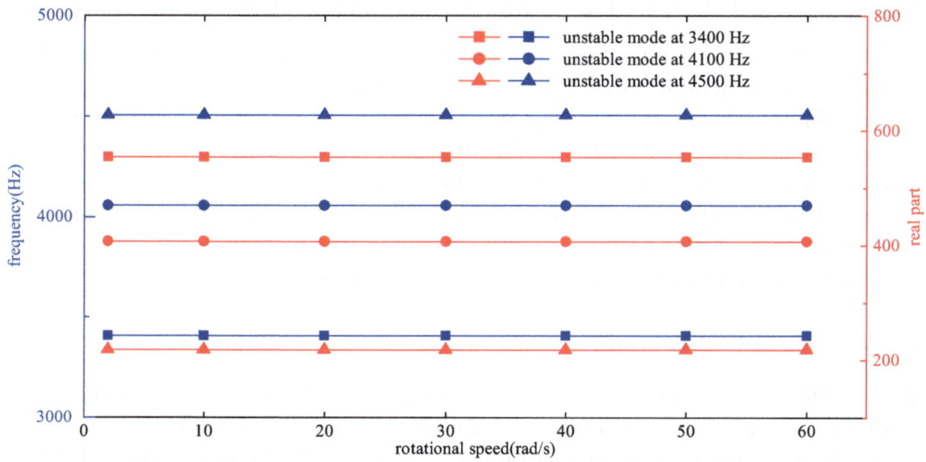

Fig. 11. Unstable modes at 3400Hz, 4100Hz and 4500 Hz under varying rotational speeds at $\mu = 0.2$

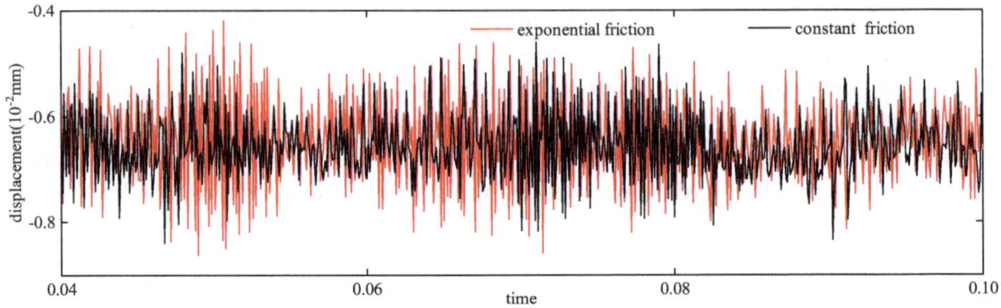

Fig. 12. Displacement of systems with constant friction and exponential friction

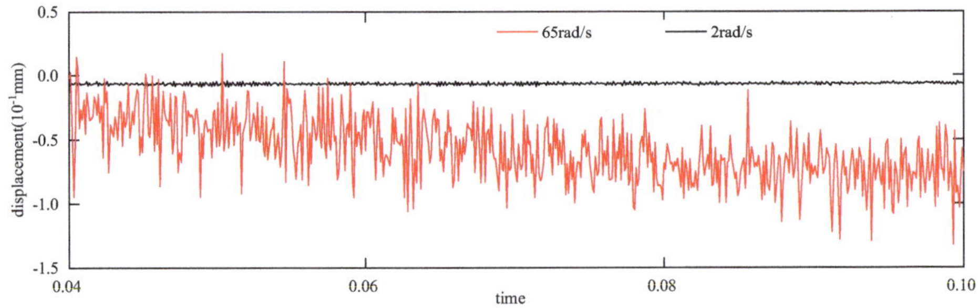

Fig. 13. Displacement of systems with exponential friction at 2 rad/s and 65 rad/s

4.3 Exponential Friction-Induced Vibration

The results of TDA shown in Fig. 12 and Fig. 13 can be analyzed as follows:

(1) The reference point on stators remains in a vibration state during braking, which may lead to structural instability and noise excitation.
(2) The vibration displacement of stators under exponential friction is generally larger than that under constant friction, indicating that exponential friction leads to stronger vibration and a higher tendency toward instability.
(3) The vibration amplitude of stators increases significantly, and vibration behavior is more complex during high-speed braking, suggesting that the initial braking speed has a notable influence on the vibration response in TDA.

4.4 Analysis of Results

CMA facilitates observing the system's deformation at characteristic frequencies and identifying unstable modes. For relatively simple cases such as constant friction models and scenarios involving extensive parameter variations, CMA is recommended as it offers sufficient accuracy with lower computational cost. However, CMA cannot reflect the actual vibration amplitude, nor can it account for the effect of rotational speed on system instability. Under complex conditions such as the exponential friction model, TDA offers distinct advantages, as it not only accounts for the influence of rotational speed but also captures the actual vibration amplitudes. Therefore, the selection between the two methods should be determined based on the specific simulation scenarios and practical engineering requirements.

5 Conclusion

A simplified aircraft FE model is established to comparatively analyze the performance of CMA and TDA in friction-induced vibration. The two methods yield consistent predictions at a friction coefficient of $\mu = 0.2$, and the validity of the results is verified. CMA tends to over-predict high-frequency unstable modes while under-predicting those at low frequencies, but it can significantly reduce computational time. In contrast, the results from TDA are more accurate and intuitive. Furthermore, CMA reveals that increased instability occurs with higher friction coefficients under constant friction, while sensitivity analysis reveals that modal characteristics remain insensitive to rotational speed. Under exponential friction conditions, TDA can capture accurate vibration responses, particularly revealing pronounced nonlinear dynamic behavior at high rotational speed. In conclusion, CMA is computationally efficient and suitable for simple conditions and parametric studies, whereas TDA offers higher accuracy and is more appropriate for complex working conditions, and the two methods should be chosen reasonably.

Acknowledgments. This project was funded by The Explorers Program of Shanghai (Basic Research Funding) (No. 24TS1402200).

References

1. Cai, C., Cheng, Y., Shuai, C.: Research on the friction noise generation mechanism and suppression method of submarine rubber-based propeller bearings—a review. Polymers **15**(16), 3352 (2023)
2. Fagan, X., Sinou, J.J., Besset, S., et al.: The influence of disks deformation on the stability analysis of an aircraft braking system. Arch. Appl. Mech. **95**(1), 13 (2025)
3. Yin, Y., Cao, J., Bao, J., et al.: Characteristics and mechanism of friction noise and vibration of disk brake under DC magnetic field. Tribol. Trans. **67**(1), 185–197 (2024)
4. Zhang, H., Qiao, J., Zhang, X.: Nonlinear dynamics analysis of disc brake frictional vibration. Appl. Sci. **12**(23), 12104 (2022)
5. Baskaran, S., Sivaprakasam, S.: Friction analysis of aircraft landing gears due to landing impact. Proc. Inst. Mech. Eng. Part J: J. Eng. Tribol. **236**(2), 274–283 (2022)
6. Wu, B.W., Qiao, Q.F., Chen, G.X., et al.: Effect of the unstable vibration of the disc brake system of high-speed trains on wheel polygonalization. Proc. Inst. Mech. Eng. Part F: J. Rail Rapid Transit **234**(1), 80–95 (2020)
7. Zhang, M., Xu, R., Nie, H.: Analysis on aircraft brake squeal problem based on finite element method. Int. J. Aerosp. Eng. **2017**(1), 3982851 (2017)
8. Zhu, Q., Chen, G.X., Wu, B.W., et al.: Effect of the material parameter and shape of brake pads on friction-induced disc brake squeal of a railway vehicle. Tribol. Trans. **64**(4), 744–752 (2021)
9. Vashishtha, G., Chauhan, S., Singh, R., et al.: A review of numerical techniques for frictional contact analysis. Lubricants **13**(1), 18 (2025)

10. Tapia, N.F., Reyes, S.I., Vassiliou, M.F., et al.: Development of a pressure-, velocity-, and acceleration-dependent phenomenological friction model using experimental data of sliding tests between 11 polymers and stainless steel. Eng. Struct. **318**, 118761 (2024)
11. Liu, W., Xiang, L., Ma, Y., et al.: Transient dynamical analysis of aircraft wheel brake based on LS-DYNA. Mech. Res. Appl. **29**(2), 15–17 (2016). (in Chinese)
12. Zhang, Z., Oberst, S., Lai, J.C.S.: On the potential of uncertainty analysis for prediction of brake squeal propensity. J. Sound Vib. **377**, 123–132 (2016)
13. He, S., Zhang, W., He, B., et al.: Vibration reduction analysis of a dry friction constrained longitudinal vibration system considering the maximum Mises stress and the magnitude and distribution of the normal load. Mech. Solids **59**(1), 345–362 (2024)

Analysis of Internal Flow Field Characteristics in the Hydraulic Retarder of Heavy-Duty Vehicle

Changshuai Zhang, Ruikai Wu(✉), Yan Wang, Yanhua Jiang, and Xin Jin

School of Mechanical Engineering, Liao Ning Technical University, Fuxin 123000, China
13203450561@163.com

Abstract. The hydraulic retarder is a crucial auxiliary braking component on heavy-duty trucks. However, the hydraulic retarder has poor braking performance over long distances and heavy loads. The internal flow field distribution directly affects the hydraulic retarder's working performance and braking torque. In this paper, we aim to analyze the influence of the structural parameters of the stator and rotor on the braking torque. The entire flow path of the hydraulic retarder was selected as the computational model. The reliability of the simulation model was verified through grid independence validation, and the accuracy of the simulated torque was validated through institutional test bench experiments. We analyze the changes of pressure field, velocity field, and turbulent kinetic energy, then explore the relationship between the internal flow field and output torque. The results indicate that the angle of inclination contributes to the torque to a certain extent. The maximum braking torque occurs at 40°. The severe flow separation occurs at 45°. Compared to the initial model, the braking torque loss at 45° is 13.7%. The turbulent kinetic energy of the 45 mm inner diameter was increased by 55.92% compared to the initial model. However, the 45 mm inner diameter results in unstable flow and a loss of 14.93% in braking torque.

Keywords: hydraulic retarder · blade inclination angle · rotor inner diameter · numerical simulation · braking torque

1 Introduction

The hydraulic retarder converts kinetic energy into heat energy through hydraulic friction to reduce the wear of mechanical brakes [1]. The hydraulic retarder's braking performance is directly affected by changes in the internal flow field. The geometric structures of the rotor and stator influence these changes to the flow field. Therefore, we analyze the internal flow field under different structural parameters of the stator and rotor.

Domestic and foreign scholars have conducted various studies on the internal flow field and operating characteristics of hydraulic retarders. The braking system must provide continuous braking force to ensure the safety and stability of heavy vehicles on long downhill slopes. Zhang, Qixiang et al. [2] studied the heat generated during braking for different durations. Wang et al. [3] designed a novel hierarchical brake actuation control (HBAC) algorithm to improve control accuracy and energy recovery efficiency.

S. Halgamuge et al. (Eds.): ICACMVE 2025, LNME, pp. 653–664, 2026.
https://doi.org/10.1007/978-981-95-6911-3_66

Li Xuebo et al. [4] proposed a new strategy for managing braking energy to optimize its distribution during downhill driving. Wei, W et al. [5] proposed a composite hysteresis characteristic compensation control strategy to enable precise braking torque control. Chen, X. Q et al. [6] proposed a deep reasoning and control method to address the problem of torque tracking control with time delays. Zhang, K et al. [7] adopted a hybrid braking control strategy that distributes braking torque under high- and low-speed conditions. DAI C et al. [8] found through experimentation that the effects of rotor-stator interaction and flow separation intensify as the fluid filling volume increases. Gao Z, W et al. [9] proposed a novel internal pump hydraulic retarder axle to address the issue of brake failure and overheating during the downhill braking of semi-trailers. Yang K et al. [10] gradually increased the braking torque by optimizing the parameters.

Wu C, X et al. [11] reduced friction and wear in the longitudinal control of heavy vehicles by developing a hydraulic retarder model with pneumatic oil flow rate control. Nahmkeo Hur et al. [12] found that an increase in the flow rate of the hydraulic retarder resulted in a decrease in the deceleration torque and flow circulation between the wheels. Zheng Hong-Peng et al. [13] used a Shear Stress Transport (SST) turbulence model to simulate braking torque more accurately. Lei Yulong et al. [14] designed a hydraulic retarder filling ratio controller with excellent performance in constant torque control and interference resistance. Liu C et al. [15] found that a decrease in viscosity can suppress vortex generation as reflected by a decrease in vorticity. Chen M et al. [16] studied the effect of the blade inclination angle on the performance of a hydraulic retarder. They found that the inclination angle affected the formation of secondary vortices and the separation flow.

The above literature uses joint simulations, experimental testing, and numerical simulations to conduct relevant studies on hydraulic retarders. The retarder's braking performance is affected by multiple factors. This study utilizes various structural parameters further to investigate a hydraulic retarder's internal flow field distribution, examining the impact of these parameters on the internal flow field.

2 Hydraulic Retarder Simulation Model Construction

2.1 Working Principle of the Hydraulic Retarder

The geometric structure of the hydrodynamic retarder mainly comprises a shell, a stator, a rotor, a working chamber, a bearing, and a bearing seat. As shown in Fig. 1. The rotor is connected to the vehicle's drive shaft, and the stator is connected to the retarder housing. When the vehicle needs to decelerate, the rotor—which rotates with the drive shaft—is obstructed by oil flowing into the working chamber from the stator. This process converts the mechanical energy of the drive shaft into the kinetic and internal energy of the oil, thereby achieving deceleration and braking. The oil then flows out at high speed from the oil outlet on the stator side.

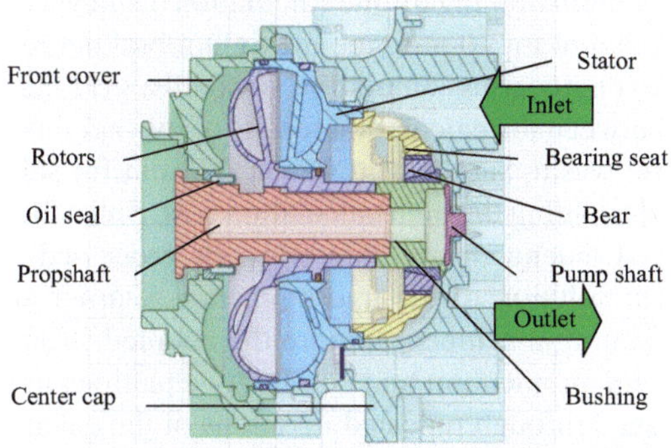

Fig. 1. Structural diagram of the hydrodynamic retarder.

2.2 Three-Dimensional Flow Channel Extraction for Hydrodynamic Retarder

The structural simplification of the rounded corners, chamfers, and other parts of the internal structure of the hydraulic retarder can avoid simulation errors. The seal's influence had been disregarded. The full-flow channel fluid domain is the computational model, as shown in Fig. 2.

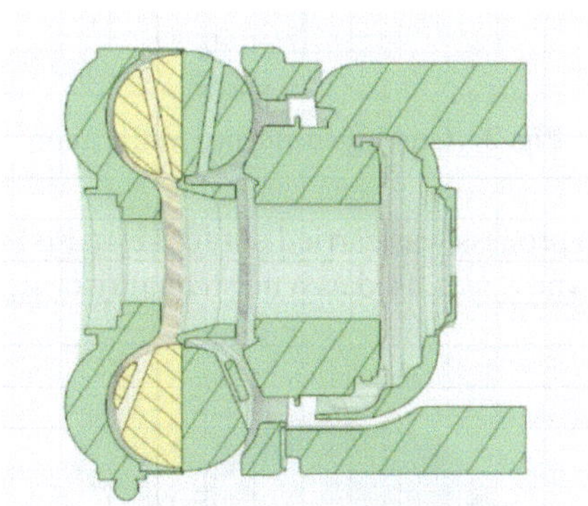

Fig. 2. Cross-section of the fluid domain computational model of the hydrodynamic retarder.

3 Numerical Simulation and Experimental Validation

3.1 Grid Delineation and Grid Independence Verification

We verified grid independence to ensure simulation accuracy and efficiency, and avoided significant effects of grid size on fluid domain simulation results. Then we set the grid size to 0.5, 1.0, 1.5, 2.0, 2.5, and 3.0 mm. The rotor speed was 1069 r/min. The inlet pressure was 2 bar, and the export quality flow rate was 1.594 kg/s. The maximum braking torque (T) was used to judge the irrelevance of the grid. As shown in Fig. 3, the output torque fluctuation was monitored when the mesh size increased from 0.5 mm to 2.0 mm. The change rate of the maximum braking torque calculation result was 1.7%. When the mesh size increased from 2.0 mm to 3.0 mm, the maximum braking torque change rate rose to 3.3%. According to engineering simulation standards, grid independence is reached when the difference between the calculation results of two neighboring grids is less than 2%. The final decision was to set the grid size to 2.0 mm for subsequent calculations, ensuring accuracy and avoiding the high calculation cost. Since the velocity and pressure fields cannot be solved independently, they must be processed simultaneously in the solution. The transient solution of the hydrodynamic retarder and the complexity of the internal flow in rotating machinery require the coupled strong coupling method to simultaneously solve the momentum and continuity equations in the solution, with the coupling results obtained through repeated iterations of the calculation.

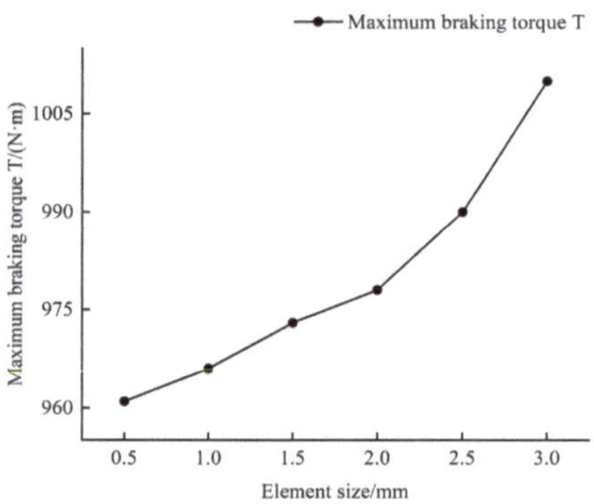

Fig. 3. Grid Independence Verification.

The mesh encryption process was carried out for key parts, such as the blade surface. As shown in Fig. 4. Skewness was used as a mesh quality criterion, with a target quality of 0.7.

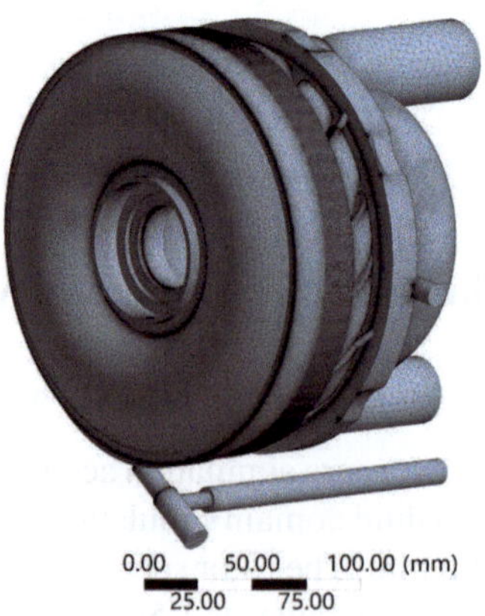

Fig. 4. Computational fluid domain mesh model.

3.2 Simulation Condition Setting

The Fluent module's pressure-based transient solution was used for the numerical calculations, and the realizable k-ω turbulence model was selected. The working medium was oil at a temperature of 80 °C, with a density of 821.9 kg/m^{3}and a dynamic viscosity of 0.020219 kg/m·s.

The remaining surfaces had wall boundary conditions. The velocity-pressure coupling algorithm had been selected as 'Coupled,' and the discretization method was second-order windward. The computational residual criterion was set to 1e-5 to ensure the reliable convergence of the velocity and pressure fields. The rotor can rotate through 1.8 revolutions in 0.1 s. The simulation step size was set to 0.001 s, and the simulation time was 0.2 s.

3.3 Model Validation

Connect the hydraulic retarder to the test bench for relevant tests, as shown in Fig. 5.

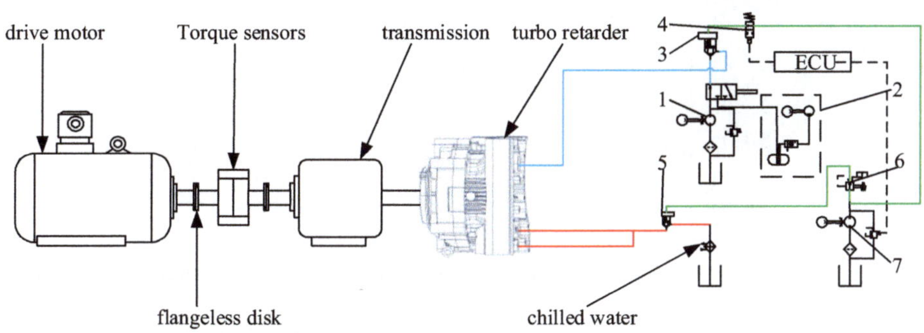

1. Main oil supply pump. 2. Pressure oil pump. 3. Main oil filling control valve. 4. Oil-filled pilot valve. 5. Main valve for controlling oil release. 6. Oil release valve. 7. Control oil pump.

Fig. 5. Retarder Test System.

The actual operating characteristics of the hydrodynamic retarder were obtained by examining the numerical simulation results. The input end of the hydrodynamic retarder was connected to the drive motor via the transmission and torque sensor; the output end connected to the hydraulic control circuit.

The input shaft's rotational speed was stabilized at 1069 rpm. The oil filling and draining control valve regulated the oil flow to maintain a pressure difference of 2.5 bar between the inlet and outlet. The oil temperature was set to 80 °C, and cooling water was circulated to cool the oil. The measured torque (T) was 965 Nm. The measured torque deviated from the numerical simulation result of 978 N·m by 1.33%, with a relatively small relative error. Repeated tests and simulations were conducted and compared to determine that the error in each data set was stabilized to within 5%. The error complied with the relevant engineering standards. The hydrodynamic retarder fluid domain computational model was accurate.

4 Fluid Retarder Flow Field Characterization

4.1 Pressure Field Analysis

The tilt angles of the stator and rotor of the initial model are both 40°, and the inner diameter of the rotor is 49 mm. The direction of rotation of the rotor is consistent with the right-hand screw rule in the positive direction of the Z-axis. The leading edge of the rotor blade is the area where it first contacts the fluid and guides the oil into the blade flow channel. The trailing edge is where the oil exits this channel.

Figure 6(a) shows the overall static pressure distribution. As the oil flows through the blade's leading edge, the fluid velocity decreases sharply near the impact point, and the fluid's kinetic energy is converted into pressure energy.

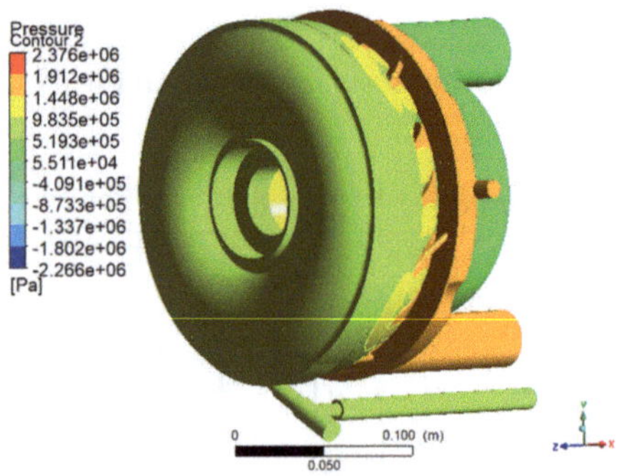

(a) Cloud view of the global pressure distribution.

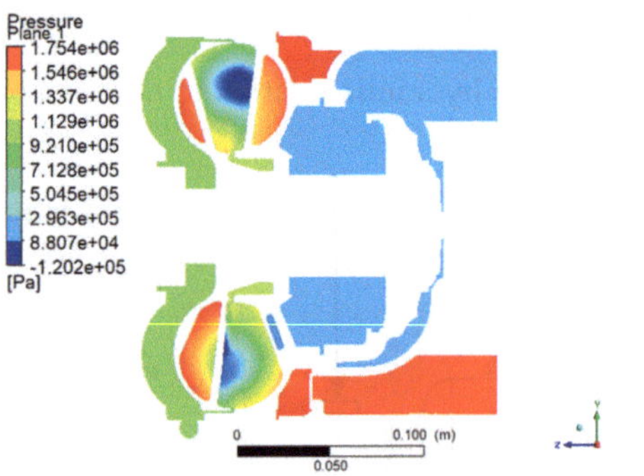

(b) Cloud view of the nodal pressure distribution.

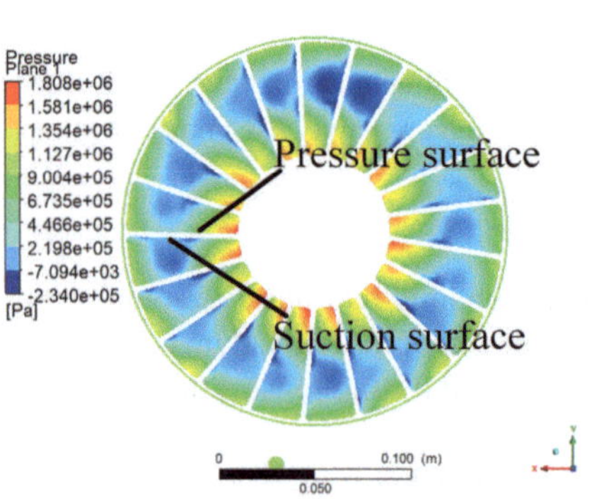

(c) Rotor cross-sectional pressure distribution.

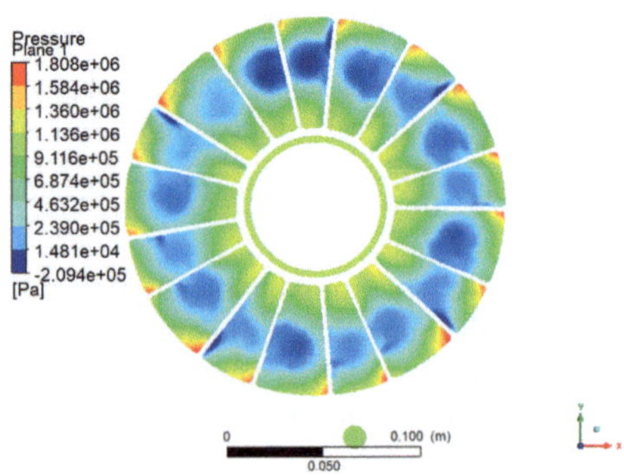

(d) Stator cross-sectional pressure distribution.

Fig. 6. Pressure Distribution of Hydraulic Retarder.

According to Bernoulli's equation:

$$P_{total} = P_{static} + \frac{1}{2}\rho v^2 \tag{1}$$

When the flow rate decreases, the static pressure increases significantly, resulting in a higher pressure at the leading edge. As shown in Fig. 6(b), the pressure gradient between the leading and trailing edges of the rotor blades causes the oil to flow from the rotor to the stator and back within the circulation circle, forming a continuous flow. In the annular arc surface region of the rotor, fluid is subjected to centrifugal force, forming a high-pressure zone. The pressure gradient from the root to the rotor blade tip also drives fluid circulation.

Figures 6(c) and (d) show the pressure at the intersection of the rotor and stator fluid domains. The difference in pressure between the pressure and suction sides of the rotor blades directly generates torque. Figure 7 illustrates how the pressure difference (expressed as ΔP) varies between the pressure and suction sides of the rotor blades at different inner diameters and blade angles. As the tilt angle of the blade increases, the braking torque first increases and then decreases. At the tilt angle of 40°, the blades obtain a greater fluid force, resulting in the braking torque reaching its maximum value. Reducing the inner diameters of the stator and rotor weakens the centrifugal pressure and significantly decreases the pressure difference on the blade surface. Consequently, the braking torque decreases sharply.

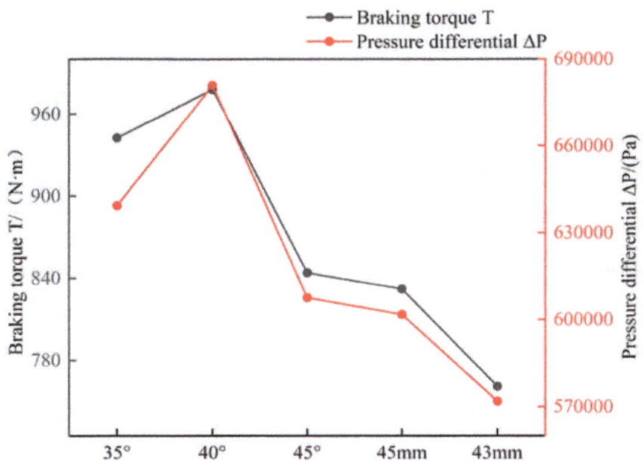

Fig. 7. Surface pressure difference on blades and braking torque under different inner diameters and blade angles.

As the inclination angle increases to 45°, severe flow separation occurs at the back of the blade, as shown in Fig. 8. The sudden change in curvature at the blade's leading edge, combined with a large angle of inclination, results in a dramatic change in pressure. The sudden change in pressure triggers a fluid acceleration and deceleration cycle, inducing vortex shedding. Ultimately, fluid separation results in a significant increase in energy loss and a corresponding decrease in torque.

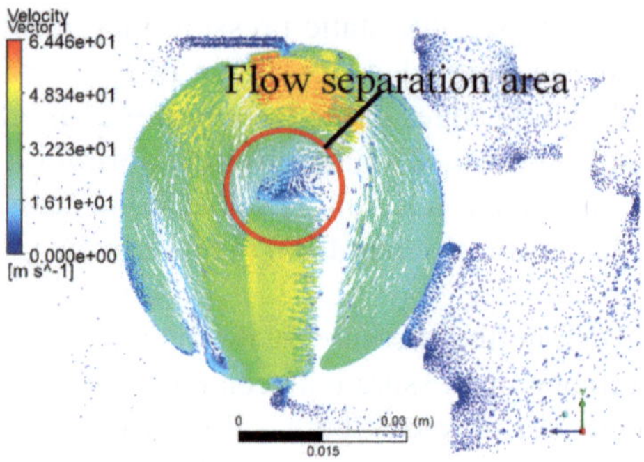

Fig. 8. Fluid separation phenomenon.

4.2 Velocity Field Analysis

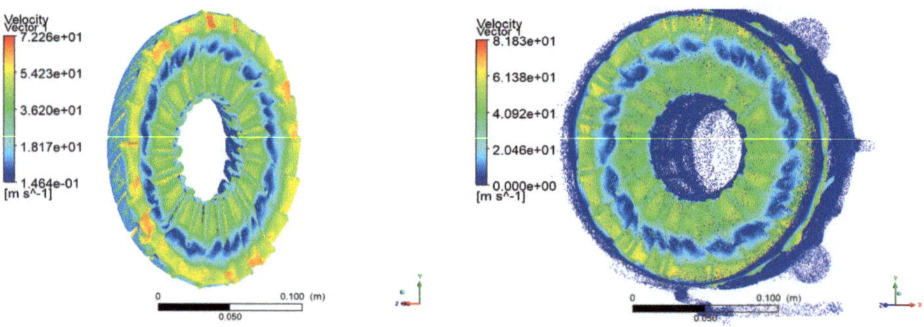

(a) Rotor side velocity vector distribution. (b) Stator side velocity vector distribution.

Fig. 9. Speed Distribution of Hydraulic Retarder.

According to Bernoulli's equation, areas of low pressure have high velocity. The areas of high pressure have low velocity. As shown in Figs. 9(a) and (b), when oil flows from the stator into the rotor, it is rapidly flung outwards towards the circumference of the circulation circle under the influence of centrifugal force. The oil fluid achieves high absolute velocity and kinetic energy. Here, the magnitude and direction of the velocity of the oil fluid serve as the 'raw material' for generating torque in the stator. Changes in the velocity vector are the root cause of braking torque. The centrifugal force exerted by the rotor results in high tangential and radial velocity at the outer edge and low velocity at the inner edge. The complex radial velocity gradient interacts with the pressure gradient.

4.3 Characteristics of Turbulent Kinetic Energy Distribution

Turbulence kinetic energy is a measure of the ability to exchange energy; the greater the turbulence kinetic energy, the more intense the turbulent pulsation of the fluid, and the greater the mixing and energy exchangeability. As shown in Fig. 10, turbulent arterial motion is strongly present in the leading-edge areas of the rotor and stator blades. The blade pressure surface at the leading edge of the rotor blade rotates with the rotor at high speed, forcing the fluid that has just flowed in from the stator side to accelerate. At the same time, the velocity direction changes, and intense energy exchange occurs. After the rotor side accelerates, the oil accelerates due to the change in velocity direction. The oil then flows back into the stator side from the leading edge of the stator blade region. At this point, the stator obstructs the high-speed fluid flow, causing its velocity to drop by 25.22%, as can be seen. Turbulent pulsation of the fluid is intense at this location.

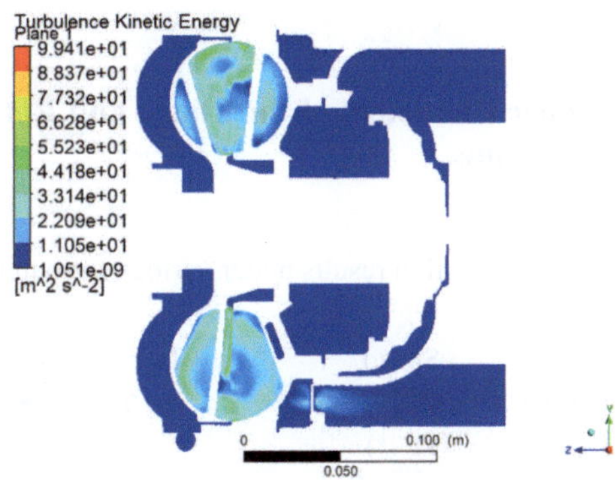

Fig. 10. Turbulent kinetic energy distribution.

Figure 11 illustrates the velocity differences and turbulent kinetic energies associated with various structural parameters. The maximum velocity values for all three inclination angles are approximately 81 m/s. However, at an inclination angle of 45°, the fluid velocity decreases significantly upon entering the stator. A large amount of kinetic energy is converted into heat energy on impact with the working surface.

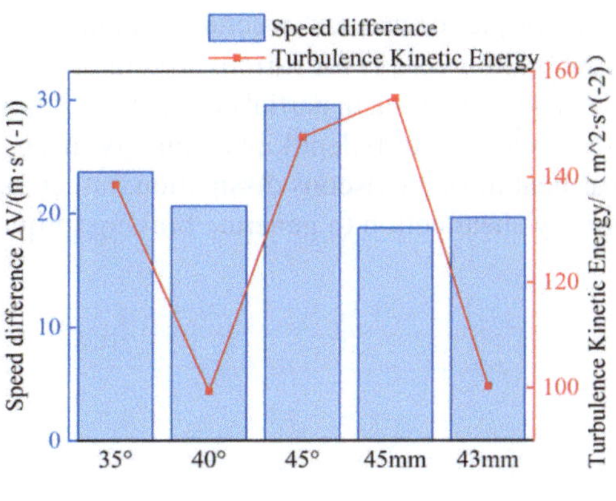

Fig. 11. Velocity and turbulent kinetic energy of different structural parameters.

The reduction in the inner diameter causes a local contraction of the flow channel. The contraction generates a highly intense alternation of high- and low-pressure zones, leading to flow separation. The separation zone occupies the flow channel, reducing the equivalent cross-sectional area. However, the backflow from the separation zone reduces the net flow rate, decreasing the overall flow velocity. The reduction in the inner diameter brings the fluid closer to the axis of rotation. Due to the constant rotor speed and conservation of angular momentum, the fluid velocity decreases. The fluid's low velocity increases the relative velocity between the fluid and the blades, enhances shear strength, and increases the rate of turbulent kinetic energy generation. However, the flow is unstable.

4.4 The Effect of the Inner Diameters of the Stator and Rotor and the Blade Angles on Braking Torque.

Table 1. Numerical simulation results under various structural parameters.

Structural parameters	Pressure differential/Pa	Speed difference/(m $\bullet s^{-1}$)	Turbulent kinetic energy/($m^2 \bullet s^{-2}$)	Braking torque T/(N·m)
40°	639300	23.68	138.6	942
35°	681000	20.64	99.41	978
45°	607600	29.55	147.6	844
45mm	601800	18.73	155.0	832
43mm	571900	19.64	100.3	761

Table 1 presents the pressure difference between the two surfaces of the blade, the velocity difference between the rotor and the stator, the maximum turbulent kinetic energy, and the braking torque for various structural parameters.

Figure 12 shows the braking torque for various structural parameters. A vortex forms at an inclination angle of 45°, resulting in partial energy loss. An inner diameter of 43 mm causes flow instability. Excessive turbulent kinetic energy means that the fluid's kinetic energy is dissipated as heat through viscous dissipation, rather than being transferred to the stator through orderly shear action to generate braking torque.

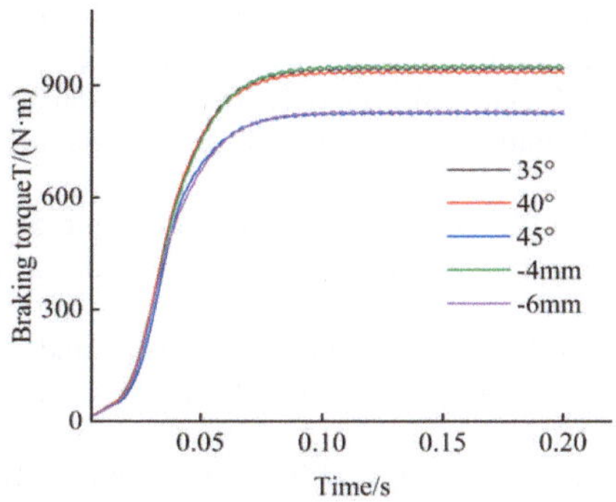

Fig. 12. Braking torque of different structural parameters.

5 Conclusion

This paper addresses the issue of poor braking performance of the hydraulic retarder over long distances and heavy loads. The following conclusions are drawn:

(1) As the blade angle increases, the braking torque first increases and then decreases. The maximum braking torque occurs at an angle of 40°. The angle contributes to the torque to a certain extent. However, when the angle increases beyond 40°, severe flow separation occurs on the back of the blade.
(2) Analysis of the velocity vector diagram shows that the small-scale vortex generated at an inclination angle of 45° is caused by a sudden change in curvature at the leading edge of the large-angle blade. Sudden changes in pressure caused cycles of acceleration and deceleration of the fluid. Compared to the initial model, brake torque loss is 13.7%.
(3) At the inner diameter of 45 mm, the turbulent kinetic energy increases by 55.92% compared to the initial model. However, changes to the inner diameter can cause flow instability and result in flow separation. The reduced inner diameter compresses the geometric flow channel, resulting in a 14.93% loss in braking torque compared to the initial model.

Acknowledgments. This study was funded by Liaoning Province Natural Fund Joint Fund Project (No. 20240318) and Liaoning Province Natural Fund Joint Fund Project (No. 20240326).

References

1. Ding, Z.W., Zhou, L., Xu, J., Tong, J.W.: Vehicle Retarders: A Review. IEEE Access **11**, 102757–102767 (2023)
2. Zhang, Q.X., et al.: The evolution of friction and wear behavior of train brake friction pairs during sustained drag braking on long steep slopes. Wear **564–565**, 205724 (2025)
3. Wang, J.C.: Hierarchical braking accurate control of electrohydraulic composite braking system for electric vehicles. ISA Trans. **158**, 594–608 (2025)

4. Li, X.B., et al.: A novel braking energy management strategy for battery electric trucks with hydraulic retarder on long downhill. Energy **322**, 135684 (2025)
5. Wei, W., Tao, T. L., Jian, H. C., Guo, M., Hu, N.P., Yan, Q.D.: On the hysteresis characteristics and compensation control strategy of a pneumatic hydrodynamic retarder. Phys. Fluids **36** (2024)
6. Chen, X.Q., Wei, W., Yan, Q.D., Yang, N.K., Huang, J.Q.: Time-delay deep Q-network based retarder torque tracking control framework for heavy-duty vehicles. IEEE Trans. Veh. Technol. **72**, 149–161 (2023)
7. Zhang, K., Shang, H.C., Xu, J., Niu, J.G., Yue, Y.G.: Testing and performance analysis of an integrated electromagnetic and hydraulic retarder for heavy-duty vehicles. Eng. Appl. Artif. Intell. **126**, 106906 (2023)
8. Dai, C., Dong, L., Zhang, X., Zhu, J.C., Lin, H.B.: Experimental and internal flow investigation on the performance of a hydraulic retarder with different liquid-filled amount and blade inclination angles. J. Therm. Sci. **31**, 923–933 (2022)
9. Gao, Z.W., Li, D.S., Liu, Z.Q., Ye, L.Z.: Design and braking stability analysis of a novel internal pump-hydraulic retarder axle for heavy articulated vehicle. Int. J. Automot. Technol. **23**, 1285–1294 (2022)
10. Yang, K.H., Liu, C.B., Wu, Q.T., Li, X.S.: Multi-objective optimization design for a hydrodynamic retarder based on CFD simulation considering cavitation effect. Proc. Inst. Mech. Eng. C J. Mech. Eng. Sci. **236**, 1443–1460 (2021)
11. Wu, C.X., Guo, X.X., Yang, B., Pei, X.F., Guo, S.J.: Hydraulic retarder torque control for heavy duty vehicle longitudinal control. Int. J. Heavy Vehicle Syst. **26** (2019)
12. Hur, N., Moshfeghi, M., Lee, W.: Flow and performance analyses of a partially-charged water retarder. Comput. Fluids **164**, 18–26 (2018)
13. Zheng, H.P., Lei, Y.L., Song, P.X.: Water medium retarders for heavy-duty vehicles: computational fluid dynamics and experimental analysis of filling ratio control method. Journal of Hydrodynamics. Ser. B **29**, 1067–1075 (2017)
14. Lei, Y.L., Song, P.X., Zheng, H.P., Fu, Y., Li, X.Z., Song, B.: Application of fuzzy logic in constant speed control of hydraulic retarder. Advances in Mechanical Engineering, 9 (2017)
15. Liu, C., Bu, WeiyaWng, Wang, T.: Numerical investigation on effects of thermophysical properties on fluid flow in hydraulic retarder. Int. J. Heat Mass Transf. **114**, 1146–1158 (2017)
16. Chen, M., Guo, X.X., Tan, G.F., Pei, X.F., Zhang, W.: Effects of blade lean angle on a hydraulic retarder. Adv. Mech. Eng. **8**, 1–9 (2016)

Topology Optimization of Excavator Arm Based on Altair Inspire

Bohao Su, Junli Sun(✉), and Dafeng Long

School of Electronic Information and Electrical Engineering, Huizhou University, Huizhou 516007, China
15395655@qq.com

Abstract. This study employs Altair Inspire for topology optimization of an excavator arm to achieve lightweight design objectives. A simplified CAD model incorporating predefined material properties, operational constraints, and critical loading conditions was established. An initial structural configuration was generated through geometry reconstruction, followed by topology optimization targeting maximum stiffness via iterative finite element simulations. Modal analysis of the baseline design identified key vibration modes, guiding structural refinements to enhance dynamic stability. The optimized model was rigorously validated through comparative assessment of strength indices (including von Mises stress and deformation) and dynamic characteristics (natural frequencies and mode shapes) against the original design. Results confirm a 71.76% mass reduction while the minimum safety factor is greater than or equal to 3.0 under JS-02 operational loads. This methodology demonstrates significant potential for improving energy efficiency and structural sustainability in heavy machinery applications without compromising safety margins. The workflow establishes a replicable framework for topology-driven optimization of complex industrial components.

Keywords: Altair Inspire · Excavator arm · Topology optimization · Strength analysis · Modal analysis

1 Introduction

Under the guidance of China's Made in China 2025 strategy (State Council Document [2015] No. 28), which mandates energy efficiency and emission reduction for sustainable development, manufacturing industries are actively transitioning toward low-carbon operations. Given the construction machinery sector's substantial environmental impact, technological advancements in this field are imperative. Consequently, lightweight design has gained considerable focus as a pivotal strategy to reduce operational energy consumption. As critical equipment in the construction machinery sector, excavators play a vital role in mining, hydraulic engineering, power generation, and construction industries. Industry reports confirm that excavators handle approximately 65% to 70% of global earthmoving operations [1], making enhanced energy efficiency of these machines imperative for achieving sustainable development.

S. Halgamuge et al. (Eds.): ICACMVE 2025, LNME, pp. 665–677, 2026.
https://doi.org/10.1007/978-981-95-6911-3_67

Extensive research has focused on lightweight design for excavator working devices. Luigi Solazzi (2010) achieved weight reduction through material substitution, replacing structural steel in the boom and arm with aluminum alloy 6061-T6. This reduced total weight from 2050 kg to 1080 kg, attaining 47.3% reduction [2]. In a later study, Solazzi (2019) replaced structural steel in the arm with a hybrid of aluminum alloy 6063-T6 and carbon fiber composite. This approach yielded 64.1% weight reduction compared to the original steel design [3]. Alternatively, Zhang et al. proposed a structure lightweight method based on adaptive approximate model combined with the genetic algorithm, achieving a 9.38% mass reduction in a certain type of hydraulic excavator boom [4]. Wang et al. implemented the Weight-Guide Method for the excavator working device, attaining 21% total weight reduction [5]. Similarly, Chen et al. combined Multilayer Perceptron (MLP) with NSGA-II algorithm to iteratively optimize the bucket structure, the optimized bucket achieves a mass reduction of 7.06% [6]. These studies optimized the working device while maintaining conventional manufacturing processes. Although this conservative strategy reduces process validation costs, it fails to fully exploit lightweight design potential due to insufficient exploration of material innovations and topology alterations.

This paper proposes a novel topology optimization framework implemented in Altair Inspire for excavator arms. The methodology targets an optimal structural configuration to achieve significant mass reduction of the working assembly, directly reducing operational energy consumption while enhancing excavation efficiency.

2 Theory of Topology Optimization

Structural optimization comprises three fundamental approaches: size optimization, shape optimization, and topology optimization [7]. Topology optimization of solid structures involves the determination of features such as the number and location and shape of holes and the connectivity of the domain. The purpose of topology optimization is to find the optimal lay-out of a structure within a specified region [8]. Among topology optimization methodologies, the Solid Isotropic Material with Penalization (SIMP) remains prevalent due to its computational efficiency and implementation simplicity. The mathematical formulation based on SIMP is defined as follows:

$$\begin{cases} \min \ c(\rho_e) = P^T u \\ s.t. \ \sum_{e=1}^{N} v_e \rho_e \leq V \\ d.v. \ 0 \leq \rho_{\min} \leq \rho_e \leq 1 \end{cases} \tag{1}$$

where c is the structural compliance; v_e is the volume of element e; ρ_e is the density of element e; V is the total design domain volume.

Altair Inspire, an integrated computer-aided engineering (CAE) platform, employs its embedded OptiStruct solver to execute topology optimization for excavator arm designs. The solver generates structurally optimal configurations while strictly complying with manufacturing constraints. Figure 1 presents the flowchart of the topology optimization process.

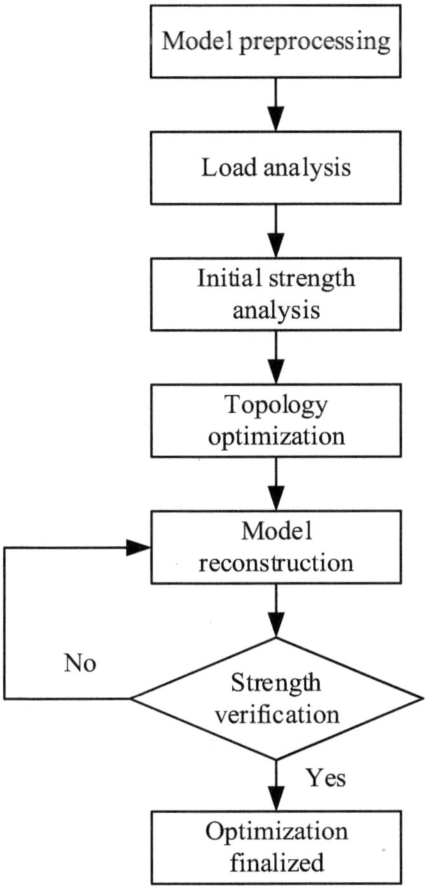

Fig. 1. Flowchart of the topology optimization process

3 Static Analysis of Working Device

3.1 Component Analysis of Working Device

The excavator's working device consists of the boom, arm, bucket, rocker link, connecting rod, and hydraulic cylinders interconnected by revolute joints. During excavation cycles, hydraulic cylinders actuate via extension/retraction motions, transmitting kinetic energy to the bucket through this articulated kinematic chain. Figure 2 presents the working device of the excavator.

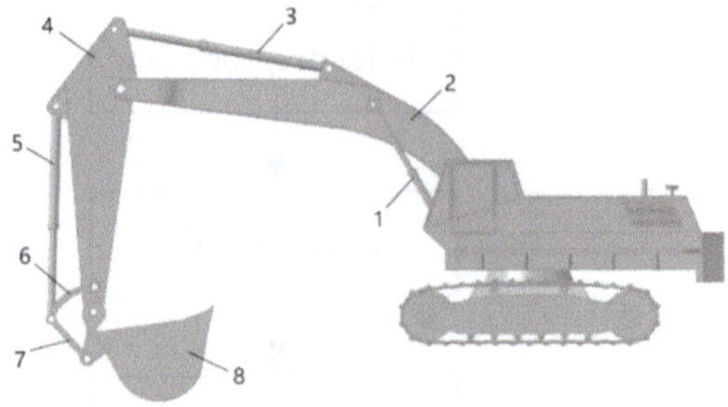

Fig. 2. Working device of the excavator: 1 – Boom hydraulic cylinder; 2 – Boom; 3 – Arm hydraulic cylinder; 4 – Arm; 5 – Bucket hydraulic cylinder; 6 – Rocker link; 7 – Connecting rod; 8 – Bucket.

Figure 3 presents the arm structure schematic, while Fig. 4 details dimensional specifications of its 3D model.

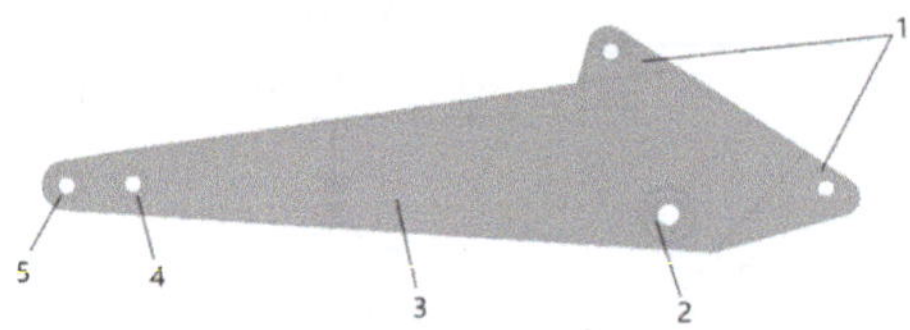

Fig. 3. Schematic diagram of the arm structure: 1 – Lug plate; 2 – Front sleeve bushing; 3 – Arm main structure; 4 – Central sleeve bushing; 5 – Rear Sleeve bushing.

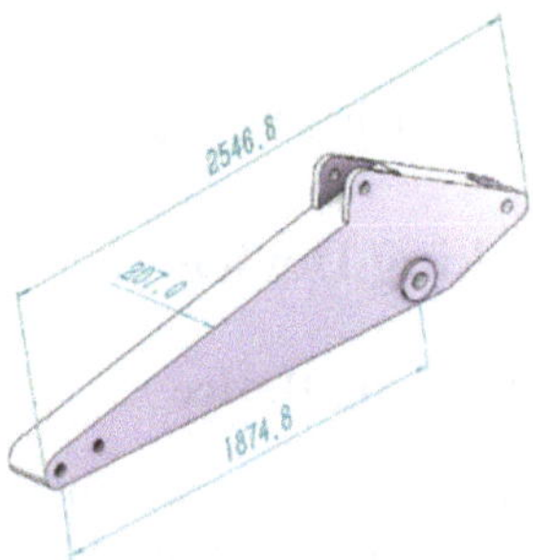

Fig. 4. Dimensions of the arm 3D model

Table 1 presents primary mechanical properties of the selected materials.

Table 1. Material properties and performance parameters

Material	Young's Modulus (GPa)	Poisson's Ratio	Density (kg/m3)	Yield Strength (MPa)
Q345B	206	0.30	7850	345

Through geometry simplification with feature suppression, the arm model was reduced to a solid baseline configuration. This idealized model has a pre-optimization mass of 1427.3 kg.

3.2 Design Objectives

According to von Mises' law, yielding begins when the (recoverable) elastic energy of distortion reaches a critical value [9]. According to this theory, the yield criterion ultimately established is [10]:

$$\sqrt{\frac{1}{2}\left[(\sigma_1-\sigma_2)^2+(\sigma_2-\sigma_3)^2+(\sigma_3-\sigma_1)^2\right]} \le [\sigma] \quad (2)$$

where σ_1, σ_2, σ_3 represent the first, second, and third principal stresses respectively, $[\sigma]$ is the allowable stress.

The allowable stress must be rigorously quantified to incorporate adequate strength reserves, especially for critical components where failure may cause catastrophic consequences, thus ensuring structural integrity. Under the mandatory requirement that the arm's minimum safety factor n shall not be less than 3.0[11], the allowable stress is derived from:

$$[\sigma] = \frac{\sigma_y}{n} \quad (3)$$

where σ_y is the yield strength of Q345B.

This defines the allowable stress as 115 MPa. The maximum von Mises stress in the optimized model shall not exceed 115 MPa.

3.3 Working Load Conditionsmass

This study implements the static stress test condition JS-02 mandated by Chinese National Standard GB/T 9141–1988. In this critical posture that maximizes hydraulic cylinder moment arms, the bucket tip, bucket-arm joint, and arm-boom joint form a collinear force transmission path. The resultant excavator configuration is illustrated in Fig. 5.

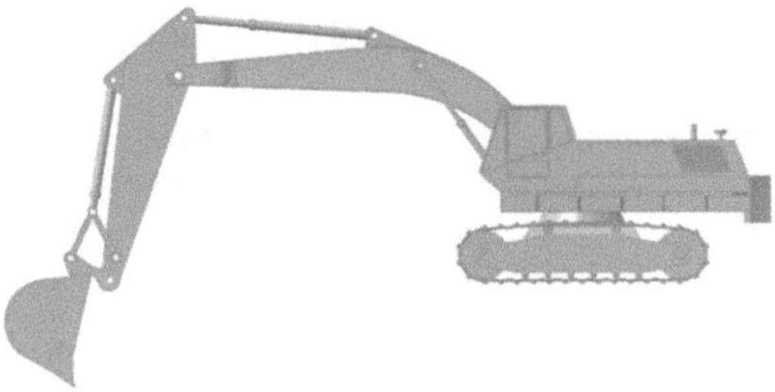

Fig. 5. Working device configuration under JS-02 critical loading condition

3.4 Static Analysis of the Working Device

Forces on the arm originate at articulation joints and propagate through the linkage mechanism from the bucket. Thus, quantifying the bucket load is imperative. This load acts perpendicular to the arm-bucket-joint-to-tip line, with magnitude derived from the bucket hydraulic cylinder thrust force via moment equilibrium. Figure 6 presents the analysis of bucket digging force.

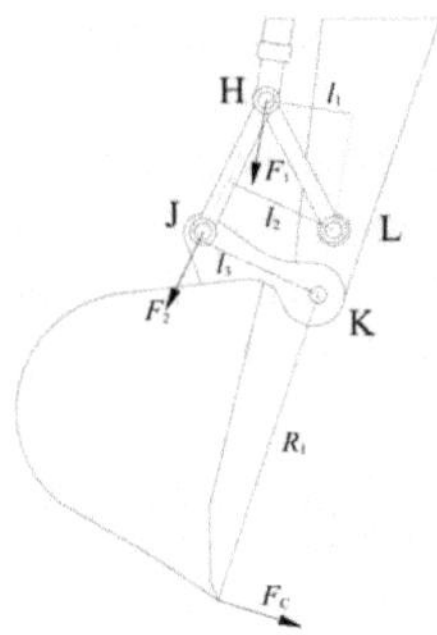

Fig. 6. Analysis of bucket digging force

The bucket digging force is calculated as follows:

$$\begin{cases} F_C = \dfrac{F_1 \times l_1 \times l_3}{l_2 \times R_1} \\ F_1 = \dfrac{P \pi D^2}{4} \end{cases} \quad (4)$$

where F_1 is the thrust force of the bucket hydraulic cylinder; F_C is the theoretical bucket digging force; l_1 is the moment arm of F_1 about pivot L; l_2 is the moment arm of F_2 about L; l_3 is the moment arm of F_2 about pivot K; R_1 is the moment arm of F_C about K; D is the piston-side diameter of the bucket hydraulic cylinder; P is the hydraulic operating pressure, valued at 28 MPa.

Through computational analysis, the bucket digging force F_C is derived as 42 kN.

4 Topology Optimization Design for the Arm

4.1 Model Preprocessing

The topology optimization pre-processing in Altair Inspire involved defining the undercarriage assembly (left/right tracks and suspension) as fixed supports with gravity along -Y, assigning the arm's design space and non-design regions to Rigid Group 1, and grouping the undercarriage/cabin/non-optimized linkages into Rigid Group 2. All linkage connections were constrained by revolute joints with supplemental locked translational joints at hydraulic cylinder interfaces, while the derived 42 kN bucket digging force was applied to the bucket tip. Figure 7 presents the joint settings, and Fig. 8 presents the designed space and load configuration.

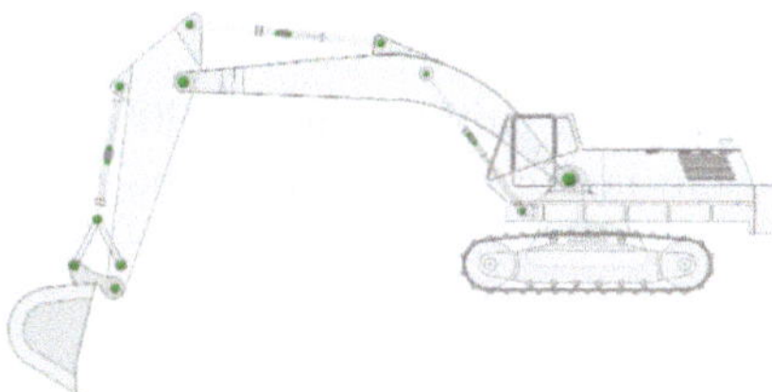

Fig. 7. Joint settings

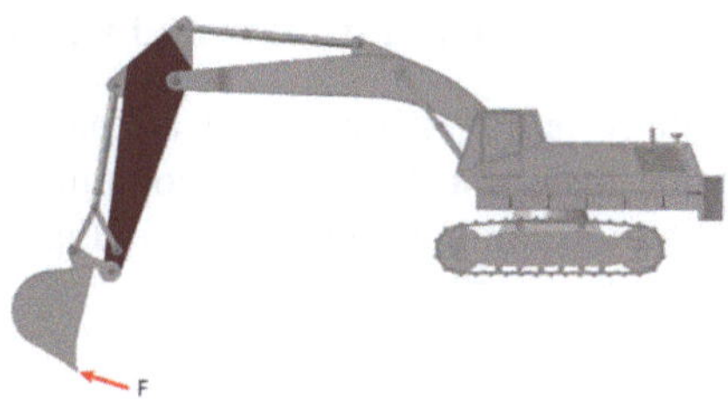

Fig. 8. Design space and load configuration

4.2 Initial Strength Analysis

Initial strength analysis of the baseline model yielded maximum displacement of 0.169 mm, maximum von Mises stress of 53.6 MPa, and minimum safety factor of 6.4, with corresponding full-field distributions presented in Figs. 9, 10 and 11.

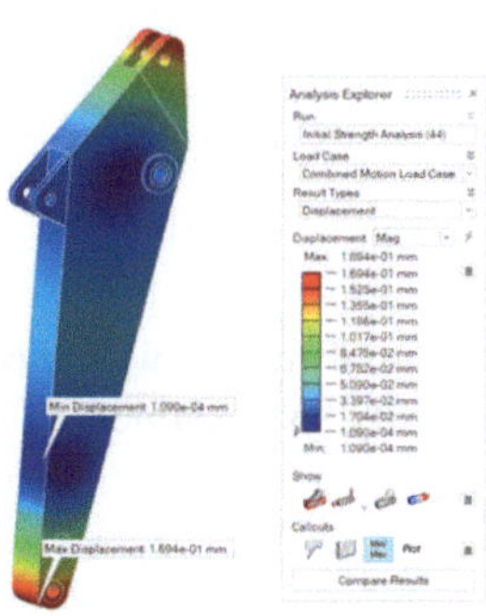

Fig. 9. Displacement contour

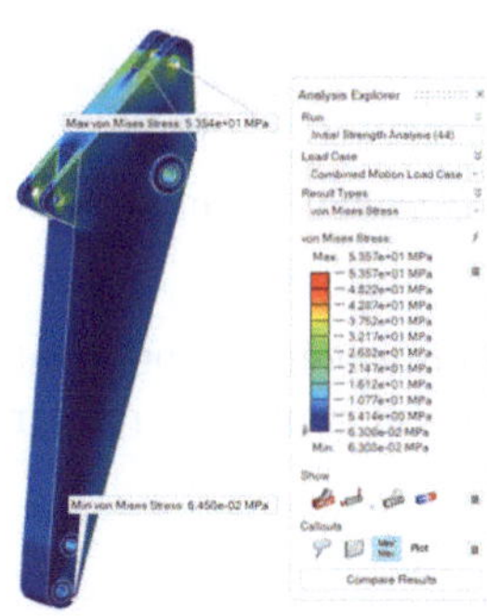

Fig. 10. Von Mises stress contour

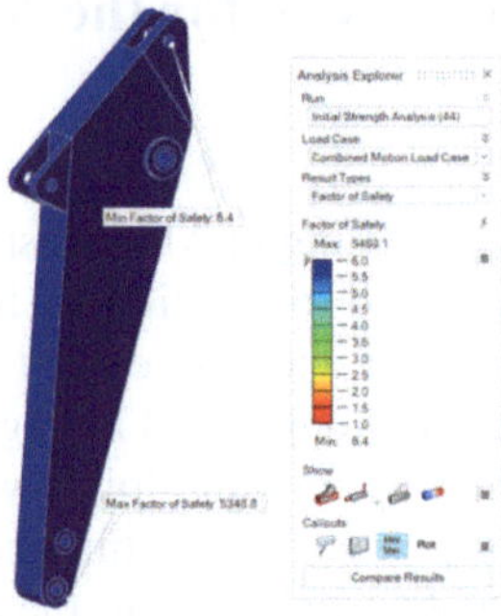

Fig. 11. Factor of safety contour

4.3 Topology Optimization and Geometry Reconstruction

Topology optimization aimed at maximizing structural stiffness under 30% design volume mass constraint. The computational setup in Fig. 12 generated the material distribution configuration shown in Fig. 13.

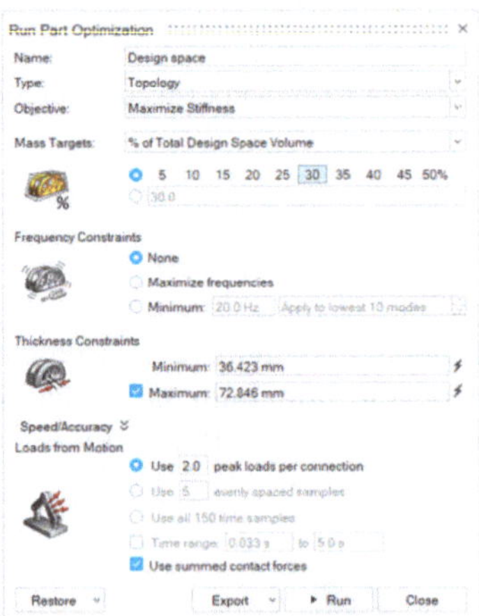

Fig. 12. Topology optimization setup

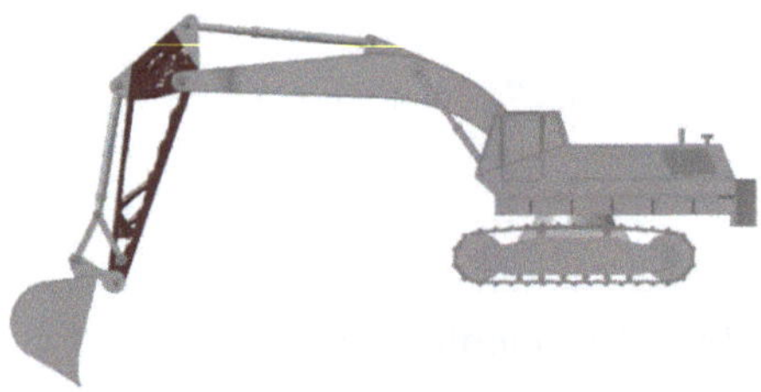

Fig. 13. Topology optimization result

Based on topology optimization results, the conceptual model was reconstructed with concurrent geometric refinement for manufacturability, as documented in Fig. 14.

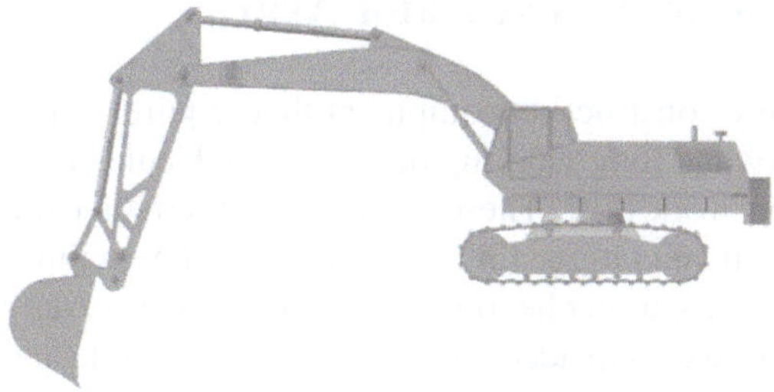

Fig. 14. Reconstructed model

4.4 Strength Analysis of the Reconstructed Model

Strength analysis of the reconstructed model yielded maximum displacement of 0.348 mm, maximum von Mises stress of 96.4 MPa, and minimum safety factor of 3.6. Full-field distributions are presented in Figs. 15, 16 and 17.

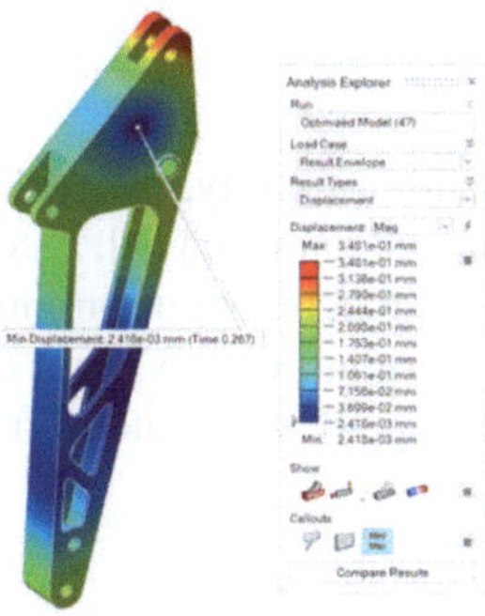

Fig. 15. Displacement contour

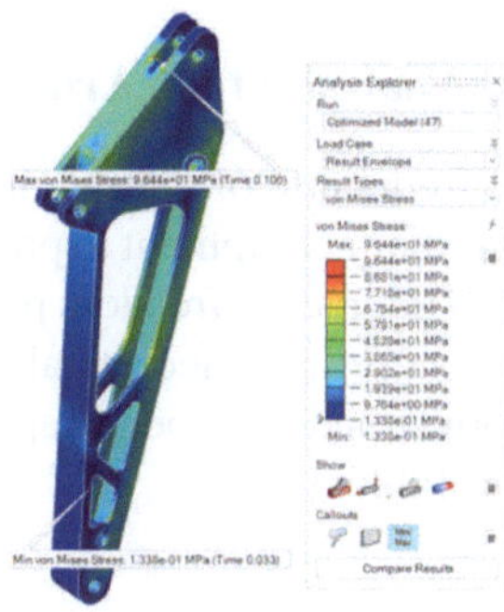

Fig. 16. Von Mises stress contour

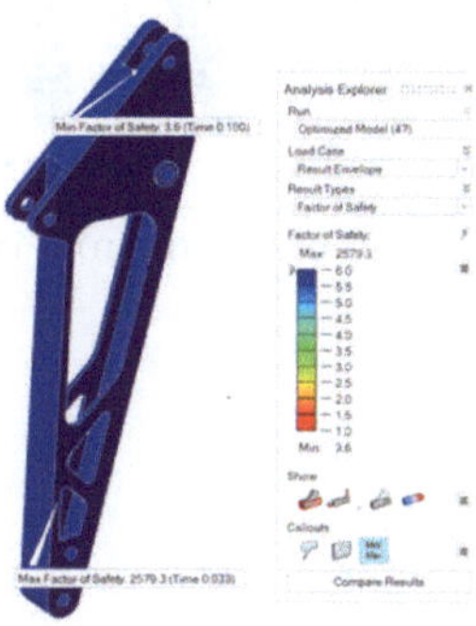

Fig. 17. Factor of safety contour

5 Modal Analysis of the Excavator Arm

The excavator is heavy construction equipment that requires sophisticated manoeuvrability due to the bulky and long working devices, which can cause significant vibration that hinders precision work [12]. These structural vibrations expose the arm to cyclic stresses, creating significant fatigue failure risks. Given the established efficacy of modal analysis in diagnosing vibration phenomena, this technique serves as a fundamental validation tool for structural design adequacy. Consequently, modal analysis was conducted on the topology-optimized arm to verify dynamic performance integrity.

5.1 Theoretical Framework

Modal properties characterize the intrinsic vibration behavior of mechanical systems. Through modal analysis, three fundamental parameters per vibration mode are derived: natural frequencies, mode shapes, and damping ratios. Modal analysis of the excavator arm can effectively prevent resonance by ensuring that the frequency of externally applied loads does not approach the arm's natural frequency [13].

5.2 Modal Analysis of the Arm

The natural frequency of a structure is a critical dynamic property, particularly when mounting auxiliary equipment to prevent resonance-induced damage [14]. A constrained system with N degrees of freedom possesses N natural frequencies. As a continuum structure, the excavator arm theoretically exhibits infinite vibration modes upon discretization. The corresponding mode shapes are illustrated in Fig. 18, with natural frequencies quantified in Table 2.

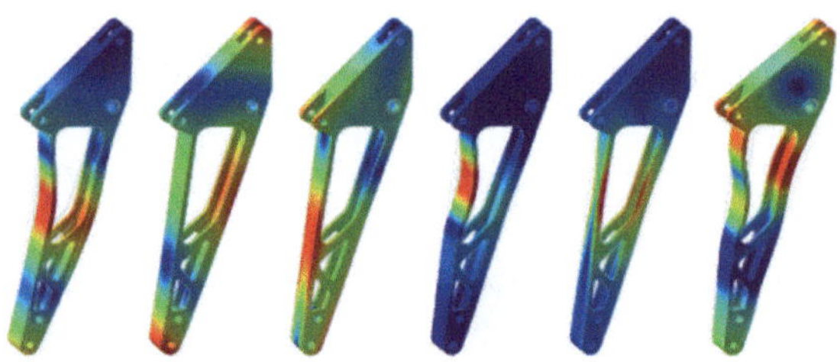

Fig. 18. Mode shapes of the arm from 1st to 6th order

Table 2. Natural frequencies of the arm for the first six modes

Mode order	Natural frequency (Hz)	Mode order	Natural frequency (Hz)
1	158.642	4	232.498
2	165.881	5	297.708
3	179.803	6	407.980

The first three natural frequencies of the arm cluster in the narrow 158–180 Hz frequency range. When operational excitation frequencies approach this range, mode coupling phenomena amplify resonant vibration amplitudes during service. To mitigate this risk, two structural modifications are proposed in Fig. 19 and Fig. 20.

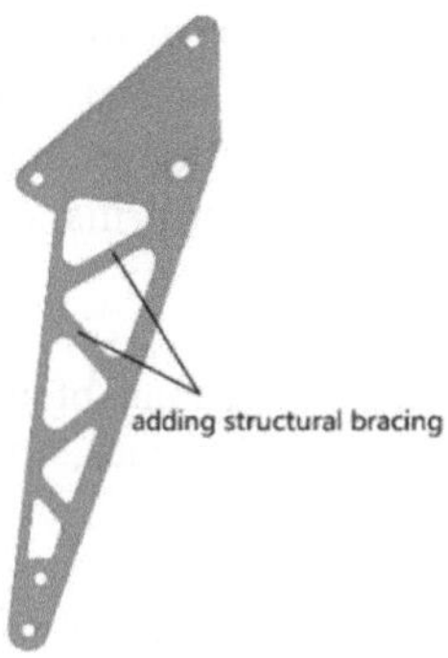

Fig. 19. Design 1 model

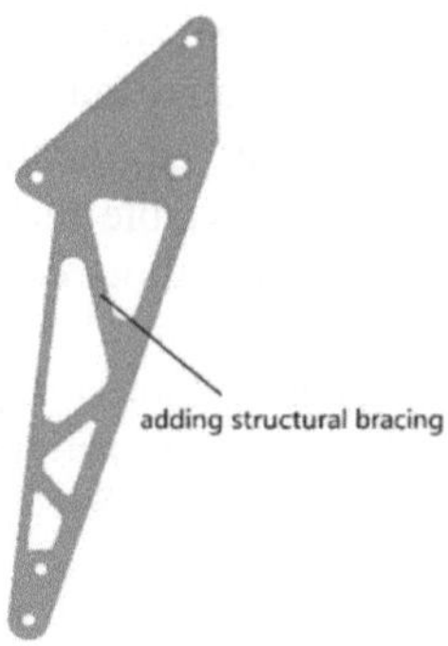

Fig. 20. Design 2 model

6 Results and Discussion

Strength analysis of both modified arm configurations was conducted using the embedded OptiStruct solver, with key performance metrics benchmarked against the baseline design in Table 3.

Table 3. Comparison of optimization designs

Element	Design 1	Design 2
Weight(kg)	405.327	403.085
Maximum displacement (mm)	0.3515	0.3316
Maximum von Mises stress (MPa)	97.47	101.60
Minimum safety factor	3.5	3.4

Benchmarked in Table 3, Design 1 achieves 71.6% mass reduction versus baseline, while Design 2 attains 71.8%. Both maintain peak von Mises stresses below the 115 MPa allowable limit. Although Design 2 exhibits 12% lower maximum displacement than Design 1, its safety factor is marginally lower. Given Design 2's superior manufacturability with simplified rib structures, it is selected as the final configuration.

7 Conclusions

Leveraging Altair Inspire's topology optimization capabilities, this study generated a lightweight excavator arm configuration achieving a 71.8% mass reduction while maintaining structural integrity. Through rigorous strength and modal analyses, the optimized design's structural integrity and dynamic characteristics were validated, with targeted modifications resolving critical performance risks identified during simulation. The finalized model demonstrates significant mass reduction in the arm component, directly contributing to enhanced operational energy efficiency and sustainability in construction machinery applications. This successful implementation establishes topology optimization as a viable methodology for developing high-performance, eco-efficient heavy equipment without compromising safety margins.

Acknowledgments. Project supported by the Key Research Program of the Higher Education Institutions of Department of Education of Guangdong Province, China (Grant No. 2024GCZX009; 2022ZDJS056; 2021ZDZX1012).

Disclosure of Interests. The authors have no competing interests to declare that are relevant to the content of this article.

References

1. Editorial Department of China Journal of Highway and Transport.: Review on China's Road Construction Machinery Research Progress: 2018. China J. Highway Transp. **31**(06), 1–164 (2018)
2. Solazzi, L.: Design of aluminium boom and arm for an excavator. J. Terramechanics **47**(4), 201–207 (2010)
3. Solazzi, L., et al.: Excavator arms: numerical, experimental and new concept design. Compos. Struct. **217**, 60–74 (2019)
4. He, Z., et al.: Lightweight design of hydraulic excavator based on adaptive approximation model. Comput. Integr. Manuf. Syst. **29**(12), 4051–4062 (2023)
5. Jian, W., et al.: Finite element analysis and optimization design of the working device of eight ton excavator. Mach. Des. Manuf. **03**, 19–23 (2022)
6. Yi-xin, C., et al.: Structural optimization design of excavator bucket based on improved depth surrogate model. J. Jilin University (Eng. Technol. Ed.), 1–12 (2024)
7. Jie, Y., et al.: Overview of discrete structure topology optimization. Sci. Technol. Eng. **20**(24), 9673–9682 (2020)
8. Bendsøe, M.P., Sigmund, O.: Topology Optimization. Springer, Heidelberg (2013)
9. Hill, R.: The Mathematical Theory of Plasticity. Oxford University Press (1998)
10. Wenguang, W.: Mechanics of Materials. 1st edn. Chemical Industry Press (2020)

11. Tongji University.: Hydraulic Excavators: Single-Bucket Type. 2nd edn. China Architecture & Building Press (1986)
12. Rameshkumar, R., et al.: Vibration analysis of dipper arm cylinder in an excavator using MATLAB-Adams co-simulations. Int. J. Veh. Struct. Syst. **16**(3), 442–445 (2024)
13. Wei, W., et al.: Modal analysis and lightweight design of hydraulic cylinders based on ANSYS. Hydraulics Pneumatics Seals **42**(07), 15–19 (2022)
14. Prasad, K.D., et al.: Natural frequencies of thin and thick metallic plates: analytical and experimental approach. Int. J. Res. Eng. Appl. Sci. **5**(6), 31–47 (2015)

The Simulation and Analysis Method for Deformation of Medium Carbon Steel Laser Quenching Considering Nonlinear Structural Characteristics

Yanglin Chen, Gangyan Li(✉), Yutong Guo, Bucheng Zhang, Chenxu Hu, and Honghui Huang

School of Mechanical and Electric Engineering, Wuhan University of Technology, Wuhan, China
gangyanli@whut.edu.cn

Abstract. Laser quenching, as one of the key technologies for surface strengthening of medium carbon steel, significantly influences the deformation of the workpiece due to the nonlinear structural characteristics during the processing. However, traditional simulation methods are mostly based on linear assumptions, making it difficult to accurately predict the quenching deformation in the unsteady heat transfer process of laser quenching, resulting in a significant deviation between simulation and experimental results. To address this, this paper proposes a simulation and analysis method for the deformation of medium carbon steel laser quenching based on nonlinear structural characteristics, considering the nonlinear high-temperature mechanical properties of medium carbon steel, the nonlinear deformation, and the nonlinear boundary constraints. The necessity of considering the nonlinear structural characteristics of medium carbon steel in laser quenching simulation is also discussed based on the simulation analysis.

Keywords: Laser Quenching · Deformation · Nonlinear Structural Characteristics · Simulation Analysis

1 Introduction

1.1 Background

Laser quenching is a heat treatment process that utilizes a high-energy photon beam to irradiate the surface of a material, rapidly raising the surface temperature (with a heating rate of 10^3–10^6 °C/s) to above the austenite transformation temperature. After the irradiation, the material cools quickly through its own heat conduction (with a cooling rate of 10^4–10^5 °C/s), resulting in a phase transformation to martensite [1]. Due to its fast heating and high cooling rates, this process produces a dense martensitic structure that significantly enhances surface hardness, wear resistance, and fatigue strength, while maintaining good toughness in the core. As one of the key processes for improving the performance of parts in modern manufacturing, laser quenching has become essential.

S. Halgamuge et al. (Eds.): ICACMVE 2025, LNME, pp. 678–691, 2026.
https://doi.org/10.1007/978-981-95-6911-3_68

With the rapid advancements in computer technology and the increasing sophistication of numerical simulation methods, researchers can now utilize various numerical simulation tools, including the finite element method (FEM), and advanced simulation software such as ANSYS and COMSOL, to conduct virtual simulations of temperature fields, stress evolution, and phase transformation mechanisms during the laser quenching process. This research approach has not only significantly reduced experimental costs but also accelerated the exploration of new materials and process parameters.

1.2 Current Research Status

With the growing application of laser heat treatment technologies, numerical simulation of the laser quenching process for medium carbon steel has emerged as a prominent research topic. Existing studies have primarily focused on the evolution of temperature fields, optimization of process parameters, and prediction of hardened layer depth. For instance, Guo et al. [2] employed ABAQUS software to simulate the temperature distribution during laser surface treatment of 65Mn steel and CrMo cast iron, respectively, and established predictive models for peak temperature and hardened layer depth. Zhao et al. [3] developed a coupled temperature-stress-phase transformation numerical model for the dual-layer laser quenching of 12Cr2Ni4A gear steel, revealing residual stress concentrations in the secondary tempering region. Zhang et al. [4] constructed thermo-mechanical and multi-physics coupling models to analyze the dynamic evolution of microstructure and stress fields in GGG70L ductile iron and 42CrMo alloy, respectively. Ma et al. [5] developed a multi-physics simulation platform for 45 steel based on critical phase transformation conditions and CCT diagrams, providing quantitative insights into the transient thermal gradients and transformation strengthening zones influenced by laser parameters.

Despite significant progress, several challenges remain in this field. First, simulation studies specifically targeting the deformation behavior of medium carbon steel during laser quenching are relatively scarce. Most existing research has emphasized temperature distribution or hardness prediction, which is insufficient for applications requiring high precision and repeatability. Second, current simulation approaches often fail to accurately reflect the nonlinear characteristics of the laser quenching process. The thermal–physical properties of medium carbon steel vary significantly with temperature during rapid heating and cooling, resulting in highly nonlinear structural responses. However, simplified models are frequently employed, which limits the accuracy of deformation prediction. Therefore, it is imperative to develop advanced nonlinear simulation methods to more precisely capture the deformation evolution during laser quenching of medium carbon steel.

2 Simulation and Analysis Method for Deformation of Medium Carbon Steel Laser Quenching Considering Nonlinear Structural Characteristics

The simulation and analysis method for deformation of medium carbon steel laser quenching proposed in this paper summarizes the characteristic nonlinear structural behavior of medium carbon steel during laser quenching, which results from the rapid scanning of the laser. The rapid temperature variation over a large area in a very short time causes significant changes in the material properties. Using the JMatPro platform in combination with phase transformation thermodynamics, the material property parameters as functions of temperature were obtained to characterize the nonlinear structural characteristics of medium carbon steel during laser quenching. During the simulation process, an APDL command flow was implemented to switch the material property parameters for the heating and cooling stages of the laser quenching process. This study effectively reflects the nonlinear structural characteristics of the quenched workpiece.

2.1 Nonlinear Structural Characteristics of Medium Carbon Steel Laser Quenching

The deformation of medium carbon steel during laser quenching differs from that observed in conventional tensile test specimens. In laser quenching, medium carbon steel is not subjected to external loads but instead experiences internal stresses caused by thermal input. The formation of these stresses is complex, involving the coupling effects of thermal expansion, solid-state phase transformation, and other factors.

The nonlinear structural characteristics of medium carbon steel during laser quenching are manifested in the complex changes in its internal physical properties and microstructure as they vary with temperature, stress, and phase transformation. These behaviors impact the material's mechanical properties, and therefore, when studying the deformation behavior of medium carbon steel during laser quenching, its nonlinear structural characteristics must be considered. Since the nonlinear structural characteristics in laser quenching are essentially caused by the changes in material property parameters, the nonlinear variation of the internal material properties can be used to fully account for the nonlinear structural characteristics of medium carbon steel during the laser quenching process.

Nonlinear Structural Characteristics of Medium Carbon Steel Induced by Temperature during laser Quenching: In order to obtain the temperature-dependent material property parameters required for the deformation simulation analysis of medium carbon steel laser quenching, this study utilizes the thermodynamic simulation platform JMatPro, based on the CalPhaD framework, to numerically analyze the temperature response characteristics of thermodynamic parameters during the phase transformation of medium carbon steel. Using the elemental composition of the S45C, thermodynamic modeling was performed with JMatPro software to obtain the evolution of thermodynamic parameters during the material's phase transformation process. The resulting temperature-dependent material property curves are shown in Fig. 1.

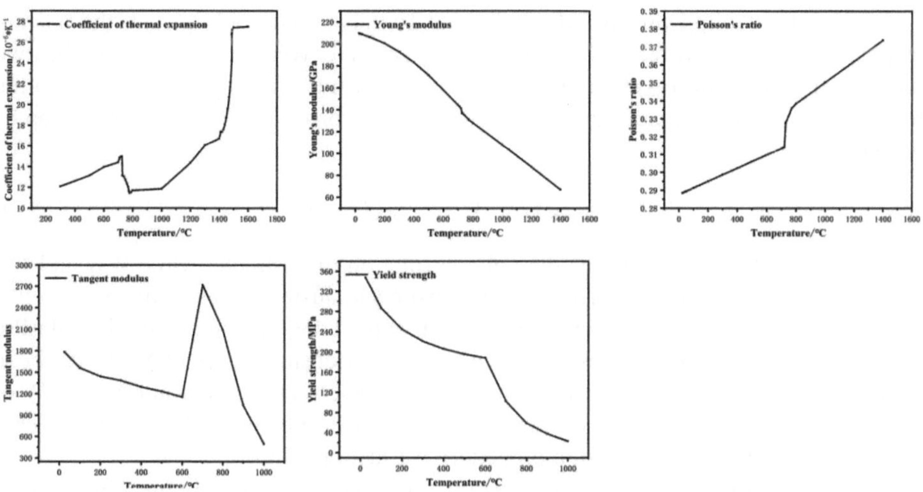

Fig. 1. Temperature-dependent material property curves of S45C.

Nonlinear Structural Characteristics of Medium Carbon Steel Induced by Phase Transformation during Laser Quenching: During the cooling process, the primary solid-state phase transformation in medium carbon steel laser quenching is the formation of martensite. While the heating stage involves the transformation from ferrite and cementite to austenite, the cooling stage is characterized by the transformation from austenite to martensite. As the phase composition has already changed during cooling, the material properties used during the heating phase—corresponding to ferrite and cementite—are no longer applicable. Furthermore, after quenching, the phase composition of medium carbon steel has been altered, making it difficult to directly determine the chemical composition of the quenched specimen. Therefore, in this study, the material properties during the cooling process are determined based on the material phase-field simulation platform JMatPro, in conjunction with thermodynamic calculations of phase transformations.

First, the determination of material properties during the cooling stage is carried out based on phase transformation kinetics. The evolution of the volume fraction can be represented by the K-M equation [6], and its mathematical expression is as follows:

$$\frac{df_M}{dT} = \alpha(1 - f_M) \tag{1}$$

In the equation, f_M represents the volume fraction of martensite; T represents the temperature, in °C; and α is an empirical coefficient, typically taken as 0.011.

Based on the temperature-dependent relationship of phase volume fraction shown in Eq. (1), the expression for the martensite phase volume fraction can be derived as shown in Eq. (2). The critical phase transformation temperature parameter can be calculated using its constitutive equation, yielding a value of 330 °C.

$$f_M(T) = 1 - \exp[-\alpha(M_S - T)] \tag{2}$$

During the cooling process, due to the extremely rapid cooling rate, the austenite phase transforms directly into martensite, and the volume fractions of the two phases satisfy the conservation condition. Based on this single-phase transformation model assumption, the macroscopic effective material property W of the system can be expressed as:

$$W = f_M(T) \cdot W_M + f_A(T) \cdot W_A \tag{3}$$

In the equation, f_A represents the volume fraction of austenite, while W_M and W_A are the material property parameters of martensite and austenite, respectively.

Next, by inputting the chemical composition of the S45C specimen into the JMatPro software, the material properties of austenite and martensite phases of S45C steel can be obtained. Substituting these values into Eq. (3), the effective material property W of the specimen treated as a mixture of austenite and martensite during the cooling process can be determined. The calculation procedure is illustrated in Fig. 2.

The relevant material property parameters for the cooling stage obtained through the above method are shown in Fig. 3.

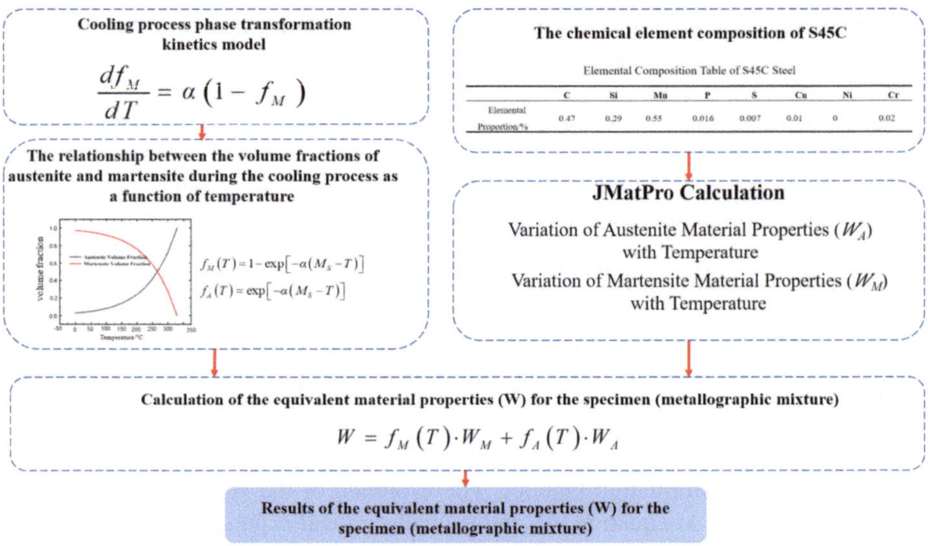

Fig. 2. Schematic diagram of the material property parameter calculation process during the cooling stage.

The nonlinear structural characteristics of medium carbon steel laser quenching are essentially due to the nonlinear changes in material property parameters. By analyzing the changes in material phase microstructures during the laser quenching process of medium carbon steel and combining phase transformation thermodynamics, the nonlinear variation of material property parameters can be more accurately reflected. In the simulation analysis of medium carbon steel laser quenching deformation, this characteristic can be accounted for by considering the nonlinear material property parameters.

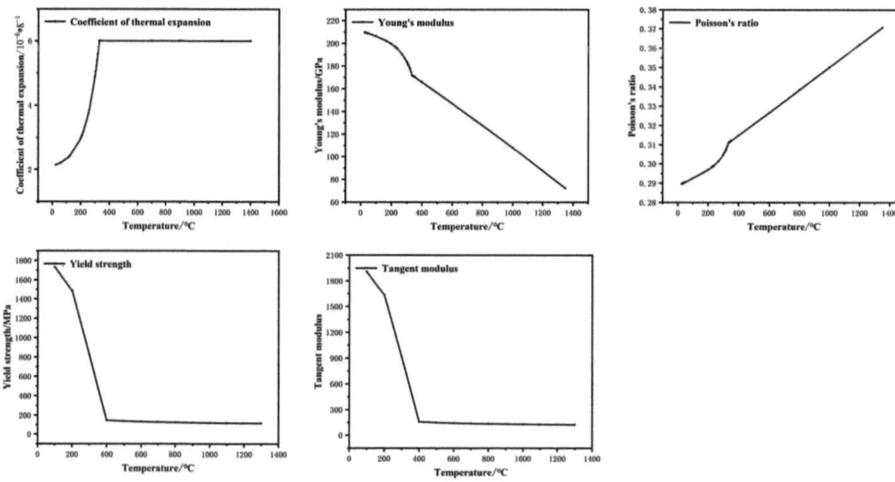

Fig. 3. Temperature-dependent material property curves for the cooling stage.

2.2 Simulation and Analysis Method for the Deformation of Medium Carbon Steel Laser Quenching Considering Nonlinear Structural Characteristics

Based on the nonlinear structural characteristics of the medium carbon steel laser quenching process described above, APDL language is used in ANSYS finite element simulation to dynamically update material parameters as a function of temperature, fully accounting for the nonlinear structural characteristics of medium carbon steel laser quenching. Figure 4 illustrates the logical framework for material property parameter changes, which enables the corresponding material property parameter database to be called during the heating and cooling stages. The partial code for implementing the substitution of material property parameters during the heating and cooling stages of medium carbon steel laser quenching is shown in Fig. 5. The substitution of material property parameters is mainly achieved using the *mpchg command.

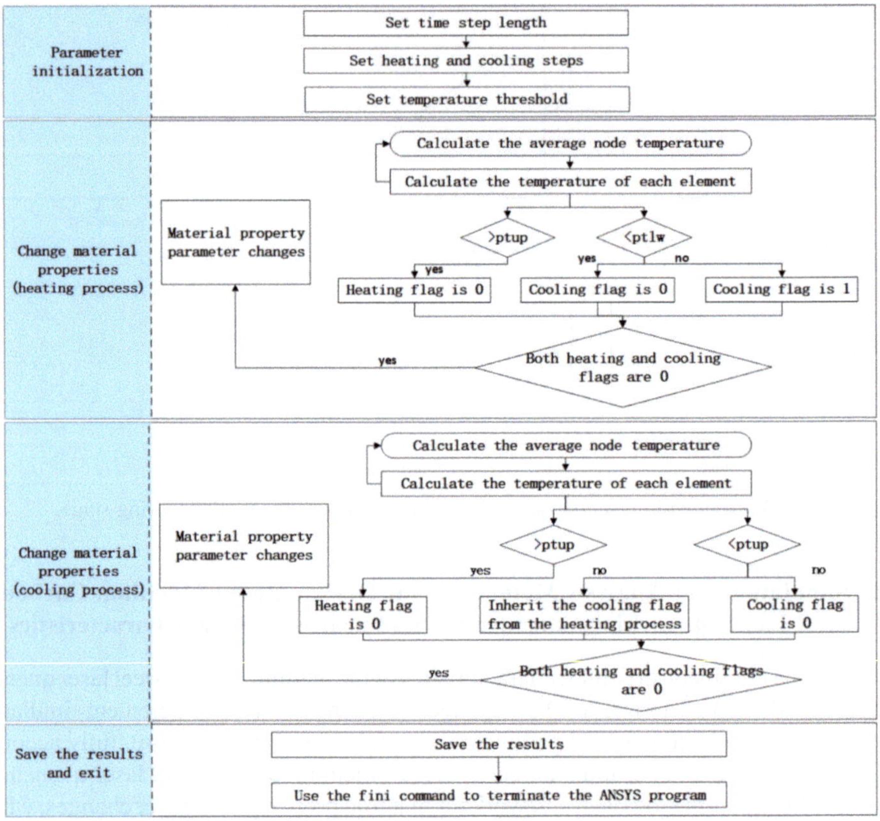

Fig. 4. Schematic diagram of material property parameter substitution logic.

```
ptup=900
ptlw=800
!...
!Heating process
*if,petemp,gt,ptup,then
mflg(j,1)=0
*endif
*if,petemp,lt,ptlw,then
mflg(j,2)=0
*else
mflg(j,2)=1
*endif
*if,mflg(j,1),eq,0,then
*if,mflg(j,2),eq,0,then
mpchg,pmat,j
!...
```

```
!...
!Cooling process
*if,petemp,gt,ptup,then
mflg(j,1)=0
*endif
*if,petemp,lt,ptup,then
mflg(j,2)=0
*endif
*if,mflg(j,1),eq,0,then
*if,mflg(j,2),eq,0,then
mpchg,pmat,j
!...

pmat=matid
!Obtaining the material
! number of this material
```

Fig. 5. Partial code for material property parameter substitution.

3 Validation of the Nonlinear Structural Characteristics of Medium Carbon Steel Laser Quenching Based on Simulation Analysis

To validate the superiority of the simulation analysis method for medium carbon steel laser quenching deformation considering nonlinear structural characteristics, this study used a medium carbon steel specimen with dimensions of 100mm × 30mm × 10mm for laser quenching experiments. Two sets of simulation control groups were established, and the nonlinear structural characteristics of medium carbon steel laser quenching were verified through a comparative analysis. Figure 6(a), (b) shows the laser quenching experimental equipment and deformation detection equipment.

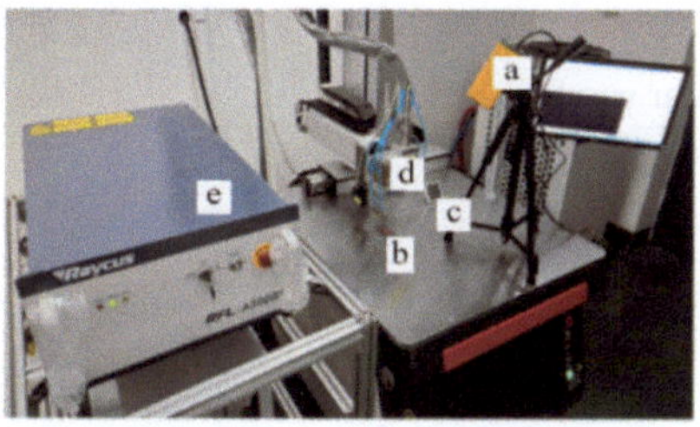

(a) laser quenching equipment

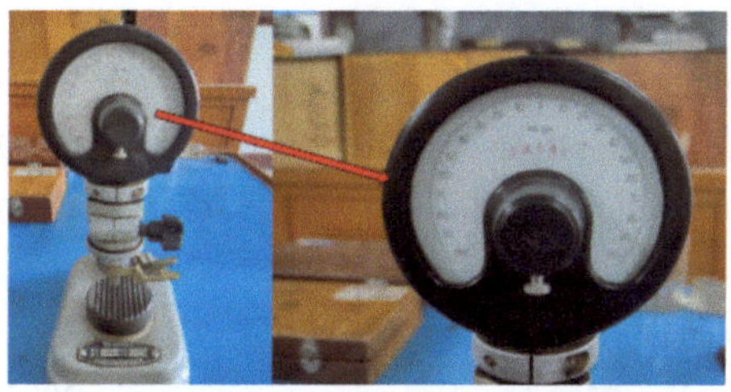

(b) deformation measuring equipment

Fig. 6. Experimental equipment.

The specific setup for the two sets of simulation experimental groups is as follows: Group A: The material property parameters related to medium carbon steel laser quenching deformation are set to fixed values, with the specific values shown in Table 1.

Table 1. Specific material property parameter settings for simulation control group A.

Material Property Parameters	Specific Values
Density/ (kg·m^{-3})	7826.6
Specific Heat/ (J·kg^{-1}·C^{-1})	471
Thermal Conductivity/ (W·m^{-1}·C^{-1})	47.86
Thermal Expansion/K^{-1}	12*10^{-6}
Young's Modulus/GPa	209
Poisson's Ratio	0.29
Yield Strength/MPa	347.4
Tangent Modulus/MPa	1781

Group B: This group does not consider the nonlinear structural characteristics of medium carbon steel laser quenching caused by the different phase transformation behaviors during heating and cooling. In the simulation, no separate heating and cooling stages are set; instead, only the material property parameters from the heating process are used for the simulation analysis. Except for the material property parameters, all other settings remain the same, as shown in Table 2.

Table 2. Basic settings in simulation analysis.

Simulation Setting	Specific Information
Laser Power/W	2200
Scanning Speed/m·min^{-1}	1
Spot Radius/mm	Rectangular spot8 × 8
Absorption Rate	0.777
Convection Coefficient/ (W·m^{-2}·°C^{-1})	133.77
Ambient Temperature/°C	22
Quenching Time/s	5.28
Cooling Time/s	2003
Heat Source Function	2.67e7 × (0.483 × exp(0.947 × (2) × (({X})^4 + {Y}^4)/0.004^4) + 0.517 × exp((-2) × 0.22 × (({X})^50 + {Y}^50)/0.004^50))

3.1 Temperature Field Comparative Analysis

As shown in Fig. 7(a), (b) and (c), the temperature field distribution contour maps illustrate the results of three comparative simulation experiments after three seconds of laser heat source loading. It can be observed that all three groups exhibit a relatively large temperature gradient in the transverse direction (i.e., along the laser scanning path), while the gradient in the longitudinal direction is relatively small. The transverse temperature gradients are generally consistent across the three groups, whereas notable differences appear in the longitudinal direction. Among them, the Simulation Control Group A exhibits the widest range of longitudinal temperature gradient variation, whereas the Simulation Experimental Group shows the narrowest range of variation in the longitudinal temperature gradient.

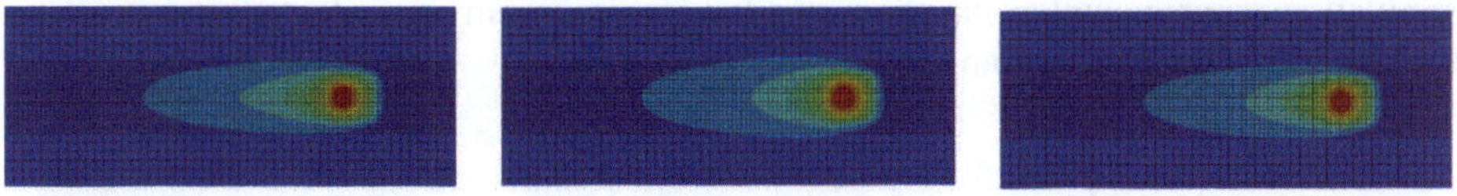

(a) Simulation control group A (b) Simulation control group B (c) Simulation experimental group

Fig. 7. Schematic of the simulated temperature field distribution under different settings.

The highest temperature results at Points A, B, and C are shown in Fig. 8. The maximum temperature results at Points A, B, and C for the three comparison simulation groups are listed in the table below.

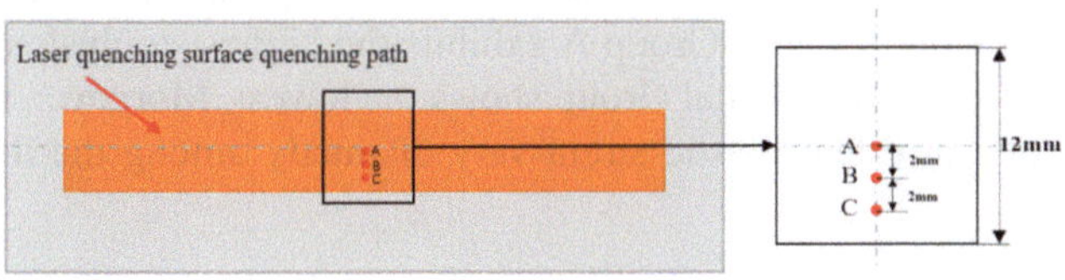

Fig. 8. Schematic of the quenching sample at Points A, B, and C.

Table 3. Maximum temperature results at Points A, B, and C under different settings.

Group Name	Maximum Temperature at Different Points/°C		
	A	B	C
Simulation Experiment Group	1283.9	1157.9	550.9
Simulation Control Group A	1145.6	1027.8	570
Simulation Control Group B	1284	1158.1	551.1

Analysis of the peak temperatures at Points A, B, and C in Table 3 indicates that the simulation control group B shows results nearly identical to those of the simulation experimental group, while Control Group A exhibits approximately 100 °C higher temperatures at Points A and B, and about 20 °C higher at Point C.

Based on the above analysis, in the laser quenching process of medium carbon steel, the maximum temperature and the temperature gradient of the thermal field differ among the three simulation groups. This is because specific heat capacity and thermal conductivity affect the amount of heat absorbed by the material surface and the rate of heat transfer.

3.2 Structural Field Comparison Analysis

The equivalent elastic strain distribution contour maps of the structural fields in the three comparative simulation experiments are shown in Fig. 9(a), (b) and (c). Among them,

Simulation Control Group A exhibits the highest maximum equivalent elastic strain and the widest distribution range of maximum equivalent elastic strain. In contrast, the simulation experimental group shows the lowest maximum equivalent elastic strain and the narrowest distribution range.

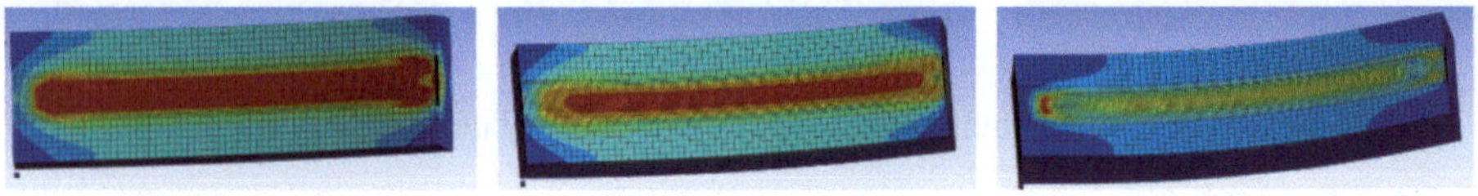

(a) Simulation control group A (b) Simulation control group B (c) Simulation experimental group

Fig. 9. Schematic diagrams of equivalent elastic deformation distribution in the simulated structural fields under different settings.

The equivalent plastic strain distribution contour maps are shown in Fig. 10(a), (b) and (c). The Simulation Control Group A exhibits the highest equivalent plastic strain, while the Simulation Experimental Group shows the lowest. Moreover, the distribution characteristics of equivalent plastic strain vary distinctly among the three simulation groups.

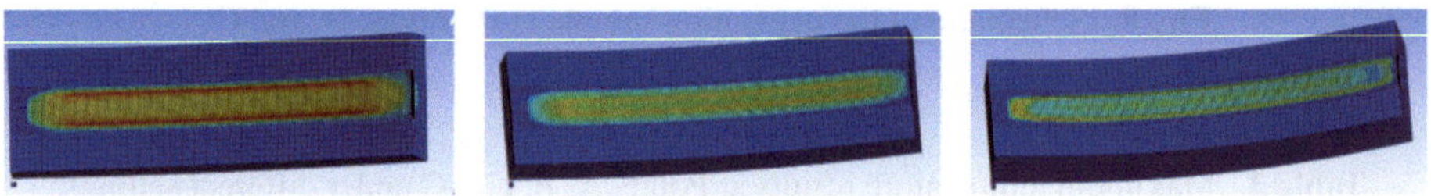

(a) Simulation control group A (b) Simulation control group B (c) Simulation experimental group

Fig. 10. Schematic diagrams of equivalent plastic deformation distribution in the simulated structural fields under different settings.

During the laser quenching process of medium-carbon steel, the focus is primarily on the deformation of the backside of the laser-quenched surface. When the laser-quenched surface is used in actual production, it is essential to ensure that the deformation along the Z-axis on the backside of the quenching does not affect the installation accuracy of the part. The deformation along the Z-axis at five different paths on the backside of the laser-quenched surface is observed as shown in Fig. 11.

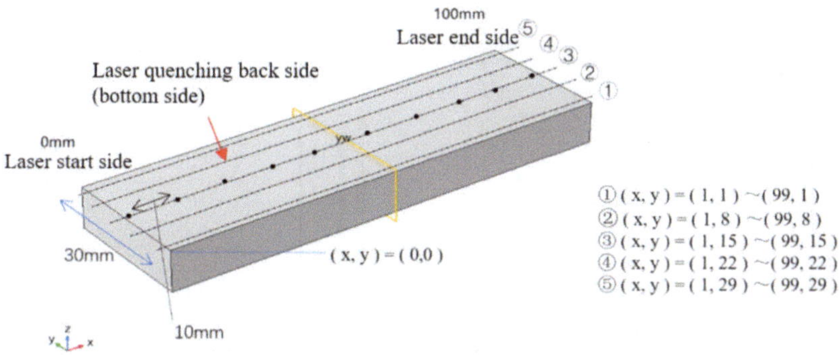

Fig. 11. Schematic Diagram of Defined Path Locations on the Backside of Laser Quenching.

The deformation along the Z-axis at different times along the five paths on the backside of the laser-quenched sample is compared between the three simulation groups and the experimental deformation data, as shown in Fig. 11. By observing the deformation curves along the Z-axis for the five paths in Fig. 12(a), (b), (c), (d) and (e), it can be seen that the simulation experiment group has the highest fitting degree with the experimental deformation data. The Z-axis deformation along the five paths is relatively smooth, with changes within 2μm for adjacent paths. In contrast, the other two simulation control groups show larger variations in Z-axis deformation for adjacent paths, with a maximum difference of 20μm. For Simulation Control Group A, the Z-axis deformation is largest at Path 3 and smallest at Path 1. For Simulation Control Group B, the situation is reversed, with the Z-axis deformation being smallest at Path 3 and largest at Path 1.

Based on the above analysis, during the laser quenching process of medium carbon steel, the magnitude and distribution of elastic and plastic deformation differ among the three simulation groups. This is because the elastic modulus and Poisson's ratio influence the extent of elastic deformation, while the coefficient of thermal expansion characterizes deformation caused by temperature and affects the degree of thermal deformation. The yield strength determines whether the material enters the yielding stage and undergoes plastic deformation, thus influencing the distribution between elastic and plastic deformation. Additionally, the tangent modulus affects the extent of plastic deformation.

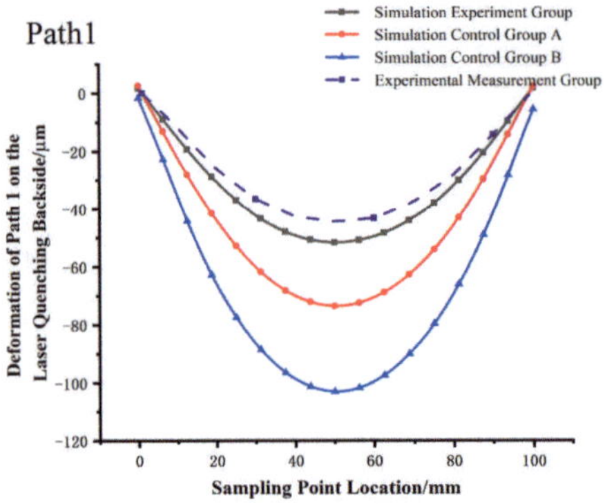

(a) comparison of deformation of path 1

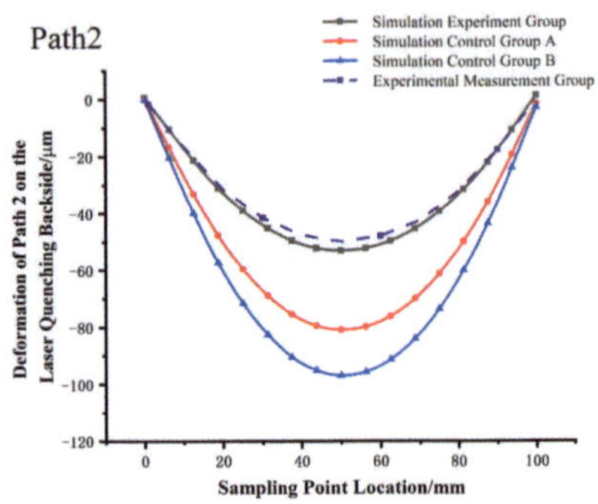

(b) comparison of deformation of path 2

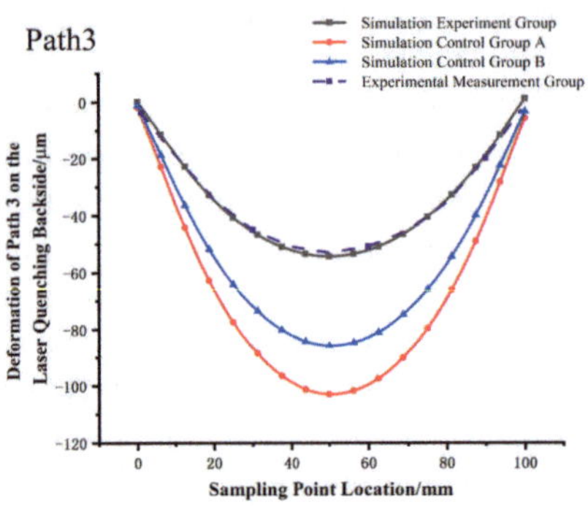

(c) comparison of deformation of path 3

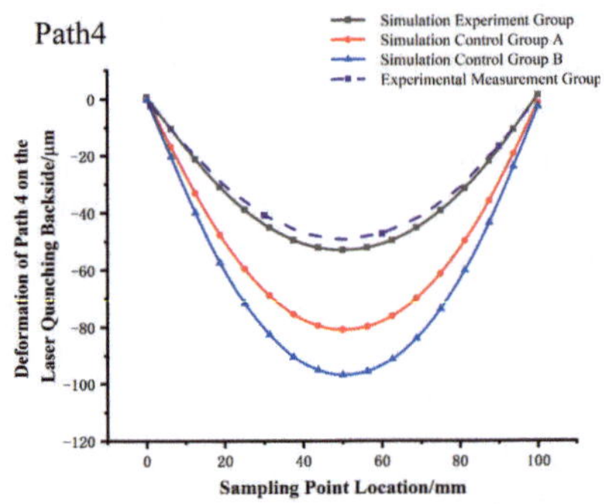

(d) comparison of deformation of path 4

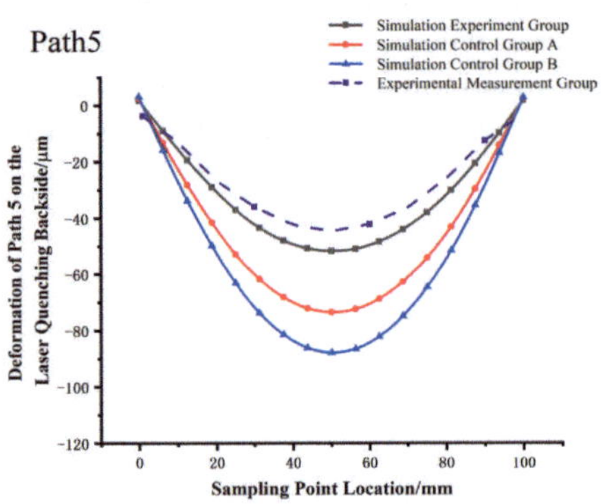

(e) comparison of deformation of path 5

Fig. 12. Comparison of simulation deformation curves at defined paths under different settings.

4 Conclusion

(1) Studies have shown that material property parameters significantly influence the simulation results and deformation distribution characteristics during the laser quenching of medium carbon steel. Considering the nonlinear structural characteristics of the material can effectively improve the accuracy of deformation prediction.
(2) The method proposed in this study is primarily applicable to relatively simple geometries, and certain limitations remain when applied to complex components.

References

1. Wu, L., Long, W., Yang, Y., et al.: Research status of laser quenching of steel. Metal Process. (Hot Processing) **05**, 20–31 (2023)
2. Guo, S.H., Li, Z.X., Cen, Y.M., et al.: Organization distribution and peak temperature model of the laser surface treatment process of 65Mn steel. Metal Heat Treatment **49**(01), 249–256 (2024)
3. Zhao, J., Li, C., Zhan, H., et al.: Study on laser quenching of aircraft gears with a rectangular beam based on a multi-field coupling mechanism. Appl. Phys. A **131**(1), 1–21 (2025)
4. Zhang, Q.L., Li, G.C., Dong, H.P., et al.: Stress and deformation research of laser-electromagnetic induction composite quenching process for 42CrMo steel. Surface Technol. **53**(21), 142–152 (2024)
5. Ma, Z.H., Adaei, X.D., Liu, J.J., et al.: Temperature field and phase transformation hardening zone in the laser quenching of 45 steel. Metal Heat Treat. **48**(09), 258–264 (2023)
6. Rong, Y.M.: Study on the mechanism of phase transformation and precipitation effects on thermo-elastic-plasticity in laser welding of high-strength steels, and mechanical properties of welds. Wuhan: Huazhong University of Science and Technology (2017)

Simulation Analysis of the Drilling Force Model for Double Chamfered Cutting Edges

Yanyu Li, Xiaoshi Zhang(✉), Liming Jiang, Cheng Fu, and Qiang Zhang

Chongqing Construction Industry (Group) Co., Ltd, Chongqing 400054, China
2306834665@qq.com

Abstract. In order to study the influence of double chamfer parameters on drilling performance, a finite element J-C thermodynamic coupling damage model was established for double chamfer TBT drilling and processing of 7075-T651 aluminum alloy using Abaqus software. Theoretical analyses were conducted on the force analysis of the ideal contour of the deformation area during the drilling process, the material flow mode of the chamfered cutting edge, the force analysis of the cutting force, and the conveniently calculated drilling force model. The torque formula and axial force formula of the drilling force model for the chamfered cutting edge of TBT drilling were obtained. Research shows that the theoretical formula of the established drilling force model for the chamfered cutting edge of TBT drilling can be applied to the calculation of the torque and axial force of the chamfered cutting edge of TBT drilling. The distribution laws of the chamfer cutting edge width and chamfer Angle on the axial force and torque of the TBT drill were obtained. It is concluded that when designing the bevel cutting edge of TBT drilling, a bevel cutting edge width of 0.1–0.2 mm and bevel angles of 30° and 60° can obtain a smaller axial force.

Keywords: TBT drilling bevel edge · Drilling force model · Torque · Axial force

1 Introduction

The TBT drill is a specialized tool for deep hole processing, featuring excellent rigidity and toughness. However, during the grinding process of TBT drills, microscopic defects such as cracks and notches may occur, affecting the service life of the cutting tools. The research background of this paper is to shape TBT drills with chamfered cutting edges during the processing and manufacturing of cutting tools, effectively limiting the occurrence of microscopic notches and prolonging the service life of the cutting tools.

At present, scholars at home and abroad have conducted extensive research on chamfered cutting edge tools. Weng et al. studied the influence law of the drilling force of milling cutter chamfering tools, introduced the cutting force model for correcting the sharp cutting edge by reference factors, and proposed the selection of tool parameters and chip separation basis for chamfered double straight edge drills, providing a theoretical basis for analyzing the chip breaking mechanism of chamfered double straight edge drills in this paper. Zhuang et al. studied the geometric chamfer edge parameters of the

S. Halgamuge et al. (Eds.): ICACMVE 2025, LNME, pp. 692–704, 2026.
https://doi.org/10.1007/978-981-95-6911-3_69

turning tool's cutting edge, and based on the cutting theory, established a theoretical prediction model of cutting force and proposed the process parameters during high-speed cutting, providing appropriate process parameters for this study. Zhuang et al. studied the slip line theory of the tool cutting process and obtained the orthogonal cutting force model of the turning tool with a chamfered cutting edge, providing a reference idea for the parameter optimization of the chamfered double straight cutting edge drill bit in this paper.

This paper studies the force analysis of the ideal contour of the deformation area during the drilling process, the material flow mode of the chamfered cutting edge, the force analysis of the cutting force, as well as the torque and axial force in the theoretical model of the drilling force with chip distribution grooves. In this paper, the Abaqus finite element analysis software is used to simulate the hole processing process of the chamfered cutting edge parameters, verifying the correctness of the theoretical model of the drilling force of the chamfered cutting edge tool proposed in this paper. Studies [1–3] provide theoretical support for the tool design of chamfered cutting edge TBT drills.

2 Establishment of the Finite Element Model

2.1 Establishment of the Geometric Model

A tool model for simulation was established in NX8.0 to study the influence of the double straight cutting edge of the chamfering in TBT drilling on the drilling performance, as shown in Fig. 1. The cross-sectional view of the parameters of the double straight cutting edge of the TBT drill chamfer is shown in Fig. 2. γ_0 is the rake Angle, α_0 is the relief Angle, b is the chamfer width of the cutting edge, and γ is the chamfer Angle of the cutting edge.

Fig. 1. Tool model.

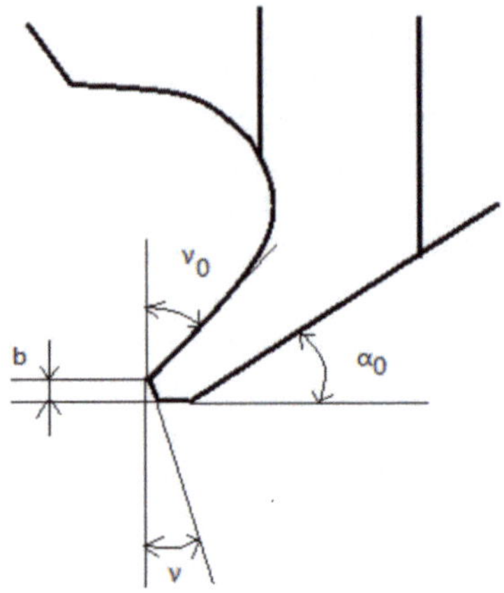

Fig. 2. Sectional view of the parameters of the beveled double straight cutting edge.

2.2 Analysis of Drilling Parameters and Cutting Force of Chamfered Double Straight Edge Drill

The schematic diagram of the chamfered double straight edge drilling of the inner hole is shown in Fig. 3 [8]. In the actual drilling process, V_f is the feed rate of 12 mm/min, V_c is the cutting speed of 22.6 m/min, f_n is the feed rate of 0.01 mm/r, a_p is the cutting depth of 1.5 mm, D_c is the tool diameter of 6 mm, n is the spindle speed of 1200 r/min, and K_r is the cutting Angle of the first cutting edge of 60°. The schematic diagram of the cutting force in drilling is shown in Fig. 4. The axial force F_f is the feed force, the torque generated is the tangential force F_c, and the radial force F_p is generated by guidance.

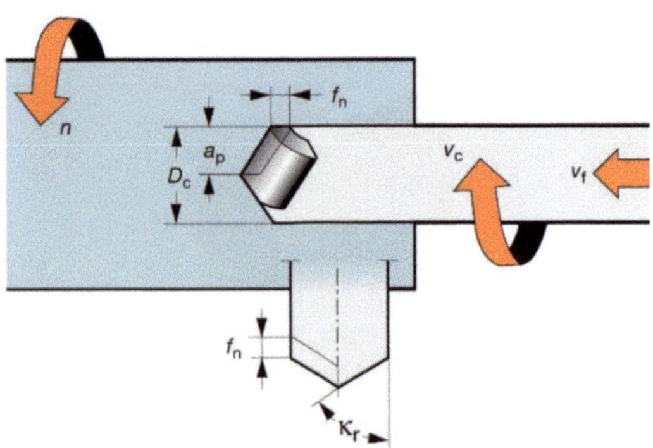

Fig. 3. Drilling parameter schematic diagram.

In Fig. 3, V_f represents the feed rate, V_c represents the cutting speed, f_n represents the feed rate, a_p represents the cutting depth of 1.5 mm, D_c represents the tool diameter, n represents the spindle speed, and K_r represents the cutting Angle of the first cutting edge.

In Fig. 4, the feed force F_f, tangential force F_c and radial force F_p are shown.

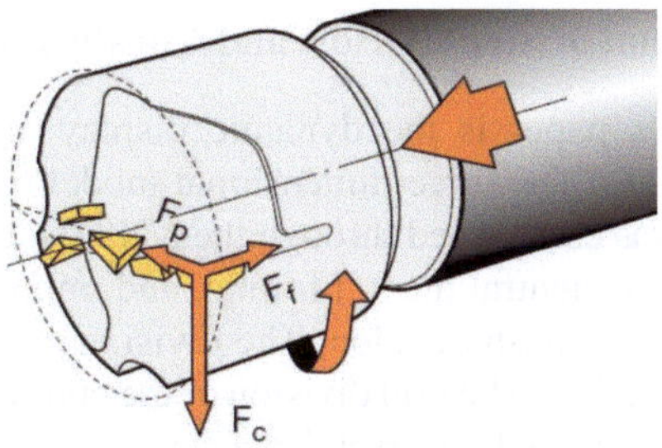

Fig. 4. Schematic diagram of cutting force.

2.3 Establishment of the Finite Element Simulation Material Model

The basis for the separation and discrimination of materials in the specific simulation is to adopt the John-Cook model separation criterion commonly used in the Abaqus finite element analysis.

$$\sigma_{\mathrm{f}} = \left[A + B(\varepsilon^{\mathrm{p}})^{\mathrm{n}}\right] \times 1 + C\ln\frac{\varepsilon^{\mathrm{p}}}{\varepsilon_0} \times \left[1 - \frac{T - T_0}{T_{\mathrm{melt}} - T_0}^{\mathrm{m}}\right]$$

where A, B, C, n, and m are material parameters as shown in Table 1, σ_{f} is the equivalent stress, ε^{p} is the plastic strain, ε_{o} is the reference strain, T_{o} is the reference temperature, and T_{melt} is the melting temperature.

Table 1. Parameters of 7075 Aluminum Alloy Materials.

A	B	C	n	m
527	575	0.017	0.72	1.61

ε^{f} is the equivalent stress when the material fails.

$$\varepsilon^{\mathrm{f}} = \left[\mathrm{d}_1 + \mathrm{d}_2(\varepsilon^{\mathrm{p}})^{\mathrm{d}_3}\right] \times 1 + \mathrm{d}_4\ln\frac{\varepsilon^{\mathrm{p}}}{\varepsilon_0} \times \left[1 - \mathrm{d}_5\frac{T - T_0}{T_{\mathrm{melt}} - T_0}^{\mathrm{m}}\right]$$

where d_1, d_2, d_3, d_4 and d_5 are failure parameters as shown in Table 2 [4–8].

Table 2. Failure Parameters of 7075 Aluminum Alloy.

d_1	d_2	d_3	d_4	d_5
0.11	0.572	−3.446	0.016	1.099

2.4 Analyze the Addition of Steps, Grids, and Constraints

The analysis step of this paper is the dynamic display analysis step with a large temperature-stress coupling. The three-dimensional model of the double straight edge bevel cutting edge drill was established through the NX8.0 three-dimensional modeling software, converted into the neutral file x_t format, and imported into the general finite element abaqus software for mesh division. The division of meshes mainly adopts the idea of local refinement meshes. The grid division of the cutting area is denser, while that of the non-cutting area is relatively coarser. This can save the limited computing space of the computer. The contact between the surfaces of the beveled double straight edge drill face and the cutting part of the blank belongs to motion contact. Self-contact is also added, which belongs to general contact. The chamfered double straight edge drill adopts the C3D4T grid type and the blank adopts the C3D8RT grid type. The C3D8RT has an hour-controlled reductive integral model, which is easy to converge during the iterative process. The addition of constraints mainly considers the addition of tool paths and tool tracks. Through the point-analytical rigid body coupling method, the tool paths and tool tracks are controlled to control the depth of cut and feed rate of the back of the double straight edge chamfer cutting edge. The rotational speed of the points is controlled to control the spindle speed and thereby control the cutting speed.

The specific approach is to adopt the idea of model segmentation in the drilling area and the non-drilling area. A cylinder is separated from the rectangular blank, and then a single-sided pattern is distributed along the edge of the cylinder to refine the mesh. This can exert the maximum computing power of the computer under limited computing resources. The addition of constraints is carried out through the point-rigid body coupling method. Specifically, a reference coordinate is established, and the reference point is coupled with the finite element model of the tool. The spindle speed is controlled by controlling the rotational speed of the control point, and the feed speed is controlled by controlling the axial speed of the control point (at this time, the speed is a constant). In the actual drilling process, the drilling feed speed is not constant. By controlling the variation relationship between displacement and time to control the feed rate, an effective and feasible method is provided for the simulation of changing the feed rate. The mesh division is shown in Fig. 5 [9–16].

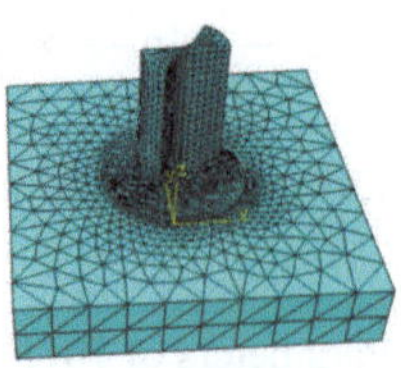

Fig. 5. The finite element model for meshing.

3 Establishment of the Theoretical Model of Drilling Force

The force analysis of the single cutting edge cutting process is now conducted. The micro-element OA' B' triangle is analyzed and subjected to the boundary OA', OB', normal force $F_b{}'$ and shear force $F_a{}'$ respectively, as well as the shear force F_S and normal force F_p of the chamfered cutting edge. According to the force equilibrium condition, the force analysis diagram can be written as:

$$F'_b = FScos\varphi 1 + F_p sin\varphi_1 + F_{S1} \tag{1}$$

$$F'_a = F_S sin\varphi_1 - F_p cos\varphi_1 + F_{p1} \tag{2}$$

In Fig. 6, OA' is L, and the width is W. It can be calculated that $L = t_1 cot(\varphi_0 - 1)$, $F_S = \tau WL$, $F_p = \sigma WL$, $F_{S1} = \tau_{S1} WL$, $F_p = \sigma_{P1} WL$. In the microelement, $\tau = \tau_{S1} = \tau_S$, and $\sigma = \sigma_{P1} = \sigma_P$. Formulas (1–2) can be rewritten as Formulas (3–4).

$$F'_b = \tau_S W t_1 cot(\varphi_0 + 1) \tag{3}$$

$$F'_a = \sigma_P W t_1 cot(\varphi_0 - 1) \tag{4}$$

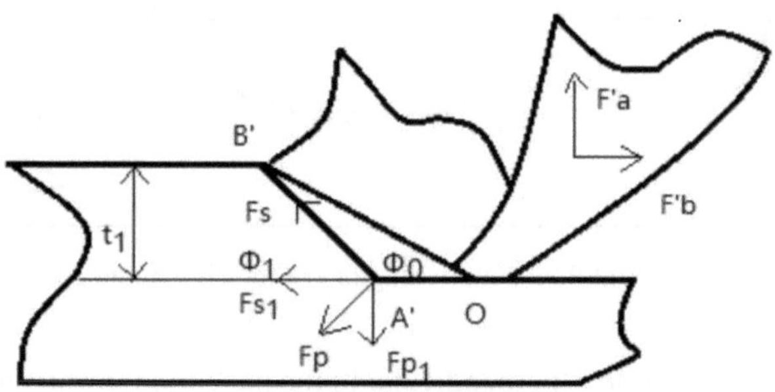

Fig. 6. Force analysis diagram of the ideal contour in the deformed area.

As shown in Fig. 7, introducing the geometric proportionality coefficient c, h = cb. According to the geometric relationship of triangle ABC, $l_1 = cbtan\gamma$, $l_2 = cbcot\alpha_0$. The static pressure in triangle ABC is denoted as p. The component force in the normal direction of the main direction of the cutting edge $F_a{}^1$ and the component force in the tangential direction $F_b{}^1$ are shown in Formulas (5, 6). Substituting l_1 and l_2 into Formulas (5, 6), we can obtain Formulas (7, 8).

$$F_b^1 = \mu F_a^1 \tag{5}$$

$$F_a^1 = pW(l_1 + l_2) \tag{6}$$

$$F_b^1 = \mu pW(cbtan\gamma + cbcot\alpha_0) \tag{7}$$

$$F_a^1 = pW(cbtan\gamma + cbcot\alpha_0) \tag{8}$$

For the resultant force in the normal direction of the main direction of the cutting edge F_a and the resultant force in the tangential direction F_b, substituting Formulas (3, 4) and (7, 8) into (9, 10), we can obtain Formulas (11, 12).

$$F_a = F_a^1 + F'_a \tag{9}$$

$$F_b = F_b^1 + F'_b \tag{10}$$

$$F_a = \sigma_P Wt_1 cot(\varphi_0 - 1) + pW(cbtan\gamma + cbcot\alpha_0) \tag{11}$$

$$F_b = \tau_S Wt_1 cot(\varphi_0 + 1) + \mu pW(cbtan\gamma + cbcot\alpha_0) \tag{12}$$

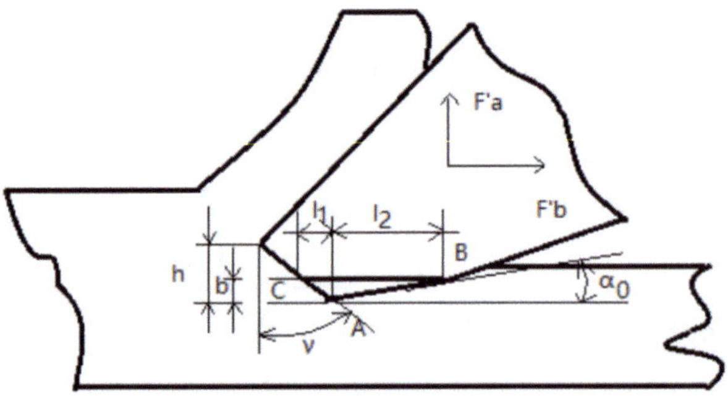

Fig. 7. The material flow pattern diagram of the chamfered cutting edge.

In Fig. 8, the F_b direction is perpendicular to the paper and inward. The infinitesimal dS between S_{i-1} and S_i is taken. There exists a formula for axial force $F = sin\varphi_3 F_a$ in the infinitesimal dS, and Torque formula $M = rF_b$. The integration of $dF = sin\varphi_3 dF_a$ and $dM = rdF_b$ in the interval $[R_1, R_2]$ is carried out by using the idea of calculus. Axial force F and Torque M can be obtained. Substituting Formulas (13, 14) and (11, 12) into (13–14), we can obtain (15–16).

$$F = \int_{R1}^{R2} sin\varphi_3 dF_a \tag{13}$$

$$M = \int_{R1}^{R2} rdF_b \tag{14}$$

$$F = (R_2 - R_1)[\sigma_P fsin\varphi_3(cot\varphi_0 - 1) + 2p(cbtan\gamma + cbcot\alpha_0)] \tag{15}$$

$$M = \left(R_2^2 - R_1^2\right)[1/2\tau_S f(cot\varphi_0 + 1) + \mu p(cbtan\gamma + cbcot\alpha_0)/sin\varphi_3] \tag{16}$$

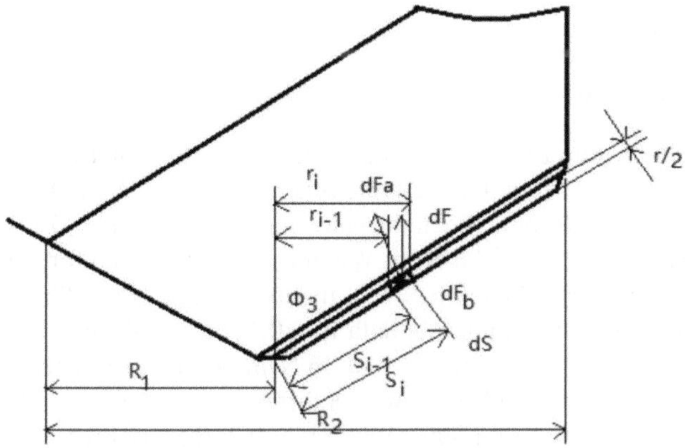

Fig. 8. Force analysis of cutting force.

Formulas (15, 16) cannot be directly calculated here yet. Since the shear Angle φ_0 and the friction coefficient μ are still unknown data, they need to be converted into known data through formulas. The deformation coefficient $\xi = t_2/t_1$ is introduced, and t_2 is the unchanged thickness of the cutting layer.

M and ξ have the following relationship (17–19). Bringing (17–18) into (19), we can obtain (20). Bringing (18) and (20) into (15–16), we can obtain (21–22).

$$\varphi_0 = \pi/4 - \beta + \gamma_0 \tag{17}$$

$$\tan\varphi_0 = \cos\gamma_0/(\xi - \sin\gamma_0) \tag{18}$$

$$\mu = \tan\beta \tag{19}$$

$$\mu = \tan\{\pi/4 - \arctan[\cos\gamma_0/(\xi - \sin\gamma_0)] + \gamma_0\} \tag{20}$$

$$F = (R_2 - R_1)[\sigma_P f\sin\varphi_3(\xi - \sin\gamma_0 + \cos\gamma_0)/\cos\gamma_0 + 2p(cb\tan\gamma + cb\cot\alpha_0)] \tag{21}$$

$$M = \left(R_2^2 - R_1^2\right)\{1/2\tau_S f(\xi - \sin\gamma_0 + \cos\gamma_0)/\cos\gamma_0 + \tan\{\pi/4 - \arctan[\cos\gamma_0/(\xi - \sin\gamma_0)] + \gamma_0\} \, p(cb\tan\gamma + cb\cot\alpha_0)/\sin\varphi_3\} \tag{22}$$

where $\sigma_P = 425$ MPa, $p = 1560$ MPa, $\tau_S = 300$ MPa, $\xi = 1.7$, $c = 0.03$, $R_1 = 3$ mm, $R_2 = 6.4$ mm, $\varphi_3 = 62.5°$, $\gamma_0 = 15°$, $\alpha_0 = 8°$. Axial force F and torque M can calculate by b which is the width of the chamfer of the cutting edge, and γ is the chamfered cutting edge.

4 Result Analysis

4.1 Model Validation

The drilling test of the chamfered cutting edge of 7075 material. Among them, the rotational speed is 2000 r/min, the feed rate is 0.2 mm/r, the depth of cut is 6.5 mm, and the tool diameter is φ12.8 mm. When the test data γ is 50° and b = 0.1, the axial force F = 884 N and the torque M = 4.56 N•m; When b = 0.2, the axial force F = 1129 N and the torque M = 6.83 N•m; When b = 0.3, the axial force F = 1386 N and the torque M = 9.09 N•m; When the simulation data b = 0.1, the axial force F = 874 N, with a difference of 1.13%, and the torque M = 4.46 N•m, with a difference of 2.19%. When b = 0.2, the axial force F = 1179 N, with a difference of 4.42%, and the torque M = 6.53 N•m, with a difference of 4.39%. When b = 0.3, the axial force F = 1346 N, with a difference of 2.88%, and the torque M = 8.79 N•m, with a difference of 3.3%. If the error between the test data and the simulation data is within 10%, it is considered that the simulation data is valid and feasible.

4.2 Data Analysis

The distribution laws of the chamfer cutting edge width and chamfer Angle on the axial force and torque of the TBT drill are explored. The following is an effective comparison between the theoretical calculation data and the simulation data. γ is divided into 30°, 40°, 50°, and 60°. When b is 0.1, 0.2, and 0.3, orthogonal simulation experiments are conducted with a total of 12 groups of data, among which 1 is γ = 30° and b = 0.1; 2 is γ = 30° and b = 0.2; 3 is γ = 30° and b = 0.3; 4 is γ = 40° and b = 0.1; 5 is γ = 40° and b = 0.2; 6 is γ = 40° and b = 0.3; 7 is γ = 50° and b = 0.1; 8 is γ = 50° and b = 0.2; 9 is γ = 50° and b = 0.3; 10 is γ = 60° and b = 0.1; 11 is γ = 60° and b = 0.2; 12 is γ = 60° and b = 0.3.

When the chamfer Angle γ = 30°, the chamfer width b = 0.1, the axial force increases from 410 N to 0.3, the axial force increases from 793 N, and the torque increases from 3.7 N•m to 4.9 N•m. When γ = 40° b = 0.1, the axial force increases from 837 N to the chamfer width b = 0.3, the axial force is 1070 N, and the torque increases from 4.5 N•m to 5.5 N•m. When γ = 50° b = 0.1, the axial force increases from 874 N to the chamfer width b = 0.3, the axial force is 1370 N, and the torque increases from 5.3 N•m to 7.0 N•m. With the increase of the chamfer width b, the axial force and torque are positively correlated, as shown in Fig. 9. For the same chamfer width b, as the chamfer Angle is within the range of 30° to 50°, the chamfer Angle is also positively correlated with the axial force and torque. When the chamfer Angle γ = 60°, b = 0.1, the axial force F = 590 N, and the torque M = 4.0 N•m; When b = 0.1, the axial force F = 625 N, and the torque M = 4.2 N•m, it can be seen from Fig. 10 that when the chamfer Angle γ = 60°, the axial force and torque decrease to γ = 30°. When designing the beveling edge of the TBT drill, a beveling edge width of 0.1 to 0.2 mm and beveling angles of 30° and 60° can obtain a relatively small axial force.

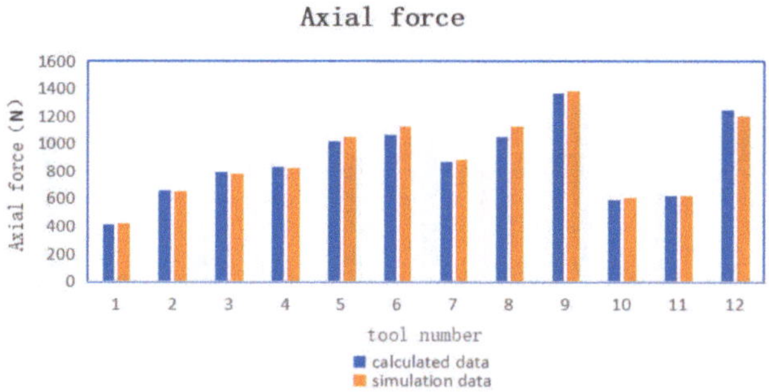

(a) Axial force

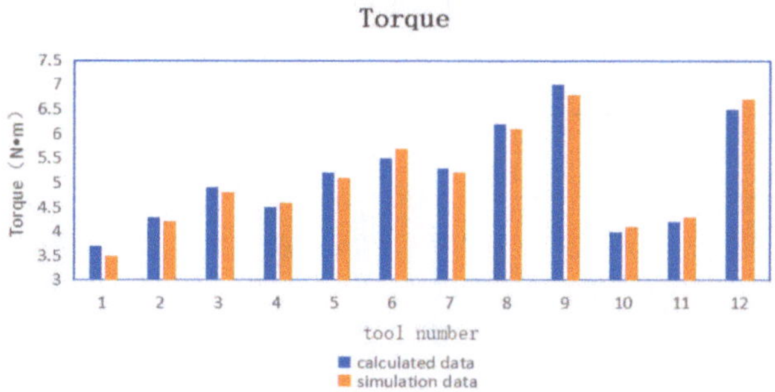

(b) Torque

Fig. 9. Comparison of calculated data and simulation data.

Figure 10 shows the relative changes in axial force and torque at γ angles of 40°, 50°, and 60° compared to γ Angle of 30°. For tools 1 to 3, the chamfering Angle is 30°; for tools 4 to 6, it is 40°; and for tools 7 to 9, it is 60°. It can be clearly seen that when the chamfer Angle $\gamma = 60°$ and b = 0.1, the calculated value of the axial force is 180 N, the simulated value is 190 N, and the calculated value of the torque is 0.3 N•m and the simulated value is 0.6 N•m. When the chamfer Angle $\gamma = 60°$ and b = 0.2, the calculated value of the axial force is 25 N, the simulated value is 30 N, the calculated value of the torque is 0.1 N•m, and the simulated value is 0.1 N•m.

A detailed analysis was conducted on the relative changes of the calculated data and the simulation data when the chamfer edge width was 0.1–0.2 mm and the chamfer Angle was 30° and 60°. For Tools 1–3, the chamfer Angle was 30°, and for Tools 4–6, it was 60°. The maximum change in torque was 5.4%, and the maximum change in axial force was 3.3%. The average value of torque change is 2.9%, the torque fluctuation value is 3.4%, the average value of axial force change is 1.6%, and the axial force fluctuation value is 1.2%. It can be seen that when the drill bit with double chamfered cutting edges is drilling, the axial force is relatively stable with a relative change of 1.2%, and the torque is relatively stable with a relative change of 3.4%, remaining stable at the technical level of 600 N axial force and 4 N•m torque shown in Fig. 11.

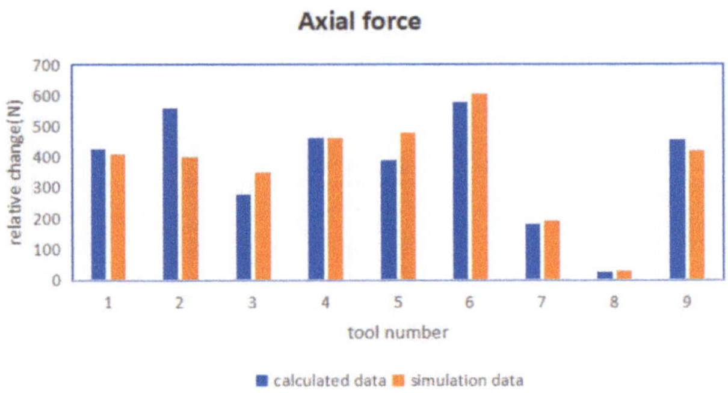

(a) Axial force

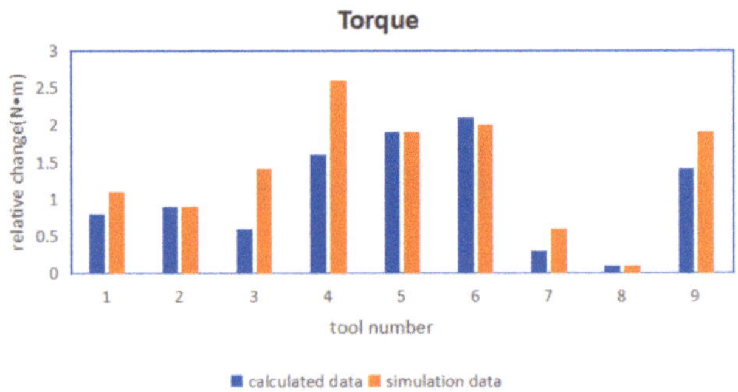

(b) Torque

Fig. 10. Comparison of Angle $\gamma = 30°$ of calculated data and simulation data.

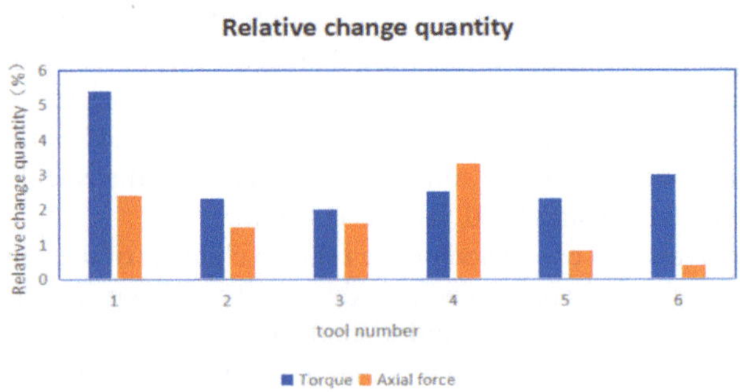

Fig. 11. Relative change quantity of calculated data and simulation data.

5 Conclusion

This paper studies the drilling force model of TBT for drilling double straight edge chamfered cutting edges. Firstly, the parameter model of the chamfered cutting edge is established and the Abaqus finite element simulation is carried out. The torque formula and axial force formula of the drilling force model for the chamfered cutting edge of TBT are obtained. The distribution laws of the chamfer cutting edge width and chamfer Angle on the axial force and torque of the TBT drill are explored. The following is an effective comparison between the theoretical calculation data and the simulation data. γ is divided into 30°, 40°, 50°, and 60°, and orthogonal simulation experiments are conducted when b is at 0.1, 0.2, and 0.3, totaling 12 groups of data. The correctness of the theoretical model of the drilling force of the beveled cutting edge tool proposed in this paper is verified. Theoretical support is provided for the tool design of chamfered cutting edge TBT drills.

a) By using the idea of calculus to integrate dF = sinφ3dFa and dM = rdFb in the interval [R1, R2], the axial force and torque of the drilling force model for the double straight bevel cutting edge of the TBT drill can be obtained.
b) When designing the beveling edge of the TBT drill, a beveling edge width of 0.1 to 0.2 mm and beveling angles of 30° and 60° can obtain a relatively small axial force.
c) As the chamfer width b increases, the axial force and torque are positively correlated. For the same chamfer width b, when the chamfer Angle is within the range of 30° to 50°, the chamfer Angle is also positively correlated with the axial force and torque.
d) The average value of torque change is 2.9%, the torque fluctuation value is 3.4%, the average value of axial force change is 1.6%, and the axial force fluctuation value is 1.2%. It can be seen that when the drill bit with double chamfered cutting edges is drilling, the axial force is relatively stable with a relative change of 1.2%, and the torque is relatively stable with a relative change of 3.4%, remaining stable at the technical level of 600 N axial force and 4 N•m torque.

References

1. He, T.: Simulation research on the drilling mechanism of small deep holes in titanium alloys. North University of China, Taiyuan (2017)
2. Zeng, T.: Research on the design of twist drills with a large length-to-diameter ratio and their application in deep Hole processing. Shandong: Shandong University (2022)
3. Wang, S.Q.: Deep Hole Machining Technology. Northwestern Polytechnical University Press, Xi'an (2003)
4. Liu, Z.F., Han, X.: Deep Hole Processing Techniques for Typical Difficult-to-Machine Materials Technique. Science Press, Beijing (2020)
5. Xiao, M.H.: Wear mechanism of high-speed cutting tools for nickel-based SuperAlloys research. Nanjing University of Aeronautics and Astronautics, Nanjing (2010)
6. Liu, Q.T.: Research on the processing technology of 17–4PH ultra-long and thin deep holes. Xi'an Shiyou University, Xi'an (2018)
7. Qian, Q., Cheng, Y., Zheng, L.H.: Deep holes based on maximum stability optimization of structural parameters of machining gun drill. Comb. Mach. Tool Autom. Mach. Technol. **6**, 155–158 (2021)
8. Zhao, G.L., Feng, P.F., Zhang, J.F.: Simulation and experimental research on ultrasonic vibration drilling process characteristics of titanium Alloy. J. Beijing Univ. Aeronaut. Astronaut. **45**(8), 1597–1605 (2019)
9. Long, J.C., Rong, Q., Zhang, W., et al.: Cooling and lubrication in deep hole machining analysis of the function of lubricating fluid. Tool Technol. **53**(3), 91–93 (2019)

10. Li, X.D., Jiang, N., Zhou, X.L., et al.: Comparison of grinding and milling surface quality of 17–4PH steam turbine blade roots. Tool Technol. **52**(1), 30–33 (2018)
11. Chen, L.H., Yu, D.G.: Zhao H Y TC4 titanium alloy deep hole drilling research on cutting mode and axial force. Mach. Tool Hydraulics **51**(7), 153–157 (2023)
12. Lu, N.: Cemented carbide based on ABAQUS finite element simulation research on tool wear mechanism. Mach. Tool Hydraulics **49**(4), 164–168 (2021)
13. Zhu, H.X., Shen, X.Q., Zhang, L.X.: Quasi-dry deep hole drilling simulation research on medium temperature field and tool wear. Mach. Des. Manuf. **2**, 193–195 (2014)
14. Cui, J., Liu, L., Xia, H.Y., et al.: Millisecond excitation based on the VOF method numerical study and process optimization of Photohole-making process. Laser Infrared **52**(2), 211–216 (2022)
15. Wang, J.: Modern deep Hole machining technology. Ha Er bin Institute of Technology Press, Harbin (2005)
16. Wang, G.L.: Theoretical model of drilling force considering the geometric parameters of the chamfering cutting edge of the shovel drill tool. Acta Ordnance Engineering 26, 44(6), 1733–1743

Research on the Thinning Rate of Conical Parts with Variable Curvature by Static Pressure Support in Single-Point Progressive Forming

Xinyu Wang, Mingshun Yang(✉), Han Du, and Haoze Zhang

School of Mechanical and Precision Instrument Engineering, Xi'an University of Technology, Xi'an 710001, China
yangmingshun@xaut.edu.com

Abstract. In single-point progressive forming, problems such as cracking are prone to occur due to uneven thickness. In this paper, the single-point progressive forming technology assisted by static pressure support is adopted. Firstly, the principle of single-point incremental forming with static pressure support is analyzed, and the work piece area is divided according to the thinning rate. Secondly, the variation of the thickness reduction rate in each work piece area under different static pressure conditions is studied through finite element simulation. The research found that the thinning rates in Zone I and Zone III remained at around ±3%, while Zone II was the main deformation area. Its thinning rate fluctuated at 105 mm, dropping from +25% to −35%, and local instability of the material might occur. The thinning rate of the end path resumed positive growth and reached 20% at 150 mm.

Keywords: Thinning rate · wall thickness distribution uniformity · forming limit · stress concentration · springback

1 Introduction

Single Point Incremental Forming (SPIF) is an emerging plastic forming method for sheet metal. It does not rely on molds. According to the concept of "layered manufacturing" in rapid prototyping, under the control of the machine tool, the tool head processes the sheet metal layer by layer along the predetermined path to form the required work piece. Single Point Incremental Forming (SPIF) is an emerging sheet metal forming technology based on the "layered manufacturing" concept from rapid prototyping. The forming tool progressively shapes the sheet material layer by layer along a predefined path to create the desired component. Compared to conventional sheet metal forming processes, SPIF eliminates the need for dedicated dies, offering shorter lead time, lower cost, and higher flexibility. These advantages make it particularly suitable for small-batch rapid manufacturing of complex curved components in the aerospace and medical industries. However, when wall thickness reduction exceeds a critical threshold, there is a significant deterioration in part quality and the potential for fracture.

S. Halgamuge et al. (Eds.): ICACMVE 2025, LNME, pp. 705–713, 2026.
https://doi.org/10.1007/978-981-95-6911-3_70

To address this technical challenge, Static Pressure Support (SPS) has been introduced into SPIF. By applying controlled static pressure to the non-forming zone of the sheet material, SPS improves material flow uniformity and optimizes thickness distribution, reducing thinning rate by 20%–30% while increasing forming limit by over 20%. Pressure regulation enables precision control of complex-curved components and effectively prevents sheet fracture caused by strain concentration, establishing a technical foundation for high-precision sheet forming.

The research on the thickness thinning mechanism and process parameters of single-point incremental forming (SPIF) has undergone years of development and has gradually formed a systematic theoretical framework. In the early studies, Pinard et al. [1] proposed magnetic field-assisted SPIF (M-SPIF), using magnetic sphere tools to regulate the forming force. Xiao et al. [2] found that low-frequency vibration could reduce the forming force of AA1050 aluminum alloy. Almadani et al. [3] developed the hot air non-contact polymer SPIF, revealing the crucial role of temperature and feed rate. Hou et al. [4] studied the forming mechanism of jute/PLA composites by heat-assisted SPIF. Szpunar et al. [5] verified the accuracy of friction stir assisted SPIF for titanium plates through numerical simulation (with an error of 3%). Naranjo et al. [6] analyzed the influence of heat-assisted SPIF (HA-SPIF) on the surface properties of Ti6Al4V. Gohil et al. [7] proposed the forming limit criterion of AA3003 in ultrasonic vibration-assisted SPIF. Bai Lang [8] developed the hydropressure-ultrasonic coupled SPIF, which significantly improved the accuracy. Lin Yunbo [9] optimized static pressure and ultrasonic parameters to reduce axial error. Nie Kaixin [10] confirmed that variable pressure bracing improves the residual stress distribution of SPIF. Jing Zhangshuai [11] clarified that ultrasonic vibration optimizes residual stress and thinning rate through the "surface/volume effect". Wang Xiaoye et al. [12] simulated and revealed the nonlinear influence of static pressure and ultrasonic waves on the forming force.

2 Single-Point Incremental Forming Principle and Geometric Model Establishment

2.1 Single-Point Incremental Forming Principle

As shown in Fig. 1, the SPS-SPIF schematic diagram, this technology uses a hydraulic system to process the sheet metal. Its working principle is as follows: The sheet metal with an initial thickness is fixed on a sealed fixture, and the sheet metal is extruded along a preset trajectory through the tool head. During the forming process, the hydraulic pump provides adjustable pressure, and the relief valve maintains a constant pressure, ensuring that the sheet metal is always subjected to a constant static pressure supporting force.

2.2 Geometric Model Establishment

Workpiece Dimensions For processing variable-curvature conical components, a hemispherical forming tool (10 mm diameter) and a circular fixture were used to shape 1060 aluminum sheets (1 mm thickness). The material properties and stress-strain characteristics are detailed in Tables 1 and 2. The forming tool and fixture were modeled as

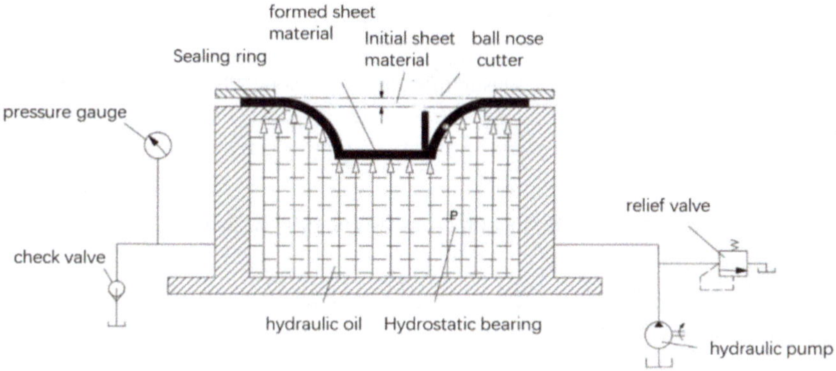

Fig. 1. SPS-SPIF schematic diagram.

analytical rigid bodies. The sheet had dimensions of 136 mm in diameter, and the fixture had an outer diameter of 140 mm and an inner diameter of 100 mm. The simulation used a stable, dynamic, explicit algorithm.

Tensile tests were conducted on the specimens in three directions: 0 °, 45 °, and 90 °. The material properties were tested using the ASTM E8 standard tensile test method, and 1060Al was stretched using an electronic tensile testing machine (see Table 1). The density was measured by the water displacement method. The elastic modulus and yield strength were determined using the initial linear segment of the stress-strain curve and the 0.2% offset. The Poisson's ratio was measured by the bidirectional strain gauge method. The data in Table 2 are the average values of three repeated experiments, as shown in Fig. 2.

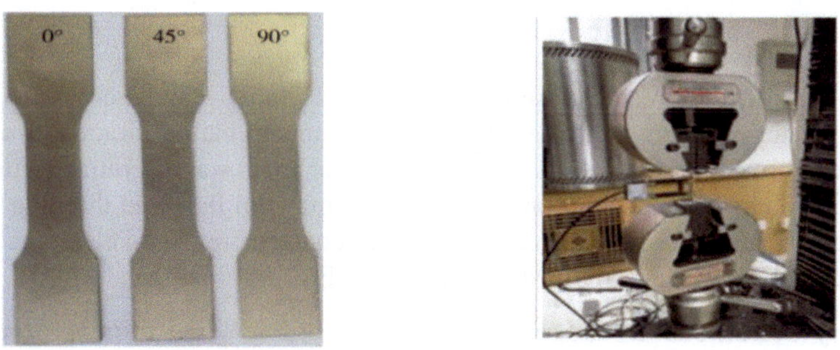

Fig. 2. Tensile specimens and HT2402 electronic tensile testing machine.

Tool Path Generation The variable-curvature conical tool path was programmatically generated using Python (see Fig. 3), enabling rapid adaptation to different target geometries through parametric code adjustments. This algorithmic approach demonstrates three key advantages over conventional UG software.

Table 1. Properties of 1060 AL material.

Material Name	Density	Elastic modulus	Songbaibi	Yield strength
1060 AL	2, 71	70	0.3	30

Table 2. Stress-strain properties of 1060 Al material.

Stress	20	44	64	77	87	91	98
Strain	0.002	0.0169	0.0507	0.0845	0.1183	0.1352	0.169

When the upper and lower pressure plates are assembled coplanar with the sheet metal, the sheet's thickness will extend downward by 1 mm due to the shell offset being referenced to the top surface. This assembly configuration will cause immediate computational errors during simulation execution.

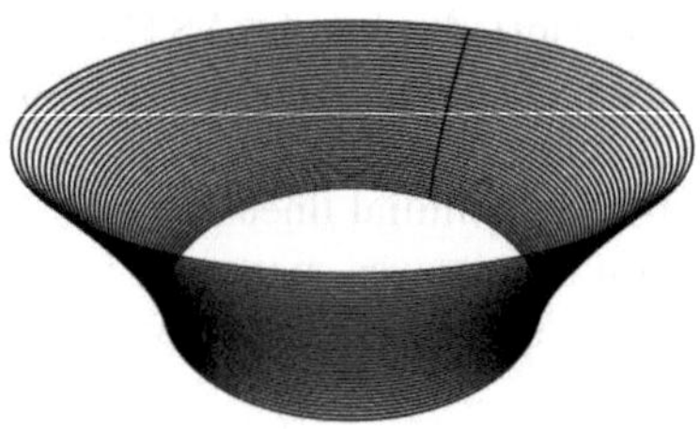

Fig. 3. Trajectory of conical parts with variable curvature.

Mesh Generation For single-point incremental forming simulations, an optimized meshing strategy was implemented to balance computational accuracy and efficiency. The critical forming region (central 100 mm diameter zone) was discretized with fine 1 × 1 mm quadrilateral elements to precisely capture material flow and thinning behavior, while the peripheral non-forming area employed coarser 2 × 2 mm elements to reduce computational cost. The grid division is shown in Fig. 4.

Boundary Conditions Boundary conditions: According to the forming principle, the upper and lower pressure plates are fixed and completely constrained. Fig. 5 shows the assembly schematic diagram of the sheet, tool head, and pressure plate. The degrees of freedom of the tool head for X/Y/Z three-direction translation and Z direction rotation are configured, with the amplitudes set for each direction. The amplitudes are substituted into the generated NC code.

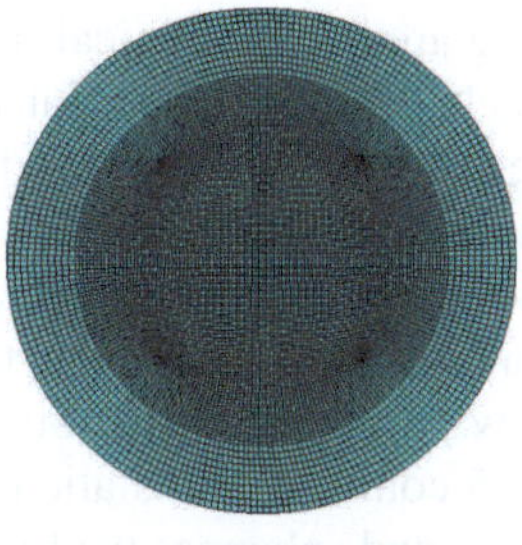

Fig. 4. Divide the grid.

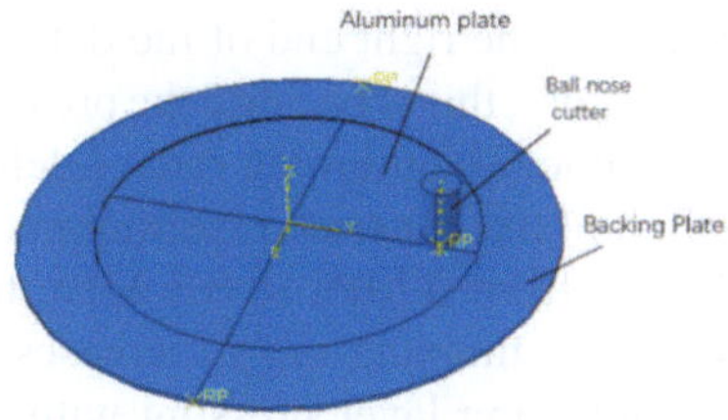

Fig. 5. Finite element model of static support single-point incremental forming.

3 Simulation of the Thinning Rate of Static Pressure Support - Single-Point Incremental Forming Parts

3.1 Thinning Partition Characteristics

The thickness characteristics of the formed part along the normal direction are shown in Fig. 6. According to the deformation mechanism, the work piece is divided into three deformation zones. The bending transition zone I is subjected to bending-tension and other loads, and the bottom stable zone III is protected by the static pressure support reaction force. These two zones have less thinning. The deformation zone II is the direct action area of the tool head and the main deformation zone, with a higher thinning rate. The definition of the wall thickness thinning rate is as follows:

$$\eta = \frac{t_0 - t}{t_0} \times 100\% \tag{1}$$

In the formula, η represents the thinning rate, t_0 denotes the initial sheet thickness, and t indicates the deformed sheet thickness

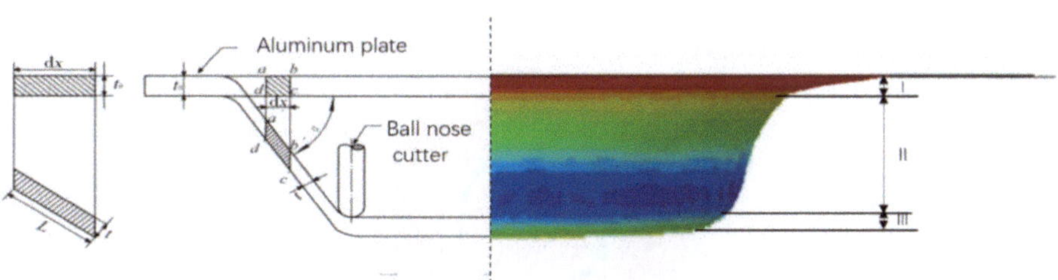

Fig. 6. Schematic diagram of SPIF thickness variation.

3.2 Simulation Result Analysis

As shown in Fig. 7, it is a cloud map of the thickness distribution of formed parts under different static pressure conditions. It can be seen from the figure that there is a darker thinning band at the right end of the deformation thinning area of the conical part with variable curvature, this is because the pressing points of each layer are concentrated along the tool path line. The thinning rate of deformation zone II will significantly reduce the forming limit of the material, resulting in cracking in this area. When the static pressure load increases from 0 MPa to 0.3 MPa, the material flows counterclockwise, and the thickness of the thinning strip increases by 15%. Concurrently, optimizing the tool head path and using layer pressing with uniform downward pressure distribution can effectively reduce the risk. The cloud image data in Fig. 5 confirms that static pressure support reduces stress concentration, improves spring back, and enhances the form limit and accuracy.

To analyze the influence law of static pressure parameters on the wall thickness thinning rate of each area, a node path was created along the center line of the inner surface of the formed part, as shown in Fig. 8.

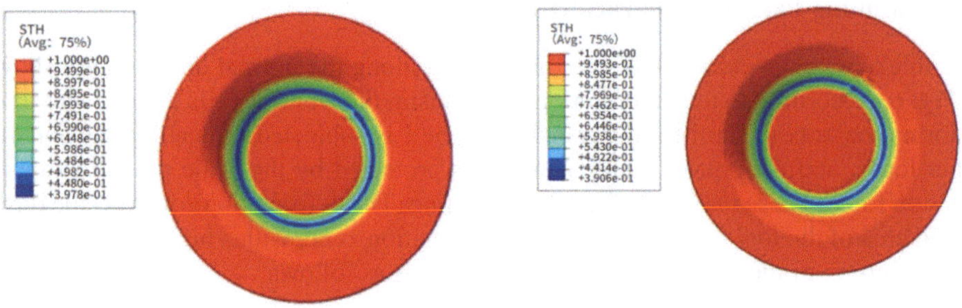

(a) (b)

Fig. 7. Cloud maps of thickness distribution under different static pressure parameters (a = 0 MPa, b = 0.03 MPa).

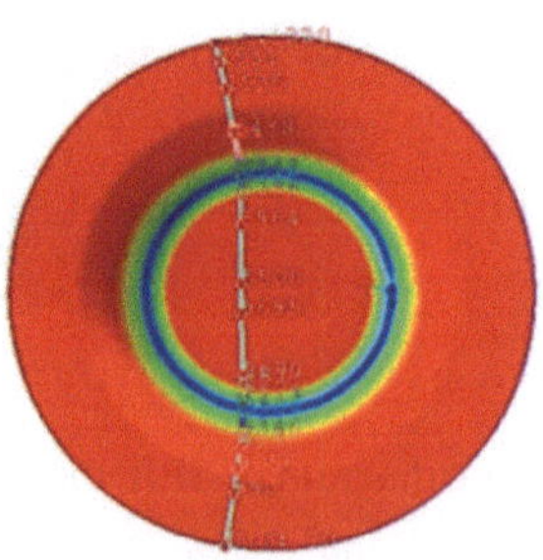

Fig. 8. Create a node path.

By analyzing each node of the forming path and using the thinning rate calculated by Formula (1), the thickness variation laws of each area of the formed parts under different static pressure parameters (0 MPa and 0.03 MPa) were analyzed, as shown in Fig. 9.

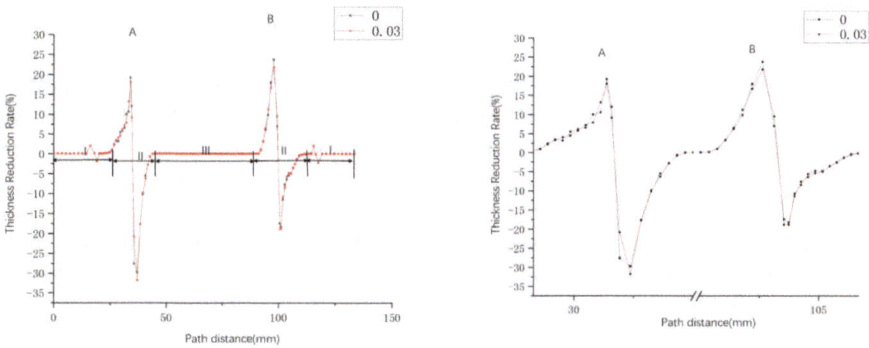

(a) The variation of the overall thinning rate under each static pressure value

(b) A magnified view of the AB area

Fig. 9. The variation curves of the thinning rate along the path nodes under each static pressure value.

According to the thickness reduction rate distribution curve (as shown in Fig. 7), a slight increase of approximately −5% occurs at 20 mm. This is attributed to the local compression caused by the contact between the tool head and the sheet, leading to material accumulation. In the range of 20–37.5 mm, the thinning rate increases significantly to 30%. In the range of 38–75 mm, the thinning rate initially decreased before showing an uptick, from 30% to 15%, and then rebounding to 25%. The thinning rates of zone I and zone III have consistently remained within ±3%, validating the characteristics of their non-formed zones. Zone II, as the main deformation area, has a thinning rate that decreases from +25% to −35% at 105 mm. Local instability or shear slip of the material may occur at this position. The thinning rate at the end of the path resumed positive growth and finally reached 20% at 150 mm. The overall variation of the wall thickness on both sides is symmetrical, but the maximum thinning rate at the starting end is approximately 8% higher than that at the ending end. This is due to the asymmetric accumulation caused by the counterclockwise flow of the material.

In Zone II, the thinning rate fluctuates. If the thinning rate is greater than 20%, the feed rate of the tool head is reduced by half. The current static pressure of 0.03 MPa improves the uniformity of the sheet. In Zone II, the use of step static pressure (e.g. 0 ~ 20 mm: 0 MPa; 20–100 mm: 0.05 MPa; 100–150 mm: 0.03 MPa) is a viable option for achieving a balanced thinning gradient. For the asymmetric stacking of plates, the single-pass trajectory can be changed to multiple passes to improve the forming accuracy.

4 Conclusion

This paper aims to address the problems of uneven wall thickness distribution and fracture of conical parts with variable curvature in single-point incremental forming (SPIF). It proposes to adopt the single-point progressive forming technology assisted by static pressure support. The finite element simulation method is employed to analyze the influence of static pressure support on the thinning rate of the work piece. The research findings indicate that:

1) The introduction of static pressure support can significantly improve the uniformity of wall thickness distribution, reduce the thinning rate, prevent part cracking, and increase the forming limit. When the static pressure value is maintained within the range of 0.2–0.3 MPa, the thinning rate can be reduced by 30–45%, while simultaneously increasing the forming limit by more than 20%.
2) Based on the simulation results, process optimization is carried out. Using a 10 mm tool head with a static pressure value of 0.25 MPa can improve forming accuracy. Optimizing the feed rate to between 800 and 1,200 mm/min will significantly reduce local stress. The layer-down spiral progressive tool path can control wall thickness deviation to within ±0.1 mm.

References

1. Pinard, D., Skelton, D.R., Gucciardi, M., et al.: Study of forming mechanics of magnetic field–assisted single point incremental forming. J. Manuf. Process. **79**, 28–34 (2022) [CrossRef]
2. Xiao, X., Oh, S.H., Kim, S.H., et al.: Effects of low-frequency vibrations on single point incremental sheet forming. Metals. **12**(2), 346 (2022) [CrossRef]
3. Almadani, M., Guner, A., Hassanin, H., et al.: Optimisation of a novel hot air contactless single incremental point forming of polymers. J. Manuf. Process. **117**, 302–314 (2024) [CrossRef]
4. Hou, C., Su, X., Peng, X., et al.: Thermal-assisted single point incremental forming of jute fabric reinforced poly (lactic acid) biocomposites. Fibers Polym. **21**, 2373–2379 (2020) [CrossRef]
5. Szpunar, M., Trzepieciński, T., Ostrowski, R., et al.: Thermo-mechanical numerical simulation of friction stir rotation-assisted single point incremental forming of commercially pure titanium sheets. Materials. **17**(13), 3095 (2024) [CrossRef]
6. Naranjo, J.A., Miguel, V., Coello, J., et al.: Tribological characterization of the heat-assisted single point incremental forming process applied to the Ti6Al4V alloy with the definition of an adhesion parameter for the tool surface. Materials. **14**(24), 7641 (2021) [CrossRef]
7. Gohil, A., Modi, B., Patel, K.: Effect of amplitude of vibration in ultrasonic vibration-assisted single point incremental forming. Mater. Manuf. Process. **37**(16), 1837–1849 (2022) [CrossRef]
8. Bai Lang. Research on the single-point incremental forming accuracy and its influencing mechanism of static pressure support - ultrasonic vibration. Xian university of science and technology, 000049 (2019). DOI: https://doi.org/10.27398/dcnki.Gxalu. [CrossRef]
9. Lin Yunbo Static pressure support - ultrasonic vibration of single point incremental forming precision research: Xi'an university of science and technology, 001881 (2020). DOI: https://doi.org/10.27398/dcnki.Gxalu. [CrossRef]
10. Nie Kaixin Pressure support single point incremental forming characteristics research: Xi'an university of science and technology, 000202 (2023). DOI: https://doi.org/10.27398/dcnki.Gxalu. [CrossRef]
11. Zhangshuai, J.: Research on the Influence of Wall Thickness of 1060 Aluminum Parts on the Single-Point Incremental Forming Characteristics of Ultrasonic Vibration, p. 001911. Xian university of science and technology (2021). https://doi.org/10.27398/dcnki.Gxalu. [CrossRef]
12. Xiaoye, W., Chunlei, L.: Simulation of single-point incremental forming force under the assistance of multiple technologies. J. Xi'an Jiaotong Univ. **54**(12), 176–184 (2020) [CrossRef]

Finite Element Analysis of the Effect of Tooth Number on Thermal Deformation of Cylindrical Milling Cutter

Zhen Ai and Yuebo Ji(✉)

School of Marine Equipment and Mechanical Engineering, Jimei University, Fujian Xiamen, China
ybji@jmu.edu.cn

Abstract. As an advanced high-performance light alloy material with exceptional ductility, electrical conductivity, and superior physical-mechanical properties, 7075 aluminum alloy has found extensive applications in critical fields such as new energy vehicles, aerospace, and shipbuilding, emerging as an ideal material to meet the stringent performance requirements of high-end manufacturing. Acknowledging the current gap in systematic research on the number of teeth in milling cutters, this study employs 7075 aluminum alloy as the workpiece material and utilizes ANSYS finite element analysis software to establish digital models of two-tooth, three-tooth, and four-tooth cylindrical milling cutters, followed by transient thermal analysis. Through parameter optimization and solution calculation, the results demonstrate that the four-tooth milling cutter exhibits significantly lower directional heat flux compared to its two-tooth and three-tooth counterparts, albeit with relatively higher outer wall temperature. Concurrently, the four-tooth milling cutter demonstrates notably smaller total deformation, along with lower instantaneous and average milling forces. These findings provide a substantial foundation for optimizing the structural design of milling cutters and enhancing their performance capabilities, offering critical insights for advancing machining technology in high-precision manufacturing domains.

Keywords: Number of milling cutter teeth · directional heat flux · total deformation

1 Introduction

In high-end manufacturing fields such as aerospace and automotive manufacturing, high-speed milling has become a core processing technique due to its advantages of high efficiency and precision [1]. As a key cutting tool, the cutting performance of cylindrical milling cutters directly impacts machining quality and production efficiency. However, under high-speed cutting conditions, heat generated by friction between the tool and workpiece, as well as chip deformation, can cause thermal deformation of the milling cutter, thereby affecting tool life, machining accuracy, and surface quality. Currently, scholars both domestically and internationally have conducted extensive research on

S. Halgamuge et al. (Eds.): ICACMVE 2025, LNME, pp. 714–722, 2026.
https://doi.org/10.1007/978-981-95-6911-3_71

the thermo-mechanical coupling characteristics of milling cutters using methods such as experimental testing, theoretical modeling, and finite element analysis, preliminarily revealing the mechanisms and variation patterns of key parameters like cutting speed and feed rate on tool thermal deformation. However, existing research predominantly focuses on cutting parameters, and systematic investigation into tool structural parameters (such as the number of teeth) remains insufficient. As a key structural element affecting milling cutter performance, changes in the number of teeth indirectly alter cutting force distribution, heat dissipation conditions, and vibration characteristics, consequently influencing the tool's thermal deformation behavior. Unfortunately, the quantitative relationship between the number of teeth and milling cutter thermal deformation is still unclear, and there is a lack of in-depth analysis of the tool's thermo-structural coupling effects under different tooth counts. This makes it difficult to effectively control thermal deformation through tooth number optimization during the tool design stage. Al7075 aluminum alloy, with its exceptional ductility, electrical conductivity, and superior physical-mechanical properties, is widely used as a high-performance light alloy material in critical fields such as new energy vehicles, aerospace, and shipbuilding, making it an ideal choice to meet the stringent performance requirements of high-end manufacturing. Therefore, this study selects Al7075 aluminum alloy as the workpiece material [2]. In-depth analysis of the influence mechanism of tooth number on the thermal deformation of cylindrical milling cutters not only helps improve the theoretical system of thermo-structural coupling for milling cutters but also provides theoretical support for the optimal structural design of cutting tools. Quantifying the thermal deformation patterns of milling cutters with different tooth numbers through finite element analysis can guide engineers to select the optimal tooth number during the design stage, achieving a balance between tool thermal stability and cutting performance, thereby reducing machining errors and enhancing production efficiency. This study, based on the finite element method, constructs thermo-structural coupling models of cylindrical milling cutters with different numbers of teeth. It systematically investigates the impact of tooth number variation on the temperature field distribution and thermal deformation patterns of milling cutters, quantifies their correlation, and proposes tooth number optimization strategies based on thermal stability. This provides new ideas and methods for solving the problem of tool thermal deformation in high-speed milling.

2 Establishing the Finite Element Model

2.1 Milling Cutter Model Construction and Properties

This study utilizes the 3D geometric modeling tools integrated within Workbench to construct digital models of two-tooth, three-tooth, and four-tooth milling cutters, respectively. These models are then imported into the Explicit Dynamics module of the Workbench platform and combined with the ANSYS finite element analysis system to establish coupled simulation models of milling cutters with different tooth counts and the workpiece. The finite element models of the various milling cutters and the workpiece are shown in Fig. 1. Their geometric parameters and meshing strategies are meticulously configured based on actual machining conditions, providing a reliable foundation for subsequent mechanical analysis of the cutting process.

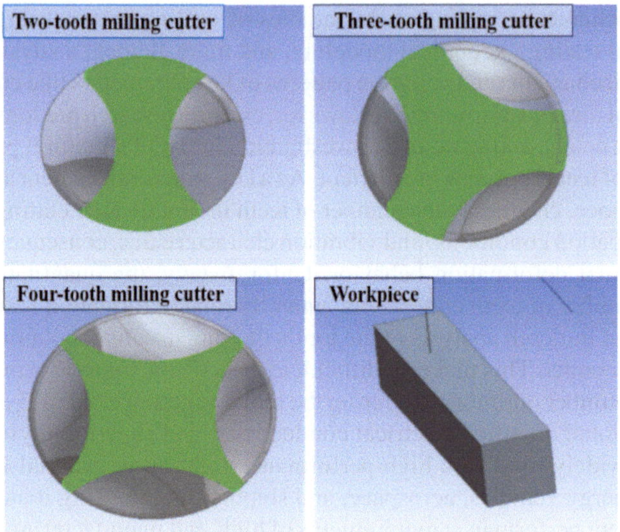

Fig. 1. Milling Cutter and Workpiece Models

His study assigns material parameters to the milling cutter and workpiece models, respectively. The physical property parameters of the titanium carbide (TiC) carbide end mill are detailed in Table 1, and the material properties of Al7075 aluminum alloy are listed in Table 2, providing fundamental data support for subsequent analysis.

Table 1. Physical Properties of Titanium Carbide (TiC) Carbide End Mill

Material	Young's Modulus (GPa)	Poisson's	Density g/cm^{-3}	Hardness (HRA)	Thermal Conductivity (W/(m.C)	Coefficient of Thermall (/°C)
Carbide Alloy	110	0.36	4.5	290	12	8.6x10^{-6}

Table 2. Physical Properties of Al7075 Aluminum Alloy

Material	Young's Modulus (GPa)	Poisson's	Density g/cm^{-3}	Thermal Conductivity (W/(m.°C))	Coefficient of Thermall (/°C)
Al7075	70	0.33	2.81	218	2.36x10^{-5}

During the simulation modeling process, considering the rigid characteristics of the milling cutter, it is set as a rigid body to exclude the influence of the cutter's own deformation during simulation. In contrast, the workpiece undergoes significant deformation during milling. Therefore, the Johnson-Cook (J-C) constitutive model is adopted to describe its mechanical behavior. The J-C model parameters for Al7075 aluminum alloy are detailed in Table 3. The model is functionally represented as: $\sigma = (A + B\varepsilon^n)\left[1 + \mathrm{C}\ln\left(\frac{\dot{\varepsilon}}{\dot{\varepsilon}_0}\right)\right]\left[1 - \left(\frac{T - T_a}{T_m - T_a}\right)^m\right]$. In Table 3, σ is the internal stress of the material, A is the initial yield stress; B is the work hardening constant; n is the strain hardening exponent; C is the strain rate sensitivity coefficient; T is the temperature; T_a is room temperature; T_m is the material melting point; m is the thermal softening exponent [3]. This constitutive equation, by coupling the effects of strain, strain rate, and temperature, provides a theoretical basis for accurately simulating the plastic deformation of the workpiece during milling.

Table 3. [3] Johnson-Cook Model Parameters for Al7075

Material	A	B	C	n	m	T_a (°C)	T_m(°C)
Al7075	546	678	0.024	0.71	1.56	20	800

2.2 Meshing

In ANSYS finite element analysis, mesh size parameters significantly impact simulation accuracy. Reasonable meshing is crucial for enhancing the accuracy and reliability of simulation results, and element size optimization is key to ensuring the validity of numerical simulations. Since the tool model is often set as a rigid body considering only heat transfer effects, appropriate mesh types need to be selected during meshing based on the physical characteristics of the components. Compared to tetrahedral meshes, hexahedral meshes, with their stable topology and high element quality, can simultaneously improve computational efficiency and simulation accuracy [4]. This differentiated mesh selection strategy can effectively optimize numerical simulation performance. As shown in Fig. 2, discrete density coupling technology is employed to realize the coupled interaction between the tool and workpiece materials. By enhancing the coupling density and uniformity in the tool-workpiece contact area and the surfaces of the base material and chips, the stress distribution and deformation characteristics during machining can be accurately captured, significantly improving the agreement between simulation results and experimental data.

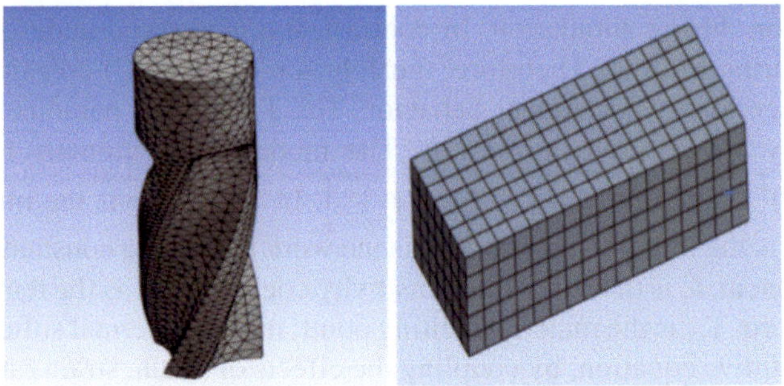

Fig. 2. Meshing of Milling Cutter and Workpiece

2.3 Defining Coordinates and Contact

The setup of milling cutter motion parameters constitutes the core of machining process simulation, where precise velocity definition is particularly critical. The feed velocity along the positive X-axis direction must be determined based on actual machining conditions, as its appropriate setting directly influences milling dynamics characteristics and simulation accuracy. In ultrasonic vibration milling simulations, it is further necessary to specify the angular velocity of the cutter's clockwise rotation about the Z-axis [5]. This parameter couples spindle rotation with feed motion. Through precise parameterization, the actual operational state of the milling cutter can be effectively simulated, ensuring the fidelity of cutting force distribution results to engineering practice [6]. The coordinate definitions for the tool and workpiece are illustrated in Fig. 3, providing fundamental support for subsequent kinematic and dynamic analyses. The definition of load contact interfaces between the tool and workpiece, as depicted in Fig. 4, enables comprehensive simulation of mechanical behaviors during milling through thorough load characterization.

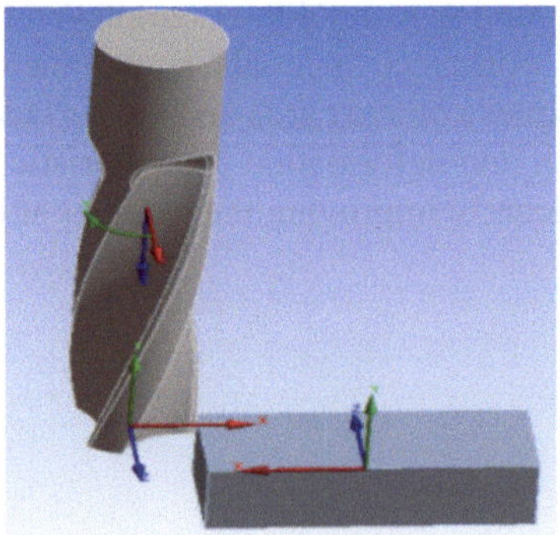

Fig. 3. Coordinates of Milling Cutter and Workpiece

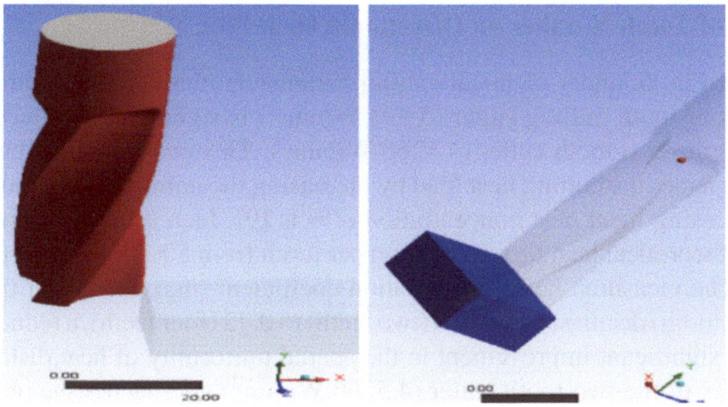

Fig. 4. Defining Contact Surfaces of Milling Cutter and Workpiece

3 Solution and Results

3.1 Effect of Tooth Number on Milling Temperature

During the milling process with titanium carbide (TiC) carbide end mills, the maximum temperature can reach 800–1000 °C. Studying Fig. 5 reveals that the minimum temperatures for two-tooth, three-tooth, and four-tooth end mills are 139.21 °C, 140.44 °C, and 144.19 °C, respectively. The surface temperature of the milling cutter increases from 904.36 °C to 906.02 °C. In short, the data in Fig. 5 clearly indicate that in titanium alloy milling, as the number of cutter teeth increases, the total heat generated per unit time increases, while the cooling time per tooth decreases. This leads to an overall rise in tool temperature (increased minimum temperature) and indirectly elevates the peak temperature at the cutting edge [7]. This significantly enhanced thermal effect is one of the main factors limiting the application of multi-tooth milling cutters in titanium alloy machining, as it can severely shorten tool life [8].

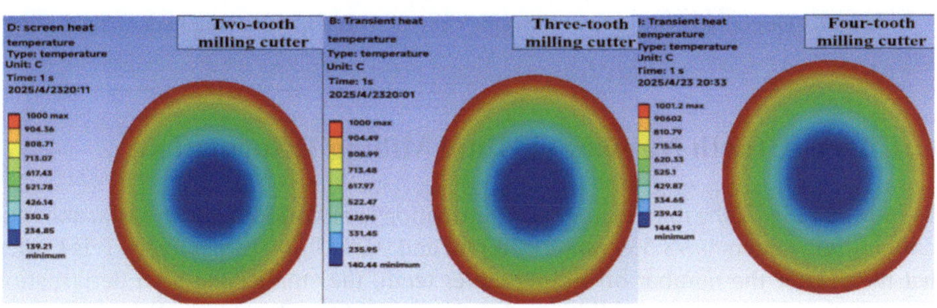

Fig. 5. Milling Temperature of Milling Cutters with Different Numbers of Teeth

3.2 Effect of Tooth Number on Directional Heat Flux

As shown in Fig. 6, under identical cutting parameters, the maximum directional heat flux of the four-tooth milling cutter (3.994 W/mm^2) is significantly reduced by 11.8% compared to the two-tooth cutter (4.5266 W/mm^2). This confirms that the multi-tooth structure disperses the cutting heat load by increasing the number of heat sources, effectively suppressing local heat concentration effects [9]. Increasing the number of teeth reduces the theoretical heat flux proportion per tooth from 50% for two teeth to 25% for four teeth. The measured heat concentration coefficient (maximum heat flux / average heat flux per tooth) decreases from 1.0 (two teeth) to 0.72 (four teeth), a reduction of 28%, indicating a significant improvement in the spatial uniformity of heat distribution. The peak heat flux of the two-tooth cutter (4.5266 W/mm^2) approaches the phase transformation threshold of carbide (>4.5 W/mm^2), while the four-tooth cutter (3.994 W/mm^2) falls outside this high-risk range. This reveals that increasing the number of teeth can reduce the risk of local thermal damage, but vigilance is required regarding the temperature rise of the substrate caused by overall heat accumulation [10]. Multi-tooth tools alleviate local heat concentration through the heat source dispersion effect (each unit increase in tooth number reduces maximum heat flux by approximately 0.5 W/mm^2), but are constrained by substrate heat accumulation resulting from reduced cooling time. For efficient titanium alloy machining, it is recommended to use a four-tooth tool combined with high-pressure internal cooling to achieve synergistic optimization of heat distribution and material removal rate.

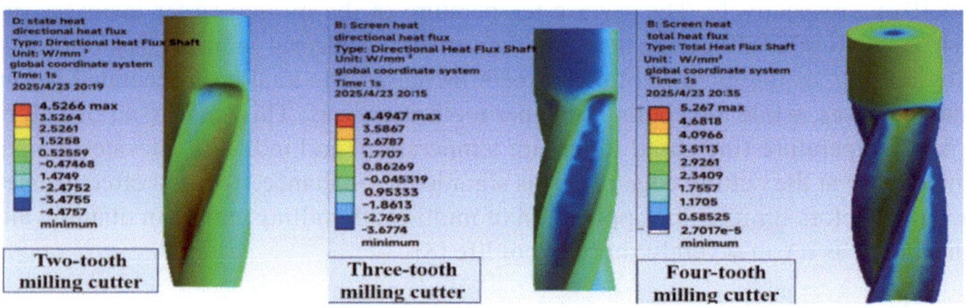

Fig. 6. Directional Heat Flux of Milling Cutters with Different Numbers of Teeth

4 Effect of Tooth Number on Total Milling Deformation

As shown in Fig. 7, the maximum total deformation for two-tooth, three-tooth, and four-tooth milling cutters is 98.539 μm, 21.38 μm, and 9.0853 μm, respectively. It is evident that the greater the number of milling cutter teeth, the smaller the total deformation, with the four-tooth cutter exhibiting significant rigidity advantages. When subjected to the same thermal load, this high rigidity characteristic enables it to effectively resist thermal deformation effects, resulting in smaller overall deformation [11, 12]. The multi-tooth structure suppresses milling deformation through a dual mechanism: Firstly, the increase in tooth number reduces the cutting force per tooth (a 50% decrease for four teeth compared to two teeth), weakening the intensity of the excitation source [13, 14]. Secondly, the increase in tool body solid ratio leads to a significant increase in cross-sectional moment of inertia (theoretical increase >40%), enhancing structural stiffness. The synergy of these two mechanisms reduces the total deformation of the four-tooth cutter sharply to 9.2% of that of the two-tooth cutter, effectively avoiding common tool deflection and chatter problems in titanium alloy machining. It should be noted that although substrate temperature rise can partially weaken material stiffness, the dominant role of structural stiffness still ensures the deformation control advantage of multi-tooth tools.

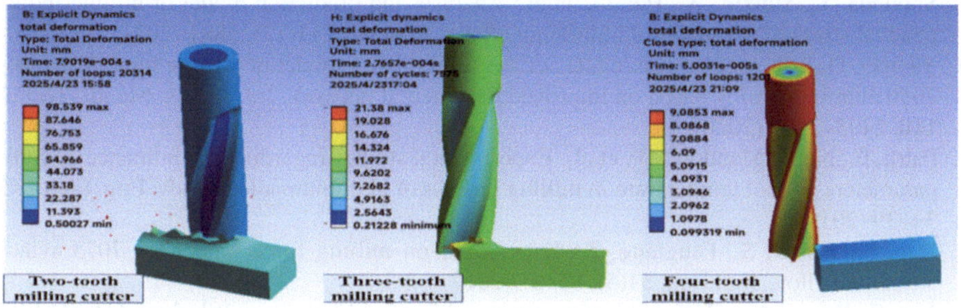

Fig. 7. Total Deformation of Milling Cutters with Different Numbers of Teeth

5 Conclusion

(1) Utilizing finite element simulation technology, this study systematically investigated the temperature variation patterns of milling cutters with different numbers of teeth during the milling process. The results indicate that compared to two-tooth and three-tooth cutters, the four-tooth cutter exhibits a higher temperature distribution during milling.

(2) Finite element simulation was performed to analyze the effect of different tooth numbers on directional heat flux during milling. Under identical cutting conditions, the directional heat flux of the four-tooth milling cutter is significantly smaller than that of the two-tooth and three-tooth cutters.

(3) Finite element simulation was performed to analyze the effect of different tooth numbers on total deformation during milling. When evaluating the total deformation of the milling cutter, it was found that the total deformation of the four-tooth cutter is significantly smaller than that of the two-tooth and three-tooth cutters.

Acknowledgments. Fujian Provincial Science and Technology Programme Funding Project No. (2022H0021)

References

1. Yunfeng, Z.O., Jianhua, Z., Dongliang, X., et al.: Simulation analysis and experimental research on milling of aluminum alloy Al2024. Manuf. Technol. Mach. Tool. **1**, 101–103 (2011)
2. Zhengxi, Z.: Research on Modeling and Simulation of Milling Force in Milling Process of Aluminum Alloy 7075-T6. Liaoning University of Science and Technology, Anshan (2021)
3. Yuanchang, X., Lei, W.: Simulation experimental research on axial force during milling with different number of milling cutter teeth. Tool Eng. **57**(08), 119–123 (2023)
4. Yaohui, Z., Peng, W., Na, L., et al.: Simulation research on optimization of milling parameters for 7075 aluminum alloy. Mach. Des. Manuf. **5**, 171–174 (2020)
5. Shouhua, Y., Yunxin, W., Hai, G., et al.: Analysis and prediction model of milling-induced residual stress of aeronautical aluminum alloys. Appl. Sci. **11**(13), 5881 (2021)
6. Yaonan, D., Xiaotao, Z., Xubing, C., et al.: A prediction model of milling force for aviation 7050 aluminum alloy based on improved RBF neural network. Int. J. Adv. Manuf. Technol. **110**, 2493–2501 (2020)
7. Patru, E., Nubia Mandora, D., et al.: Experimental studies regarding the influence of cutting parameters on tool temperature in milling process of aluminum alloys. Adv. Eng. Forum. **34**, 34–39 (2019)
8. Dan, L., Huilai, S., Fangfang, Z.: Experiment on milling force model for 7075 aviation aluminum alloy. Tool Eng. **51**(5), 85–87 (2017)
9. Yue, S.: Research on Thermal Deformation and Control Methods in Whirlwind Hard Milling of Ball Screws. Shandong Jianzhu University (2022). https://doi.org/10.27273/d.cnki.gsajc.2022.000684
10. Yi, Z.: Simulation of Surface Topography and Roughness Prediction in Ball-End Milling. Harbin University of Science and Technology (2014)
11. Ping, Z.: Modeling and Analysis of Tool Wear and Machining Error in Virtual NC Milling. Harbin University of Science and Technology (2012)
12. Na, Z.: Simulation and Research on Surface Topography in High-Speed CNC Milling. Lanzhou University of Technology (2011)
13. Xiulin, S.: Research on Key Technologies of Physical Simulation for Virtual NC Milling. Harbin Engineering University (2011)
14. Yunping, Y., Na, Z., Dingxiang, W., et al.: Influence of tool eccentricity and deformation on milling force of ball-end milling cutter. J. Lanzhou Univ. Technol. **37**(01), 33–36 (2011)

Uncertainty-Incorporated Robust Optimization of Welding Process Parameters for Fatigue Life of Welded Structures of Mining Wide-Body Dump Truck

Guandong Lu[1], Meng Zhou[1], Weiguo Gao[1](✉), Lingtao Weng[1], Ketong Dong[1], Jianying Zhao[2], and Dawei Zhang[1]

[1] Tianjin Key Laboratory of Equipment Design and Manufacturing Technology, School of Mechanical Engineering, Tianjin University, Tianjin 300354, China
lgd060017@163.com
[2] Sany Heavy Equipment Co., Ltd., Shenyang 110000, China

Abstract. To address fatigue cracking in localized welded structures of mining wide-body dump truck frames during service, this study proposes a robust fatigue life optimization methodology incorporating uncertainties in welding process parameters. First, a high-fidelity surrogate model is established leveraging extensive numerical simulations, with welding process parameters as inputs and minimum structural fatigue life as the response. Subsequently, parameter uncertainties are introduced within an interval framework to formulate a multi-objective optimization model. This model aims to simultaneously maximize the minimum structural fatigue life and minimize its robustness index. Ultimately, optimization conducted via the NSGA-II algorithm delivers a 48.4% enhancement in the minimum fatigue life of welded structures and a 26.3% improvement in its robustness.

Keywords: Welding Fatigue Life · Surrogate Model · Uncertainty Robust Optimization

1 Introduction

Large-scale engineering machinery is typically deployed in extreme environments and harsh operating conditions. Taking the mining wide body dump truck as an example, this core equipment for stripping and transportation in open-pit mines operates fully loaded on uneven and extreme road surfaces year-round. Its frame, serving as the primary load-bearing structure, incorporates numerous welded structures [1]. Should fatigue cracking occur within its service life, it can directly trigger major downtime accidents. Consequently, optimizing the fatigue life and robustness of welded structures is particularly critical.

G. Lu and M. Zhou—These authors contributed equally to this work.

S. Halgamuge et al. (Eds.): ICACMVE 2025, LNME, pp. 723–734, 2026.
https://doi.org/10.1007/978-981-95-6911-3_72

Welding process parameters, specifically core variables including current, voltage, and welding speed [2], significantly govern welding quality, thereby directly determining the fatigue life of welded structures. Current optimization of welding process parameters is predominantly achieved through orthogonal experimental design [3–5]. However, the search space is strictly confined to the discrete domain spanned by predefined experimental matrices, rendering it difficult to transcend local optimum boundaries and consequently incapable of achieving genuine global optimization.

In practical engineering system optimization design, the uncertainty of design parameters is universally present [6]. Specifically for welding process parameters, their uncertainty primarily stems from real-time fluctuations in welding current and voltage, along with systematic errors in welding speed control. Traditional deterministic optimization methods treat design parameters as fixed constants [7]. This idealized assumption frequently induces significant deviations between optimized results and actual performance.

To address the aforementioned challenges, this study focuses on local welded structures of mining wide-body dump truck frames that are prone to fatigue cracking under actual operating conditions. We propose a robust fatigue life optimization methodology integrating uncertainties in welding process parameters, achieving synchronous improvement in both fatigue life and its robustness of welded structures.

2 Numerical Simulation of Fatigue Life of Welded Structures

This study employs Ansys finite element simulation to computationally determine the fatigue life of welded structures. Compared to traditional physical testing for fatigue life acquisition, the finite element numerical approach significantly shortens optimize cycles and reduces testing costs while predicting structural fatigue life with relatively high accuracy. Although discrepancies persist between simulation results and actual performance, the finite element method establishes a reliable data foundation for subsequent optimization design processes through iterative simulations. Consequently, this study conducts analysis grounded in theoretical principles and finite element simulation.

2.1 Theoretical Basis

Gaussian Heat Source Model The welding heat source model constitutes a fundamental prerequisite for numerical simulation of welding processes. Gaussian distributed heat source model can ensure satisfactory temperature field analysis while maintaining computational efficiency. Eagar and Tsai [8] characterized the heat flux density distribution of the arc heat source on the workpiece surface using a two-dimensional Gaussian function. The heat flux density function of the welding heat source can be expressed as follows:

$$Q_{gauss}(r) = q_{\max} \exp\left(-\frac{3r^2}{r_0^2}\right) \tag{1}$$

$$q_{\max} = \frac{3}{\pi r_0^2} \eta U I \tag{2}$$

where r denotes the distance from any point in the heat source to the arc center; q_{max} is the maximum heat flux density at the heat source center, r_0 is the effective heating radius of the heat source; η is the welding thermal efficiency, U is the welding electrode voltage, and I is the welding current.

Fatigue Life Analysis Fatigue refers to a progressive failure process wherein materials accumulate irreversible damage in localized regions under cyclic loading, initiating microscopic cracks that propagate with increasing load cycles, ultimately culminating in complete structural fracture.

The S-N curve, alternatively known as the Wöhler curve, serves as the foundation for stress-based fatigue analysis. It characterizes the relationship between stress amplitude and the number of cycles to failure under cyclic loading, thereby quantifying the material's fatigue performance [9].

The logarithmic formulation is the most commonly used mathematical model for characterizing S-N curves, expressed as follows:

$$\lg N = \lg C - b \lg(\Delta\sigma) \tag{3}$$

where $\Delta\sigma$ denotes the stress amplitude, N is the number of cycles to failure, b is the fatigue strength exponent, and C is the fatigue strength coefficient.

2.2 Finite Element Simulation of Welding Fatigue Life

This study focuses on a local welded structure within the frame of a mining wide-body dump truck that is prone to fatigue cracking under actual operating conditions, with the specific geometry illustrated in Fig. 1.

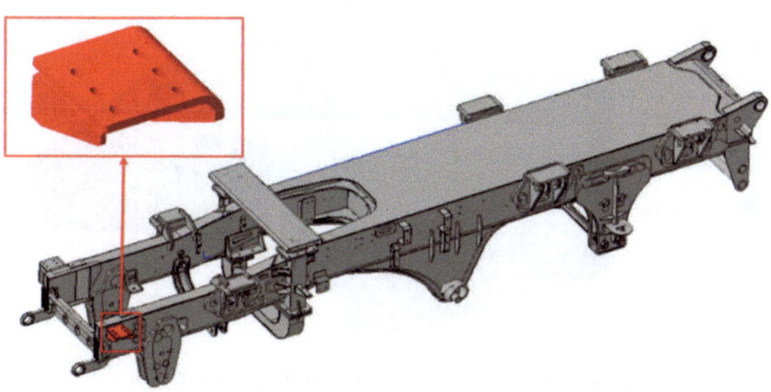

Fig. 1. Selection of local welding structures.

Establishment of Finite Element Model We imported the target structure into ANSYS Workbench and suppressed minor fillets to streamline the model geometry. Solid weld beads were then created at joint locations, with material properties assigned from the nCode material library (Struct Steel). Following mesh parameter configuration, the structure was discretized with local refinement near weld zones, yielding a final mesh of 30,935 elements and 58,312 nodes.

We applied a moving Gaussian heat source along the weld path using ANSYS Parametric Design Language (APDL) to simulate transient thermal input during the welding process. The Gaussian heat source parameters are specified in Table 1.

Table 1. Gaussian heat source parameter setting.

Gaussian heat source parameters	value
Welding electrode voltage(V)	30
Welding current(A)	300
Welding speed(cm/min)	45
Welding thermal efficiency	0.8
Effective heating radius of heat source(mm)	1

Building upon the finite element model, we performed sequentially coupled transient thermal and static structural analyses to obtain stress and strain data generated during the welding thermal process. The results were saved in *.rst file format, with the resulting welding residual stress contour plot shown in Fig. 2.

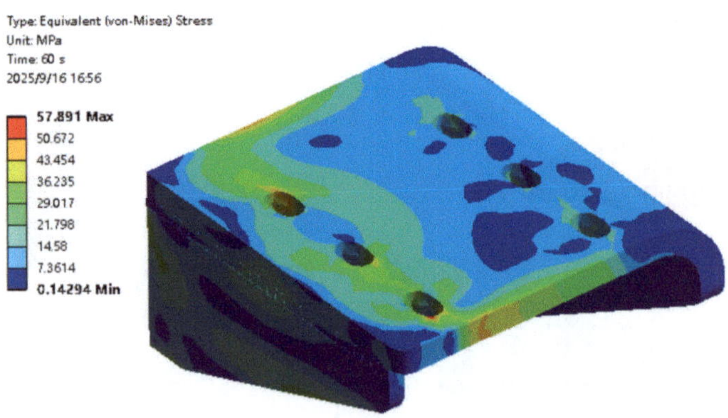

Fig. 2. The welding residual stress contour plot.

The finite element results in *.rst format were imported into ANSYS nCode DesignLife for structural fatigue life assessment. The material model employed Struct Steel. The reference S-N curve for fully reversed loading ($R = -1$) is shown in Fig. 3. Fatigue boundary conditions were configured as: a 20 kN sinusoidal cyclic load (10 Hz) applied parallel to the x-axis at the interface between the structure and main frame, serving as the load spectrum for subsequent fatigue calculations. After completing module configuration in nCode and executing the computation, the fatigue life contour plot of the welded structure was obtained as shown in Fig. 4.

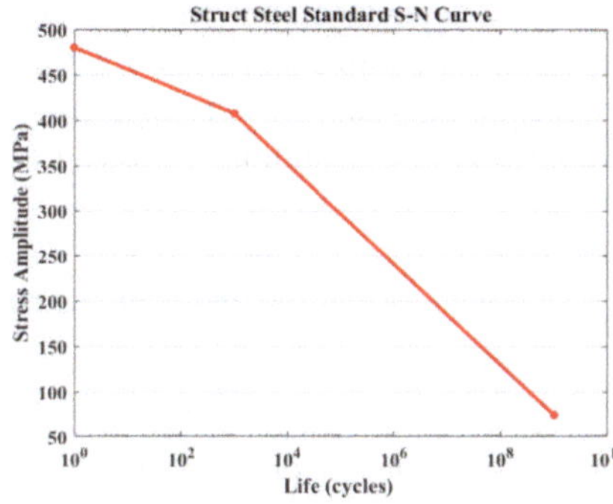

Fig. 3. Struct steel standard S-N curve.

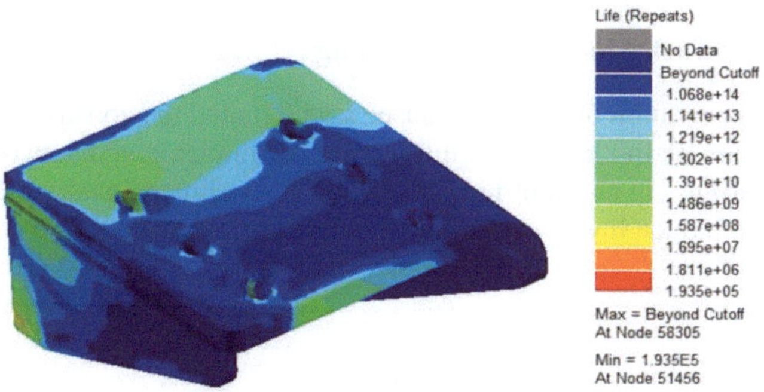

Fig. 4. Fatigue life contour plot.

3 BP Neural Network-Based Surrogate Modeling

Although the preceding chapter has completed numerical simulations of welded structure fatigue life using the finite element method, driving optimization iterations directly with this FE model would incur prohibitive computational costs. To address this challenge, this study introduces a backpropagation (BP) neural network to construct a high-fidelity surrogate model, thereby enabling rapid fatigue life approximation with acceptable computational expenditure.

3.1 Sample Data Generation

Employing surrogate models for optimization requires sufficient accuracy to ensure reliable results. Model precision depends on both approximation capability and training data quality. This study adopts Latin Hypercube Sampling (LHS) to generate training data within the design space. LHS's stratified uniform distribution [10] maximizes space-filling with minimal samples, reducing clustering/void effects while enhancing coverage uniformity.

Using welding electrode voltage, current, and velocity as design variables, we sampled 200 points via Latin Hypercube Sampling (LHS) within prescribed optimization ranges (Table 2). The sampled input variables were then processed through batch simulations on ANSYS Workbench, with minimum fatigue life extracted as the output response.

Table 2. Sample design variable sampling range.

Design variable name	Sample value range
Welding electrode voltage(V)	[26, 36]
Welding current(A)	[260, 340]
Welding velocity (cm/min)	[35, 55]

3.2 Neural Network Modeling and Training

Following the generation of training datasets, we initiated the construction and training of the neural network model. The architecture of the implemented network is illustrated in Fig. 5, featuring: (1) an input layer receiving three-dimensional variables, (2) three fully-connected hidden layers performing nonlinear transformations.

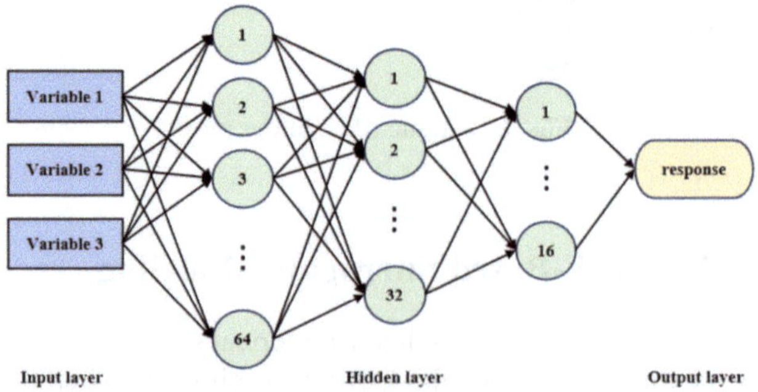

Fig. 5. Schematic diagram of neural network topology.

During network training, we employed backpropagation (BP) algorithm with an 80–20 train-validation split for performance evaluation. In the forward pass, 3D inputs sequentially propagated through hidden layers (64–32-16 neurons), undergoing linear transformation followed by ReLU activation and 20% dropout for enhanced robustness. During backpropagation, L2 weight decay ($\lambda = 1 \times 10^{-5}$) was incorporated into MSE loss gradient computation, with BP-updated parameters constrained to prevent overfitting. Layer normalization parameters were dynamically saved to maintain data distribution consistency between training and test sets.

3.3 Accuracy Verification of Surrogate Model

As shown in Fig. 6, the mean squared error (MSE) loss curves of both training and validation sets demonstrate characteristic convergence trends during 100 training epochs. The training loss (red curve) decreases steadily with iterations, indicating effective parameter optimization. The validation loss (blue curve) exhibits rapid initial decline before stabilizing at a low plateau, confirming the model's generalization capability.

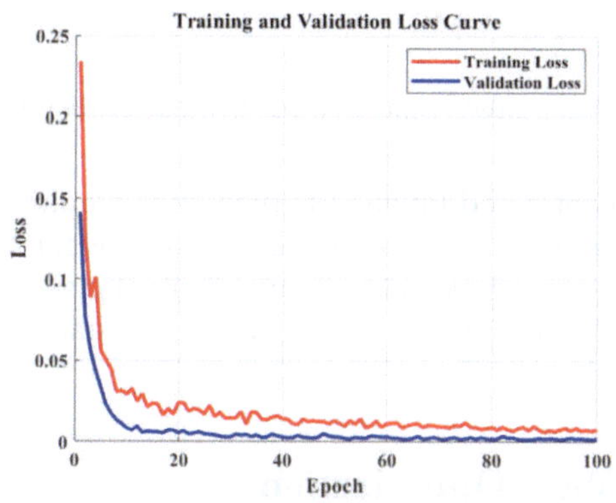

Fig. 6. Training and validation loss curve.

The prediction accuracy of the surrogate model was evaluated using the coefficient of determination (R^2), calculated as [11]:

$$R^2 = 1 - \frac{\sum_{i-1}^{n} (t_i - p_i)^2}{\sum_{i=1} (t_i - T)^2} \tag{4}$$

where n denotes the sample size of the validation set; t_i is the true output response values in the validation data, p_i is the predicted output response values, and T is the mean of p_i. An R^2 value closer to 1 indicates higher predictive accuracy of the surrogate model. The trained model achieved $R^2 = 0.98$, demonstrating excellent accuracy and successful fitting. Fig. 7 presents the parity plot comparing true versus predicted output responses for the validation set.

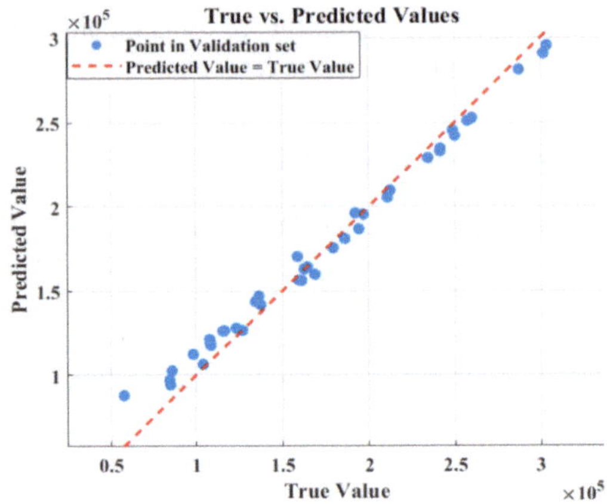

Fig. 7. Parity plot of true vs. predicted output responses for validation set.

A single finite element-based fatigue life prediction requires approximately 7 min, whereas the trained surrogate model completes predictions for 1,000 input data sets in merely 1.0651 s. This approach boosts computational efficiency by hundreds of times, dramatically reduces optimization iteration time, and substantially shortens the overall design cycle.

4 Uncertainty Robust Optimization

Building upon the high-precision BP neural network surrogate model established in the preceding chapter, this study conducts fatigue life optimization within the welding process parameter design space. The developed model enables instantaneous approximation of minimum fatigue life responses at any design variable configuration across the optimization domain, thereby circumventing computational bottlenecks inherent to finite element analysis and providing a time-efficient, reliable response evaluation mechanism for optimization iterations.

4.1 Construction of Interval Uncertainty Optimization Model

Optimization design involves multiple sources of uncertainty. In this study, the primary uncertainties originate from real-time fluctuations in welding current/voltage and systematic errors in welding speed control. To facilitate mathematical modeling, we quantify these uncertainties in three design variables (electrode voltage, current, and welding speed) as uniform interval disturbances around nominal values — specifically, $\pm 5\%$ variations from their baseline settings.

To assess the impact of these variations, we employ Monte Carlo sampling to generate 10,000 points within the defined intervals. The mean output response is treated as the nominal minimum fatigue life for each design variable's baseline value, while the standard deviation serves as the robustness metric. This approach transforms the uncertainty-aware optimization model into a deterministic formulation, expressed mathematically as follows:

$$\begin{cases} \max Fatigue(X_i), & i = 1, 2, 3 \\ \min Fatigue\_Robustness(X_i), & i = 1, 2, 3 \end{cases}$$
$$X = (X_1, X_2, X_3) = (U, I, v) \tag{5}$$
$$s.t. \quad Fatigue(X_i) > 1 \times 10^4$$

where X_1, X_2, X_3 denote the three design variables (welding voltage(U), current(I), and velocity(v)).

The first optimization model maximizes the nominal fatigue life value, while the second minimizes the robustness metric.

Constraint: Fatigue life $>10^4$ cycles.

4.2 Optimization Based on NSGA-II

Building upon the interval uncertainty model, this study employs the NSGA-II (Non-dominated Sorting Genetic Algorithm II) to solve the optimization problem. The workflow of NSGA-II-based optimization proceeds as follows [12,13, 1) Population Initialization. (2) Fitness Evaluation. (3) Genetic Operations. (4) Non-dominated Sorting and Crowding Distance Calculation. (5) Iterative Convergence.

The parameter configuration for the NSGA-II algorithm in this study is specified in Table 3. The resulting Pareto frontier obtained after algorithmic computation is presented in Fig. 8.

Table 3. The parameter configuration for the NSGA-II algorithm.

NSGA-II algorithm parameter name	Value
Number of iterations	40
Population size	200
Crossover probability	0.7
Mutation probability	0.2
Mutation rate	0.05

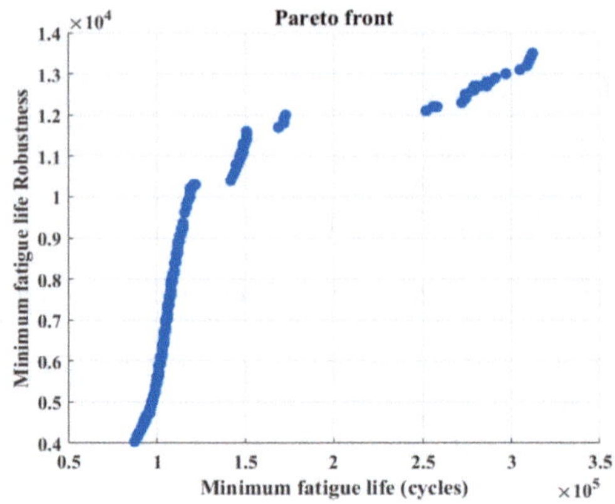

Fig. 8. Pareto frontier diagram of optimization results

A solution demonstrating significant fatigue life improvement was selected from the Pareto frontier as the final optimized result. Using the surrogate model, this optimized solution was compared with the pre-optimization parameters in terms of minimum fatigue life and robustness predictions, with comparative data summarized in Table 4. Relative to the pre-optimization design, the optimized solution achieved a 48.4% enhancement in minimum fatigue life and a 26.3% improvement in robustness, demonstrating substantial optimization effectiveness.

Table 4. Comparison table of parameters before and after optimization.

	U (V)	I (A)	v (cm/min)	M F L(cycles)	Robustness
Before	30	300	45	1.88E+05	1.71E+04
After	26.94	265.27	49.3	2.79E+05	1.26 E+04
Percentage improvement	\	\	\	48.4%	26.3%

5 Conclusion

This study focuses on a fatigue-prone local welded structure in a mining wide-body dump truck frame under operational conditions, proposing a robust fatigue life optimization method that incorporates welding process parameter uncertainties. By developing a BP neural network surrogate model for rapid and accurate minimum fatigue life prediction, we introduced parameter uncertainties within an interval framework to perform robustness-oriented optimization. The final implementation achieved a 48.4% improvement in minimum fatigue life and a 26.3% enhancement in robustness, demonstrating significant optimization effectiveness.

Acknowledgments. This research was supported by a project from the National Key Research and Development Program of China (2022YFB3402004).

References

1. Ma, X., Gu, Z., Mi, C., et al.: Weak ternary interval evaluation of fatigue life and its application to mining dump truck. Fatigue Fract. Eng. Mater. Struct. **43**(10), 2441–2454 (2020)
2. Qingzhi, L.: Research on the influence of welding process parameter optimization on weld quality Mold manufacturing. Mold Manuf. **25**(06), 168–170 (2025)
3. Zeyong, Y., Yisheng, Y., Chengrui, Z., et al.: Laser spiral spot welding process parameters optimization research. Trans. China Weld. Inst. **25**(07), 1–8 (2025)
4. Yao, C., Cong, Z., Xubing, C., et al.: Optimization of laser soldering welding parameters based on VIKOR method. Laser J. **46**(06), 196–202 (2025)
5. Uday, K.N., Rajamurugan, G.: Influence of process parameters and its effects on friction stir welding of dissimilar aluminium alloy and its composites–a review. J. Adhes. Sci. Technol. **37**(5), 767–800 (2023)
6. Xiaopeng, N., Shunpeng, Z., Jiewei, G., et al.: Fatigue reliability analysis and optimization Design of Turbine Blade Disks under multi-source uncertainties. J. Propul. Technol. **43**(02), 228–236 (2022)
7. Hai, R., Jianzhong, J., Zaiqi, Y., et al.: Multidisciplinary design optimization of a vehicle body for lightweight based on a neural network surrogate model. J. Automot. Saf. Energy. **16**(01), 32–42 (2025)
8. Eagar, T.W., Tsai, N.S.: Temperature fields produced by traveling distributed heat sources. Weld. J. **62**(12), 346–355 (1983)
9. Jinbao, Z., Zheng, H., Jinle, Z., et al.: Review of advances in fatigue S-N curve prediction models. Sci. Technol. Eng. **23**(13), 5390–5411 (2023)
10. Deru L, Yang L, et al.: Naiwei L, Fatigue reliability analysis of orthotropic steel bridge deck based on LHS-kriging method. Chin. J. Comput. Mech. 35(04), 408–416 (2018)
11. Bo, Y., Ao, Z., Ronghua, S., et al.: Reliability analysis of cylindrical Shell structures based on BPNN optimized by genetic algorithm. Eng. Mech. **42**(S1), 16–22 (2025)
12. Meng, Z., Guandong, L., Ketong, D., et al.: Multi-objective genetic algorithm for suspension system performance based on adaptive interval congestion. In: Technical Committee on Process Control, Chinese Association of Automation. The 36th Chinese Process Control Conference (2025)
13. Ketong, D., Weiguo, G., Meng, Z., et al.: Optimization of cutting performance of cantilever-type Roadheader based on improved seagull optimization algorithm. In: Technical Committee on Process Control, Chinese Association of Automation. The 36th Chinese Process Control Conference (2025)

Research on Water Jet Polishing Process of Translucent Molds for Automotive Interior Parts

Lei Chen[1] and Xiaomin Cheng[2(✉)]

[1] Ningbo Joyson Quin Automotive Systems Co., Ltd., Ningbo 315100, Z.J, China
[2] Hangzhou City University, Hangzhou 310015, Z.J, China
chengxiaomin@hzcu.edu.cn

Abstract. The molding molds for multi-layer composite transparent inserts of automotive interior parts need to meet requirements such as light transmittance, complex cavity structure, and high reliability, posing severe challenges to surface precision and greenization level in their manufacturing process. Traditional processing technologies have problems such as high energy consumption, heavy pollution, and difficulty in adapting to complex structures, which cannot meet the manufacturing needs of such molds. Focusing on the greenization of the entire mold processing process, this paper explores the optimization paths of key technologies including high-efficiency cutting, low-temperature nitriding heat treatment, and abrasive water jet polishing (AWJP). It studies the process parameters and experimental devices of AWJP for polishing complex inner cavities of transparent molds. The experimental results show that the surface roughness of the mold first decreases and then increases with the increase of jet pressure, with an optimal pressure range. The polishing effect is the best when the polishing time is 60–90 min and the jet angle is around 45°. Different abrasive grain sizes have significant impacts on the polishing effect: W5 abrasive can achieve lower roughness but with lower efficiency, while W12 abrasive has a higher removal rate but slightly higher roughness. With 2–30 MPa jet pressure and W12-W5 diamond abrasives as core parameters, high-precision finishing of complex cavity surfaces can be achieved through the flexible impact and grinding effect of gas-liquid-solid three-phase flow.

Keywords: Automobile Interior Parts · Water Jet · Polishing Process

1 Introduction

As an important part of automobiles, automotive interior parts involve multiple links such as material processing, molding, and assembly in their production process, and are key to realizing the "greenization of the production process" in the automotive industry chain [1]. Molds for automotive interior parts are core tools for producing interior parts, and their manufacturing process includes multiple high-energy-consumption and high-emission links such as material cutting, heat treatment, surface processing, and assembly

S. Halgamuge et al. (Eds.): ICACMVE 2025, LNME, pp. 735–743, 2026.
https://doi.org/10.1007/978-981-95-6911-3_73

[2]. The greenization of interior part mold manufacturing focuses on reducing resource consumption and environmental load throughout the entire mold manufacturing process through material innovation, process optimization, energy upgrading, waste recycling, and digital management and control [3], while ensuring mold precision and service life. In some high-end models, instrument panels mostly use transparent interior parts, which are not only aesthetically pleasing but also energy-saving, and there is a trend of using transparent interior parts in ordinary models. This puts forward higher requirements for the surface finish, dimensional stability, and microstructures of interior part molding molds, such as high-precision surfaces driven by light transmittance, complex cavity designs caused by multi-layer composites, and precise positioning required by insert integration, while also meeting the strict standards of the automotive industry for reliability and durability. Therefore, compared with ordinary molds, they have significantly improved design complexity, processing precision, and material adaptability. Traditional processing relies on a multi-step process of "rough milling → finish milling → grinding → electrical discharge machining", which has high energy consumption and much waste, making it difficult to meet the manufacturing requirements of transparent interior part molds.

In recent years, research on the application of abrasive water jet polishing technology in precision mold processing has made many progresses. In 2022, Li et al. proposed a process parameter optimization model based on jet characteristics through research on water jet polishing of complex aerospace parts [4]; in 2023, Wang's team deeply analyzed the influence mechanism of abrasive types and grain sizes on polishing effects for surface finishing of mold steel [5]; Shuaikang C.'s team studied the thermo-mechanical coupling mechanism and surface integrity in alloy abrasive water jet processing [6]; Simon S.'s team conducted a full-factor experiment on the surface roughness of chrome-nickel steel after water jet processing [7]. These studies provide important references for the water jet polishing process of transparent molds for automotive interior parts.

2 Greenization of Mold Processing Technology

The greenization of mold processing technology refers to a sustainable development model that minimizes resource consumption, reduces environmental pollution to the lowest level, and balances processing efficiency and product quality throughout the entire life cycle of mold design, manufacturing, use, and recycling through technological innovation, process optimization, material upgrading, and management innovation. Its core goal is to break the traditional processing model of "high energy consumption, high pollution, and low efficiency" and build a green manufacturing system of "reduction, reuse, and resource utilization" [8].

2.1 High-Efficiency Cutting Process to Reduce Processing Energy Consumption

In traditional wet cutting methods, the cost of cutting fluid accounts for nearly a quarter of the total processing cost. Volatilization of cutting fluid produces VOCs pollution, and it is prone to cause thermal adhesion between tools and materials. If no cutting fluid is needed during cutting, the costs of purchasing, filtering, and treating waste cutting fluid can be saved [9]. Using ultra-fine grain cemented carbide tools and high-speed spindles to realize dry cutting of interior part molds can improve processing efficiency by 35% and reduce carbon emissions by 22% [10].

2.2 Changing Heat Treatment Process to Reduce Processing Steps

Automotive interior part molds require heat treatment. The traditional heat treatment method is "high-temperature quenching + multiple tempering". If the low-temperature nitriding + rapid cooling process is adopted: ion nitriding is carried out at 380–420 °C, and nitrogen atoms penetrate into the mold surface through glow discharge, forming a nitrided layer with a hardness of over 1000HV. The treatment time is shortened by 35%, energy consumption is reduced by 30%, and mold deformation is reduced, thereby lowering the energy consumption of subsequent straightening processing.

2.3 Abrasive Water Jet Polishing to Replace Traditional Chemical Polishing to Improve Quality and Efficiency

Automotive interior part molds have high surface requirements, and the cavity generally requires a roughness of Ra ≤ 0.02 μm, so polishing is necessary. The traditional chemical polishing process uses a mixture of nitric acid and hydrofluoric acid, which is highly corrosive, difficult to treat waste liquid, and has high pollution problems. Abrasive Water Jet Polishing (AWJP) is a low-energy-consumption surface processing technology that flexibly polishes the mold surface by driving and controlling abrasives with water jets, without chemical pollution, and improves polishing efficiency [3].

3 Greenization of Polishing Process for Transparent Molds of Interior Parts

The multi-layer composite transparent insert molds for automotive interior parts have complex shapes, and the edge matching between molded parts and assembly components has high requirements. Manual polishing relies on the experience of technicians, with low efficiency and poor quality consistency; rigid tools for mechanical polishing such as grinding wheels and cloth wheels cannot adapt to complex curved surfaces, easily causing scratches or over-cutting, and are difficult to reach small parts; electrolytic polishing is sensitive to material composition, and electrolyte is easy to accumulate in inner cavity corners, causing local over-corrosion. AWJP can adaptively bend with the shape of the inner cavity, with flexible jet characteristics and non-contact processing mode, which can easily enter complex inner cavities and solve the above pain points [8].

3.1 Principle of Abrasive Water Jet Polishing

The inner cavity surface of automotive interior part molds has high requirements, especially for microstructures, which need to achieve low surface roughness, good surface precision, and long service life, putting forward higher requirements for processing technology. Abrasive water jet polishing of mold inner cavities is a surface finishing technology based on high-pressure fluid dynamics [9]. After high-pressure water is mixed with abrasive particles, a high-speed solid-liquid mixed jet is formed. Under pressure, abrasive particles move forward and rotate, scraping and grinding the inner cavity surface of the mold, removing surface peaks slightly, and reducing surface roughness.

3.2 Abrasive Water Jet Polishing Process and Device

The polishing effect of AWJP is mainly reflected in reducing the surface roughness of the mold cavity through micro-removal of the material surface. The experimental process parameters selected for polishing interior part molds in this study are as follows: the higher the jet pressure, the greater the abrasive kinetic energy and the higher the material removal rate, but excessively high pressure easily causes surface scratches. Therefore, the working pressure of the water jet is 2–30 MPa; the grain size of diamond abrasives is W12-W5; the jet distance is 5–50 mm, and the jet angle is in the range of 30°-60°; the jet time is determined according to the final processing roughness, and the jet trajectory is controlled by a numerical control program with reciprocating scanning to ensure polishing uniformity.

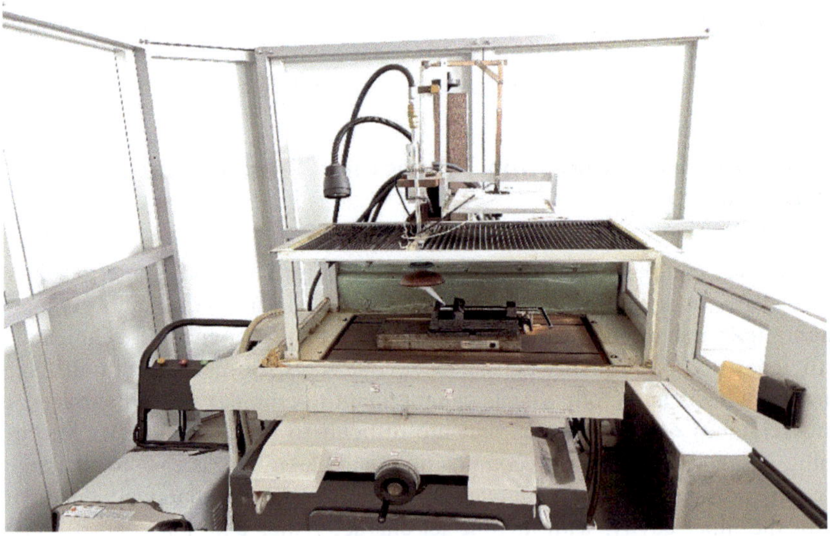

Fig. 1. Polishing Experimental Device.

The polishing experimental device is shown in Fig. 1, consisting of X and Y axis motion platforms, a polishing experimental chamber, a jet system, a pressure boosting system, and a filtering system. A cylindrical jet is used, so that abrasives can obtain greater kinetic energy under the same flow rate, resulting in high polishing efficiency. The pressure boosting system boosts water to a preset pressure, forming a high-speed water flow through the nozzle. Abrasives are mixed with high-pressure water through a mixing device to form a gas-liquid-solid three-phase flow. The mixed jet is sprayed onto the inner cavity surface of the mold along a preset path. Abrasive particles impact surface micro-protrusions at high speed, removing materials through mechanisms such as plastic deformation, brittle fracture, or fatigue spalling, realizing surface flattening and finishing. The filtering system includes a booster pump, a pressure gauge, a clean water tank, a sedimentation tank, a water storage tank, multi-stage filters, a separator, an abrasive recovery device, a clean water recovery device, etc., which filters, screens, and separates water, abrasives, and removed materials. Water and abrasives are reused multiple times to reduce costs and achieve green production.

3.3 Polishing Experiment on Complex Inner Cavities of Transparent Molds for Interior Parts

Influence of Jet Pressure on Inner Cavity Roughness. The influence of jet pressure on the inner cavity roughness of the mold is shown in Fig. 2, where the abscissa represents jet pressure (MPa) and the ordinate represents surface roughness Ra value (μm). When the jet pressure increases, the surface roughness Ra value of the mold changes: the Ra value at the corresponding position of the mold first decreases and then increases, with the minimum roughness and the best polishing effect at a certain pressure. When the pressure continues to increase, the energy contained in the abrasive particles is too large, the impact on the cavity surface increases, the surface removal increases, the surface roughness increases, and the polishing effect deteriorates.

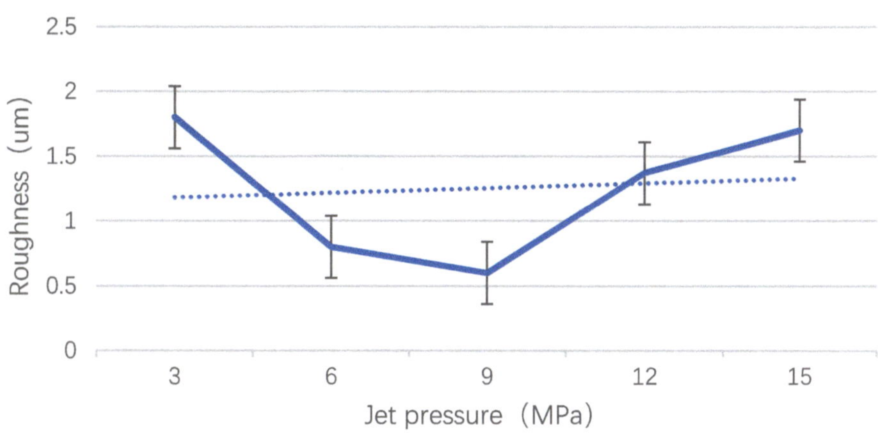

Fig. 2. The impact of pressure on the roughness of the mold cavity.

Influence of Abrasive Grain Size on Polishing Effect. Polishing experiments were carried out with two types of diamond abrasives, W12 and W5, under fixed parameters: jet pressure 18 MPa, jet angle 45°, polishing time 45 min, and jet distance 25 mm. The experimental results are shown in Fig. 3. When using W12 abrasive, the surface roughness Ra value of the mold finally stabilizes at 0.03 μm; when using W5 abrasive, the surface roughness Ra value can be reduced to 0.02 μm. In terms of material removal rate, due to the larger particle size, W12 abrasive removes more materials than W5 abrasive in the same time, with the material removal amount being about 1.8 times after multi-stage filtration, sorting, abrasive recovery, and separation. However, the larger W12 abrasive particles easily obtain more energy, which is more likely to cause greater impact on the polished surface, leading to deeper scratches and pits on the polished surface. W5 abrasive particles are fine, which can act more uniformly on surface micro-protrusions, achieving more precise grinding and thus lower roughness. However, the polishing efficiency of W5 abrasive is relatively low, and it takes about 1.4 times longer than W12 abrasive to achieve the same surface quality.

The used abrasives were filtered and recovered, and recycled abrasives meeting the grain size requirements were selected. Polishing experiments were repeated with the same process parameters. For W12 recycled abrasives, the surface roughness Ra value after polishing was basically stable in the first 3 uses, which was not much different from the polishing effect of new abrasives. The Ra value increased slightly in the 4th use and further increased in the 5th use. The reason is that during multiple uses of recycled abrasives, some particles will break or wear, resulting in smaller particle size or uneven distribution, thus affecting the polishing effect. However, in general, W12 recycled abrasives can maintain good polishing quality consistency within 3 uses, and W5 recycled abrasives within 2 uses, which can meet actual production needs, further reflecting the greenization and economy of the abrasive water jet polishing process.

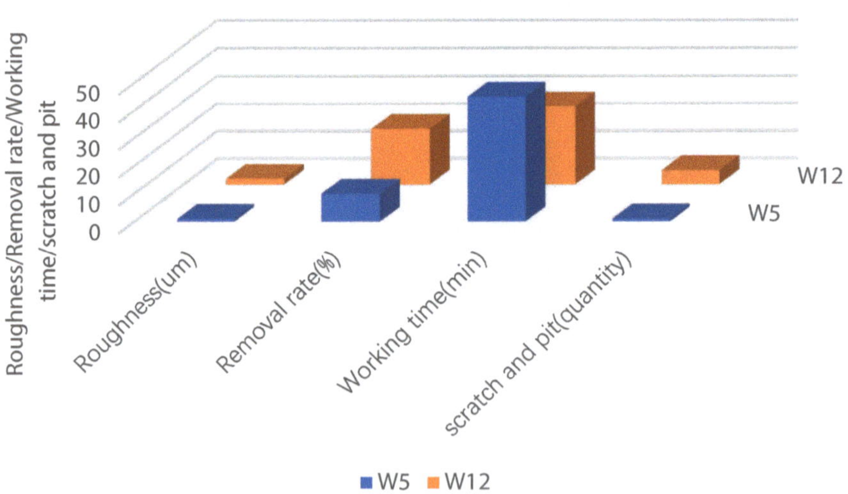

Fig. 3. The impact of abrasive material on the roughness of the mold cavity.

Combined Influence of Jet Angle and Polishing Time on Inner Cavity Roughness. The combined influence of jet angle and polishing time on the inner cavity roughness of the mold is shown in Fig. 3, where the abscissa represents polishing time (min), the ordinate represents surface roughness Ra value (μm), and the depth coordinate represents jet angle (°). First, polishing experiments were carried out with a fixed jet angle, and the roughness values of different parts of the mold inner cavity were detected after polishing for 30 min, 60 min, 90 min, 120 min, and 150 min, and the average value was taken as the experimental result. Then, the above experiments were repeated by rotating the jet angle, and Fig. 4 was obtained after synthesis. The blue part in Fig. 4 has smaller roughness, corresponding to a polishing time of 60–90 min. That is, when the jet angle is in the range of 30°-60°, the polishing effect improves with the increase of polishing time, with the best effect in the 60–90 min period, and deteriorates after 90 min. The polishing effect is the best when the jet angle is around 45°

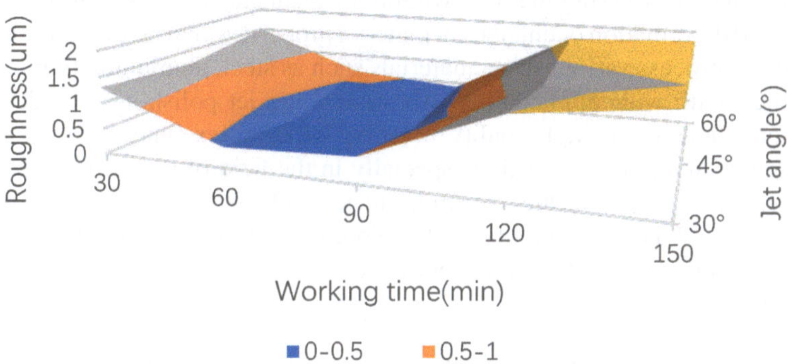

Fig. 4. Influence of the combined effect of injection angle and polishing time on the roughness of the mold cavity.

3.4 Analysis of Polishing Effect

To visually confirm the reduction of surface roughness after polishing, an optical 3D surface profiler was used to observe the inner cavity surface of the mold before and after polishing. Figure 5 shows the undulation images of the mold inner cavity surface before and after polishing, with a sampling range of 0.5mm*0.5mm. The left image of Fig. 5 shows the inner cavity surface before polishing, with surface micro-protrusions between 55–65 μm; the right image shows the inner cavity surface after polishing, with surface micro-protrusions between 40–50 μm. The surface after polishing has smaller undulations and lower roughness than that before polishing.

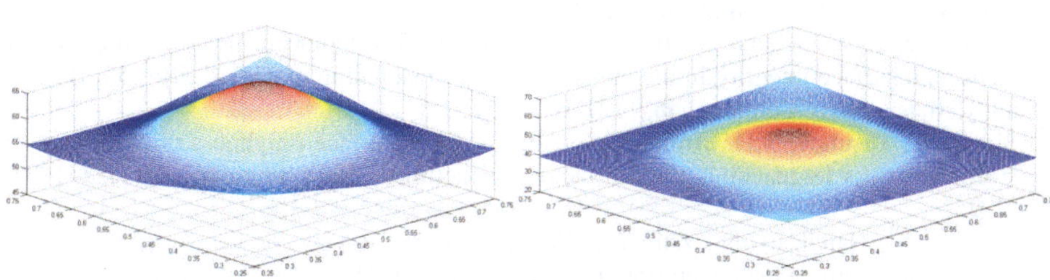

Fig. 5. Polishing effect.

4 Conclusions and Discussion

Different abrasive grain sizes have significant impacts on the polishing effect: W5 abrasive can achieve lower surface roughness but with relatively lower efficiency; W12 abrasive has a higher material removal rate but slightly higher surface roughness after polishing. Optical images intuitively confirm the reduction of surface roughness after polishing, with the microstructure becoming more flat and smooth.

When using abrasive water jet polishing, the jet can bend with the shape of the inner cavity, without tool accessibility restrictions, and flexibly adapt to complex inner cavities; it is a non-contact processing method, without mechanical stress or heat-affected zone, avoiding mold deformation or changes in material properties, and no thermal/mechanical damage; it can process various mold materials such as steel, aluminum, titanium alloys, and ceramics, with wide applicability. Abrasive water jet polishing provides an ideal solution for efficient and high-quality finishing of complex inner cavities of molds, and has significant application value especially in the field of precision molds such as automobiles. In the future, further research can be conducted on the recycling process of abrasives to extend their service life, and explore the optimal combination of polishing process parameters for molds of different materials to expand the application range of this technology.

Acknowledgments. This study was funded by Major Scientific and Technological Special Projects of Zhejiang Province and Ningbo City (grant number 2024Z149, 2024C01039,2023Z032 and 2022Z071).

References

1. Laiyong, P.: The Impact of low-carbon policies on the green and sustainable development of automotive interior parts manufacturing enterprises. Sales Mark. **36**, 34–36 (2024)
2. Smith, J., Johnson, L.: Advanced manufacturing techniques for automotive interior molds. Int. J. Automot. Technol. **23**(4), 987–1001 (2022)
3. www.gov.cn. (2022) https://www.gov.cn/gongbao/content/2022/content_5717004.htm.
4. Li, M., Zhang, H.: Parameter optimization model for water jet polishing of complex aerospace parts. J. Manuf. Sci. Eng. **144**(6), 061005 (2022)
5. Wang, Q., Liu, J.: Influence mechanism of abrasive type and size on polishing effect of die steel. Mater. Sci. Eng. A **862**, 144321 (2023)
6. Shuaikang, C., Wenchuan, L., Jiren, T., Mengyan, F.: Thermomechanical coupling mechanisms and surface integrity in the abrasive waterjet machining of Ti-5Al-5Mo-5V-1Fe-1Cr alloy. Mater. Today Commun. **46**, 112770 (2025)
7. Simon, S., Yusubov, N.D., Amirli, S.F., Amirov, F.G.: Surface roughness of chromonickel steel after water jet machining: a full factorial experiment. Russ. Eng. Res. **45**(3), 341–345 (2025)
8. Zhifeng, L., Haichuan, N., Zhiqiang, W., Qingdi, K.: Review of key technologies and applications in green manufacturing. Electro-Mech. Eng. **41**(3), 20–36 (2025)
9. Qingcheng, L.: Research on efficient processing technology of mechanical manufacturing and automation. J. Int. Nat. Sci. Stud. **2**(1), 43–45 (2025)
10. Brown, R., Davis, S.: Green polishing technologies for automotive mold manufacturing. J. Clean. Prod. **310**, 127563 (2021)

Effect Law of Cavity Structure Parameters on Asphalt Foaming Behavior

Fang Guo, Haiying Cheng(✉), Yongzhen Deng, and Zhihao Song

Key Laboratory of Road Construction Technology and Equipment of MOE, School of Construction Machinery, Chang'an University, Xi'an 710064, China
chychdktz@163.com

Abstract. The foaming process of asphalt is a complex multiphase flow mixing and heat transfer behavior. Analyzing this behavior from a macroscopic perspective often overlooks the details of multiphase flow, while microscopic analysis overly focuses on theoretical studies at the molecular level and neglects the overall characteristics of multiphase flow. Therefore, this paper adopts the lattice Boltzmann method at the mesoscopic scale to analyze the asphalt foaming behavior. Taking the asphalt foaming chamber as the research object and using the average pressure and average flow velocity of the chamber as evaluation indicators, a numerical simulation analysis of the asphalt foaming behavior inside the chamber is conducted, with comparative verification through asphalt foaming experiments. The study explores the influence of structural parameters on the average pressure and average flow velocity of the chamber, and analyzes how different structural parameters of the chamber affect the asphalt foaming behavior. The results show that within a certain range, the smaller the inlet and outlet sizes, the greater the outlet pressure and velocity of the chamber. This leads to more sufficient heat exchange between hot asphalt and water inside the chamber, thereby increasing the expansion rate and half-life of the foamed asphalt. The volume of the chamber has a significant impact on its average pressure; increasing the chamber volume can promote heat exchange between hot asphalt and water, thus enhancing the expansion rate of the foamed asphalt.

Keywords: Lattice Boltzmann Method · Foamed Asphalt · Cavity Structure Parameters

1 Introduction

Hot asphalt and room temperature water in a specific volume of foam cavity direct contact heat exchange, the water rapidly into water vapor, the volume increases resulting in a rapid increase in the pressure in the cavity, and then the water vapor is pressed into the asphalt continuous phase to form a fine foam, and from the foam cavity outlet, the fine foam under the action of internal and external pressure difference continues to increase the volume, the formation of foam asphalt, its viscosity is significantly reduced. Mixing the foamed asphalt with aggregate at this time can effectively improve the mixing performance of the mixture [1]. Foamed bitumen can be used in warm mix technology

S. Halgamuge et al. (Eds.): ICACMVE 2025, LNME, pp. 744–755, 2026.
https://doi.org/10.1007/978-981-95-6911-3_74

and cold recycling technology in road works to ensure that aggregates are coated at lower temperatures [2]. Warm mix technology and cold recycling technology can effectively solve the problem of high pollution caused by hot mix asphalt paving roads [3]. Tests suggested that the WMA binders will experience equal or less aging over the service life of a pavement structure as compared to HMA binders [4]. However, the application of foamed asphalt requires professional equipment to prepare foamed asphalt, which is based on the analysis of factors affecting the foaming behavior of asphalt. In order to analyze the foaming behavior of asphalt, scholars have analyzed it from multiple scale levels.

At first, scholars numerically simulated the asphalt foaming behavior using computational fluid dynamics methods (CFD) on a macroscopic scale. They analyzed the effects of asphalt temperature, asphalt flow rate, and water dosage on the expansion ratio of foamed asphalt during the base asphalt foaming process, optimized the design of the foaming chamber structure, studied the asphalt viscosity range suitable for asphalt foaming, and compared the impacts of different chamber pressures on the half-life and expansion ratio of foamed asphalt [5–7]. In recent years, scholars began to use Molecular Dynamics (MD) methods to analyze the foaming behavior of asphalt from a microscopic scale, and explain the interaction between asphalt and water from a molecular level [8–10]. The asphaltic molecular model commonly used in this method is the 12-component molecular model proposed by Li [8]. The molecular dynamics method studies the interaction between asphalt molecules and water molecules from the molecular level, which has theoretical guiding significance for the analysis of dynamic behavior of foamed asphalt. Mesoscopic scale is the scale between macro and micro. The Lattice Boltzmann method (LBM) is the most commonly used method for studying multiphase flow at mesoscopic scale [11]. LBM can easily and accurately describe the complex interactions between different phases and different components [12]. In LBM, fluid is regarded as consisting of a series of fluid particles larger than microscopic molecules but infinitely small on the macro level, and an evolutionary mechanism conforming to certain physical laws is constructed to allow these fluid particles to evolve, so as to obtain simulation results consistent with physical laws, which takes into account both microscopic nature and mesoscopic characteristics. Therefore, many scholars began to study the formation and breaking process of bubbles in the foaming behavior from the mesoscopic scale 13, 15].

The analysis of asphalt foaming behavior on the macro scale provides a method for the design and optimization of asphalt foaming device, and the study on the interaction between asphalt molecules and water molecules on the micro scale reveals the foaming mechanism of asphalt more deeply. However, because asphalt foaming behavior occurs in a cavity with a specific volume and involves complex problems such as the movement, deformation and crushing of the phase interface in multiphase flow, the flow characteristics of asphalt in the cavity become extremely complicated, which makes it difficult to simulate the foaming behavior of asphalt intuitively by macro and micro methods. The mesoscale numerical simulation method can accurately simulate the complex problems such as the movement, deformation and fragmentation of the multiphase interface in the process of asphalt foaming, and effectively simulate the changes of pressure and flow rate in the cavity during the process of asphalt foaming. Therefore, LBM will be used in this paper to analyze the foaming behavior of asphalt cavity from the mesoscopic scale, in order to explore the influence of the structural parameters of asphalt foam cavity on the pressure and flow rate in the foam cavity.

2 Numerical Simulation and Experimental Scheme Design of Asphaltic Foaming at Mesoscale

2.1 Numerical Simulation of Flow Field Inside Asphalt Foaming Cavity

The internal flow field of asphalt foam cavity was numerically simulated based on X-Flow, a computational fluid dynamics software. X-Flow is developed based on lattice Boltzmann method and adopts particle-based Lagrange method to realize meshless division of fluid region and the surface complexity of the model is no longer limited. In addition, X-Flow adopts the large eddy simulation turbulence model (LES) as a flow solver, which is easy to predict the flow pattern under complex characteristics, and is suitable for the complex movement inside the asphalt foam cavity.

Material Property Definition. The flow medium for the numerical simulation of asphalt foaming is two materials, namely asphalt and water, whose material properties need to be defined in X-Flow, as shown in Table 1.

Table 1. Related parameters of water and asphalt.

Parameter	Water	Asphalt
Reference density/ $(kg \cdot m^{-3})$	998.3	1019
Operating temperature/(K)	288.15	433
Dynamic viscosity/$(Pa \cdot s)$	0.001	0.22
Thermal conductivity/$(W \cdot (m \cdot K)^{-1})$	0.58	0.699
Specific heat capacity/$(J \cdot (kg \cdot K)^{-1})$	4182	2621

Processing of Boundary Conditions. The import and export interface and boundary conditions of the model should be determined before setting the calculation parameters. Figure 1 shows the foamed asphalt cavity model. The inner surface of the model is set as the wall, and the standard rebound format is used as the wall boundary conditions. In Fig. 1, the upper part of the model is asphalt and water inlet, and the lower part is foam asphalt outlet. The boundary conditions of the inlet and outlet of the flow field adopt non-equilibrium rebound boundary.

The standard rebound format assumes a reversal of velocity after a particle collides with a wall, as shown in Fig. 2 [15].

According to the collision-migration process of lattice Boltzmann and the particle velocity dispersion direction shown in Fig. 3, the particle distribution function of the standard rebound scheme can be expressed as

$$f_{\bar{i}}(x_b, t) = f_i(x_f, t) \tag{1}$$

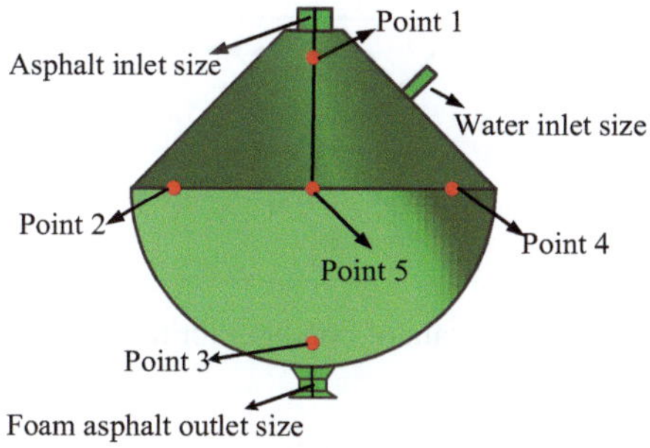

Fig. 1. Model setup in X-Flow.

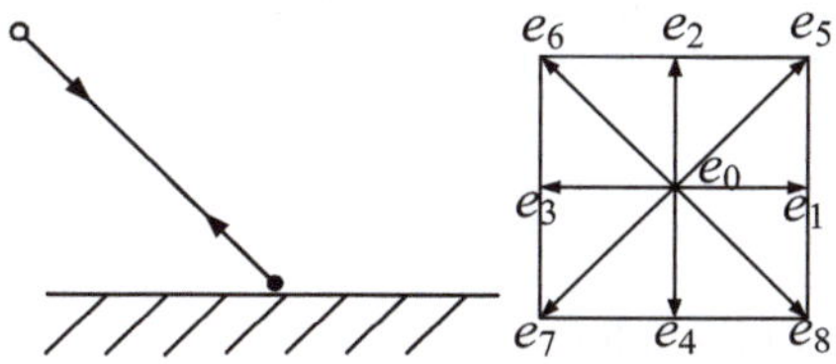

Fig. 2. Schematic diagram of standard rebound format and particle velocity dispersion.

where: i is the direction pointing to the wall; $\bar{i}$ is the opposite direction of i; x_b is the solid wall node; x_f is the fluid node, $x_f = x_b - e_i\delta_t$.

The non-equilibrium rebound boundary is shown in Fig. 3 where the particles migrate to the boundary node where the distribution function $f_0, f_1, f_3, f_4, f_7, f_8$ is known, while the distribution function f_2, f_5, f_6 and density are unknown. Taking velocity as an example, the macroscopic velocity of the fluid at this point is known, but the macroscopic density is not known. Based on the relationship between the macroscopic quantity and the distribution function, the following relationship can be derived:

$$\begin{cases} f_2 + f_5 + f_6 = \rho - (f_0 + f_1 + f_3 + f_4 + f_7 + f_8) \\ f_2 + f_5 + f_6 = \rho u_y + (f_4 + f_7 + f_8) \\ f_5 - f_6 = \rho u_x - (f_3 + f_7 - f_8 - f_1) \end{cases} \tag{2}$$

There are only 3 independent equations in the above equations but 4 unknowns. To make the system of equations solvable, a set of relational expressions needs to be added to close the system. This is based on the non-equilibrium rebound relationship proposed by Zou et al. [16].

$$f_2 - f_2^{eq} = f_4 - f_4^{eq} \tag{3}$$

In this case, there are 4 independent equations with 4 unknowns, and all unknown distribution functions on the boundary nodes can be solved. This idea can be used to solve both pressure and density boundary conditions.

Calculation of Relevant Parameters. The mass flow rate at the velocity inlet is determined by Eq. (4).

$$\dot{m} = \rho VA \tag{4}$$

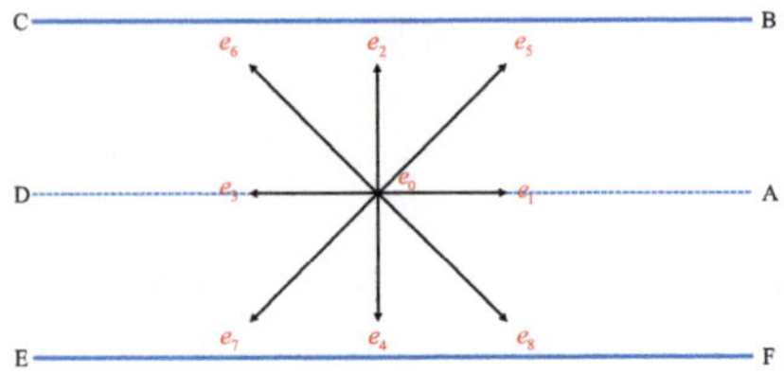

Fig. 3. Schematic diagram of non-equilibrium rebound boundary.

where, $\dot{m}$ is the flow rate; ρ is the fluid density; V is the fluid velocity; A is the area of cross section.

The asphalt inlet and water inlet are set as velocity inlets with a circular shape. The velocity magnitude is calculated based on the flow rate and inlet diameter, as follows.

$$V = \frac{4\dot{m}}{\rho \pi D^2} \tag{5}$$

where, D is the diameter of the inlet.

Calculation Accuracy. To ensure simulation accuracy while controlling computational time, it is necessary to select an appropriate grid scale and perform local refinement for complex areas. The overall grid scale is set to 0.0008 m, the encryption method near the wall is selected, and the grid scale at the wall surface, asphalt entrance and foam asphalt exit is 0.0002 m. The water inlet size is small, and its grid scale is 0.0001 m. Considering that the asphalt injection time in the actual asphalt foaming process is 5 s, the simulation time is set to 5 s and the frequency is 80 Hz.

Through the numerical simulation of asphalt foaming behavior, the stability coefficients of the simulation results all range from 0.1 to 0.3, meeting the convergence requirements.

2.2 Design of Test Scheme

To verify the validity of the mesoscale numerical simulation results of asphalt foaming, a foaming test was conducted using 70# asphalt, with its parameters listed in Table 2. During the test, the asphalt temperature was 160 °C. To measure the flow field parameters inside the cavity, pressure sensors and temperature sensors were installed inside the cavity to measure the pressure and temperature in the cavity, respectively. Figure 4(a) is the asphalt foaming device used in the test, and Fig. 4(b) is the change of the volume in the process of asphalt foaming. Considering the significant influence of water proportion on asphalt foaming behavior and the difficulty in measuring the flow velocity in the asphalt foaming process, the water proportion was used as the adjustment variable in the test, and the water proportion of 3%, 4% and 5% were used for three repeated tests to obtain the average value, and the pressure in the cavity was used as the main evaluation index. During the test, the aggregate mass was 25 kg, the mixing ratio was 3%, and the foaming time was 5 s.

The bitumen foaming test process is as follows:

Table 2. Indicators of 70# matrix asphalt.

Type of asphalt	Needle penetration at 25 °C/0.1 mm	10 °C elongation/cm	Softening point/°C	60°C power viscosity/*pa · s*
70#	71.5	38	47.6	187.5

(1) Prepare test materials and test equipment, including matrix asphalt, water, asphalt foam testing machine, measuring scale, measuring barrel, etc.;
(2) Pour the asphalt into the asphalt bucket of the foaming machine, and heat the asphalt to the test temperature;
(3) Record foaming time, asphalt flow, water proportion and asphalt temperature, and adjust the asphalt flow and water flow to the values required for the test;
(4) Start the foaming test, weigh the quality of the ejected asphalt, and record the pressure change inside the cavity during the foaming process of the asphalt, and repeat three times to take the average value of the pressure data.
(5) Adjust the proportion of foaming water and repeat the above test.

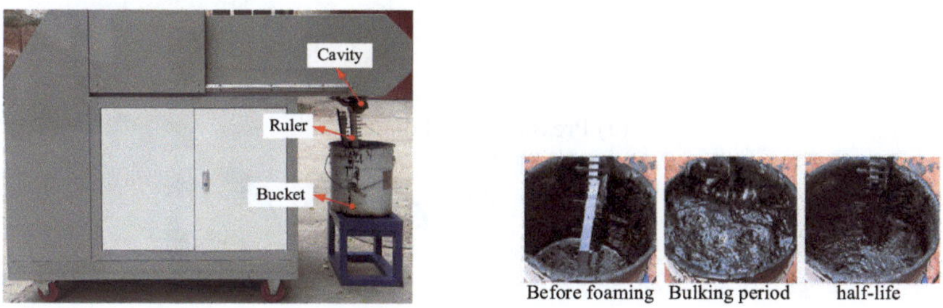

(a) (b)

Fig. 4. Asphalt foaming test.

3 Results and Analysis

3.1 Verification of Numerical Simulation Results of Flow Field in Asphalt Foam Cavity

Based on the aggregate particles of the same mass used in the asphalt foaming test, the flow field simulation of asphalt foaming was carried out with three water ratios of 3%, 4% and 5%. The Angle between the asphalt inlet and the y axis is 180°, and the Angle between the water inlet and the y axis is 45°. The inlet flow rates of asphalt and water could be obtained by Eq. (5).

In order to analyze the average pressure and flow velocity inside the cavity during the asphalt foaming process, the flow velocity and pressure were collected from 5 sampling points as shown in Fig. 1, and their average values were taken for subsequent analysis.

The flow field distribution formed during the foaming process was simulated, and the flow velocity and pressure distribution inside the cavity were obtained, as shown in Fig. 5. Figure 5(a) showed the pressure states inside the foaming cavity at different times. The pressure at the asphalt inlet was low, while the pressure at the bottom was high. The average pressure inside the cavity was higher than the external atmospheric pressure, under which the foamed asphalt was ejected from the cavity. In addition, the pressure in the area where asphalt experienced turbulence was low. The generation of turbulence promoted the mixing of asphalt and water, which was beneficial to improving the foaming effect of asphalt. In Fig. 5(b), asphalt entered the foaming cavity from the inlet and, during the flow process, was affected by the water flow from the right, causing its flow path to shift to the left. When the asphalt reached the bottom of the cavity, part of the asphalt was directly ejected through the outlet, and another part, after coming into contact with the bottom of the cavity, flowed back along the cavity wall to the left side of the cavity. After water entered the cavity, it moved toward the outlet of the foaming cavity and, during the movement, fully mixed with asphalt. In the mixing process, heat exchange occurred between water and hot asphalt, causing the water to phase change into water vapor, and the volume of the asphalt expanded, with the final result that it was ejected from the cavity.

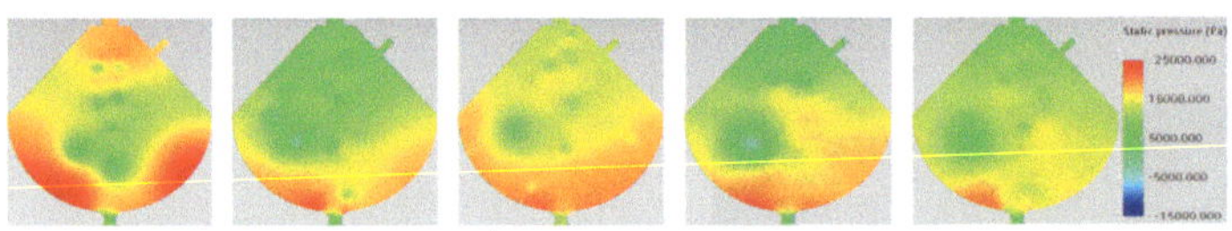

(a) Pressure cloud diagram

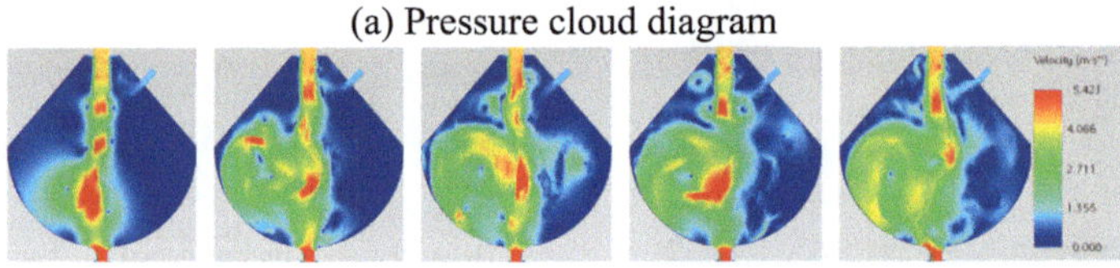

(b) Stream velocity maps

Fig. 5. Cavity pressure and flow velocity cloud.

In order to verify the reference-ability of the simulation model, pressure data were collected in the foam cavity, and the test results were compared with the simulation results, as shown in Table 3.

Table 3. Cavity pressure simulation results and test results.

Water proportion (%)	Measured pressure/MPa				Simulated pressure/MPa	Relative error/%
	three measurements			average value		
3	0.0538	0.0824	0.0856	0.0739	0.0759	2.7
4	0.0888	0.1098	0.1074	0.1020	0.1126	10.4
5	0.1686	0.1712	0.1586	0.1661	0.1839	10.7

It could be seen from Table 3 that as the proportion of water increased, the difference between the experimental results and the simulation results gradually grew. This was because during the test, the asphalt exchanged heat with the outside through the cavity shell, resulting in partial heat loss. However, heat loss was not considered in the simulation. Therefore, compared with the simulation results, the average pressure of the cavity in the experiment was lower than the simulated pressure. Under the conditions where the water proportions were 3%, 4%, and 5% respectively, the relative errors between the simulation results and the test results of the cavity pressure were all less than 11%, indicating that the established numerical model could meet the requirements of engineering analysis.

3.2 Influence of Asphalt Foaming Cavity Structure Parameters on Asphalt Foaming Behavior

The structural parameters of the cavity (such as the size of outlet and inlet and the volume of the cavity) affect the asphalt foaming effect by affecting the velocity of asphalt in the foaming cavity, the formation of pressure in the cavity and the mixing effect of two-phase flow. Therefore, this paper mainly analyzes the influence laws of three structural parameters, namely the cavity inlet, outlet and volume, on the asphalt foaming behavior. By increasing the parameters to 1.3 times their original values or reducing them to 0.7 times their original values, the changes in the foaming behavior inside the foaming cavity are observed.

Influence of Cavity Inlet Size. Figure 6 show the curves of central pressure variation and central velocity variation under different inlet sizes. It can be observed from Fig. 8 that during the first 2 s of the foaming process, the central pressure gauge of the cavity exhibits large amplitude changes and high frequency fluctuations. This phenomenon can be explained by the results of Fig. 5. In the first 2 s of foaming process, because the asphalt has not completely filled the cavity, the asphalt flow in the cavity is relatively intense, resulting in a large amplitude of pressure fluctuation. As the bitumen flowing into the cavity gradually increases and fills the whole cavity, the central pressure in the cavity tends to be stable and the fluctuation amplitude decreases significantly.

Coefficient of Variation (CV) is a dimensionless indicator used in statistics to measure relative dispersion, which is the ratio of standard deviation to the mean. The coefficient of variation is used here as an evaluation index of the fluctuation degree of the central pressure and central flow velocity of the cavity during the asphalt foaming process. Under different structural parameters, the data of central pressure and central velocity of 401 groups of cavity were collected respectively, and the coefficient of variation was calculated. It can be obtained that when the inlet size of asphalt is 0.7D, D and 1.3D, the coefficient of variation of the central pressure is 0.1925, 0.3046 and 0.2877 respectively. As the inlet size increases, the central pressure of the asphalt in the cavity gradually decreases, and its coefficient of variation showed a trend of first increasing and then decreasing.

On the premise that the asphalt inlet flow remains unchanged, the inlet velocity decreases with the increase of the asphalt inlet size, which results in the decrease of the center velocity inside the cavity. When the asphalt inlet size is 0.7D, D and 1.3D, the coefficient of variation of the velocity in the center of the cavity is 0.3148, 0.3804 and 0.5012 respectively. With the increase of the asphalt inlet size, the central flow velocity in the cavity gradually decreases, and its fluctuation degree gradually increases. Meanwhile, the central flow velocity and central pressure in the cavity exhibit a fluctuating pattern where one increases as the other decreases.

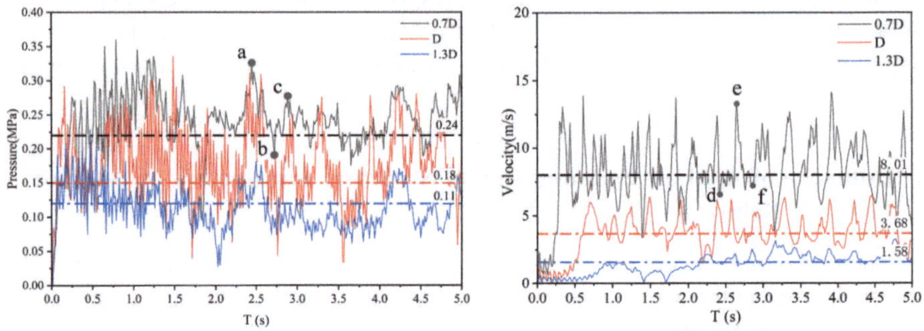

(a) variation of center pressure (b) variation of center flow rate

Fig. 6. Variation of center pressure and center flow rate with different inlet sizes.

As can be seen from Fig. 6: When the flow velocity at the center of the cavity increases from d to e, an excessively high flow velocity is unfavorable for heat exchange between water and hot asphalt, and the expansion rate of foamed asphalt also decreases, which causes the pressure at the center of the cavity to drop from point a to point b. During the process where the central flow velocity of the cavity decreases from point e to point f, the heat exchange time between the hot asphalt and water in the cavity increases, allowing more heat in the hot asphalt to participate in the phase change of water; at the same time, the gradually decreasing central flow velocity of the cavity increases the flow velocity gradient inside the cavity, intensifies the turbulence of asphalt in the cavity, and further promotes the mixing of water and hot asphalt, thereby increasing the expansion rate of foamed asphalt. This causes the central pressure inside the cavity to increase from b to point c, and the foamed asphalt is ejected from the cavity at a higher speed under the higher pressure, which in turn makes the central flow velocity in the cavity start to increase again. It can be concluded from this that the central flow velocity and central pressure show a fluctuation trend with a phase difference of 180°.

Influence of Cavity Outlet Size. Figure 7 shows the curves of outlet pressure variation and outlet velocity variation under different outlet sizes. As can be seen from Fig. 7(a), when the flow rate of hot asphalt injected into the foam cavity is constant and the size of the cavity outlet decreases, the momentum flux of the foam asphalt decreases. Thus, more pressure needs to be applied to maintain the normal ejection of the foam asphalt, resulting in an increase in the internal pressure of the cavity. However, high or low chamber pressure during the foaming process is not conducive to the normal foaming of asphalt. This shows that in order to obtain high quality foamed asphalt, the outlet size of foamed asphalt should be controlled within a certain range, so as to avoid the negative impact of too high or too low cavity pressure on the quality of foamed asphalt.

When the outlet size of foam asphalt is 0.7d, d and 1.3d, the coefficient of variation of outlet pressure is 0.1982, 0.2992 and 0.3395 respectively, and the coefficient of variation of outlet flow rate is 0.1161, 0.1232 and 0.1476 respectively. The increase of the coefficient of variation indicates that the increase of the outlet size of foam asphalt will aggravate the fluctuation of pressure and flow rate. In a certain range, the smaller the outlet size, the greater the outlet pressure and velocity of the cavity, the more sufficient the heat exchange between hot asphalt and water inside the cavity, and the higher the expansion rate and half-life of the foam asphalt.

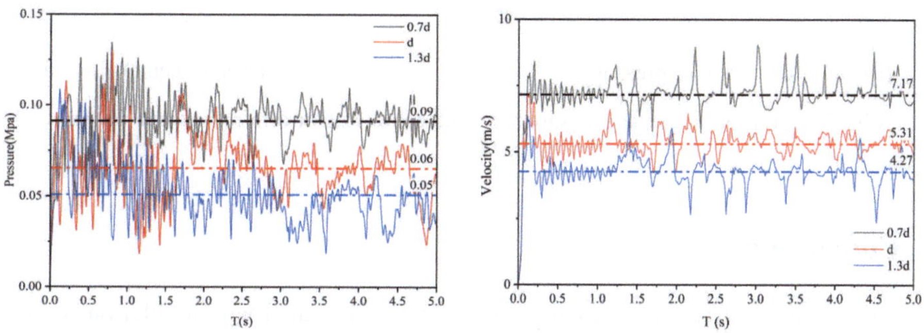

(a) variation of outlet pressure (b) variation of outlet flow rate

Fig. 7. Variation of outlet pressure and outlet flow rate with different outlet sizes.

Influence of Cavity Volume. As can be seen from Fig. 8, with the increase of the volume of the cavity, the central pressure of the cavity presents a decreasing trend. Although a larger cavity volume is not conducive to increasing the expansion rate of foamed asphalt, it is conducive to prolonging the half-life of foamed asphalt. When the volume of the cavity is 0.7 V, V and 1.3 V, the coefficient of variation of the central flow velocity of the cavity is 0.3434, 0.4064 and 0.4597, and the coefficient of variation of the central pressure of the cavity is 0.2187, 0.2540 and 0.3886. This indicates that although the increase of the volume of the cavity has little influence on the central velocity of the cavity, it significantly increases the fluctuation degree of the central pressure and the central velocity of the cavity.

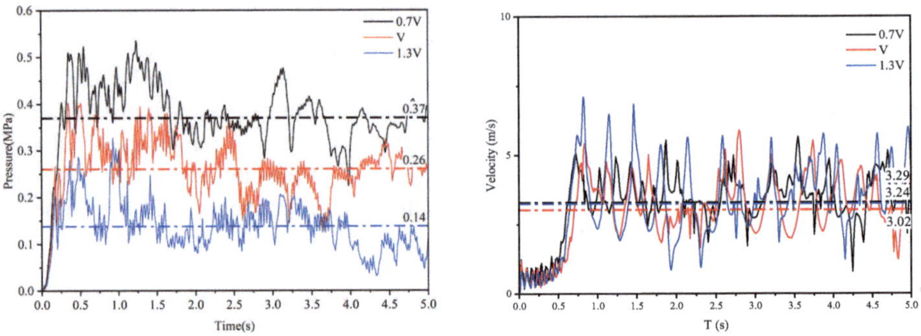

(a) variation of center pressure (b) variation of center flow rate

Fig. 8. Variation of center pressure and center flow rate or different cavity volumes.

4 Conclusion

(1) The average intracavity pressure based on the asphalt foam simulation model is compared with the test result, and the error is less than 11%, which indicates that the asphalt foam simulation analysis has a certain reference.
(2) During the asphalt foaming process, the center velocity and center pressure of the cavity fluctuated periodically, and the phase difference between the two was 180°.
(3) With the increase of the size of asphalt inlet and outlet, the mean value and fluctuation range of the velocity and pressure inside the cavity showed a decreasing trend. The volume of the cavity has little influence on the velocity in the cavity, but with the increase of the volume of the cavity, the average pressure in the cavity shows a decreasing trend.

Acknowledgments. This work was study was funded by the National Natural Science Foundation of China (grant number 52375230), the Natural Science Basic Research Program of Shaanxi Province of China (grant number 2022 JZ-26).

Disclosure of Interests. The authors have no competing interests to declare that are relevant to the content of this article.

References

1. Cheng, H., Wei, F., Huang, Z.: The structural parameters optimization of asphalt foaming cavity by response surface analyzing. AMSE JOURNALS-2016-Series: Modelling B **15**(1), 134–149 (2016)
2. Li, Q., Song, S., Wang, J.: Characterization of residual moisture distribution in cold-recycled mixtures with foamed asphalt using hyperspectral technology. J. Mater. Civ. Eng. **34**(11), 04022288 (2022)

3. Behnood, A., Karimi, M., Cheraghian, G.: Coupled effects of warm mix asphalt (WMA) additives and rheological modifiers on the properties of asphalt binders. Clean. Eng. Technol. 100028 (2020)
4. Abu, Q., Nazzal, M., Abbas, A.: Micromechanical and chemical characterization of foamed warm-mix asphalt aging. J. Mater. Civ. Eng. **30**(9), 04018213 (2018)
5. Liu, F., Wang, A., Fu, Z.: Evaluation for VOF simulation of flow field in asphalt foaming chamber based on volume fraction. J. Test. Eval. **46**(4), 1391–1398 (2018)
6. Han, Y., Ding, Z., Wu, Y.: Influence of foam condition on foam expansion rate of base asphalt. J. Traffic Transp. Eng. **19**(3), 1671–1637 (2019)
7. Cheng, H., Luo, Z., Seliverstov, N.: A study of foam bitumen preparation for effective recycling of pavement layers. Sustainability **14**(15), 9375 (2022)
8. Li, D., Greenfield, M.: Chemical compositions of improved model asphalt systems for molecular simulations. Fuel **115**, 347–356 (2014)
9. Yao, H., Liu, J., Wang, Y.: Mechanism exploration of the foamed asphalt binder using the molecular dynamics (MD) method. J. Clean. Prod. **374**, 134015 (2022)
10. Liang, S., Liang, C., Li, M.: Investigation of nanoscale interfacial bonding properties in foamed asphalt cold recycled mixtures under chloride salt erosion. Case Stud. Constr. Mater. **21**, e03390 (2024)
11. Li, Q., Luo, K., Kang, Q.: Lattice Boltzmann methods for multiphase flow and phase-change heat transfer. Prog. Energy Combust. Sci. **52**, 62–105 (2016)
12. Haussmann, M., Ries, F., Jeppener-Haltenhoff, J.: Evaluation of a near-wall-modeled large eddy lattice Boltzmann method for the analysis of complex flows relevant to IC engines. Computation **8**(2), 43 (2020)
13. Cubeddu, A., Rauh, C., Ulrich, V.: Simulations of bubble growth and interaction in high viscous fluids using the lattice Boltzmann method. Int. J. Multiph. Flow **93**, 108–114 (2017)
14. Lin, G., Geng, C., Zhang, L.: Improved multicomponent multiphase lattice Boltzmann model for physical foaming simulation. J. Appl. Fluid Mech. **15**(4), 1125–1136 (2022)
15. Shen, L., Wang, Z., Li, S.: Numerical simulation for seepage field of soil based on mesoscopic structure reconfiguration technology. Rock Soil Mech. **36**(11), 3307–3314 (2015)
16. Zou, Q., He, X.: On pressure and velocity boundary conditions for the lattice Boltzmann BGK model. Phys. Fluids **9**(6), 1591–1598 (1997)

Author Index

S. Halgamuge et al. (Eds.): ICACMVE 2025, LNME, pp. 757–760, 2026.
https://doi.org/10.1007/978-981-95-6911-3

K

L

M

P

Q

R

S

MIX
Papier aus verantwortungsvollen Quellen
Paper from responsible sources
FSC® C105338

If you have any concerns about our products,
you can contact us on
ProductSafety@springernature.com

In case Publisher is established outside the EU,
the EU authorized representative is:
Springer Nature Customer Service Center GmbH
Europaplatz 3, 69115 Heidelberg, Germany

Printed by Libri Plureos GmbH
in Hamburg, Germany